T0188813

Communications in Computer and Information Science 854

Commenced Publication in 2007
Founding and Former Series Editors:
Alfredo Cuzzocrea, Xiaoyong Du, Orhun Kara, Ting Liu, Dominik Ślęzak,
and Xiaokang Yang

More information about this series at http://www.springer.com/series/7899

Jesús Medina · Manuel Ojeda-Aciego
José Luis Verdegay · David A. Pelta
Inma P. Cabrera · Bernadette Bouchon-Meunier
Ronald R. Yager (Eds.)

Information Processing and Management of Uncertainty in Knowledge-Based Systems

Theory and Foundations

17th International Conference, IPMU 2018
Cádiz, Spain, June 11–15, 2018
Proceedings, Part II

 Springer

Editors
Jesús Medina
Universidad de Cádiz
Cádiz, Cadiz
Spain

Manuel Ojeda-Aciego
Universidad de Málaga
Málaga, Málaga
Spain

José Luis Verdegay
Universidad de Granada
Granada
Spain

David A. Pelta
Universidad de Granada
Granada
Spain

Inma P. Cabrera
Universidad de Málaga
Málaga, Málaga
Spain

Bernadette Bouchon-Meunier
LIP6
Université Pierre et Marie Curie, CNRS
Paris
France

Ronald R. Yager
Iona College
New Rochelle, NY
USA

ISSN 1865-0929   ISSN 1865-0937   (electronic)
Communications in Computer and Information Science
ISBN 978-3-319-91475-6   ISBN 978-3-319-91476-3   (eBook)
https://doi.org/10.1007/978-3-319-91476-3

Library of Congress Control Number: 2018944294

Printed on acid-free paper

This Springer imprint is published by the registered company Springer International Publishing AG
part of Springer Nature
The registered company address is: Gewerbestrasse 11, 6330 Cham, Switzerland

To Lotfi A. Zadeh

Preface

These are the proceedings of the 17th International Conference on Information Processing and Management of Uncertainty in Knowledge-Based Systems, IPMU 2018. The conference was held during June 11–15, in Cádiz, Spain.

The IPMU conference is organized every two years with the aim of bringing together scientists working on methods for the management of uncertainty and aggregation of information in intelligent systems. Since 1986, the IPMU conference has been providing a forum for the exchange of ideas between theoreticians and practitioners working in these areas and related fields.

This IPMU edition held special meaning since one of its co-founders, Lotfi A. Zadeh, passed away on September 6, 2017. To pay him a well-deserved tribute, and in memory of his long relationship with IPMU participants, a special plenary panel was organized to discuss the scientific legacy of his ideas. Renowned researchers and Lotfi's good friends made up the panel: it was chaired by Ronald Yager, while Bernadette Bouchon-Meunier, Didier Dubois, Janusz Kacprzyk, Rudolf Kruse, Rudolf Seising, and Enric Trillas acted as panelists. Besides this, a booklet of pictures with Lotfi Zadeh and friends was compiled and distributed at the conference.

Following the IPMU tradition, the Kampé de Fériet Award for outstanding contributions to the field of uncertainty and management of uncertainty was presented. Past winners of this prestigious award were Lotfi A. Zadeh (1992), Ilya Prigogine (1994), Toshiro Terano (1996), Kenneth Arrow (1998), Richard Jeffrey (2000), Arthur Dempster (2002), Janos Aczel (2004), Daniel Kahneman (2006), Enric Trillas (2008), James Bezdek (2010), Michio Sugeno (2012), Vladimir N. Vapnik (2014), and Joseph Y. Halpern (2016). In this 2018 edition, the award was given to Glenn Shafer (Rutgers University, Newark, USA) for his seminal contributions to the mathematical theory of evidence and belief functions as well as to the field of reasoning under uncertainty. The so-called Dempster–Shafer theory, an alternative to the theory of probability, has been widely applied in engineering and artificial intelligence.

The program consisted of the keynote talk of Glenn Shafer, as recipient of the Kampé de Feriet Award, five invited plenary talks, two round tables, and 30 special sessions plus a general track for the presentation of the 190 contributed papers that were authored by researchers from more than 40 different countries. The plenary presentations were given by the following distinguished researchers: Gloria Bordogna (IREA CNR – Institute for the Electromagnetic Sensing of the Environment of the Italian National Research Council), Lluis Godo (Artificial Intelligence Research Institute of the Spanish National Research Council, Barcelona, Spain), Enrique Herrera-Viedma (Department of Computer Science and Artificial Intelligence, University of Granada, Spain), Natalio Krasnogor (School of Computing Science at Newcastle University, UK), and Yiyu Yao (Department of Computer Science, University of Regina, Canada).

The conference followed a single-blind review process, respecting the usual conflict-of-interest standards. The contributions were reviewed by at least three reviewers. Moreover, the conference chairs further checked the contributions in those cases were conflicting reviews were obtained. Finally, the accepted papers are published in three volumes: Volumes I and II focus on "Theory and Foundations," while Volume III is devoted to "Applications."

The organization of the IPMU 2018 conference was possible thanks to the assistance, dedication, and support of many people and institutions. In particular, this renowned international conference owes its recognition to the great quality of the contributions. Thank you very much to all the participants for their contributions to the conference and all the authors for the high quality of their submitted papers. We are also indebted to our colleagues, members of the Program Committee, and the organizers of special sessions on hot topics, since the successful organization of this international conference would not have been possible without their work. They and the additional reviewers were fundamental in maintaining the excellent scientific quality of the conference. We gratefully acknowledge the local organization for the efforts in the successful development of the multiple tasks that a great event like IPMU involves.

We also acknowledge the support received from different areas of the University of Cádiz, including the Department of Mathematics, the PhD Program in Mathematics, the Vice-Rectorate of Infrastructures and Patrimony, and the Vice-Rectorate for Research; the International Global Campus of Excellence of the Sea (CEI·Mar) led by the University of Cádiz and composed of institutions of three different countries; the European Society for Fuzzy Logic and Technology (EUSFLAT); and the Springer team who managed the publication of these proceedings. Finally, J. Medina, M. Ojeda-Aciego, J. L. Verdegay, I. Cabrera, and D. Pelta acknowledge the support of the following research projects: TIN2016-76653-P, TIN2015-70266-C2-P-1, TIN2014-55024-P, TIN2017-86647-P, and TIN2017-89023-P (Spanish Ministery of Economy and Competitiveness, including FEDER funds).

June 2018   Jesús Medina
                                              Manuel Ojeda-Aciego
                                                    Irina Perfilieva
                                              José Luis Verdegay
                                        Bernadette Bouchon-Meunier
                                                  Ronald R. Yager
                                                  Inma P. Cabrera
                                                  David A. Pelta

Organization

General Chair

Jesús Medina   Universidad de Cádiz, Spain

Program Chairs

Manuel Ojeda-Aciego   Universidad de Málaga, Spain
Irina Perfilieva   University of Ostrava, Czech Republic
José Luis Verdegay   Universidad de Granada, Spain

Executive Directors

Bernadette   LIP6 - Université Pierre et Marie Curie, CNRS, Paris,
  Bouchon-Meunier   France
Ronald R. Yager   Iona College, USA

Sponsors and Publicity Chair

Martin Stepnicka   University of Ostrava, Czech Republic

Special Sessions Chair

Inma P. Cabrera   Universidad de Málaga, Spain

Publication Chair

David A. Pelta   Universidad de Granada, Spain

Local Organizing Committee

María José   Universidad de Cádiz, Spain
  Benítez-Caballero
María Eugenia Cornejo   Universidad de Cádiz, Spain
Juan Carlos Díaz-Moreno   Universidad de Cádiz, Spain
David Lobo   Universidad de Cádiz, Spain
Óscar Martín-Rodríguez   Universidad de Granada, Spain
Eloísa Ramírez-Poussa   Universidad de Cádiz, Spain

International Advisory Board

Anne Laurent
Benedetto Matarazzo
Christophe Marsala
Enric Trillas
Eyke Hüllermeier
Giulianella Coletti
João Paulo Carvalho
José Luis Verdegay
Julio Gutierrez-Rios
Laurent Foulloy
Llorenç Valverde

Lorenza Saitta
Luis Magdalena
Manuel Ojeda-Aciego
Maria Amparo Vila
Maria Rifqi
Marie-Jeanne Lesot
Mario Fedrizzi
Miguel Delgado
Olivier Strauss
Salvatore Greco
Uzay Kaymak

Program Committee

Michal Baczynski, Poland
Rafael Bello-Pérez, Cuba
Jim Bezdek, USA
Isabelle Bloch, France
Ulrich Bodenhofer, Austria
Bernadette Bouchon-Meunier, France
Humberto Bustince, Spain
Inma P. Cabrera, Spain
Joao Carvalho, Portugal
Giulianella Coletti, Italy
Oscar Cordon, Spain
Inés Couso, Spain
Keeley Crockett, UK
Fabio Cuzzolin, UK
Bernard De Baets, Belgium
Guy De Tré, Belgium
Sébastien Destercke, France
Antonio Di Nola, Italy
Didier Dubois, France
Fabrizio Durante, Italy
Francesc Esteva, Spain
Juan C. Figueroa-García, Colombia
Sylvie Galichet, France
Lluis Godo, Spain
Fernando Gomide, Brazil
Gil González-Rodríguez, Spain
Michel Grabisch, France
Przemysław Grzegorzewski, Poland
Lawrence Hall, USA

Francisco Herrera, Spain
Enrique Herrera-Viedma, Spain
Ludmilla Himmelspach, Germany
Janusz Kacprzyk, Poland
Uzay Kaymak, The Netherlands
James Keller, USA
Laszlo Koczy, Hungary
Vladik Kreinovich, USA
Tomas Kroupa, Italy
Rudolf Kruse, Germany
Christophe Labreuche, France
Weldon A. Lodwick, USA
Jean-Luc Marichal, Luxembourg
Trevor Martin, UK
Sebastian Massanet, Spain
Gaspar Mayor, Spain
Jesús Medina, Spain
Radko Mesiar, Slovakia
Ralf Mikut, Germany
Enrique Miranda, Spain
Javier Montero, Spain
Jacky Montmain, France
Serafín Moral, Spain
Zbigniew Nahorski, Poland
Pavel Novoa, Ecuador
Vilem Novak, Czech Republic
Hannu Nurmi, Finland
Manuel Ojeda-Aciego, Spain
Gabriella Pasi, Italy

Witold Pedrycz, Canada
David A. Pelta, Spain
Irina Perfilieva, Czech Republic
Fred Petry, USA
Vincenzo Piuri, Italy
Olivier Pivert, France
Henri Prade, France
Marek Reformat, Canada
Daniel Sánchez, Spain
Mika Sato-Ilic, Japan
Ricardo C. Silva, Brazil

Martin Stepnicka, Czech Republic
Umberto Straccia, Italy
Eulalia Szmidt, Poland
Settimo Termini, Italy
Vicenc Torra, Sweden/Spain
Linda van der Gaag, The Netherlands
Barbara Vantaggi, Italy
José L. Verdegay, Spain
Thomas Vetterlein, Austria
Susana Vieira, Portugal
Slawomir Zadrozny, Poland

Additional Reviewers

Jesús Alcalá-Fernández
José Carlos R. Alcantud
Svetlana Asmuss
Laszlo Aszalos
Mohammad Azad
Cristobal Barba Gonzalez
Gleb Beliakov
María J. Benítez-Caballero
Pedro Bibiloni
Alexander Bozhenyuk
Michal Burda
Ana Burusco
Camilo Alejandro Bustos Téllez
Francisco Javier Cabrerizo
Yuri Cano
Andrea Capotorti
J. Manuel Cascón
Dagoberto Castellanos
Francisco Chicano
María Eugenia Cornejo
Susana Cubillo
Martina Dankova
Luis M. de Campos
Robin De Mol
Yashar Deldjoo
Pedro Delgado-Pérez
Juan Carlos Díaz
Susana Díaz
Marta Disegna
Alexander Dockhorn
Paweł Drygaś

Talbi El Ghazali
Javier Fernández
Joao Gama
José Gámez
M. D. García Sanz
José Luis García-Lapresta
José García Rodríguez
Irina Georgescu
Manuel Gómez-Olmedo
Adrián González
Antonio González
Manuel González-Hidalgo
Jerzy Grzymala Busse
Piotr Helbin
Daryl Hepting
Michal Holcapek
Olgierd Hryniewicz
Petr Hurtik
Atamanyuk Igor
Esteban Induráin
Vladimir Janis
Andrzej Janusz
Sándor Jenei
Pascual Julian-Iranzo
Aránzazu Jurío
Katarzyna Kaczmarek
Martin Kalina
Gholamreza Khademi
Margarita Knyazeva
Martins Kokainis
Galyna Kondratenko

Oleksiy Korobko
Piotr Kowalski
Oleksiy Kozlov
Anna Król
Sankar Kumar Roy
Angelica Leite
Tianrui Li
Ferenc Lilik
Nguyen Linh
Hua Wen Liu
Bonifacio Llamazares
David Lobo
Marcelo Loor
Ezequiel López-Rubio
Gabriel Luque
Rafael M. Luque-Baena
M. Aurora Manrique
Nicolás Marín
Ricardo A. Marques-Pereira
Stefania Marrara
Davide Martinetti
Víctor Martínez
Raquel Martínez España
Miguel Martínez-Panero
Tamás Mihálydeák
Arnau Mir
Katarzyna Miś
Miguel A. Molina-Cabello
Michinori Nakata
Bac Nguyen
Joachim Nielandt
Wanda Niemyska
Juan Miguel Ortiz De Lazcano Lob
Sergio Orts Escolano
Iván Palomares
Esteban José Palomo
Manuel Pegalajar-Cuéllar
Barbara Pękala
Renato Pelessoni
Tomasz Penza
Davide Petturiti
José Ramón Portillo

Cristina Puente
Eloísa Ramírez-Poussa
Ana Belén Ramos Guajardo
Jordi Recasens
Juan Vicente-Riera
Rosa M. Rodríguez
Luis Rodríguez-Benítez
Estrella Rodríguez-Lorenzo
Maciej Romaniuk
Jesús Rosado
Clemente Rubio-Manzano
Pavel Rusnok
Hiroshi Sakai
Sancho Salcedo-Sanz
José María Serrano
Ievgen Sidenko
Gerardo Simari
Julian Skirzynski
Andrzej Skowron
Grégory Smits
Marina Solesvik
Anna Stachowiak
Sebastian Stawicki
Michiel Stock
Andrei Tchernykh
Jhoan S. Tenjo García
Luis Terán
Karl Thurnhofer Hemsi
S. P. Tiwari
Joan Torrens
Gracian Trivino
Matthias Troffaes
Shusaku Tsumoto
Diego Valota
Matthijs van Leeuwen
Sebastien Varrette
Marco Viviani
Pavel Vlašánek
Yuriy Volosyuk
Gang Wang
Anna Wilbik
Andrzej Wójtowicz

Special Session Organizers

Stefano Aguzzoli	University of Milan, Italy
José M. Alonso	University of Santiago de Compostela, Spain
Michal Baczynski	University of Silesia, Poland
Isabelle Bloch	University Paris-Saclay, France
Reda Boukezzoula	LISTIC- Université de Savoie Mont-Blanc, France
Humberto Bustince	Universidad Pública de Navarra, Spain
Inma P. Cabrera	Universidad de Málaga, Spain
Tomasa Calvo	Universidad de Alcalá, Spain
Ciro Castiello	University of Bari Aldo Moro, Italy
Juan Luis Castro	University of Granada, Spain
Yurilev Chalco-Cano	Universidad de Tarapacá, Chile
Davide Ciucci	University of Milano-Bicocca, Italy
Didier Coquin	LISTIC- Université de Savoie Mont-Blanc, France
Pablo Cordero	University of Málaga, Spain
Rocío de Andrés Calle	University of Salamanca, Spain
Bernard De Baets	Ghent University, Belgium
Juan Carlos de la Torre	University of Cádiz, Spain
Graçaliz Dimuro	Institute of Smart Cities and Universidade Federal do Rio Grande, Brazil
Enrique Domínguez	University of Málaga, Spain
Bernabe Dorronsoro	University of Cádiz, Spain
Krzysztof Dyczkowski	Adam Mickiewicz University in Poznań, Poland
Ali Ebrahimnejad	Islamic Azad University, Iran
Tommaso Flaminio	Artificial Intelligence Research Institute (CSIC), Barcelona, Spain
Pilar Fuster-Parra	Universitat de les Illes Balears, Spain
M. Socorro García Cascales	Polytechnic University of Cartagena, Spain
Brunella Gerla	University of Insubria, Italy
Juan Gómez Romero	University of Granada, Spain
Teresa González-Arteaga	University of Valladolid, Spain
Balasubramaniam Jayaram	Indian Institute of Technology Hyderabad, India
László Kóczy	Budapest University of Technology and Economics, Hungary
Yuriy Kondratenko	Petro Mohyla Black Sea National University, Ukraine
Weldon Lodwick	University of Colorado, USA
Vincenzo Loia	University of Salerno, Italy
Nicolás Madrid	University of Málaga, Spain
Luis Magdalena	Universidad Politécnica de Madrid, Spain
María J. Martín-Bautista	University of Granada, Spain
Sebastián Massanet	University of the Balearic Islands, Spain
Corrado Mencar	University of Bari Aldo Moro, Italy
Radko Mesiar	University of Technology, Slovakia
Enrique Miranda	University of Oviedo, Spain
Ignacio Montes	University of Oviedo, Spain

Contents – Part II

Rough and Fuzzy Similarity Modelling Tools

Tri-partitions and Uncertainty

Fuzzy Methods in Data Mining and Knowledge Discovery

Fuzzy Analysis of Sentiment Terms for Topic Detection Process in Social Networks

Karel Gutiérrez-Batista[✉][iD], Jesús R. Campaña[iD], Maria-Amparo Vila[iD], and Maria J. Martin-Bautista[iD]

Department of Computer Science and Artificial Intelligence, ETSIIT, University of Granada, 18071 Granada, Spain
{karel,jesuscg,vila,mbautis}@decsai.ugr.es

Abstract. The aim of this paper is to analyze the influence of sentiment-related terms on the automatic detection of topics in social networks. The study is based on the use of an ontology, to which the capacity to gradually identify and discard sentiment terms in social network texts is incorporated, as these terms do not provide useful information for detecting topics. To detect these terms, we have used two resources focused on the analysis of sentiments. The proposed system has been assessed with real data sets of the social networks Twitter and Dreamcatcher in English and Spanish respectively.

Keywords: Topic detection · Hierarchical clustering · Fuzzy sets Sentiment terms

1 Introduction

The growth and popularity of social networks has made them one of the main sources of unstructured textual data, so it is expected that organizations and researchers employ time and resources studying them. However, the great amount of texts, along with their lack of structure, hinders automatic processing and analysis at a large scale, so it is practical to group the texts beforehand according to the addressed topic. In this sense, it is particularly useful, to detect automatically the main topics that are being addressed and are relevant to users.

On the other hand, the detection of topics depends to a great extent on the origin of the texts. In social networks, users tend to express ideas, feelings and opinions on a certain topic in a colloquial manner, so there is a high frequency of terms that express sentiments or opinions related to certain products or services. These terms are a great source of information in tasks such as sentiment analysis, opinion mining or recommendation systems, but not for topic detection.

In order to solve the above problem, this research aims to gradually analyze the influence of sentiment or opinion terms on the automatic detection of topics in social media texts. Our proposal is based on data mining, lexical resources for

© Springer International Publishing AG, part of Springer Nature 2018
J. Medina et al. (Eds.): IPMU 2018, CCIS 854, pp. 3–14, 2018.
https://doi.org/10.1007/978-3-319-91476-3_1

sentiment analysis and a multilingual knowledge base. This new approach allows us to identify and gradually discard sentiment terms with the aim of improving the automatic detection of topics in social media texts.

To achieve the task, we applied a filter to detect sentiment terms during the semantic processing stage. To this end, we used the resources SentiWordNet 3.0 [1] and SenticNet 3 [2]. The main advantage of using these two resources is that they both assign a polarity to every sense or concept, which allowed us to carry out an analysis establishing various polarity thresholds (α-cuts). These values allow us the possibility of deleting only the terms which are above a certain polarity threshold. This is extremely interesting, as sometimes the analyzed texts have a large quantity of opinion terms and many texts are therefore automatically rejected, thus affecting topic detection negatively.

Sentiment analysis has been addressed using various techniques. In the literature there are various studies where we can observe that the main focuses to take into account when analyzing sentiments are based on Machine learning techniques [3–5] and the use of a opinion lexicon (sentiment lexicon) [6,7]. There is also a series of studies which combine other techniques related with the subject, such as the use of ontologies [8] and part-of-speech tagging [9].

From a fuzzy logic approach, in [10] the author analyzes the progress that has been made and the challenges there are in the field of fuzzy sentiment analysis. Opinion words are fuzzy by nature, so it is interesting to combine current sentiment analysis techniques with fuzzy logic [10]. In order to take advantage of fuzzy logic, we need to know the polarity of opinion words, provided by opinion lexicons such as SentiWordNet 3.0 and SenticNet 3.

This is the case in [11], where membership and non-membership values of sentiment words are calculated using SentiWordNet. Once calculated, a fuzzy set operator is applied to classify the term as positive, negative or neutral. In [12] the authors put forward a system using fuzzy functions to adjust the polarities of SentiWordNet with the aim of emulating the effect of various linguistic hedges. Before applying the function, an extraction of features is performed applying a system based on rules. Finally, texts are classified using a K-Means algorithm.

In [13], fuzzy logic is used to model concept polarities and the uncertainty associated with them regarding different domains. A knowledge graph is built on the basis of the resources WordNet and SenticNet. The graph is then exploited using a graph-propagation algorithm that propagates sentiment information learnt from labelled data sets. The graph has two levels; the first represents the semantic relationships between concepts, whilst the second contains the links between the concept membership functions and the different domains.

Although in this study we do not make sentiment analysis specifically, opinion words are taken into account as they have to be identified and then gradually discarded to be able to analyze their influence in the automatic detection of topics. In Sect. 3.3 we describe the fuzzy model to take into account for the goals of our research.

The rest of the paper is structured as follows: Sect. 2 briefly describes the proposal for detecting topics automatically. Section 3 describes in detail the process for identifying and discarding sentiment terms gradually from the textual data of social networks. Section 4 presents and analyzes the experimental results. Finally, in Sect. 5 the conclusions derived from the analysis are presented.

2 Overview of Our Topic Detection Approach

In this section, we describe the different stages of our approach to automatically detect topics in social networks.

2.1 Syntactic Preprocessing

The data used for this first stage are the texts extracted from social media. These texts are processed syntactically and different filters are applied to aid in their automatic processing.

First of all, grammatical categories labelling and entities recognition were carried out. This was done using the Stanford Part-of-Speech (POS) Tagger [14] and Stanford Named Entity Recognition (NER) [15] tools, respectively. The first tool assigns grammatical categories to the words of a text (noun, adjective, verb, etc.). The second identifies words within a text representing names or things, such as people, companies or places, on the basis of various provided models.

A token filter was then applied to delete all punctuation as it hinders automatic text processing. Subsequently, filters are applied to delete empty words, terms that are not identified as nouns by the POS, words identified as proper nouns by the entity recognizer and words that are not included in the external knowledge base Multilingual Central Repository 3.0 (MCR 3.0) [16], as none of these provide useful information to detect contexts and topics.

2.2 Semantic Preprocessing

Once the above filters have been applied, the texts are ready to be semantically processed. The main goal of semantic preprocessing is homogenizing the syntactic representation of the concepts in the text.

To achieve this, each term is replaced with its corresponding labels from the WordNet Domains taxonomy [17] of the knowledge base MCR 3.0. To determine the set of labels of a term, it is necessary to know the sense of the term, as each sense has a set of labels of WordNet Domains. For this reason, it is necessary to disambiguate all the terms to know their real sense before replacing them with their labels.

2.3 Hierarchical Clustering

Once the texts have been homogenized, the hierarchical clustering of the texts is carried out on the basis of the labels of WordNet Domains. To do this, a weight matrix relating the labels to the documents has to be created on the basis of the texts obtained in the previous phase.

When the weight matrix has been created, the hierarchical clustering algorithm is run to create text clusters on the basis of the semantics of WordNet Domains. Each text is associated to a cluster (context).

2.4   Cluster Labeling

Finally, the clusters (contexts) are labelled. During this process each cluster is assigned the most relevant set of labels of the texts belonging to that cluster. To this end, the Chi-Square (X^2) technique is used. This is one of the most widely used techniques for (*Differential cluster labeling*) [18] and one of the techniques showing best results.

3   Sentiment Influence in Topic Detection

In the following section we will describe in detail the process we used to detect and discard sentiment terms (positive or negative) in social network texts. This is the main contribution of this research. The main reason to discard these terms is that we aim to detect the main topics addressed in social network texts, where users usually express their feelings or opinions on a subject.

These terms, as well as empty words, are very frequent in this kind of texts so, when applying certain data mining techniques, they can disguise the real topics of texts. This is why, it is important to detect and discard the noise that sentiment terms provoke.

To detect sentiment terms, the resources SentiWordNet 3.0 [1] and SenticNet 3 [2] have been used, which are based on WordNet [19] and allow us to determine if a term in a certain context expresses any sentiment. The first step is to disambiguate the term in order to know its real sense and determine if it expresses sentiment.

In general, when sentiment terms are discarded, regardless of the lexical resource that has been used, the most frequent terms belonging to certain topic are more related to it. Also, many texts which have been filtered for sentiment can then not be processed as all the terms have been discarded. This is mainly because most texts of social networks only express an opinion on a certain topic. For this reason, we have decided to establish α-cuts to discard only the terms above certain polarity thresholds, in order to increase the final number of texts used to detect topics.

It is important to highlight that the idea of applying a filter to detect sentiment terms is totally new, as it allows the separation of terms with relevant information to detect topics, from sentiment-related terms, which are useful for many computational studies, but not for the goals of this study.

3.1   SentiWordNet 3.0

This is a lexical resource created specifically for tasks focused on classifying sentiments, as well as applications based on opinion mining [1]. It is an improved version of WordNet 1.0 [20] and is publicly available for research purposes. SentiWordNet 3.0 is the result of assigning to all the senses (synsets) of WordNet two numerical values, which indicate the polarity value (positive and negative), these values are in the range [0, 1] [1].

Figures 1 and 2 show histograms of positive and negative polarities in SentiWordNet, respectively. As can be observed, most senses have polarities in the range [0.0, 0.2].

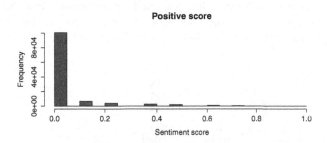

Fig. 1. Positive polarities in SentiWordNet 3.0

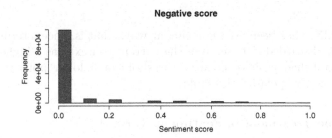

Fig. 2. Negative polarities in SentiWordNet 3.0

In our case, once each term is disambiguated, we determine the positive and negative values assigned in SentiWordNet to the corresponding sense of each of the analyzed terms. If the sense has a positive value higher that the established positive polarity threshold or a negative value higher than the established negative polarity threshold, the term is totally discarded and is not taken into account for the subsequent analysis of the main topics addressed in texts.

3.2   SenticNet 3

SenticNet 3 [2] is one of most relevant lexical resources regarding sentiment analysis. It is a semantic and sentiment resource based on WordNet and is used mainly to carry out concept-level sentiment analysis. This trait allows us to infer the semantics and emotions present in natural language opinions, so it is a very useful tool to carry out sentiment analysis based on the characteristics of products and services [2].

In other words, SenticNet 3.0, goes further than assessing the opinion on a certain element, offering the chance to compare each of the characteristics of the element. It is composed of 30,000 expressions with a polarity between $[-1, 1]$.

Figure 3 shows the histogram of the polarities of SenticNet 3. As in SentiWordNet, most senses have polarities in the range $[0.0, 0.2]$. But, unlike in SentiWordNet, where each sense has assigned positive and negative values, in SenticNet, in order to determine the polarity of a sense, it was necessary to calculate the average of the polarities of each of the terms of a sense.

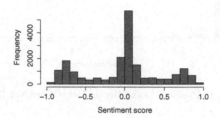

Fig. 3. Sentiment score histogram of SenticNet

SentiWordNet has been used in a similar way. That is, once a term is disambiguated, it is discarded if its sense in the current context has a higher polarity than the established positive polarity threshold or a lower polarity than the established negative polarity threshold.

3.3 Flexible Treatment of Sentiment Terms

We now present the polarity model to construct fuzzy sets of sentiment terms. This model is the basis of our proposal and allows us to gradually analyze the influence of sentiment terms when detecting topics.

SentiWordNet Case. Let $\mathcal{V}$ be the fuzzy set of terms in SentiWordNet: $\forall t \in \mathcal{V}$ $\exists \alpha_1, \alpha_2 \in [0, 1]$ such that α_1 is the intensity of t representing positive sentiment and α_2 the intensity of t representing negative sentiment.

We define $\hat{F}$ as the fuzzy set of terms derived from $\mathcal{V}$ with the membership function:

$$\forall t \in \mathcal{V}, \mu_{\hat{F}(t)} = \max\left(\alpha_1, \alpha_2\right). \tag{1}$$

therefore, for each term we reflect the intensity of its representation as a sentiment, regardless of it being positive or negative.

SenticNet Case. Let $\mathcal{W}$ be the fuzzy set of terms included in SenticNet: $\forall t \in \mathcal{W}$ $\exists \alpha_1 \in [-1, 1]$ such that α_1 is the intensity of t representing positive or negative sentiment. So that t can only have one polarity.

We define $\hat{S}$ as the fuzzy set of terms derived from $\mathcal{W}$ with the membership function:

$$\forall t \in \mathcal{W}, \mu_{\hat{S}(t)} = \quad \bmod \; \alpha_1. \tag{2}$$

obviously, for this case, we also reflect the intensity of t representing a sentiment, regardless of it being positive or negative.

The topics can now be obtained considering the α-cuts of the $\hat{F}$ and $\hat{S}$ sets, instead of discarding all the sentiment terms (in that case, the α-cut would be 0). This increases the flexibility of the process, including a larger amount of documents without deteriorating the Silhouette Coefficient of resulting clusters.

4 Experiments

We will now demonstrate experimentally the validity of our proposal. We should highlight that we did not have any previous information on the topics present in the texts (categories or labels). Therefore, we selected the Silhouette Coefficient [21] as a quality measure. It is an unsupervised measure that can determine the number of clusters for which the clustering algorithms offer best results.

The data sets we used for the experiment have been extracted from the social networks Twitter and Dreamcatchers. The first, is one of the most popular social networks and one of the most used in research. On the other hand, Dreamcatchers has been developed using a collaborative approach and the database that supports it is available. The data selected from Twitter and Dreamcatchers is in English and Spanish respectively, demonstrating the validity of the proposal, regardless of the language.

4.1 Data Sets

Eight data sets (Table 1) belonging to Twitter and Dreamcatchers have been selected. We only included the graphs of data sets 4 and 8 and a summary of the most relevant results of the other data sets, due to space constraints. The amount of documents is varied in order to validate the proposal, regardless of the number of processed texts.

The data from Twitter were downloaded using Sentiment140[1], which is in CSV format and has six fields, one of which is the text of the tweets that was used in this research. The reason why the data were selected is that they are directed at sentiment analysis.

On the other hand, we also used the Dreamcatchers database with 61 tables in total. The gathered information refers to personal and membership data of users and the interactions they carry out in their profiles and with other users. Mainly, the *posts* and their *comments*, the *ideas* and their *dreams* and the *chat* provide textual information.

---

[1] http://www.sentiment140.com/.

Table 1. Description of used data sets

Data set	Number of documents	Source	Number of different terms	Total number of terms
Data set 1	1000	Twitter	1665	4875
Data set 2	2000	Twitter	2417	9542
Data set 3	5000	Twitter	3189	12915
Data set 4	10000	Twitter	4597	25634
Data set 5	1000	Dreamcatchers	808	3138
Data set 6	2000	Dreamcatchers	1284	6806
Data set 7	5000	Dreamcatchers	1661	8851
Data set 8	10000	Dreamcatchers	2218	17141

4.2 Evaluation

In this section we explain how our proposal to detect topics automatically applying the filter to discard sentiment terms was assessed, using the different resources and for the different selected α-cuts. To assess the results of the hierarchical clustering algorithm and then compare the different employed resources, the Silhouette Coefficient has been used as a goodness measure.

Experimentation has been carried out using the clustering method Complete-link, establishing the number of clusters (17, 25, 40, 60, 80, 100 and 120) and seven polarity α-cuts (0, 0.2, 0.3, 0.4, 0.5, 0.7 and 1). We experimented with these numbers of clusters as in previous experiments, we realised that for values of clusters lower than 17 and higher than 120, the performance of the algorithms is lower than with the values indicated above. In all cases, the cosine distance was used as measure of similarity.

4.3 Results and Discussion

In Table 2 the statistics referring to the number of sentiment terms discarded for each data set and for each opinion lexicon are shown. In this case, all the opinion terms have been discarded (the α-cut is 0) and the SenticNet 3 is the resource with which more terms have been discarded.

As mentioned previously, it is worth studying how our proposal behaves when instead of discarding all the sentiment terms, only the terms above a certain polarity threshold are discarded. This analysis is of great importance, as opinion terms are very frequent, especially in social media texts and when we discard them many texts will not be taken into account when detecting topics. That is why, on some occasions, it may be convenient to discard only some of the opinion terms in order for more of the texts to remain to detect topics.

Table 2. Number of sentiment-related terms discarded by data set

Dataset	SentiWordNet	%	SenticNet	%
Data set 1	2904	59.57	4086	83.83
Data set 2	5775	60.52	7949	83.31
Data set 3	1696	13.13	9160	70.9
Data set 4	3281	12.8	18141	70.77
Data set 5	2081	66.32	2137	68.1
Data set 6	4147	60.93	4592	67.47
Data set 7	1314	14.85	5122	57.87
Data set 8	2533	14.78	10101	58.93

Figure 4(a) and (b) show the values of the Silhouette Coefficient for data sets 4 and 8 respectively. In both cases, the hierarchal clustering method Complete-link and the lexical resource SentiWordNet have been used. The number of clusters (17, 25, 40, 60, 80, 100 and 120) and seven polarity thresholds (0, 0.2, 0.3, 0.4, 0.5, 0.7 and 1) have been established, where 0 means all opinion terms are discarded and 1 means not discarding any term. As can be observed for each data set, from 60 clusters onwards, which is when the values of the Silhouette Coefficient stabilise, the best results are obtained for the threshold 0.

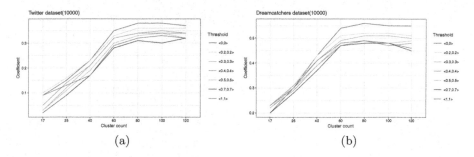

Fig. 4. (a) Silhouette Coefficient for data set 4 using SentiWordNet, (b) Silhouette Coefficient for data set 8 using SentiWordNet

Figure 5(a) and (b) analyze the relationship of the Silhouette Coefficient and the number of documents that remain after applying the filter that discards opinion terms, according to the different selected thresholds for each data set. In these graphs we can observe, that although the Silhouette Coefficient always has the highest value when the threshold is 0, for this same value the number of documents that remain after applying the filter that discards opinion terms is always the lowest. As expected, the lower the α-cut, the more sentiment terms are discarded and the more documents are discarded and are not taken into account to detect topics. It is therefore useful to achieve a consensus (equilibrium point) between both parameters to improve the topic detection process. This point is obviously the intersection between both lines.

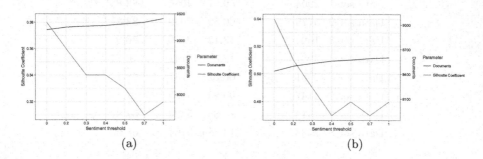

(a)   (b)

Fig. 5. Relationship between the Silhouette Coefficient and the number of documents that remain after discarding the sentiment-related terms for different α-cuts (a) data set 4 using SentiWordNet, (b) data set 8 using SentiWordNet

As in the above examples, Fig. 6(a) and (b) show the Silhouette Coefficient for data sets 4 and 8, respectively. In this case, all parameters are the same except the lexical resource used to discard opinion terms, which is SenticNet3. In this case also from 60 groups onwards, which is when the values of the Silhouette Coefficient stabilise, the best results are obtained for the threshold 0.

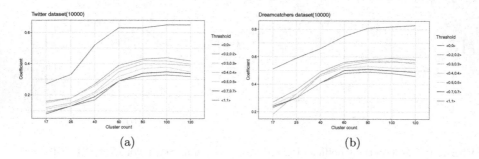

(a)   (b)

Fig. 6. (a) Silhouette Coefficient for data set 4 using SenticNet, (b) Silhouette Coefficient for data set 8 using SenticNet

Figure 7(a) and (b) analyze the relationship between the Silhouette Coefficient and the number of documents that remain after applying the filter that discards opinion terms, according to the different selected thresholds for each data set. The conclusions are the same as for the previous case.

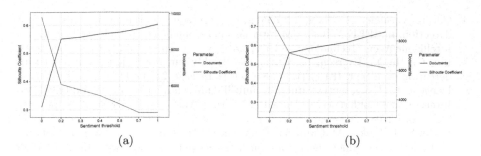

Fig. 7. Relationship between the Silhouette Coefficient and the number of documents that remain after discarding the sentiment-related terms for different α-cuts (a) data set 4 using SenticNet, (b) data set 8 using SenticNet

5   Conclusions

In this study, a new approach for the automatic detection of topics in text from social networks is developed. For such purpose, a filter has been incorporated during the semantic preprocessing of texts, allowing us to detect and discard sentiment-related terms, as these terms do not constitute relevant information to detect topics. To this end, SentiWordNet 3.0 and SenticNet 3 are used.

The experiments carried out with Twitter and Dreamcatchers, allow us to demonstrate the viability of the proposal. With the aim of establishing a comparison, we have experimented with various polarity thresholds.

For each data set, a study was carried out to establish a consensus between the value of the Silhouette Coefficient and the number of texts to analyze in order to detect topics, as when discarding all the sentiment-related terms a great amount of documents can be discarded and will not be part of the topic detection process. To this end, seven different thresholds were established. The conclusion was that in order to obtain an equilibrium between the two analyzed parameters, the threshold varies in the range [0.0, 0.3] depending on the case.

Acknowledgements. This research was partially supported by the Andalusian Government (Junta de Andalucía) under projects P11-TIC-7460 and P10-TIC-6109.

References

1. Baccianella, S., Esuli, A., Sebastiani, F.: Sentiwordnet 3.0: an enhanced lexical resource for sentiment analysis and opinion mining. In: LREC, vol. 10, pp. 2200–2204 (2010)
2. Cambria, E., Olsher, D., Rajagopal, D.: Senticnet 3: a common and common-sense knowledge base for cognition-driven sentiment analysis. In: AAAI Conference on Artificial Intelligence (2014)
3. Nadali, S., Murad, M., Abdul Kadir, R.: Sentiment classification of customer reviews based on fuzzy logic, vol. 2, pp. 1037–1044 (2010)

4. Cai, K., Spangler, S., Chen, Y., Zhang, L.: Leveraging sentiment analysis for topic detection. In: IEEE/WIC/ACM International Conference on Web Intelligence and Intelligent Agent Technology, WI-IAT 2008, vol. 1, pp. 265–271, December 2008
5. Lin, C., He, Y., Everson, R., Ruger, S.: Weakly supervised joint sentiment-topic detection from text. IEEE Trans. Knowl. Data Eng. **24**(6), 1134–1145 (2012)
6. Denecke, K.: Using sentiwordnet for multilingual sentiment analysis. In: IEEE 24th International Conference on Data Engineering Workshop, ICDEW 2008, pp. 507–512, April 2008
7. Poria, S., Gelbukh, A., Cambria, E., Yang, P., Hussain, A., Durrani, T.: Merging senticnet and wordnet-affect emotion lists for sentiment analysis. In: 2012 IEEE 11th International Conference on Signal Processing, vol. 2, pp. 1251–1255, October 2012
8. Kontopoulos, E., Berberidis, C., Dergiades, T., Bassiliades, N.: Ontology-based sentiment analysis of Twitter posts. Expert Syst. Appl. **40**(10), 4065–4074 (2013)
9. Prabowo, R., Thelwall, M.: Sentiment analysis: a combined approach. J. Inf. **3**(2), 143–157 (2009)
10. Rajnish, R.: Fuzzy aspects in sentiment analysis and opinion mining. Int. J. Innov. Res. Sci. Eng. Technol. **5**, 7750–7755 (2016)
11. Yadav, S., Tayal, D.K.: Word level sentiment analysis using fuzzy sets. Int. J. Adv. Sci. Technol. **54**, 73–78 (2015)
12. Rahmath P.H., Ahmad, T.: Fuzzy based sentiment analysis of online product reviews using machine learning techniques. Int. J. Comput. Appl. **99**(17), 9–16 (2014)
13. Dragoni, M., Tettamanzi, A.G.B., da Costa Pereira, C.: Propagating and aggregating fuzzy polarities for concept-level sentiment analysis. Cogn. Comput. **7**(2), 186–197 (2015)
14. Toutanova, K., Klein, D., Manning, C.D., Singer, Y.: Feature-rich part-of-speech tagging with a cyclic dependency network. In: Proceedings of the 2003 Conference of the North American Chapter of the Association for Computational Linguistics on Human Language Technology, NAACL 2003, vol. 1, pp. 173–180. Association for Computational Linguistics, Stroudsburg (2003)
15. Finkel, J.R., Grenager, T., Manning, C.: Incorporating non-local information into information extraction systems by gibbs sampling. In: Proceedings of the 43rd Annual Meeting on Association for Computational Linguistics, ACL 2005, pp. 363–370. Association for Computational Linguistics, Stroudsburg (2005)
16. Agirre, A.G., Laparra, E., Rigau, G.: Multilingual central repository version 3.0. In: Proceedings of the Eight International Conference on Language Resources and Evaluation (LREC 2012), Istanbul, Turkey. European Language Resources Association (ELRA), May 2012
17. Magnini, B., Cavaglia, G.: Integrating subject field codes into WordNet. In: LREC. European Language Resources Association (2000)
18. Manning, C.D., Raghavan, P., Schütze, H.: Introduction to Information Retrieval. Cambridge University Press, New York (2008)
19. Navigli, R., Ponzetto, S.P.: BabelNet: the automatic construction, evaluation and application of a wide-coverage multilingual semantic network. Artif. Intell. **193**, 217–250 (2012)
20. Andrea, E., Fabrizio, S.: SENTIWORDNET: a publicly available lexical resource for opinion mining. In: Proceedings of the 5th Conference on Language Resources and Evaluation (LREC 2006), pp. 417–422 (2006)
21. Rousseeuw, P.: Silhouettes: a graphical aid to the interpretation and validation of cluster analysis. J. Comput. Appl. Math. **20**(1), 53–65 (1987)

Fuzzy Association Rules Mining Using Spark

Carlos Fernandez-Bassso[1], M. Dolores Ruiz[2]([✉]), and Maria J. Martin-Bautista[1]

[1] Computer Science and A.I. Department, CITIC-UGR, University of Granada,
Granada, Spain
{cjferba,mbautis}@decsai.ugr.es
[2] Computer Engineering Department, University of Cádiz, Cádiz, Spain
mariadolores.ruiz@uca.es

Abstract. Discovering new trends and co-occurrences in massive data is a key step when analysing social media, data coming from sensors, etc. Traditional Data Mining techniques are not able, in many occasions, to handle such amount of data. For this reason, some approaches have arisen in the last decade to develop parallel and distributed versions of previously known techniques. Frequent itemset mining is not an exception and in the literature there exist several proposals using not only parallel approximations but also Spark and Hadoop developments following the MapReduce philosophy of Big Data.

When processing fuzzy data sets or extracting fuzzy associations from crisp data the implementation of such Big Data solutions becomes crucial, since available algorithms increase their execution time and memory consumption due to the problem of not having Boolean items. In this paper, we first review existing parallel and distributed algorithms for frequent itemset and association rule mining in the crisp and fuzzy case, and afterwards we develop a preliminary proposal for mining not only frequent fuzzy itemsets but also fuzzy association rules. We also study the performance of the proposed algorithm in several datasets that have been conveniently fuzzyfied obtaining promising results.

Keywords: Big data algorithms · Fuzzy frequent itemset
Fuzzy association rules · Data Mining · Apriori

1 Introduction

The vast amounts of data generated, stored and analysed by companies, and by extension by private users, has given rise to a new phenomenon known as Big Data. Every day millions of tweets are published on Twitter, countless messages are sent via messaging apps and multitudes of comments are generated in online shops. In addition to this, every year more buildings are summed to the "smart sensored" fashion to collect data and use it in their daily performance to be more efficient. The necessity of constantly extracting information from all the gathered

© Springer International Publishing AG, part of Springer Nature 2018
J. Medina et al. (Eds.): IPMU 2018, CCIS 854, pp. 15–25, 2018.
https://doi.org/10.1007/978-3-319-91476-3_2

data is a fact, and the Big Data philosophy using MapReduce framework enables it. In particular, Data Mining techniques are currently under development to benefit from this new framework [3,5,11,12].

One important technique, often employed for exploratory analysis, is that of association rules. They have the form of implications $A \rightarrow B$, which represent the joint co-occurrence of A and B. However, many of the data to be analysed have a nature that is difficult to represent, such as natural language texts. Beside this, when discovering associations between numerical variables we have to take care when data is discretized since final results can vary a lot depending on how the ranges are defined [17]. To better represent this kind of data, Fuzzy Sets theory [26] has proved to be a good option, having as a result fuzzy databases where we can search for fuzzy association rules [6].

In this paper we propose a solution to perform this analysis. There are various methods for mining fuzzy association rules which enable us to analyse and extract interesting information from datasets. These methods run into problems when they are used to analyse vast amounts of data, becoming less efficient at processing and analysis. To this end, we propose a new technique for mining fuzzy association rules that enables the processing of big amounts of data. We have implemented it using Spark [8] which enable faster memory operations than Hadoop [25] since it allows in-memory computations. The results of our experiments show that this method improves the efficiency of the algorithm, with respect to traditional techniques, in terms of time and memory when the number of transactions increases. However when the number of items increases, the algorithm does not offer in all cases significant efficiency improvements in terms of execution time compared to traditional methods. Nonetheless, thanks to the fact that the Big Data techniques offer greater memory capacity, substantial improvements can be achieved when memory problems arise in the generation of the item combinations to be analysed in massive datasets.

The paper is organized as follows: Sect. 2 reviews the literature to show how Big Data technologies can improve existent Data Mining algorithms. Section 3 introduces the measures and methods employed to mine fuzzy association rules. Next section presents the BDFARE algorithm developed for mining fuzzy association rules employing Big Data technologies. Section 5 shows the experiments and results obtained, prior to concluding the paper in Sect. 6.

2 Preliminary Concepts and Related Work

In the literature there are several approaches for mining frequent itemsets using Big Data techniques. The most famous framework, called *MapReduce* was designed by Google in 2003. Since then, there have been proposed several ways to perform association rule analysis with some minor changes. The MapReduce framework bases in two different functions to distribute the computation. On one hand, the *Map()* function transforms data into *(key, value)* pairs according to some criteria, and on the other hand, the *Reduce()* function aggregates the lists of key-value pairs sharing the same key to obtain a piece of processed data.

There are two different frameworks for distributed processing of data: Hadoop and Spark. Some of the methods that are already implemented in these platforms include RandomForest and K-means (clustering) within the official Spark MLlib library [8]. These methods have achieved substantial improvements compared to traditional forms of implementation in the sense that we can now make full use of our cluster, obtaining thus substantial improvements compared to traditional techniques and more scalable algorithms. In particular, in Spark it is included the PFP (Parallel FP-Growth) which is a distributed version of FP-Growth to obtain the frequent itemsets of higher level exceeding the minimum support threshold [18].

In addition to this, we can find in the literature other proposals for mining frequent itemsets using MapReduce techniques for the non-fuzzy case. We can highlight some approaches implementing Apriori extensions using hadoop: [9,10,19]. As mentioned, Spark accelerates the performance versus Hadoop implementations since it makes computations in memory. In addition to this, the algorithms proposed in [9,10,19] search directly in the data the itemsets instead of using other data structures, e.g. trees, hash tries or hash tables, which can decrease time execution as concluded in a study made in [24]. Spark frameworks for Apriori extensions can be found in [21,22]. The R-Apriori and YAFIM algorithms proposed in [21,22] respectively, are very similar to the non-fuzzy phase for each α-cut of our approach but they make a loop to search k-itemsets inside the distributed process using a hash tree while we make the MapReduce for every k-itemset using a hash table.

To the best of our knowledge there is only one work presenting how to discover fuzzy association rules employing the MapReduce framework. This work [14] is based on an extension to the fuzzy case of the Count Distribution algorithm [13,15].

3 Fuzzy Association Rules

Association rules were formally defined for the first time by Agrawal et al. [1]. The problem consists in discovering implications of the form $X \to Y$ where X, Y are subsets of items from $I = i_1, i_2, ..., i_m$ fulfilling that $X \cap Y = \emptyset$ in a database formed by a set of n transactions $D = t_1, t_2, ..., t_n$ each of them containing subsets of items from I. X is usually referred as the antecedent and Y as the consequent of the rule.

The problem of discovering association rules is divided into two sub-tasks:

- Finding all the sets above the minimum support threshold, where support is provided by the percentage of transactions in the set. These sets are known as frequent sets.
- On the basis of the frequent sets are found, rules are discovered as those exceeding the minimum threshold for confidence or another measurement of interestingness generally given by the user.

However, the nature of the data can be diverse and can come described in numerical, categorical, imprecisely, etc. In the case of numerical elements, a first approximation could be to categorise them so that, for example, the height of a person may be given by a range to which it belongs, as for instance [1.70, 1.90]. However, depending on how these intervals are defined, the obtained results may vary a lot. To avoid this, the use of linguistic labels such as "high" represented by a fuzzy set is a good option to represent the height of a person having at the same time a meaningful semantic to the user. Beside this, we may also have a dataset with imprecise knowledge where ordinary crisp methods cannot be directly applied.

To deal with this kind of data we introduce the concept of fuzzy transaction and fuzzy association rule defined in [4,6].

Definition 1. *Let I be a set of items. A fuzzy transaction, t, is a non-empty fuzzy subset of I in which the membership degree of an item $i \in I$ in t is represented by a number in the range [0, 1] and denoted by $t(i)$.*

By this definition a crisp transaction is a special case of fuzzy transaction. We denote by $\tilde{D}$ a database consisting in a set of fuzzy transactions. For an itemset, $A \subseteq I$, the degree of membership in a fuzzy transaction t is calculated as the minimum of the membership degree of all its items:

$$t(A) = \min_{i \in A} t(i). \tag{1}$$

Then, a fuzzy association rule $A \to C$ is satisfied in $\tilde{D}$ if and only if $t(A) \leq t(C)$ $\forall t \in \tilde{D}$, that is, the degree of satisfiability of C in $\tilde{D}$ is greater than or equal to the degree of satisfiability of A for all fuzzy transactions t in in $\tilde{D}$.

Using this model the support and confidence measures are defined using a semantic approach based on the evaluation of quantified sentences as proposed in [4,6]. Using the GD-method [6] and the quantifier $Q_M(x) = x$ the support of a fuzzy rule $A \to B$ results:

$$FSupp(A \to B) = \sum_{\alpha_i \in \Lambda(A \cap B)} (\alpha_i - \alpha_{i-1}) \frac{|(A \cap B)_{\alpha_i}|}{|\tilde{D}|} \tag{2}$$

where $\Lambda(A \cap B) = \{\alpha_1, \alpha_2, \ldots, \alpha_p\}$ with $\alpha_i > \alpha_{i+1}$ and $\alpha_{p+1} = 0$ is a set of α-cuts. In the previous formula, by abuse of notation, we consider the associated fuzzy set to a set of items, i.e. X_{α_i} represents the α-cut of the fuzzy set derived from the itemset X, which is the fuzzy set with membership degree $\mu_X(t) = t(X) = \min_{i \in X} t(i)$ where $X \subset I$. Note that the elements of the fuzzy set derived from X are the transactions.

Analogously the confidence is computed as follows:

$$FConf(A \to B) = \sum_{\alpha_i \in \Lambda(A \cap B)} (\alpha_i - \alpha_{i-1}) \frac{|(A \cap B)_{\alpha_i}|}{|A_{\alpha_i}|} \tag{3}$$

Employing support and confidence measures and setting appropriated thresholds for them fuzzy association rules can be discovered. In [7] it is proposed a parallelization of the computation of $FSupp$ and $FConf$ using a set of predefined α-cuts. Note that considering a sufficiently dense set of α-cuts in the unit interval, the obtained measure will be a good approximation of the real measure that should consider every $\alpha \in [0, 1]$ appearing in the dataset. This is the main idea of our proposal for mining fuzzy association rules using MapReduce. Firstly, the algorithm developed using MapReduce is applied repeatedly for every α-cut. The final step consists in applying a MapReduce phase which aggregates the results using the previous formulas for support and confidence.

4   BDFARE Algorithm

Traditional algorithms have some problems when dealing with large amounts of data as they make multiple scans of the whole database. This means that the execution time increases with the number of transactions. In our research we have used Spark to improve Apriori algorithm for mining fuzzy association rules as follows.

The data is stored in a Big Data architecture, for which we used Hadoop (which allows for replication through its HDFS - Hadoop Distributed File System) and enables distributed processing using Spark.

This new algorithm, called BDFARE (Big Data Fuzzy Association Rules Extraction) algorithm, is based on two phases. The first one involves loading the dataset and calculating how often each item appears in the set of transactions using the Map and Reduce functions. In Fig. 1 we can see an example of first phase with the value of $\alpha = 0.5$. In this example, the $Map()$ function, in particular it is used the $FlatMap()$ function, returns only the items with support higher than 0.5 (column in the middle only contains items with membership values ≥ 0.5). After that, a re-counting of the obtained items is done using the $Reduce()$ function.

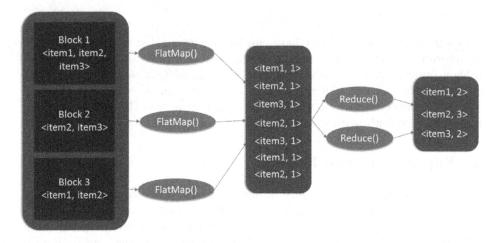

Fig. 1. First phase of BDFARE

This is followed by the second phase, in which we extract the size-k itemsets (see Fig. 2). For this task we used a function that returns the candidate k-itemsets from the frequent items. As a result, a list of pairs of the type $<itemset, degree\_of\_membership>$ is saved in a global variable representing the candidate list. This enables a faster access to the list. After this, the $map()$ function returns, as in the previous phase, the counts of the candidates, having a pair $<itemset, 1>$ when the membership value of the itemset is higher than the value of the α considered in that iteration. In this way, the algorithm calculates in each step all the cardinalities necessary for the computation of the final support and confidence measures. These two phases will be repeated until no new k-itemsets can be found.

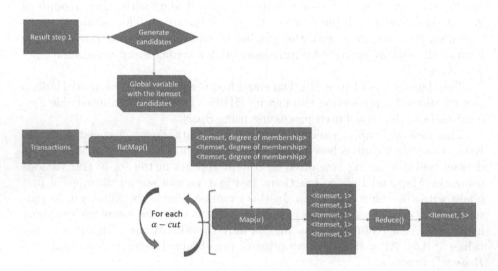

Fig. 2. Second phase of BDFARE

As mentioned in Sect. 2, we have employed a hash table as the central data structure to accelerate the searching of itemsets. If we use instead of the hash table a linear search at each node it will result in an increase of time. In [24] a comparison between the different data structures was made concluding that the hash table outperformed among the three data structures (hash tree, trie and hash table) for both real-life and synthetic datasets.

5   Experiments and Results

In order to check the performance of our proposal we have carried out several experiments to compare running time using the non-distributed version of Apriori for discovering fuzzy association rules with the distributed proposal using BDFARE algorithm. Our aim is to study the behaviour of the algorithm with and without Big Data techniques. To this end, we applied the algorithms in several fuzzy transactional datasets and we study their running time according to different parameters: the number of items and the number of transactions of the datasets.

Three different datasets have been considered from the UCI machine learning repository[1] where some attributes have been conveniently fuzzyfied as described in [23]. The German dataset consists of transactions about credits offered by a german bank. Three variables were fuzzyfied: amount of the credit, its duration and the age of the person who owns the credit. The Autompg dataset consists of several attributes about cars. In this case, the continuous attributes were fuzzified using the following linguistic labels: low, medium and high. The Bank dataset contains data about marketing campaigns of a Portuguese banking institution. In this dataset we have fuzzified their continuous attributes by defining a suitable fuzzy partition according to the semantics of the attribute (description of the fuzzy sets employed can be found in [23]).

The final datasets used in the experiments have been replicated in order to obtain larger datasets to prove the performance of the algorithm in extreme situations. Their original size can be found in Table 1. Actually, there is not any large fuzzy dataset available in open data repositories, but we plan to apply the algorithm to data collected during a time period from sensored buildings. These sensors give numerical values from a continuous scale that are very close among them (e.g. 25° and 25.2°). In this case, the building operators are more interested in obtaining patterns relating temperatures such as "warm", "cold", "very cold", etc. that can be represented by convenient fuzzy sets instead of using intervals that may divide very close data in two different intervals.

Table 1. Datasets

Fuzzy database	Transactions	Items
German	1000	79
Autompg	398	39
Bank	45211	112

5.1 Results

The experimental evaluation have been made in an computer architecture consisting of a cluster comprised of three processing units with Intel Xeon processors with 4 cores at 2.2 GHz, while the traditional algorithm was executed on one of these clusters. We have performed several experiments with different thresholds values, but we show here the results for minimum support equal to 0.2 and minimum confidence equal to 0.8. The important thing here is to set the same thresholds for distributed and non-distributed approaches since we are

---

[1] http://archive.ics.uci.edu/ml/.

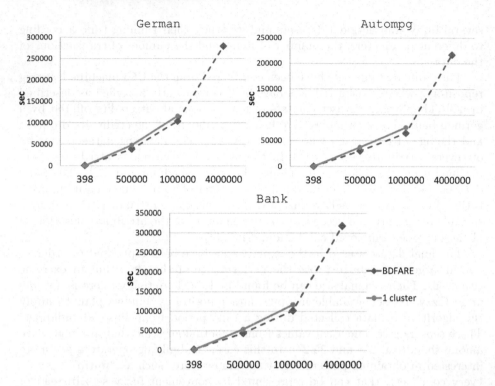

Fig. 3. Performance of BDFARE vs non-distributed algorithm when the quantity of transactions increases

more interested here in observing the performance (time and memory) of both approaches depending on the number of transactions and the number of items.

We began by studying the behaviour of the algorithms (distributed and non-distributed) when the number of transactions increases. To this end, they were run as subsets of the whole datasets in order to observe the behaviour with regard to the number of transactions. Figure 3 shows the behaviour of the algorithm when the quantity of transactions increases in each of the datasets (the number of transactions has been replicated till obtaining four millions). It can be observed that the traditional algorithm performed worst than the BDFARE algorithm, as expected.

To be exhaustive, with the datasets consisted of 0.5 million and 1 million transactions, we achieved time savings in average of 12% and 18% respectively compared to the non-distributed version. We can also see in this graph that the traditional algorithm did not complete its execution for the dataset containing 4 million transactions due to an error resulting from the lack of memory. By contrast, BDFARE completed the execution thanks to the use of Big Data techniques, which allow us to use the memory of the three nodes, thereby obtaining a higher capacity.

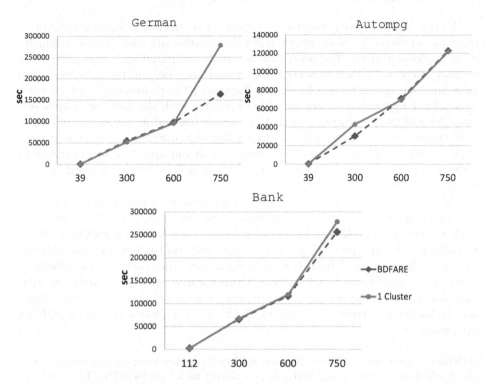

Fig. 4. Performance of BDFARE vs non-distributed algorithm when the quantity of items increases

Figure 4 shows the running time when the number of items increases in the three datasets. In this graph we observe that the execution time using BDFARE does not offer substantial improvements in some cases compared to the traditional algorithm. This is because the Apriori algorithm explores all possible item combinations and in each of these explorations it consults the dataset. Additionally, with a small number of items BDFARE does not achieve always more efficient executions due to the time consumed in the planning of the jobs, necessary when distributing data. However when the number of transactions increases, the performance of the BDFARE algorithm tends to improve the non-distributed approach.

6   Conclusions and Future Research

As we have seen, this paper has focused on the study of one of the most commonly used techniques in data mining, association rules, which allows to extract co-occurrence patterns from datasets. The algorithms proposed traditionally for mining association rules fail when analysing massive datasets because the process results in very high computational costs and its efficiency decreases when the dataset grows.

To this end we have presented an extension to Apriori algorithm for mining fuzzy association rules using Spark, a Big Data framework which enables MapReduce implementations. The algorithm has been compared with non-distributed version of the algorithm showing improvements not only in terms of execution time and but also in terms of memory, improving the processing capacity, since some of the experiments were not able to process with the non-distributed version. An additional advantage is that our algorithm and its performance can be easily improved even further just by expanding our processing system with more clusters (computation nodes). This allows to scale our approach in external data centers or in cloud systems such as AWS (Amazon Web Services), another great advantage of Big Data technology.

As regards future research, we want to implement more efficient approaches [16] that have been proved that performs quite well in the non-distributed case such as the Apriori-TID [2], FP-Growth [20] or ECLAT [27] algorithms. Additionally, we plan to apply the presented approach and the new implementations to sensor data collected from several buildings in order to study the efficiency patterns relating indoor and outdoor temperatures, HVAC (Heating, ventilation, air-conditioning) set points and energy consumptions. In this case, fuzzy sets are suitable to represent understandable value ranges for the users, building operators, etc.

Acknowledgements. The research reported in this paper was partially supported by the Andalusian Government (Junta de Andalucía) under projects P11-TIC-7460 and the Spanish Ministry for Economy and Competitiveness by the project grant TIN2015-64776-C3-1-R.

References

1. Agrawal, R., Imielinski, T., Swami, A.: Mining associations between sets of items in large databases. In: ACM-SIGMOD International Conference on Data, pp. 207–216 (1993)
2. Agrawal, R., Srikant, R.: Fast algorithms for mining association rules in large databases. In: Proceedings of the Twentieth International Conference on Very Large Databases, Santiago, Chile, pp. 487–499 (1994)
3. Anastasiu, D.C., Iverson, J., Smith, S., Karypis, G.: Big data frequent pattern mining. In: Aggarwal, C.C., Han, J. (eds.) Frequent Pattern Mining, pp. 225–259. Springer, Cham (2014). https://doi.org/10.1007/978-3-319-07821-2_10
4. Berzal, F., Delgado, M., Sánchez, D., Vila, M.A.: Measuring accuracy and interest of association rules: a new framework. Intell. Data Anal. **6**(3), 221–235 (2002)
5. del Río, S., López, V., Benítez, J.M., Herrera, F.: On the use of MapReduce for imbalanced big data using random forest. Inf. Sci. **285**, 112–137 (2014). Processing and Mining Complex Data Streams
6. Delgado, M., Marín, N., Sánchez, D., Vila, M.A.: Fuzzy association rules: general model and applications. IEEE Trans. Fuzzy Syst. **11**(2), 214–225 (2003)
7. Delgado, M., Ruiz, M.D., Sánchez, D., Serrano, J.M.: A formal model for mining fuzzy rules using the RL representation theory. Inf. Sci. **181**(23), 5194–5213 (2011)
8. Meng, X., et al.: MLlib: machine learning in apache spark. arXiv preprint: abs/1505.06807 (2015)

9. Farzanyar, Z., Cercone, N.: Accelerating frequent itemset mining on the cloud: a MapReduce-based approach. In: IEEE 13th International Conference on Data Mining Workshops, pp. 592–598 (2013)
10. Farzanyar, Z., Cercone, N.: Efficient mining of frequent itemsets in social network data based on MapReduce framework. In: Proceedings of the 2013 International Conference on Advances in Social Networks Analysis and Mining (ASONAM 2013), pp. 1183–1188 (2013)
11. Fernández, A., Carmona, C.J., del Jesus, M.J., Herrera, F.: A view on fuzzy systems for big data: progress and opportunities. Int. J. Comput. Intell. Syst. **9**, 69–80 (2016)
12. Fernandez-Basso, C., Ruiz, M.D., Martin-Bautista, M.J.: Extraction of association rules using big data technologies. Int. J. Des. Nat. Ecodyn. **11**(3), 178–185 (2016)
13. Gabroveanu, M., Cosulschi, M., Constantinescu, N.: A new approach to mining fuzzy association rules from distributed databases. Ann. Univ. Bucharest **54**, 3–16 (2005)
14. Gabroveanu, M., Cosulschi, M., Slabu, F.: Mining fuzzy association rules using MapReduce technique. In: International Symposium on INnovations in Intelligent SysTems and Applications, INISTA, pp. 1–8 (2016)
15. Gabroveanu, M., Iancu, I., Cosulschi, M., Constantinescu, N.: Towards using grid services for mining fuzzy association rules. In: Proceedings of the 1st East European Workshop on Rule-Based Applications, RuleApps, pp. 507–513 (2007)
16. Hipp, J., Güntzer, U., Nakhaeizadeh, G.: Algorithms for association rule mining - a general survey and comparison. ACM SIGKDD Explor. Newsl. **2**(1), 58–64 (2000)
17. Hüllermeier, E., Yi, Y.: In defense of fuzzy association analysis. IEEE Trans. Syst. Man Cybern. Part B Cybern. **37**(4), 1039–1043 (2007)
18. Li, H., Wang, Y., Zhang, D., Zhang, M., Chang, E.Y.: PFP: parallel FP-growth for query recommendation. In: Proceedings of the 2008 ACM Conference on Recommender Systems, pp. 107–114. ACM (2008)
19. Li, N., Zeng, L., He, Q., Shi, Z.: Parallel implementation of Apriori algorithm based on MapReduce. In: Proceedings of the 2012 13th ACIS International Conference on Software Engineering, Artificial Intelligence, Networking and Parallel/Distributed Computing, SNPD 2012, pp. 236–241. IEEE Computer Society, Washington, D.C. (2012)
20. Pei, J., Yin, Y., Mao, R., Han, J.: Mining frequent patterns without candidate generation: a frequent-pattern tree approach. Data Mining Knowl. Discov. **8**(1), 53–87 (2004)
21. Qiu, H., Gu, R., Yuan, C., Huang, Y.: YAFIM: a parallel frequent itemset mining algorithm with spark. In: 2014 IEEE International Parallel & Distributed Processing Symposium Workshops (IPDPSW), pp. 1664–1671. IEEE (2014)
22. Rathee, S., Kaul, M., Kashyap, A.: R-Apriori: an efficient Apriori based algorithm on spark. In: Proceedings of the PIKM 2015, Melbourne, VIC, Australia. ACM (2015)
23. Ruiz, M.D., Sánchez, D., Delgado, M., Martin-Bautista, M.J.: Discovering fuzzy exception and anomalous rules. IEEE Trans. Fuzzy Syst. **24**(4), 930–944 (2016)
24. Singh, S., Garg, R., Mishra, P.K.: Performance analysis of Apriori algorithm with different data structures on hadoop cluster. Int. J. Comput. Appl. **128**(9), 45–51 (2015)
25. White, T.: Hadoop: The Definitive Guide, 4th edn. O'Reilly, Sebastopol (2015)
26. Zadeh, L.A.: Fuzzy sets. Inf. Control **8**, 338–353 (1965)
27. Zaki, M.J.: Scalable algorithms for association mining. IEEE Trans. Knowl. Data Eng. **12**(3), 372–390 (2000)

A Typology of Data Anomalies

Ralph Foorthuis[(⊠)] [iD]

UWV, La Guardiaweg 116, 1040 HG Amsterdam, The Netherlands
ralph.foorthuis@uwv.nl

Abstract. Anomalies are cases that are in some way unusual and do not appear to fit the general patterns present in the dataset. Several conceptualizations exist to distinguish between different types of anomalies. However, these are either too specific to be generally applicable or so abstract that they neither provide concrete insight into the nature of anomaly types nor facilitate the functional evaluation of anomaly detection algorithms. With the recent criticism on 'black box' algorithms and analytics it has become clear that this is an undesirable situation. This paper therefore introduces a general typology of anomalies that offers a clear and tangible definition of the different types of anomalies in datasets. The typology also facilitates the evaluation of the functional capabilities of anomaly detection algorithms and as a framework assists in analyzing the conceptual levels of data, patterns and anomalies. Finally, it serves as an analytical tool for studying anomaly types from other typologies.

Keywords: Anomalies · Outliers · Deviants · Typology · Data analysis
Classification · Pattern recognition · Exploratory analytics · Machine learning
Data mining

1 Introduction

Anomalies are cases that are in some way unusual and do not appear to fit the general patterns present in the dataset [1–3]. Such cases are often also referred to as *outliers*, *novelties* or *deviant observations* [3, 4]. Anomaly detection (AD) is the process of analyzing the dataset to identify these deviant cases. Anomaly detection can be used for various goals, such as fraud detection, data quality analysis, security scanning, process and system monitoring, and data cleansing prior to training statistical models [1–4].

Several ways to distinguish between different kinds of anomalies have been presented in the literature. These conceptualizations, however, are either only relevant for specific situations or too abstract to provide a clear and concrete understanding of anomalies (see Sects. 2 and 4). This paper therefore presents a *typology of anomalies* that offers a theoretical and tangible *understanding* of the nature of different types of anomalies, assists researchers with *evaluating* the functional capabilities of their anomaly detection algorithms, and as a framework aids in *analyzing*, i.a., the conceptual levels of data and anomalies. A preliminary version has been presented briefly in [1, 5] to evaluate an unsupervised non-parametric AD algorithm. This paper extends that initial typology and discusses its theoretical properties in more depth.

© Springer International Publishing AG, part of Springer Nature 2018
J. Medina et al. (Eds.): IPMU 2018, CCIS 854, pp. 26–38, 2018.
https://doi.org/10.1007/978-3-319-91476-3_3

A clear understanding of the types of anomalies that can be encountered in datasets is relevant for several reasons. First, it is important in statistics, data science, machine learning, analytics and knowledge-based systems to have a fundamental and tangible understanding of anomalies, of the various anomaly types that exist, and of their defining characteristics. In this context, the typology presented here not only helps in theoretically understanding the nature of data and (deviations from) patterns, but also provides a functional evaluation framework that enables researchers to demonstrate which anomaly type(s) their AD algorithms are able to detect. Second, with the recent criticism on 'opaque' and 'black box' analytics methods that may result in unfair outcomes [6, 7], it has become clear that it is undesirable to have algorithms and analysis results that lack transparency and cannot be interpreted meaningfully. This is especially true for AD algorithms, as these may be used to identify and act on 'suspicious' cases. Although the typology presented here does not make the algorithms themselves more transparent, a clear understanding of (the types of) anomalies and their properties helps in making the results of data analyses understandable and transparent. Third, even if statistical and machine learning algorithms are functionally transparent and understandable, the implementations of these algorithms – and the knowledge-based systems they are part of – may be done poorly or simply fail due to overly complex real-world settings [8, 9]. The results of data analyses conducted in practice may thus prove to be incorrect and unpredictable. A deep understanding of anomalies is therefore needed to determine whether detected cases indeed constitute true anomalies. This is especially relevant for unsupervised AD algorithms, as these are often not used with known labelled data. Finally, the *no free lunch* theorem, which posits that no single algorithm will show superior performance in all problem domains, also holds for anomaly detection [10–12]. Individual AD algorithms are generally not able to detect all types of anomalies and will differ in their performance. In addition, more complex algorithms do not necessarily perform better than relatively simple ones. The typology assists researchers in making transparent which algorithms are able to detect what types of anomalies to what degree.

This paper proceeds as follows. Section 2 discusses related research. Section 3 presents the typology of anomalies. Section 4 discusses the properties of the typology and compares it with other research. Finally, Sect. 5 is for conclusions.

2 Related Work

The literature acknowledges various ways to distinguish between types of anomalies. In [3] a distinction is made between a *weak outlier* (noise that can be attributed to the statistical variation of a random variable) and a *strong outlier* (a true anomaly that may be generated by a mechanism different from the one generating the normal cases). The general typology presented in [13] differentiates between three types. A *point anomaly* refers to one or several individual cases that are deviant with respect to the rest of the data. A *contextual anomaly* appears normal at first sight, but is deviant when an explicitly mentioned context is taken into account [also see 14]. For example, when a measured temperature value is not low in general, but only in the summer season. Finally, a *collective outlier* refers to a group of data points that belong together and, as a group, deviates from the rest of the data. This requires 'dependent data', in which individual cases (rows) are related by e.g. a time, space or identification attribute. A unique sequence of events is an example of such a collective outlier.

Several more specific and concrete typologies are also known, especially from time series or sequence analysis. In [15] several within-sequence types are presented. An *additive outlier* in this context is an isolated spike during a short period, whereas a *transitory change outlier* is a spike that requires some time to disappear. A *level shift outlier* is a sudden but structural change to a higher or lower value level, whereas an *innovational outlier* may show shifts in both the trend and the seasonal pattern. The taxonomy presented in [16] focuses on between-sequences anomalies and makes a distinction between *isolated outliers, shift outliers, amplitude outliers* and *shape outliers*. Even more types can be acknowledged, such as *deviant cycle anomalies*. Figure 5 in the Discussion section illustrates several of these types. Another example of a specific typology is known from regression analysis, in which it is common to distinguish between *outliers, high-leverage points* and *influential points* [2, 17].

The above mentioned typologies, summarized in Fig. 1, are either too general and too abstract to provide a clear and concrete understanding of anomaly types, or feature well-defined types that are only relevant for a specific purpose (such as time series analysis or regression modeling). This paper therefore presents a typology of anomalies that is general but still offers a meaningful, tangible and useful description of the various types of anomalies. Such a typology has significant value for both practitioners and researchers. It seems that a similar typology, grounded in the fundamental dimensions of data types and attribute relationships, has not been published before (note that many typologies regarding anomaly detection *techniques* do exist).

Ref.	G/S	Anomaly types	Use of Explicit Dimensions
[1,5,18]	G	Extreme value anomaly, Multidimensional numerical anomaly, Sparse class anomaly, Multidimensional mixed data anomaly	Nature of data (c.f. Types of data), Number of interacting attributes (c.f. the Cardinality of relationship)
[3]	G	Weak outlier, Strong outlier	None (random noise is a candidate)
[13]	G	Point anomaly, Contextual anomaly, Collective anomaly	None
[2, 17]	S	Outliers, High-leverage points, Influential points	None
[15]	S	Additive outlier, Transitory change outlier, Level shift outlier, Innovational outlier	None
[16]	S	Isolated outliers, Shift outliers, Amplitude outliers, Shape outliers	None

Fig. 1. Overview of typologies and anomaly types (G/S refers to general vs. specific)

3 Typology of Anomalies

This section presents the general typology of anomalies that offers a theoretical, detailed and concrete understanding of the types of anomalies that can be encountered in datasets. It also gives researchers a tool to evaluate which types of anomalies can be detected by a given AD algorithm, assists in analyzing the conceptual levels of patterns and anomalies, and aids in studying anomaly types from other typologies.

The typology uses two dimensions, each of which describes a fundamental aspect of the nature of data, to distinguish between anomaly types. The first dimension represents the **types of data** involved in describing the behavior of the cases. This refers to the data types of the attributes (i.e. variables) that are involved in the anomalous character of a deviant case and thus have to be handled appropriately during the analysis in order for it to be detected. The data types are:

- *Continuous*: The variables that capture the anomalous behavior are all numeric in nature. Examples of such variables are age, height and temperature.
- *Categorical*: The variables that capture the anomalous behavior all represent codes or class values. This includes binomial and multinomial attributes. Examples of such variables are gender, country, color and animal species.
- *Mixed*: The variables that capture the anomalous behavior are both continuous and categorical in nature. At least one attribute of each type is present.

Not all data types acknowledged in so-called multimodal objects are included here. The reason is that e.g. video, audio and free text anomalies can generally be reduced to class- or number-based deviations, or require a very specific analysis.

| | | Types of Data | | |
		Continuous attributes	Categorical attributes	Mixed attributes
Cardinality of Relationship	**Univariate** Described by individual attributes (independence)	Type I Extreme value anomaly	Type II Rare class anomaly	Type III Simple mixed data anomaly
	Multivariate Described by multidimensionality (dependence)	Type IV Multidimensional numerical anomaly	Type V Multidimensional rare class anomaly	Type VI Multidimensional mixed data anomaly

Fig. 2. The typology of anomalies

The second dimension is the **cardinality of relationship** and represents how the various attributes relate to each other when describing anomalous behavior. These attributes, individually or jointly, are responsible for the deviant character of the case:

- *Univariate*: Except for being part of the same set, no relationship between the variables exists to which the anomalous behavior of the deviant case can be attributed. To detect the anomaly, its attributes can therefore be analyzed separately – i.e. the analysis can assume independence between the variables.
- *Multivariate*: The deviant behavior of the anomaly lies in the relationships between its variables. The anomaly can thus not be detected by studying the individual

attributes separately. Variables need to be analyzed jointly in order to take into account their relations, i.e. combinations of values. 'Relationships' should be interpreted broadly here, including (partial) correlations, interactions, collinearity, as well as associations between attributes of different data types.

The preliminary typology presented in [1, 5, 18] is summarized in the first row of Fig. 1. The typology presented in this paper is an updated and extended version. All data types are now treated separately, yielding six basic anomaly types. In addition, the terminology is updated. The new typology is depicted in Fig. 2. The types are illustrated in Figs. 3 and 4 (note: the reader might want to zoom in on a digital screen to see colors, patterns and data points in detail). Figures 3A, B and 4A are simulated datasets, while Fig. 4B depicts real-world data from the Polis Administration, an official register of income data in the Netherlands [1]. The six types of anomalies, which follow naturally and objectively from the two dimensions, are described below.

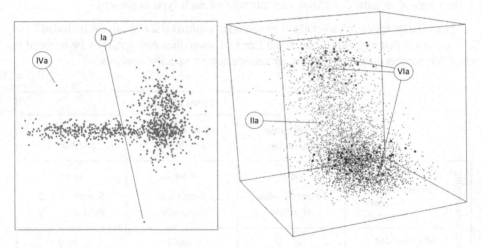

Fig. 3. (A) Set with two numerical variables; (B) Set with three numerical attributes and two categorical attributes (color and size) (Color figure online)

I. **Extreme value anomaly:** A case with an extremely high, low or otherwise rare value for one or multiple individual numerical attributes [cf. 3, 19]. As such cases deviate w.r.t. one or more individual attributes, their anomalous nature does not rely on relationships between attributes. However, the more attributes take on an extremely high, low or rare numerical value, the more anomalous the case is. The two cases with label *Ia* in Fig. 3A are examples, as are the *Ib* cases in Fig. 4B. Traditional univariate statistics typically offers methods to detect this type, e.g. by using a measure of central tendency plus or minus 3 times the standard deviation or the median absolute deviation [3, 13, 17]. These cases are literally 'outliers', as they lie in an isolated region of the numerical space. However, note that this type includes rare cases [cf. 19] and that such low-density values can also be located in the middle of the value range.

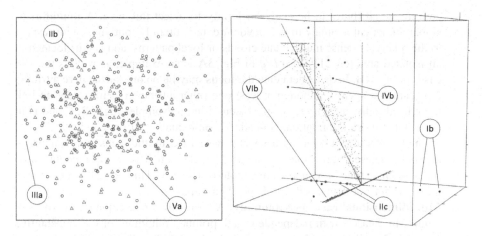

Fig. 4. (A) Set with two numerical attributes and two categorical attributes (color and shape); (B) Polis Administration set with one categorical and three numerical attributes, and large dots representing the 30 most extreme anomalies detected by SECODA [1] (Color figure online)

II. **Rare class anomaly:** A case with an uncommon class value for one or multiple categorical variables. Cases of this type are anomalous w.r.t. one or more individual attributes, so the deviant nature of rare class anomalies does not rely on relationships between attributes. However, like Type I cases, the more attributes take on a rare class value, the more anomalous the case is. The research in [20] deals with this type of anomaly. Case *IIa* in Fig. 3B is a rare class anomaly, being the only green data point in the set. Case *IIb* in Fig. 4A, the only square class, is another example. The rare red and orange colors of Fig. 4B's *IIc* points make for rare class anomalies as well.

III. **Simple mixed data anomaly:** A case that is both a Type I and Type II anomaly, i.e. with at least one extreme value and one rare class. This anomaly type deviates with regard to multiple data types. This requires deviant values for at least two attributes, each anomalous in its own right. These can thus be analyzed separately; analyzing the attributes jointly is unnecessary because the case is not anomalous in terms of a combination of values. However, similar to the other univariate anomaly types, the more attributes take on a rare value, the more anomalous the case in question is. Case *IIIa* in Fig. 4A is an example. Case *IIa* in Fig. 3B would be a Type III anomaly if it had been positioned to the extreme left (at the location of label 'IIa').

IV. **Multidimensional numerical anomaly:** A case that does not conform to the general patterns when the relationship between multiple continuous attributes is taken into account, but which does not have extreme values for any of the individual attributes that partake in this relationship. The anomalous nature of a case of this type lies in the deviant or rare combination of its continuous attribute values, and as such hides in multidimensionality. It therefore requires several continuous attributes to be analyzed jointly to detect this type. A multidimensional numerical anomaly in independent data is literally 'outlying' with respect to the relatively dense multivariate clouds or local patterns, and is thus located in an isolated area [cf. 21]. Case *IVa* in Fig. 3A is an example, as well as the *IVb* cases in Fig. 4B. In dependent data the focus may lie on one substantive attribute (e.g. 'amount spent'), although at least one other attribute is still needed to link the related individual cases. See the discussion on examples Fig. 5B, C and D in Sect. 4 for more information. So-called 'contextual' [13] or 'conditional' [14] anomalies should be seen as a special case of a multidimensional numerical anomaly. These require that the respective contextual or environmental attributes, such as time or location, are denoted explicitly. This explicit denotation is allowed, but not demanded, for Type IV anomalies.

V. **Multidimensional rare class anomaly:** A case with a rare combination of class values. In datasets with independent data points a minimum of two substantive categorical attributes needs to be analyzed jointly to discover a multidimensional rare class anomaly. An example is this curious combination of values from three attributes used to describe dogs: 'MALE', 'PUPPY' and 'PREGNANT'. Another example is case *Va* in Fig. 4A, as it is the only red circle in the set. When dealing with dependent data (see Sects. 2 and 4), the anomaly can also be a deviant combination of class values of a single substantive attribute, but from multiple related cases. Again, an additional attribute, such as time, is still required to link these dependent cases. An example of such a type V anomaly in dependent data is the deviant phase-sequence in Sect. 4.

VI. **Multidimensional mixed data anomaly:** A case with a class or a combination of classes that in itself is not rare in the dataset as a whole, but is only rare in its local pattern or neighborhood (numerical area). The anomalous nature of a case of this type lies in the deviant relationship between its continuous and categorical attributes. As with the other multivariate anomalies, such cases hide in multidimensionality and thus multiple attributes need to be jointly taken into account to identify them. As a matter of fact, multiple data types need to be used, as anomalies of this type per definition are comprised of both numerical and categorical attributes. Cases *VIa* in Fig. 3B are illustrations of a multidimensional mixed data anomaly in independent data, as they are points with a color rarely seen in their respective neighborhoods. This also holds for the *VIb* cases in Fig. 4B, being blue points in an otherwise pink local pattern (or vice versa). Type VI cases can also take the form of second- or higher-order anomalies, with categorical values that are not rare (not even in their neighborhood), but are rare in their combination in that specific area. Here is another way to look at this: a first-order Type VI anomaly can be seen as a *rare class anomaly* in its local neighborhood, while a second- or higher-order Type VI anomaly can be seen as a *multidimensional rare class anomaly* in its local neighborhood. More examples of the different types of anomalies will be provided in the Discussion section. Additional illustrations (including those of higher-order anomalies) can be found in [1].

4   Discussion

The typology presented here offers a clear and tangible definition of the different types of anomalies. As the various figures show, these types lend themselves to be clearly illustrated by visual plots. In addition to providing a clear *understanding* of the different kinds of anomalies that exist, the typology can be used to *evaluate* AD algorithms. This is a relevant contribution because most research publications do not make it very clear which types can be detected by the anomaly detection algorithms presented, even though it is clear that many of those algorithms are incapable of identifying all types [1, 11]. It is therefore advised that researchers use the typology to provide clear insight into the functional capabilities of their AD algorithms by explicitly stating which anomaly type(s) can be detected. This also gives due acknowledgment of the *no free lunch* theorem in an AD context [cf. 10, 11, 12].

Evaluation of Algorithms. Using the typology for algorithm evaluation has more implications than merely stating which types can be detected, since the typology is ideally also used to *create test sets*. AD studies often evaluate algorithms by treating (a sample of) one class in existing datasets as anomalies [22]. However, this is a questionable practice because these classes may actually represent true patterns rather than true deviants, and may be very similar to other classes in the dataset. This latter situation can indeed be observed for several classes in the real-world Polis dataset. Moreover, there is no guarantee that all anomaly types will be present in such a test set. A better approach for creating AD test sets would therefore be to take the typology presented here and insert several instances of each anomaly type in a simulated or real-world dataset. This ensures that the different types of anomalies are present in the set and a thorough evaluation of the algorithm can thus be conducted. Researchers should at least aim to include the most important types, based on the domain or the problem being studied. See [1] for an example of an evaluation.

Local vs. Global Anomalies. The typology presented here also offers a natural way to distinguish between *local* and *global* anomalies. It follows from the typology that the three univariate anomaly types are global anomalies, as these are unusual w.r.t. an individual attribute (possibly several individual attributes, but each attribute is anomalous in its own right). They are anomalous with regard to the entire dataset. When taking all the set's cases into account, extreme value anomalies will always have an extremely low, high or rare value for the given attribute. Rare class anomalies and simple mixed data anomalies likewise have an extremely rare value for the given attribute(s), without any condition and regardless of the other attributes. The three multivariate anomaly types, on the other hand, are only deviant given the categorical condition or the specific numerical area the case in question is located in. This is the result of the fact that the anomalous nature of the case lies in the combination of its attribute values. This is clearly illustrated by the *VIb* cases in Fig. 4B, where the blue cases are not anomalous because they are blue (which is a very normal color in the set), but because they seem to be misplaced between the pink cases. A similar argument holds true for the pink *VIb* cases. In short, the three multivariate anomaly types are situational and therefore local. The three univariate anomaly types represent global anomalies, as they are anomalous regardless of the values of other attributes.

Other Typologies. The typology presented here can be used both for *clarifying* more abstract typologies and for positioning the anomalies of more specific typologies in a *broader framework*. The typology presented in [13] is very general in nature, yielding rather abstract anomaly types. This is made clear by the fact that, given some assumptions, all six anomaly types of this paper's typology can manifest themselves as a point anomaly. A *point anomaly* is simply an individual case that deviates from the rest of the data [13]. The example is given of a very high 'amount spent' in a dataset with credit card transactions. This is exactly what the *extreme value anomaly* in the typology of Fig. 2 is. Another example in [13] concerns isolated cases that are described by two numerical dimensions, of which none has an extremely high or low value. This is therefore this paper's *multidimensional numerical anomaly*. The explanation in [13] does not explicitly state whether point anomalies can also be comprised of categorical data. However, if this type is interpreted in a broad sense, then *rare class anomalies* are also point anomalies because these are unique or rare data points and there is no need for dependent data or an explicitly denoted context (see below). A similar argument can then be made for *multidimensional rare class anomalies*, *multidimensional mixed data anomalies* and *simple mixed data anomalies*, which renders point anomalies a very broad and abstract type indeed. The typology presented in Fig. 2 is thus helpful, and even needed, to obtain a more concrete understanding of how point anomalies can manifest themselves.

The *contextual anomaly* in [13] is only deviant in a specific and explicitly specified context, such as a certain location or time period. This requires relationships between variables, making this a *multidimensional numerical anomaly* (and possibly a *multidimensional rare class anomaly* or *multidimensional mixed data anomaly*) for which the analyst has explicitly specified the contextual variables before running the analysis.

Finally, the *collective anomaly* in [13] refers to a group of cases that, as a combined whole, shows deviant behavior. An example is when individual cases are not deviant in themselves, but only as a group of cases that represents a deviant sequence. The set of red underlined classes in the following phase-sequence can therefore be regarded as such an anomaly:

phase1, phase2, phase3, phase1, phase2, phase3, **phase1. phase3**, phase1, phase2, phase3

In terms of this study's typology this is a *multidimensional rare class anomaly*, in which the combination (sequence) of classes deviates from the regular pattern (the cycle 'phase1, phase2, phase3'). As one can see from the example, the anomaly is comprised of multiple cases in a set with dependent data (related points or rows). If relevant, however, one could abstract from the original individual points and declare the anomaly at the group level (i.e. the cycle), turning this into a *rare class anomaly*.

Collective anomalies in sequence data can be described in more detail both by the typology presented in this paper and by the specific typologies from time series analysis [15, 16]. Additional examples are shown in Fig. 5 and will be discussed below. In time series analysis, the left red spike of Fig. 5A is an *additive outlier*, the right spike a *transitory change outlier* that takes some time to disappear [15]. In terms of this paper's typology, both are *extreme value anomalies* as they have an extremely high respectively low value for a single attribute. The isolated spike in Fig. 5B also constitutes an *additive outlier*. However, this is not an *extreme value anomaly*, as it deviates from the local pattern without exhibiting extreme values from a global perspective. This is therefore a *multidimensional numerical anomaly*, as it requires two numerical attributes to identify the anomaly in the local pattern. Interestingly, the typology of Fig. 2, albeit in principle more abstract than a specific typology dedicated to time series anomalies, is thus able to distinguish between instances of one and the same time series type. The typology's ability to make this more specific distinction is due to its fundamental dimensions: data types and cardinality of relationship.

The red transition of Fig. 5C constitutes a *level shift outlier*. This can be regarded as a collective anomaly because no individual point is anomalous – the deviation lies in the sudden level shift of the sequence. The deviant behavior can only be detected by taking into account both the time and the value variable, making this a *multidimensional numerical anomaly*. However, by first determining the difference between two consecutive cases, this change point detection problem can be turned into a simple search for extreme values. Given the transformed dataset that results from this operation, this would thus be an *extreme value anomaly*, representing a deviant transition – which is now a single case – rather than (a group of) cases from the original set.

The red part of Fig. 5D is another example of a collective anomaly, in the form of a *deviant cycle anomaly*. As this involves numerical data of which the red part – which does not feature extremely high, low or rare individual values – is found to be anomalous when taking into account the entire sequence, this is a *multidimensional numerical anomaly*. Re-ordering the sequence in a random fashion (i.e. discarding the time attribute) would make the anomaly 'disappear', leaving neither a *deviant cycle anomaly* nor a *multidimensional numerical anomaly*. It is similar to Fig. 5B's anomaly, except for the fact that Fig. 5D's anomaly represents a group (cycle) of related cases instead of a single data point. Note that collective anomalies such as those in Fig. 5D can usually only be discovered in sets with dependent data and by specialized algorithms [cf. 13, 22]. As with Fig. 5C, the example of Fig. 5D can be transformed from a *multidimensional numerical anomaly* into a simpler type. The individual cycles could first be detected and classified, after which a *rare class anomaly* can be denoted.

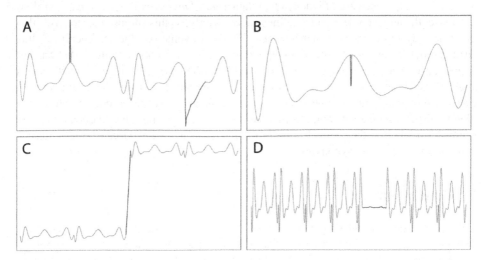

Fig. 5. Four time series with time on the horizontal axis and the anomalies in red (Color figure online)

Conceptual Levels. The examples of Fig. 5C, D and the phase-sequence make clear that one can *abstract from* the original micro-level data points to view the data and anomalies at a higher and somewhat simpler conceptual level. The grouping variables, such as time, location and identification attributes, are generally used for this. In Fig. 5D (and the phases example) the sequence data were also analyzed at the level of a cycle, transforming a Type IV (and Type V) anomaly into a *rare class anomaly*. In terms of the typology of Fig. 2 this implies a change from a multivariate anomaly to a univariate one. In terms of [13] this changes the anomaly from a collective to a point anomaly. In addition to the change in conceptual perspective, this may involve a different AD algorithm or a transformation of the dataset. There can be several reasons to change the conceptual level of a dataset and its anomalies. First, the goal of the analysis may imply a certain conceptual focus. The aim may be to detect anomalous individual data points (e.g. logged events such as login attempts) or aggregated entities (e.g. entire user sessions comprised of multiple actions). A second reason to change the level of a dataset is the fact that some sets may be too big to process. The data reduction obtained by transforming the dataset into a set with aggregated cases may be required to make the analysis more manageable. A third reason concerns the AD algorithms the analyst has at his or her disposal. An advanced algorithm to analyze dependent data may simply not be available, meaning that the analyst first needs to transform the dataset to a format that is suitable for the algorithms at hand.

Terminology. To conclude this discussion, it is worthwhile to re-assess the synonyms mentioned at the beginning of this paper. As stated, the terms *anomaly*, *outlier*, *novelty* and *deviant* are often treated as having an identical meaning. However, in light of the typology and discussion presented here, several of these terms should be defined more clearly. The term *outlier*, from a traditional statistical perspective, refers to observations that literally lie outside the general patterns or dense data clouds. In other words, such cases lie in a numerically isolated region of the space. Given this typology, the term *outlier* can thus best be reserved for *extreme value anomalies, simple mixed data anomalies* and, in the case of independent data, *multidimensional numerical anomalies*. Likewise, the term *novelty* can be defined more strictly, as this should refer to cases that in some way represent new and hitherto unknown events or objects. Therefore, this term can best be applied to situations in which a case represents something that has not happened or been detected before. This could be the case in change point detection analysis, such as in the time series of Fig. 5. Alternatively, a *novelty* could refer to cases discovered with unsupervised or one-class anomaly detection, in which the identified case is not a data point from a pattern that the algorithm has learned before by training on labelled data. Finally, the terms *anomaly* and *deviant* can be regarded as general terms and true synonyms.

5 Conclusion

This paper has presented a general typology of anomalies that offers a concrete understanding of the different anomalies one can encounter in datasets. The typology can also be used to evaluate AD algorithms in a more transparent way. In particular, researchers can utilize it to create test sets that will be used in the evaluation and should report explicitly which types of anomalies can be detected by a given AD algorithm. Furthermore, as a result of its fundamental dimensions, the typology can be used both for clarifying existing typologies that are more abstract in nature [e.g. 13] and for studying the anomalies of specific typologies [e.g. 15] through a more general lens. For some specific, dedicated typologies, this study's typology can even provide deeper insight by proposing meaningful sub-divisions within existing types. Finally, the typology clearly distinguishes between local and global anomalies, and can be used as a framework to analyze the conceptual levels of data and anomalies.

Remarks. The data examples and the R code to analyze them can be downloaded from www.foorthuis.nl. The author thanks Emma Beauxis-Aussalet for her valuable remarks.

References

1. Foorthuis, R.: SECODA: segmentation- and combination-based detection of anomalies. In: Proceedings of the 4th IEEE International Conference on Data Science and Advanced Analytics (DSAA 2017), Tokyo, Japan, pp. 755–764 (2017). https://doi.org/10.1109/dsaa.2017.35
2. Izenman, A.J.: Modern Multivariate Statistical Techniques: Regression, Classification, and Manifold Learning. Springer, New York (2008). https://doi.org/10.1007/978-0-387-78189-1
3. Aggarwal, C.C.: Outlier Analysis. Springer, New York (2013). https://doi.org/10.1007/978-1-4614-6396-2
4. Pimentel, M.A.F., Clifton, D.A., Clifton, L., Tarassenko, L.: A review of novelty detection. Sig. Process. **99**, 215–249 (2014)
5. Foorthuis, R.: Anomaly detection with SECODA. Poster Presentation at the 4th IEEE International Conference on Data Science and Advanced Analytics (DSAA), Tokyo (2017)
6. Mittelstadt, B.D., Allo, P., Taddeo, M., Wachter, S., Floridi, L.: The ethics of algorithms: mapping the debate. Big Data Soc. 3(2), 2053951716679679 (2016)
7. Ziewitz, M.: Governing algorithms: myth, mess, and methods. Sci. Technol. Hum. Values **41**(1), 3–16 (2016)
8. Sculley, D., et al.: Hidden technical debt in machine learning systems. In: Proceedings of NIPS 2015, vol. 2, pp. 2503–2511 (2015)
9. Breck, E., Cai, S., Nielsen, E., Salib, M., Sculley, D.: What's your ML test score? A rubric for ML production systems. In: Proceedings of NIPS 2016 (2016)
10. Clarke, B., Fokoué, E., Zhang, H.H.: Principles and Theory for Data Mining and Machine Learning. Springer, New York (2009). https://doi.org/10.1007/978-0-387-98135-2
11. Janssens, J.H.M.: Outlier selection and one-class classification. Ph.D. Thesis, Tilburg University (2013)
12. Rokach, L., Maimon, O.: Data Mining With Decision Trees: Theory and Applications, 2nd edn. World Scientific Publishing, Singapore (2015)
13. Chandola, V., Banerjee, A., Kumar, V.: Anomaly detection: a survey. ACM Comput. Surv. **41**(3), 15 (2009)
14. Song, X., Wu, M., Jermaine, C., Ranka, S.: Conditional anomaly detection. IEEE Trans. Knowl. Data Eng. **19**(5), 631–645 (2007)
15. Kaiser, R., Maravall, A.: Seasonal outliers in time series. Universidad Carlos III de Madrid, working paper number 99-49 (1999)

16. Hubert, M., Rousseeuw, P., Segaert, P.: Multivariate functional outlier detection. Stat. Methods Appl. **24**(2), 177–202 (2015)
17. Chatterjee, S., Hadi, A.: Regression Analysis by Example, 4th edn. Wiley, Hoboken (2006)
18. Foorthuis, R.: The SECODA Algorithm for the Detection of Anomalies in Sets with Mixed Data. www.foorthuis.nl. 18 Nov 2017
19. Embrechts, P.: Extreme value theory: potential and limitations as an integrated risk management tool. Deriv. Use Trading Regul. **6**(1), 449–456 (2000)
20. Koufakou, A., Ortiz, E., Georgiopoulos, M., Anagnostopoulos, G., Reynolds, K.: A scalable and efficient outlier detection strategy for categorical data. In: Proc of ICTAI (2007)
21. Ben-Gal, I.: Outlier detection. In: Maimon, O., Rockach, L. (eds.) Data Mining and Knowledge Discovery Handbook. Kluwer Academic Publishers, Boston (2005). https://doi.org/10.1007/0-387-25465-X_7
22. Goldstein, M., Uchida, S.: A comparative evaluation of unsupervised anomaly detection algorithms. PLoS ONE **11**(4), e0152173 (2016)

IF-CLARANS: Intuitionistic Fuzzy Algorithm for Big Data Clustering

Hechmi Shili[1,2]([✉]) and Lotfi Ben Romdhane[2]

[1] Faculty of Sciences, University of Monastir, Monastir, Tunisia
hechmi.shili@fsm.rnu.tn
[2] Modeling of Automated Reasoning Systems (MARS),
Research Laboratory LR17ES05, Higher Institute of Computer Science and Telecom
(ISITCom), University of Sousse, Sousse, Tunisia

Abstract. Clustering method is one of the most important and basic technique for data mining which aims to group a collection of samples into clusters based on similarity. Clustering Big datasets has always been a serious challenge due to its high dimensionality and complexity. In this paper, we propose a novel clustering algorithm which aims to introduce the concept of intuitionistic fuzzy set theory onto the framework of CLARANS for handling uncertainty in the context of mining Big datasets. We also suggest a new scalable approximation to compute the maximum number of neighbors. Our experimental evaluation on real data sets shows that the proposed algorithm can obtain satisfactory clustering results and outperforms other current methods. The clusters quality was evaluated by three well-known metrics.

Keywords: IF-CLARANS · Clustering algorithm · Big data
Intuitionistic fuzzy set

1 Introduction

In the digital world today, according to unpreceded progress and development of the internet and online world technologies such as big and powerful data servers, the amounts of data have increased exponentially not only according to the size but also in terms of variety and complexity data sets. Nevertheless, the treatment of this large amount of data becomes a major challenge.

Classical data mining, like classification and clustering, approaches are not sufficient for analyzing such data. In fact, big data is fundamentally different from other data because of the three V's volume, velocity, and variety [9].

- The volume of data stored today is exploding. In the year 2000, 800,000 petabytes (PB) of data were stored in the world and by 2020 it is anticipated to hit 35 zettabytes.
- The velocity is the speed at which the data is generated, captured and shared.
- Variety means the diversity of the data sources and data formats.

© Springer International Publishing AG, part of Springer Nature 2018
J. Medina et al. (Eds.): IPMU 2018, CCIS 854, pp. 39–50, 2018.
https://doi.org/10.1007/978-3-319-91476-3_4

Clustering big data is a very challenge to data mining because large volumes and different varieties must be taken into account. A broad collection of clustering methods has been proposed for clustering big data, see [9] for a recent review. Uncertainty caused by the overlapping nature of the various partitions is a very challenging problem in cluster analysis, and several algorithms have been proposed for it. Fuzzy set theory is one of the most popular tools to handle, with greater flexibility, uncertainty and imprecision in intelligent systems. Fuzzy models allows us to represent patterns or members in a vague or ambiguous way. In fact, the concept of fuzzy membership μ, lying in $[0, 1]$, allows a pattern to simultaneously belong in more than one partition when, in conventional techniques, it is assumed that each object belongs to exactly one cluster. Fuzzy methods are less sensitive to local minima than crisp ones because of the fuzzy updating at each iteration [12].

In fuzzy set theory, the membership of an object to a fuzzy set is a real number between zero and one. But in reality, it may not always be certain that the degree of non-membership of an element in a fuzzy set is equal to 1 minus the membership degree because there may be some hesitation degree. Therefore, a generalization of fuzzy sets was introduced by Atanassov [15] as intuitionistic fuzzy sets (IFS) which incorporated the degree of hesitation called hesitation margin (and is defined as 1 minus the sum of membership and non-membership degrees respectively).

In this paper, a new clustering algorithm called Intuitionistic Fuzzy CLARANS (IF-CLARANS) is proposed. The concept of intuitionistic fuzzy membership (like IFCM) is incorporated onto the framework of CLARANS for handling uncertainty in the context of mining large data. The goodness of clustering is evaluated using the Xie-Beni (XB) cluster validity index [5], the partition entropy (PE) [11] and the accuracy of the partitioning results [16]. Algorithms like FCM [12], IFCM [17] and CLARANS [2] are used for comparative study.

The rest of this paper is organized as follows. Section 2 is devoted to some preliminaries in order to make the paper as self-contained as possible. The proposed model is introduced in Sect. 3. Section 4 presents the main results of the paper. Finally, the paper ends with some conclusions and references.

2   Preliminaries

2.1   Original Fuzzy C-Means (FCM)

Fuzzy C-means clustering (FCM) is an objective function-based clustering originally developed by Dunn in 1971 and further improved by Bezdek [12]. Unlike the traditional k-means algorithm, which partitions a set of data into predetermined c clusters and each element in the data set belongs to only one exact cluster, FCM admits partial belongingness (membership) of a data to more than a single cluster, and member-ship grades quantify the degree to which this element belongs to different clusters. Given a finite collection of data set containing N elements, $X = x_1, x_2, \ldots, x_N \subset R^q$, the FCM method splits the data into c

clusters by minimizing a certain objective function. The algorithm returns cluster centers (prototypes) $r_1, r_2, \ldots, r_c$ and a fuzzy partition matrix $U = [u_{ij}]$,
$u_{ij} \in [0, 1]$, $i = 1, 2, \ldots, N$; $j = 1, 2, \ldots, c$. An ij^{th} entry of U, U_{ij}, indicates
to which extent the element x_i belongs to the jth cluster. The FCM algorithm
minimizes the following objective function:

$$f = \sum_{i=1}^{N} \sum_{j=1}^{c} U_{ij}^m d_{ij}^2 \tag{1}$$

$d_{ij} = \|x_i - r_j\|$ is the Euclidean distance between x_i and r_j.

In Eq. (1), the fuzziness exponent (fuzziness coefficient) m assumes values
greater than 1, while the value $m = 2$ is the most commonly used. Different
values of m control the shape of membership functions produced by the FCM
algorithm. Higher value of m leads to spike-like membership functions while the
values close to 1 produce more Boolean-like shapes of membership functions. The
parameter exhibits some influence on the performance of the FCM algorithm and
can be subject to optimization. The iterative updates of the partition matrix and
the prototypes are realized as follows [12]:

$$U_{ij} = \frac{1}{\sum_{k=1}^{c} (\frac{d_{ij}}{d_{ik}})^{\frac{2}{m-1}}}, \tag{2}$$

where i is an integer in range $[1, N]$ and j is an integer in range $[1, c]$

$$r_j = \frac{\sum_{i=1}^{N} U_{ij}^m x_i}{\sum_{i=1}^{N} U_{ij}^m} \tag{3}$$

Where j is an integer in range $[1, c]$
The algorithm is terminated once the following condition is satisfied:

$$max \left\{ \left| U_{ij}^{k+1} - U_{ij}^k \right| \right\} \prec \epsilon \tag{4}$$

Where ϵ is a certain non negative threshold value whereas k denotes the
index of the successive iteration of the algorithm.

2.2 Intuitionistic Fuzzy Sets (IFSs)

Let X be a universe of discourse. In [13], Zadeh introduced the concept of fuzzy
set:

$$F = \{(x, \mu_F(x)) | x \in X\} \tag{5}$$

whose basic component is only a membership degree $\mu_F(x)$ with the non-membership degree being $1 - \mu_F(x)$. However, in real-life situations, when a person is asked to express his/her preference degree to an object, there usually exists an uncertainty or hesitation about the degree, and there is no means to incorporate the uncertainty or hesitation in a fuzzy set [14]. To solve this issue, Atanassov [15] generalized Zadeh's fuzzy set to intuitionistic fuzzy set (IFS) by adding an uncertainty (or hesitation) degree. IFS is defined as follows:

$$A = \{(x, \mu_A(x), v_A(x)) | x \in X\} \tag{6}$$

which is characterized by a membership degree $\mu_A(x)$ and a non-membership degree $v_A(x)$, where:

$$\mu_A : X \to [0, 1], x \in X \to \mu_A(x) \in [0, 1], \tag{7}$$

$$v_A : X \to [0, 1], x \in X \to v_A(x) \in [0, 1], \tag{8}$$

and it holds the condition $0 \le \mu_A(x) + v_A(x) \le 1$.
when $v_A(x) = 1 - \mu_A(x)$ for every x in set A, then the set A becomes a fuzzy set.

For all intuitionistic fuzzy sets, Atanassov [15] also indicated a hesitation degree, $\pi_A(x)$, which arises due to lack of knowledge in defining the membership degree of each element x in set A and is given by:

$$\pi_A(x) = 1 - \mu_A(x) - v_A(x) \tag{9}$$

It is evident that $0 \le \pi_A \le 1$ for each $x \in A$.
Due to the hesitation degree, the membership values lie in the interval

$$[\mu_A(x), \mu_A(x) + \pi_A(x)]$$

3 The Proposed Algorithm

3.1 Our Contribution

This work deals with the problem of clustering large datasets. The main contributions of the proposed algorithm called IF-CLARANS are as follows:

1. The concept of Atanassov's intuitionistic fuzzy set theory is used onto the framework of CLARANS for handling uncertainty in the context of mining large data.
2. We suggest a new scalable approximation to compute the maximum number of neighbors in CLARANS algorithm.

3.2   CLARANS Algorithm

CLARANS [2] considers two parameters *numlocal*, representing the number of iterations for the algorithm, and *maxneighbor*, the number of adjacent nodes in the graph G that need to be searched up to convergence. These parameters are provided as input at the beginning. While CLARA [1] compared very few neighbors corresponding to a fixed small sample, CLARANS uses random search to generate neighbors by starting from an arbitrary node and randomly checking maxneighbor neighbors [2]. If a neighbor represents a better partition, the process continues with this new node. Otherwise a local minimum is found, and the algorithm restarts until *numlocal* local minima are obtained.

The clustering process searches through a graph $G_{N,c}$, where node v^q is represented by a set of c medoids $m_1^q, \ldots, m_c^q$ of the clusters. Two nodes are said to be neighbors if they differ by only one medoid, and are connected by an edge. More formally, two nodes $v^1 = m_1^1, \ldots, m_c^1$ and $v^2 = m_1^2, \ldots, m_c^2$ are said to be neighbors if and only if the cardinality of the intersection of v^1 and v^2 is given as $card(v^1 \bigcap v^2) = c - 1$. Hence each node in the graph has $c * (N - c)$ neighbors. For each node $v^q \in G_{N,c}$ we assign a cost function:

$$J_c^q = \sum_{i=1}^{N} \sum_{j=1}^{c} d_{ij}^q, \tag{10}$$

where d_{ij}^q denotes the dissimilarity measure of the ith object x_i from the jth cluster medoid m_j^q in the qth node. The aim is to determine that set of c-medoids $m_1^*, \ldots, m_c^*$ at node v^*, for which the corresponding cost is the minimum as compared to all other nodes in the graph.

The main steps of the algorithm are outlined as follows:

```
CLARANS algorithm
begin
1)   Input parameters numlocal and maxneighbor. Initialize i to 1,
and mincost to a large number.
2)   Set current to an arbitrary node in G_{n,k}.
3)   Set j to 1.
4)   Consider a random neighbor S of current, and based on cost
function (Equation (14)), calculate the cost differential of the
two nodes.
5)   If S has a lower cost set current to S, and go to 3).
6) Otherwise, increment j by 1. If j  <= maxneighbor, go to 4).
7)   Otherwise, when j > maxneighbor, compare the cost of current
with mincost. If the former is less than mincost, set mincost to
the cost of current and set bestnode to current.
8)   Increment  i by 1. If i > numlocal, output bestnode and halt.
 Otherwise, go to 2).
end.
```

Note that *maxneighbor* is computed as:

$$maxneighbor = p\%(c * (N - c)) \tag{11}$$

with p being provided as input by the user. Typically, $1.25 \leq p \leq 1.5$ [2].

3.3 Intuitionistic Fuzzy CLARANS (IF-CLARANS)

In order to incorporate intuitionistic fuzzy property in conventional CLARANS algorithm, the clusters centers are updated in the following manner.

Hesitation degree is initially calculated using

$$\pi_A(x) = 1 - \mu_A(x) - (1 - \mu_A(x)^\alpha)^{1/\alpha} \tag{12}$$

Where $\alpha \succ 0$ is the Yager's coefficient [24].

The intuitionistic fuzzy membership values are obtained as follows:

$$U_{ik}^* = U_{ik} + \pi_{ik} \tag{13}$$

where $U_{ik}^*(U_{ik})$ denotes the intuitionistic (conventional) fuzzy membership of the kth data in i^{th} class.

The objective function is defined as follow:

$$J^q = \sum_{\substack{N \\ i=1}} \sum_{j=1}^{c} (U_{ij}^{*q})(d_{ij}^q)^2, \tag{14}$$

Here the distance component is weighted by the corresponding membership value $d_{ij}^{*q} = (x_i, v_j^*)$. This is used in Steps 1, 4, and 7, of the algorithm CLARANS.

Fuzzy partitioning is carried out through an iterative optimization of the cost function defined in Eq. (14), with the membership at node v^q being computed as defined in Eq. (13).

Where $v_i^* = \dfrac{\sum_{k=1}^{n} U_{ij}^* x_k}{\sum_{k=1}^{n} U_{ij}^*}$

3.4 Estimating the Maximum Number of Neighbors Examined

In CLARANS, the process of finding k medoids from n objects is viewed abstractly as a search for a minimum through a certain graph $G_{n,k}$. Each node in the graph has $c * (N - c)$ neighbors. At each step, PAM [1] searches and examines all of the neighbors of the current node in its search for a minimum cost solution. The main improvement made by the CLARANS algorithm compared to the PAM algorithm is that it performs the search only in a subset of neighboring nodes by the definition of the *maxneighbor* parameter (Eq. 11). This reduces the number of searches, and hence greatly reduces the local minimum search time [2]. But the value of *maxneighbor* turns up to be very high when N is sufficiently large ($N \geq 10,000$). By Step 6 of the CLARANS algorithm, this increases the computational burden. We, therefore, define a new scalable approximation of the *maxneighbor* parameter expressed as:

$$maxneighbor = \frac{c^2 log_2(N/c)}{2log_2(c)} \qquad (15)$$

By introducing this new approximation of *maxneighbor*, the number of neighbors to be examined at each iteration is increased reducing the computational burden of the algorithm to reach at the best solution, which is not worse than that obtained using *maxneighbor* in Eq. (11). We are, as a result, able to eliminate the user-defined parameter p while also reducing computational time of the algorithm. This new equation has been employed in this article for experiments involving large datasets ($N \geq 10,000$). Note that *maxneighbor*, is gently increasing and thus able to model larger clusters better.

The choice of the logarithmic function with respect to the linear and power function is justified by the fact that, the slope of the curve actually becomes more gentle as the value of N increases, which is more suitable for large datasets.

4   Experimental Results

4.1   Experimental Environments

In this part, the experimental environments are described such as,

Experimental Tools: In all experiments we use MATLAB program as a powerful tool to compute clusters and Intel Core i7-7700, 3.6 GHz, 16 GB RAM/3 TB HDD running windows 10. The experimental results are taken as the average values after 10 runs.

Cluster Validity Measurement: The results are expressed in terms of indices XB [5], Partition Entropy (PE) [11] and the accuracy measure [16].

$$XB = \frac{\sum_{i=1}^{N}\sum_{j=1}^{c} U_{ij}^m d_{ij}}{N * \min_{i,j} d'(U_i, U_j)^2} \qquad (16)$$

where U_{ij} is the membership of pattern x_i to cluster c_j, d_{ij} denotes the dissimilarity measure of the ith object x_i from the jth cluster medoid and d' denotes the dissimilarity measure of the ith and j^{th} cluster medoids. Minimization of XB is indicative of better clustering, particularly in case of fuzzy data.

$$PE(c) = -\frac{1}{N}\sum_{i=1}^{N}\sum_{j=1}^{c} U_{ij} log_2 U_{ij} \qquad (17)$$

where $0 \leq PE(c) \leq log_2c$. In general, we find an optimal c^* by solving $min_{2 \leq c \leq n-1} PE(c)$ to produce a best clustering performance for the data set X [11].

Experimental Datasets: We present results on five well-known datasets (Spambase, connect-4, Nursery, covtype and FARS) which are available from the UCI Machine Learning Repository (available at: http://archive.ics.uci.edu/ ml/). All attributes/variables of these datasets are used concurrently for the best evaluation of the algorithms. The basic information of the five real data sets that we will use in our experiments is illustrated in Table 1.

Table 1. Real data sets summary.

Dataset	# instances	#attributes	#clusters
Spambase	4597	57	2
connect-4	67,557	126	3
Nursery	12690	8	5
covtype	581,012	54	7
FARS	100968	29	8

4.2   The Comparison of Clustering Quality

We compared the solution results of the IF-CLARANS algorithm with those of the FCM algorithm [12], the IFCM algorithm [17] and the CLARANS algorithm [2]. The clustering quality results of the different clustering algorithms, in terms of the XB and PE, are shown in Table 2. The results show that our algorithm yields reasonably good results in all datasets. The XB and PE measures value results of the IF-CLARANS method are much lower than those of the FCM, IFCM and the CLARANS methods, which means that by using the proposed method, the best quantitative evaluation results available have been achieved.

The clustering efficiency is also experimented using the accuracy measure. Good clustering corresponds to higher values of the accuracy that represents the average of well clustered elements in their corresponding classes. The obtained results are shown in Table 3. Our method gives the highest clustering accuracy in the majority of the tests, except for the Spambase dataset, which is the smallest dataset used in the experiments, where IFCM algorithm achieved higher accuracy.

We also measured the execution time of CLARANS and IF-CLARANS algorithms. The experiments are conducted on the five real data sets listed in Table 1 when the number of instances is very large (N reaches $581,012$ instances). the aim of this study is to investigate the behavior of the two algorithms when operating on large data sets. Figure 1 lists the average execution time of our proposed

Table 2. Clustering performance of the involved clustering algorithms on real data for $c = c^*$

Dataset	Algorithm	maxneighbor	XB	PE
Spambase	FCM	-	0.41	0.61
	IFCM	-	0.26	0.39
	CLARANS	$p\%c*(N-c) = 91$	0.37	0.46
	IF-CLARANS	$c^2log_2(N/c)/2log_2(c) = 44$	0.22	0.33
connect-4	FCM	-	0.38	0.51
	IFCM	-	0.23	0.41
	CLARANS	$p\%c*(N-c) = 2026$	0.34	0.48
	IF-CLARANS	$c^2log_2(N/c)/2log_2(c) = 82$	0.24	0.42
Nursery	FCM	-	0.35	0.47
	IFCM	-	0.43	0.49
	CLARANS	$p\%c*(N-c) = 634$	0.28	0.44
	IF-CLARANS	$c^2log_2(N/c)/2log_2(c) = 121$	0.18	0.40
covtype	FCM	-	0.36	0.43
	IFCM	-	0.24	0.36
	CLARANS	$p\%c*(N-c) = 40670$	0.28	0.39
	IF-CLARANS	$c^2log_2(N/c)/2log_2(c) = 285$	0.21	0.35
FARS	FCM	-	0.28	0.41
	IFCM	-	0.26	0.36
	CLARANS	$p\%c*(N-c) = 8076$	0.22	0.35
	IF-CLARANS	$c^2log_2(N/c)/2log_2(c) = 290$	0.19	0.35

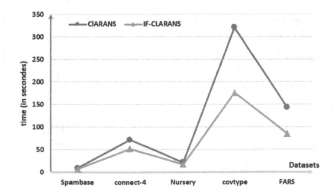

Fig. 1. Running time comparison of CLARANS and IF-CLARANS for different datasets.

algorithm and the CLARANS algorithm. As the graph shows, our algorithm outperforms CLARANS on all datasets in execution time, especially as the number of objects increase. In fact, for the covtype dataset (which is the largest one)

Table 3. Clustering performance in terms of accuracy for different number of clusters.

Dataset	Algorithm	#clusters						
		2	3	4	5	6	7	8
Spambase	FCM	52.3	21.2	23.4	9.8	11.3	15.4	17.6
	IFCM	**56.4**	24.1	25.7	12.1	11.5	18.3	14.2
	CLARANS	42.8	12.4	25.2	10.1	9.7	11.8	17.4
	(maxneighbor)	*(92)*	*(137)*	*(183)*	*(229)*	*(175)*	*(321)*	*(367)*
	IF-CLARANS	47.2	17.4	21.4	14.3	11.5	11.1	10.3
	(maxneighbor)	*(45)*	*(60)*	*(81)*	*(106)*	*(133)*	*(163)*	*(196)*
connect-4	FCM	21.2	48.3	18.7	15.4	14.2	16.7	13.2
	IFCM	23.4	46.8	17.4	16.5	13.4	17.4	15.5
	CLARANS	18.7	41.2	17.3	20.2	18.3	15.3	16.6
	(maxneighbor)	*(1351)*	*(2026)*	*(2702)*	*(3378)*	*(4053)*	*(4728)*	*(5404)*
	IF-CLARANS	19.8	**48.7**	18.2	21.1	19.6	17.8	18.1
	(maxneighbor)	*(60)*	*(82)*	*(112)*	*(148)*	*(187)*	*(231)*	*(278)*
Nursery	FCM	11.2	14.6	19.1	42.2	21.3	17.5	10.3
	IFCM	13.3	14.8	21.2	43.7	24.5	18.9	12.8
	CLARANS	15.3	18.5	23.4	45.2	25.7	21.2	17.9
	(maxneighbor)	*(254)*	*(381)*	*(507)*	*(634)*	*(761)*	*(888)*	*(1014)*
	IF-CLARANS	17.8	13.1	20.6	**52.6**	18.1	23.5	20.5
	(maxneighbor)	*(51)*	*(68)*	*(93)*	*(122)*	*(154)*	*(189)*	*(227)*
covtype	FCM	11.5	15.1	19.4	21.3	25.4	41.8	27.8
	IFCM	13.4	17.8	20.1	19.7	21.3	43.4	22.2
	CLARANS	15.4	17.3	19.7	24.3	23.7	49.2	30.1
	(maxneighbor)	*(11620)*	*(17430)*	*(23240)*	*(29050)*	*(34860)*	*(40670)*	*(46480)*
	IF-CLARANS	14.6	22.4	18.7	19.1	15.4	**53.8**	22.8
	(maxneighbor)	*(73)*	*(100)*	*(137)*	*(181)*	*(231)*	*(285)*	*(344)*
FARS	FCM	12.5	15.4	17.6	11.8	21.1	14.9	39.4
	IFCM	17.3	11.3	18.6	11.4	17.3	18.3	41.2
	CLARANS	14.3	14.3	17.9	15.4	21.1	17.2	44.7
	(maxneighbor)	*(1019)*	*(2028)*	*(3038)*	*(4004)*	*(5122)*	*(6170)*	*(8076)*
	IF-CLARANS	20.1	18.1	14.6	15.8	19.7	17.3	**47.5**
	(maxneighbor)	*(63)*	*(85)*	*(117)*	*(154)*	*(196)*	*(241)*	*(291)*

when the number of instances is equal to $581,012$, the running time of our algorithm was $176\ sec$ compared to $320.8\ sec$ for running CLARANS Algorithm on the same database. The main cause of this big difference may be attributed to the effect of the *maxneighbor* choice. However, for CLARANS algorithm equation (Eq. 11) is used to calculate *maxneighbor* and it number corresponding to the covtype dataset is equal to 40670. CLARANS algorithm takes a long time

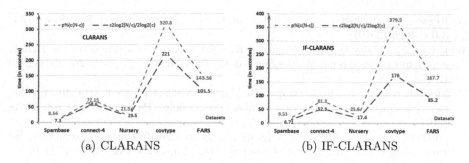

Fig. 2. Effect of maxneighbors choice on the clustering running time for different datasets.

to examine all this number of neighbors while for our algorithm, using equation (Eq. 15), the number of neighbors to examine is equal to *285* for the same base.

To see how effective was our method of selecting the number of *maxneighbor*, we compared the execution times of our algorithm and CLARANS algorithm using the modified expression for maxneighbor given by Eq. (15) and that of the CLARANS algorithm Eq. (11).

We can observe from Fig. 2 that, for both algorithms, the execution time improves by a significant amount when the modified expression for *maxneighbor* is used. This is clearly seen for the case of very large datasets with high dimensionality (such as covtype dataset described in Table 1).

5   Conclusion

In this paper, we introduced a new algorithm to intuitionistic fuzzy clustering, called IF-CLARANS that uses intuitionistic fuzzy set theory onto the framework of CLARANS for handling uncertainty. In order to evaluate the performance of our algorithm, we used five real data sets. based on the criteria of the solution quality, the experimental results show that IF-CLARANS has better quality level of clustering and performance that outperforms some of the best-known approaches.

References

1. Aboubi, Y., Drias, H., Kamel, N.: BAT-CLARA: BAT-inspired algorithm for clustering LARge applications. In: 8th IFAC Conference on Manufacturing Modelling, Management and Control, MIM 2016, vol. 49, pp. 243–248 (2016)
2. Ng, R.T., Han, J.: CLARANS: a method for clustering objects for spatial data mining. IEEE Trans. Knowl. Data Eng. **14**, 1003–1016 (2002)
3. Lorbeer, B., et al.: Variations on the clustering algorithm BIRCH. Big Data Res. 2214–5796 (2017)
4. Lathiya, P., Rani, R.: Improved CURE clustering for big data using Hadoop and Mapreduce. In: International Conference on Inventive Computation Technologies (ICICT), Coimbatore, India, pp. 1–5 (2016)

5. Rezaee, B.: A cluster validity index for fuzzy clustering. Fuzzy Sets Syst. **161**, 3014–3025 (2010)
6. Dutta, M., Mahanta, A.K., Pujari, A.K.: QROCK: a quick version of the ROCK algorithm for clustering of categorical data. Pattern Recogn. Lett. **26**, 2364–2373 (2005)
7. Mahesh Kumar, K., Rama Mohan Reddy, A.: A density based algorithm for discovering clusters in large spatial databases with noise. Pattern Recogn. **58**, 39–48 (2016)
8. Ankerst, M., et al.: OPTICS: ordering points to identify clustering structure. In: Proceedings of ACM SIGMOD Conference on Management of Data. ACM Press, Philadelphia (1999)
9. Saxena, A., Prasad, M., Gupta, A., Bharill, N., Patel, O.P., Tiwari, A., Er, M.J., Ding, W., Lin, C.-T.: A review of clustering techniques and developments. Neurocomputing **267**, 664–681 (2017)
10. Berkhin, P.: Survey of Clustering Data Mining techniques. Accrue Software Inc., San Jose (2000)
11. Yu, H., Zhi, X., Fan, J.: Image segmentation based on weak fuzzy partition entropy. Neurocomputing **168**, 994–1010 (2015)
12. Bezdek, J.C. (ed.): Pattern Recognition with Fuzzy Objective Function Algorithms. Plenum Press, New York (1981)
13. Zadeh, L.A.: Outline of a new approach to the analysis of complex systems and decision processes. IEEE Trans. Syst. Man Cybern. **3**, 28–44 (1973)
14. Deschrijver, G., Cornelis, C., Kerre, E.E.: On the representation of intuitionistic fuzzy t-norms and t-conorms. IEEE Trans. Fuzzy Syst. **12**, 45–61 (2004)
15. Yuan, X., Li, H., Zhang, C.: The theory of intuitionistic fuzzy sets based on the intuitionistic fuzzy special sets. Inf. Sci. **277**, 284–298 (2014)
16. Halkidi, M., Gunopulos, D., Vazirgiannis, M., et al.: A clustering framework based on subjective and objective validity criteria. ACM Trans. Knowl. Disc. Data **1**(4), 1–25 (2008)
17. Zhang, H.-M., Xu, Z.-S., Chen, Q.: On clustering approach to intuitionistic fuzzy sets. Control Decis. **22**, 882 (2007)
18. Dhillon, I., Guan, Y., Kulis, B.: Kernel k-means: spectral clustering and normalized cuts. In: Proceeding of KDD, Proceedings of 10th ACM SIGKDD International Conference on Knowledge Discovery and Data Mining, pp. 551–556 (2004)
19. Dhillon, I., Modha, D.: Concept decompositions for large sparse text data using clustering. Mach. Learn. **42**(1–2), 143–175 (2001)
20. de Amorim, R.C., Mirkin, B.: Minkowski metric, feature weighting and anomalous cluster initializing in k-means clustering. Pattern Recogn. **45**(3), 1061–1075 (2012)
21. Pelleg, D., Moore, A.W.: X-means: extending k-means with efficient estimation of the number of clusters. In: Proceedings of 17th International Conference on Machine Learning, pp. 727–734. Morgan Kaufmann (2000)
22. Cai, X., Nie, F., Huang, H.: Multi-view k-means clustering on big data. In: Rossi, F. (ed.) Proceedings of 23rd International Joint Conference on Artificial Intelligence, IJCAI 2013. IJCAI/AAAI (2013)
23. Mahesh Kumar, K., Rama Mohan Reddy, A.: An efficient k-means clustering filtering algorithm using density based initial cluster centers. Inf. Sci. **418**, 286–301 (2017)
24. Klir, G.J., Yuan, B.: Fuzzy Sets and Fuzzy Logic Theory and Applications. Prentice Hall of India Private Limited, New Delhi (2002)

Semi-supervised Fuzzy c-Means Variants: A Study on Noisy Label Supervision

Violaine Antoine[1]([⊠]) and Nicolas Labroche[2]

[1] Clermont Auvergne University, UMR 6158, LIMOS,
63006 Clermont-Ferrand, France
violaine.antoine@uca.fr
[2] University of Tours, EA 6300, LIFAT, Tours, France

Abstract. Semi-supervised clustering algorithms aim at discovering the hidden structure of data sets with the help of expert knowledge, generally expressed as constraints on the data such as class labels or pairwise relations. Most of the time, the expert is considered as an oracle that only provides correct constraints. This paper focuses on the case where some label constraints are erroneous and proposes to investigate into more detail three semi-supervised fuzzy c-means clustering approaches as they have been tailored to naturally handle uncertainty in the expert labeling. In order to run a fair comparison between existing algorithms, formal improvements have been proposed to guarantee and fasten their convergence. Experiments conducted on real and synthetical datasets under uncertain labels and noise in the constraints show the effectiveness of using fuzzy clustering algorithm for noisy semi-supervised clustering.

Keywords: Fuzzy clustering · Label constraints
Semi-supervised clustering · Noise

1 Introduction

Semi-supervised clustering algorithms are part of exploratory data analysis. They intend to extract the underlying structure of datasets by grouping similar objects together with the help of some partial external knowledge usually provided as pairwise constraints [1], e.g. must-link/cannot-link constraints between pairs of objects that indicate if two objects must (or not) be in the same cluster, or labels constraints [2], that specify explicitly the class labels for some objects. These approaches can lead clustering algorithms towards a better definition of the existing structures in the data, or at least to a definition that better fits the needs of the final user. For clustering algorithms that are directly derived from the optimization of an objective function, like k-means and its variants, various methods have been proposed by adding a penalty term [2–4] or by learning a proper metric [2,5] that adapts the topology so that less constraints are violated.

However, all these methods heavily depend on the quality of the provided expert knowledge. Even in the best case, where only correct constraints are provided to the algorithms, it has been shown that improperly chosen constraints

© Springer International Publishing AG, part of Springer Nature 2018
J. Medina et al. (Eds.): IPMU 2018, CCIS 854, pp. 51–62, 2018.
https://doi.org/10.1007/978-3-319-91476-3_5

can deteriorate performances [6]. Hence, solutions have been proposed to evaluate the quality or the utility of constraints prior to clustering to avoid such problem [7,8]. But, to the best of our knowledge, no work has directly tackled the problem of semi-supervised clustering when the expert does not provide relevant constraints.

This paper shows that, in this context of erroneous or uncertain expert labeling, it is possible to use the natural property of fuzzy clustering algorithm to handle uncertainty in constraints to maintain good clustering performances. For the sake of clarity, this paper is restricted to label constraints since they are more general than pairwise constraints. The study is also limited to variants of fuzzy c-means (FCM) that include a term to penalize the solution when label constraints are not respected. As such, we discard more complex FCM algorithms as the kernel-based [9] or those that determine the number of clusters [2,10].

Without loss of generality, we consider label constraints expressed as a fuzzy membership matrix $\tilde{\mathbf{U}} = (\tilde{u}_{ik})$ that indicates to which extent each object i is supposed to be assigned to the cluster k according to the expert. In this case, an object does not necessarily have constraints and these constraints may not be completely certain, i.e. $0 \leq \sum_k \tilde{u}_{ik} \leq 1$. Table 1 illustrates such matrix $\tilde{\mathbf{U}}$ with 4 objects and 3 clusters and introduces the vocabulary that will be used in the experiments.

Table 1. Example of a constraint membership matrix. Object o_1 represents the traditional seed constraint with a crisp assignment to a single cluster. Object o_4 is not constrained. Objects o_2 and o_3 show the expressiveness brought by fuzzy representation of constraints with certain or uncertain/single or multi-labels.

	c_1	c_2	c_3	$\sum_k \tilde{u}_{ik}$	Explanations
o_1	1	0	0	1	Single-label and certain constraint
o_2	0	0.3	0	0.3	Single-label and uncertain constraint
o_3	0	0.5	0.5	1	Double-label and certain constraint
o_4	0	0	0	0	Unconstrained object

A comparative review on semi-supervised fuzzy c-means algorithms with label contraints has already been performed in [11]. However, their objective is not to evaluate the ability of the algorithms to deal with erroneous or noisy expert labels and the soundness of optimization techniques is not discussed, as a strict copy of the original algorithms is employed. In this paper, we consider modified algorithms to conduct a fair comparison that only involves penalty term employed in FCM for the constraints. To this aim, we ensure and fasten the convergence of the optimization and we introduce the Mahalanobis distance when it is not already achieved, as a specific and adaptive distance for each cluster is beneficial for some datasets.

The rest of the paper is then organized as follows. Semi-supervised clustering algorithms and their modifications are presented Sects. 2 and 3. Experiments on raw, uncertain and noisy labels are introduced Sect. 4 and a conclusion is available Sect. 5.

2   Semi-supervised Clustering Algorithms

Let $\mathbf{X} = \{\mathbf{x}_i, \ldots \mathbf{x}_n\}$ be a dataset composed of n objects such that $\mathbf{x}_i \in \mathbb{R}^p$ is the feature vector representing the object i. The clusters are defined by centroids $\mathbf{V} = \{\mathbf{v}_1, \ldots \mathbf{v}_c\}$ and d_{ik}^2 corresponds to the squared Euclidean distance between the object $\mathbf{x}_i$ and the centroid $\mathbf{v}_k$. The standard fuzzy c-means algorithm minimizes the intraclass inertia by alternatively optimizing the degrees of membership $\mathbf{U} = (u_{ik})$ and the centroids $\mathbf{V}$ [12,13]. The objective function is the following:

$$J_{FCM}(\mathbf{U}, \mathbf{V}) = \sum_{i=1}^{n} \sum_{k=1}^{c} u_{ik}^m d_{ik}^2, \tag{1}$$

where $m > 1$ is a fixed value that controls the degree of fuzziness for the partition and u_{ik} should satisfy:

$$\sum_{k=1}^{c} u_{ik} = 1; \quad u_{ik} > 0 \quad \forall i \in \{1 \ldots n\}, \forall k \in \{1 \ldots c\}. \tag{2}$$

Gustafson and Kessel have proposed a variant of FCM that use a specific Mahalanobis distance for each cluster [13]. The distance between an object $\mathbf{x}_i$ and a cluster k becomes $d_{ik}^2 = (\mathbf{x}_i - \mathbf{v}_k)^T \mathbf{S}_k (\mathbf{x}_i - \mathbf{v}_k)$, where $\mathbf{S}_k$ is the norm-inducing matrix of the cluster k. The matrices $\mathbf{S}_1 \ldots \mathbf{S}_c$ are defined as fuzzy covariance matrices and enable to detect the optimal geometrical shapes of the clusters.

sfcm is a famous algorithm that add a penalty term in the objective function of FCM to take into account uncertain labels [10] and for which an extension with Mahalanobis distance already exists [2]. The proposed objective function minimizes the following criteria such that constraints (2) are respected:

$$J_{sfcm}(\mathbf{U}, \mathbf{V}) = \sum_{i=1}^{n} \sum_{k=1}^{c} u_{ik}^m d_{ik}^2 + \alpha \sum_{i=1}^{n} \sum_{k=1}^{c} (u_{ik} - \tilde{u}_{ik} \mathbf{b}_i)^m d_{ik}^2, \tag{3}$$

where $m > 1$ must be an even number, $\alpha \in \mathbb{R}^+$ is a coefficient controlling the tradeoff between the objective function of FCM and the constraints, $\tilde{\mathbf{U}} = (\tilde{u}_{ik})$ is a partition given by an analyst and $\mathbf{b}_i$ is such that $b_i = 1$ if $\mathbf{x}_i$ is constrained and $b_i = 0$ otherwise.

This paper proposes a simple correction of the update equation of the prototypes $\mathbf{V}$ that is similar to what is proposed in [2]:

$$\mathbf{v}_k = \frac{\sum_{i=1}^{n} \alpha \left(u_{ik}^m + (u_{ik} - \tilde{u}_{ik} \mathbf{b}_i)^m \right) \mathbf{x}_i}{\sum_{i=1}^{n} \alpha u_{ik}^m + (u_{ik} - \tilde{u}_{ik} \mathbf{b}_i)^m}, \forall k \in \{1 \ldots c\}. \tag{4}$$

ssfcm is the first of the two semi-supervised FCM algorithms proposed in [14]. It minimizes the following objective function:

$$J_{ssfcm}(\mathbf{U}, \mathbf{V}) = \sum_{i=1}^{n} \sum_{k=1}^{c} |u_{ik} - \tilde{u}_{ik}|^m d_{ik}^2, \tag{5}$$

with $m \geq 1$ and such that constraints (2) are respected.

The algorithm ssfcm has no coefficient to set for some tradeoff between the inherent structure of the data and the constraints. Thus, the optimization is straightforward and the convergence ensured. However, it enforces a total respect of the constraints and consequently may not be able to deal efficiently with noisy or erroneous constraints.

In our test, we have proposed an extension of ssfcm with a Mahalanobis distance following the approach of Gustafson and Kessel [13] to make possible a fair comparison with the other algorithms when ellipsoidal clusters are to be found. Learning a Mahalanobis distance comes down to defining a $(p \times p)$ matrix $\mathbf{S}_k$ for each cluster k and minimizing the objective function (5) with the respect to $\mathbf{U}$, $\mathbf{V}$ and $\mathbf{S} = (\mathbf{S}_1 \dots \mathbf{S}_c)$. In order to avoid trivial solution consisting of $\mathbf{S}_k$ with only zeros that would minimize the objective function, a constant volume $\rho_k > 0$ is assigned to each cluster k:

$$det(\mathbf{S}_k) = \rho_k, \forall k \in \{1 \dots c\} \tag{6}$$

The constrained optimization problem is solved by introducing c Lagrange multipliers λ_k in J_{ssfcm}:

$$\mathcal{L} = J_{ssfcm}(\mathbf{U}, \mathbf{V}, \mathbf{S}) - \sum_{k=1}^{c} \lambda_k(\rho_k - det(\mathbf{S}_k)). \tag{7}$$

Setting the derivative of the Lagragian function to 0 leads to the following result:

$$\mathbf{S}_k = \rho_k \det(\mathbf{\Sigma}_k)^{\frac{1}{p}} \mathbf{\Sigma}_k^{-1},$$

$$\mathbf{\Sigma}_k = \sum_{i=1}^{n} \sum_{k=1}^{c} |u_{ik} - \tilde{u}_{ik}|^m (\mathbf{x}_i - \mathbf{v}_k)^T (\mathbf{x}_i - \mathbf{v}_k).$$

esfcm is an entropy regularized FCM [14] with the following objective function:

$$J_{esfcm} = \sum_{i=1}^{n} \sum_{k=1}^{c} u_{ik} d_{ik}^2 + \lambda^{-1} \sum_{i=1}^{n} \sum_{k=1}^{c} (|u_{ik} - \tilde{u}_{ik}|) \log(|u_{ik} - \tilde{u}_{ik}|), \tag{8}$$

such that $\lambda \in \mathbb{R}^+$ and constraints (2) are respected. In order to minimize this objective function, the authors remove the absolute value and replace it by new constraints $u_{ik} \geq \tilde{u}_{ik} \; \forall i \in \{1 \dots n\}, \; \forall k \in \{1 \dots c\}$ so that the function is still convex. As for ssfcm, the update equation of u_{ik} depends on $\tilde{u}_{ik}$ which may limit the way esfcm deals with erroneous constraints. Finally, enriching esfcm with a Mahalanobis distance is similar to what is performed for FCM.

3   Mapping Function

One common problem when evaluating semi-supervised clustering algorithms based on random initial centers such as FCM, is that the label assigned randomly to these centers may not coincide with the labels used to express the constraints. The problem can be solved by taking as initial centers the barycenter computed with the constrained labeled objects [15]. However, this solution is inappropriate when there exists clusters without labels or when the constraints set is noisy.

As an exemple, let us consider a dataset with 4 objects and for each object the following constraints labels: x_1 and x_2 in cluster 1, x_3 in cluster 2 and x_4 in cluster 3. Figure 1 presents a dataset with the previous constraints and some initial centers named $v_{1,2,3}$. It is obvious to observe that there exists a mismatch between the clusters, more particularly their centers labels, and the labels of the constrained objects. For instance, $\mathbf{x}_4$ should be in the class 3 but is assigned to cluster 1. In this case, the convergence of the algorithm to a solution where the centroid $\mathbf{v}_3$ is close to $\mathbf{x}_4$ is too expensive compared to a solution where some constraints are violated which in turn leads to poor results.

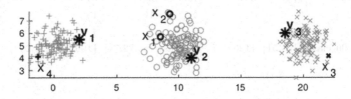

Fig. 1. Dataset with three clusters. Symbols '+', 'o', 'x' correspond to the real classes whereas stars represent centroids.

To this aim, our mapping function simply considers all pairing of labels between the one provided by clusters centers and the one provided by the constraints and each time performs the complete clustering. The pairing that is finally kept is the one that minimizes the objective function.

4   Experiments

This section is devoted to the comparison of sfcm, esfcm, ssfcm as well as skmeans when possible for several real-world and synthetic datasets. The skmeans algorithm is a semi-supervised clustering method that uses labeled data to improve a traditional k-means algorithm [15]. We use it as a baseline to show the interest of using fuzzy approaches in the case of uncertain or noisy supervision. We also implicitly compare our approach to a traditional FCM as it corresponds to sfcm without constraints. First, a study of the λ parameter for esfcm is conducted as its behavior highly depends on this parameter. Next, experiments are carried out to represent different scenarios where expert annotation can induce errors in constrained algorithms. In the case of single constraint, the membership degree provided by the expert can either be wrong (error in the chosen class label), uncertain (low membership constraint while 1 was expected) or both at the same time. Finally, in our multi-label scenario, we deal with the case where the expert may hesitate between two class labels to annotate one object.

4.1 Experimental Settings

We selected six well-known datasets from the UCI repository[1]: Glass, Ionosphere, Iris, Letters, Vehicle, Wine and a synthetic dataset generated with Gaussians: GaussK6. Characteristics are available in Table 2. For the Letters dataset, only the three letters I, J, L are kept as done in [16]. GaussK6 contains 2 overlapped classes. This dataset, as well as Wine, is suitable for a Euclidean distance whereas the other datasets offer better results with the Mahalanobis distance.

Table 2. Description of the datasets.

Name	GaussK6	Glass	Ionosphere	Iris	Letters	Vehicle	Wine
n	1200	214	351	150	227	846	178
p	2	8	33	4	16	18	13
c	6	2	2	3	3	4	3
Class sizes	200/class	{163,51}	{126,225}	50/class	{81,72,74}	{199,217,218,212}	{59,71,48}

In order to obtain a fair comparison between the algorithms, the same constraints and the same centers initializations have been tested at each experiment. An experiment consists in 100 trials where 1 trial executes 5 different initializations of the centers. The final result selected is the one with the minimal objective function.

In our experiments, our objective is to see how fuzzy clustering algorithms may help reaching better performances than crisp clustering algorithms when dealing with uncertain/noisy labels. However, in the end, we are interested in solving the crisp clustering problem since a decision has to be made about the class memberships of the objects. For this reason, the evaluation of the accuracy is calculated with the Adjusted Rand Index (ARI) [17] rather than a specific index related to fuzzy clustering. Moreover, ARI measures the similarity between two crisp partitions by taking into account the possibility that the obtained clustering is observed by chance. For fuzzy clustering algorithms, hard partition was determined by assigning objects to the cluster with the maximum membership value provided by the final fuzzy partition.

The modified partition coefficient (MPC) [18] has also been calculated to choose the λ parameter. This validity index measures the fuzziness of a partition: a crisp partition corresponds to a 1 value and a total fuzzy partition to a 0 value.

---

[1] Available at http://archive.ics.uci.edu/ml.

4.2 Choice of Parameters

For all experiments the exponent m controlling the fuzziness of the final partition is set to 2.

The α parameter is set in such a way that two terms of the objective function have the same importance. Then, it gives a balance between the search for an underlying structure and the respect of the constraints. It is left to future work to study the influence of this parameter.

The λ parameter is more complicated to set, as it plays a key role on the behavior of esfcm even without constraints. Thus, experiments were conducted on esfcm with no partial supervision to set the value of λ. Various λ values have been tested and both MPC and ARI measures have been calculated.

As a result, we noticed that the MPC value is increasing as the λ value increases. This behavior is easily explained by the fact that MPC measures the fuzziness of a partition and λ acts as a fuzzy controller of the final partition. Thus, setting a MPC value close to 0.8 ensures us to obtain a partition neither too crisp nor too fuzzy. Nonetheless, we have also observed that the MPC and ARI measures are not totally correlated, particularly when a Mahalanobis distance is used. Experiments reported in Table 3 show that, in general, a good accuracy is reached when MPC is around 0.8.

Table 3. λ values used in esfcm for the average MPC measure around 0.8 and the average corresponding ARI.

	GaussK6	Glass	Ionosphere	Iris	LettersIJL	Vehicle	Wine
λ	0.14	3.24	2.35	4	0.125	2.5	0.31
MPC	0.80	0.81	0.80	0.80	0.81	0.79	0.80
ARI	0.81	0.37	0.08	0.67	0.22	0.16	0.90

4.3 Comparative Experiments

Several experiments are reported in this section depending on the presence or not of constraints and on the quality of constraints ranging from single-label (un)certain constraints with added noise, to multi-label (un)certain constraints to simulate expert annotation errors.

No Constraint. First, the algorithms are executed without constraints to establish a comparative baseline for each dataset. Table 4 illustrates the average ARI and its 95% confidence interval for skmeans, sfcm, ssfcm and esfcm without constraints, i.e. k-means, FCM and FCM with an entropy regularization. Since skmeans has only the possibility to use a Euclidean distance, it cannot be compared to algorithms employing a Mahalanobis distance, hence the missing values in Table 4.

The sfcm and ssfcm algorithms without constraints, which correspond to FCM, outperform most of the time esfcm and skmeans. Low values of ARI are still visible, for example with the Vehicle dataset or the Ionosphere dataset. It means that the global structure of the data is difficult to detect and requires background knowledge to help its discovery.

Since we observed that the confidence interval remains stable when constraints are introduced, their values are not presented in the next tables.

Table 4. No constraint: average ARI and 95% confidence intervals for each algorithm and each dataset.

Dataset	skmeans	sfcm	ssfcm	esfcm
GaussK6	0.80 ± 0.1	**0.91 ± 0.1**	**0.91 ± 0.1**	0.78 ± 0.1
Glass	/	**0.48 ± 0.1**	**0.48 ± 0.1**	0.41 ± 0.2
Ionosphere	/	**0.46 ± 0.0**	**0.46 ± 0.0**	0.10 ± 0.1
Iris	/	**0.75 ± 0.0**	**0.75 ± 0.0**	0.68 ± 0.1
Letters	/	0.21 ± 0.1	0.21 ± 0.1	**0.22 ± 0.1**
Vehicle	/	0.06 ± 0.0	0.06 ± 0.0	**0.16 ± 0.0**
Wine	0.82 ± 0.2	**0.90 ± 0.0**	**0.90 ± 0.0**	**0.90 ± 0.0**

Single Labels. In this experiment, we assume that each constraint is expressed on a single cluster label with a specific membership value μ. Table 5 describes, for all the datasets, the performances of the algorithms when $\mu = 1$ (like in any traditional crisp seed-based semi supervised clustering) or $\mu = 0.5$. Figure 2(a) depicts the evolution of the ARI varying with the percentage of constraints. Results with $\mu = 0.2$ are similar to those with $\mu = 0.5$ and thus are not reported.

As expected, when provided contraints are correct, adding constraints enables the clustering algorithms to improve their accuracies and a membership on the constraint label equal to 1 achieves better results than a membership equal to 0.5.

As a general manner, esfcm and sfcm outperform the ssfcm algorithm although ssfcm holds better results than esfcm without constraints. As a matter of fact for ssfcm, constraints are not taken into account to compute the new centers, reducing indirectly its capacity to take into account an harmonious solution encompassing both constrained and unconstrained objects.

Single Labels with Noise. Noise effect is studied by randomly modifying the labels of 20% of the constrained objects so as to produce erroneous annotations. In the end, 6% of the constraints are incorrect, 24% have the correct label and the rest is unconstrained. Table 6 and Fig. 2(b) presents, with the same parametrization as before, the results with noisy constraints.

These results show that as a general manner, noisy sets of labels generate lower quality solutions compared to labels constraints without noise. However,

Table 5. Single label constraints: average ARI for each algorithm and each dataset containing 30% of single label constraints with membership $\mu = 1$ or $\mu = 0.5$.

Dataset	$\mu = 1$				$\mu = 0.5$		
	skmeans	sfcm	ssfcm	esfcm	sfcm	ssfcm	esfcm
GaussK6	0.99	0.99	0.99	0.99	0.99	0.99	0.99
Glass	/	**0.75**	0.56	0.74	0.61	0.47	**0.65**
Ionosphere	/	**0.59**	0.50	0.46	**0.59**	0.56	0.44
Iris	/	**0.92**	0.82	**0.92**	0.87	0.83	**0.92**
Letters	/	0.69	0.39	**0.73**	0.48	0.33	**0.69**
Vehicle	/	0.48	0.13	**0.53**	0.31	0.13	**0.42**
Wine	**0.93**	**0.93**	0.91	**0.93**	0.92	0.92	**0.93**

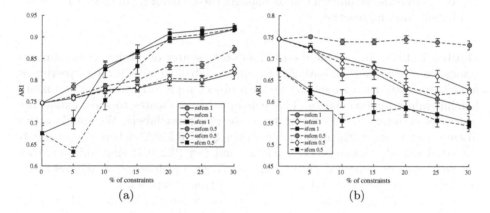

(a)   (b)

Fig. 2. Average ARI and 95% confidence intervals on the Iris dataset as a function of the percentage of (a) not noisy (b) noisy constraints for sfcm, ssfcm and esfcm. Continuous lines represent constraints with membership $\mu = 1$ and dotted lines constraints with $\mu = 0.5$.

Table 6. Single label constraints with noise: average ARI for each algorithm and each dataset containing 30% of single label constraints with membership $\mu = 1$ or $\mu = 0.5$. Here 20% of the constraints are mislabeled.

Dataset	$\mu = 1$				$\mu = 0.5$		
	skmeans	sfcm	ssfcm	esfcm	sfcm	ssfcm	esfcm
GaussK6	0.84	**0.85**	**0.85**	0.84	**0.98**	0.84	0.84
Glass	/	**0.55**	0.35	0.51	**0.67**	0.33	0.43
Ionosphere	/	**0.39**	0.34	0.23	**0.49**	0.38	0.26
Iris	/	0.59	**0.63**	0.55	**0.73**	0.62	0.54
Letters	/	**0.45**	0.28	0.43	0.38	0.25	**0.42**
Vehicle	/	0.31	0.09	**0.38**	0.21	0.09	**0.33**
Wine	**0.75**	**0.75**	0.74	**0.75**	**0.87**	0.75	0.75

the sfcm algorithm is still able to reach a better accuracy than FCM (when there is no constraint). Indeed, sfcm can adjust to which extent it will respect the constraints. Thus, sfcm has a flexibility to ignore some constraints if it enables to keep a coherent overall structure. Inversely, esfcm and ssfcm force the total respect of the constraints, leading to a drop in performances in the presence of noise.

The sfcm algorithm with noisy labels has a better accuracy than FCM in two situations. The first situation happens when the overall structure of a dataset is difficult to retrieve without constraints. It is for example the case for Vehicle or Letters, where the ARI without constraints is low. Consequently, the constraints, even a little noisy, enable to lead the algorithm towards a totally different solution, improving the accuracy. In the second situation, when constraints are uncertain (i.e. with membership strictly below 1), it let sfcm more degrees of freedom to make a choice amongst the constraints in order to preserve a coherent overall structure.

Double Labels. In real-life use-case, an other source of erroneous annotations comes from an expert hesitating between two labels. The following experiment models such situation by setting for each object a pair of constraints on membership values for two classes. This pair of values indicates to some extent the degree of certainty of the expert for these two class labels. We simulate two distinct cases: one with membership values $\xi = (0.5, 0.5)$ where the expert is sure that one of the two labels is correct and $\xi = (0.2, 0.2)$ that indicates that the choice of the expert is not certain. As Glass and Ionosphere datasets only contains two classes, they are discarded from this experiment.

Table 7 and Fig. 3 illustrate the results of both experimentations. Most of the time, the sfcm algorithm outperforms esfcm and ssfcm. While sfcm works better with membership values set to $\xi = (0.5, 0.5)$, esfcm and ssfcm often achieves higher accuracies with lower membership values. Indeed, sfcm has the ability to violate constraints when the solution gets too far from a coherent choice for an overall structure, whereas esfcm and ssfcm are directly incorporating the constraints membership values in the fuzzy partition.

Table 7. Double labels constraints: average ARI for each algorithm and each dataset with 30% of constraints with either $\xi = (0.5, 0.5)$ or $\xi = (0.2, 0.2)$.

Dataset	$\xi = (0.5, 0.5)$			$\xi = (0.2, 0.2)$		
	sfcm	ssfcm	esfcm	sfcm	ssfcm	esfcm
GaussK6	**0.99**	0.89	0.88	**0.99**	0.95	0.98
Iris	**0.85**	0.71	0.67	**0.80**	0.73	0.71
Letters	**0.63**	0.32	0.55	0.38	0.30	**0.55**
Vehicle	0.43	0.10	**0.44**	0.22	0.11	**0.34**
Wine	**0.92**	0.79	0.81	**0.92**	**0.92**	0.91

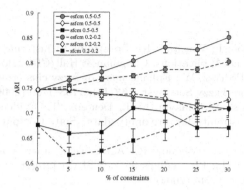

Fig. 3. Double labels constraints: average ARI and 95% confidence intervals as a function of the percentage of constraints for sfcm, ssfcm and esfcm on the Iris dataset. Continuous lines represent constraints $\xi = (0.5, 0.5)$ while dotted lines corresponds to $\xi = (0.2, 0.2)$.

5 Conclusion

In this paper, we propose to use fuzzy algorithms to handle erroneous or uncertain expert annotations for the semi-supervised clustering problem. For the sake of clarity, we restrict our study to three main fuzzy semi-supervised algorithms. In order to make the comparison fair, each algorithm has been either corrected or improved with Mahalanobis distance to ensure comparable performances on all our test datasets. Moreover, we propose a first mapping function that solves the mismatch problem that may occur between labels defined by the initial cluster centers and labels defined in the constraints set. This mapping function although fully functional needs to be optimized, eventually based on a Hungarian algorithm.

Several scenarios are introduced to represent the variety of causes of annotation errors by an expert: either a wrong label, a low confidence in the chosen label or an hesitation between two labels.

We observed that sfcm reaches the more stable results with a good accuracy and esfcm obtains high accuracies only when labels constraints are certain. The ssfcm algorithm often does not achieve good performances. Such results can be explained by the fact that sfcm allows to violate constraints in the final solution whereas esfcm and ssfcm prohibit this behavior.

In our opinion, the major interest of fuzzy semi-supervised algorithms is their ability to handle constraints with a degree of certainty. In case of noise, lowering the labels confidence enables to keep a good improvement of the accuracy when compared to unsupervised clustering algorithm. A perspective is to investigate the addition of labels constraints in other soft clustering algorithms, that generates for instance possibilistic partitions.

References

1. Basu, S., Davidson, I., Wagstaff, K.: Constrained Clustering: Advances in Algorithms, Theory, and Applications. Chapman & Hall/CRC, Boca Raton (2008)
2. Bouchachia, A., Pedrycz, W.: Enhancement of fuzzy clustering by mechanisms of partial supervision. Fuzzy Sets Syst. **157**(13), 1733–1759 (2006)
3. Antoine, V., Quost, B., Masson, M.H., Denœux, T.: Evidential clustering with instance-level constraints for proximity data. Soft. Comput. **18**(7), 1321–1335 (2014)
4. Basu, S., Banerjee, A., Mooney, R.: Active semi-supervision for pairwise constrained clustering. In: Proceedings of 2004 SIAM Interernational Conference on Data Mining, pp. 333–344 (2004)
5. Bilenko, M., Basu, S., Mooney, R.J.: Integrating constraints and metric learning in semi-supervised clustering. In: Proceedings of 21st ICML (2004)
6. Wagstaff, K.L.: When is constrained clustering beneficial, and why. In: AAAI (2006)
7. Vu, V., Labroche, N., Bouchon-Meunier, B.: Boosting clustering by active constraint selection. In: Proceedings of 2010 19th ECAI, pp. 297–302 (2010)
8. Vu, V., Labroche, N., Bouchon-Meunier, B.: An efficient active constraint selection algorithm for clustering. In: 20th ICPR, pp. 2969–2972 (2010)
9. Zhang, D., Tan, K., Chen, S.: Semi-supervised kernel-based fuzzy c-means. In: Pal, N.R., Kasabov, N., Mudi, R.K., Pal, S., Parui, S.K. (eds.) ICONIP 2004. LNCS, vol. 3316, pp. 1229–1234. Springer, Heidelberg (2004). https://doi.org/10.1007/978-3-540-30499-9_191
10. Pedrycz, W., Waletzky, J.: Fuzzy clustering with partial supervision. IEEE Trans. Syst. Man Cybern. Part B Cybern. **27**(5), 787–795 (1997)
11. Lai, D., Garibaldi, J.: A comparison of distance-based semi-supervised fuzzy c-means clustering algorithms. In: IEEE International Conference on Fuzzy Systems (FUZZ-IEEE), pp. 1580–1586 (2011)
12. Bezdek, J.: Pattern Recognition with Fuzzy Objective Function Algorithms. Advanced Applications in Pattern Recognition. Springer, New York (1981). https://doi.org/10.1007/978-1-4757-0450-1
13. Gustafson, D., Kessel, W.: Fuzzy clustering with a fuzzy covariance matrix. In: IEEE Conference on Decision and Control Including the 17th Symposium on Adaptive Processes, pp. 761–766 (1979)
14. Endo, Y., Hamasuna, Y., Yamashiro, M., Miyamoto, S.: On semi-supervised fuzzy c-means clustering. In: IEEE International Conference on Fuzzy Systems (FUZZ-IEEE), pp. 1119–1124 (2009)
15. Basu, S., Banerjee, A., Mooney, R.: Semi-supervised clustering by seeding. In: Proceedings of 19th International Conference on Machine Learning (ICML), pp. 27–34 (2002)
16. Basu, S., Bilenko, M., Banerjee, A., Mooney, R.: Probabilistic Semi-supervised Clustering with Constraints, pp. 71–98. MIT Press, Cambridge (2006)
17. Rand, W.: Objective criteria for the evaluation of clustering methods. J. Am. Stat. Assoc. **66**(336), 846–850 (1971)
18. Dave, R.: Validating fuzzy partitions obtained through c-shells clustering. Pattern Recogn. Lett. **17**(6), 613–623 (1996)

Towards a Hierarchical Extension of Contextual Bipolar Queries

Janusz Kacprzyk[1,2(✉)] and Sławomir Zadrożny[1,2]

[1] Systems Research Institute, Polish Academy of Sciences,
ul. Newelska 6, 01-447 Warsaw, Poland
{kacprzyk,zadrozny}@ibspan.waw.pl
[2] WIT – Warsaw School of Information Technology,
ul. Newelska 6, 01-447 Warsaw, Poland

Abstract. We are concerned with the bipolar database queries in which the query is composed of a necessary (required) and optional (desired) part connected with a non-conventional aggregation operator "and possibly", combined with context, as, for instance, in the query "find houses which are cheap and – with respect to other houses in town – possibly close to a railroad station". We deal with a multivalued logic based interpretation of bipolar queries. We assume that the human user, usually a database novice, tends to use general terms in the queries in natural language, which do not directly relate to attributes, and via a question and answer process these terms are "decoded" using a concept hierarchy that at the end involves terms directly related to attribute values. We propose a novel extension of our contextual hierarchical bipolar database query in which the original query is considered a level 0 query at the bottom of the precisiation hierarchy, then its required and optional parts are assumed to be bipolar queries themselves, with an account of context. This makes it possible to further precisiate the user's intentions/preferences. A level 1 of precisiation is obtained, and the process is continued so far as it is necessary for the user to adequately reflect his/her intentions/preferences as to what is sought. The new concept is demonstrated on an intuitively appealing real estate example which will serve the role of both an illustration of the idea of our approach and of a real example.

Keywords: Database query · Bipolar query · Context · Fuzzy logic
User intention · User preference

1 Introduction

The paper is concerned with a very important issue in database querying context, as well as in information retrieval, Web search, etc., in that the human user is mostly a database novice for whom the only fully natural way of articulation and communication is natural language, with its inherent imprecision. Therefore, questions, queries, requests, etc. posed originally by the humans in

© Springer International Publishing AG, part of Springer Nature 2018
J. Medina et al. (Eds.): IPMU 2018, CCIS 854, pp. 63–74, 2018.
https://doi.org/10.1007/978-3-319-91476-3_6

natural language should be "rephrased" into another form suitable for the computer. In our work we deal with database querying in which the user tends to formulate queries to a database involving human specific terms and relations that are not directly applicable in a querying language exemplified by the SQL. For instance, in a simple real estate example that will be used here, a "natural" human query may be "find all houses that are comfortable and well located", and it is obvious that "comfortable" and "well located" are imprecise terms that may not relate directly to database attributes, and should be further precisiated, or "decoded". Moreover, both conditions present in the query may not be of the same importance, meant in a subtle way as that the user may be happy with a house just comfortable if at a given location there are no houses both "comfortable" and "well located". However, if it turns out that such an "ideal" house exists at a given location, then the user is not anymore interested in houses just "comfortable" but not "well located".

Thus, in the proposed approach we aim at providing a database user with a high flexibility in forming a query via taking into account the fact that conditions may be satisfied to a degree and that they may be of a different importance, in the above mentioned sense, as well with a possibility to express his or her preferences at a higher level, i.e., characterizong the data sought using terms not necessarily directly represented in a database.

Our point of departure is an important new direction in fuzzy querying which – by extending straightforward approaches that have been used since the late 1970s, and even some extended approaches based on fuzzy logic with linguistic quantifiers (cf. Kacprzyk and Ziółkowski [21], Kacprzyk et al. [22]) – tries to explicitly take into account *bipolarity* in human judgments and intention/preference articulation. Further, we adopt our extension of the concept of bipolarity including the notion of the context. Finally, we combine that with our concept of hierarchical bipolar queries and propose a novel approach to the flexible querying of databases.

In Sect. 2 we briefly remind the essence of one of possible approaches to bipolarity in querying. Next, in Sect. 3 the concept of the context, as used in our approach to bipolar queries, is briefly discussed. Then, in Sect. 4 we will present the new hierarchical contextual bipolar queries.

2   Bipolar Fuzzy Queries

The term "bipolar query", first introduced by Dubois and Prade [8–11], boils down practically to the distinguishing of two types of query conditions which express negative and positive user preferences. Essentially, as proved by many psychological, cognitive, etc. experiments, a human being in his/her assessments is usually tempted to use some sort of a *bipolar scale* in the sense that while trying to provide an opinion or testimony about some objects, values of their features, capabilities, etc. he or she would rather assess and evaluate them through:

- some degree of being *negative*, i.e., to be rejected,
- some degree of being *positive*, i.e., to be accepted.

Here we follow the main line of reasoning in the formalization of bipolarity in judgments and evidence via fuzzy logic and possibility theory due to Dubois and Prade and their collaborators, e.g.: Benferhat et al. [1], Dubois and Prade [8, 9,11], cf. also Dubois and Prade [12], De Tré et al. [6], Dziedzic et al. [13,20], Hadjali et al. [15], Lietard and Rocacher [24], Lietard et al. [26], Matthé et al. [27], to name a few.

The first issue is to properly choose a scale to one's positive/negative evaluation. In practice two such scales are used (cf. Grabisch et al. [14]):

- *bipolar univariate* and
- *unipolar bivariate,*

with the former assuming one scale with three main levels of, respectively, negative, neutral and positive evaluation, gradually changing from one end of the scale to another, usually represented by $[-1, 1]$, while the latter assuming two independent scales for a positive and negative evaluation, usually represented by $[0, 1]$. The latter will be used here.

Then, we adopt the approach to bipolar queries which assigns a different semantics to the negative and positive evaluations. Namely, the objects (here: tuples) admitting negative evaluation are rejected and positive evaluation contributes to the overall evaluation of an object only if it is non-rejected objects. This is a bit oversimplified view on the semantics of the bipolar queries we adopt as the positive/negative evaluations are assumed to be gradual and their combination is more peculiar but this serves well the purpose of presenting the essence of our new proposal. Moreover, even if the role of positive evaluations seems to be somehow secondary from what was said above, they are however equally important as negative evaluations in case when there exist non-rejected objects admitting positive evaluations. Such a bipolar semantics may be best described via introducing a special aggregation operator "and if possibly" which is briefly described in what follows.

A prototypical example of a bipolar query considered in this paper is:

$$C \text{ and possibly } P \tag{1}$$

exemplified by "find a house which is inexpensive (C) and possibly close to public transportation (P)" which is basically meant as that the above query is satisfied by a tuple t only if either one of the two conditions holds:

1. it satisfies (to a high degree) both conditions C and P, or
2. it satisfies just C and there is no tuple in the whole database which satisfies both conditions.

The idea of such a query has been introduced in the seminal paper of Lacroix and Lavency [23] who proposed the use of a query (C, P) with two categories of conditions: C which is required (mandatory), and P which expresses just mere preferences (desires).

The semantics of Lacroix and Lavency's approach is provided by the "and possibly" operator meant as sketched above, i.e., if at least one tuple in given database satisfies both mandatory and desired condition then the "and possibly" operator is interpreted as the standard conjunction, otherwise only the mandatory condition is taken into account while evaluating the tuples.

Such an aggregation operator has been later proposed independently by Dubois and Prade [7] in default reasoning and by Yager [28,29] in multicriteria decision making, cf. also Bordogna and Pasi [2] in information retrieval.

Bipolar queries with the "and possibly" operator my be also seen as a special case of Chomicki's [5] *queries with preferences* based on an extra relational algebra operator called the *winnow*, proposed by Chomicki. Such an operator is associated with a preference relation on the universe of tuples and returns as a result those tuples which are non-dominated with respect to this preference relation. More on how a bipolar query (C, P) may be expressed using a query with preferences may be found, e.g., in Zadrożny and Kacprzyk [32]; cf. also (4).

While discussing the bipolarity, this semantics may well be explained as follows. The unipolar bivariate scale is assumed, and a special interpretation in which the negative and positive assessments are considered to correspond to the *required* and *desired* conditions, i.e. the negative assessment is identified with the degree to which the required condition is <u>not</u> satisfied as, e.g., if a house sought has to be cheap (the required condition), then its negative assessment corresponds to the degree to which it is not cheap. The desired condition directly corresponds to the positive assessment.

Lacroix and Lavency [23] consider only the crisp (nonfuzzy) conditions C and P. Then, a bipolar query (C, P) can be processed via the "first select using C then order using P" strategy, i.e., by finding tuples satisfying C and, second, choosing from among them those satisfying P, if any. Some fuzzifications of the original Lacroix and Lavency's approach are proposed by Zadrożny [30], and Zadrożny and Kacprzyk [31], which will be used here; for some other approaches, see Bosc et al. [3,4], or Lietard et al. [25], etc.

Consider the general form of the bipolar query (1) and denote, as previously, by C the complement of the negative assessment (e.g., "price is cheap") and by P the positive assessment (e.g., located "near a railroad station"). Then, the semantics of the bipolar query (1) may be formally expressed as:

– a tuple t belongs to the answer set of the query (1) if it satisfies (notice that $P(t)$ and $C(t)$ are binary predicates):

$$C(t) \ and \ possibly \ P(t) \equiv C(t) \wedge \exists s(C(s) \wedge P(s)) \Rightarrow P(t) \qquad (2)$$

– and if there are tuples satisfying both P and C, then (2) boils down to $C \wedge P$ while otherwise it boils down to C alone.

Basically, the fuzzification of the above mentioned concepts of a bipolar query, can be done in the following ways (cf. Zadrożny and Kacprzyk [32]):

– by a direct fuzzification of (2); notice that the notation is the same as for (2) but now the predicates are fuzzy:

$$C(t) \ and \ possibly \ P(t) \equiv C(t) \wedge \exists s \ (C(s) \wedge P(s)) \Rightarrow P(t) \qquad (3)$$

- by a direct fuzzification of the winnow operator (cf. Chomicki [5]) and applying it with a preference relation based on $P(\cdot)$, i.e., t is preferred to s iff $P(t)$ and $\neg P(s)$ (this will not be dealt with in more detail since it is not the main focus of this paper):

$$C(t) \text{ and possibly } P(t) \equiv C(t) \wedge \neg \exists s \, ((C(s) \wedge P(s) \wedge \neg P(t))) \qquad (4)$$

- by using our fuzzy version of the winnow operator (cf. Zadrożny and Kacprzyk [32]) and applying it with a preference relation based on $P(\cdot)$ as above:

$$C(t) \text{ and possibly } P(t) \equiv C(t) \wedge \forall s \, (C(s) \Rightarrow (\neg P(s) \vee P(t))) \qquad (5)$$

and, clearly, these forms are equivalent in the classic Boolean logic.

The above logic formulas can then be precisiated in the sense that the one can choose a specific form of the conjunction and disjunction, i.e. a t-norm and t-conorm (often called an s-norm), and the negation, which form so-called De Morgan Triples $(\wedge, \vee, \neg)$ that comprise a t-norm operator $\wedge$, a t-conorm (s-norm) operator $\vee$ and a negation operator $\neg(x \vee y) = \neg x \wedge \neg y$, which then give rise to specific S-implications and R-implications; Clearly a type of a fuzzy logic formal system should be specified. For details, cf. Zadrożny and Kacprzyk [32].

3 Contextual Bipolar Queries

The general form of a bipolar query can be extended in various ways but for our purposes its extension into a *contextual bipolar query* proposed by Zadrożny. Kacprzyk and Dziedzic [33,34] is relevant.

We start, as before, with the bipolar query meant within the required/desired semantics and, the query "C and possibly P" is satisfied by a tuple t only if either of two conditions holds: (1) it satisfies (of course, possibly to a high degree) both conditions C and P, or (2) it satisfies C and there is no tuple in the whole database which satisfies both conditions.

However, a natural extension of the "if possibly" in (1) is that the satisfaction of both C and P can be meant in a certain *context*. For instance, a person planning to visit a few regions (towns, ...) of a given country can be looking for a cheap and comfortable hotel(s) but maybe this wish cannot be fulfilled in case of some regions because, for instance, a given region can just be very expensive. The inclusion of a *context* can therefore be natural making it possible to constrain the check for the possibility of satisfying both conditions to a suitable subset of tuples. This implies the concept of a *contextual bipolar query* introduced by Zadrożny. Kacprzyk and Dziedzic [33,34] which may be exemplified by:

Find *cheap* and possibly *−with respect to the hotels located in the same region− comfortable* hotels $\qquad (6)$

to be meant as to be satisfied by a hotel if:

1. it is cheap (to a high degree) and is comfortable (to a high degree), or
2. it is cheap (to a high degree) and there is no other hotel   (7) located in the same region which is both cheap and comfortable.

The new "and possibly + context" operator may be formalized as follows. The context is identified with a part of the database defined by an additional binary predicate W, i.e.,

$$Context(t) = \{s \in R : W(t,s)\} \tag{8}$$

where R denotes the whole database (relation).

The "and possibly + context" operator has three arguments:

$$C \text{ and possibly } P \text{ with respect to } W \tag{9}$$

where the predicates C and P should be interpreted, as previously, as representing the required and desired conditions, respectively, while the predicate W denotes the context.

Then, the formula (9) is interpreted as:

$$C(t) \text{ and possibly } P(t) \text{ with respect to } W \equiv$$
$$C(t) \wedge \exists s(W(t,s) \wedge C(s) \wedge P(s)) \Rightarrow P(t) \tag{10}$$

In our example, C and P represent the properties of "cheap" and "comfortable", respectively, while W denotes the relation of being "located in the same region", i.e., $W(t,s)$ is true if both tuples represent hotels located in the same region; W can also be fuzzy.

A relation expressed by the context, i.e. predicate W, defines basically a partition (in a broad sense) of the set of tuples, crisp or fuzzy. This can be formalized in various ways, for instance by specifying an equivalence, similarity, etc. relation, an ordering, or even a modal logic based interpretation. For a lack of space, for details we refer to our papers, e.g. Zadrożny. Kacprzyk and Dziedzic [33,34].

4   Hierarchical Contextual Bipolar Queries

Now, by following the argument for the *hierarchical bipolar queries* proposed by Kacprzyk and Zadrożny [16] (cf. also a related concept of a compound bipolar query, cf. Kacprzyk and Zadrożny [17–19]), we will present how the degrees of truth for particular tuples are calculated for the *contextual bipolar queries* in a hierarchical context. We will follow, to make our short presentation easier comprehensible, the real estate example presented in Sect. 1.

Suppose that a customer looks for houses that are *"financially advantageous and possibly well located"* but is interested in houses in different parts of the town and is well aware that in some of them meeting both conditions is perfectly possible while in some other this may not be the case, so that a contextual bipolar query should be more appropriate, that is, a potential customer can pose the initial query as:

find a house that is *financially advantageous* (C_0) *and possibly*
– *with respect to other houses located in the same part of town*   (11)
(W) – *is well located* (P_0)

to be meant to that a house satisfies this query if:

1. it is financially advantageous and is well located (to a high degree), or
2. it is financially advantageous (to a high degree) and there is no other house *located in the same part of town* which is both financially advantageous and well located,

and notice the lower index 0 associated with the predicates C and P indicates that they correspond to the initial (zero) level of the query, and we also assume – for clarity of presentation – that the context W is the same for all levels. Clearly, one can also assume different contexts for different levels which does not change the essence of the approach.

Then, using the interpretation of bipolar queries due to (2), used throughout the rest of this paper for simplicity, (9) is interpreted as:

$$C_0(t) \text{ and possibly } P_0(t) \text{ with respect to } W \equiv$$
$$C_0(t) \wedge \exists s(W(t,s) \wedge C_0(s) \wedge P_0(s)) \Rightarrow P_0(t) \quad (12)$$

where t, as previously, denotes a tuple (here: representing a house).

Then, since the real estate agent tries to "decode" the intention/preference of the customer, he or she is requested to more specifically formulate the required condition (here "financially advantageous") and desired condition (here "good location"), in context W. For instance, the customer can say that (notice that the upper indexes of the C and P conditions will correspond to the level of concept hierarchy, i.e. generality, at which a concept in question is considered):

– the extent of the predicate C_0, representing the required condition "a financially advantageous house", may be equated with the answer set of the following query:

find a house that is *inexpensive* $(C_1^{C_0})$ *and possibly* – *with respect to other houses located in the same part of town* (W) –   (13)
in a *modern building* $(P_1^{C_0})$

which, similarly as for (12), yields

$$C_1^{C_0}(t) \text{ and possibly } P_1^{C_0}(t) \text{ with respect to } W \equiv$$
$$C_1^{C_0}(t) \wedge_o \exists s(W(t,s) \wedge C_1^{C_0}(s) \wedge P_1^{C_0}(s)) \Rightarrow P_1^{C_0}(t) \quad (14)$$

- the extent of the predicate P_0, representing the desired condition "well located"', may be equated with the answer set of the following query:

find a house that is *in an affluent part of town* $(C_1^{P_0})$ *and possibly – with respect to other houses located in the same part of town* (W) *– is close to a recreational area* $(P_1^{P_0})$ (15)

which, similarly as for (12), yields

$$C_1^{P_0}(t) \text{ and possibly } P_1^{P_0}(t) \text{ with respect to } W \equiv$$
$$C_1^{P_0}(t) \wedge_o \exists s(W(t,s) \wedge C_1^{P_0}(s) \wedge P_1^{P_0}(s)) \Rightarrow P_1^{P_0}(t) \quad (16)$$

and these can be viewed as the first level query formulations (intentions/preferences).

Again, these condition can also be viewed to be too general, not directly related to database attribute values, and one can proceed further:

- for the second level formulation of the first level required condition $C_1^{C_0}$, i.e. "inexpensive":

find a house that has a *low price* $(C_2^{C_1^{C_0}})$ *and possibly – with respect to other houses located in the same part of town* (W) *– has a good bank loan offer* $(P_2^{C_1^{C_0}})$ (17)

which, similarly as for (14), yields

$$C_2^{C_1^{C_0}}(t) \text{ and possibly } P_2^{C_1^{C_0}}(t) \text{ with respect to } W \equiv$$
$$C_2^{C_1^{C_0}}(t) \wedge_o \exists s(W(t,s) \wedge C_2^{C_1^{C_0}}(s) \wedge P_2^{C_1^{C_0}}(s)) \Rightarrow P_2^{C_1^{C_0}}(t) \quad (18)$$

- for the second level formulation of the first level desired condition $P_1^{C_0}$, i.e. "modern building":

find a house that has an *intelligent energy management* $(C_2^{P_1^{C_0}})$ *and possibly – with respect to other houses located in the same part of town* (W) *– has fast lifts* $(P_2^{P_1^{C_0}})$ (19)

which, similarly as for (16), yields

$$C_2^{P_1^{C_0}}(t) \text{ and possibly } P_2^{P_1^{C_0}}(t) \text{ with respect to } W \equiv$$
$$C_2^{P_1^{C_0}}(t) \wedge_o \exists s(W(t,s) \wedge C_2^{P_1^{C_0}}(s) \wedge P_2^{P_1^{C_0}}(s)) \Rightarrow P_2^{P_1^{C_0}}(t) \quad (20)$$

– for the second level formulation of the first level required condition $C_1{}^{P_0}$, i.e. "an affluent part of town":

> find a house that is in a *quiet zone* $(C_2{}^{C_1{}^{P_0}})$ *and possibly* (21)
> *– with respect to other houses located in the same part of town*
> *(W) – is close to the business district* $(P2{}^{C_1{}^{P_0}})$

which, similarly as for (20), yields

$$C_2{}^{C_1{}^{P_0}}(t) \ and \ possibly \ P_2{}^{C_1{}^{P_0}}(t) \ with \ respect \ to \ W \equiv$$
$$C_2{}^{C_1{}^{P_0}}(t) \wedge_o \exists s(W(t,s) \wedge C_2{}^{C_1{}^{P_0}}(s) \wedge C_2{}^{C_1{}^{P_0}}(s)) \Rightarrow C_2{}^{C_1{}^{P_0}}(t) \quad (22)$$

– for the second level formulation of the first level required condition $P_1{}^{P_0}$, i.e. "close to a recreational area":

> find a house that is *close to a park* $(C_2{}^{P_1{}^{P_0}})$ *and possibly* (23)
> *– with respect to other houses located in the same part of town*
> *(W) – is close to a lake* $(P2{}^{P_1{}^{P_0}})$

which, similarly as for (22), yields

$$C_2{}^{P_1{}^{P_0}}(t) \ and \ possibly \ P_2{}^{P_1{}^{P_0}}(t) \ with \ respect \ to \ W \equiv$$
$$C_2{}^{P_1{}^{P_0}}(t) \wedge_o \exists s(W(t,s) \wedge C_2{}^{P_1{}^{P_0}}(s) \wedge P_2{}^{P_1{}^{P_0}}(s)) \Rightarrow P_2{}^{P_1{}^{P_0}}(t) \quad (24)$$

and, if necessary, one can continue until the conditions involve attributes in the database.

Notice that the context condition in the above queries (i.e. "with respect to other houses located in the same part of the town") is assumed the same for all queries and conditions but, in general, we can also employ local contexts which will not be considered here.

5 Concluding Remarks

We have further extended two classes of our works: first, on contextual bipolar queries to databases in which the query is composed of a required and desired part connected with a non-conventional aggregation operator "and possibly", with context, as, for instance, in the query "find houses which are cheap and – with respect to other houses in town – possibly close to a railroad station" assuming a multivalued (fuzzy) logic based perspective. Second, our hierarchical bipolar database query in which, basically, the original query is considered a level 0 query at the bottom of the precisiation hierarchy, then its required and desired parts are assumed to be bipolar queries themselves, in this paper under a specified context. This makes it possible to further precisiate the user's intentions/preferences that involve initially natural language terms the human user cannot initially present via values of attributes. A level 1 of precisiation is obtained, and the process is continued as far as it is necessary for the user to adequately reflect his/her intentions/preferences as to what is sought. The new concept has been illustrated on an intuitively appealing real estate example. Notice that, for simplicity and clarity, this example – used both for the illustration of our approach and as an application example – is shown in a general way using linguistic information only.

Acknowledgment. This work has been partially supported by Project 691249, RUC-APS: Enhancing and implementing Knowledge based ICT solutions within high Risk and Uncertain Conditions for Agriculture Production Systems (www.ruc-aps.eu), funded by the EU under H2020-MSCA-RISE-2015.

References

1. Benferhat, S., Dubois, D., Kaci, S., Prade, H.: Modeling positive and negative information in possibility theory. Int. J. Intell. Syst. **23**, 1094–1118 (2008)
2. Bordogna, G., Pasi, G.: Linguistic aggregation operators of selection criteria in fuzzy information retrieval. Int. J. Intell. Syst. **10**(2), 233–248 (1995)
3. Bosc, P., Pivert, O., Mokhtari, A., Lietard, L.: Extending relational algebra to handle bipolarity. In: Shin, S.Y., Ossowski, S., Schumacher, M., Palakal, M.J., Hung, Ch.-Ch. (eds.) Proceedings of 2010 ACM Symposium on Applied Computing (SAC), pp. 1718–1722. ACM, Sierre (2010)
4. Bosc, P., Pivert, O.: On four noncommutative fuzzy connectives and their axiomatization. Fuzzy Sets Syst. **202**, 42–60 (2012)
5. Chomicki, J.: Preference formulas in relational queries. ACM Trans. Database Syst. **28**(4), 427–466 (2003)
6. De Tré, G., Zadrożny, S., Bronselaer, A.: Handling bipolarity in elementary queries to possibilistic databases. IEEE Trans. Fuzzy Sets **18**(3), 599–612 (2010)
7. Dubois, D., Prade, H.: Default reasoning and possibility theory. Artif. Intell. **25**(2), 243–257 (1988)
8. Dubois, D., Prade, H.: Bipolarity in flexible querying. In: Carbonell, J.G., Siekmann, J., Andreasen, T., Christiansen, H., Motro, A., Legind Larsen, H. (eds.) FQAS 2002. LNCS (LNAI), vol. 2522, pp. 174–182. Springer, Heidelberg (2002). https://doi.org/10.1007/3-540-36109-X_14
9. Dubois, D., Prade, H.: Handling bipolar queries in fuzzy information processing. In: Galindo, J. (ed.) Handbook of Research on Fuzzy Information Processing in Databases, pp. 97–114. Information Science Reference, New York (2008)
10. Dubois, D., Prade, H.: An introduction to bipolar representations of information and preference. Int. J. Intell. Syst. **23**, 866–877 (2008)
11. Dubois, D., Prade, H.: An overview of the asymmetric bipolar representation of positive and negative information in possibility theory. Fuzzy Sets Syst. **160**(10), 1355–1366 (2009)
12. Dubois, D., Prade, H.: Modeling and if possible "and or at least": different forms of bipolarity in flexible querying. In: Pivert, O., Zadrożny, S. (eds.) Flexible Approaches in Data, Information and Knowledge Management. SCI, vol. 497, pp. 3–19. Springer, Cham (2014). https://doi.org/10.1007/978-3-319-00954-4_1

13. Dziedzic, M., Kacprzyk, J., Zadrozny, S.: Contextual bipolarity and its quality criteria in bipolar linguistic summaries. Tech. Trans.: Autom. Control **4-AC**, 117–127 (2014)
14. Grabisch, M., Greco, S., Pirlot, M.: Bipolar and bivariate models in multicriteria decision analysis: descriptive and constructive approaches. Int. J. Intell. Syst. **23**, 930–969 (2008)
15. Hadjali, A., Kaci, S., Prade, H.: Database preference queries a possibilistic logic approach with symbolic priorities. Ann. Math. Artif. Intell. **63**(3–4), 357–383 (2011)
16. Kacprzyk, J., Zadrozny, S.: Hierarchical bipolar fuzzy queries: towards more human consistent flexible queries. In: IEEE International Conference on Fuzzy Systems (FUZZ-IEEE 2013), pp. 1–8 (2013)
17. Kacprzyk, J., Zadrozny, S.: Compound bipolar queries: combining bipolar queries and queries with fuzzy linguistic quantifiers. In: Proceedings of 8th Conference of the European Society for Fuzzy Logic and Technology (EUSFLAT-2013), University of Milano-Bicocca, Milan, Italy, 11–13 September 2013, pp. 848–855. Atlantis Press (2013)
18. Kacprzyk, J., Zadrożzny, S.: Compound bipolar queries: the case of data with a variable quality. In: Proceedings of IEEE-2017, IEEE International Conference on Fuzzy Systems, Naples, Italy, 9–12 July 2017, pp. 1–6. IEEE Press (2017)
19. Kacprzyk, J., Zadrożny, S.: Compound bipolar queries: a step towards an enhanced human consistency and human friendliness. In: Matwin, S., Mielniczuk, J. (eds.) Challenges in Computational Statistics and Data Mining. SCI, vol. 605, pp. 93–111. Springer, Cham (2016). https://doi.org/10.1007/978-3-319-18781-5_6
20. Kacprzyk, J., Zadrożny, S., Dziedzic, M.: A novel view of bipolarity in linguistic data summaries. In: Kóczy, L.T., Pozna, C.R., Kacprzyk, J. (eds.) Issues and Challenges of Intelligent Systems and Computational Intelligence. SCI, vol. 530, pp. 215–229. Springer, Cham (2014). https://doi.org/10.1007/978-3-319-03206-1_16
21. Kacprzyk, J., Ziółkowski, A.: Database queries with fuzzy linguistic quantifiers. IEEE Trans. Syst. Man Cybern. SMC **16**, 474–479 (1986)
22. Kacprzyk, J., Zadrożny, S., Ziółkowski, A.: FQUERY III+: a "human consistent" database querying system based on fuzzy logic with linguistic quantifiers. Inf. Syst. **14**(6), 443–453 (1989)
23. Lacroix M., Lavency P.: Preferences: putting more knowledge into queries. In: Proceedings of 13 International Conference on Very Large Databases, pp. 217–225 (1987)
24. Lietard, L., Rocacher, D.: On the definition of extended norms and co-norms to aggregate fuzzy bipolar conditions. In: Carvalho, J.P., Dubois, D., Kaymak, U., da Costa Sousa, J.M. (eds.) Proceedings of Joint 2009 International Fuzzy Systems Association World Congress and 2009 European Society of Fuzzy Logic and Technology Conference, Lisbon, Portugal, 20–24 July 2009, pp. 513–518 (2009)
25. Lietard, L., Rocacher, D., Bosc, P.: On the extension of SQL to fuzzy bipolar conditions. In: Proceedings of NAFIPS-2009 Conference, pp. 1–6 (2009)
26. Lietard, L., Tamani, N., Rocacher, D.: Fuzzy bipolar conditions of type "or else". In: FUZZ-IEEE, pp. 2546–2551. IEEE (2011)
27. Matthé, T., De Tré, G., Zadrożny, S., Kacprzyk, J., Bronselaer, A.: Bipolar database querying using bipolar satisfaction degrees. Int. J. Intell. Syst. **26**(10), 890–910 (2011)
28. Yager, R.: Higher structures in multi-criteria decision making. Int. J. Man Mach. Stud. **36**, 553–570 (1992)

29. Yager, R.: Fuzzy logic in the formulation of decision functions from linguistic specifications. Kybernetes **25**(4), 119–130 (1996)
30. Zadrożny, S.: Bipolar queries revisited. In: Torra, V., Narukawa, Y., Miyamoto, S. (eds.) MDAI 2005. LNCS (LNAI), vol. 3558, pp. 387–398. Springer, Heidelberg (2005). https://doi.org/10.1007/11526018_38
31. Zadrożny, S., Kacprzyk, J.: Bipolar queries and queries with preferences. In: Proceedings of DEXA 2006, pp. 415–419 (2006)
32. Zadrożny, S., Kacprzyk, J.: Bipolar queries: an aggregation operator focused perspective. Fuzzy Sets Syst. **196**, 69–81 (2012)
33. Zadrożny, S., Kacprzyk, J., Dziedzic, M., De Tré, G.: Contextual bipolar queries. In: Jamshidi, M., Kreinovich, V., Kacprzyk, J. (eds.) Advance Trends in Soft Computing. SFSC, vol. 312, pp. 421–428. Springer, Cham (2014). https://doi.org/10.1007/978-3-319-03674-8_40
34. Zadrożny, S., Kacprzyk, J., Dziedzic, M.: Contextual bipolar queries. In: Proceedings of IFSA/EUSFLAT 2015, Gijon, Spain, pp. 1266–1273. Atlantis Press (2015)

Towards an App Based on FIWARE Architecture and Data Mining with Imperfect Data

Jose M. Cadenas[✉], M. Carmen Garrido, and Cristina Villa

Department of Information and Communications Engineering,
University of Murcia, Murcia, Spain
{jcadenas,carmengarrido,cristina.villa}@um.es

Abstract. In this work, the structure for the prototype construction of an application that can be framed within ubiquitous sensing is proposed. The objective of application is to allow that a user knows through his mobile device which other users of his environment are doing the same activity as him. Therefore, the knowledge is obtained from data acquired by pervasive sensors. The FIWARE infrastructure is used to allow to homogenize the data flows.

An important element of the application is the Intelligent Data Analysis module where, within the *Apache Storm* technology, a Data Mining technique will be used. This module identifies the activity carried out by mobile device user based on the values obtained by the different sensors of the device.

The Data Mining technique used in this module is an extension of the Nearest Neighbors technique. This extension allows the imperfect data processing, and therefore, the effort that must be made in the data preprocessing to obtain the minable view of data is reduced. It also allows us to parallelize part of the process by using the *Apache Storm* technology.

Keywords: Intelligent Data Analysis · Fuzzy k nearest neighbors
Imperfect data · Activity recognition

1 Introduction

With the technology development in recent years, more specifically in the Internet of Things area (IoT), nowadays large amounts of data of a very diverse nature are generated. Due to this, one of the investigation fields that is taking more relevancy inside the Intelligent Data Analysis is Smart Cities. It intends to make use of the data that are generated to provide services of higher quality, with greater efficiency, more sustainable and where citizens have an active participation.

In this sense, one of the most interesting topics is the city contextualization and its inhabitants, so that the environment can be perceived and the services

© Springer International Publishing AG, part of Springer Nature 2018
J. Medina et al. (Eds.): IPMU 2018, CCIS 854, pp. 75–87, 2018.
https://doi.org/10.1007/978-3-319-91476-3_7

can be adapted according to the actual state of them. In this way, it is necessary to understand the citizens day to day based on the activities that their inhabitants carry out and their way of interacting with the environment.

Inside this framework, mobility and the use of transport have become two major sectors to pay attention to. Therefore, lots of works are found, as [18, 24, 26], where information about points and the most important connections of a city, the social use of those zones and the traffic conditions can be obtained from the trajectory marked by the GPS sensors in the taxis. This information can help to make decisions when choosing a route and even serves as a basis for urbanism according to the needs or characteristics of each place. Following this approach, in [2], several elements such as GPS sensors located in buses, the stop geolocation and the smart cards of the users are used to improve the management of the bus lines of a city classifying zones based on four categories: residential, work, nightlife and personal. Making use of these cards, in [4] points of user origin-destination are used together with each trajectory timestamps to understand the use of the lines depending on the area where the main stations are located (residential, work, etc.) and the time slots. In this way, several lines and links can be proposed to the user avoiding crowds at peak times.

However, to achieve a real service personalization, it is necessary to collect data directly on people [13, 25], working within the Ubiquitous Sensing scope. Because of this, the use of smartphones as ubiquitous devices and provided with diverse sensors constitutes an essential tool within this field. Currently, smartpthones incorporate accelerometers, pedometers, heart rate monitors, and global positioning systems, among many other sensors, which make them ideal for the activity recognition task.

So, data generated by GSM and WIFI signals are used in several works, as in [16, 21], to detect several activities (walk, drive, stay at home/be still, ride a bicycle) by means of the distances between the reception towers and using decision trees to determine the activity. In [12, 23], accelerometers located on smartphones were used to obtain the data and then different studies of classification techniques were performed for the carried out activities detection.

Activity recognition from a user can have many applications in different areas. For example, it can be used to monitor and diagnose patients who need more continuous control, [14, 22], or to help detecting falls of elderly sending alerts for a faster intervention [1]. It can also be used to adapt the devices to the activity realized on a certain moment, such as increasing the music volume or sending calls to voicemail directly if the user is running [11] or, as in [15], increasing the letter size of the devices and, thereby, to improve readability when the user is walking.

However, the human activity recognition takes great importance in the sports domain. For people who want to monitor themselves when they perform physical exercise is useful to be aware of the time and intensity of it, the calories burned, their activity history, etc. This explains the success of applications like [7, 17] that give the user all this information and the opportunity to plan their own routines.

That is why the aim of this work is to create an application that identifies the activity that a user is doing in real time, as well as its location, to be able to recommend points of the city where there are more people doing the same kind of activity as him to improve the realization of it and the social skills.

This paper is structured as follows. In Sect. 2, the proposed social application is presented, illustrating its main objective and its general design. In Sect. 3, two application elements are presented: the Intelligent Data Analysis module and the data preprocessing. These elements are preliminarily evaluated in Sect. 4. Finally, in Sect. 5 the conclusions and future works are presented.

2 Developing a Social Application from Smartphone Sensors

In this paper, the initial prototype of an application is proposed. The application can be framed within ubiquitous sensing since its main objective is the extraction of knowledge from data acquired by pervasive sensors [19]. More specifically, the application carries out the human activity recognition from smartphone sensors aimed to create a system to social activity monitoring. This application is named Social Jogging (*SJogg*).

The objective of *SJogg* application is to allow that a user knows through his smartphone, which other users of his environment are doing the same activity. In this way the user will be able to carry out this activity in a group way, improving the performance of the same as well as his social skills. There are several studies that indicate that doing group activities has the advantage of greater motivation, improving performance and avoiding the boredom of doing it alone and without interacting with others. Figure 1 shows the objective of *SJogg* application illustratively.

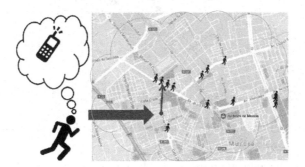

Fig. 1. *SJogg* application objective

The starting point of the application development is a dataset obtained from sensors that are currently being incorporated into mobile devices, such as mobile phones or music players. On these data, an Intelligent Data Analysis (IDA) process is performed to model the activity carried out by the mobile device user, based on the values obtained by the different sensors. From this model and with the values of the sensors at a given moment, the application must be able to identify the activity that the user is doing and suggest him what other application users are doing the same activity in a nearby environment.

To implement the application, the FIWARE infrastructure is used. FIWARE allows us to homogenize the data flows in order to make the ETL process (Extract, Transform and Load) easier. FIWARE [8] is part of the FI-PPP (Future Internet Public-Private Partnership), the main reference in the European Union in terms of the construction and implementation of Future Internet policies. Its objective is to become a general purpose tool by standardizing the most common phases, allowing that new sources of data can be added in a comfortable and agile way.

A whole architecture scheme used in the application design is shown in Fig. 2. In this scheme, the workflow begins with the data collected by the user's mobile device sending. *Accelerometer* data are sent in packets of X data collected every Y seconds to the *App Server* (where the data preprocessing will be performed) together with the user's *GPS* coordinates. The data, in the appropriate format, are sent to the IDA module, where using the *Apache Storm* technology, a Data Mining (DM) technique will be used.

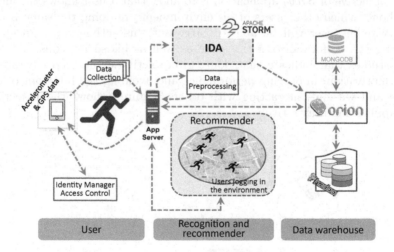

Fig. 2. Integration of the *SJogg* application in the FIWARE architecture

Once the data have been analyzed, *IDA* module will generate an output composed by the feasible activities recognized. This information will serve both to update the user information in the temporary database (*MongoDB*) and to ask to the *Context Orion Broker* the information about which users are performing the same activity in an environment close to the user. The information is retrieved and sent to the *Recommender* module, which will finally generate the information that is provided to the user.

On the other hand, from time to time the application will ask the user to label his activity. This information will be sent to the *Orion Context Broker* which, using the connector that *Cygnus* provides for *Hadoop*, will register it in that framework. In this way, a new dataset is generated allowing to update the model when the sensors change their technology and a new model is needed.

The next section focuses on IDA module to explain the steps followed since the input of information until obtaining the information that will be provided as output.

3   Intelligent Data Analysis Module and Data Preprocessing

This section focuses on IDA module and the preprocessing data that are needed in the *SJogg* application. The technique used in IDA module is based on k Nearest Neighbors technique, that has been used in many applications that solve problems of daily life obtaining good results. In this paper, an extension of this technique is used. The main is not only to use a technique with a high accuracy but also to allow the imperfect data processing. This possibility will reduce the effort that must be made in the data preprocessing to obtain the minable view of data. It also allows us to parallelize part of the process by using the *Apache Storm* technology.

3.1   Data Collection and Preprocessing

The starting point of the IDA module of *SJogg* application is a set of labeled data (which will be the system model) obtained from different users through of accelerometers of their mobile devices while performing activities of daily life (such as walking, jogging, etc.). When the collection of the labeled data is carried out, it is interesting that the user can provide the label of the activity he is performing in a more natural way. So, the labeling/classification of certain instances in an imprecise way is allowed when performing an activity that is not totally determined, for example, when the activity is to walk fast. In this case, it would be more natural to label these instances in an imprecise way as {0.5/jogging, 0.5/walking} or {jogging, walking}.

The accelerometer measures the acceleration that occurs in the axes of the three space dimensions. In addition, it can detect the orientation of the device and the Earth gravity. So, placing the device in the trouser pocket and making this the reference system, we have to:

- Axis x: measures the value of the acceleration on the horizontal axis.
- Axis y: measures the value of the acceleration on the vertical axis.
- Axis z: measures the value of the acceleration that occurs in the forward movement.

These data are collected with a temporary frequency that results in collecting large amounts of instances. A first processing of these data is their grouping in certain time intervals. Thus, groups of m vectors (x, y, z) are made for the measures obtained in the readings. As a second step, the following information, used in various works as in [12], is obtained for each group:

- $\bar{x}, \bar{y}, \bar{z}$ indicating the means of the accelerations in the axes x, y, z, respectively.
- s_x, s_y, s_z indicating the standard deviation for each axis.
- da_x, da_y, da_z indicating the mean absolute deviation for each axis.
- $resultant = \frac{\sum_i^m \sqrt{(x_i^2 + y_i^2 + z_i^2)}}{m}$ indicating the average resultant acceleration.

Therefore, several datasets based on the obtained information are considered (some experiments are performed to select the specific dataset that will be incorporated into the *SJogg* application):

- Dataset with 10 attributes, $\bar{x}, \bar{y}, \bar{z}, s_x, s_y, s_z, da_x, da_y, da_z, resultant$, and also the *class* attribute indicating the performed activity (**SJoggDS-11** dataset).
- Dataset with 7 attributes, $F_x, F_y, F_z, da_x, da_y, da_z, resultant$, and the *class* attribute indicating the performed activity. Attributes F_x, F_y, F_z indicate fuzzy (triangular) values obtained as follows:
  ○ $[\bar{x} - s_x, \bar{x}, \bar{x} + s_x]$, $[\bar{y} - s_y, \bar{y}, \bar{y} + s_y]$ and $[\bar{z} - s_z, \bar{z}, \bar{z} + s_z]$ (**SJoggDS-F-8** dataset).
- Dataset with 7 attributes, $I_x, I_y, I_z, da_x, da_y, da_z, resultant$, and the *class* attribute indicating the performed activity. I_x, I_y and I_z indicate interval-valued attributes obtained as follows:
  ○ $[\bar{x} - s_x, \bar{x} + s_x]$, $[\bar{y} - s_y, \bar{y} + s_y]$ and $[\bar{z} - s_z, \bar{z} + s_z]$ (**SJoggDS-I-8** dataset).

In addition, during the data collection the sensors can have some operation error producing the loss of some values. This situation can lead to some instances having missing values and it would be interesting not to have to discard these instances.

Therefore, after the data collection and preprocessing, instances in the minable view of the application are expressed by imperfect values (interval values, fuzzy values, missing values and imprecise classes). In order to carry out the DM process it is necessary the use of techniques that can deal with this kind of data.

3.2   Data Mining

To approach the DM process from imperfect data is necessary a technique that can work with these data and can be used in the IDA module of *SJogg* application. Next, the kNN$_{imp}$ technique proposed in [3], and potentially useful for our purposes is described.

kNN$_{imp}$ - A Classifier From Imperfect Data. The classification/identification process is carried out with the kNN$_{imp}$ technique that can work with imperfect data. Due to this technique lacks of learning phase, the model will be formed by the set of available data.

The technique supports nominal and numerical attributes. Nominal attributes can be expressed by crisp, crisp subset, fuzzy subset and missing values. Numerical attributes can be expressed by crisp, interval, fuzzy and missing values. In addition, the technique can provide as output an imprecise class if the weights of the most important classes differ by less than an amount that is specified by an external parameter. The classification or identification of the activity carried out by an application user is obtained using the Algorithm 1.

Algorithm 1. kNN_{imp} technique

Input Dataset E, Instance $\mathbf{z}$, Value k ($1 \leq k \leq |E|$), Values U_D and U_I ($U_D, U_I \in [0, 1]$)
Let $KIMP_z$ be the set of the k nearest instances of $\mathbf{z}$ according to $d_{imp}(\cdot, \cdot)$
Calculate imperfection weight ($p(\mathbf{x}^j)$) and distance weight ($q(\mathbf{x}^j)$) for all $\mathbf{x}^j \in KIMP_z$
if (degree of imperfection of $KIMP_z$) $\leq U_I$ **then**
    Aggregate the information of each neighbor in order to obtain possible class values for the instance $\mathbf{z}$ using *AggreN* and *AggreF* functions
    Calculate the set of output classes z_n using U_D
    Output z_n
else
    Output Classification is not performed
end if

As Algorithm 1 shows, kNN_{imp} technique computes the set $KIMP_z$ that contains the k instances $\mathbf{x}^j \in E$ which are the nearest to z according to the measure $d_{imp}(\mathbf{x}^j, \mathbf{z})$. Then, for each instance $\mathbf{x}^j \in KIMP_z$, two weights are calculated depending on its degree of imperfection ($p(\cdot)$) and its distance to $\mathbf{z}$ ($q(\cdot)$). Furthermore, the overall degree of imperfection in $KIMP_z$ is measured, if it is too high, the classification is not performed. To establish the maximum degree of imperfection, kNN_{imp} uses the parameter U_I. This parameter plays an important role when the dataset can contain instances with high imperfection degree. If $KIMP_z$ passes the imperfection check, the functions *AggreN* and *AggreF* obtain the set of possible weighted classes taking into account the k nearest neighbors. The class with the highest score is chosen as output, together with other classes whose score is similar to the highest. To assess if a class should be included in the final output, kNN_{imp} uses the threshold U_D.

Next, some kNN_{imp} elements are briefly commented (for more depth, see [3]): Distance/dissimilarity measures, contribution of neighbors to the classification, controlling the output class similarity, aggregation methods for classification and the process to obtain the accuracy.

Distance/Dissimilarity Measures. In order to calculate the nearest neighbors, the technique uses a measure (distance/dissimilarity) which computes the distance between two instances and can work with/without imperfect data coming from numerical and nominal attributes. The measure is defined as $d_{imp}(\mathbf{x}, \mathbf{x}') = \sqrt{\frac{\sum_{i=1}^{n-1} f(x_i, x_i')^2}{n-1}}$ where $f(x_i, x_i')$ is defined by two functions $f_1(x_i, x_i')$ and $f_2(x_i, x_i')$. $d_{imp}(\mathbf{x}, \mathbf{x}')$ is a heterogeneous function defined from different functions, $f_1(\cdot, \cdot)$ and $f_2(\cdot, \cdot)$, on different kinds of attributes (numerical and nominal respectively) where $f_1(\cdot, \cdot)$ and $f_2(\cdot, \cdot)$ are normalized fuzzy distance or dissimilarity measures.

Contribution of Neighbors to the Classification.

- Weights based on distance: $q(\mathbf{x}) = 1 - d_{imp}(\mathbf{x}, \mathbf{z})$ with $\mathbf{x} \in KIMP_z$
- Weights based on imperfection: $p(\mathbf{x}) = 1 - imp(\mathbf{x})$ with $\mathbf{x} \in KIMP_z$ and $imp(\cdot) : E \to [0, 1]$ defined as $imp(\mathbf{x}) = \frac{1}{n} \sum_{i=1}^{n} g(x_i)$ where $g(\cdot) : \Omega_{x_i} \to [0, 1]$ measures the imperfection of the value in the attribute x_i.

Controlling the Output Class Similarity. The kNN_{imp} technique exploits the definition of a similarity value between possible classes, defined as $sim(\omega_M, \omega_i) = \frac{\mu(\omega_M) - \mu(\omega_i)}{\mu(\omega_M)}$, to perform the classification of an instance. The minimum $sim(\omega_M, \omega_i)$ necessary to consider that the classes ω_M and ω_i are possible outputs is controlled by the threshold $0 \le U_D \le 1$. Thus, let us assume that ω_c is the class having the highest membership degree $\mu(\omega_c)$ to classify an instance. If there are other classes with very close membership degrees to $\mu(\omega_c)$, all these classes could be returned as possible instance classification. The role of U_D threshold is to define how close to ω_c must be a class to be considered an output class. Thus, the threshold U_D allows the output of kNN_{imp} technique is multivalued.

Aggregation Methods for Classification. The aggregation methods defined for kNN_{imp} technique are composed of the two functions $AggreN(\cdot)$ and $AggreF(\cdot)$. These two functions provide high flexibility to this technique, allowing choose them according to the classification problem. In [3], different aggregation methods are defined. Next, two aggregation methods are described because they are used later in the experiments. Method WM_{CV} uses $AggreN() = WCVEN()$ and $AggreF() = CV()$, and method SM_{CV} uses $AggreN() = SVEN()$ and $AggreF() = CV()$.

- $SVEN()$ returns a vote of 1 to the class of x^j with the highest membership degree and 0 to the other classes:

$$SVEN(i, x_n^j, p(\mathbf{x^j}), q(\mathbf{x^j})) = \begin{cases} 1 & \text{if } i = \arg \max_{h=1,\ldots,I} \mu^j(\omega_h) \\ 0 & \text{otherwise} \end{cases}$$

- $WCVEN()$ returns the score assigned by a neighbor (x^j) to each class value $(i = 1, \ldots, I)$ and is defined as follows:

$$WCVEN(i, x_n^j, p(\mathbf{x^j}), q(\mathbf{x^j})) = \mu^j(\omega_i) \cdot p(\mathbf{x^j}) \cdot q(\mathbf{x^j})$$

that is, the weight assigned by a neighbor to each class value is determined by the weight of that value in x_n^j ($\mu^j(\omega_i)$), by the weight of x^j according to its distance ($q(\mathbf{x^j})$) and by the weight of x^j according to its imperfection ($p(\mathbf{x^j})$).

- $CV()$ provides as output the fuzzy set $\{\mu(\omega_i)/\omega_i\}$ composed of ω_i with $\mu(\omega_i) > 0$.

$$\mu(\omega_i) = \frac{\sum_{j=1}^{k} AggreN(i, x_n^j, p(\mathbf{x^j}), q(\mathbf{x^j}))}{\sum_{j=1}^{k} \sum_{i=1}^{I} AggreN(i, x_n^j, p(\mathbf{x^j}), q(\mathbf{x^j}))}$$

4 Preliminary Evaluation of the IDA Process

As a preliminary evaluation of the kNN$_{imp}$ technique in the context of the *SJogg* application, an available public dataset is used. This dataset has the necessary characteristics for the proposed application.

4.1 Dataset Description

The selected dataset was generated as part of the work done by the WISDM team for the activity recognition in [12]. For the experiments, data obtained from the accelerometers located on the smartphones are used. The acceleration values in the axes x, y, z have been grouped in time intervals of 10 s, where each group contains a total of 150 instances. From these groupings, the complete preprocessing is performed as described in the Sect. 3.1, obtaining the different datasets with 11 attributes (SJoggDS-11 dataset) and 8 attributes (SJoggDS-F-8 and SJoggDS-I-8 datasets) with 7430 instances each one.

4.2 Parameter Configuration

On the datasets described in the previous section, four DM techniques have been applied: for datasets without imperfect values, a classic kNN technique (IBk) and a decision tree (J48) provided by Weka package [9]; and a fuzzy decision tree (FID3.5) [10] have been applied. For datasets with imperfect values, the kNN$_{imp}$ technique has been applied. The parameter configuration for the techniques is shown in Table 1.

Table 1. Parameters of techniques

J48$_{Weka}$	IBk$_{Weka}$	FID3.5	kNN$_{imp}$
M=2	k=1	Fuzzy Top-down	For numerical attributes: similarity proposed in [5]
(best M)	(best k)	discretization	For nominal attributes: similarity proposed in [20]
	For others, the default values		Imperfection function: Power of fuzzy sets [6]

Also, since the class inferred by kNN$_{imp}$ technique, using the aggregation methods defined above and the threshold U_D, can be a set, it is necessary to define how the accuracy in classification is measured. Algorithm 2 shows this process where the considered dataset is a subset of reserved instances as test dataset (E_{test}).

Algorithm 2. Classification Accuracy

Input Dataset E_{test}, Class value inferred to each $\mathbf{z} \in E_{test}$ $(class_{kNN_{imp}}(\mathbf{z}))$
$Suc,\ SucErr=0;$
for all $\mathbf{z}$ in E_{test} **do**
  if $class_{kNN_{imp}}(\mathbf{z}) = class(\mathbf{z})$ **then** $Suc = Suc + 1$
  else **if** $(class_{kNN_{imp}}(\mathbf{z}) \bigcap class(\mathbf{z})) \neq \emptyset$ **then** $SucErr = SucErr + 1$
end for
$Acc_{min} = \frac{Suc}{|E_{test}|},\ Acc_{max} = \frac{Suc+SucErr}{|E_{test}|}$
Output $[Acc_{min}, Acc_{max}]$

In the definition of the upper bound of this interval, those cases where the class value of a test instance is not the same but it is included in the inferred class value are considered as success. Note that situations in which the two values of interval are equal will be denoted with a single value Acc.

4.3 Preliminary Results

The results obtained when performing a 10-fold cross validation with the different datasets are shown in Table 2. In addition, for kNN_{imp} technique, the k values, the combination method and U_D that obtain the best results are shown.

Table 2. Averaged accuracies (%)

Datasets	Techniques			
	$J.48_{Weka}$	IBk_{Weka}	FID3.5	kNN_{imp}
SJoggDS-11	91.79	96. 96	78.84	–
SJoggDS-F-8	–	–	–	$[94.12,97.87]_{k=2;(SMcv,U_D=0);(WMcv,U_D=0.05)}$
SJoggDS-I-8	–	–	–	$[94.02,97.89]_{k=2;(SMcv,U_D=0);(WMcv,U_D=0.05)}$

Table 3 shows the confusion matrix of kNN_{imp} technique (with $k = 2$, WMcv and $U_D = 0.05$) and JoggDS-F-8 dataset. The matrix shows (when is compared with the confusion matrix of IBk_{Weka}) that some successes have a second nearest neighbor with a different class to the correct one. In addition, some errors have a second nearest neighbor with the correct class. These successes/errors are included in the upper bound of the accuracy interval (therefore, this upper bound is higher). This is an important result because the technique's output indicates that some activities have a similar behavior to others, as for example {Jogging, Uptairs} or {Walking, Uptairs}.

These preliminary results show the importance of expressing the information in a more adequate way to its true nature, avoiding important loss of information it contains. Expressions of the information in a more appropriate or concrete way

Table 3. Confusion matrix for kNN$_{imp}$

a	b	c	d	e	ab	ac	ad	ae	bc	bd	be	cd	ce	de	←classified as
2187	3	18	1	1	1	13	0	0	3	0	0	0	0	0	a=Jogging
0	2768	19	0	0	0	3	0	0	62	0	0	0	0	0	b=Walking
8	86	1304	3	3	3	46	1	0	147	0	0	2	2	1	c=Upstairs
0	0	2	404	0	0	0	0	0	1	0	0	1	0	2	d=Sitting
0	0	2	0	330	0	0	0	0	0	0	0	0	0	3	e=Standing

allows us to improve the accuracy results obtained, but it is necessary to have DM techniques that can deal with this kind of information.

4.4 Building the IDA Module

Based on the obtained results, the SJoggDS-F-8 dataset and the kNN$_{imp}$ technique are used with the parameter values defined by $U_D = 0.05, k = 2$ and WM_{CV}.

The mobile device of an application user will collect data from the accelerometer and GPS for 10 s. These data are sent to the application server. On the server, data will be transformed into the set of 7 attributes described in Sect. 3.1. These 7 attributes will be the input of the IDA module that holds the kNN$_{imp}$ technique. The execution of the technique in Apache Storm will be carried out by partitioning the dataset SJoggDS-F-8, which constitutes the model, in smaller subsets distributed in the different system nodes. Each node provides the two nearest neighbors of the local subset and a final node is responsible for obtaining the two global nearest neighbors and providing the output class. With this value and the user GPS coordinates, Context Orion Broker obtains the GPS coordinates of the users that perform the same or similar activity. This information is downloaded to the application server and, through the recommender module, the output that will be sent to the user is generated by means of a text message or on a city map.

5   Conclusion and Future Works

This paper shows the importance of data mining techniques development, that allow working with data that express the true nature of them or an information very close to said nature. Therefore, it is important to continue working on the data mining technique extension that currently obtain good results when addressing daily life problems so that they can perform other type of data treatment, maintaining their performance. This is one of the work lines that can be addressed. In this way, there will be several possibilities to implement the IDA module in order to select the one that works best in each domain. Since the preliminary results obtained in this work indicate that kNN$_{imp}$ obtains good results in the social sport domain and the technique model is the set of examples, it is interesting to deep in the study and design of instance selection techniques that can also work from imperfect data. In this way, a reduced dataset and the commented technique parallelization, makes kNN_{imp} a suitable real-time response technique. On the other hand, although this work focuses on the human activity recognition applied to social sport, the following objective is to carry out an activity recognition in other domains such as healthcare, transport and other types of social activities.

Acknowledgement. Supported by the project TIN2017-86885-R (AEI/FEDER, UE) granted by the Ministry of Economy, Industry and Competitiveness of Spain (including ERDF support).

References

1. de la Concepción, M.Á.Á., Morillo, L.M.S., García, J.A.Á., González-Abril, L.: Mobile activity recognition and fall detection system for elderly people using Ameva algorithm. Pervasive Mob. Comput. **34**, 3–13 (2017). https://doi.org/10.1016/j.pmcj.2016.05.002
2. Barth, R.S., Galante, R.M.: Passenger density and flow analysis and city zones and bus stops classification for public bus service management. In: Proceedings of the Brazilian Symposium on Databases, Salvador, Brazil, pp. 217–222 (2016)
3. Cadenas, J.M., Garrido, M.C., Martínez, R., Muñoz, E., Bonissone, P.P.: A fuzzy K-nearest neighbor classifier to deal with imperfect data. Soft. Comput. **22**, 18 (2017). https://doi.org/10.1007/s00500-017-2567-x
4. Ceapa, I., Smith, C., Capra, L.: Avoiding the crowds: understanding tube station congestion patterns from trip data. In: Proceedings of the ACM SIGKDD International Workshop on Urban Computing, New York, 134–141 (2012)
5. Chen, S.M.: New methods for subjective mental workload assessment and fuzzy risk analysis. Cybern. Syst. **27**(5), 449–472 (1996). https://doi.org/10.1080/019697296126417
6. DeLuca, A., Termini, S.: A definition of a nonprobabilistic entropy in the setting of fuzzy sets theory. Inf. Control **20**(4), 301–312 (1972). https://doi.org/10.1016/S0019-9958(72)90199-4
7. Fitbit: The fitness app for everyone, San Francisco, CA, Fitbit App. https://www.fitbit.com/es/app
8. FIWARE Community. To build an open sustainable ecosystem around public, royalty-free and implementation-driven software platform standards that will ease the development of new Smart Applications in multiple sectors everyone. https://www.fiware.org
9. Hall, M., Frank, E., Holmes, G., Pfahringer, B., Reutemann, P., Witten, I.H.: The WEKA data mining software: an update. SIGKDD Explor. **11**(1), 1–18 (2009)
10. Janikow, C.Z.: FID3.5: one of the FID programs originally proposed in "Fuzzy decision trees: issues and methods". IEEE Trans. Man Syst. Cybern. **28**(1), 1–14 (1998). http://www.cs.umsl.edu/janikow/fid/index.html
11. Kozina, S., Gjoreski, H., Gams, M., Luštrek, M.: Efficient activity recognition and fall detection using accelerometers. In: Botía, J.A., Álvarez-García, J.A., Fujinami, K., Barsocchi, P., Riedel, T. (eds.) EvAAL 2013. CCIS, vol. 386, pp. 13–23. Springer, Heidelberg (2013). https://doi.org/10.1007/978-3-642-41043-7_2

12. Kwapisz, J.R., Weiss, G.M., Moore, S.A.: Activity recognition using cell phone accelerometers. SIGKDD Explor. Newsl. **12**(2), 74–82 (2010). https://doi.org/10. 1145/1964897.1964918
13. Lara, O.D., Labrador, M.A.: A survey on human activity recognition using wearable sensors. IEEE Commun. Surv. Tutor. **15**(3), 1192–1209 (2013)
14. Lau, S.L., König, I., David, K., Parandian, B., Carius-Düssel, C., Schultz, M.: Supporting patient monitoring using activity recognition with a smartphone. In: Proceedings of the 7th International Symposium on Wireless Communication Systems, York, UK, pp. 810–814 (2010)
15. Mashita, T., Shimatani, K., Iwata, M., Miyamoto, H., Komaki, D., Hara, T., Kiyokawa, K., Takemura, H., Nishio, S.: Human activity recognition for a content search system considering situations of smartphone users. In: Proceedings of the IEEE Virtual Reality Workshops, Costa Mesa, CA, pp. 1–2 (2012)
16. Mun, M., Estrin, D., Burke, J., Hansen, M.: Parsimonious mobility classification using GSM and WiFi traces. In: Proceedings of the Fifth Workshop on Embedded Networked Sensors, Charlottesville, Virginia, USA (2008)
17. NikeFuel: A universal way to measure movement, Beaverton, OR, NikeFuel App. https://secure-nikeplus.nike.com/plus/what_is_fuel
18. Pan, G., Qi, G., Wu, Z., Zhang, D., Li, S.: Land-use classification using taxi GPS traces. IEEE Trans. Intell. Transp. Syst. **14**(1), 113–123 (2013)
19. Perez, A.J., Labrador, M.A., Barbeau, S.J.: G-sense: a scalable architecture for global sensing and monitoring. IEEE Netw. **24**(4), 57–64 (2010). https://doi.org/ 10.1109/MNET.2010.5510920
20. Santini, S., Jain, R.: Similarity is a geometer. Multimed. Tools Appl. **5**(3), 277–306 (1997). https://doi.org/10.1023/A:1009651725256
21. Sohn, T., Varshavsky, A., LaMarca, A., Chen, M.Y., Choudhury, T., Smith, I., Consolvo, S., Hightower, J., Griswold, W.G., Lara, E.: Mobility detection using everyday GSM traces. In: Proceedings of the 8th International Conference on Ubiquitous Computing, Orange County, CA, pp. 212–224 (2006)
22. Tryon, W.W., Tryon, G.S., Kazlausky, T., Gruen, W., Swanson, J.M.: Reducing hyperactivity with a feedback actigraph: initial findings. Clin. Child Psychol. Psychiatry **11**(4), 607–617 (2006). https://doi.org/10.1177/1359104506067881
23. Weiss, G.M., Lockhart, J.W.: The impact of personalization on smartphone-based activity recognition. In: Proceedings of the AAAI 2012 Workshop on Activity Context, Toronto, CA, pp. 98–104 (2012)
24. Yuan, J., Zheng, Y., Xie, X., Sun, G.: T-drive: enhancing driving directions with taxi drivers' intelligence. IEEE Trans. Knowl. Data Eng. **25**(1), 220–232 (2013)
25. Zheng, Y., Li, Q., Chen, Y., Xie, X., Ma, W.-Y.: Understanding mobility based on GPS data. In: Proceedings of the 10th International Conference on Ubiquitous Computing, Seoul, Korea, pp. 312–321 (2008)
26. Zheng, Y., Liu, Y., Yuan, J., Xie, X.: Urban computing with taxicabs. In: Proceedings of the 13th International Conference on Ubiquitous Computing, Beijing, China, pp. 89–98 (2011)

A Fuzzy Close Algorithm for Mining Fuzzy Association Rules

Régis Pierrard[1,2]([✉]), Jean-Philippe Poli[1], and Céline Hudelot[2]

[1] CEA, LIST, Data Analysis and System Intelligence Laboratory,
91191 Gif-sur-Yvette Cedex, France
{regis.pierrard,jean-philippe.poli}@cea.fr
[2] Mathematics Interacting with Computer Science, CentraleSupélec,
Paris-Saclay University, 91190 Gif-sur-Yvette, France
celine.hudelot@centralesupelec.fr

Abstract. Association rules allow to mine large datasets to automatically discover relations between variables. In order to take into account both qualitative and quantitative variables, fuzzy logic has been applied and many association rule extraction algorithms have been fuzzified.

In this paper, we propose a fuzzy adaptation of the well-known Close algorithm which relies on the closure of itemsets. The Close-algorithm needs less passes over the dataset and is suitable when variables are correlated. The algorithm is then compared to other on public datasets.

Keywords: Fuzzy data mining · Fuzzy closure operator
Frequent itemsets mining · Fuzzy logic · Association rules

1 Introduction

Extracting association rules from data has been one of the main tasks in data mining for years. It relies on the extraction of frequent itemsets. In order to deal with both quantitative and qualitative variables, some algorithms have used the fuzzy set theory. Fuzzy logic provides tools to manage the vagueness inherent in both the natural language and the knowledge itself. Different fuzzy association rule mining algorithms have already been developed to handle this kind of data.

Because datasets are nowadays getting bigger and bigger, the way these fuzzy association rule mining algorithms manage huge databases is essential. Some algorithms store a big amount of data while some others need to perform many database passes.

There exist several crisp association rule mining algorithms that do not store a lot of data or need only a limited number of database passes. However, most of them do not have a fuzzy counterpart. In this paper, we propose an algorithm that uses the fuzzy set theory and the fuzzified version of the Close mining algorithm [1] to extract frequent itemsets from data with a reduced number of database passes.

© Springer International Publishing AG, part of Springer Nature 2018
J. Medina et al. (Eds.): IPMU 2018, CCIS 854, pp. 88–99, 2018.
https://doi.org/10.1007/978-3-319-91476-3_8

The rest of the paper is organized as follows. Section 2 reviews related work. In Sect. 3, we present the fuzzy set framework and we describe the algorithm. Section 4 presents the experimental results we got and Sect. 5 concludes the paper.

2 Related Works

2.1 Fuzzy Association Rule Mining

The first fuzzy association rule mining algorithms were based on the Apriori algorithm [2]. It consists in two main steps. First, finding the frequent itemsets and second, generating fuzzy rules based on the previously extracted frequent itemsets. In order to find the frequent itemsets, it first scans the whole database to extract frequent itemsets that contain only one item (1-itemsets). An itemset is said to be frequent when the support of this itemset in the database, i.e. the number of occurrences, is larger than a user-specified minimum support threshold. After that first step, frequent 1-itemsets are used to generate candidate 2-itemsets. Frequent 2-itemsets are extracted computing their support. The process continues until no more candidate can be generated. It requires n database passes, where n is the size of the maximum length frequent itemset. Once frequent itemsets have been mined, every candidate association rule is generated. An association rule is valid when its confidence is larger than a user-specified minimum confidence threshold. For a frequent itemset I and an association rule $I_1 \Rightarrow I_2$ such as $I_1 \subset I$, $I_2 \subset I$ and $I_1 \cap I_2 = \varnothing$, the confidence of this association rule is its number of occurrences among the occurrences of I. All candidate association rules are generated to find the most confident ones. Many fuzzy association rule mining algorithms rely on the Apriori algorithm. The F-APACS algorithm [3] first converts data into linguistic terms using the fuzzy set theory. A statistical analysis is performed to automatically set both the minimum support threshold and the minimum confidence threshold. The FDTA algorithm [4] proposes another way of converting quantitative data into linguistic terms. AprioriTid [18] is an improved version of FDTA. Kuok et al. [5] proposed a different approach to handle quantitative databases for generating fuzzy association rules.

A completely different way of mining fuzzy frequent itemsets relies on a frequent-pattern tree structure. The generic framework is as follows. The first step consists in fuzzifying data, if necessary. Then, the tree is constructed and the final step is the mining of fuzzy frequent itemsets based on the previously constructed tree. Papadimitriou and Mavroudi [6] proposed an algorithm called fuzzy frequent pattern tree (FFPT). Non frequent 1-itemsets are removed from the database and each transaction is sorted according to the membership value of its frequent 1-itemsets. Then, the tree is constructed by handling each transaction one by one. Since transactions are sorted by membership values, several different paths may represent the same itemset. As a consequence, a few useless tree nodes are generated. The compressed fuzzy frequent pattern tree (CFFPT) algorithm solves this problem by using a global sorting strategy [7]. However, this solution leads to attaching an array to each node. Lin et al. [8] proposed the upper bound fuzzy frequent pattern tree algorithm (UBFFPT). It estimates the upper bound membership values of frequent itemsets to avoid attaching an array to each node. This algorithm requires four database passes to build the tree. Then, the tree is parsed several times to generate all candidate frequent itemsets. Depending on the database, the tree can be long and have a large amount of nodes. An ultimate database pass is performed to compute the support of every candidate frequent itemset.

2.2   The Close Algorithm

Pasquier et al. [1] proposed the Close algorithm. This algorithm handles non-fuzzy databases. It uses a closure operator to find closed itemsets. Those itemsets have interesting properties that benefit the mining of frequent itemsets. Since there are often less frequent closed itemsets than frequent itemsets, the search space is smaller, the computation is less costly and the number of database passes is reduced. The algorithm relies on the following properties [1]:

1. all subsets of a frequent itemset are frequent;
2. all supersets of an infrequent itemset are infrequent;
3. all closed subsets of a frequent closed itemset are frequent;
4. all closed supersets of an infrequent closed itemset are infrequent;
5. the set of maximal frequent itemsets is identical to the set of maximal frequent closed itemsets;
6. the support of a frequent itemset I which is not closed is equal to the support of the smallest frequent closed itemset containing I.

   The algorithm goes through three phases to generate association rules. First, it generates all frequent closed itemsets from the database. Then, it derives all frequent itemsets from the previously generated frequent closed itemsets. The final step consists in generating all confident association rules.

3   Fuzzified Close Algorithm

3.1   Fuzzy Sets

Zadeh introduced the fuzzy set theory [9]. In a universe X, a fuzzy set F is characterized by a mapping $\mu_F : X \rightarrow [0, 1]$. This mapping specifies in what extent each $x \in X$ belongs to F and it is called *the membership function of F*. If F is a non-fuzzy set, $\mu_F(x)$ is either 0, i.e. x is not a member of F, or 1, i.e. x is a member of F. The set of all fuzzy sets in a universe X is written F^X.

   The *kernel* of a fuzzy set F is a non-fuzzy set defined as

$$\ker(F) = \{x \in X | \mu_F(x) = 1\}. \tag{1}$$

   A binary fuzzy relation can be defined the same way as a fuzzy set. Given two universes X and Y, a binary fuzzy relation $\mathcal{R}$ is a mapping defined as

$$\mathcal{R} : X \times Y \rightarrow [0, 1]. \tag{2}$$

It assigns a degree of relationship to any $(x, y) \in X \times Y$.

3.2   Formal Concept Analysis

Formal Concept Analysis (FCA) [10–12] provides a framework for analyzing the relationship between a set of objects and a set of attributes. A database with fuzzy values can be represented by a triplet $\langle \mathcal{O}, \mathcal{A}, \mathcal{R} \rangle$ with $\mathcal{O}$ a finite set of objects, $\mathcal{A}$ a finite set of attributes and $\mathcal{R}$ a binary fuzzy relation defined as $\mathcal{R} : \mathcal{O} \times \mathcal{A} \to [0,1]$. This triplet is called a formal fuzzy context.

In the following, a fuzzy set of attributes (objects) is a fuzzy set whose membership function is defined as $\mu : \mathcal{A} \to [0,1]$ ($\mu : \mathcal{O} \to [0,1]$).

Operators $\uparrow$ and $\downarrow$ can then be defined [13]. Let X be a fuzzy set of objects and Y be a fuzzy set of attributes. $\uparrow$ and $\downarrow$ are defined as follows:

$$\forall a \in \mathcal{A}, \mu_{X^\uparrow}(a) = \bigwedge_{o \in O} \left(\mu_X(o) \to \mathcal{R}(o, a) \right), \tag{3}$$

$$\forall o \in \mathcal{O}, \mu_{Y^\downarrow}(o) = \bigwedge_{a \in A} \left(\mu_Y(a) \to \mathcal{R}(o, a) \right). \tag{4}$$

$X^\uparrow$ is a fuzzy set of attributes and $Y^\downarrow$ is a fuzzy set of objects. In the next section, the composition of these two functions is written $\uparrow\downarrow$.

We use the Lukasiewicz implication operator defined as

$$a \to b = \min(1 - a + b, 1). \tag{5}$$

The Lukasiewicz implication is compatible with the implication from classical logic.

3.3   Fuzzy Closure Operator

The closure operator cannot be the same in the fuzzified version of the algorithm. It still takes as an argument a crisp set, which we call a *generator*, and also returns a crisp set. However, the relation $\mathcal{R}$ between objects and attributes is no longer crisp. That is why this operator needs to be modified.

Definition 1. *A fuzzy closure operator in a universe X is defined as $h : F^X \to F^X$ and satisfies the following conditions:*

$$\forall I \subset F^X, I \subset h(I), \tag{6}$$

$$\forall I \subset F^X, h(h(I)) = h(I), \tag{7}$$

$$\forall I, J \subset F^X, I \subset J \Rightarrow h(I) \subset h(J). \tag{8}$$

For any formal fuzzy context $\langle \mathcal{O}, \mathcal{A}, \mathcal{R} \rangle$, for a fuzzy set of attributes Y, $\uparrow\downarrow$ is a fuzzy closure operator [12,14]. The fuzzy closure of Y by $\uparrow\downarrow$ is $Y^{\uparrow\downarrow}$, which is a fuzzy set of attributes.

In our case, the closure operator takes a crisp set of items (or attributes) as a generator. Let I be a crisp set of items. It can be turn into a fuzzy set to be used by the fuzzy closure operator as follows:

$$\forall a \in \mathcal{A}, \mu_I(a) = \begin{cases} 1, & \text{if } a \in I \\ 0, & \text{otherwise} \end{cases}. \tag{9}$$

As for the set the closure operators returns, it also has to be a crisp set. The fuzzy closure operator ↑↓ returns a fuzzy set F. We can get a crisp set of items I using the kernel function as follows:

$$I = \ker(F). \tag{10}$$

This operator is still a closure operator. In the following, this closure operator is written h such as $h : \mathcal{P}(\mathcal{A}) \to \mathcal{P}(\mathcal{A})$. One can interpret the result of this closure operator as the set of attributes that are shared by all the objects that have all the attributes from the generator.

3.4   Support and Confidence

The support of an itemset and the confidence of a rule are computed as stated in [17]. Both of them are numbers between 0 and 1.

The following property states that an itemset and its closure have the same support. This will be used in our algorithm.

Proposition 1. $\forall I \in \mathcal{P}(\mathcal{A}), \text{support}(h(I)) = \text{support}(I)$.

3.5   Algorithm Description

The proposed fuzzy association rule mining approach integrates concepts from both the fuzzy set theory and the Close algorithm [1]. It does not tackle the fuzzification of the database. This task has been addressed in the previously mentioned articles [3–5]. Besides, the generation of all confident association rules is the same as in the Apriori algorithm [2].

FCC_i refers to the set of triplets associated with all the frequent closed candidate itemsets whose generator's size is i. FC_i refers to the set of triplet associated with all the frequent closed itemsets whose generator's size is i. Each triplet is under the following form:

$$(\text{generator}, \text{closure}, \text{support}).$$

Thus, in the remainder of this article, for any $p \in FCC_i$ or FC_i, p.generator refers to the generator linked to p, p.closure is its closure and p.support is its support. FCC_i.generators refers to the set of all generators in FCC_i. FCC_i.closures and FCC_i.supports are defined the same way.

Algorithm 1 below describes the process. On line 1, FCC_1 is initialized with every item from the set of attributes $\mathcal{A}$. On line 5, for each generator in FCC_i, the *generateClosures* function provides the corresponding closure and support. This function is detailed below. Then, on lines 6 to 9, the set of candidate closed itemsets FCC_i is pruned to get the set of frequent closed itemsets FC_i. New generators, whose size is $i + 1$, are generated on line 11 using the *generateGenerators* function. This function is described below. The whole process will last until no new generators can be generated. The output is the set of all frequent closed itemsets that will be used to generate all frequent itemsets.

The *generateClosures* function is stated as shown in Algorithm 2 below. This function has been designed to compute the closures and the supports of the generators in FCC_i performing only one database pass. For each object $o \in \mathcal{O}$, for each element $p \in FCC_i$, the contribution k to the support and $\mu_{p\downarrow}(o)$ are computed looping over the items in p.generator (from line 10 to line 13). Then, for each attribute $a \in \mathcal{A}$, the membership function $\mu_{p\uparrow\downarrow}$ of the fuzzy closure is updated (line 15). When the last object is reached and there is no more update to the membership function, the kernel of the fuzzy closure is computed (from line 16 to line 20).

This *generateGenerators* is exactly the same as in the Close algorithm. This function generates all the potential generators of size $i + 1$ from the generators in FC_i. In order to get one potential generator, two generators from FC_i that have the same $i - 1$ first elements are combined. Then, this set of potential generators is pruned to avoid useless computations. In particular, if one of the new generators is included in the closure of one of the former generators, then it is pruned.

Overall, the whole algorithm, i.e. Algorithm 1, needs one database pass per iteration. That is the same as the algorithms based on the Apriori algorithm. However, the total number of iterations is usually smaller with the close algorithm because there are often less frequent closed itemsets than frequent itemsets.

After this phase, all the frequent closed itemsets are used to find all the frequent itemsets. This new phase is exactly the same as in the original Close algorithm. The first step consists in splitting the set of all frequent closed itemsets according to their size. Then, these new sets L_i are browsed in descending order of size to generate all frequent itemsets of size $i - 1$. The process will finish when the set of frequent 1-itemsets is completed.

3.6   Example

For the sake of comprehension, we apply in this section the algorithm on a small database D, shown in Table 1. D contains five objects (1 to 5) and five items (A to E). The minimum support is equal to 0.4 (40%).

Table 1. The fuzzy database D

Objects	Items				
	A	B	C	D	E
1	0.8	0.1	0.9	0.8	0
2	0	0.3	0.2	0	0.9
3	1	0.7	0.7	1	0.6
4	0	0.2	0	0.2	1
5	0.9	0.6	0.8	1	0.9

The pruning of FCC_1 leads to removing {B} since its support is smaller than the minimum support threshold. The other elements from FCC_1 are kept to generate FC_1. This corresponds to line 5 to line 10 in Algorithm 1. FCC_1 and FC_1 are shown in Table 2.

Table 2. FCC_1 on the left and FC_1 on the right. {B} is pruned from FCC_1 to FC_1 because it is not frequent.

Generator	Closure	Support
{A}	{AD}	0.54
{B}	{B}	0.38
{C}	{C}	0.52
{D}	{D}	0.60
{E}	{E}	0.68

Generator	Closure	Support
{A}	{AD}	0.54
{C}	{C}	0.52
{D}	{D}	0.60
{E}	{E}	0.68

Then, on line 11, FCC_2 is generated. {AD} is not a generator in FCC_2 because it is included in the closure of {A}. FC_2 is then generated. {CD} and {AC} have the same closure, so only one of them is kept. FCC_2 and FC_2 are shown in Table 3.

Table 3. FCC_2 on the left and FC_2 on the right.

Generator	Closure	Support
{AC}	{ACD}	0.46
{AE}	{ADE}	0.30
{CD}	{ACD}	0.46
{CE}	{CE}	0.32
{DE}	{DE}	0.34

Generator	Closure	Support
{AC}	{ACD}	0.46

FC_2 contains only one element, that is why FCC_3 is empty. That is the end of the first phase, which corresponds to Algorithm 1. FC is returned. It is shown in Table 4.

The second phase consists in deriving frequent itemsets from frequent closed itemsets. The longest closed itemset contains three items. That is why three different sets are generated for deriving frequent itemsets: L_3, L_2 et L_1. Bold itemsets are itemsets which have been derived from a bigger closed itemset. These three sets are shown in Table 5.

Table 4. FC

Closure	Support
{AD}	0.54
{C}	0.52
{D}	0.60
{E}	0.68
{ACD}	0.46

Table 5. Deriving frequent itemsets. Bold lines refer to derived itemsets. From left to right: L_3, L_2 and L_1.

Itemset	Support		Itemset	Support		Itemset	Support
{ACD}	0.46		{AD}	0.54		{C}	0.52
			{AC}	**0.46**		{D}	0.60
			{CD}	**0.46**		{E}	0.68
						{A}	**0.54**

4 Experimental Results

In order to compare our algorithm to the fuzzy version of Apriori and to UBFFPT, we have implemented these algorithms. As our implementations of the algorithms may not be fully optimized, our results do not show any execution time. The metric that we used is the number of database passes. It allows to directly compare the fuzzy version of Apriori to our algorithm.

4.1 Datasets

We used three different datasets. The first one is the *mushroom* dataset [15]. It contains 8124 examples (objects). The number of attributes is 22. Those are all categorical attributes, so the final binary dataset contains 119 attributes. To fuzzify it, zeros were replace by a uniform random number in $[0, 0.5]$ and ones were replace by a uniform random number in $[0.5, 1]$.

The two other datasets come from the 2017 Civil Service People Survey [16]. Those are surveys that only contain numbers in $[0, 1]$. One dataset, that is called *benchmark scores*, contains 9 examples. Attributes have been pruned to avoid missing values for a final amount of 87 attributes. The other dataset is called *all organisation scores*. After filtering missing values, the dataset contains 93 examples and 84 attributes.

Algorithm 1. Close algorithm

---

input : A fuzzy formal context $\langle \mathcal{O}, \mathcal{A}, \mathcal{R} \rangle$
          A minimum support threshold $S \in [0; 1]$
output: All frequent closed itemsets and their support

1 generators in $FCC_1 \leftarrow \{1\text{-itemsets}\}$
2 **for** $(i \leftarrow 1;\ FCC_i.\text{generators} \neq \varnothing;\ i\text{++})$ **do**
3 | closures in $FCC_i \leftarrow \varnothing$
4 | supports in $FCC_i \leftarrow 0$
5 | $FCC_i \leftarrow \text{generateClosures}(FCC_i)$
6 | **forall** candidate closed itemsets $c \in FCC_i$ **do**
7 | | **if** $c.\text{support} \geq minsupport$ **then**
8 | | | $FC_i \leftarrow FC_i \cup \{c\}$
9 | | **end**
10 | **end**
11 | $FCC_{i+1} \leftarrow \text{generateGenerators}(FC_i)$
12 **end**
13 $FC \leftarrow \bigcup_{j=1}^{i-1} \{FC_j.\text{closures}, FC_j.\text{supports}\}$
14 **return** FC

4.2 Results and Discussion

Results are shown in Fig. 1. For the mushroom dataset, we can observe that our algorithm makes at best one less database pass than the fuzzy version of Apriori. This is due to the fact that data are not highly correlated and are sparse. That means that most frequent itemsets are closed. As a consequence, with the cost of computing closures, our algorithm should not be expected to outperform Apriori and UBFFPT on such a dataset.

Observations are different with the two other datasets. We can see that the lower the minimum support threshold, the larger the difference between the number of database passes of both algorithms. These data come from surveys, whose data are usually highly correlated and dense. Our algorithm takes advantage of this using the closure operator. Thus, most generators are much shorter than their closures. That explains the lower amount of database passes.

The UBFFPT algorithm needs 4 database passes to construct its tree and to extract frequent itemsets. Besides, frequent pattern mining algorithms, such as UBFFPT, spend most of their time traversing the tree. For highly correlated data, as in the benchmark dataset, our algorithm has an edge on these algorithms. Moreover, it consumes less memory than Apriori, which generates many candidates at each iteration, and than UBFFPT, which browses all the paths to the currently studied item[1] to generate candidates.

---

[1] One item is usually represented by several nodes in the tree.

Algorithm 2. generateClosures function

---

input : The set of candidate closed itemsets FCC_i
output: Updated FCC_i after the computation of closures and supports

1   $n \leftarrow 0$
2   **forall** $p \in FCC_i$ **do**
3 | numbers in $\mu_{p\uparrow\downarrow}{}^a \leftarrow 1$
4   **end**
5   **forall** objects $o \in \mathcal{O}$ **do**
6 | n++
7 | **forall** $p \in FCC_i$ **do**
8 | $k \leftarrow 1$
9 | $\mu_{p\downarrow}{}^b \leftarrow 1$
10 | **forall** attributes $i \in p$.generator **do**
11 | $k \leftarrow \min\left(k, \mathcal{R}(o,i)\right)$
12 | $\mu_{p\downarrow} \leftarrow \min\left(\mu_{p\downarrow}, 1, \mathcal{R}(o,i)\right)$
13 | **end**
14 | **forall** attributes $i \in \mathcal{A}$ **do**
15 | $\mu_{p\uparrow\downarrow,i} \leftarrow \min\left(\mu_{p\uparrow\downarrow,i}, 1, 1 + \mathcal{R}(o,i) - \mu_{p\downarrow}\right)$
16 | **if** $n = Card(\mathcal{O})$ **then**
17 | **if** $\mu_{p\uparrow\downarrow,i} = 1$ **then**
18 | p.closure $\leftarrow p$.closure $\cup \{i\}$
19 | **end**
20 | **end**
21 | **end**
22 | p.support $\leftarrow p$.support $+ k$
23 | **end**
24   **end**
25   **return** FCC_i

---

[a] $\mu_{p\uparrow\downarrow}$ is a vector corresponding to the membership function of the fuzzy closure $p^{\uparrow\downarrow}$.
[b] $\mu_{p\downarrow}$ is a fuzzy number that corresponds to $\mu_{p\downarrow}(o)$.

Also, the first iteration of generating closures in our algorithm can bring valuable insight. Indeed, if most 1-itemsets are closed, then the data is likely to be weakly correlated and another algorithm may perform better. However, if the proportion of closed 1-itemsets is low, the data is likely to be highly correlated and our algorithm will then compute all the frequent itemsets in few database passes.

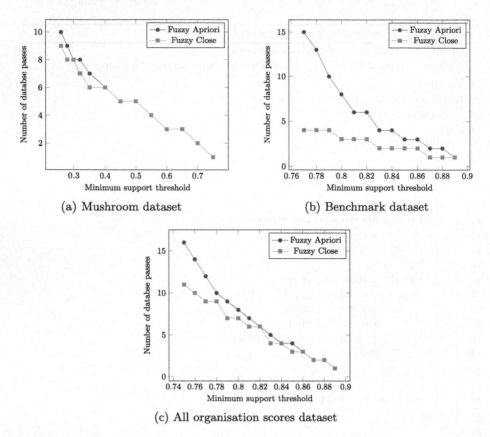

(a) Mushroom dataset   (b) Benchmark dataset

(c) All organisation scores dataset

Fig. 1. Plots showing the number of database passes relatively to the minimum support threshold for the three datasets.

5   Conclusion

In this paper, we introduced a new fuzzy association rule mining algorithm inspired by the Close algorithm. Our goal was to make it able to mine frequent itemsets from data in a reduced number of database passes and without storing too much data.

It relies on a closure operator that is able to process fuzzy data while both taking as an argument and returning a crisp set. This new closure operator is based on a fuzzy closure operator of whom we take the kernel. The closure is the set of items that are shared by all the objects that include the generator. That is why it is very efficient with highly correlated data.

The algorithm finds the set of all the closed frequent itemsets. This set is sufficient to extract all the frequent itemsets. As it is usually a smaller set than the set of all the frequent itemsets, the search space is also smaller.

We have tested our algorithm on three different datasets. We have shown that this approach outperforms other algorithms when dealing with correlated and dense data, which are the kind of data that can be found in surveys, census dataset or in some classification datasets. It needs less database passes and stores a small amount of data to extract all the frequent itemsets.

References

1. Pasquier, N., Bastide, Y., Taouil, R., Lakhal, L.: Efficient mining of association rules using closed itemset lattices. Inf. Syst. **24**(1), 25–46 (1999)
2. Agrawal, R., Srikant, R.: Fast algorithms for mining association rules. In: Proceedings of the 20th International Conference on Very Large Data Bases, pp. 487–499 (1994)
3. Chan, K.C.C., Au, W.H.: An effective algorithm for discovering fuzzy rules in relational databases. In: Proceedings of the 1998 IEEE International Conference on Fuzzy Systems, pp. 1314–1319 (1998)
4. Hong, T.P., Kuo, C.S., Chi, S.C.: Mining association rules from quantitative data. Intell. Data Anal. **3**, 363–376 (1999)
5. Kuok, C.M., Fu, A., Wong, M.H.: Mining fuzzy association rules in databases. SIGMOD Rec. **27**(1), 41–46 (1998)
6. Papadimitriou, S., Mavroudi, S.: The fuzzy frequent pattern tree. In: Proceedings of the 9th WSEAS International Conference on Computers, pp. 3:1–3:7 (2005)
7. Lin, C.W., Hong, T.P., Lu, W.H.: An efficient tree-based fuzzy data mining approach. Int. J. Fuzzy Syst. **12**(2), 150–157 (2010)
8. Lin, C.W., Hong, T.P., Lu, W.H.: A two-phase fuzzy mining approach. In: International Conference on Fuzzy Systems (2010)
9. Zadeh, L.A.: Fuzzy sets. Inf. Control **8**(3), 338–353 (1965)
10. Wille, R.: Restructuring lattice theory: an approach based on hierarchies of concepts. In: Rival, I. (ed.) Oredered Sets, pp. 445–470. Reidel, Dordrecht (1982)
11. Ganter, B., Wille, R.: Formal Concept Analysis: Mathematical Foundations. Springer, Berlin (1999). https://doi.org/10.1007/978-3-642-59830-2
12. Belohlavek, R.: Fuzzy Relational Systems: Foundations and Principles. Kluwer/ Plenum, New York (2002). https://doi.org/10.1007/978-1-4615-0633-1
13. Belohlavek, R.: Concept lattices and order in fuzzy logic. Ann. Pure Appl. Log. **128**(1), 277–298 (2004)
14. Gerla, G.: Fuzzy Logic: Mathematical Tools for Approximate Reasoning. Kluwer, Dordrecht (2001). https://doi.org/10.1007/978-94-015-9660-2
15. Lichman, M.: UCI machine learning repository. University of California, School of Information and Computer Sciences, Irvine (2013)
16. Contains public sector information licensed under the Open Government Licence v3.0 (2017)
17. Delgado, M., Marin, N., Sanchez, D., Vila, M.-A.: Fuzzy association rules: general model and applications. IEEE Trans. Fuzzy Syst. **11**(2), 214–225 (2003)
18. Hong, T.-P., Kuo, C.-S., Wang, S.-L.: A fuzzy AprioriTid mining algorithm with reduced computational time. Appl. Soft Comput. **5**(1), 1–10 (2004)

Datil: Learning Fuzzy Ontology Datatypes

Ignacio Huitzil[1(✉)], Umberto Straccia[2], Natalia Díaz-Rodríguez[3],
and Fernando Bobillo[1]

[1] I3A, University of Zaragoza, Zaragoza, Spain
{ihuitzil,fbobillo}@unizar.es
[2] ISTI-CNR, Pisa, Italy
straccia@isti.cnr.it
[3] U2IS, ENSTA ParisTech and Inria FLOWERS, Paris, France
natalia.diaz@ensta-paristech.fr

Abstract. Real-world applications using fuzzy ontologies are increasing in the last years, but the problem of fuzzy ontology learning has not received a lot of attention. While most of the previous approaches focus on the problem of learning fuzzy subclass axioms, we focus on learning fuzzy datatypes. In particular, we describe the *Datil* system, an implementation using unsupervised clustering algorithms to automatically obtain fuzzy datatypes from different input formats. We also illustrate the practical usefulness with an application: semantic lifestyle profiling.

Keywords: Fuzzy ontologies · Machine learning · Lifestyle profiling

1 Introduction

Ontologies can nowadays be considered a standard for knowledge representation. An ontology is an explicit and formal specification of the concepts, individuals and relationships that exist in some area of interest, created by defining axioms that describe the properties of these entities [20]. Ontologies can provide semantics to data, making knowledge maintenance, information integration, and reuse of components easier. The current standard language for ontology representation is OWL 2 (Web Ontology Language) [24].

Classical ontologies are not appropriate to deal with imprecise and vague knowledge, inherent to several real world domains. Fuzzy ontologies [3,21] extend classical ontologies with elements of fuzzy set theory and fuzzy logic [25]. Although fuzzy ontologies have been successfully used in several real-world applications [9–11,13] and some methodologies [1] and tools [5] supporting their development are available, the cold start problem is still common. It is difficult for ontologists who are experts in a domain but not familiar with fuzzy logic to develop fuzzy ontologies. To overcome this problem, *fuzzy ontology learning techniques* are necessary, but unfortunately there is not been a lot of research in this direction.

J. Medina et al. (Eds.): IPMU 2018, CCIS 854, pp. 100–112, 2018.
https://doi.org/10.1007/978-3-319-91476-3_9

The few exceptions are focused on learning fuzzy subclass axioms [4, 15–17, 22]; the three latter references are implemented in the *FuzzyDL-Learner* tool.[1] Starting from the (possibly partial) membership of individuals to classes, it is possible to automatically compute some partial inclusions between (possibly fuzzy) concepts. Fuzzy concept descriptions can be based on fuzzy datatypes.

Because the focus is on learning fuzzy subclass axioms, these approaches usually restrict to a simple case: a uniform partition of the domain into fuzzy datatypes. Instead, we propose to compute fuzzy datatypes from existing real data using clustering algorithms.

In some cases, only attribute values are available and there are no data about the membership to classes. In such cases, we can still learn the fuzzy datatypes, use them to build some preliminary fuzzy subclass axioms to populate the classes, and learn a more complete set of fuzzy subclass axioms using existing approaches.

The main contribution of this paper is the description of the *Datil* system, an implementation of an automatic fuzzy datatype learning algorithm for fuzzy ontologies supporting different input and output formats. We also discuss how to integrate this learning step into existing approaches for fuzzy subclass learning and illustrate it with a real-world use case: semantic lifestyle profiling, i.e., the automatic classification of the lifestyle of people given their digital footprints. The ultimate aim is to help tasks such as long-term human behavior classification and thus improve virtual coaching or customize lifestyle recommendation and intervention programs from free form non-labelled sensor data.

The rest of this paper is organized as follows. Section 2 provides some background on fuzzy ontologies and clustering algorithms. Next, Sect. 3 describes the Datil tool. Then, Sect. 4 illustrates the usefulness of the system with a use case on semantic lifestyle profiling. Finally, Sect. 5 sets out some conclusions and addresses some ideas for future work.

2   Background

2.1   Fuzzy Ontologies

Fuzzy ontologies extend classical (crisp) ontologies by considering several notions of fuzzy set theory and fuzzy logic [3, 21]. Before going into the details, let us briefly recall the elements of an ontology:

- *Individuals* denote domain elements or objects. For example, john and mary.
- *Datatypes* denote elements that do not belong to the represented domain, but rather to a different domain that is already structured and whose structure is already known to the machine. Data values can be numerical values, textual, or dates, among many other possibilities.
- *Concepts* or classes denote unary predicates and are interpreted as sets of individuals, such as Human. Concept can be simple (atomic) or complex, built up using different types of concept constructors depending on the expressivity of the ontology language.

---

[1] www.umbertostraccia.it/cs/software/FuzzyDL-Learner.

– *Properties* or roles denote binary predicates relating a pair of elements. There are two types of properties: *object properties* (or abstract roles) link a pair of individuals, whereas *data properties* (or concrete roles) relate an individual with a data value. For instance, isFriendOf relates two human individuals and is an object property, while hasAge links an individual with an integer number and is a data property.

– *Axioms* are formal statements involving these ontology elements, like a recipe that defines how to combine the previous ingredients to represent the knowledge of some particular domain. The available types of axioms depend on the expressivity of the ontology language, but some typical types are:

- *Concept assertions* state the membership of an individual to a class. For example, the fact that john belongs to the concept of Human people.
- *Object property assertions* describe the relation between two individuals. For instance, one can state that john and mary are related via hasChild.
- *Data property assertions* describe the relation between an individual and a data value. For example, it is possible to express that John's age is 18 by relating john and the number 18 via hasAge.
- *Subclass* axioms, stating that a concept is more specific (a subclass) of another one. For example, Woman is more specific than Human.

The interested reader can find a complete list of the OWL 2 elements in [24].

In fuzzy ontologies, the elements of an ontology are extended in such a way that concepts, relations, datatypes, and axioms are fuzzy. In particular:

– *Fuzzy concepts* and *fuzzy properties* are interpreted as fuzzy sets of individuals and fuzzy binary relations, respectively. For example, YoungHuman can contain the fuzzy set of young people.

– *Fuzzy axioms* express statements that are not either true or false but hold to some degree. For example, we can state that john belongs to the concept of YoungHuman with at least degree 0.9, meaning that he is rather young.

– *Fuzzy datatypes* generalize crisp values by using a fuzzy membership function. For example, instead of considering the crisp value 18, now it is possible to consider about18. The former datatype is incompatible with the value 17.99, whereas the latter one is not. Some popular membership functions, commonly used to define fuzzy datatypes are the trapezoidal, the triangular, the left-shoulder, and the right-shoulder, depicted in Fig. 1.

Although there is not an standard fuzzy ontology language, *Fuzzy OWL 2* [5] is a popular choice. The language extends OWL 2 ontologies with OWL 2 annotations encoding fuzzy information using a XML-like syntax. The key idea of this representation is to start with an OWL 2 ontology created as usual, with a classical ontology editor. Then, it is possible to annotate the elements to represent the features of the fuzzy ontology that OWL 2 cannot directly encode. In particular, it is possible to annotate fuzzy axioms by adding a degree of truth, to represent fuzzy datatypes, and to define specific elements of fuzzy ontologies

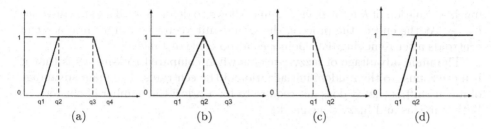

Fig. 1. (a) Trapezoidal; (b) Triangular; (c) Left-shoulder; (d) Right shoulder functions.

(such as fuzzy modifiers or aggregated concepts). There is a Protégé plug-in making the syntax of the annotations transparent to the users.[2]

For practical reasons, the range of the fuzzy datatypes is usually restricted to an interval $[k_1, k_2]$, e.g., in the *fuzzyDL* reasoner and Fuzzy OWL 2 [5,6].

2.2 Clustering

This section recaps three well-known unsupervised clustering algorithms, namely k-means, fuzzy c-means, and mean-shift. These learning algorithms cluster a collection of n real data values (or points) denoted x_j into a set of classes or clusters C_i described by means of their centroids (one per cluster) c_i.

K-means groups a set of data into k clusters [18]. The algorithm starts by computing randomly the k initial centroids c_i. Then, it repeats two steps: first, each point x_j is assigned to its nearest cluster, denoted $C(x_j)$, according to the Euclidean distance: $C(x_j) = C_k$ if $\arg \min_i ||x_j - c_i||^2 = k$. Second, the centroids are updated: $c_i = (\sum_{x_j \in C_i} x_j)/|C_i|$. The algorithm aims at minimizing a squared error function and finishes when a stopping criteria is met (typically, after a total number of iterations or when there are no further changes in the centroids).

Fuzzy c-means [2] is an extension where every point can belong to several clusters with different degrees of membership. To this end, the algorithm considers c fuzzy clusters and a matrix of membership degrees μ, where $\mu_{ij} \in [0, 1]$ denotes the membership degree of the datum x_j to the i-th cluster. The positions of the centroids are computed as $c_i = (\sum_{j=1}^{n} \mu_{ij}^m x_j)/\sum_{j=1}^{n} \mu_{ij}^m$, where $m \geq 1$ is a parameter indicating a fuzziness degree. The membership degrees are then updated as $\mu_{ij} = \left(\sum_{k=1}^{c} \frac{||x_j - c_i||^{2/(m-1)}}{||x_j - c_k||^{2/(m-1)}} \right)^{-1}$.

Mean-shift [7,8] is widely used in clustering but also in image segmentation. It seeks modes or local maxima of density in a feature space by computing a mean-shift vector $m(x)$. The algorithm defines a window around each point, computes the mean of the data points in the window and then shifts its center to the mean. It uses a Gaussian Kernel K_g to keep track of the nearest neighbors of each x_i according to a bandwidth or window size h. To compute the bandwidth we use the rule of thumb proposed in [23]. This rule can be used to compute a

---

[2] http://www.umbertostraccia.it/cs/software/FuzzyOWL.

quick estimation of h for a given K_g, and allows to define a local seeking distance $l = \frac{h}{2}$. At the end of the process, the mean-shift vector converges into a set of centroids after removing data points at a too close distance.

The main advantage of fuzzy c-means when compared to k-means is that it is more robust to the random initialization of the centroids. The main advantage of mean-shift is that it does not require to fix a priori the number of clusters, as both k-means and fuzzy c-means do.

3   The Datil System

Overview. Datil[3] (DATatypes with Imprecision Learner) is a software that automatically learns fuzzy datatypes for fuzzy ontologies from different types of inputs. Datil implements several unsupervised clustering algorithms: k-means, fuzzy c-means, and mean-shift (see Sect. 2.2). The tool is publicly available.[4].

For each data property in an ontology with a numerical range (or with assertions involving numerical values), Datil collects an array of real numbers corresponding to the values of the property for different individuals. A clustering algorithm provides a set of centroids from these array of values. These centroids are used as the parameters to build fuzzy membership functions partitioning the domain, as illustrated in Fig. 2.

Learning the fuzzy datatypes. Assuming that $k \geq 2$, Datil creates the following datatypes using a set of centroids $\{c_1, c_2, \ldots, c_k\}$:

- a left-shoulder function with parameters c_1 and c_2,
- a right-shoulder function with parameters c_{k-1} and c_k, and
- $k - 2$ triangular functions, where the i-th triangular function has parameters c_i, c_{i+1}, and c_{i+2}.

As already mentioned, the fuzzy datatypes are defined over a range $[k_1, k_2]$ and not over $(-\infty, \infty)$. For each data property d, Datil uses several strategies to compute such k_1, k_2 for all the fuzzy datatypes defined for d:

- First, checking if the range of dp is of the form [>= k1, <= k2], where >= and <= denote xsd:minInclusive and xsd:maxInclusive OWL 2 facets, respectively, that constrain the possible values of an OWL 2 numerical datatype.
- Otherwise, it computes the minimum (min) and the maximum (max) of the array of real numbers formed by the values of dp and defines $k1 = min - \sigma$ and $k2 = max + \sigma$ for some σ. In the case of mean-shift, $\sigma = h/2$.

For small numbers of clusters ($k \leq 7$), Datil automatically provides readable names for the fuzzy datatype labels. For example, for a data property SkinTemperature, the tool can create 7 fuzzy datatypes VeryVeryLowSkinTemperature, VeryLowSkinTemperature, LowSkinTemperature, NeutralSkinTemperature,

---

[3] Dátil is the Spanish for the date fruit.
[4] http://webdiis.unizar.es/~ihvdis/Datil.

HighSkinTemperature, VeryHighSkinTemperature and VeryVeryHighSkinTemperature. If the number of clusters was 6, NeutralSkinTemperature would be omitted. For an arbitrary number of clusters $k > 7$, label names are formed by concatenating the name of the data property dp and an integer number (the order of the fuzzy datatype according to an increasing value of the smaller centroid).

If the clustering algorithm provides a unique centroid c, Datil only creates one triangular function with parameters $c - \sigma$, c, and $c + \sigma$.

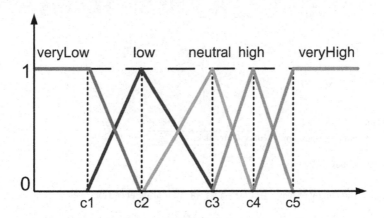

Fig. 2. Some fuzzy membership functions built from the centroids.

Input formats. Datil supports 3 possible input formats: .owl, .fdl, and .csv.

- .owl format correspond to ontologies in the standard language OWL 2. Files can be classical ontologies but also fuzzy ontologies in Fuzzy OWL 2; in the latter case the annotations with the fuzzy information are discarded. Datil restricts itself to data property assertions and range restrictions. A semantic reasoner is used to retrieve both explicit and implicit axioms.
- .fdl is the own syntax of the fuzzyDL reasoner to define fuzzy ontologies [6]. As in previous case, Datil restricts itself to data property assertions and range restrictions and does not consider any fuzzy information (not even the degree of truth of the assertions).
- .csv (Comma-Separated Values) format consists of large data (numbers and text) in plain text. Each record (row) in the file contains one or more fields (columns) separated by commas. In this case, the clustering algorithm takes as an input all the values for a given column. Typically, the first line of the file is special and contain the column names, which should correspond to datatype properties from an ontology.

Output format. Datil supports 2 possible output formats: .owl, and .fdl. The output is a fuzzy ontology with some fuzzy datatype definitions that can be represented as OWL 2 annotations (as specified in Fuzzy OWL 2) or as fuzzyDL

(.fdl) axioms. If the output is a .fdl file, apart from the definition of the fuzzy datatypes, Datil adds further axioms required by fuzzyDL reasoner (functional and range data property axioms).

If the input was an ontology (.owl or .fdl), the output is an extension with the new elements. If the input was a .csv file, the output ontology is created from scratch, and the user can import it later on from another ontology file.

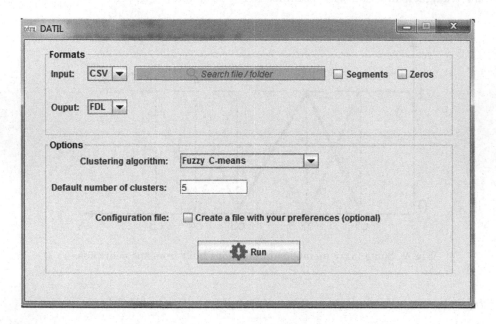

Fig. 3. Snapshot of the main user interface of Datil.

Dependencies. Datil is implemented in Java and uses some external libraries:

- *OWL API*[5] [14] is an ontology API to manage OWL 2 ontologies in Java applications and provides a common interface to interact with DL reasoners. It can be considered as a de facto standard, as the most recent versions of most of the semantics tools and reasoners use the OWL API to load and process OWL 2 ontologies.
- *HermiT*[6] is an OWL 2 ontology reasoner [12]. It completely supports the language and implements several optimization techniques. HermiT is implemented in Java, and is accessible through several interfaces, including the OWL API. We use it to retrieve all the data property assertions, not only those explicitly represented in the ontology but also the implicit ones.

---

[5] http://owlapi.sourceforge.net.

[6] http://www.hermit-reasoner.com.

- *Java-ML* (Java Machine Learning Library)[7] is a collection of machine learning algorithms and a common Java interface for those algorithms. Although Java-ML provides an implementation of k-means, we have implemented our own version the algorithm. However, we do use its Java data structures in all of our clustering algorithms.
- *fuzzyDL*[8] is a fuzzy ontology reasoner [6]. It supports a fuzzy extension of a significant fragment of OWL 2 and supports a notable plethora of reasoning services. The possible input formats are Fuzzy OWL 2, its own syntax in FDL format, and a Java API. We use fuzzyDL to translate fdl fuzzy ontologies into Fuzzy OWL 2.

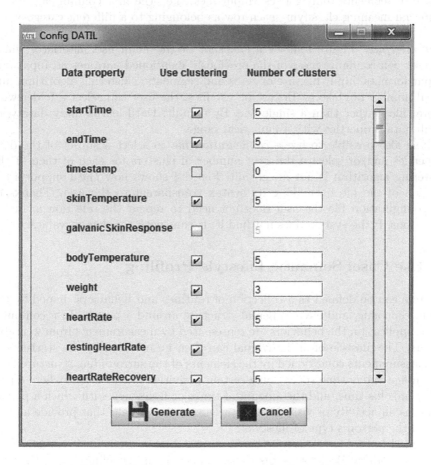

Fig. 4. GUI to create a configuration file in Datil.

---

7 http://java-ml.sourceforge.net.
8 http://www.umbertostraccia.it/cs/software/fuzzyDL/fuzzyDL.html.

Configuration options. Datil requires several parameters:

- The input and output formats.
- The input file. The output file is not a parameter; Datil uses the same filename (with a different filename extension if there is a format change).
- The selected clustering algorithm.
- The number of clusters (only for k-means and fuzzy c-means) for all the properties, or a different number for each property.
- The properties for which to learn the fuzzy datatypes.
- Use of zeros (only for .csv files): zero values can be either taken into account or skipped (in practice, they are often used just to represent empty data).
- Use of segments (only for .csv input files), i.e., the first column can have a special meaning classifying each row as belonging to a different category.

User interface. Figure 3 shows a snapshot of the main user interface, where the user can configure most of the previously mentioned parameters: input and output formats, input file, use of zeros and segments, clustering algorithm, and global number of clusters. In case of .csv files, the user can select a folder with several files rather than a single one. By default, Datil learns fuzzy datatypes for all data properties with a numerical range.

It is also possible to use a *configuration file* to select a subset of the data properties and/or select a different number of clusters for each of them if the clustering algorithm is not mean-shift. Figure 4 shows how Datil supports the creation of the file by making its syntax transparent to the user. Thanks to the configuration file the user does not need to repeat the selection in future executions. If the system does not find it, it runs with the default values.

4   Use Case: Semantic Lifestyle Profiling

Lifestyle can be defined as a collection of routines and behaviors shaped by the social, economic, and environmental structure around a person. In a computational application the behaviors are represented by measurements from wearable sensors. The lifestyle of an individual can then be modeled by the statistics of the measurements conditioned by the elements of the surrounding structure. The percentage of day time spent doing certain activities, the locations where a person spends his time, and the amount of times or frequency with which a person performs an activity or visits a place, are examples of data that provide a good idea of the person's type of lifestyle.

A model of a lifestyle can be based on matching a predefined semantic template to the data. We propose discovering these templates blindly from the data using machine learning techniques [9]. Semantically meaningful and interpretable models to better understand the underlying statistics of individual lifestyle patterns of people is not a trivial task because of the variability of the individuals. Even if technology allows for a broad spectrum of sensors, it is not straightforward to choose the most appropriate data acquisition, data imputation and data fusion techniques [19]. Semantics can enhance data-driven processes and improve accuracy and precision of recognition in human activities [10, 11].

To serve application development in Ambient Intelligence scenarios ranging from activity monitoring in smart homes to active healthy aging or lifestyle profiling, we have developed a fuzzy ontology that allows to describe the lifestyle of a user given its digital footprints such as wrist-born activity trackers, GPS and mobile phone applications[9]. The ontology includes information such as height, weight, locations, cholesterol, sleep, activity levels, activity energy expenditure, heart rate, or stress levels, among many other aggregated features.

In order to populate the ontology, the only information that we have used are real data obtained from digital traces such as sleep and activity sensors and other wearable devices. In particular, we used 40 records of volunteers of middle age living in the Eindhoven area (The Netherlands) [9]. These data were provided by a private company and are confidential (little details are thus given in this paper for privacy reasons). However, we would like to point out that this scenario is a typical case where we do not have data about the membership to classes but we do know the values of several data properties.

The next step is to learn some fuzzy datatypes for each of the 68 data properties and 14 day segments by using Datil. If we use k-means with $k = 5$ fuzzy datatypes, we end up trying to learn 4760 fuzzy datatypes, although for some combinations of data property and day segment there were no data. For instance, one of the learnt fuzzy datatypes for the data property HighSkinTemperature restricted to the day segment day AtWork is HighSkinTemperatureAtWork, defined (in Fuzzy OWL 2 Manchester syntax) as follows:

```
Datatype: HighSkinTemperatureAtWork
   Annotations:
     fuzzyLabel "<fuzzyOwl2 fuzzyType="datatype">
        <Datatype type="triangular" a="22.54" b="27.97" c="30.22" />
        </fuzzyOwl2>"

     EquivalentTo:
       (xsd:double[>= "-6.71"^^xsd:double] and xsd:double[<= "37.97"^^xsd:double])
```

Note that indeed the fuzzy datatypes add more knowledge, in the sense that we can make new inferences. For example, two individuals with slightly different skin temperatures at work, even if such values are different than the center of the triangular function 27.97, would be compatible with the fuzzy datatype HighSkinTemperatureAtWork with different degrees of truth.

To compute the categorization of a person into some lifestyle pattern, we propose the following process:

1. Build a crisp ontology with the features of interest, using domain data scientists with diet and specialists that monitor cardiac disease patients. At this point, experts should identify lifestyle patterns like MediterranLuncher.
2. Populate it with data property assertions obtained from wearable devices.
3. Learn some fuzzy datatypes from the data property assertions using Datil.

---

[9] https://github.com/NataliaDiaz/Ontologies.

4. Define some preliminary rules (concept equivalence or fuzzy subclass axioms) with the help of an expert. Some of the concepts will have complex definitions, being defined in terms of the learnt datatypes. For example, one can define the concept of MediterranLuncher from the (late) time and the (high) duration of the lunch breaks.

```
(define-concept MediterraneanLuncher (g-and
  (some StartTimeFixedAfternoon HighStartTimeFixedAfternoon)
  (some ActivityDurationFixedAfternoon HighActivityDurationFixedAfternoon)
)
```

5. Ask a fuzzy semantic reasoner to retrieve all the instances of each of the defined concepts together with the degrees of membership.
6. Represent it as fuzzy concept assertions and add them to the fuzzy ontology.
7. Run a learning algorithm starting from memberships to fuzzy classes (Fuzzy DL-Learner) to get the final rules. Final rules are complex definitions of lifestyle pattern concepts, similar to the preliminary rules but automatically derived from the real data.

5   Conclusions and Future Work

This paper has presented Datil, a novel tool which is able to learn fuzzy datatypes automatically. Datil supports different input and output formats, allowing both enriching existing fuzzy ontologies and helping in the extension of crisp ontologies to the fuzzy case. Three well-known clustering algorithms are implemented to compute a set of centroids from available data, and then the fuzzy datatypes are defined after them. We have also discussed how fuzzy datatype learning has been applied to a real-world application: semantic lifestyle profiling. In order to characterize the lifestyle of people given their digital footprints, we start from numerical data obtained from different sensors and use Datil to cluster them into fuzzy datatypes interpretable by human users.

Fuzzy datatype learning is a complementary technique to other approaches for fuzzy ontology learning. In particular, our implementation could be used to extend Fuzzy DL-Learner, a system learning fuzzy subclass axioms. Other ideas for future work include the implementation of more sophisticated clustering algorithms. Last but not least, we plan to apply Datil to learn fuzzy datatypes in some more real-world domains; this practical experience will surely provide more ideas to extend our tool.

Acknowledgment. I. Huitzil was partially funded by Universidad de Zaragoza - Santander Universidades (Ayudas de Movilidad para Latinoamericanos - Estudios de Doctorado). N. Díaz-Rodríguez acknowledges AAPELE.eu EU COST Action IC1303 and EU Erasmus+ Funding; part of her work was done during internship at Philips Research. I. Huitzil and F. Bobillo were partially supported by the projects TIN2016-78011-C4-3-R and CUD2017-17. Special thanks are due to Aki Härmä and Rim Helaoui (Philips Research) for their invaluable help with lifestyle real data.

References

1. Alexopoulos, P., Wallace, M., Kafentzis, K., Askounis, D.: IKARUS-Onto: a methodology to develop fuzzy ontologies from crisp ones. Knowl. Inf. Syst. **32**(3), 667–695 (2012)
2. Bezdek, J.C.: Pattern Recognition with Fuzzy Objective Function Algorithms. Advanced Applications in Pattern Recognition, 2nd edn. Plenum Press, New York (1987)
3. Bobillo, F., Cerami, M., Esteva, F., García-Cerdaña, À., Peñaloza, R., Straccia, U.: Fuzzy description logics. In: Cintula, P., Fermüller, C., Noguera, C. (eds.) Handbook of Mathematical Fuzzy Logic Volume III, Studies in Logic, Mathematical Logic and Foundations, vol. 58, pp. 1105–1181. College Publications (2015). chapter XVI
4. Bobillo, F., Ruiz, M.D., Gómez-Romero, J., Sánchez, D.: On the application of data mining techniques to graded ontology building. In: Actas del XVIII Congreso Español sobre Tecnologías y Lógica Fuzzy (ESTYLF 2016), pp. 142–143 (2016)
5. Bobillo, F., Straccia, U.: Fuzzy ontology representation using OWL 2. Int. J. Approx. Reason. **52**(7), 1073–1094 (2011)
6. Bobillo, F., Straccia, U.: The fuzzy ontology reasoner fuzzyDL. Knowl.-Based Syst. **95**, 12–34 (2016)
7. Cheng, Y.: Mean shift, mode seeking, and clustering. IEEE Trans. Pattern Anal. Mach. Intell. **17**(8), 790–799 (1995)
8. Comaniciu, D., Meer, P.: Mean shift: a robust approach toward feature space analysis. IEEE Trans. Pattern Anal. Mach. Intell. **24**, 603–619 (2002)
9. Díaz-Rodríguez, N., Härmä, A., Helaoui, R., Huitzil, I., Bobillo, F., Straccia, U.: Couch potato or gym addict? Semantic lifestyle profiling with wearables and knowledge graphs. In: Proceedings of the 6th NIPS Workshop on Automated Knowledge Base Construction (AKBC 2017), December 2017
10. Díaz-Rodríguez, N., León-Cadahía, O., Pegalajar-Cuéllar, M., Lilius, J., Delgado, M.: Handling real-world context-awareness, uncertainty and vagueness in realtime human activity tracking and recognition with a fuzzy ontology-based hybrid method. Sensors **14**(10), 18131–18171 (2014)
11. Díaz-Rodríguez, N., Pegalajar-Cuéllar, M., Lilius, J., Delgado, M.: A fuzzy ontology for semantic modelling and recognition of human behaviour. Knowl.-Based Syst. **66**, 46–60 (2014)
12. Glimm, B., Horrocks, I., Motik, B., Stoilos, G., Wang, Z.: HermiT: an OWL 2 reasoner. J. Autom. Reason. **53**(3), 245–269 (2014)
13. Gómez-Romero, J., Bobillo, F., Ros, M., Molina-Solana, M., Ruiz, M.D., Martín-Bautista, M.J.: A fuzzy extension of the semantic building information model. Autom. Constr. **57**, 202–212 (2015)
14. Horridge, M., Bechhofer, S.: The OWL API: a Java API for OWL ontologies. Semant. Web J. **2**(1), 11–21 (2011)
15. Iglesias, J., Lehmann, J.: Towards integrating fuzzy logic capabilities into an ontology-based inductive logic programming framework. In: Proceedings of the 11th International Conference on Intelligent Systems Design and Applications (ISDA 2011), pp. 1323–1328 (2011)
16. Lisi, F.A., Straccia, U.: A logic-based computational method for the automated induction of fuzzy ontology axioms. Fundam. Inform. **124**(4), 503–519 (2013)
17. Lisi, F.A., Straccia, U.: Learning in description logics with fuzzy concrete domains. Fundam. Inform. **140**(3–4), 373–391 (2015)

18. Lloyd, S.P.: Least squares quantization in PCM. IEEE Trans. Inf. Theory **28**(2), 129–137 (1982)
19. Pires, I.M., Garcia, N.M., Pombo, N., Flrez-Revuelta, F.: From data acquisition to data fusion: a comprehensive review and a roadmap for the identification of activities of daily living using mobile devices. Sensors **16**(2), 184 (2016)
20. Staab, S., Studer, R. (eds.): Handbook on Ontologies. IHIS. Springer, Heidelberg (2004). https://doi.org/10.1007/978-3-540-92673-3
21. Straccia, U.: Foundations of Fuzzy Logic and Semantic Web Languages. CRC Studies in Informatics Series. Chapman & Hall, New York (2013)
22. Straccia, U., Mucci, M.: pFOIL-DL: learning (fuzzy) $\mathcal{EL}$ concept descriptions from crisp OWL data using a probabilistic ensemble estimation. In: Proceedings of the 30th Annual ACM Symposium on Applied Computing (SAC-15), Salamanca, Spain, pp. 345–352. ACM (2015)
23. Turlach, B.A.: Bandwidth selection in kernel density estimation: a review. CORE and Institut de Statistique (1993)
24. W3C OWL Working Group: OWL 2 Web Ontology Language: Document Overview (2008). http://www.w3.org/TR/owl2-overview
25. Zadeh, L.A.: Fuzzy sets. Inf. Control **8**(3), 338–353 (1965)

Fuzzy Transforms: Theory and Applications to Data Analysis and Image Processing

Axiomatic of Inverse Lattice-Valued F-transform

Jiří Močkoř[(✉)]

Institute for Research and Applications of Fuzzy Modeling,
Centre of Excellence IT4Innovations, University of Ostrava,
30. dubna 22, 701 03 Ostrava 1, Czech Republic
Jiri.Mockor@osu.cz
http://irafm.osu.cz/

Abstract. Axioms of two versions of inverse fuzzy transformation systems are introduced, and it is proved that a transformation function satisfies these axioms if and only if it is an upper or lower inverse lattice-valued F-transform with respect to a fuzzy partition. Categories of inverse transformation systems are introduced, and it is proved that these categories are isomorphic to the category of spaces with fuzzy partitions.

1   Introduction

Fuzzy transform (or F-transform) is a method that is successfully used in signal and image processing [2], compression [3,6,10,21], numerical solutions of ordinary and partial differential equations [5,20,22], data analysis [12] and many other applications. The main strategy of fuzzy transform techniques is to transform an original space of functions into a special space of functions where various computations are simpler. Inverse transformations back to the original spaces yield either the original functions or their approximations (see, e.g., [1]). These techniques, which were specially developed for fuzzy sets, were introduced by I. Perfilieva in a series of papers [10–16,18] with various levels of generalization. Originally, the notion of a fuzzy transform (or F-transform) was introduced for real-valued functions and fuzzy sets with values in the Łukasiewicz algebra. In the same paper [10], a notion of F-transform that is based on a residuated lattice in the interval $[0, 1]$ is also introduced as a generalization of the original definition. Moreover, in paper [14], the notions of direct and inverse F-transforms were extended to the case of Q-valued functions on a space that is defined by Q-valued fuzzy partitions, where Q is a complete residuated lattice.

The basic strategy of the F-transform is to approximate values of real-valued functions that are defined on a set X by values of these functions that are calculated in the neighbourhoods of elements of a fuzzy partition that is defined on X. Fuzzy transform can also be characterized as a method for converting

This research was partially supported by the project 18-06915S provided by the Grant Agency of the Czech Republic.

J. Medina et al. (Eds.): IPMU 2018, CCIS 854, pp. 115–126, 2018.
https://doi.org/10.1007/978-3-319-91476-3_10

complicated and huge fuzzy sets to smaller and simpler fuzzy sets. From that point of view, a one-dimensional lattice-valued fuzzy transform can be described as a map $F : Q^X \to Q^Y$, where X is a "large" set (the original universe of fuzzy sets), Y is a "smaller" set (a new, simpler universe), and Q is an appropriate complete lattice.

The ground structure for F-transform constructions is the *space with a fuzzy partition*, which is represented by a set X (a domain of real- or lattice-valued functions) and a system $\mathcal{A} = \{A_y : y \in Y\}$ of fuzzy sets in X, which is called a *fuzzy partition* on X. In most definitions of F-transform methods, fuzzy partitions must satisfy additional properties. The original definition of F-transform of real-valued functions (see [13]) describes a fuzzy partition as a finite system of fuzzy sets $A_i, i = 1, \ldots, n$ that are defined in a real interval $X = [a, b]$ and satisfy the following properties with respect to a finite set of real points x_k, where $a = x_0 < \cdots < x_n = b$, for each $k = 1, \ldots, n$:

1. $x_k \in core(A_k) = \{x \in X : A_k(x) = 1\}$,
2. $Supp A_k = (x_{k-1}, x_{k+1})$,
3. A_k is continuous,
4. A_k is strictly increasing on $[x_{k-1}, x_k]$ and strictly decreasing on $[x_k, x_{k+1}]$,
5. $\sum_{k=1}^{n} A_k(x) = 1$, for each $x \in [a, b]$.

The original definition of a fuzzy partition is not suitable for lattice-valued functions. For that case, Perfilieva [11] introduced the notion of a lattice-valued fuzzy partition that is defined for lattice-valued functions, which is a natural generalization of a fuzzy partition that is defined by a lattice-valued similarity relation. The category **SpaceFP** of lattice-valued fuzzy partitions was introduced in [7], where principal properties of that category were described. Both real-valued and lattice-valued F-transforms are originally defined by a formula with fuzzy sets using a given fuzzy partition. This ad hoc approach, which is often used in the theory of fuzzy sets, alters the properties of such objects. Many basic properties of these objects need to be derived from the specific shape of the formula that defines them, instead of being explicitly described by axioms of the objects. In [8], we introduced a new structure (without the notion of a fuzzy partition) that we call *a fuzzy transformation system*. This structure has been proven equivalent to a lattice-valued F-transform with respect to a fuzzy partition. As we mentioned, an approximation of the original function can be derived by inverse F-transform, which was also introduced by Perfilieva [13]. As in the case of a direct F-transform, an inverse F-transform is also defined by an ad hoc formula that is based on the explicit use of a fuzzy partition.

To eliminate as much as possible the explicit definition of the inverse F-transform based on fuzzy partitions, we introduce in this article a new axiomatically defined structure called *inverse fuzzy transformation*. This axiomatic definition focuses exclusively on the properties of the inverse transformation function $Q^Y \to Q^X$ and does not use any fuzzy partitions. We show that analogous to the direct F-transform, the inverse transformation system is isomorphic to the inverse F-transform with respect to a fuzzy partition.

2   Upper and Lower F-transforms

Since Zadeh's original paper [23] was published, the notion of a "fuzzy set" has changed significantly and is now more general. The first important modification concerns the value set: instead of the real-number interval $I = [0, 1]$, more general lattice structures Q are considered. Among these lattice structures, *complete residuated lattices* play an important role, (see, e.g., [9]) and are also called *strictly two-sided commutative quantales* (see [19]). Such a structure is defined as follows: $Q = (Q, \wedge, \vee, \otimes, \rightarrow, 0, 1)$ such that $(Q, \wedge, \vee)$ is a complete lattice with bottom element 0 and top element 1, $(Q, \otimes, 1)$ is a commutative monoid, and $\rightarrow$ is a binary operation that is adjoint to $\otimes$, i.e.,

$$\alpha \otimes \beta \leq \gamma \text{ iff } \alpha \leq \beta \rightarrow \gamma.$$

Recall that a negation of an element a in Q is defined by $\neg a = a \rightarrow 0$. A residuated lattice Q is called a (commutative, integral) Girard monoid [4] if, for arbitrary $a \in Q$, it satisfies the double negation law $\neg\neg a = a$.

For a residuated lattice Q, a Q-fuzzy set in a crisp set X is a map $f : X \rightarrow Q$. An F-transform, in a form that was introduced by Perfilieva [14], is based on the so-called fuzzy partitions on the crisp set.

Definition 1 [14]. *Let X be a set. A system $\mathcal{A} = \{A_\lambda : \lambda \in \Lambda\}$ of normal Q-fuzzy sets in X is a fuzzy partition of X if $\{core(A_\lambda) : \lambda \in \Lambda\}$ is a partition of X. A pair $(X, \mathcal{A})$ is called a space with a fuzzy partition.*

If $(X, \mathcal{A})$ is a space with a fuzzy partition, then by $|\mathcal{A}|$ we denote the index set of $\mathcal{A}$. The category **SpaceFP** of spaces with fuzzy partitions is defined in [7].

Definition 2. *The category **SpaceFP** is defined by*

1. *Fuzzy partitions $(X, \mathcal{A})$ as objects,*
2. *Morphisms $(g, \sigma) : (X, \mathcal{A}) \rightarrow (Y, \mathcal{B})$ such that*
   (a) $g : X \rightarrow Y$ is a map,
   (b) $\sigma : |\mathcal{A}| \rightarrow |\mathcal{B}|$ is a map such that

$$\forall \lambda \in |\mathcal{A}|, x \in X \quad A_\lambda(x) \leq B_{\sigma(\lambda)}(g(x)).$$

3. *The composition of morphisms in **SpaceFP** is defined by $(h, \tau) \circ (g, \sigma) = (h \circ g, \tau \circ \sigma)$.*

If $(g, \sigma) : (X, \mathcal{A}) \rightarrow (Y, \mathcal{B})$ is a morphism, then for any $\lambda \in |\mathcal{A}|$, $g(core(A_\lambda)) \subseteq core(B_{\sigma(\lambda)})$, which follows from the definition.

Objects of the category **SpaceFP** represent ground structures for a fuzzy transform, which was proposed by Perfilieva [13] and, in the case where it is applied to Q-valued functions with Q-valued partitions, in [14]. Two variants of lattice-valued fuzzy transforms are defined – lower and upper F-transforms.

Definition 3 [14]. *Let $(X, \mathcal{A})$ be a space with a fuzzy partition $\mathcal{A} = \{A_\lambda : \lambda \in |\mathcal{A}|\}$.*

1. *An upper F-transform with respect to the space $(X, \mathcal{A})$ is a function $F^{\uparrow}_{X,\mathcal{A}}$: $Q^X \to Q^{|\mathcal{A}|}$ that is defined by*

$$f \in Q^X, \lambda \in |\mathcal{A}|, \quad F^{\uparrow}_{X,\mathcal{A}}(f)(\lambda) = \bigvee_{x \in X} (f(x) \otimes A_\lambda(x)).$$

2. *A lower F-transform with respect to the space $(X, \mathcal{A})$ is a function $F^{\downarrow}_{X,\mathcal{A}}$: $Q^X \to Q^{|\mathcal{A}|}$ that is defined by*

$$f \in Q^X, \lambda \in |\mathcal{A}|, \quad F^{\downarrow}_{X,\mathcal{A}}(f)(\lambda) = \bigwedge_{x \in X} (A_\lambda(x) \to f(x)).$$

In [8], we introduced *upper and lower transformation systems* on a set X as special maps $Q^X \to Q^Y$ and proved that these upper and lower transformation systems are equivalent to upper or lower F-transforms, respectively, with respect to a fuzzy partition $\mathcal{A}$ that is defined on X, such that $|\mathcal{A}| = Y$. Moreover, we proved that the categories of upper and lower transformation systems are isomorphic to the category **SpaceFP**.

3   Axiomatic Definition of Inverse F-transforms

It is natural to ask the following question: can we reconstruct a function by its F-transform? As expected, in general, precise reconstruction is not possible because we lose information when passing to the F-transform. However, for any lattice-valued F-transform $F : Q^X \to Q^{|\mathcal{A}|}$ with respect to a fuzzy partition $\mathcal{A}$, there exists an inverse lattice-valued F-transform $H : Q^{|\mathcal{A}|} \to Q^X$ such that $HF(f)$ is an approximation of an original function $f \in Q^X$. The inverse lower and upper lattice-valued F-transforms were also defined by Perfilieva [13] and the proximity between the original function f and the function $HF(f)$ was investigated in [13]. In what follows, by Q, we denote a complete residuated lattice.

Definition 4. *Let $(X, \mathcal{A})$ be a space with a fuzzy partition $\mathcal{A} = \{A_\lambda : \lambda \in |\mathcal{A}|\}$.*

1. *An upper inverse F-transform with respect to $(X, \mathcal{A})$ is a function $H^{\uparrow}_{X,\mathcal{A}}$: $Q^{|\mathcal{A}|} \to Q^X$ that is defined by*

$$(\forall p \in Q^{|\mathcal{A}|}, x \in X) \quad H^{\uparrow}_{X,\mathcal{A}}(p)(x) = \bigwedge_{\lambda \in |\mathcal{A}|} A_\lambda(x) \to p(\lambda).$$

2. *A lower inverse F-transform with respect to $(X, \mathcal{A})$ is a function $H^{\downarrow}_{X,\mathcal{A}}$: $Q^{|\mathcal{A}|} \to Q^X$ that is defined by*

$$(\forall p \in Q^{|\mathcal{A}|}, x \in X) \quad H^{\downarrow}_{X,\mathcal{A}}(p)(x) = \bigvee_{\lambda \in |\mathcal{A}|} A_\lambda(x) \otimes p(\lambda).$$

In this section, we define inverse transformation systems on a set X as maps $Q^Y \to Q^X$ that satisfy specified axioms and we prove that the only inverse transformation systems that satisfy the axioms are upper and lower inverse F-transforms with respect to fuzzy partitions. Moreover, we show that the categories of upper or lower inverse transformation systems are isomorphic to the category **SpaceFP** of spaces with fuzzy partitions.

In what follows, we use the following notation: Let X be a set and $\alpha \in Q$. Then, $\underline{\alpha}_X \in Q^X$ is the constant function with value α. By $\chi_{X,\{x\}}$ we denote the characteristic function $X \to Q$ of a singleton $\{x\}$ in a set X. If there is no risk of misunderstanding, instead of $\chi_{X,\{x\}}$ we use only $\chi_{\{x\}}$.

The notion of a lower inverse transformation system is defined as follows:

Definition 5. *A system* (X, Y, u, T) *is called a (Q-valued)* **lower inverse transformation system** *if*

1. *X, Y are sets;*
2. *$u : X \twoheadrightarrow Y$ is a surjective map;*
3. *$T : Q^Y \to Q^X$ is a map such that*
   (a) *For each $\{s_i : i \in J\} \subseteq Q^Y$, $T(\bigvee_{i \in J} s_i) = \bigvee_{i \in J} T(s_i)$;*
   (b) *For each $s \in Q^Y, \alpha \in Q$, $T(\underline{\alpha}_Y \otimes s) = \underline{\alpha}_X \otimes T(s)$;*
   (c) *For each $y \in Y, x \in X$, $T(\chi_{Y,\{y\}})(x) = 1 \Leftrightarrow u(x) = y$.*

Our goal is to prove that *lower inverse transformation systems* are precisely *lower inverse F-transforms*, which are defined in Definition 4.

Theorem 1. *Let X, Y be sets, $u : X \to Y$ a surjective map and $T : Q^Y \to Q^X$ a map. The following statements are equivalent:*

(1) (X, Y, u, T) is a lower inverse transformation system;
*(2) There exists $(X, \mathcal{A}) \in$ **SpaceFP** such that $|\mathcal{A}| = Y$, $u(x) = y$ iff $x \in core(A_y)$ and $T = H^{\downarrow}_{X,\mathcal{A}}$.*

Proof. For the proof of the theorem, we recall that according to [17], each function $s \in Q^Y$ can be represented in the form

$$s = \bigvee_{y \in Y} \underline{s(y)}_Y \otimes \chi_{\{y\}}. \tag{1}$$

$(1) \Rightarrow (2)$. For $y \in Y$, we set $A_y = T(\chi_{\{y\}}) \in Q^X$. Since u is a surjective map, it follows from property (c) that $\mathcal{A} = \{A_y : y \in Y\}$ is a fuzzy partition in the set X, $|\mathcal{A}| = Y$ and $u(x) = y$ iff $x \in core(A_y)$. We show that for each $s \in Q^Y$, $T(s) = H^{\downarrow}_{X,\mathcal{A}}(s)$. According to Eq. (1), for each $x \in X$, we have

$$T(s)(x) = T(\bigvee_{y \in Y} \underline{s(y)}_Y \otimes \chi_{\{y\}})(x) = \bigvee_{y \in Y} T(\underline{s(y)}_Y \otimes \chi_{\{y\}})(x) =$$

$$\bigvee_{y \in Y} \underline{s(y)}_X(x) \otimes T(\chi_{\{y\}})(x) = \bigvee_{y \in Y} s(y) \otimes A_y(x) = H^{\downarrow}_{X,\mathcal{A}}(s)(x).$$

(2) $\Rightarrow$ (1). We show that $(X, |\mathcal{A}|, u, H_{X,\mathcal{A}}^{\downarrow})$ is a lower transformation system, where $u(x) = \alpha$ iff $x \in core(A_\alpha)$. It is clear that $H_{X,\mathcal{A}}^{\downarrow}$ satisfies conditions (a) and (b). For $\alpha \in |\mathcal{A}|$, we have

$$H_{X,\mathcal{A}}^{\downarrow}(\chi_{|\mathcal{A}|,\{\alpha\}})(x) = \bigvee_{\beta \in |\mathcal{A}|} A_\beta(x) \otimes \chi_{|\mathcal{A}|,\{\alpha\}}(\beta) = A_\alpha(x) \otimes \chi_{|\mathcal{A}|,\{\alpha\}}(\alpha) = A_\alpha(x),$$

and $H_{X,\mathcal{A}}^{\downarrow}$ satisfies condition (c). ∎

Any lower inverse transformation system (X, Y, u, T) has the following property:

$$(\forall s \in Q^Y, x \in X) \quad T(s)(x) \geq s(u(x)).$$

From relation (1) and properties from Definition 5, it follows that

$$T(s)(x) = \bigvee_{y \in Y} s(y) \otimes T(\chi_{Y,\{y\}})(x) \geq s(u(x)) \otimes T(\chi_{Y,\{u(x)\}})(x) = s(u(x)).$$

We introduce an upper variant of an inverse transformation system.

Definition 6. *A system* (X, Y, w, S) *is called a (Q-valued)* **upper inverse transformation system** *if*

1. *X, Y are sets,*
2. *$w : X \twoheadrightarrow Y$ is a surjective map,*
3. *$S : Q^Y \to Q^X$ is a map such that*
   (a) For each $\{s_i : i \in J\} \subseteq Q^Y$, $S(\bigwedge_{i \in J} s_i) = \bigwedge_{i \in J} S(s_i)$;
   (b) For each $s \in Q^Y, \alpha \in Q$, $S(\underline{\alpha}_Y \to s) = \underline{\alpha}_X \to S(s)$;
   (c) For each $x \in X, y \in Y$, $S(\neg\chi_{Y,\{y\}})(x) = 0 \Leftrightarrow w(x) = y$.

We want to show that upper inverse transformation systems are precisely upper inverse F-transforms, which were introduced in Definition 4.

Theorem 2. *Let X, Y be sets, $w : X \to Y$ a surjective map and $S : Q^Y \to Q^X$ a map. Let the double negation law hold in Q. Then, the following statements are equivalent:*

(1) (X, Y, w, S) is an upper inverse transformation system;
*(2) There exists $(X, \mathcal{A}) \in$ **SpaceFP** such that $|\mathcal{A}| = Y$, $w(x) = y$ iff $x \in core(A_y)$ and $S = H_{X,\mathcal{A}}^{\uparrow}$.*

Proof. For the proof of Theorem 2, we need the following representation of a function $\neg s$, where $s \in Q^Y$, which was proved in [8]:

$$\neg s = \bigwedge_{y \in Y} s(y)_Y \to \neg\chi_{\{y\}}. \tag{2}$$

(1) $\Rightarrow$ (2). For $y \in Y$, we set $A_y = \neg S(\neg \chi_{\{y\}})$. According to (3c), $core(A_y) = \{x \in X : w(x) = y\}$ and A_y is a normal fuzzy set in X. Moreover, it follows that $\mathcal{A} = \{A_y : y \in Y\}$ is a fuzzy partition in X and $|\mathcal{A}| = Y$. We show that $S(s) = H^\uparrow_{X,\mathcal{A}}$. Let $s \in Q^Y, x \in X$. Then, according to (2), we have

$$s = \neg\neg s = \bigwedge_{y \in Y} \underline{\neg s(y)}_Y \to \neg \chi_{\{y\}},$$

and we obtain

$$S(s)(x) = S(\neg\neg s)(x) = S(\bigwedge_{y \in Y} \underline{\neg s(y)}_Y \to \neg \chi_{\{y\}})(x) =$$

$$\bigwedge_{y \in Y} S(\underline{\neg s(y)}_Y \to \neg \chi_{\{y\}})(x) = \bigwedge_{y \in Y} \underline{\neg s(y)}_X(x) \to S(\neg \chi_{\{y\}})(x) =$$

$$\bigwedge_{y \in Y} \underline{\neg s(y)}_X(x) \to \neg\neg S(\neg \chi_{\{y\}})(x) = \bigwedge_{y \in Y} \neg S(\neg \chi_{\{y\}})(x) \to s(y) =$$

$$\bigwedge_{y \in Y} A_y(x) \to s(y) = H^\uparrow_{X,\mathcal{A}}(s)(x).$$

(2) $\Rightarrow$ (1). To prove that $(X, |\mathcal{A}|, w, H^\uparrow_{X,\mathcal{A}})$ is an upper transformation system, we need to prove only properties (b) and (c). The proof of the property (a) is trivial. For $\alpha \in Q, s \in Q^{|\mathcal{A}|}$, we have

$$H^\uparrow_{X,\mathcal{A}}(\underline{\alpha}_Y \to s)(x) = \bigwedge_{\beta \in |\mathcal{A}|} A_\beta(x) \to (\underline{\alpha}_Y \to s)(\beta) =$$

$$\bigwedge_{\beta \in |\mathcal{A}|} (A_\beta(x) \otimes \alpha) \to s(\beta) = \bigwedge_{\beta \in |\mathcal{A}|} (\alpha \otimes A_\beta(x)) \to s(\beta) =$$

$$\bigwedge_{\beta \in |\mathcal{A}|} \alpha \to (A_\beta(x) \to s(y)) = \bigwedge_{\beta \in |\mathcal{A}|} (\alpha \to (A_\beta(x) \to s(y))) =$$

$$\alpha \to (\bigwedge_{\beta \in |\mathcal{A}|} (A_\beta(x) \to s(y)) = \underline{\alpha}_X(x) \to H^\uparrow_{X,\mathcal{A}}(s)(x) =$$

$$(\underline{\alpha}_X \to H^\uparrow_{X,\mathcal{A}}(s))(x),$$

and $H^\uparrow_{X,\mathcal{A}}$ satisfies property (b). In addition, for $\alpha \in |\mathcal{A}|, x \in X$, we have

$$\neg H^\uparrow_{X,\mathcal{A}}(\neg \chi_{|\mathcal{A}|,\{\alpha\}})(x) = \neg \bigwedge_{\beta \in |\mathcal{A}|} A_\beta(x) \to (\neg \chi_{|\mathcal{A}|,\{\alpha\}})(\beta) =$$

$$\neg \bigwedge_{\beta \in |\mathcal{A}|} A_\beta(x) \otimes \chi_{|\mathcal{A}|,\{\alpha\}}(\beta) \to 0 = \neg(A_\alpha(x) \to 0) = A_\alpha(x).$$

Hence,

$$H^\uparrow_{X,\mathcal{A}}(\neg \chi_{|\mathcal{A}|,\{\alpha\}})(x) = 0 \Leftrightarrow A_\alpha(x) = 1 \Leftrightarrow w(x) = \alpha,$$

and $H^{\uparrow}_{X,\mathcal{A}}$ satisfies property (c).   ■

If Q satisfies the double negation law, any upper inverse transformation system (X, Y, w, S) has the following property:

$$(\forall s \in Q^Y, x \in X) \quad S(s)(x) \leq s(w(x)).$$

From the above proof, it follows that

$$S(s)(x) = \bigwedge_{y \in Y} \neg S(\neg \chi_{\{y\}})(x) \rightarrow s(y) \leq \neg S(\neg \chi_{\{w(x)\}})(x) \rightarrow s(w(x)) =$$

$$1_Q \rightarrow s(w(x)) = s(w(x)).$$

In [8], we introduced the categories **FTrans**$^{\uparrow}$ and **FTrans**$^{\downarrow}$ of upper and lower transformation systems and proved that these categories are isomorphic to the category **SpaceFP**. We prove analogous results for the categories of upper and lower *inverse* transformation systems. For any sets X, Y and a map $f : X \rightarrow Y$, by $f^{\rightarrow}$ we denote a Zadeh's extension $Q^X \rightarrow Q^Y$ of f, i.e., for $t \in Q^X, y \in Y$,

$$f^{\rightarrow}(t)(y) = \bigvee_{x \in X, f(x)=y} t(x).$$

Analogously, $f^{\leftarrow} : Q^Y \rightarrow Q^X$ is an extension of f that is defined by

$$f^{\leftarrow}(s)(x) = s(f(x)),$$

for each $s \in Q^Y, x \in X$.

Definition 7. *The category* **ITrans**$^{\downarrow}$ *of lower inverse transformation systems is defined as follows:*

1. *Objects are lower inverse transformation systems;*
2. *Morphisms* $(f, \sigma) : (X, Y, u, T) \rightarrow (X_1, Y_1, u_1, T_1)$ *are defined as follows:*
   (a) $f : X \rightarrow X_1$ *and* $\sigma : Y \rightarrow Y_1$ *are maps;*
   (b) In the diagram

$$\begin{array}{ccc} Q^Y & \xrightarrow{\;T\;} & Q^X \\ {\scriptstyle \sigma^{\rightarrow}}\downarrow & & \downarrow{\scriptstyle f^{\rightarrow}} \\ Q^{Y_1} & \xrightarrow{\;T_1\;} & Q^{X_1}, \end{array}$$

   we have $T_1 \circ \sigma^{\rightarrow} \geq f^{\rightarrow} \circ T;$
   (c) Morphisms are composed component-wise.

It can be verified easily that **ITrans**$^{\downarrow}$ is a category.

Theorem 3. *The categories* **SpaceFP** *and* **ITrans**$^{\downarrow}$ *are isomorphic.*

Proof. Let $(f, \sigma) : (X, Y, u, T) \rightarrow (X_1, Y_1, u_1, T_1)$ be a morphism in **ITrans**$^\downarrow$. Then, the functor $K : \mathbf{ITrans}^\downarrow \rightarrow \mathbf{SpaceFP}$ is defined by

$$K(X, Y, u, T) = (X, \mathcal{A} = \{T(\chi_{Y, \{y\}}) : y \in Y\}), \quad K(f, \sigma) = (f, \sigma).$$

From the proof of Theorem 1, it follows that $(X, \mathcal{A}) \in \mathbf{SpaceFP}$. We show that $(f, \sigma) : (X, \mathcal{A}) \rightarrow (X, \mathcal{A}_1)$ is a morphism in the category **SpaceFP**. For $x \in X$, we have

$$T(\chi_{Y, \{y\}})(x) \leq \bigvee_{x' \in X, f(x') = f(x)} T(\chi_{Y, \{y\}})(x') = f^\rightarrow T(\chi_{Y, \{y\}})(f(x)) \leq$$

$$T_1 \circ \sigma(\chi_{Y, \{y\}})(f(x)) = T_1(\chi_{Y_1, \{\sigma(y)\}})(f(x)).$$

It is clear that K is a functor. Let the functor K^{-1} be defined by

$$K^{-1}(X, \mathcal{A}) = (X, |\mathcal{A}|, u, H_{X, \mathcal{A}}^\downarrow)$$

where $u : X \rightarrow |\mathcal{A}|$ is such that $u(x) = \lambda \Leftrightarrow x \in core(A_\lambda)$. From the proof of Theorem 1, it follows that $K^{-1}(X, \mathcal{A}) \in \mathbf{ITrans}^\downarrow$. Let $(f, \sigma) : (X, \mathcal{A}) \rightarrow (Y, \mathcal{B})$ be a morphism in **SpaceFP**. Then, $K^{-1}(f, \sigma) = (f, \sigma)$ is a morphism $K^{-1}(X, \mathcal{A}) \rightarrow K^{-1}(Y, \mathcal{B})$ in **ITrans**$^\downarrow$. Let $s \in Q^{|\mathcal{A}|}, y \in Y$. Then, we have

$$f^\rightarrow H_{X, \mathcal{A}}^\downarrow(s)(y) = \bigvee_{x, f(x) = y} H_{X, \mathcal{A}}^\downarrow(s)(x) = \bigvee_{x, f(x) = y} \bigvee_{\alpha \in |\mathcal{A}|} A_\alpha(x) \otimes s(\alpha) \leq$$

$$\bigvee_{x, f(x) = y} \bigvee_{\alpha \in |\mathcal{A}|} B_{\sigma(\alpha)}(f(x)) \otimes s(\alpha) \leq$$

$$\bigvee_{\alpha \in |\mathcal{A}|} B_{\sigma(\alpha)}(y) \otimes \bigvee_{\alpha', \sigma(\alpha') = \sigma(\alpha)} s(\alpha') \leq \bigvee_{\beta \in |\mathcal{B}|} B_\beta(y) \otimes \bigvee_{\alpha, \sigma(\alpha) = \beta} s(\alpha) =$$

$$\bigvee_{\beta \in |\mathcal{B}|} B_\beta(y) \otimes \sigma^\rightarrow(s)(\beta) = H_{Y, \mathcal{B}}^\downarrow \sigma^\rightarrow(s)(y),$$

and (f, σ) is a morphism in **ITrans**$^\downarrow$. From the proof of Theorem 1, it follows directly that $K.K^{-1} = K^{-1}.K = id$. ∎

Definition 8. *The category* **ITrans**$^\uparrow$ *of upper inverse transformation systems is defined as follows:*

1. *Objects are upper inverse transformation systems;*
2. *Morphisms* $(f, \sigma) : (X, Y, w, S) \rightarrow (X_1, Y_1, w_1, S_1)$ *are defined as follows:*
   (a) $f : X \rightarrow X_1$ *and* $\sigma : Y \rightarrow Y_1$ *are maps;*
   (b) In the diagram

$$\begin{array}{ccc} Q^Y & \xrightarrow{S} & Q^X \\ \sigma^\leftarrow \uparrow & & \uparrow f^\leftarrow \\ Q^{Y_1} & \xrightarrow{S_1} & Q^{X_1}, \end{array}$$

we have $S \circ \sigma^\leftarrow \geq f^\leftarrow \circ S_1;$

(c) Morphisms are composed component-wise.

It is clear that **ITrans**$^\uparrow$ is a category.

Theorem 4. *Let the double negation law hold in Q. Then, categories* **SpaceFP** *and* **ITrans**$^\uparrow$ *are isomorphic.*

Proof. The functor $L : \textbf{ITrans}^\uparrow \to \textbf{SpaceFP}$ is defined by

$$L(X, Y, w, S) = (X, \mathcal{A}), \quad L(f, \sigma) = (f, \sigma),$$

where $(f, \sigma) : (X, Y, w, S) \to (X_1, Y_1, w_1, S_1)$ is a morphism in **ITrans**$^\uparrow$ and $(X, \mathcal{A})$ is a space with a fuzzy partition such that $S = H^\uparrow_{X,\mathcal{A}}$. According to Theorem 2, $\mathcal{A} = \{A_y = \neg S(\neg \chi_{Y,\{y\}}) : y \in Y\}$. We prove that $(f, \sigma) : (X, \mathcal{A}) \to (X_1, \mathcal{A}_1)$ is a morphism in **SpaceFP**. For each $y \in Y$, we have

$$\sigma^\leftarrow(\neg \chi_{Y_1, \{\sigma(y)\}}) \leq \neg \chi_{Y, \{y\}}.$$

Using that relation, for each $x \in X$ we obtain

$$\neg A^1_{\sigma(y)}(f(x)) = S_1(\neg \chi_{Y_1, \{\sigma(y)\}})(f(x)) = f^\leftarrow \circ S_1(\neg \chi_{Y_1, \{\sigma(y)\}})(x) \leq$$
$$S \circ \sigma^\leftarrow(\neg \chi_{Y_1, \{\sigma(y)\}})(x) \leq S(\neg \chi_{Y, \{y\}})(x) = \neg A_y(x),$$

and it follows that $A_y(x) \leq A^1_{\sigma(y)}(f(x))$. Hence, L is a functor. The functor $L^{-1} : \textbf{SpaceFP} \to \textbf{ITrans}^\uparrow$ is defined by

$$L^{-1}(X, \mathcal{A}) = (X, |\mathcal{A}|, u, H^\uparrow_{X,\mathcal{A}}), \quad L^{-1}(f, \sigma) = (f, \sigma),$$

where $(f, \sigma) : (X, \mathcal{A}) \to (Y, \mathcal{B})$ is a morphism in **SpaceFP**. According to Theorem 2, $L^{-1}(X, \mathcal{A}) \in \textbf{ITrans}^\uparrow$ and $(f, \sigma) : L^{-1}(X, \mathcal{A}) \to L^{-1}(Y, \mathcal{B})$ is a morphism in **ITrans**$^\uparrow$. Let $t \in Q^{|\mathcal{B}|}, x \in X$. Then, we have

$$f^\leftarrow H^\uparrow_{Y,\mathcal{B}}(t)(x) = H^\uparrow_{Y,\mathcal{B}}(t)(f(x)) = \bigwedge_{\beta \in |\mathcal{B}|} B_\beta(f(x)) \to t(\beta) \leq$$

$$\bigwedge_{\alpha \in |\mathcal{A}|} B_{\sigma(\alpha)}(f(x)) \to t(\sigma(\alpha)) \leq \bigwedge_{\alpha \in |\mathcal{A}|} A_\alpha(x) \to t(\sigma(\alpha)) =$$

$$H^\uparrow_{X,\mathcal{A}}(t.\sigma)(x) = H^\uparrow_{X,\mathcal{A}} \sigma^\leftarrow(t)(x).$$

It follows that L and L^{-1} are mutually inverse functors.   ∎

4   Conclusions

The paper attempts to eliminate the *ad hoc* construction of the inverse F-transform, which is frequently used in fuzzy set theory and applications. In the paper, we investigated a lattice-valued version of the inverse F-transform, which was introduced by Perfilieva [13], that is based on a new ground structure, namely, *spaces with fuzzy partitions*, which generalizes classically defined fuzzy partitions. We defined a new structure (X, Y, v, H), which is called a Q-upper (or lower, respectively) inverse transformation system on X, and for these structures, we introduced a system of natural axioms. We proved that unique models of these structures that satisfy these axioms are lattice-valued versions of lower and upper inverse F-transforms, which further justifies the appropriateness of *ad hoc* formulas of inverse F-transforms. We also introduced categories **ITrans**$^\uparrow$ and **ITrans**$^\downarrow$ of Q-upper and Q-lower inverse transformation systems and proved that these categories are isomorphic to the category of spaces with fuzzy partitions **SpaceFP**. (For lower transformation systems, this holds if the double negation law holds in Q.)

References

1. Bede, B., Rudas, I.J.: Approximation properties of fuzzy transforms. Fuzzy Sets Syst. **180**(1), 20–40 (2011). https://doi.org/10.1016/j.fss.2011.03.001
2. Di Martino, F., et al.: An image coding/decoding method based on direct and inverse fuzzy tranforms. Int. J. Approx. Reason. **48**, 110–131 (2008). https://doi.org/10.1016/j.ijar.2007.06.008
3. Gaeta, M., Loia, V., Tomasiello, S.: Cubic B-spline fuzzy transforms for an efficient and secure compression in wireless sensor networks. Inf. Sci. **339**, 19–30 (2016). https://doi.org/10.1016/j.ins.2015.12.026
4. Höhle, U.: Commutative residuated monoids. In: Höhle, U., Klements, P. (eds.) Nonclassical Logic and Their Applications to Fuzzy Subsets, pp. 53–106. Kluwer Academic Publishers, Dordrecht (1995). https://doi.org/10.1007/978-94-011-0215-5-5
5. Khastan, A., Perfilieva, I., Alijani, Z.: A new fuzzy approximation method to Cauchy problem by fuzzy transform. Fuzzy Sets Syst. **288**, 75–95 (2016). https://doi.org/10.1016/j.fss.2015.01.001
6. Loia, V., Tomasiello, S., Vaccaro, A.: Using fuzzy transform in multi-agent based monitoring of smart grids. Inf. Sci. **388/389**, 209–224 (2017). https://doi.org/10.1016/j.ins.2017.01.022
7. Močkoř, J.: Spaces with fuzzy partitions and fuzzy transform. Soft. Comput. **13**(21), 3479–3492 (2017). https://doi.org/10.1007/s00500-017-2541-7
8. Močkoř, J.: Axiomatic of lattice-valued F-transform. Fuzzy Sets Syst. (2017). https://doi.org/10.1016/j.fss.2017.08.008
9. Novák, V., Perfilijeva, I., Močkoř, J.: Mathematical Principles of Fuzzy Logic. Kluwer Academic Publishers, Boston (1991)
10. Perfilieva, I.: Fuzzy transforms and their applications to image compression. In: Bloch, I., Petrosino, A., Tettamanzi, A.G.B. (eds.) WILF 2005. LNCS (LNAI), vol. 3849, pp. 19–31. Springer, Heidelberg (2006). https://doi.org/10.1007/11676935_3
11. Perfilieva, I.: Fuzzy transforms: a challange to conventional transform. In: Hawkes, P.W. (ed.) Advances in Image and Electron Physics, vol. 147, pp. 137–196. Elsevier Academic Press, San Diego (2007). https://doi.org/10.1016/S1076-5670(07)47002-1
12. Perfilieva, I., Novak, V., Dvořak, A.: Fuzzy transforms in the analysis of data. Int. J. Approx. Reason. **48**, 36–46 (2008). https://doi.org/10.1016/j.ijar.2007.06.003

13. Perfilieva, I.: Fuzzy transforms: theory and applications. Fuzzy Sets Syst. **157**, 993–1023 (2006). https://doi.org/10.1016/j.fss.2005.11.012
14. Perfilieva, I., Singh, A.P., Tiwari, S.P.: On the relationship among F-transform, fuzzy rough set and fuzzy topology. In: Proceedings of IFSA-EUSFLAT, pp. 1324–1330. Atlantis Press, Amsterdam (2015). https://doi.org/10.1007/s00500-017-2559-x
15. Perfilieva, I., Valasek, R.: Fuzzy transforms in removing noise. Adv. Soft Comput. **2**, 221–230 (2005). https://doi.org/10.1007/3-540-31182-3-19
16. Perfilieva, I., De Baets, B.: Fuzzy transforms of monotone functions with application to image compression. Inf. Sci. **180**, 3304–3315 (2010). https://doi.org/10.1016/j.ins.2010.04.029
17. Rodabaugh, S.E.: Powerset operator based foundation for point-set lattice theoretic (poslat) fuzzy set theories and topologies. Quaest. Math. **20**(3), 463–530 (1997). https://doi.org/10.1080/16073606.1997.9632018
18. Perfilieva, I., Kreinovich, V.: Fuzzy transforms of higher order approximate derivatives: a theorem. Fuzzy Sets Syst. **180**(1), 55–68 (2011). https://doi.org/10.1016/j.fss.2011.05.005
19. Rosenthal, K.I.: Quantales and Their Applications. Pitman Research Notes in Mathematics, vol. 234. Longman/Burnt Mill, Harlow (1990)
20. Štěpnička, M., Valašek, R.: Numerical solution of partial differential equations with the help of fuzzy transform. In: Proceedings of the FUZZ-IEEE 2005, Reno, Nevada, pp. 1104–1009 (2005)
21. Tomasielbo, S., Loia, V., Gaeta, M.: Quasi-consensus in second-order multi-agent systems with sampled data through fuzzy transform. J. Uncertain Syst. **10**, 243–250 (2017)
22. Tomasiello, S.: An alternative use of fuzzy transform with application to a class of delay differential equations. Int. J. Comput. Math. (2016). https://doi.org/10.1080/00207160.2016.1227436
23. Zadeh, L.A.: Fuzzy sets. Inf. Control **8**, 338–353 (1965). https://doi.org/10.1016/S0019-9958(65)90241-X

Why Triangular Membership Functions are Often Efficient in F-transform Applications: Relation to Probabilistic and Interval Uncertainty and to Haar Wavelets

Olga Kosheleva[ID] and Vladik Kreinovich[(⊠)][ID]

University of Texas at El Paso, El Paso, TX 79968, USA
{olgak,vladik}@utep.edu

Abstract. Fuzzy techniques describe expert opinions. At first glance, we would therefore expect that the more accurately the corresponding membership functions describe the expert's opinions, the better the corresponding results. In practice, however, contrary to these expectations, the simplest – and not very accurate – triangular membership functions often work the best. In this paper, on the example of the use of membership functions in F-transform techniques, we provide a possible theoretical explanation for this surprising empirical phenomenon.

Keywords: F-transform · Triangular membership functions
Probabilistic uncertainty · Interval uncertainty · Haar wavelets

1   Formulation of the Problem

Practical Problem: Need to Find Trends in Observations. In many practical situations, we analyze how a certain quantity x changes with time t. For example, we may want to analyze how an economic characteristic changes with time:

- we want to analyze the trends,
- we want to know what caused these trends, and
- we want to make predictions and recommendations based on this analysis.

To perform this analysis, we observe the values $x(t)$ of the desired quantity at different moments of time t. Often, however, the observed values themselves do not provide a good picture of the corresponding trends, since the observed values contain some random (noise-type) factors that prevent us from clearly seeing the trends.

For economic characteristics such as the stock market value, on top of the trend – in which we are interested – there are always day-by-day and even hour-by-hour fluctuations. For physical measurements, a similar effect can be caused

© Springer International Publishing AG, part of Springer Nature 2018
J. Medina et al. (Eds.): IPMU 2018, CCIS 854, pp. 127–138, 2018.
https://doi.org/10.1007/978-3-319-91476-3_11

by measurement uncertainty, as a result of which the measured values $x(t)$ differ from the clear trend by a random measurement error – error that differs from one measurement to another.

How can we detect the desired trend in the presence of such random noise?

F-transform Approach to Solving This Problem: A Brief Reminder. One of the successful approach for solving the above trend-finding problem comes from the F-transform idea; see, e.g., [13,14,16–19].

One of the ideas behind F-transform comes from the fact that what we really want is not just a *quantitative* mathematical model, we want a good *qualitative* understanding of the corresponding trend – and of how this trend changes with time. For example, we want to be able to say that the stock market first somewhat decreases, then rapidly increases, etc. In other words, we want these trends to be described in terms of time-localized natural-language properties.

Once we selected these properties, we can use fuzzy logic techniques (see, e.g., [1,6,9,12,15,23]) to describe these properties in computer-understandable terms, as time-localized membership functions $x_1(t)$, ..., $x_n(t)$. Time-localized means that when we analyze the process $x(t)$ on a wide time interval $[\underline{T}, \overline{T}]$:

- the first membership function $x_1(t)$ is different from 0 only on a narrow interval $[\underline{T}_1, \overline{T}_1]$, where $\underline{T}_1 = \underline{T}$;
- the second membership function $x_2(t)$ is different from 0 only on a narrow interval $[\underline{T}_2, \overline{T}_2]$, where $\underline{T}_2 \leq \overline{T}_1$;
- etc.

so that the whole range $[\underline{T}, \overline{T}]$ is covered by the corresponding ranges $[\underline{T}_i, \overline{T}_i]$.

Once we have these functions $x_i(t)$, then, as a good representation of the original signal's trend, it is reasonable to consider, e.g., linear combinations

$$x_a(t) = \sum_{i=1}^{n} c_i \cdot x_i(t) \tag{1}$$

of these functions as the desired reconstruction for the no-noise signal.

This approach has indeed led to many successful applications.

In Many Practical Applications, Triangular Membership Functions Work Well. Which membership functions should we use in this approach? At first glance, since the objective of a membership function is to capture the expert reasoning, we may expect that the more adequately these functions capture the expert reasoning, the more adequate will be our result. From this viewpoint, we expect complex membership functions to work the best.

Somewhat surprisingly, however, in many practical applications, the simplest possible triangular membership functions work the best, i.e., functions of the type

$$x_i(t) = \max\left(1 - \frac{|x - c|}{w}, 0\right)$$

that:

- linearly rise from 0 to 1 on the interval $[c - w, c]$, and then
- linearly decrease from 1 to 0 on the interval $[c, c + w]$.

Why? The above empirical fact needs explanation: why triangular membership functions work so well?

What we do in this paper. In this paper, we provide a possible explanation for this empirical phenomenon.

2   Analysis of the Problem and the Main Ideas Behind Our Explanation

What is a Trend: Discussion. As we have mentioned earlier, we are interested *not* so much in predicting the moment-by-moment values of the corresponding quantity $x(t)$ – these values contains random fluctuations. What we are interested in is the *trend*. So, to analyze this problem in precise terms, we need to understand what we mean by a trend.

A trend may mean increasing or decreasing, decreasing fast vs. decreasing slow, etc. In the ideal situation, in which we do not have any random fluctuations, all these properties can be easily described in terms of the time derivative $x'(t) \overset{\text{def}}{=} \dfrac{dx}{dt}$ of the corresponding process.

From this viewpoint, understanding the trend means reconstructing the *derivative* $x'(t)$ of the observed process based on its random-fluctuation-corrupted observed values.

What is F-transform from this Viewpoint. We are interested in the trend, so once we have applied the F-transform technique and obtained the desired no-boise expression (1), what we really want is to use its derivative

$$x'_a(t) = \sum_{i=1}^{n} c_i \cdot x'_i(t). \tag{2}$$

If we denote the derivatives $x'_i(t)$ of the membership functions by $e_i(t)$, the formula (2) then means that we approximate the derivative $e(t) \overset{\text{def}}{=} x'(t)$ of the original signal by a linear combination of the functions $e_i(t)$:

$$e(t) \approx e_a(t) = \sum_{i=1}^{n} c_i \cdot e_i(t). \tag{3}$$

In these terms, we approximate the original derivative by a function from a linear space spanned by the functions $e_i(t)$. In this sense, selecting the functions $x_i(t)$ means selecting the proper linear space – i.e., the proper functions $e_i(t)$.

For Computational Convenience, It Makes Sense to Select an Orthonormal Basis. What is important is the linear space.

Each linear space can have many possible bases. From the computational viewpoint, it is often convenient to use orthonormal bases, i.e., bases for which:

– we have $\int e_i^2(t)\,dt = 1$ for all i, and
– we have $\int e_i(t) \cdot e_j(t)\,dt = 0$ for all $i \neq j$.

Thus, without losing generality, we can assume that the basis $e_i(t)$ is orthonormal.

Comment. For the typically used equally spaced triangular functions on intervals $[\underline{T}_i, \overline{T}_i] = [\underline{T} + (i-1) \cdot h, \underline{T} + (i+1) \cdot h]$, for some $h > 0$, the corresponding derivatives $e_i(t)$ are indeed orthogonal, i.e., we indeed have $\int e_i(t) \cdot e_j(t)\,dt = 0$ for all $i \neq j$, but, in general, we have

$$\int e_i^2(t)\,dt = 2h \cdot \left(\frac{1}{h}\right)^2 = \frac{2}{h} \neq 1.$$

However, it is easy to transform this basis into an orthonormal one without changing the corresponding linear space: namely, it is sufficient to consider the new functions $e_i^*(t) = \sqrt{\dfrac{h}{2}} \cdot e_i(t)$.

Mathematical Analysis of the Problem. Once we know the original function $e_a(t)$ and we have selected the basis $e_i(t)$, what are the parameters c_i that provide the best approximation?

We start with a tuple $e \overset{\text{def}}{=} (e(t_1), e(t_2), \ldots)$ that contains all observations – to be more precise, numerical derivatives $e(t_k) = \dfrac{x(t_{k+1}) - x(t_k)}{t_{k+1} - t_k}$ based on these observations. Once we have an approximating function $e_a(t)$, we can form a similar tuple based on the approximating values: $e_a \overset{\text{def}}{=} (e_a(t_1), e_a(t_2), \ldots)$ It is reasonable to select the coefficients c_i for which the new tuple is the closest to the original one, i.e., for which the distance

$$\sqrt{(e_a(t_1) - e(t_1))^2 + (e_a(t_2) - e(t_2))^2 + \ldots}$$

between the tuples e_a and e is the smallest possible. Since the square $z \to z^2$ is a monotonic function, minimizing the distance is equivalent to minimize the square of the distance, i.e., the quantity

$$(e_a(t_1) - e(t_1))^2 + (e_a(t_2) - e(t_2))^2 + \ldots$$

In most practical situations, measurements are performed at regular intervals, so this sum is proportional to the corresponding integral

$$\int (e_a(t) - e(t))^2\,dt.$$

So, we want to find the values c_i for which this integral attains its smallest possible value. Since we assumed that the basis is orthonormal, the optimal coefficients c_i can be simply obtained as

$$c_i = \int e(s) \cdot e_i(s) \, ds. \tag{4}$$

Thus, the representation (3) takes the form

$$e(t) \approx e_a(t) = \sum_{i=1}^{n} e_i(t) \cdot \left(\int e(s) \cdot e_i(s) \, ds \right). \tag{5}$$

We Want to Select the Functions $e_i(t)$ for Which the Noise has the Least Effect on the Result. The whole purpose of this analysis is to eliminate the noise – or at least to decrease its effect. From this viewpoint, it is reasonable to select the functions $e_i(t)$ for which the effect of the noise on the reconstructed signal $e_{a(t)}$ is as small as possible.

According to the formula (5), the function $e_a(t)$ is the sum of n values

$$v_i(t) \stackrel{\text{def}}{=} e_i(t) \cdot \left(\int e(s) \cdot e_i(s) \right) ds. \tag{6}$$

Thus, it is desirable to make sure that the effect of noise on each of these values v_i is as small as possible.

Noise $n(t)$ means that instead of the original function $e(t)$, we have a noise-infected function $e(t) + n(t)$. If we use this noisy function instead of the original function $e(t)$, then, instead of the original value $v_i(t)$, we get a new value

$$v_i^{\text{new}}(t) = e_i(t) \cdot \left(\int (e(s) + n(s)) \cdot e_i(s) \, ds \right). \tag{7}$$

The difference $\Delta v_i(t) = v_i^{\text{new}}(t) - v_i(t)$ between the new and the original values is thus equal to

$$\Delta v_i(t) = e_i(t) \cdot \left(\int n(s) \cdot e_i(s) \, ds \right). \tag{8}$$

What Noises $n(t)$ Should We Consider? In principle, in different situations, we can have different types of noise, with different statistical characteristics.

- In some cases, we know the probability distribution of the noise, i.e., we have the case of *probabilistic uncertainty*.
- In other ases, we do now know the probabilities of different noise values; the only information that we have is an upper bound Δ on the value of the noise: $|n(t)| \leq \Delta$; see, e.g., [5,8,11,20]. In this case, $e(t) + n(t) \in [e(t) - \Delta, e(t) + \Delta]$, i.e., we have an *interval uncertainty*.
- In practice, we often have *partial* information about the probabilities.

What We Do in this Paper. We show that in both extreme cases – when we have full knowledge of the probabilities and when we do not know probabilities – the optimal membership functions $x_i(t)$ are triangular.

The fact that the same family of membership functions is optimal in both extreme cases explains why such membership functions are indeed often the most efficient in applications of F-transform techniques.

3   Case of Interval Uncertainty: Precise Formulation of The Problem and Its Solution

Analysis of the Problem. The difference $\Delta v_i(t)$ depends on time t and on the noise $n(t)$. To make sure that we reconstruct the trend correctly, it makes sense to require that for all possible moments of time t and for all possible noises $n(t)$, this difference does not exceed a certain value – and this value should be as small as possible. In other words, we would like to minimize the worst-case value of this difference:

$$J_{\text{int}}(e_i) \stackrel{\text{def}}{=} \max_{t,n(t)} \left| e_i(t) \cdot \left(\int n(s) \cdot e_i(s)\, ds \right) \right|. \tag{9}$$

So, we arrive at the following mathematical problem.

Definition 1. *Let us assume that we are given:*

- *the value $\Delta > 0$, and*
- *an interval $[\underline{T}_i, \overline{T}_i]$.*

We consider functions $e_i(t)$ defined on the given interval for which $\int e_i^2(t) = 1$. For each such function $e_i(t)$, we define its degree of noise-dependence *as the value*

$$J_{\text{int}}(e_i) = \max_{t,n(t)} \left| e_i(t) \cdot \left(\int n(s) \cdot e_i(s)\, ds \right) \right|, \tag{10}$$

where the maximum is taken:

- *over all moments of time $t \in [\underline{T}_i, \overline{T}_i]$, and*
- *over all functions $n(t)$ for which $|n(t)| \leq \Delta$ for all t.*

We say that the function $e_i(t)$ is optimal *if its degree of noise-dependence is the smallest possible.*

Proposition 1. *A function $e_i(t)$ is optimal if and only if $|e_i(t)| = $ const for all t.*

Discussion. We usually consider membership functions $x_i(t)$ which:

- first increase, and
- then decrease.

For such functions $x_i(t)$, the derivative $e_i(t) = x_i'(t)$ is:

- first positive, and
- then negative.

Thus, for the optimal function, we:

- first have $e_i(t)$ equal to a positive constant c, and
- then equal to minus this same constant.

By integrating this piece-wise constant function, we conclude that the function $x_i(t)$:

- first linearly increases,
- then linearly decreases with the same slope,

i.e., that $x_i(t)$ is a triangular membership function.

   Thus, *we have indeed explained why triangular membership functions are often efficient in F-transform applications.*

Comment. The piece-wise constant functions described above are well-known: they are known as *Haar wavelets*; see, e.g., [3,7,10,22]. These functions indeed form a basis, and often, by using this basis to approximate signals and images, practitioners get very good results.

   From this viewpoint, the use of triangular membership functions in F-transform techniques is equivalent to using Haar wavelets to approximate the corresponding trend. Since Haar wavelets are known to be practically efficient, it is not surprising that F-transform techniques using triangular membership functions are practically efficient as well.

Proof of Proposition 1. The desired objective function J_{int} is the largest value of the quantity

$$q(t, n(t)) \stackrel{\text{def}}{=} \left| e_i(t) \cdot \left(\int n(s) \cdot e_i(s) \, ds \right) \right| = |e_i(t)| \cdot \left| \int n(s) \cdot e_i(s) \, ds \right| \qquad (11)$$

over all possible values of t and $n(t)$:

$$J_{\text{int}} = \max_{t, n(t)} q(t, n(t)). \qquad (12)$$

This double maximum can be equivalently described as

$$J_{\text{int}} = \max_{n(t)} Q(n(t)), \qquad (13)$$

where we denoted

$$Q(n(t)) \stackrel{\text{def}}{=} \max_t q(t, n(t)). \qquad (14)$$

Once the noise function $n(t)$ is fixed, the value

$$q(t, n(t)) \qquad (15)$$

is proportional to $|e_i(t)|$. Thus, the maximum of $q(t, n(t))$ over t is attained when $|e_i(t)|$ is the largest:

$$Q(n(t)) = \max_t q(t, n(t)) = \left(\max_t |e_i(t)|\right) \cdot \left|\int n(s) \cdot e_i(s)\, ds\right|, \qquad (16)$$

i.e.,

$$Q(n(t)) = \left(\max_t |e_i(t)|\right) \cdot F(n(t)), \qquad (17)$$

where we denoted

$$F(n(t)) \stackrel{\text{def}}{=} \left|\int n(s) \cdot e_i(s)\, ds\right|. \qquad (18)$$

The first factor in the formula (17) is a positive constant not depending on the noise $n(t)$. So, to find the largest value of $Q(n(t))$, we need to find the largest possible value of $F(n(t))$:

$$J_{\text{int}} = \max_{n(t)} Q(n(t)) = \left(\max_t |e_i(t)|\right) \cdot \max_{n(t)} F(n(t)). \qquad (19)$$

The absolute value of the sum does not exceed the sum of absolute values, so

$$F(n(t)) = \left|\int n(s) \cdot e_i(s)\, ds\right| \le \int |n(s) \cdot e_i(s)|\, ds = \int |n(s)| \cdot |e_i(s)|\, ds. \qquad (20)$$

For each s, we have $|n(s)| \le \Delta$, hence

$$F(n(t)) \le \Delta \cdot \int |e_i(s)|\, ds. \qquad (21)$$

On the other hand, for $n(s) = \Delta \cdot \text{sign}(e_i(s))$, we have

$$n(s) \cdot e_i(s) = \Delta \cdot \text{sign}(e_i(s)) \cdot e_i(s) = \Delta \cdot |e_i(s)|. \qquad (22)$$

Hence, for this particular noise, we have

$$F(n(t)) = \left|\int \Delta \cdot |e_i(s)|\, ds\right| = \Delta \cdot \int |e_i(s)|\, ds. \qquad (23)$$

So, the upper bound in the inequality (21) is always attained, hence

$$\max_{n(t)} F(n(t)) = \Delta \cdot \int |e_i(s)|\, ds. \qquad (24)$$

Substituting the expression (24) into the formula (19), we conclude that

$$J_{\text{int}} = \left(\max_t |e_i(t)|\right) \cdot \Delta \cdot \int |e_i(s)|\, ds. \qquad (25)$$

We want to find a function $e_i(t)$ for which this expression is the smallest possible. To find this $e_i(t)$, it is convenient to take into account that both e_i-dependent factors in the formula (25) correspond to known norms of the function $e_i(t)$ (see, e.g., [4]):

– the expression $\max\limits_{t} |e_i(t)|$ is the L^∞-norm $\|e_i\|_{L^\infty}$, and
– the expression $\int |e_i(s)|\, ds$ is the L_1-norm $\|e_i\|_{L^1}$.

Thus, we have

$$J_{\text{int}} = \Delta \cdot \|e_i\|_{L^\infty} \cdot \|e_i\|_{L^1}. \tag{26}$$

We consider the functions $e_i(t)$ for which $\int e_i^2(t)\, dt = 1$. This property can also be described in terms of a standard norm: namely, it can be described as $\|e_i\|_{L^2} = 1$, where

$$\|e_i(t)\|_{L^2} \overset{\text{def}}{=} \sqrt{\int e_i^2(t)\, dt}. \tag{27}$$

There is a known inequality connecting these three norms: Hölder's inequality (see, e.g., [4]):

$$\|f\|_{L^2}^2 \leq \|f\|_{L^1} \cdot \|f\|_{L^\infty}, \tag{28}$$

for which it is known that the equality is attained if and only if $|f(t)|$ is constant – wherever it is different from 0.

In our case, this inequality implies that

$$J_{\text{int}} = \Delta \cdot \|e_i\|_{L^\infty} \cdot \|e_i\|_{L^1} \geq \Delta \cdot \|e_i\|_{L^2}^2 = \Delta \cdot 1 = \Delta, \tag{29}$$

and that the smallest possible value Δ is attained when $|e_i(t)|$ is constant. This is exactly what we wanted to prove.

Comment. It should be mentioned that the ideas of this proof are similar to the ideas from our paper [2].

4   Case of Probabilistic Uncertainty: Precise Formulation of The Problem and Its Solution

Analysis of the Problem. We consider the case when for each moment t, we know the probability distribution of the corresponding noise $n(t)$. Let us select a reasonable model.

Since we do not have any reason to assume that the characteristics of noise change with time, it makes sense to assume that all the variables $n(t)$ corresponding to different moments t are identically distributed.

Since we do not have any reason to assume that the positive values of the noise are more probable than the negative values, it makes sense to assume that the distribution is symmetric, and that, as a result, its mean value is 0.

Finally, since we do not have any reason to assume that the noises $n(t)$ and $n(t')$ corresponding to different moments of time are correlated, it makes sense to assume that these noises are independent, i.e., that we have a *white noise*.

Under these assumptions, the difference $\Delta v_i(t)$ – as expressed by the formula (8) – is a linear combination of the large number of independent variables $n_i(s)$. Thus, due to the Central Limit Theorem (see, e.g., [21]), we can conclude that the difference $\Delta v_i(t)$ is normally distributed.

A normal distribution is uniquely determined by its mean and variance. Since the mean value of each $n_i(s)$ is 0, the mean of $\Delta v_i(t)$ is also 0. The variance of the sum of independence random variables is equal to the sum of the variances. Since the integral is nothing else but the limit of the sums, for the variance $\sigma^2(t)$ of $\Delta v_i(t)$, we get the following formula:

$$\sigma_i^2(t) = e_i^2(t) \cdot \sigma^2 \cdot \int e_i^2(s)\, ds, \tag{30}$$

where σ characterizes the standard deviation of each noise value $n(s)$. Since we have transformed the vectors $e_i(t)$ into an orthonormal base, we have $\int e_i^2(s)\, ds = 1$ and thus, $\sigma_i^2(t) = \sigma^2 \cdot e_i^2(t)$.

This variance depends on the time t. Similarly to the interval case, it is reasonable to minimize the worst-case value of the variance, i.e., to minimize the value $\max_t(\sigma^2 \cdot e_i^2(t))$. Since σ^2 is a constant, minimizing this value is equivalent to minimizing the quantity $\max_t e_i^2(t)$.

So, we arrive at the following mathematical problem.

Definition 2. *Let us assume that we are given an interval $[\underline{T}_i, \overline{T}_i]$. We consider functions $e_i(t)$ defined on the given interval for which $\int e_i^2(t) = 1$. For each such function $e_i(t)$, we define its* degree of noise-dependence *as the value*

$$J_{\mathrm{prob}}(e_i) = \max_t e_i^2(t), \tag{31}$$

where the maximum is taken over all moments of time $t \in [\underline{T}_i, \overline{T}_i]$. We say that the function $e_i(t)$ is optimal *if its degree of noise-dependence is the smallest possible.*

Proposition 2. *A function $e_i(t)$ is optimal if and only if $|e_i(t)| = \mathrm{const}$ for all t.*

Comment. In the previous section, we have already shown that this implies that the original membership function $x_i(t)$ is triangular.

Proof of Proposition 2. It is known that, in general,

$$\int_a^b f(t)\, dt \leq (b - a) \cdot \max_s f(s), \tag{32}$$

and that the equality happens only if $f(t) = \max_s f(s)$ for almost all t. In particular, this means that

$$\int_{\underline{T}_i}^{\overline{T}_i} e_i^2(t)\, dt \leq (\overline{T}_i - \underline{T}_i) \cdot \max_t e_i^2(t), \tag{33}$$

and the equality is attained only if $|e_i(t)| = \mathrm{const}$.

Since we consider orthonormalized functions $e_i(t)$, the left-hand side of the inequality (33) is equal to 1. Thus, we can conclude that

$$\max_t e_i^2(t) \geq \frac{1}{\overline{T}_i - \underline{T}_i},$$

and the equality is attained if and only if $|e_i(t)| = $ const. So, the minimum of the functional (31) is indeed attained when $|e_i(t)| = $ const. The proposition is proven.

Acknowledgments. This work was supported in part by the US National Science Foundation grant HRD-1242122.

The authors are greatly thankful to the anonymous referees for valuable suggestions.

References

1. Belohlavek, R., Dauben, J.W., Klir, G.J.: Fuzzy Logic and Mathematics: A Historical Perspective. Oxford University Press, New York (2017)
2. Brito, A.E., Kosheleva, O.: Interval + image = wavelet: for image processing under interval uncertainty, wavelets are optimal. Reliab. Comput. **4**(3), 291–301 (1998)
3. Chui, C.: An Introduction to Wavelets. Academic Press, San Diego (1992)
4. Edwards, R.E.: Functional Analysis: Theory and Applications. Dover, New York (2011)
5. Jaulin, L., Kiefer, M., Dicrit, O., Walter, E.: Applied Interval Analysis. Springer, London (2001). https://doi.org/10.1007/978-1-4471-0249-6
6. Klir, G., Yuan, B.: Fuzzy Sets and Fuzzy Logic. Prentice Hall, Upper Saddle River (1995)
7. Mallat, S.: A Wavelet Tour of Signal Processing: The Sparse Way. Academic Press, Burlington (2008)
8. Mayer, G.: Interval Analysis and Automatic Result Verification. De Gruyter, Berlin (2017)
9. Mendel, J.M.: Uncertain Rule-Based Fuzzy Systems: Introduction and New Directions. Springer, Cham (2017). https://doi.org/10.1007/978-3-319-51370-6
10. Meyer, Y.: Wavelets Algorithms and Applications. SIAM, Philadelphia (1993)
11. Moore, R.E., Kearfott, R.B., Cloud, M.J.: Introduction to Interval Analysis. SIAM, Philadelphia (2009)
12. Nguyen, H.T., Walker, E.A.: A First Course in Fuzzy Logic. Chapman and Hall/CRC, Boca Raton (2006)
13. Novak, V., Perfilieva, I., Holcapek, M., Kreinovich, V.: Filtering out high frequencies in time series using F-transform. Inf. Sci. **274**, 192–209 (2014)
14. Novak, V., Perfilieva, I., Kreinovich, V.: F-transform in the analysis of periodic signals. In: Inuiguchi, M., Kusunoki, Y., Seki, M. (eds.), Proceedings of the 15th Czech-Japan Seminar on Data Analysis and Decision Making under Uncertainty CJS'2012, Osaka, Japan, 24–27 September 2012 (2012)
15. Novák, V., Perfilieva, I., Močkoř, J.: Mathematical Principles of Fuzzy Logic. Kluwer, Boston (1999)
16. Perfilieva, I.: Fuzzy transforms: theory and applications. Fuzzy Sets Syst. **157**, 993–1023 (2006)

17. Perfilieva, I.: F-transform. In: Kacprzyk, J., Pedrycz, W. (eds.) Springer Handbook of Computational Intelligence, pp. 113–130. Springer, Heidelberg (2015). https://doi.org/10.1007/978-3-662-43505-2_7
18. Perfilieva, I., Danková, M., Bede, B.: Towards a higher degree F-transform. Fuzzy Sets Syst. **180**(1), 3–19 (2011)
19. Perfilieva, I., Kreinovich, V., Novak, V.: F-transform in view of trend extraction. In: Inuiguchi, M., Kusunoki, Y., Seki, M. (eds.) Proceedings of the 15th Czech-Japan Seminar on Data Analysis and Decision Making under Uncertainty CJS'2012, Osaka, Japan, 24–27 September 2012 (2012)
20. Rabinovich, S.G.: Measurement Errors and Uncertainty: Theory and Practice. Springer, Berlin (2005). https://doi.org/10.1007/0-387-29143-1
21. Sheskin, D.J.: Handbook of Parametric and Nonparametric Statistical Procedures. Chapman and Hall/CRC, Boca Raton (2011)
22. Vetterli, M., Kovacevic, J.: Wavelets and Subband Coding. Prentice Hall, Englewood Cliffs (1995)
23. Zadeh, L.A.: Fuzzy sets. Inf. Control **8**, 338–353 (1965)

Enhanced F-transform Exemplar Based Image Inpainting

Pavel Vlašánek[✉]

Centre of Excellence IT4Innovations, Institute for Research
and Applications of Fuzzy Modeling, University of Ostrava,
30. dubna 22, 701 03 Ostrava, Czech Republic
pavel.vlasanek@osu.cz

Abstract. This paper focuses on a completion of the partially damaged image. There are a variety of techniques to deal with this task. Our contribution belongs to the group of exemplar based image inpainting techniques which process the image what was separated to the many small regions. The regions are called patches and the task of inpainting becomes the task of searching for the most suitable patch from the undamaged part of the image to replace the partially damaged one. Our novelty is in processing based on fuzzy mathematics and new filling order prioritization function.

Keywords: Image inpainting · Fuzzy transform · Patches

1 Introduction

The image inpainting deals with region completion after object removal. The object can be building, tourist, electric wires, inscription, etc. There are many techniques which are based on many several principles.

The first ones were based on diffusion. Let us mention technique by Ogden et al. [1] who proposed to use Gaussian pyramids. Important contribution in this field was given by Bertalmio et al. [2] who established the term *image inpainting*. The idea behind was to propagate change in information (authors used derivative of Laplacian) along the edges. Some authors focus on the relation between image processing and physics. One proposition like that is to take the image inpainting as a problem of motion of viscous fluids [3]. The common idea behind all of the mentioned techniques is in the assumption that pixels surrounding the damaged area should be spread inward it and blended eventually.

Another group of techniques is known as exemplar based [4]. The new pixels in the damaged area are not derived one-by-one from the surrounding pixels but from bigger continuous regions known as patches. These patches are taken from the known region of the image and their parts are used for filling in the damaged region. Significant improvement was given by Criminisi et al. [5]. Their technique is described more in the following text.

© Springer International Publishing AG, part of Springer Nature 2018
J. Medina et al. (Eds.): IPMU 2018, CCIS 854, pp. 139–150, 2018.
https://doi.org/10.1007/978-3-319-91476-3_12

Our contribution is based on fuzzy mathematics. We propose new way of proper patch selection which takes color and gradient to the consideration. Their estimations are given by F-transform applied to the input image in a different color model and gray-scale representation. Furthermore, the proposed algorithm contains a new way of filling order prioritization.

The structure of this contribution is as follows. The preliminaries and state of the art are given in Sect. 2. Section 3 describes the theory of F-transform. The novel algorithm is described in Sect. 4, supported by examples in Sect. 5. The paper is concluded in Sect. 6.

2   Preliminaries and State of the Art

Let us fix the following notation and use it throughout the paper. Image I is a 2D function such as $I : [0, M] \times [0, N] \to [0, 255]$, where $M + 1$ denotes the image width, $N + 1$ denotes the image height, and $[0, 255]$ denotes the pixel intensity. We denote $[0, M]_{\mathbb{Z}} = \{0, 1, 2, \ldots, M\}$, $[0, N]_{\mathbb{Z}} = \{0, 1, 2, \ldots, N\}$ and $[0, 255]_{\mathbb{Z}} = \{0, 1, 2, \ldots, 255\}$. Image I is assumed to be partially known on the area Φ and unknown (damaged) on the area Ω. The border between these areas is denoted by $\delta\Omega$ and assumed to be unknown. The notation is illustrated in Fig. 1.

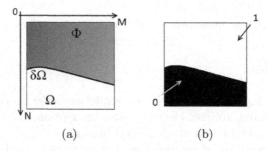

(a)   (b)

Fig. 1. (a) Two areas where image I is defined (Φ) and undefined (Ω); (b) mask S.

In exemplar based inpainting, a rectangular patch $\Psi \in \Phi$ is chosen to partially fill in region Ω. We denote $\Psi_\phi(\cdot)$ for a rectangular patch centered at pixel ϕ. If pixel ω is unknown (belongs to $\delta\Omega \cup \Omega$), then the exemplar based inpainting consists of finding a patch from the known area, e.g., $\Psi_\phi(\cdot)$, which is the most similar to $\Psi_\omega(\cdot)$ according to

$$\arg \min_\phi d(\Psi_\omega, \Psi_\phi). \tag{1}$$

Let us illustrate importance of filling order. The basic strategy is *onion-peel*. The patches centered at the border pixels $\delta\Omega$ are processed in one direction

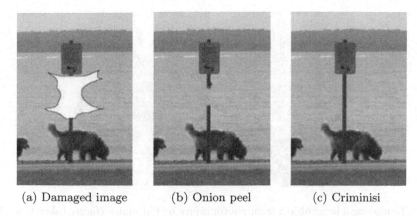

(a) Damaged image   (b) Onion peel   (c) Criminisi

Fig. 2. Illustration of the filling order importance. Onion peel way, where pixels on the border (red) are processed one by one from left to right, leads to wrong reconstruction. Better way is to process certain regions first to follow the structures. Figure taken from [5]. (Color figure online)

(e.g. from left to right) one by one. After every replacement, the border $\delta\Omega$ is updated. This solution leads to unsatisfying results frequently as can be seen in Fig. 2.

A certain development was proposed in [5]. The onion peel is replaced by a more sophisticated version. The patches from the unknown area have priorities depending on their locations. The ones that are centered on the edges and moreover contain more undamaged pixels are processed before those in the flat regions. The authors of [5] proposed the following priority function

$$P(\omega) = C(\omega)D(\omega); \omega \in \Phi,$$

where the constituent functions are defined as follows

$$C(\omega) = \frac{\sum_{e \in \Psi_\omega \cap (\Phi)} C(e)}{|\Psi_\omega|}, D(\omega) = \frac{|\nabla I_\omega^\perp \cdot n_\omega|}{\alpha}.$$

In the expression for $C(\omega)$, function $C(e)$ is such that $C(e) = 0; \forall e \in \Omega \cap \delta\Omega$ and $C(e) = 1; \forall e \in \Phi$. Thus, the priority is based on the gradient and number of the known pixels. Function $D(\omega)$ represents a strength of isophotes at ω. A larger value of P at w corresponds to a higher priority of processing for patch Ψ_w. An application of this technique is shown in Fig. 3.

In this paper, we propose to apply the F-transform [6] technique. This technique establishes a correspondence between an original object (image) and a set of orthogonal projections on elements of a fuzzy partition (details are in Sect. 3). Each projection is a feature vector of the image with respect to a partition element. The expected advantage of our proposal is to simplify the similarity (1) by similarity between feature vectors of the corresponding patches. We propose to take average intensities of the color information and average gradients. The

Fig. 3. Removing a large object from photography by Criminisi (figure taken from [5]).

details are given in the following sections. We also propose to simplify the filling order evaluation function.

3   F-transform

The F-transform technique was used for many tasks in image processing. Let us mention diffusion based image inpainting [7], compression [8], edge detection [9], image fusion [10] etc.

The definition of the F-transform [6] and the notion of a fuzzy partition[1] is as follows. Fuzzy sets (*basic functions*) $A_0, \ldots, A_m, 1 < m < M$, which are identified with their membership functions $A_0, \ldots, A_m : [0, M] \to [0, 1]$, establish a *fuzzy partition* of $[0, M]$ with nodes $0 = x_0 < x_1 < \cdots < x_m = M$, if the following conditions are fulfilled:

(1) $A_k : [0, M] \to [0, 1]$, $A_k(x_k) = 1$;
(2) $A_k(x) = 0$ if $x \notin (x_{k-1}, x_{k+1})$, $k = 0, \ldots, m$;
(3) $A_k(x)$ is continuous on $[0, M]$;
(4) $A_k(x)$ strictly increases on $[x_{k-1}, x_k]$ and strictly decreases on $[x_k, x_{k+1}]$, where $k = 1, \ldots, m$;
(5) $\sum_{k=0}^{m} A_k(x) = 1$, $x \in [0, M]$.

We say that the fuzzy partition $A_0, \ldots, A_m$ is an *h-uniform fuzzy partition* if nodes $x_k = hk$, $k = 0, \ldots, m$ are equidistant, $h = M/m$, and two additional properties are met:

(6) $A_k(x_k - x) = A_k(x_k + x)$, $x \in [0, h]$, $k = 0, \ldots, m$;
(7) $A_k(x) = A_{k-1}(x - h)$, $k = 1, \ldots, m$, $x \in [x_{k-1}, x_{k+1}]$.

---

[1] For the sake of simplicity, we consider this notion for a one dimensional function. Extension to two dimensions is straight forward.

Parameter h will be referred to as a *radius*. The radius determines width of the basic functions therefore the number of covered pixels with respect to particular axis. Triangular shaped generating function A and the corresponding h-uniform fuzzy partition is as follows

$$A_0(x) = \begin{cases} 1 - \frac{(x-x_1)}{h_1} & \text{if } x \in [x_1, x_2], \\ 0 & \text{otherwise,} \end{cases}$$

$$A_k(x) = \begin{cases} \frac{(x-x_{k-1})}{h_{k-1}} & \text{if } x \in [x_{k-1}, x_k], \\ 1 - \frac{(x-x_k)}{h_k} & \text{if } x \in [x_k, x_{k+1}], \\ 0 & \text{otherwise,} \end{cases}$$

$$A_m(x) = \begin{cases} \frac{(x-x_{m-1})}{h_{m-1}} & \text{if } x \in [x_{m-1}, x_m], \\ 0 & \text{otherwise.} \end{cases}$$

We need to distinguish between damaged and undamaged regions. For that purpose, the binary mask S is used such as

$$S(x, y) = \begin{cases} 0 & \text{if } I(x, y) \in \Omega \cup \delta\Omega \\ 1 & \text{if } I(x, y) \in \Phi. \end{cases}$$

3.1   F-Transform of Various Degrees

The F^0-transform of an image is given by a corresponding matrix of components. Let the fuzzy partition of $[0, M]$ and $[0, N]$ be given by basic functions $A_0, \ldots, A_m : [0, M] \to [0, 1]$ and $B_0, \ldots, B_n : [0, N] \to [0, 1]$, respectively.

We call the $m \times n$ matrix of real numbers $\mathbf{F^0_{mn}}[I] = (F^0_{kl})$ the *(discrete) F^0-transform* of image I with respect to $\{A_0, \ldots, A_m\}$ and $\{B_0, \ldots, B_n\}$, if for all $k = 0, \ldots, m; l = 0, \ldots, n$,

$$F^0_{kl} = \frac{\sum_{y=0}^{N} \sum_{x=0}^{M} I(x, y) A_k(x) B_l(y) S(x, y)}{\sum_{y=0}^{N} \sum_{x=0}^{M} A_k(x) B_l(y) S(x, y)}.$$

The elements F^0_{kl} are called the *components of the F^0-transform*. In image processing, the F^0-transform components determine an average intensity value of the respective area.

Let us recall the (direct) F^1-transform [9]. We say that matrix $\mathbf{F^1_{mn}}[I] = (F^1_{kl})$, $k = 0, \ldots, m$, $l = 0, \ldots, n$ is the F^1-transform of I with respect to $\{A_k \times B_l \mid k = 0, \ldots, m, l = 0, \ldots, n\}$, and F^1_{kl} is the corresponding F^1-transform component.

The F^1-transform components of I are linear polynomials in the form of

$$F^1_{kl}(x, y) = c^{00}_{kl} + c^{10}_{kl}(x - x_k) + c^{01}_{kl}(y - y_l),$$

where the coefficients are

$$c_{kl}^{00} = \frac{\sum_{y=0}^{N} \sum_{x=0}^{M} I(x,y) A_k(x) B_l(y) S(x,y)}{\sum_{y=0}^{N} \sum_{x=0}^{M} A_k(x) B_l(y) S(x,y)},$$

$$c_{kl}^{10} = \frac{\sum_{y=0}^{N} \sum_{x=0}^{M} I(x,y)(x - x_k) A_k(x) B_l(y) S(x,y)}{\sum_{y=0}^{N} \sum_{x=0}^{M} (x - x_k)^2 A_k(x) B_l(y) S(x,y)}, \qquad (2)$$

$$c_{kl}^{01} = \frac{\sum_{y=0}^{N} \sum_{x=0}^{M} I(x,y)(y - y_l) A_k(x) B_l(y) S(x,y)}{\sum_{y=0}^{N} \sum_{x=0}^{M} (y - y_l)^2 A_k(x) B_l(y) S(x,y)}.$$

In image processing, the F^1-transform components stands for the average intensity and the average gradients in x and y direction of the appropriate region.

4   Novel Algorithm

The proposed algorithm is partially based on the previous research published in [11] and more deeply in [12]. The novelty and great improvement of the new algorithm is in change of the patch comparison measure, new completion order function, strong impact to follow the edges and extension to color images. Let us describe the algorithm more deeply.

As previously mentioned, inpainting deals with unknown region completion. Image I with a designated unknown region Ω (labeled by mask S) are the inputs to this task. Region Ω must be replaced (filled in) using patches from the known area Φ, where $\Phi = [0, M] \times [0, N] \setminus \delta\Omega \cup \Omega$.

The known area Φ of image I is partitioned to many overlapped patches. Each patch has a rectangular shape with a predefined size[2]. The number of patches q is determined by their size. The patches $\Psi_\phi^i(\cdot)$ where $i = 1, \ldots, q$ are stored in the database D_Ψ, which will be used later for the reconstruction.

The unknown area Ω is surrounded by the border $\delta\Omega$. The border has significant importance because its pixels $I(x,y) \in \delta\Omega$ give the center points of the patches with the most known pixels inside. To achieve high quality reconstruction, it is necessary to start with regions determined by them [5]. We propose to find the border $\delta\Omega$ using mathematical morphology [13]. First step is to apply the operator *erosion* defined as follows

$$S \ominus T = \{z \in [0, M] \times [0, N] | T_z \subseteq S\},$$

where T is a structuring element and z is a translation vector. For the purpose of border determination, we consider square-shaped structuring element of the size 3×3. This operation is applied to the mask S thus the values of the structuring element T must be the same like the values of the mask pixels which belong to the damaged region $\Omega \cup \delta\Omega$. The border is determined as follows

$$\delta\Omega = S - (S \ominus T).$$

---

[2] The size depends on the size and distribution of the damaged regions.

It is crucial to prioritize filling order. Criminisi et al. [5] addressed this issue and proposed a solution described in Sect. 2. It consists in prioritization based on gradient and number of known pixels in the processed patch.

We propose a different way. Our idea comes from assumption that regions with strong edges should be processed first. The difference in comparison with the Criminisi way lays in their estimation. We propose to compute image Laplacian in all known region Φ such as

$$\Delta I = \frac{\partial^2}{\partial x^2} + \frac{\partial^2}{\partial y^2}.$$

Illustration can be seen in Fig. 4.

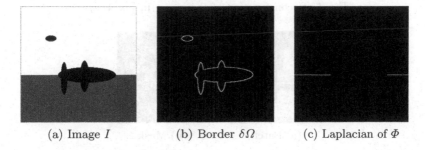

(a) Image I (b) Border $\delta\Omega$ (c) Laplacian of Φ

Fig. 4. Illustration of the border of image I and Laplacian of the respective region Φ.

The computation is fast to process and effective to find the edges. As a next step, we use variance of Laplacian. We assume that flat regions without any strong edges or high contrast texture have low Laplacian variance. Therefore, the task of filling order prioritization becomes a task of searching for a pixel $(x, y) \in \delta\Omega$ whose 3×3 region around has a maximal variance of Laplacian such as

$$(x, y) : \arg \max_{(x,y) \in \delta\Omega} Var(\Delta I(x, y)_{3\times3}),$$

where $Var(\Delta I(x, y)_{3\times3})$ stands for variance of Laplacian in the 3×3 neighborhood centered around pixel (x, y). Let us use ω for the chosen pixel. The result of the reconstruction of one particular example is shown in Fig. 5.

We take the pixel ω from the border and create the patch $\Psi_\omega(\cdot)$, whose center is at this pixel. Let s be the mask of the damaged (unknown) part in $\Psi_\omega(\cdot)$ as shown in Fig. 6. We apply s to the each patch in the database D_Ψ and create a new database D_Ψ^d, where each patch $\Psi_\phi^d(\cdot)$ is partially damaged by s. The goal is to find the closest patch $\Psi_\phi^d(\cdot)$ to patch $\Psi_\omega^d(\cdot)$ from D_Ψ^d.

To measure the closeness, we propose to transform image I to the two different color representations. Gray-scale for gradient computation and HSL for average color computation. The formula using them both is as follows

Fig. 5. Reconstruction of the image in Fig. 4. a)

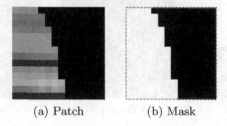

(a) Patch   (b) Mask

Fig. 6. Patch with unknown area determined by the mask. Mask is shown with added border.

$$d(\Psi_1, \Psi_2) = \frac{\mathrm{RMSE}^{00}(\Psi_1, \Psi_2) + \mathrm{RMSE}^{01}(\Psi_1, \Psi_2) + \mathrm{RMSE}^{10}(\Psi_1, \Psi_2)}{3}.$$

where coefficients Ψ_i^{00}, Ψ_i^{01} and Ψ_i^{10}, $i = 1, 2$ are the corresponding F^1-transform coefficients in (2). Namely, Ψ_i^{00} stands for c^{00} in the *hue* channel of the *HSL* image. Coefficients Ψ_i^{01} and Ψ_i^{10} stands for average gradient in the gray-scale image, therefore c^{01} and c^{10}. Operator *RMSE* stands for *round mean square error*. The patch searching process and effect of the selected measure $d(\Psi_1, \Psi_2)$ are illustrated in Fig. 7.

The number of feature vector components is determined by the radius h of the selected generating function. The exact value of h differs from one application to another. A smaller value of h corresponds to a larger number of components in a feature vector and greater computation time. A larger h leads to a faster computation but with a higher risk of a wrong assignment.

After we choose patches $\Psi_\omega(\cdot)$ and $\Psi_\phi^d(\cdot)$, the partially known $\Psi_\omega(\cdot)$ can be replaced. The patch $\Psi_\phi(\cdot)$, which corresponds to $\Psi_\phi^d(\cdot)$, is selected from the database D_Ψ and used. The usage indicates that we must replace the unknown part of $\Psi_\omega(\cdot)$ by a respective part of $\Psi_\phi^d(\cdot)$. The illustration is in Fig. 8.

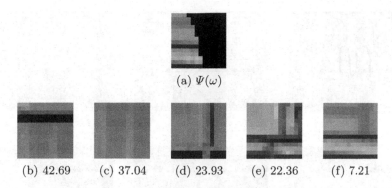

(a) $\Psi(\omega)$

(b) 42.69   (c) 37.04   (d) 23.93   (e) 22.36   (f) 7.21

Fig. 7. The damaged patch and values of the novel closeness measure to find the most similar one. As can be seen, the lowest value belongs to option (f) so its region identified with the damaged region of (a) is used for replacement. Result is in Fig. 8.

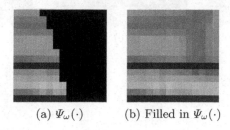

(a) $\Psi_\omega(\cdot)$ (b) Filled in $\Psi_\omega(\cdot)$

Fig. 8. (a) Patch centered around the pixel from $\delta\Omega$; (b) Patch filled in by appropriate $\Psi_\phi^d(\cdot)$.

5 Examples

Let us demonstrate our algorithm and compare it with another technique [5]. Figure 9 (a) is from a database used in [14]. Figure 9 (b) and (c) were chosen to demonstrate the feature of our novel reconstruction to follow the edges. Results can be seen in Fig. 9.

Let us magnify the details to highlight the differences in comparison with the conventional technique. This can be seen in Fig. 10. Image in Fig. 10 (d) demonstrates the ability of our novel algorithm to preserve and follow complicated structures. The details shows better edge completion, the most visible in the window regions, than using Criminisi technique. Images (e) and (f) shows higher quality of the reconstruction when it comes to follow edges and patterns as well.

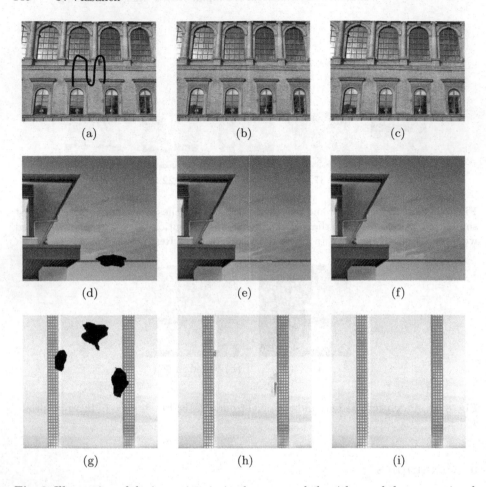

(a)   (b)   (c)

(d)   (e)   (f)

(g)   (h)   (i)

Fig. 9. Illustration of the image inpainting by our novel algorithm and the conventional one. (Left column) damaged images; (middle column) reconstruction using [5]; (right column) reconstruction using the proposed algorithm. For both algorithms, the same patch size was used.

Our current implementation of F-transform, which is used for implementation of the proposed algorithm, is available as a part of the OpenCV framework[3]. The F-transform technique is known as very fast with processing time in milliseconds on a non high end PC. Details with analysis can be seen in [15].

---

[3] Module **fuzzy**, which is included in **opencv_contrib** available at https://github. com/itseez/opencv_contrib.

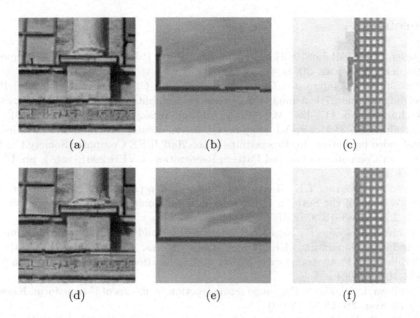

(a)   (b)   (c)

(d)   (e)   (f)

Fig. 10. Details of the Fig. 9. First row ((a), (b), (c)) are details of the reconstruction using Criminisi [5] technique and bottom row ((d), (e), (f)) using our novel algorithm.

6 Conclusion

The paper describes an exemplar based inpainting algorithm using the F^1-transform technique. We propose to represent regions of image (patches) by feature vectors composed of F^1-transform components. The patches which are partially damaged, considered as unknown, are replaced by ones taken as a known. Important contributions presented in our technique are the patch representation, measure used for decision of the patch similarity and filling order evaluation.

The patch representation is composed from average color of the *hue* channel in HSL image representation and average gradients in x and y direction. The gradients are computed from the gray-scale representation of the input image. The filling prioritization function is based on assumption that variance of Laplacian of the flat regions is minimal. Our goal is to reconstruct regions with edges or high contrast textures first. For that purpose, our prioritization is based on maximal variance of Laplacian.

We described the benefits of our novel technique and illustrated the advantages of our filling order prioritization. The technique is demonstrated on the several images and visually compared with another one.

Acknowledgment. This work was supported by The Ministry of Education, Youth and Sports from the National Programme of Sustainability (NPU II) project IT4Innovations excellence in science - LQ1602.

References

1. Ogden, J.M., Adelson, E.H., Bergen, J.R., Burt, P.J.: Pyramid-based computer graphics. RCA Eng. **30**(5), 4–15 (1985)
2. Bertalmio, M., Sapiro, G., Caselles, V., Ballester, C.: Image inpainting. In: Proceedings of the 27th Annual Conference on Computer Graphics and Interactive Techniques, pp. 417–424. ACM Press/Addison-Wesley Publishing Co. (2000)
3. Bertalmio, M., Bertozzi, A.L., Sapiro, G.: Navier-stokes, fluid dynamics, and image and video inpainting. In: Proceedings of the 2001 IEEE Computer Society Conference on Computer Vision and Pattern Recognition, CVPR 2001, vol. 1, pp. I-355. IEEE (2001)
4. Efros, A.A., Leung, T.K.: Texture synthesis by non-parametric sampling. In: The Proceedings of the Seventh IEEE International Conference on Computer Vision, vol. 2, pp. 1033–1038. IEEE (1999)
5. Criminisi, A., Pérez, P., Toyama, K.: Region filling and object removal by exemplar-based image inpainting. IEEE Trans. Image Process. **13**(9), 1200–1212 (2004)
6. Perfilieva, I.: Fuzzy transforms: theory and applications. Fuzzy Sets Syst. **157**(8), 993–1023 (2006)
7. Perfilieva, I., Vlašánek, P.: Image reconstruction by means of F-transform. Knowl.-Based Syst. **70**, 55–63 (2014)
8. Hurtik, P., Perfilieva, I.: A hybrid image compression algorithm based on jpeg and fuzzy transform. In: 2017 IEEE International Conference on Fuzzy Systems (FUZZ-IEEE), pp. 1–6. IEEE (2017)
9. Perfilieva, I., Hodáková, P., Hurtík, P.: Differentiation by the F-transform and application to edge detection. Fuzzy Sets Syst. **288**, 96–114 (2016)
10. Vajgl, M., Hurtik, P., Perfilieva, I., Hodáková, P.: Image composition using f-transform. In: 2014 IEEE International Conference on Fuzzy Systems (FUZZ-IEEE), pp. 1112–1117. IEEE (2014)
11. Vlašánek, P., Perfilieva, I.: Image reconstruction by the patch based inpainting. In: Carvalho, J.P., Lesot, M.-J., Kaymak, U., Vieira, S., Bouchon-Meunier, B., Yager, R.R. (eds.) IPMU 2016. CCIS, vol. 610, pp. 588–598. Springer, Cham (2016). https://doi.org/10.1007/978-3-319-40596-4_49
12. Vlašánek, P., Perfilieva, I.: Patch based inpainting inspired by the F1-transform. Int. J. Hybrid Intell. Syst. **13**(1), 39–48 (2016)
13. Serra, J.: Image Analysis and Mathematical Morphology. Academic Press Inc., Cambridge (1983)
14. Tiefenbacher, P., Bogischef, V., Merget, D., Rigoll, G.: Subjective and objective evaluation of image inpainting quality. In: 2015 IEEE International Conference on Image Processing (ICIP), pp. 447–451. IEEE (2015)
15. Perfilieva, I., Hurtik, P., Di Martino, F., Sessa, S.: Image reduction method based on the f-transform. Soft Comput. **21**(7), 1847–1861 (2017)

A Novel Approach to the Discrete Fuzzy Transform of Higher Degree

Linh Nguyen$^{(\boxtimes)}$ and Michal Holčapek

Institute for Research and Applications of Fuzzy Modelling, NSC IT4Innovations,
University of Ostrava, 30. dubna 22, 701 03 Ostrava 1, Czech Republic
{Linh.Nguyen,Michal.Holcapek}@osu.cz

Abstract. In this paper, we propose a new approach to the discrete fuzzy transform of higher degree based on the piecewise constant representation of discrete functions and the application of the continuous fuzzy transform. We show how a given discrete function can be reconstructed by using the discrete higher degree fuzzy transform and how convenient the latter is computed by the novel approach. Finally, we illustrate and compare the proposed technique with the original discrete fuzzy transform of higher degree.

Keywords: Continuous fuzzy transform · Discrete fuzzy transform
F-transform · Fuzzy partition

1 Introduction

The fuzzy transform (F-transform, for short) has been introduced by Perfilieva in [10]. In this paper, Perfilieva proposed both the continuous F-transform for continuous (and later locally integrable) functions and the discrete F-transform for discrete functions defined over sets of finite points. The continuous F-transform has been extended to higher degrees in [11] to improve its ability to approximate functions whose domains are connected subsets of the real line. The continuous higher degree F-transform has been reformulated for discrete functions by Holčapek and Tichý in [4]. It is well known that the F-transform consists of two phases, namely, direct and inverse. The direct phase transforms a (locally integrable or discrete) function into a set of its local approximations, which are called the direct F-transform components and are determined with respect to basic functions, i.e., fuzzy sets, that form a fuzzy partition of the domain of the given function. On the other hand, the inverse phase provides an approximate reconstruction of the original function from its direct fuzzy transform components. Due to its good reconstruction ability, low computational complexity and successful reduction of noise, the F-transform has become a popular alternative in various fields of application, e.g., data analysis, time series analysis, image processing, non-parametric regression, numerical solution of differential equations, (see, e.g., [3,5,7,9,13]).

© Springer International Publishing AG, part of Springer Nature 2018
J. Medina et al. (Eds.): IPMU 2018, CCIS 854, pp. 151–162, 2018.
https://doi.org/10.1007/978-3-319-91476-3_13

In practice, all applications of F-transform are designed in a discrete form; therefore, the importance of discrete (higher degree) F-transform increases, when one solves practical tasks. Besides a discretization of integral formulas used in the computation of the continuous F-transform (see, e.g., [12]), it seems to be reasonable to apply the discrete F-transform of higher degree in the form as has been introduced in [4]. Recall that if p is a real function defined on a finite set $D = \{t_i \in [a, b] \mid i = 1, \ldots, n, \; t_i < t_{i+1}\}$, then the k-th component of the direct F-transform of degree m ($m \in \mathbb{N}$) of p with respect to a fuzzy partition $\{A_k \mid k = 1, \ldots, \ell\}$ of D is a polynomial

$$F_k^m[p](t) = C_{k,0} + C_{k,1}(t - c_k) + \cdots + C_{k,m}(t - c_k)^m,$$

where c_k denotes the node of the k-th basic function, and

$$(C_{k,0}, \ldots, C_{k,m})^T = \left(\mathbf{X}_k^T \mathbf{A}_k \mathbf{X}_k\right)^{-1} \mathbf{X}_k^T \mathbf{A}_k \mathbf{Y} \tag{1}$$

with

$$\mathbf{X}_k = \begin{bmatrix} 1 & t_1 - c_k & \cdots & (t_1 - c_k)^m \\ \vdots & \vdots & \vdots & \vdots \\ 1 & t_n - c_k & \cdots & (t_n - c_k)^m \end{bmatrix},$$

$\mathbf{A}_k = \mathrm{diag}\{A_k(t_1), \ldots, A_k(t_n)\}$ and $\mathbf{Y} = (p(t_1), \ldots, p(t_n))^T$. The reconstruction, providing an approximation of the original function p, is then given by the linear like combinations

$$F^m[p](t_i) = \sum_{k=1}^{\ell} F_k^m[p](t_i) \cdot A_k(t_i). \tag{2}$$

One of the disadvantages of the presented approach is the setting of a fuzzy partition to ensure the invertibility of $\mathbf{X}_k^T \mathbf{A}_k \mathbf{X}_k$ for any k. Indeed, each setting of a fuzzy partition has to control that the number of elements in which each basic function gives a non-zero value is greater then or equal to $m + 1$. Another disadvantage of the presented approach is the computation of the inverse matrix in (1) for each k, if discrete functions are defined over non-uniformly distributed elements and the nodes of a fuzzy partition do not coincide with some of them. Obviously, these updates make the computation of discrete higher degree F-transform more time consuming, especially, if the number of basic functions is large, which appears in the case of higher dimensions.

The recent theory of the higher degree F-transform is mostly developed in the continuous design [1,2,8]. Particularly, we proposed in [1] an efficient approach to the computation of the direct higher degree F-transform components based on various bases of polynomials. The aim of this paper is to introduce a novel (alternative) approach to the computation of a discrete higher degree F-transform with the use of benefits of the continuous design, which can overcome the mentioned disadvantages. To justify the usefulness of our approach, we analyze the quality of reconstruction of the original discrete function provided by the proposed novel approach. Furthermore, we compare our novel approach with the original approach to the computation of discrete F-transform. For the purpose of this paper, we restrict ourselves to discrete functions defined over integers. Nevertheless, the proposed technique can be simply modified for discrete functions defined over sets with uniformly distributed elements.

The paper is structured as follows. The next section provides a brief introduction to the continuous F-transform of higher degree. In Sect. 3, we introduce the novel approach to the discrete higher degree F-transform and discuss the reconstruction of discrete functions. An illustration and comparison of the novel approach and the original approach is presented in Sect. 4. The last section is a conclusion.

2   Preliminaries

Let $\mathbb{N}$, $\mathbb{Z}$, $\mathbb{R}$ and $\mathbb{C}$ denote the set of natural numbers, integers, reals and complex numbers, respectively.

2.1   Fuzzy Partition

Fuzzy partition is a fundamental concept of the theory of F-transform of higher degree. In this paper, we restrict our analysis to a particular type of fuzzy partitions, which is called a simple uniform fuzzy partition. This type of fuzzy partition consists of fuzzy sets, determined by a generating function and uniformly spread along the real line.

Definition 1. *A real-valued function $K : \mathbb{R} \to [0,1]$ is said to be a* generating function *if it is continuous, even, non-increasing on $[0,1]$ and vanishing outside of $(-1,1)$.*

Basic examples of generating functions that are frequently used in applications of F-transform are the triangle and raised cosine functions.

Example 1. The functions $K^{tr}, K^{rc} : \mathbb{R} \to [0,1]$ defined by

$$K^{tr}(t) = \max(1 - |t|, 0) \quad \text{and} \quad K^{rc}(t) = \begin{cases} \frac{1}{2}(1 + \cos(\pi t)), & -1 \le t \le 1; \\ 0, & \text{otherwise,} \end{cases}$$

for any $t \in \mathbb{R}$, are called the *triangle* and *raised cosine* generating functions, respectively.

Definition 2. *Let K be a generating function, let h and r be positive real constants, and let $t_0 \in \mathbb{R}$. For any $k \in \mathbb{Z}$, let*

$$A_k(t) = K\left(\frac{t - t_0 - c_k}{h}\right),$$

where $c_k = kr$. The set $\mathcal{A} = \{A_k \mid k \in \mathbb{Z}\}$ is said to be a simple uniform fuzzy partition *of the real line determined by the quadruplet (K, h, r, t_0) if, for any $t \in \mathbb{R}$, there exists $k \in \mathbb{Z}$ such that $A_k(t) > 0$. The parameters h, r and t_0 are called the* bandwidth, shift *and* central node *of the fuzzy partition $\mathcal{A}$, respectively.*

Since the setting of the central node has no effect on the theoretical results concerning the fuzzy transform, for the sake of simplicity, we restrict our investigation to the simple uniform fuzzy partitions with $t_0 = 0$. Moreover, we omit the reference to $t_0 = 0$ in the quadruplet (K, h, r, t_0) and simply write (K, h, r).

2.2 Continuous F-transform of Higher Degree

Let $L^2_{loc}(\mathbb{R})$ be a set of all complex-valued functions that are square Lebesgue integrable on any closed subinterval of $\mathbb{R}$. As we have mentioned in Introduction, the F-transform of higher degree consists of two phases: *direct* and *inverse*. In what follows, we briefly recall their definitions in the form presented in [1,8].

Definition 3. *Let $f \in L_{loc}(\mathbb{R})$, $m \in \mathbb{N}$, and let $\mathcal{A}$ be a simple uniform fuzzy partition of $\mathbb{R}$ determined by the triplet (K, h, r). The direct continuous fuzzy transform of degree m (F^m-transform) of f with respect to $\mathcal{A}$ is the family*

$$F^m_{\mathcal{A}}[f] = \{F^m_k[f] \mid k \in \mathbb{Z}\}$$

where, for any $k \in \mathbb{Z}$,

$$F^m_k[f](t) = C_{k,0} + C_{k,1}\left(\frac{t - c_k}{h}\right) + \ldots + C_{k,m}\left(\frac{t - c_k}{h}\right)^m, \quad t \in [c_k - h, c_k + h],$$

such that

$$(C_{k,0}, C_{k,1}, \ldots, C_{k,m})^T = (\mathcal{Z}_m)^{-1} \cdot \mathcal{Y}_{m,k} \tag{3}$$

with $\mathcal{Z}_m = (Z_{ij})$ is the $(m+1) \times (m+1)$ invertible matrix defined by

$$Z_{ij} = \int_{-1}^{1} t^{i+j-2} K(t)\, dt, \quad i, j = 1, \ldots, m+1,$$

and $\mathcal{Y}_{m,k} = (Y_{k,1}, \ldots, Y_{k,m+1})^T$ is defined by

$$Y_{k,\ell} = \int_{-1}^{1} f(h \cdot t + c_k) \cdot t^{\ell-1} K(t)\, dt, \quad \ell = 1, \ldots, m+1. \tag{4}$$

The polynomial $F^m_k[f]$ is called the k-th component of the direct continuous F^m-transform of f.

Note that the k-th component $F^m_k[f]$ receives the interval $[c_k - h, c_k + h]$ as its support, so $F^m_k[f](t)$ is not defined for $t \notin [c_k - h, c_k + h]$. From the linearity property that holds for the Lebesgue integral, it is easy to see that the direct F^m-transform satisfies the linearity property, i.e.,

$$F^m_{\mathcal{A}}[a \cdot f + b \cdot g] = a \cdot F^m_{\mathcal{A}}[f] + b \cdot F^m_{\mathcal{A}}[g],$$

for any $a, b \in \mathbb{C}$ and $f, g \in L^2_{loc}(\mathbb{R})$. Moreover, the direct F^m-transform naturally preserves polynomials up to degree m. This fact is stated in the following lemma.

Lemma 1. *Let P be a polynomial of degree n, $n \in \mathbb{N}$, and let $F_\mathcal{A}^m[P] = \{F_k^m[P] \mid k \in \mathbb{Z}\}$ be the direct F^m-transform of P with respect to a simple uniform fuzzy partition of $\mathbb{R}$. If $m \geq n$, then, for any $k \in \mathbb{Z}$, it holds that*

$$F_k^m[P](t) = P(t), \quad t \in [c_k - h, c_k + h].$$

Proof: See [1] or [8].

Definition 4. *Let $f \in L_{loc}^2(\mathbb{R})$, and let $\mathcal{A} = \{A_k \mid k \in \mathbb{Z}\}$ be a simple uniform fuzzy partition of $\mathbb{R}$ determined by the triplet (K, h, r). Let the family $F_\mathcal{A}^m[f] = \{F_k^m[f] \mid k \in \mathbb{Z}\}$ be the direct F^m-transform of f with respect to $\mathcal{A}$. The inverse continuous fuzzy transform of degree m (F^m-transform) of f with respect to $F_\mathcal{A}^m[f]$ and $\mathcal{A}$ is defined by*

$$\hat{f}_\mathcal{A}^m(t) = \frac{\sum_{k \in \mathbb{Z}} F_k^m[f](t) \cdot A_k(t)}{\sum_{z \in \mathbb{Z}} A_k(t)}, \quad t \in \mathbb{R}. \tag{5}$$

By the linearity property of the direct F^m-transform, one can simply demonstrate that the inverse F^m-transform also satisfies the linearity property. Additionally, the inverse F^m-transform approximates the original function f, where the quality of the approximation is controlled mainly by the setting of the bandwidth parameter h. The details can be found in [1,8,11].

In the next part, we assume that a simple uniform fuzzy partition $\mathcal{A}$ determined by a triplet (K, h, r) is fixed. Moreover, if no confusion can appear, we simply write the (direct or inverse) F^m-transform of a function f, whereas the reference to a simple uniform fuzzy partition $\mathcal{A}$ determined by a triplet (K, h, r) is omitted.

3   Higher Degree Fuzzy Transform for Discrete Functions

3.1   Novel Definitions of the Discrete F^m-transform

Let $p : \mathbb{Z} \to \mathbb{C}$ be a complex-valued discrete function defined on the set of all integers.[1] Note that we chose the set of integers for a simple description of our approach, but the same idea can be applied also for an arbitrary discrete set. Let us extend the function p to a piecewise constant function $\bar{p}$ defined on the real line $\mathbb{R}$ as follows. For any $t \in \mathbb{R}$, we define

$$\bar{p}(t) = p(z) \text{ if and only if } t \in [z - 1/2, z + 1/2).$$

An example of the extension of discrete function p is depicted in Fig. 1. Since piecewise constant functions belong to the linear space $L_{loc}^2(\mathbb{R})$, the continuous F^m-transform, which has been defined in the previous section, can be directly applied to them.

The introduced conversion from the discrete to the continuous space is the core of our novel approach to the higher degree fuzzy transform for discrete functions. The following definition introduces the direct F^m-transform of a discrete function.

---

[1] Complex-valued functions are frequently used in analysis of stochastic processes or signal processing (see [14]).

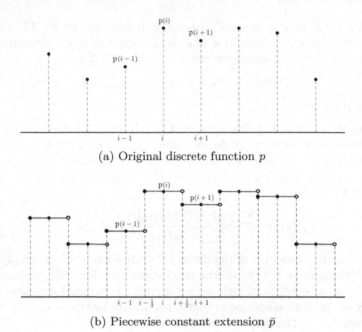

(a) Original discrete function p

(b) Piecewise constant extension $\bar{p}$

Fig. 1. The extension of a discrete function.

Definition 5. *Let $p : \mathbb{Z} \to \mathbb{C}$ be a complex-valued discrete function, $m \in \mathbb{N}$, and let $\mathcal{A}$ be a simple uniform fuzzy partition of $\mathbb{R}$ determined by the triplet (K, h, r). The direct discrete F^m-transform of p with respect to $\mathcal{A}$ is defined as follows:*

$$\mathrm{F}_{\mathcal{A}}^m[p] = \{F_k^m[\bar{p}] \mid k \in \mathbb{Z}\}, \tag{6}$$

where $\bar{p}$ is the extension of p defined above.

By the previous definition, the components of direct discrete F^m-transform of a function p with respect to a fuzzy partition can be simply computed using the product of matrices introduced in Definition 3. As a consequence of our conversion to the continuous case the verification of the invertibility of matrices in (1) is no longer required. Moreover, in spite of using integrals in the computation, the speed of the computation of the novel approach is completely comparable with the original. Indeed, the matrix $\mathcal{Z}_m$ in formula (3) is fixed. Moreover, assuming that the parameters h and r of a given fuzzy partition are natural numbers, the components of vector $\mathcal{Y}_{m,k} = (Y_{k,1}, \ldots, Y_{k,m+1})^T$ can be simply obtained as the product $Y_{k,\ell} = \mathcal{P}_k \cdot \mathcal{I}_\ell$, where

$$\mathcal{P}_k = (p(c_k - h), p(c_k - h + 1), \ldots, p(c_k), \ldots, p(c_k + h - 1), p(c_k + h))$$

and $\mathcal{I}_\ell = (I_{\ell,1}, \ldots, I_{\ell,2h+1})^T$ is determined by

$$I_{\ell,j} = \int_{a_{j-1}}^{a_j} t^{\ell-1} K(t)\, dt, \quad j = 1, \ldots, 2h + 1,$$

where $a_0 = -1, a_{2h+1} = 1$ and $a_j = -1 + \frac{2(j-1)+1}{2h}$, $j = 1, \ldots, 2h$. Since the matrix $\mathcal{I}_\ell$ is independent on the choice of k and the indefinite integral of $t^{\ell-1} K(t)$ can be found for the standard generating functions, the derivation of vector $\mathcal{Y}_{m,k}$ can be obtained in a very short time, which consequently accelerates the computation of the direct discrete F^m-transform.

The inverse F^m-transform of a discrete function is analogously defined as in Definition 4. The only difference is the use of $\mathbb{Z}$ instead of $\mathbb{R}$ as the domain of reconstructed functions.

Definition 6. *Let* $p : \mathbb{Z} \to \mathbb{C}$ *be a complex-valued discrete function, and let* $\mathcal{A} = \{A_k \mid k \in \mathbb{Z}\}$ *be a simple uniform fuzzy partition of* $\mathbb{R}$ *determined by the triplet* (K, h, r). *Let the family* $F^m_{\mathcal{A}}[p] = \{F^m_k[p] \mid k \in \mathbb{Z}\}$ *be the direct discrete* F^m-*transform of* p *with respect to* $\mathcal{A}$. *The inverse discrete* F^m-*transform of* p *with respect to* $F^m_{\mathcal{A}}[p]$ *and* $\mathcal{A}$ *is determined as follows:*

$$\hat{p}^m_{\mathcal{A}}(z) = \frac{\sum_{k \in \mathbb{Z}} F^m_k[p](z) \cdot A_k(z)}{\sum_{k \in \mathbb{Z}} A_k(z)}, \quad z \in \mathbb{Z}. \tag{7}$$

Obviously, the direct and inverse discrete F^m-transform preserves the linearity property similarly to the continuous case.

3.2   Estimation of the Reconstruction Error

Let $p : \mathbb{Z} \to \mathbb{C}$ be a discrete complex-valued function. We use $||c||$ to denote the size of the complex number c, i.e. $||c|| = \sqrt{c \cdot \bar{c}}$, where $\bar{c}$ is the complex conjugate of c. Let $z_0 \in \mathbb{Z}$ and $\delta > 0$. Then, the value

$$\omega_{z_0}(p, \delta) = \sup \{ ||\bar{p}(z_0) - \bar{p}(z_0 + \varepsilon)|| \mid \varepsilon \in \mathbb{R}, |\varepsilon| \le \delta \}, \tag{8}$$

provides us a measure of how much the function values $p(z)$ differ from each other in a neighborhood of z_0. Obviously, formula (8) imitates the definition of modulus of continuity. If p is bounded, then we define

$$\omega(p, \delta) = \sup \{ \omega_z(p, \delta) \mid z \in \mathbb{Z} \}$$

which measures the changes in the shape of function p with respect to the parameter δ.

Below, we consider the quality of reconstruction of a given function by the proposed discrete F^m-transform. First, we have to evaluate how the components of the direct F^m-transform are locally close (i.e., close in specific neighborhoods) to the original function.

Theorem 1. *Let* $p : \mathbb{Z} \to \mathbb{C}$ *be a complex-valued discrete function. Let* $\mathcal{A} = \{A_k \mid k \in \mathbb{Z}\}$ *be a simple uniform fuzzy partition of* $\mathbb{R}$ *determined by the triplet* (K, h, r), *and let* $F^m_{\mathcal{A}}[p]$ *be the direct discrete* F^m-*transform of* p *with respect to* $\mathcal{A}$. *Then, for any* $k \in \mathbb{Z}$, *it holds that*

$$||p(z) - F^m_k[p](z)|| \le \omega_z(p, 2h) \cdot \Theta(m, K),$$

for any $z \in Z$ such that $z \in [c_k - h, c_k + h]$, where

$$\Theta(m, K) = \sum_{j,\ell=1}^{m+1} |V_{j\ell}| \cdot \int_{-1}^{1} |t|^{\ell-1} K(t) \, dt$$

with $(V_{j\ell})_{j,\ell=1,\ldots,m+1} = (\mathcal{Z}_m)^{-1}$.

Proof: Let $k \in \mathbb{Z}$, and let $z \in Z$ such that $z \in [c_k - h, c_k + h]$. Since $F_k^m[c](t) = c$ holds for any complex-valued constant function c, we obtain

$$\|p(z) - F_k^m[p](z)\| = \|F_k^m[p(z)](z) - F_k^m[\overline{p}](z)\| =$$

$$\left\| (C_{k,0} - D_{k,0}) + \cdots + (C_{k,m} - D_{k,m}) \left(\frac{t - c_k}{h} \right)^m \right\| \leq$$

$$\sum_{j=0}^{m} \|C_{k,j} - D_{k,j}\| \cdot \left| \frac{t - c_k}{h} \right|^j \leq \sum_{j=0}^{m} \|C_{k,j} - D_{k,j}\|, \quad (9)$$

where we used $|(t - c_k)/h| \leq 1$, and $(C_{k,0}, \ldots, C_{k,m})$ and $(D_{k,0}, \ldots, D_{k,m})$ are determined by

$$(C_{k,0}, \ldots, C_{k,m})^T = (\mathcal{Z}_m)^{-1} \cdot \mathcal{Y}_{m,k}, \quad (10)$$

$$(D_{k,0}, \ldots, D_{k,m})^T = (\mathcal{Z}_m)^{-1} \cdot \mathcal{W}_{m,k} \quad (11)$$

with $\mathcal{Y}_{m,k} = (Y_{k,j})_{j=1,\ldots,m+1}$ and $\mathcal{W}_{m,k} = (W_{k,j})_{j=1,\ldots,m+1}$, which are the column matrices given by

$$Y_{k,j} = \int_{-1}^{1} p(z) \cdot t^{j-1} K(t) \, dt, \quad \text{and } W_{k,j} = \int_{-1}^{1} \overline{p}(h \cdot t + c_k) \cdot t^{j-1} K(t) \, dt.$$

From (10) and (11), we obtain

$$(C_{k,0} - D_{k,0}, \ldots, C_{k,m} - D_{k,m})^T = (\mathcal{Z}_m)^{-1} \cdot (\mathcal{Y}_{m,k} - \mathcal{W}_{m,k}).$$

Hence, for any $j = 0, \ldots, m$, we find that

$$\|C_{k,j} - D_{k,j}\| \leq \sum_{\ell=1}^{m+1} |V_{j+1\ell}| \cdot \|Y_{z\ell} - W_{z\ell}\|$$

$$= \sum_{\ell=1}^{m+1} |V_{j+1\ell}| \cdot \int_{-1}^{1} \|p(z) - \overline{p}(h \cdot t + c_k)\| \cdot |t|^{\ell-1} K(t) \, dt$$

$$= \sum_{\ell=1}^{m+1} |V_{j+1\ell}| \cdot \int_{-1}^{1} \|\overline{p}(z) - \overline{p}(h \cdot t + c_k)\| \cdot |t|^{\ell-1} K(t) \, dt$$

$$\leq \omega_z(p, 2h) \cdot \sum_{\ell=1}^{m+1} |V_{j+1\ell}| \cdot \int_{-1}^{1} |t|^{\ell-1} K(t) \, dt,$$

where we used $||\overline{p}(z) - \overline{p}(h \cdot t + c_k)|| \leq \omega_z(p, 2h)$. By this inequality and (9), we obtain

$$||p(z) - F_k^m[p](z)|| \leq \omega_z(p, 2h) \cdot \sum_{j=0}^{m} \sum_{\ell=1}^{m+1} |V_{j+1\ell}| \cdot \int_{-1}^{1} |t|^{\ell-1} K(t)\, dt$$

$$\leq \omega_z(p, 2h) \cdot \sum_{j,\ell=1}^{m+1} |V_{j\ell}| \cdot \int_{-1}^{1} |t|^{\ell-1} K(t)\, dt$$

$$= \omega_z(p, 2h) \cdot \Theta(m, K),$$

and the proof is finished.   □

The upper bound of the error of reconstruction of a function by its inverse discrete F^m-transform is established in the following theorem.

Theorem 2. *Let $p : \mathbb{Z} \to \mathbb{C}$ be a complex-valued discrete function. Let $\mathcal{A} = \{A_k \mid k \in \mathbb{Z}\}$ be a simple uniform fuzzy partition of $\mathbb{R}$ determined by the triplet (K, h, r). Let $\hat{p}_{\mathcal{A}}^m$ be the inverse discrete F^m-transform of p with respect to $\mathrm{F}_{\mathcal{A}}^m[p]$ and $\mathcal{A}$. Then,*

$$||p(z) - \hat{p}_{\mathcal{A}}^m(z)|| \leq \omega_z(p, 2h) \cdot \Theta(m, K), \tag{12}$$

for any $z \in \mathbb{Z}$, where $\Theta(m, K)$ is defined in Theorem 1.

Proof: Let $\mathrm{F}_{\mathcal{A}}^m[p] = \{F_k^m[p] \mid k \in \mathbb{Z}\}$ be the direct F^m-transform of p with respect to $\mathcal{A}$. For any $z \in \mathbb{Z}$, we have

$$||p(z) - \hat{p}_{\mathcal{A}}^m(z)|| = \left|\left| p(z) - \frac{\sum_{k \in \mathbb{Z}} F_k^m[p](z) \cdot A_k(z)}{\sum_{k \in \mathbb{Z}} A_k(z)} \right|\right|$$

$$= \left|\left| \frac{\sum_{k \in \mathbb{Z}} (p(z) - F_k^m[p](z)) \cdot A_k(z)}{\sum_{k \in \mathbb{Z}} A_k(z)} \right|\right| \leq \frac{\sum_{k \in \mathbb{Z}} |p(z) - F_k^m[p](z)| \cdot A_k(z)}{\sum_{k \in \mathbb{Z}} A_k(z)}.$$

It follows from Theorem 1 that

$$||p(z) - F_k^m[p](z)|| \leq \omega_z(p, 2h) \cdot \Theta(m, K),$$

for any $k \in \mathbb{Z}$. Consequently,

$$||p(z) - \hat{p}_{\mathcal{A}}^m(z)|| \leq \frac{\sum_{k \in \mathbb{Z}} \omega_z(p, 2h) \cdot \Theta(m, K) \cdot A_k(z)}{\sum_{k \in \mathbb{Z}} A_k(z)} = \omega_z(p, 2h) \cdot \Theta(m, K)$$

and the proof is finished.   □

The following corollary is a straightforward consequence of Theorem 2 in the case that the original discrete function is bounded.

Corollary 1. *Let the assumptions of Theorem 2 be satisfied. If the function p is bounded, then*

$$||p(z) - \hat{p}_{\mathcal{A}}^m(z)|| \leq \omega(p, 2h) \cdot \Theta(m, K),$$

where $\Theta(m, K)$ is defined in Theorem 1.

It is easy to see from (12) that for the bandwidth $h \leq 1/4$, we obtain $\omega_z(p, 2h) = 0$ for any $z \in \mathbb{Z}$; hence, the reconstruction of p is ideal. Note that, in practice, the bandwidth is chosen higher than $1/4$ with respect to specific tasks.

4   Illustration Examples

In this section, we illustrate the novel approach to the discrete F^m-transform on functions representing two time series. Particularly, we compare the inverse F^m-transform functions obtained by the newly defined discrete F^m-transform with that provided by the original approach presented in [4]. For the comparison, we use the MAPE and the time-consumption in computation for both approaches.

First, we consider the time series data "Monthly closings of the Dow–Jones industrial index, Aug. 1968–Aug. 1981" stored on the website http://datamarket.com that form a discrete function p with the domain $\{1, 2, \ldots, 291\}$. In Fig. 2, we display the inverse discrete F^2- and F^5-transform of p, obtained by the proposed approach, with respect to the simple uniform fuzzy partition determined by the triplet $(K^{tr}, 20, 10)$. Note that the results obtained by the novel and the standard approach are negligible; particularly, the differences between them are MAPE $= 2.1137 \times 10^{-5}$ for the F^2-transform and MAPE $= 5.7637 \times 10^{-5}$ for the F^5-transform. In Fig. 3, we depict the newly defined inverse discrete

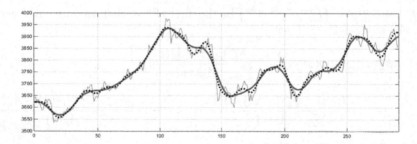

Fig. 2. The inverse F^2- (red line) and F^5-transform (dotted black line). (Color figure online)

F^3-transform of p with respect to varying fuzzy partitions $\mathcal{A}_1$, $\mathcal{A}_2$ and $\mathcal{A}_3$ determined by the triplets $(K^{tr}, 20, 10)$, $(K^{tr}, 10, 5)$ and $(K^{tr}, 1, 1)$, respectively. Among others, the results presented in Figs. 2 and 3 demonstrate the fact, which is well-known for the continuous F^m-transform, saying that a better reconstruction of the original function may be attained either by shortening the length of bandwidth h, if it is possible, or by enlarging the degree of the F-transform.

Below, we compare the time-consumption of the novel and the original approach in computation of inverse F^m-transform. For the comparison, we choose a long time series "Daily minimum temperatures in Melbourne, Australia, 1981–1990" with the dimension $d = 3650$, stored on the website http://datamarket.com. The both approaches to the discrete F^m-transform are programmed by

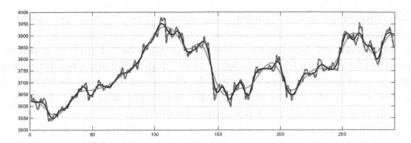

Fig. 3. The inverse F^3-transform functions (green, black and dashed red lines) with respect to $\mathcal{A}_1$, $\mathcal{A}_2$ and $\mathcal{A}_3$, respectively. (Color figure online)

Matlab 2014 on the notebook with CPU Intel Core^{TM} i5-3320M, Ram 8Gb and OS Windows 10. The computation time is also measured by Matlab. The considered simple uniform fuzzy partitions were determined by triplets $(K^{tr}, h, h/2)$, where h varies from 20 to 100 with the step 20. From the results depicted in Table 1, one can see that the computational times for both approaches are low and very similar, which is a consequence of our restriction to discrete functions defined over uniformly distributed elements (integers). This restriction actually enables us to optimize and speed up the algorithm of the original approach. Moreover, one can see that the computation time depends on the size of the bandwidth.

Table 1. Time-consumption (second) in computation of inverse F^m-transform.

F^m-transform \ Bandwidth		$h = 20$	$h = 40$	$h = 60$	$h = 80$	$h = 100$
F^0	Original	0.11479	0.07451	0.06506	0.05151	0.04940
	Novel	0.11826	0.07386	0.05214	0.04461	0.04081
F^1	Original	0.14210	0.08974	0.07528	0.06187	0.05556
	Novel	0.14657	0.08249	0.06088	0.05216	0.04426
F^2	Original	0.16912	0.09855	0.08097	0.06433	0.06021
	Novel	0.16797	0.09163	0.07021	0.05520	0.04861
F^3	Original	0.27800	0.16698	0.12056	0.09538	0.08498
	Novel	0.28627	0.16790	0.11868	0.09233	0.07524

5   Conclusions

In this paper, we introduced a novel approach to the discrete fuzzy transform of higher degree. We analyzed the quality of the reconstruction of an original discrete function provided by the inverse discrete F^m-transform. We compared the novel approach with the original one proposed in [4] on two examples of time series. The restriction to the discrete functions defined over uniformly distributed elements (integers) results in similar computation times for both approaches. The computation times are low in all tests and naturally depend on sizes of the bandwidth. Intuitively, the novel approach should provide better time consumption in computation for discrete functions defined over non-uniformly distributed values, because, in contrast to the novel approach, the original one needs to recompute the inverse matrices to get the respective F-transform components. A verification of our conjecture is a subject of our future research.

Acknowledgments. This work was supported by the project LQ1602 IT4Innovations excellence in science. The additional support was also provided by the Czech Science Foundation through the project of No. 16-09541S.

References

1. Holčapek, M., Nguyen, L., Tichý, T.: Polynomial alias higher degree fuzzy transform of complex-valued functions. Fuzzy Sets Syst. (2017). https://doi.org/10.1016/j.fss.2017.06.011
2. Hodáková, P.: Fuzzy (F)transform of functions of two variables and its applications in image processing. Ph.D. thesis, University of Ostrava (2015)
3. Holčapek, M., Tichý, T.: A probability density function estimation using F-transform. Kybernetika **46**(3), 447–458 (2010)
4. Holčapek, M., Tichý, T.: Discrete fuzzy transform of higher degree. In: IEEE International Fuzzy Systems Conference Proceedings, pp. 604–611 (2014)
5. Khastan, A., Perfilieva, I., Alijan, Z.: A new fuzzy approximation method to Cauchy problems by fuzzy transform. Fuzzy Sets Syst. **288**, 75–79 (2016)
6. Kokainis, M., Asmuss, S.: Continuous and discrete higher-degree F-transforms based on B-splines. Soft Comput. **21**(13), 3615–3639 (2017)
7. Di Martino, F., Loia, V., Sessa, S.: Fuzzy transforms method and attribute dependency in data analysis. Inf. Sci. **180**(4), 493–505 (2010)
8. Nguyen, L., Holčapek, M., Novák, V.: Multivariate fuzzy transform of complex-valued functions determined by monomial basis. Soft Comput. 3641–3658 (2017)
9. Novák, V., Štěpnička, M., Dvořák, A., Perfilieva, I., Pavliska, V., Vavříčková, L.: Analysis of seasonal time series using fuzzy approach. Int. J. Gen. Syst. **39**, 305–328 (2010)
10. Perfilieva, I.: Fuzzy transforms: theory and applications. Fuzzy Sets Syst. **157**, 993–1023 (2006)
11. Perfilieva, I., Daňková, M., Bede, B.: Towards a higher degree F-transform. Fuzzy Sets Syst. **180**, 3–19 (2011)
12. Perfilieva, I., Hodáková, P., Hurtik, P.: Differentiation by the F-transform and application to edge detection. Fuzzy Sets Syst. **288**, 96–114 (2016)
13. Perfilieva, I., Hurtik, P., Di Martino, F., Sessa, S.: Image reduction method based on the F-transform. Soft Comput. **21**(7), 1847–1861 (2017)
14. Yaglom, A.M.: An Introduction to the Theory of Stationary Random Functions, vol. 13. Prentice-Hall, Englewood Cliffs (1962). Revised English ed. Translated and edited by Silverman, R.A.

Lattice-Valued F-Transforms as Interior Operators of L-Fuzzy Pretopological Spaces

Irina Perfilieva[1(✉)], S. P. Tiwari[2], and Anand P. Singh[2]

[1] Institute for Research and Applications of Fuzzy Modeling, University of Ostrava,
30. dubna 22, Ostrava, Czech Republic
irina.perfilieva@osu.cz
[2] Indian Institute of Technology (ISM), Dhanbad, Dhanbad 826004, India
sptiwarimaths@gmail.com, anandecc@gmail.com

Abstract. The focus is on two spaces with a weaker structure than that of a fuzzy topology. The first one is a fuzzy pretopological space, and the second one is a space with an L-fuzzy partition. For a fuzzy pretopological space, we prove that it can be determined by a Čech interior operator and that the latter can be represented by a reflexive fuzzy relation. For a space with an L-fuzzy partition, we show that a lattice-valued $F^\downarrow$-transform is a strong Čech-Alexandrov fuzzy interior operator. Conversely, we found conditions that guarantee that a given L-fuzzy pretopology determines the L-fuzzy partition and the corresponding $F^\downarrow$-transform operator.

Keywords: Lattice F-transform · Fuzzy pretopological space
Fuzzy partition · Čech fuzzy interior operator

1 Introduction

This contribution is focused on various spaces with fuzzy structures and their corresponding operators. We deal with L-fuzzy objects (sets, relations, etc.) where L is a complete residuated lattice [2,3] and consider L-fuzzy topological, pretopological spaces and spaces with L-fuzzy partitions. These spaces propose an abstract approach to the notion of "closeness" and naturally arise in connection with applications to image and data analysis, time series, decision making, etc. All these spaces are introduced axiomatically, i.e. with the help of structure characterizing properties. Therefore, a natural problem is to give examples of the proposed structure or to propose a tool which produces such a structure. In the case of topological spaces, a tool can be an interior (closing) operator. However, the latter is again defined axiomatically, see [4].

In this contribution, we consider two spaces with weaker structures than that of a fuzzy topology. The first one is a fuzzy pretopological space, introduced in [13] with the purpose to find a good candidate for the extensional topological hull of fuzzy topological spaces. The second one is a space with an L-fuzzy partition

© Springer International Publishing AG, part of Springer Nature 2018
J. Medina et al. (Eds.): IPMU 2018, CCIS 854, pp. 163–174, 2018.
https://doi.org/10.1007/978-3-319-91476-3_14

introduced in [8] with the purpose to connect lattice-based (fuzzy) F-transforms with L-fuzzy rough sets operators. For a fuzzy pretopological space, we prove that it can be characterized by a Čech interior operator and that the latter can be represented by a reflexive fuzzy relation. For a space with an L-fuzzy partition, we show that the corresponding to the lattice-valued $F^\downarrow$-transform operator is a strong Čech-Alexandrov fuzzy interior operator. We discuss the opposite correspondence and prove a conditional result. This result is similar, but stronger than that proved in [9].

A deep insight into these structures from the categorical and topological viewpoints can be found in [5,9,13].

The structure of the paper is as follows. Section 2 contains preliminary information. In Sect. 3, we remind the notion of fuzzy pretopological space and introduce the Čech fuzzy interior operator and its modifications. The relationship between fuzzy pretopology, topology and their interior operators is discussed in Sect. 4. In Sect. 5, we remind notions of L-fuzzy partitions and lattice-based F-transforms. Moreover, we prove results about relational representation of lattice-based F-transforms. Section 6 contains main results of this contribution where we discuss the relationship between the direct $F^\downarrow$-transform and L-fuzzy pretopology. We show that every direct $F^\downarrow$-transform uniquely determines a strong Alexandrov L-fuzzy pretopological space, and that the corresponding $F^\downarrow$-operator is a strong Čech-Alexandrov fuzzy interior operator. The opposite correspondence is discussed at the end.

2   Preliminaries

In this section, we recall some basic concepts and properties related to residuated lattices, L-fuzzy sets, L-fuzzy relations, and L-fuzzy pretopological spaces. We refer to [1–3] for more details about residuated lattices.

Definition 21 [1]. *A **residuated lattice** L is an algebra $(L, \wedge, \vee, *, \rightarrow, 0, 1)$ such that*

(i) $(L, \wedge, \vee, 0, 1)$ *is a bounded lattice with the least element 0 and the greatest element 1;*
(ii) $(L, *, 1)$ *is a commutative monoid; and*
(iii) $\forall a, b, c \in L;$
$$a * b \leq c \iff a \leq b \rightarrow c.$$

A residuated lattice $(L, \wedge, \vee, *, \rightarrow, 0, 1)$ is **complete** if it is complete as a lattice.

The following unary operations of **negation** $\neg$ and **binarization** Δ will be used in the sequel:
$$\neg a = a \rightarrow 0,$$
$$\Delta(a) = \begin{cases} 1, & \text{if } a = 1, \\ 0, & \text{otherwise.} \end{cases}$$

Definition 22 [3]. *A residuated lattice L is called an integral, commutative Girard-monoid [3], if its negation is involutive, i.e.,*

$$\neg\neg a = a.$$

In the sequel, we will be using many standard properties of residuated lattices and complete integral commutative Girard monoids without particular references to their source. Most (if not all of them) of them came from [1–3].

Throughout this paper, we work with some fixed complete residuated lattice $L = (L, \wedge, \vee, *, \to, 0, 1)$.

For a nonempty set X, L^X denotes the collection of all L-fuzzy sets (L-valued functions) of X. Also, for all $a \in L$, $\mathbf{a}(x) = a$ is a constant fuzzy set on X. Furthermore, for all $A \in L^X$, the *core*(A) is a set of all elements $x \in X$, such that $A(x) = 1$. A fuzzy set $A \in L^X$ is called *normal*, if *core*$(A) \neq \emptyset$. Fuzzy set $S_{\{y\}} \in L^X$ is called a *singleton*, if it has the following form

$$S_{\{y\}}(x) = \begin{cases} 1, & \text{if } x = y, \\ 0, & \text{otherwise.} \end{cases}$$

The following are induced basic relations and operations of intersection $\cap$, union $\cup$, multiplication $*$, implication $\to$ and negation $\neg$ on L^X.

$$A = B \iff (\forall x)(A(x) = B(x)), \, A \leq B \iff (\forall x)(A(x) \leq B(x)),$$
$$(A \cap B)(x) = A(x) \wedge B(x), \quad (A \cup B)(x) = A(x) \vee B(x),$$
$$(A * B)(x) = A(x) * B(x), \quad (A \to B)(x) = A(x) \to B(x),$$
$$(\neg A)(x) = \neg A(x).$$

Under the assumption about completeness of L, we may consider an intersection and a union of an arbitrary family of fuzzy sets.

Definition 23. *Let X be a nonempty set. A* **fuzzy relation** *R on X is a fuzzy subset of $X \times X$, i.e. $R \in L^{X \times X}$. A fuzzy relation R is called*

(i) **reflexive** *if $\forall \, x \in X$, $R(x, x) = 1$,*
(ii) **transitive** *if $\forall \, x, y, z \in X$, $R(x, y) * R(y, z) \leq R(x, z)$.*

A reflexive and transitive fuzzy relation R is called a **fuzzy preorder**.

3 Fuzzy Pretopology and Čech Interior Operator

Let a complete residuated lattice L be fixed. In this section, we remind the notion of L-fuzzy pretopological space as it has been introduced in [13] and discuss how it can be generated by a corresponding interior operator. Moreover, we show that this operator can be represented by a reflexive fuzzy relation.

Definition 31 [10,13]. *An L-**fuzzy pretopology** on X is a collection of functions $\tau_X = \{p_x : L^X \to L \mid x \in X\}$ such that for all $A, B \in L^X$, $a \in L$, and $x \in X$,*

(i) $p_x(\mathbf{a}) = a$,
(ii) $p_x(A) \le A(x)$,
(iii) $p_x(A \cap B) = p_x(A) \wedge p_x(B)$.

The pair (X, τ_X) is called an L-**fuzzy pretopological space**. Moreover, we say that an L-fuzzy pretopological space (X, τ_X) is

(iv) **strong**, if $p_x(\mathbf{a} \to A) = a \to p_x(A)$,
 (v) **Alexandrov**, if $p_x(\bigcap\{A_j : j \in J\}) = \wedge\{p_x(A_j) : j \in J\}$,
(vi) **topological**, if $p_x(y \mapsto p_y(A)) = p_x(A)$.

Let us remark that if (X, τ_X) is an L-fuzzy pretopological space, $p_x \in \tau_X$ and $A \in L^X$, then $p_x(A)$ is a degree of the property "x belongs to the interior of A".

Let collection of functions $\tau_X = \{p_x : L^X \to L \mid x \in X\}$ be an L-fuzzy pretopology on X. With every $A \in L^X$, we associate the fuzzy set $\phi_A \in L^X$ such that for all $x \in X$, $\phi_A(x) = p_x(A)$. Obviously, $\phi : L^X \to L^X$ is an operator on L^X such that $\phi(A) = \phi_A$.

Definition 32. The map $i : L^X \to L^X$ is called a Čech fuzzy interior operator[1], if for every $\mathbf{a}, A \in L^X$, it fulfills

1. $i(\mathbf{a}) = \mathbf{a}$,
2. $i(A) \le A$,
3. $i(A \cap B) = i(A) \wedge i(B)$.

We say that a Čech fuzzy interior operator $i : L^X \to L^X$ is

4. **strong**, if for all $a \in L$, $i(\mathbf{a} \to A) = \mathbf{a} \to i(A)$,
5. **Čech-Alexandrov**, if $i(\bigcap_{j \in J} A_j) = \bigwedge_{j \in J} i(A_j)$.

It is easy to see from Definitions 31 and 32 that the following claim is valid.

Proposition 31. Collection of functions $\tau_X = \{p_x : L^X \to L \mid x \in X\}$ is an L-fuzzy pretopology on X, if and only if the map $i_\tau : L^X \to L^X$ such that for all $x \in X$, $i_\tau(A)(x) = p_x(A)$, is a Čech fuzzy interior operator. Moreover, if L-fuzzy pretopology $\tau_X = \{p_x : L^X \to L \mid x \in X\}$ is strong and Alexandrov, then the map i_τ is a strong Čech-Alexandrov interior operator.

Remark 31. In [12], a map $i : L^X \to L^X$ that enjoys properties 4 and 5 (named above as strong and Čech-Alexandrov) has been called an L-**fuzzy lower approximation operator**. In the sequel, we will use some results from [12] reformulated in the language of pretopological spaces.

Our next goal is to show that a fuzzy pretopology on X can be represented by a collection of lower approximations of fuzzy sets on X with respect to a reflexive fuzzy relation.

---

[1] Čech interior operator differs from Kuratowski interior operator by the absence of the idempotency.

Theorem 31. *Let* $(L, \wedge, \vee, *, \rightarrow, 0, 1)$ *be a complete, integral, commutative, Girard monoid. Then* $i : L^X \rightarrow L^X$ *is a strong Čech-Alexandrov fuzzy interior operator if and only if there exists a reflexive fuzzy relation* R_i *on* X *such that for every* $A \in L^X$,

$$i(A) = \underline{R_i}(A), \tag{1}$$

where $\underline{R_i}(A)$ *is a lower approximation of* A, *given by*

$$\underline{R_i}(A)(x) = \bigwedge_{y \in X} (R_i(x, y) \rightarrow A(y)). \tag{2}$$

Proof: $\Rightarrow$ By Remark 31, i is a fuzzy lower approximation operator. Therefore by [12], it can be represented by the following fuzzy relation

$$R_i(x, y) = \neg i(\neg S_{\{y\}}(x)), \tag{3}$$

where $S_{\{y\}}$ is a singleton, so that (1)+(2) holds true. Moreover, because interior operator i is anti-extensive, i.e. $i(A) \leq A$, then by [12], relation R_i is reflexive.

$\Leftarrow$ Vice versa, let (1)+(2), where R_i is an arbitrary reflexive fuzzy relation, holds trues. In [12], it has been proved that $\underline{R_i}$ is a lower approximation operator, i.e. enjoys properties 4 and 5. Because relation R_i is reflexive, $\underline{R_i}$ is obviously anti-extensive. Finally, property 1 easily follows from the reflexivity of R_i, i.e. for all $x \in X$,

$$\underline{R_i}(\mathbf{a})(x) = \bigwedge_{y \in X} (R_i(x, y) \rightarrow \mathbf{a}(y)) = \bigwedge_{y \in X} (R_i(x, y) \rightarrow a) = \bigvee_{y \in X} R_i(x, y) \rightarrow a = a.$$

$\square$

Remark 32. *On the basis of Theorem 31, we claim that a strong Čech-Alexandrov fuzzy interior operator* i *can be represented by a reflexive fuzzy relation* R_i *(3), if* i *is expressed in accordance with (1)+(2).*

Corollary 31. *Let* $(L, \wedge, \vee, *, \rightarrow, 0, 1)$ *be a complete, integral, commutative, Girard monoid. Then collection of functions* $\tau_X = \{p_x : L^X \rightarrow L \mid x \in X\}$ *is a strong Alexandrov* L-*fuzzy pretopology on* X, *if and only if there exists a reflexive fuzzy relation* R_i *on* X *such that for every* $x \in X$, $A \in L^X$,

$$p_x(A) = \underline{R_i}(A)(x).$$

Proof: By Proposition 31, a strong Alexandrov L-fuzzy pretopology τ_X is determined by the corresponding to it strong Čech-Alexandrov fuzzy interior operator i_τ. By Theorem 31, the latter can be represented by a reflexive fuzzy relation R_i.

$\square$

4 L-Fuzzy Pretopology, Topology and Their Interior Operators

Let us remind [4,14] that an L-**fuzzy topology** τ on a nonempty set X is a family of fuzzy sets in X which is closed under arbitrary suprema and finite infima and contains all constant fuzzy sets. The fuzzy sets in τ are called **open**. An L-fuzzy topology is called Alexandrov, if it is closed under arbitrary infima. In many papers (see e.g., [8,12,14]), it has been shown that Alexandrov L-fuzzy topology can be generated by a Kuratowski (fuzzy) interior operator, and the latter can be represented by a fuzzy preorder (reflexive and transitive) relation. In the majority (if not in all) of papers, this result is referred to [12] where the proof is not given.

In the preceding section, we showed (Theorem 31 and its Corollary 31) that any strong Alexandrov L-fuzzy pretopology can be generated by a Čech-Alexandrov interior operator, and that the latter can be represented by a reflexive fuzzy relation. In the below given theorem, we show the relationship between the Čech and Kuratowski interior operators and fuzzy relations that are used for their representation. We give a proof with all necessary technical details.

Theorem 41. *Let* $(L, \wedge, \vee, *, \rightarrow, 0, 1)$ *be a complete, integral, commutative, Girard monoid and* $i : L^X \rightarrow L^X$ *a strong Čech-Alexandrov fuzzy interior operator. Then the following assertions are equivalent*

(a) i *is a Kuratowski (fuzzy) interior operator,*
(b) *the corresponding to* i *fuzzy relation* R_i *is transitive.*

Proof: $(a) \Rightarrow (b)$ Let a strong Čech-Alexandrov fuzzy interior operator $i : L^X \rightarrow L^X$ be a Kuratowski (fuzzy) interior operator, i.e. for every $A \in L^X$, it fulfills

$$i(i(A)) = i(A). \tag{4}$$

Let R_i be the fuzzy relation (3) that corresponds to i. Then by (1), for every $A \in L^X$,

$$i(i(A)) = \underline{R_i}(\underline{R_i}(A)),$$

or after easy transformations,

$$i(i(A)) = \bigwedge_{y \in X} (R^2(x, y) \rightarrow A(y)),$$

where $R^2(x, y) = \bigvee_{t \in X} (R(x, t) * R(t, y))$. By (4),

$$\bigwedge_{y \in X} (R^2(x, y) \rightarrow A(y)) = \bigwedge_{y \in X} (R(x, y) \rightarrow A(y)).$$

Because R is reflexive and therefore, $R \leq R^2$, the above given equality is equal to

$$\bigwedge_{y \in X} (R^2(x, y) \rightarrow A(y)) \geq \bigwedge_{y \in X} (R(x, y) \rightarrow A(y)).$$

The latter is equivalent (see [2], Theorem 5.1.16) to

$$\bigwedge_{y \in X} (R(x,y) \to A(y)) \to (R^2(x,y) \to A(y)) = 1.$$

Therefore, for all $x, y \in X$,

$$R(x,y) \to A(y) \leq R^2(x,y) \to A(y), \text{ or}$$

$R^2 \leq R$, i.e. R is transitive.

$(b) \Rightarrow (a)$ The proof is similar to the given above. $\qquad\square$

5 L-Fuzzy Partition and Lattice-Based F-transform

The aim of this section is to remind notions of L-fuzzy partitions and lattice-based F-transforms, [8]. Moreover, we prove results about relational representation of lattice-based F-transforms.

In the literature, several notions of fuzzy partitions have been introduced and studied. Most of them connect this notion with a finite collection of fuzzy sets that are defined on the set of reals $\mathbb{R}$ or its Cartesian product. Here we recall the concept of fuzzy partition recently introduced in [8]. It has been used in [5] for the construction of the category of spaces with L-fuzzy partitions.

Definition 51. *A collection Π_X of normal fuzzy sets $\{A_\xi : \xi \in \Xi\}$ in X is an L-**fuzzy partition** of X, if the corresponding collection of ordinary sets $\{core(A_\xi) : \xi \in \Xi\}$ is a partition of X. A pair (X, Π_X), where Π_X is an L-valued fuzzy partition of X, is called a **space with an L-fuzzy partition**.*

Let $\Pi_X = \{A_\xi : \xi \in \Xi\}$ be an L-fuzzy partition of X. With this partition we associate the following surjective index-function $\ell_\Pi : X \to \Xi$:

$$\ell_\Pi(y) = \xi \iff y \in core(A_\xi). \tag{5}$$

Then Π_X is uniquely represented by the reflexive L-fuzzy relation R_Π on X, such that

$$R_\Pi(x,y) = A_\xi(x), \text{ where } \ell_\Pi(y) = \xi. \tag{6}$$

This claim follows from the fact that (6) uniquely specifies $R_\Pi(x,y)$ for every couple $(x,y) \in X \times X$. Let us remark that the opposite claim is not always true, i.e. not every reflexive fuzzy relation represents an L-fuzzy partition. Below, we give the corresponding criterion.

Proposition 51. *Let R be a reflexive L-fuzzy relation on X. Then, R represents the L-fuzzy partition Π_X of X, if and only if*

1. the binary relation $\Delta(R)$ is an equivalence on X;
2. if $z \in [y]_R$, where $[\cdot]_R$ is an equivalence class of $\Delta(R)$, then $R(\cdot, y) = R(\cdot, z)$.

Let us denote $\mathcal{R}_{\mathbf{FP}}$ the family of reflexive L-fuzzy relations on X that fulfill Proposition 51.

Definition 52. *Let $A \in L^X$ and $\Pi_X = \{A_\xi : \xi \in \Xi\}$ be an L-fuzzy partition of X. Then,*

(i) the **direct** $F^\uparrow$**-transform** *of A with respect to L-fuzzy partition $\Pi_X = \{A_\xi : \xi \in \Xi\}$ is a collection of lattice elements $\{F_\xi^\uparrow(A) : \xi \in \Xi\}$, where*

$$F_\xi^\uparrow(A) = \bigvee_{x \in X} (A(x) * A_\xi(x)),$$

(ii) the **direct** $F^\downarrow$**-transform** *of A with respect to L-fuzzy partition $\Pi_X = \{A_\xi : \xi \in \Xi\}$ is a collection of lattice elements $\{F_\xi^\downarrow(A) : \xi \in \Xi\}$, where*

$$F_\xi^\downarrow(A) = \bigwedge_{x \in X} (A_\xi(x) \to A(x)).$$

We denote by $\mathbf{F}^\uparrow[A] = \{F_\xi^\uparrow(A) : \xi \in \Xi\}$, the direct $F^\uparrow$-transform of A and $F_\xi^\uparrow(A)$ its ξ-th component. Similarly, $\mathbf{F}^\downarrow[A] = \{F_\xi^\downarrow(A) : \xi \in \Xi\}$ and $F_\xi^\downarrow(A)$ are the direct $F^\downarrow$-transform of A and its ξ-th component, respectively.

Lemma 51. *Let L-fuzzy partition $\Pi_X = \{A_\xi : \xi \in \Xi\}$ of X with the index-function ℓ_Π be represented by fuzzy relation R_Π. Then for every $A \in L^X$ and*

(i) for every $y \in X$,

$$F_{\ell_\Pi(y)}^\uparrow(A) = \bigvee_{x \in X} A(x) * R_\Pi(x, y)); \tag{7}$$

(ii) for every $x \in X$,

$$F_{\ell_\Pi(x)}^\downarrow(A) = \bigwedge_{y \in X} (R_\Pi^T(x, y) \to A(y)). \tag{8}$$

where $R_\Pi^T(x, y) = R_\Pi(y, x)$.

Proof:

(i) Let $y \in X$ and $\ell_\Pi(y) = \xi$. By (5), $y \in core(A_\xi)$ so that $R_\Pi(x, y) = A_\xi(x)$. Therefore, (7) directly follows from Definition 52, part (i).

(ii) Let $x \in X$ and $\ell_\Pi(x) = \xi$. By (5), $x \in core(A_\xi)$ so that $R_\Pi^T(x, y) = R_\Pi(y, x) = A_\xi(y)$. Therefore, (8) directly follows from Definition 52, part (ii). □

With a fuzzy partition Π_X and corresponding to it direct $F^\uparrow$ and $F^\downarrow$-transforms we associate two operators $F_\Pi^\uparrow : L^X \to L^X$ and $F_\Pi^\downarrow : L^X \to L^X$, such that for each $A \in L^X$,

$$F_\Pi^\uparrow(A)(y) = \bigvee_{x \in X} A(x) * R_\Pi(x, y)), \tag{9}$$

and

$$F_\Pi^\downarrow(A)(x) = \bigwedge_{y \in X} (R_\Pi^T(x, y) \to A(y)), \tag{10}$$

where fuzzy relation R_Π represents fuzzy partition Π_X in the sense of (6).

The following properties of the operators $F_\Pi^\uparrow : L^X \to L^X$ and $F_\Pi^\downarrow : L^X \to L^X$ have been proved in [6,8]. All of them are formulated below for arbitrary $x, y \in X$, $\mathbf{a}, A, A_i \in L^X$.

$$A \leq F_\Pi^\uparrow(A), \quad F_\Pi^\downarrow(A) \leq A, \tag{11}$$

$$F_\Pi^\uparrow(\mathbf{a}) = \mathbf{a}, \quad \mathbf{F}_\Pi^\downarrow(\mathbf{a}) = \mathbf{a}, \tag{12}$$

$$F_\Pi^\uparrow(A \cup B) = F_\Pi^\uparrow(A) \cup F_\Pi^\uparrow(B), F_\Pi^\downarrow(A \cap B) = F_\Pi^\downarrow(A) \cap F_\Pi^\uparrow(B), \tag{13}$$

$$F_\Pi^\uparrow(\bigcup_{i \in I} A_i) = \bigcup_{i \in I} F_\Pi^\uparrow(A_i), F_\Pi^\downarrow(\bigcap_{i \in I} A_i) = \bigcap_{i \in I} F_\Pi^\uparrow(A_i), \tag{14}$$

$$F_\Pi^\uparrow(\mathbf{a} * A) = a * F_\Pi^\uparrow(A), \quad F_\Pi^\downarrow(\mathbf{a} \to A) = a \to F_\Pi^\downarrow(A). \tag{15}$$

6 $F^\downarrow$-Transform as a Čech Fuzzy Interior Operator

In this section, we discuss the relationship between the direct $F^\downarrow$-transform and L-fuzzy pretopology. We show that every direct $F^\downarrow$-transform uniquely determines a strong Alexandrov L-fuzzy pretopological space, and that the corresponding $F^\downarrow$-operator is a strong Čech-Alexandrov fuzzy interior operator. We discuss the opposite correspondence and prove a conditional result. The relationship between the direct $F^\uparrow$-transform and L-fuzzy co-pretopology can be obtained using dualities between the residuated lattice operations. We will not discuss it in this contribution.

Proposition 61. *Let (X, Π_X) be a space with an L-fuzzy partition, R_Π corresponding fuzzy relation and $F_\Pi^\downarrow : L^X \to L^X$ corresponding $F^\downarrow$-operator. Then the pair (X, τ_Π) where $\tau_\Pi = \{F_\Pi^\downarrow(\cdot)(x) : L^X \to L \mid x \in X\}$ is such that for every $A \in L^X$ and $x \in X$,*

$$F_\Pi^\downarrow(A)(x) = \bigwedge_{y \in X} (R_\Pi^T(x, y) \to A(y)),$$

is a strong Alexandrov L-fuzzy pretopological space.

Proof: Let $\Pi_X = \{A_\xi : \xi \in A\}$ be an L-fuzzy partition of X with the index-function ℓ_Π where for every $x \in X$, the value $\ell_\Pi(x)$ determines unique partition element $A_{\ell_\Pi(x)}$, such that $x \in core(A_{\ell_\Pi(x)})$. For every $x \in X$, $A \in L^X$, we claim that

$$F_\Pi^\downarrow(A)(x) = F_{\ell_\Pi(x)}^\downarrow(A),$$

where the right-hand side is the $\ell_\Pi(x)$-th $F^\downarrow$-transform component of A. Indeed, for a particular $x \in X$, $F_{\ell_\Pi(x)}^\downarrow(A)$ is computed in accordance with (8) and by this, coincides with $F_\Pi^\downarrow(A)(x)$.

Let us verify that the collection $\tau_\Pi = \{F_\Pi^\downarrow(\cdot)(x) : L^X \to L \mid x \in X\}$ where for every $A \in L^X$, $F_\Pi^\downarrow(A)(x) = F_{\ell_\Pi(x)}^\downarrow(A) = \bigwedge_{y \in X}(R_\Pi^T(x,y) \to A(y))$ is a strong Alexandrov L-fuzzy pretopological space. This requires to verify properties (i)–(v) from Definition 31. It is easy to see that they immediately follow from the properties (11) – (15) of the $F_\Pi^\downarrow$ operator. □

Corollary 61. *Let (X, Π_X) be a space with an L-fuzzy partition, R_Π corresponding fuzzy relation and $F_\Pi^\downarrow : L^X \to L^X$ corresponding $F^\downarrow$-operator. Then the $F^\downarrow$-operator is a strong Čech-Alexandrov fuzzy interior operator i so that for every $A \in L^X$,*

$$i(A) = F_\Pi^\downarrow(A). \tag{16}$$

Proof: The assertion easily follows from the representation (10) and Theorem 31. □

Below, we discuss under which conditions a given L-fuzzy pretopology determines an L-fuzzy partition and the corresponding to it $F^\downarrow$-transform operator. Our result has an existential form and refers to the problem of solvability of a system of fuzzy relation equations.

Theorem 61. *Let $(L, \wedge, \vee, *, \to, 0, 1)$ be a complete, integral, commutative, Girard monoid and collection of functions $\tau_X = \{p_x : L^X \to L \mid x \in X\}$ be a strong Alexandrov L-fuzzy pretopology on X. With every $A \in L^X$, we associate fuzzy set $\phi_A \in L^X$ such that for all $x \in X$, $\phi_A(x) = p_x(A)$. Assume that the following auxiliary system*

$$\bigwedge_{y \in X}(R^T(x,y) \to A(y)) = \phi_A(x), \ A \in L^X, \tag{17}$$

of fuzzy relations equations has solution R_Π in $\mathcal{R}_{\mathbf{FP}}$.

Then, fuzzy relation R_Π represents the L-fuzzy partition Π of X and the $F^\downarrow$-transform operator $F_\Pi^\downarrow$ such that for every $x \in X$, the value of a function $p_x \in \tau_X$ at any $A \in L^X$ coincides with

$$p_x(A) = F_\Pi^\downarrow(A)(x). \tag{18}$$

Proof: Let the assumptions be fulfilled and $\tau_X = \{p_x : L^X \to L \mid x \in X\}$ be a strong Alexandrov L-fuzzy pretopology on X. By Proposition 31, τ_X determines the strong Čech-Alexandrov fuzzy interior operator $i_\tau : L^X \to L^X$ such that for all $x \in X$, $i_\tau(A)(x) = p_x(A)$. With every $A \in L^X$, we associate fuzzy set $\phi_A \in L^X$ such that for all $x \in X$,

$$\phi_A(x) = p_x(A). \tag{19}$$

Thence, $\phi_A = i_\tau(A)$. By Theorem 31, there exists a reflexive fuzzy relation R_τ on X such that for every $A \in L^X$, $i_\tau(A) = \underline{R_\tau}(A)$, where $\underline{R_\tau}(A)(x) = \bigwedge_{y \in X}(R_\tau(x,y) \to A(y))$.

On the basis of the above reasoning, the following system of fuzzy relation equations

$$\bigwedge_{y \in X} (R(x,y) \to A(y)) = \phi_A(x), \tag{20}$$

is solvable with respect to R. In particular, R_τ is one possible solution to (20). Let $\mathcal{R}_\tau$ denote a solution set of (20) and $\mathcal{R}_\tau^T$ the set of the corresponding transposed fuzzy relations. By the assumption, $\mathcal{R}_\tau^T \cap \mathcal{R}_{\mathbf{FP}} \neq \emptyset$ so that there exists fuzzy relation, say $R_\Pi \in \mathcal{R}_\tau$ such that

$$R_\Pi^T \in \mathcal{R}_\tau^T \cap \mathcal{R}_{\mathbf{FP}}.$$

By Proposition 51, fuzzy relation R_Π^T represents fuzzy partition Π_X of X in the sense of Definition 51.

By (19) and 20, for all $x \in X$ and $A \in L^X$,

$$\bigwedge_{y \in X} (R_\Pi^T(x,y) \to A(y)) = \phi_A(x),$$

which by Proposition 61, is equal to $F_\Pi^\downarrow(A)(x)$, where $F_\Pi^\downarrow : L^X \to L^X$ is the $F^\downarrow$-transform operator that corresponds to partition Π_X.

Finally, for all $x \in X$ and $A \in L^X$,

$$p_x(A) = F_\Pi^\downarrow(A)(x),$$

which confirms (18).   □

7   Conclusion

We were focused on the two spaces with fuzzy structures and their characterization by corresponding operators. The first one is a fuzzy pretopological space, and the second one is a space with an L-fuzzy partition. We showed that

- a fuzzy pretopology can be characterized by a Čech interior operator;
- in a complete, integral, commutative, Girard monoid, a strong Čech-Alexandrov fuzzy interior operator can be represented by a reflexive fuzzy relation and vice versa;
- a strong Čech-Alexandrov fuzzy interior operator represented by fuzzy relation R is a Kuratowski fuzzy interior operator if and only if R is a fuzzy preorder;
- for a space with an L-fuzzy partition, a lattice-valued $F^{\downarrow}$-transform is a strong Čech-Alexandrov fuzzy interior operator;
- under certain conditions, a given L-fuzzy pretopology determines an L-fuzzy partition and the corresponding to it $F^{\downarrow}$-transform operator.

Acknowledgement. The work was supported by the project "LQ1602 IT4Innovations excellence in science" and by the Grant Agency of the Czech Republic (project "New approaches to aggregation operators in analysis and processing of data").

References

1. Blount, K., Tsinakis, C.: The structure of residuated lattices. Int. J. Algebra Comput. **13**, 437–461 (2003)
2. Hájek, P.: Metamathematics of Fuzzy Logic. Kluwer Academic Publishers, Boston (1998)
3. Höhle, U.: Commutative residuated monoids. In: Höhle, U., Klement, P. (eds.) Non-classical Logics and Their Applications to Fuzzy Subsets, pp. 53–106. Kluwer Academic Publishers (1995)
4. Lowen, R.: Fuzzy topological spaces and fuzzy compactness. J. Math. Anal. Appl. **56**, 621–633 (1976)
5. Močkoř, J., Holčapek, M.: Fuzzy objects in spaces with fuzzy partitions. Soft Comput. https://doi.org/10.1007/s00500-016-2431-4
6. Perfilieva, I.: Fuzzy transforms: theory and applications. Fuzzy Sets Syst. **157**, 993–1023 (2006)
7. Perfilieva, I.: Fuzzy transforms: a challenge to conventional transforms. Adv. Image Electron Phys. **147**, 137–196 (2007)
8. Perfilieva, I., Singh, A.P., Tiwari, S.P.: On the relationship among F-transform, fuzzy rough set and fuzzy topology. Soft Comput. **21**, 3513–3523 (2017)
9. Perfilieva, I., Singh, A.P., Tiwari, S.P.: On F-transforms, L-fuzzy partitions and L-fuzzy pretopological spaces. In: IEEE Symposium on Foundations of Computational Intelligence: 2017 IEEE Symposium Series on Computational Intelligence (SSCI) Proceedings, 27 November 2017, Honolulu, USA, pp. 154–161 (2017)
10. Qiao, J., Hua, B.Q.: A short note on L-fuzzy approximation spaces and L-fuzzy pretopological spaces. Fuzzy Sets Syst. **312**, 126–134 (2017)
11. Singh, A.P., Tiwari, S.P.: Lattice F-transform for functions in two variables. J. Fuzzy Set Valued Anal. **3**, 185– 195 (2016)
12. She, Y.H., Wang, G.J.: An axiomatic approach of fuzzy rough sets based on residuated lattices. Comput. Math. Appl. **58**, 189–201 (2009)
13. Zhang, D.: Fuzzy pretopological spaces, an extensional topological extension of FTS. Chin. Ann. Math. **3**, 309–316 (1999)
14. Lai, H., Zhang, D.: Fuzzy preorder and fuzzy topology. Fuzzy Sets Syst. **157**, 1865–1885 (2006)

Modified F-transform Based on B-splines

Martins Kokainis[1(✉)] and Svetlana Asmuss[1,2]

[1] Department of Mathematics, University of Latvia,
Zellu 25, Riga 1002, Latvia
{martins.kokainis,svetlana.asmuss}@lu.lv
[2] Institute of Mathematics and Computer Science, University of Latvia,
Raina bulvaris 29, Riga 1459, Latvia

Abstract. The aim of this paper is to improve the F-transform technique based on B-splines. A modification of the F-transform of higher degree with respect to fuzzy partitions based on B-splines is done to extend the good approximation properties from the interval where the Ruspini condition is fulfilled to the whole interval under consideration. The effect of the proposed modification is characterized theoretically and illustrated numerically.

Keywords: Fuzzy transform · B-splines · Extrapolation
Approximation error

1 Introduction

The paper deals with the technique of fuzzy transform (F-transform or F^0-transform) introduced in 2001 [13] (see also the key paper [10]) and generalized to the case of higher degree in 2011 [12] by I. Perfilieva with co-authors. As it was shown in [12], the ordinary F-transform with constant components can be extended to the F-transform with degree-m polynomial components, $m \geq 0$ (the F^m-transform).

Properties of F-transforms significantly depend on basic functions which form a fuzzy partition. There is a number of papers dealing with fuzzy transforms with respect to a fuzzy partition with specially designed basic functions including splines (see, e.g., [1,4]). We focus on F^m-transforms with respect to a spline-based fuzzy partition introduced in [5] and further investigated in [9]. Previously we have proved that for the composite F^m-transform w.r.t. the fuzzy partition based on B-splines of degree $2k-1$ the following approximation error estimation

$$\left| f^{(n)}(t) - (F_m[f])^{(n)}(t) \right| = O(h^{r+1-n}), \ t \in [\hat{a}, \hat{b}] \subset [a, b], \tag{1}$$

holds for functions $f \in C^{r+1}[a, b]$, where $0 \leq n \leq r \leq \min\{2m+1, 2k-1\}$ and h is the parameter of the corresponding uniform crisp partition. The result $O(h^{2m+2})$ is achieved using B-splines of degree $2m + 1$ (or greater), when we consider the approximation error for functions from $C^{2m+2}([a, b])$ (see also generalizations of

© Springer International Publishing AG, part of Springer Nature 2018
J. Medina et al. (Eds.): IPMU 2018, CCIS 854, pp. 175–186, 2018.
https://doi.org/10.1007/978-3-319-91476-3_15

this result for the multivariate setting [6–8]). This improves the best estimation $O(h^{m+1})$ of approximation error which is known for the F^m-transform w.r.t. an arbitrary uniform fuzzy partition with the same parameter h.

The error estimation (1) holds on the interval where the spline-based fuzzy partition fulfills the Ruspini condition. This interval is smaller than the initial interval $[a, b]$ and does not contain the first $2k-1$ and the last $2k-1$ subintervals of the corresponding crisp partition. The aim of this contribution is to introduce a technique which allows to extend good approximation properties of the F^m-transform based on B-splines to the whole interval $[a, b]$. Such modification of the proposed technique is very important for applications and allows to use the advantages of spline-based F^m-transforms in numerical methods. Applying F^m-transforms in numerical methods, e.g., to solve differential equations with initial or boundary conditions, it is important to guarantee high quality of approximation on the whole interval under consideration and especially near the boundary points.

2   Preliminaries

By $[n..m]$ (for integers n, m with $n \le m$) we denote the set $\{n, n+1, \ldots, m\}$. Let $\mathbb{P}_l$ stand for the space of univariate polynomials of degree at most l. By $\|u\|$, where $u \in C[a, b]$, we denote the usual supremum norm of u.

2.1   Generalized Fuzzy Partition

Fix an interval $[a, b] \subset \mathbb{R}$ and a positive integer $N \in \mathbb{N}$. Let $h > 0$ and $h' > h/2$. Suppose that $t_0, \ldots, t_N$ are h-equidistant nodes s.t. $a < t_0 < \ldots < t_N < b$. Let $E_i := (t_i - h', t_i + h')$ and $\bar{E}_i$ be the closure of E_i. Furthermore, suppose that $\bigcup_{j=0}^{N} \bar{E}_j = [a, b]$.

Definition 1 (see, e.g., [11]). *Fuzzy sets* $A_0, \ldots, A_N : [a, b] \to [0, 1]$ *are said to constitute a generalized* (h, h')*-uniform fuzzy partition (FP for short) of* $[a, b]$ *if the following conditions are satisfied:*

- $A_i(t) > 0$ *if* $t \in E_i$, *and* $A_i(t) = 0$ *if* $t \in [a, b] \setminus E_i$, $i \in [0..N]$;
- A_i *is continuous on* $\bar{E}_i$, $i \in [0..N]$;
- $\sum_{j=0}^{N} A_j(t) > 0$ *for all* $t \in (a, b)$, $i \in [0..N]$;
- $A_i(t_i - t) = A_i(t_i + t)$ *for all* $t \in [0, h']$, $i \in [0..N]$;
- $A_i(t) = A_{i+1}(t + h)$ *for all* $t \in \bar{E}_i$ *and* $i \in [0..N-1]$.

Then there is a function $A : [-H, H] \to \mathbb{R}$ *(called the generating function of the partition), where* $H = h'/h$, *s.t. for all* $i \in [0..N]$ *and* $t \in \bar{E}_i$, $A_i(t) = A\left(\frac{t-t_i}{h}\right)$.

Furthermore, if there is an interval $I \subset [a, b]$ *such that* $\sum_{i=0}^{N} A_i(t) = 1$ *for all* $t \in I$, *the fuzzy partition is said to fulfill the Ruspini condition on* I.

2.2   Discrete F^m-transform

Here we recall the definition of the discrete F^m-transform. We consider the discrete transform, although the construction and related results in Sects. 3 and 4 can be extended also to the case of the integral transform. Throughout the rest of this section, let an interval $[a, b] \subset \mathbb{R}$ and its generalized fuzzy partition $\mathbb{A} = \{A_0, \ldots, A_N\}$ be fixed. Fix also an integer $m \geq 0$.

Let $\Delta = \{z_1, \ldots, z_L\} \subset [a, b]$ be a discrete set of the interval $[a, b]$ and $f : \Delta \to \mathbb{R}$. Denote $y_j = f(z_j)$, for each $j \in [1 .. L]$, and let $\boldsymbol{y} = (y_1, y_2, \ldots, y_L)^T$ be a column vector containing the values of the function f.

For each $i \in [0 .. N]$ define matrices

$$\mathbf{X}_i = \begin{pmatrix} 1 & z_1 - t_i & \ldots & (z_1 - t_i)^m \\ & & \ddots & \\ 1 & z_L - t_i & \ldots & (z_L - t_i)^m \end{pmatrix}, \quad \mathbf{A}_i = \mathrm{diag}(A_i(z_1), \ldots, A_i(z_L)).$$

Definition 2. *We say that the set Δ is sufficiently dense in the fuzzy partition $\mathbb{A}$ w.r.t. m if the matrix $\mathbf{X}_i^T \mathbf{A}_i \mathbf{X}_i$ is invertible for each $i \in [0 .. N]$.*

If the set Δ is sufficiently dense (in the fuzzy partition $\mathbb{A}$ w.r.t. m), then the discrete direct F^m-transform can be defined:

Definition 3 [3]. *The vector $F_m^{\rightarrow}[f]$ is the discrete direct F^m-transform of function f w.r.t. the fuzzy partition $\mathbb{A}$, if the ith component $F_{m,i}^{\rightarrow}[f]$, $i \in [0 .. N]$, of this vector is the polynomial*

$$F_{m,i}^{\rightarrow}[f](t) = \sum_{j=0}^{m} \beta_j^{(i)} (t - t_i)^j, \quad t \in \mathbb{R},$$

where $\beta^{(i)} = \left(\mathbf{X}_i^T \mathbf{A}_i \mathbf{X}_i \right)^{-1} \mathbf{X}_i^T \mathbf{A}_i \boldsymbol{y}$.

One can define the inverse F^m-transform, which is applied to a tuple of polynomials and is defined as a linear combination of these polynomials. Composing the direct and inverse F^m-transforms one obtains the composite F^m-transform. For the purposes of this paper, we define only the latter.

Definition 4. *Let $f : \Delta \to \mathbb{R}$. Suppose that the direct F^m-transform of f w.r.t. the fuzzy partition $\mathbb{A}$ is $F_m^{\rightarrow}[f] = (F_{m,0}^{\rightarrow}[f], \ldots, F_{m,N}^{\rightarrow}[f]) \in \mathbb{P}_m^{N+1}$. Then the function*

$$F_m[f](t) = \frac{\sum_{i=0}^{N} F_{m,i}^{\rightarrow}[f](t) \, A_i(t)}{\sum_{i=0}^{N} A_i(t)}, \quad t \in (a, b),$$

is called the (composite) F^m-transform of f w.r.t. the fuzzy partition $\mathbb{A}$.

2.3   Fuzzy Partition Based on B-splines

Central B-splines [14] are even B-splines that have 1-equidistant knots. For a fixed degree the central B-spline is unique (up to a constant factor). The properties of B-splines and construction of a fuzzy partition using central B-spline as the generating function are described in more details in [5].

Definition 5. *The central B-spline of degree* $2k - 1$, *denoted by* ϕ_{2k-1}, *is the unique piecewise polynomial function satisfying the following requirements: (1) for each* $i \in [-k..k-1]$ *the restriction of* ϕ_{2k-1} *to* $[i, i+1]$ *is a polynomial of degree at most* $2k - 1$; *(2)* $\phi_{2k-1} \in C^{2k-2}(\mathbb{R})$; *(3)* $\phi_{2k-1}(t) = 0$ *if* $t \notin (-k, k)$; *(4)* $\int_{\mathbb{R}} \phi_{2k-1}(t) \, dt = 1$.

Fix $N, k \in \mathbb{N}$ such that $N \geq 4k - 1$; let $A = \phi_{2k-1}$. Let an interval $[a, b]$ be fixed; denote $h = (b - a)/N$ and define h-equidistant nodes $t_i = a + hi$, $i \in [0..N]$.

Define the basic functions $A_i(t) := A\left(\frac{t-t_i}{h}\right)$, $i \in [k..N-k]$. Then the basic functions A_k, ..., A_{N-k} form a generalized (h, hk)-uniform fuzzy partition of $[a, b]$ (let us denote it by $\mathbb{A}_0$). We refer to $\mathbb{A}_0$ as the FPB (FP based on B-splines), or, more specifically, FPB(k, N). We also always implicitly assume that its parameters satisfy $N \geq 4k - 1$.

It is well-known that the composite F^m-transform is exact for polynomials of degree $\leq m$. The main advantage of the FPB is that the composite transform is exact for polynomials of degree $2m + 1$ (as long as it does not exceed the respective spline degree).

Let the discrete set Δ consists of the basic nodes t_i, $i \in [0..N]$ (this set is sufficiently dense w.r.t. $\mathbb{A}_0$ iff $m \leq 2k - 2$). Let $F_m[f]$ stand for the discrete composite F^m-transform of $f : \Delta \to \mathbb{R}$ w.r.t. the fuzzy partition $\mathbb{A}_0$.

Theorem 1. *Let* r, m *be non-negative integers s.t.* $r \leq \min\{2m + 1, 2k - 1\}$ *and* $m \leq 2k - 2$. *Suppose that* $f \in \mathbb{P}_r$; *then* $f(t) = F_m[f](t)$ *for all* $t \in [\hat{a}, \hat{b}]$. *Here* $[\hat{a}, \hat{b}] := [t_{2k-1}, t_{N-2k+1}]$ *is the interval where the FPB* $\mathbb{A}_0$ *satisfies the Ruspini condition.*

3   Extended FPB and Modified F-transform

Suppose we are given a fuzzy partition $A_0, A_1, \ldots, A_N$ of an interval $[a, b]$; furthermore, let $\sum_{j=0}^{N} A_j(t) \leq 1$ for all $t \in [a, b]$. If $\sum_{j=0}^{N} A_j(t) = 1$ for all $t \in [a, b]$, then the fuzzy partition fulfills the Ruspini condition on the whole interval $[a, b]$. Otherwise, one might introduce additional basic functions (i.e., *extend* the given FP) to have the Ruspini condition fulfilled everywhere on $[a, b]$. When the fuzzy partitions is FPB(k, N), there is a natural extension consisting of $2(2k - 1)$ additional basic functions.

3.1   eFPB: An Extension of FPB

Throughout the rest of this section we introduce the following notation and concepts, in addition to those described in Sect. 2.3:

- $\mathbb{A}_0$: the FPB(k, N) of $[a, b]$ consisting of the basic functions $A_k, \ldots, A_{N-k}$.
- $t_i = a + ih$, with $i \in \{-2k+1, \ldots, -2, -1, N+1, N+2, \ldots, N+2k-1\}$: additional nodes outside the interval $[a, b]$.
- $\tilde{a} := t_{-2k+1} = a - (2k-1)h$ and $\tilde{b} := t_{N+2k-1} = b + (2k-1)h$: the first and the last of the additional nodes.
- $A_i(t) = A((t-t_i)/h)$, with $i \in [-k+1 \ldots k-1]$ and $i \in [N-k+1 \ldots N+k-1]$: additional $2(2k-1)$ basic functions extending the original FPB $\mathbb{A}_0$.
- $\mathbb{A}$: the fuzzy partition of $[a, b]$, formed by the basic functions $A_{-k+1}|_{[a,b]}, \ldots,$ $A_{N+k-1}|_{[a,b]}$.
- $\bar{\mathbb{A}}$: the FPB$(k, N+4k-2)$ of the interval $[\tilde{a}, \tilde{b}]$, formed by the basic functions $A_{-k+1}, \ldots, A_{N+k-1}$.

From now on, we will refer to the fuzzy partition $\mathbb{A}$ as the (k, N)-eFPB of $[a, b]$. Notice that $\mathbb{A}$ of $[a, b]$ satisfies the Ruspini condition on the whole $[a, b]$. Furthermore, $\bar{\mathbb{A}}$ (i.e., the FPB of $[\tilde{a}, \tilde{b}]$) satisfies the Ruspini condition on the interval $[a, b]$.

We illustrate these concepts in Fig. 1. Consider $[a, b] = [0, 1]$ and its FPB$(2, 7)$, i.e., the FPB based on cubic B-splines, consisting of basic functions $A_2, \ldots, A_5$ (which are plotted in Fig. 1 with blue, solid lines), with basic nodes $t_i = i/7$, $i \in [0 \ldots 7]$. This FPB is denoted by $\mathbb{A}_0$.

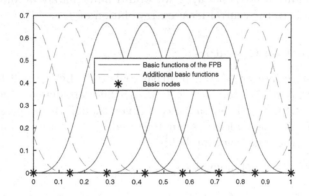

Fig. 1. FPB $\mathbb{A}_0$ of $[0, 1]$ (with blue lines) and the corresponding eFPB $\mathbb{A}$ (Color figure online)

Introducing additional 6 basic functions (plotted in Fig. 1 with red, dashed lines) yields the $(2, 7)$-eFPB $\mathbb{A}$, which fulfills the Ruspini condition on the whole interval $[a, b]$.

The additional basic functions A_{-1}, A_0, A_1 and A_6, A_7, A_8 are supported in the wider interval $[\tilde{a}, \tilde{b}] = [-3/7, 10/7]$, and $\mathbb{A}$ is obtained when these functions are restricted to $[a, b] = [0, 1]$. However, when we view the unrestricted functions (in Fig. 2 we depict them outside $[a, b]$ with green dash-dot lines), they form an FPB$(2, 13)$ of the interval $[\tilde{a}, \tilde{b}]$ (which is denoted by $\bar{\mathbb{A}}$). In Fig. 2 we also illustrate the additional basic nodes outside $[a, b]$.

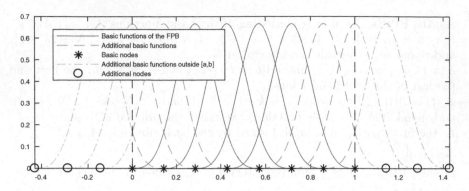

Fig. 2. Unrestricted basic functions of $\bar{\mathbb{A}}$ (the FPB of $[\tilde{a}, \tilde{b}]$) (Color figure online)

The reason we defined eFPB $\mathbb{A}$ is to obtain an analogue of Theorem 1. However, performing the usual F^m-transform w.r.t. the fuzzy partition $\mathbb{A}$ does not produce this result. Moreover, when performing the discrete F^m-transform, the partition $\mathbb{A}$ is unlikely to be sufficiently dense. If only the function f was defined on the interval $[\tilde{a}, \tilde{b}]$, then we could use $\bar{\mathbb{A}}$, i.e., the FPB of $[\tilde{a}, \tilde{b}]$, instead of $\mathbb{A}$, and perform the usual F^m-transform of f with respect to $\bar{\mathbb{A}}$. Since for this partition the Ruspini condition holds on $[a, b]$, we would immediately obtain the desired result. This leads to the central idea: extrapolation of f.

3.2　Extrapolation Operators

Fix a nonnegative integer M and positive reals δ, δ'. Let $\epsilon_0 > 0$ be such that $\delta\epsilon_0 = b - a$. Suppose that for every $\epsilon \in (0, \epsilon_0)$ there are defined linear bounded operators $\mathfrak{E}_{\epsilon,1} : C[a, a + \delta\epsilon] \to C[a - \delta'\epsilon, a]$ (even though these operators depend on δ, δ' and M, this dependence is not reflected in the notation for the sake of simplicity) such that the following holds:

1. $\mathfrak{E}_{\epsilon,1} f = f$ for every polynomial $f \in \mathbb{P}_M$;
2. $(\mathfrak{E}_{\epsilon,1} f)(a) = f(a)$;
3. the family $\{\mathfrak{E}_{\epsilon,1}\}_\epsilon$ is uniformly bounded, i.e., there exists a constant $C_\star > 0$ (independent of ϵ) such that $\|\mathfrak{E}_{\epsilon,1}\| \le C_\star$ for all ϵ.

Let there be a family of operators $\mathfrak{E}_{\epsilon,2} : C[b - \delta\epsilon, b] \to C[b, b + \delta'\epsilon]$ with analogous properties; in fact, one can define $\mathfrak{E}_{\epsilon,2}$ via $\mathfrak{E}_{\epsilon,1}$, by taking

$$(\mathfrak{E}_{\epsilon,2} f)(b + t) = (\mathfrak{E}_{\epsilon,1} g)(a - t), \quad t \in [0, \delta'\epsilon], \tag{2}$$

where $g \in C[a, b]$ is defined by $g(a + t) = f(b - t)$, $t \in [0, b - a]$. Define an operator $\mathfrak{E}_\epsilon : C[a, b] \to C[a - \delta'\epsilon, b + \delta'\epsilon]$ as follows:

$$\mathfrak{E}_\epsilon f(t) = \begin{cases} \mathfrak{E}_{\epsilon,1} f(t), & a - \delta'\epsilon \le t < a, \\ f(t), & a \le t \le b, \\ \mathfrak{E}_{\epsilon,2} f(t), & b < t \le b + \delta'\epsilon. \end{cases} \tag{3}$$

We call the operator $\mathfrak{E}_\epsilon$ an order-M extrapolator (with parameters δ, δ', for functions $f \in C[a,b]$).

The parameter δ controls the length of the interval where the values of f are used for the extrapolation; the parameter δ' controls how far we wish to extend the function f. The parameter ϵ allows to 'scale' the extrapolation, i.e., for smaller ϵ we consider f only in a small vicinity of a (or b) and extrapolate its values to another small interval outside $[a,b]$.

Remark 1. We intend to set $\delta' = 2k - 1$ and $\epsilon = h$, then $a - \delta'\epsilon = \tilde{a}$ and $\mathfrak{E}_h$ maps $f \in C[a,b]$ to a function $\tilde{f} \in C[\tilde{a}, \tilde{b}]$.

Construction of an Order-M Extrapolator. There are several ways to define operators $\mathfrak{E}_{\epsilon,1}$; arguably the simplest of them is to use the Lagrange interpolating polynomial. Suppose that $\delta = M$, $\delta' = 2k - 1$ and $\epsilon M < b - a$. Let $t_i = a + i\epsilon$, $i \in [0..M]$ (this notation is consistent with our previous use of t_i in light of Remark 1).

Fix a function $f \in C[a,b]$ and denote $y_i = f(t_i)$. Let $p \in \mathbb{P}_M$ be the (unique) polynomial satisfying $p(t_i) = y_i$ for all $i \in [0..M]$, and let $\mathfrak{E}_{\epsilon,1}(f) = p$. Operators $\mathfrak{E}_{\epsilon,2}$ and $\mathfrak{E}_\epsilon$ are defined via (2) and (3). It is easy to check that $\mathfrak{E}_{\epsilon,1}$ possesses the necessary properties (the norm of the operator depends on M, but not ϵ, thus we have the uniform boundedness, the remaining properties follow trivially). Thus $\mathfrak{E}_\epsilon$ is a valid order-M extrapolator.

However, the Lagrange interpolation, combined with the equispaced points t_i, exhibits poor numerical properties: the 2-norm condition number of the matrix used to find the polynomial (matrix $\mathbf{T}_{M,M}$ in the notation below) grows exponentially with M [2] and actual computations in floating point arithmetic become difficult even for relatively small values of M.

Our preferred route is to use least squares fitting instead of interpolation. Fix an integer $R \geq M$; set $\delta = R$, $\delta' = 2k-1$ and let $\epsilon < (b-a)/R$. Denote $t_i = a+i\epsilon$, $y_i = f(t_i)$, $i \leq R$, and $\boldsymbol{t} := (t_0, t_1, \ldots, t_R)$, $\boldsymbol{y} := f(\boldsymbol{t}) = (y_0, y_1, \ldots, y_R)$.

Consider the following task: find a polynomial $p \in \mathbb{P}_M$ which minimizes the quantity $\|\boldsymbol{y} - p(\boldsymbol{t})\|_2$. While p can be expressed in any basis of $\mathbb{P}_M$, we choose the Chebyshev polynomial basis. Express $p = \sum_{l=0}^{M} c_l T_l$, where T_l is the scaled lth Chebyshev polynomial of the first kind, i.e., $T_l(t) = \cos(l \cos^{-1} \tau)$, where $\tau := 2(t - t_0)/(t_R - t_0) - 1 \in [-1, 1]$, for all $t \in [t_0, t_R]$.

Let $\boldsymbol{c} = (c_0, \ldots, c_M)^T$ be the vector of the unknown coefficients, then it satisfies the normal equations $\mathbf{T}_{R,M}^T \mathbf{T}_{R,M} \boldsymbol{c} = \mathbf{T}_{R,M}^T \boldsymbol{y}$. Here $\mathbf{T}_{R,M}$ stands for the $(R + 1) \times (M + 1)$ matrix containing $T_l(t_i)$ as its (i,l)-th entry. The least squares solution to this problem is $\boldsymbol{c} = \left(\mathbf{T}_{R,M}^T \mathbf{T}_{R,M}\right)^{-1} \mathbf{T}_{R,M}^T \boldsymbol{y}$, which gives p. In [2, Theorem 8] authors show that for $M \leq 0.5\sqrt{R}$ the condition number of $\mathbf{T}_{R,M}^T \mathbf{T}_{R,M}$ is linear in M, thus the normal equations provide a practical way to compute $\boldsymbol{c}$ and $p(t)$.

We let $\mathfrak{E}_{\epsilon,1}(f) = p$. Again, it is easy to check that $\mathfrak{E}_{\epsilon,1}$ admits the necessary properties and the corresponding $\mathfrak{E}_\epsilon$ is a valid order-M extrapolator.

3.3 Modification of F^m-transform

Let a (k, N)-eFPB $\mathbb{A}$ be fixed, with parameter $h = (b - a)/N$. Consider also the respective FPB of the extended interval $[\tilde{a}, \tilde{b}]$, denoted by $\bar{\mathbb{A}}$. Fix a family of order-M extrapolators $\{\mathfrak{E}_\epsilon\}_\epsilon$, with parameters $\delta > 0$ and $\delta' = 2k - 1$. We shall call the direct (composite) F^m-transform (w.r.t. fuzzy partition $\bar{\mathbb{A}}$ of $[\tilde{a}, \tilde{b}]$) of the extrapolant $\mathfrak{E}_h f$ a *modified F^m-transform*:

Definition 6. *Let $f \in C[a, b]$ and $\mathfrak{E}_h$ be an extrapolator of the family $\{\mathfrak{E}_\epsilon\}_\epsilon$ described above. Denote $\tilde{f} = \mathfrak{E}_h f$ and let $\overrightarrow{F_m}[\tilde{f}] = (\overrightarrow{F_{m,-k+1}}[\tilde{f}], \dots, \overrightarrow{F_{m,N+k-1}}[\tilde{f}])$ be the direct F^m-transform of $\tilde{f}$ w.r.t. $\bar{\mathbb{A}}$.*

We call this vector the direct modified F^m-transform *(or direct $\tilde{F}^m$-transform for short) of f, based on the extrapolator $\mathfrak{E}_h$, w.r.t. the fuzzy partition $\mathbb{A}$, and denote it by $\overrightarrow{\tilde{F}_m}[f] = (\overrightarrow{\tilde{F}_{m,-k+1}}[f], \dots, \overrightarrow{\tilde{F}_{m,N+k-1}}[f])$.*

The (composite) $\tilde{F}^m$-transform is defined as the usual inverse F^m-transform applied to the vector $\overrightarrow{\tilde{F}_m}[f]$ and denoted by $\tilde{F}_m[f]$.

This term is justified by the fact that this transform coincides with the usual direct F^m-transform for most entries (i.e., for all $i \in [k .. N - k]$) and we only modify the behavior of the transformation near the endpoints of $[a, b]$ (i.e., extend the corresponding basic functions and the function f outside $[a, b]$). Since we construct $\tilde{f}$ from the values of f in $[a, b]$, one can think of $\overrightarrow{F_m}[\tilde{f}]$ as a single-step transformation of f, instead of a two-step procedure (i.e., first constructing an extrapolant and then applying the direct F^m-transform).

Notice that the modified F^m-transform depends on the particular choice of the family of extrapolators. For the sake of simplicity, we shall not reflect this dependence in the notation $\overrightarrow{\tilde{F}_m}$.

Some immediate observations:

- $\tilde{F}_m$ is a linear operator, by the linearity of $\mathfrak{E}_h$ and the usual F^m-transform.
- We have $F_{m,i}^\rightarrow[\tilde{f}] = F_{m,i}^\rightarrow[f]$ for all $i \in [k .. N - k]$, since the corresponding basic function A_i is supported in $(t_{i-k}, t_{i+k}) \subset [a, b]$, where $f \equiv \tilde{f}$.
- If the modified F^m-transform is based on an order-$(2m+1)$ extrapolator and $f \in \mathbb{P}_{2m+1}$, then $F_m[\tilde{f}] \equiv f$ on $[a, b]$ (assuming k to be large enough, i.e., $k \geq m - 1$), since $[a, b]$ is the interval where the Ruspini condition holds for the considered FPB of $[\tilde{a}, \tilde{b}]$. This is due to Theorem 1 and the fact that an order-M extrapolant of any $f \in \mathbb{P}_M$ coincides with f. It immediately yields the following:

Theorem 2. *Let r, m be non-negative integers s.t. $r \leq \min\{2m + 1, 2k - 1\}$. In the discrete case we also assume $m \leq 2k - 2$, so that Δ is sufficiently dense. Assume the extrapolator order to be $M \geq 2m + 1$. Then every $f \in \mathbb{P}_r$ satisfies $f(t) = \tilde{F}_m[f](t)$ for all $t \in [a, b]$.*

4   Approximation of Polynomials and Smooth Functions

Throughout this section, we fix an interval $[a, b] \subset \mathbb{R}$, a family of order-$(2k - 1)$ extrapolators $\mathfrak{E}_\epsilon$, and use the same notation as in Sect. 3. Furthermore, F_m will stand for (either the integral or the discrete) F^m-transform w.r.t. $\mathbb{A}_0$ (unless stated otherwise), and $\tilde{F}_m$ for the modified transform (based on $\mathfrak{E}_h$) w.r.t. $\mathbb{A}$.

We shall sometimes use notation $f(\tau) \in \mathbb{P}_l[\tau]$ to underline that the expression f is a polynomial (of degree $\leq l$) w.r.t. the variable τ. Also, by convention, $\mathbb{P}_{-l} := \{0\}$ for $l > 0$ stands for the set containing only the zero function.

We proceed by briefly sketching a claim about the usual F^m-transform of polynomials w.r.t. the fuzzy partition $\mathbb{A}_0$.

Theorem 3. *Let r, m be non-negative integers satisfying $2m + 1 < r \leq 2k - 1$. Suppose that $f \in \mathbb{P}_r$; then the difference $f - F_m[f]$, when restricted to the interval $[\hat{a}, \hat{b}]$ where $\mathbb{A}_0$ satisfies the Ruspini condition, is a polynomial of degree at most $r - (2m + 2)$.*

Proof (sketch of). Express $f - F_m[f] = \sum_{l=m+1}^{r} \sum_{i=0}^{N} c_{i,l} \mathrm{P}_{i,l}(t) A_i(t)$, where $\mathrm{P}_{i,l}(t) = \mathrm{P}_l((t - t_i)/h)$ and $\mathrm{P}_l \in \mathbb{P}_l$ are the orthogonal basis polynomials corresponding to the inner product associated with the weight function A (for precise definitions and properties of P_l, see [9]).

Thus it suffices to show that in the interval $[\hat{a}, \hat{b}]$ for all l the function $\sum_{i=0}^{N} c_{i,l} \mathrm{P}_{i,l}(t) A_i(t)$ is in $\mathbb{P}_{r-2l}$. Using the fact [9, Lemma 3] that $c_{i,l} = q_l(i)$ for some $q \in \mathbb{P}_{r-l}$, it suffices to show that for all $p \in \mathbb{P}_n$, $n \leq 2k - 1 - l$, and all $\tau \in [0, 1]$ we have $\sum_{i=1-k}^{k} p(i) \mathrm{P}_l(\tau - i) A(\tau - i) \in \mathbb{P}_{n-l}[\tau]$.

By [9, Lemma 19], it is equivalent to $I(\tau) := \int_{-k}^{k} p(x + \tau) \mathrm{P}_l(x) A(x) \, \mathrm{d}x \in \mathbb{P}_{n-l}[\tau]$. Express $p(x + \tau)$ as $\sum_{j=0}^{n} p_j(x) q_{n-j}(\tau)$, where $\deg p_j \leq j$, $\deg q_j \leq j$. Letting $\alpha_j := \int_{-k}^{k} p_j(x) \mathrm{P}_l(x) A(x) \, \mathrm{d}x$, we obtain that the integral $I(\tau)$ equals to the sum $\sum_{j=0}^{n} \alpha_j q_{n-j}(\tau)$. Moreover, $\alpha_j = 0$ for $j \leq l - 1$ by [9, Lemma 4], therefore $I(\tau) = \sum_{j=l}^{n} \alpha_j q_{n-j}(\tau) \in \mathbb{P}_{n-l}[\tau]$, concluding the argument. $\square$

Taking into account that for $f \in \mathbb{P}_{2k-1}$ the modified F^m-transform, based on order-$(2k - 1)$ extrapolators, is equivalent to the usual F^m-transform w.r.t. the fuzzy partition $A_{-k+1}, \ldots, A_{N+k-1}$ of $[\tilde{a}, \tilde{b}]$, we can conclude from Theorems 1 and 3 that.

Theorem 4. *Let r, m be non-negative integers and suppose $r \leq 2k - 1$. Let $f \in \mathbb{P}_r$; then $f - \tilde{F}_m[f]$ is a polynomial of degree at most $r - (2m + 2)$.*

This somewhat technical result has a consequence (which can be shown by a simple induction on r) which becomes important when applying the modified F-transform to solve numerically boundary value problems via the collocation method.

Corollary 1. *For every $p \in \mathbb{P}_r$, $r \leq 2k - 1$, there is a polynomial $f \in \mathbb{P}_r$ such that $\tilde{F}_m[f] = p$.*

Remark 2. When approximating smooth functions with their F^m-transforms, we shall fix m, k and consider the F^m-transforms of f w.r.t. a sequence of (k, N)-FPBs (with $N \to \infty$). In this regime, let a sequence of FPBs be fixed and denote by $F_N^m[f]$ the F^m-transform of f w.r.t. the (k, N)-FPB. For each N we can view

F_N^m as an operator mapping $C[a, b]$ into itself. Then it is easy to show that these operators are uniformly bounded.

Theorem 5. *Let non-negative integers r, m satisfy $r \leq \min\{2m + 1, 2k - 1\}$. Then for all $f \in C^{r+1}[a, b]$ it holds that $\left\| f - \tilde{F}_m[f] \right\| = O(h^{r+1})$.*

Proof. Fix any $f \in C^{r+1}[a, b]$ and $N \geq 4k-1$. As usual, we let $h = (b-a)/N$ and $\tilde{a} = a - (2k-1)h$, $\tilde{b} = b + (2k-1)h$; furthermore, $\bar{A}$ stands for the FPB$(k, N + 4k - 2)$ of $[\tilde{a}, \tilde{b}]$.

Assume $r \geq 1$. Apply the rth order Taylor's formula with the integral form of the remainder, with a as the center of expansion. Let $p \in \mathbb{P}_r$ be the corresponding Taylor polynomial for f, then

$$f(x) = p(x) + \int_a^b f^{(r+1)}(t)\, K_t(x)\, dt, \quad K_t(x) := \begin{cases} \frac{(x-t)^r}{r!}, & t \leq x \leq b \\ 0, & a \leq x < t. \end{cases}$$

Now, $(f - \tilde{F}_m[f])(x) = \int_a^b f^{(r+1)}(t) \left(K_t - \tilde{F}_m[K_t] \right)(x)\, dt$, by Theorem 2 and the linearity of $\tilde{F}_m$. Notice that the extrapolant $\mathfrak{E}_h(f)$ is not required to be differentiable, since Taylor's formula is applied only to f. Hence we can upper-bound $\left\| f - \tilde{F}_m[f] \right\|$ by $\left\| f^{(r+1)} \right\| \cdot \sup_{x \in [a,b]} \int_a^b \left| K_t(x) - \tilde{F}_m[K_t](x) \right| dt$, or, since $K_t(x) - \tilde{F}_m[K_t](x) = 0$ unless $|x - t| \leq \max\{2kh, \delta h\}$,

$$\left\| f - \tilde{F}_m[f] \right\| \leq \left\| f^{(r+1)} \right\| \cdot h \max\{2k, \delta\} \cdot \sup_{x,t \in [a,b]} \left| K_t(x) - \tilde{F}_m[K_t](x) \right|. \quad (4)$$

Let $F_m[g]$ stand for the usual F^m-transform of a function $g \in C[\tilde{a}, \tilde{b}]$ w.r.t. $\bar{A}$. To estimate the maximum in (4), denote $\hat{K}_t := \mathfrak{E}_h(K_t)$, then we need to estimate the difference $K_t - \tilde{F}_m[K_t] = K_t - F_m[\hat{K}_t]$. Even though K_t is formally defined only in $[a, b]$, we can naturally extend its definition to the whole $[\tilde{a}, \tilde{b}]$: $K_t(x) = 0$ for $x \in [\tilde{a}, t]$ and $K_t(x) = (x - t)^r/r!$ for $x \in [t, \tilde{b}]$.

We recall the following fact about the ordinary F^m-transform of K_t which follows from [9, Lemma 11]: if $|x - t| > 2kh$, then $K_t(x) - F_m[K_t](x) = 0$, otherwise this difference is of order $O(h^r)$ (because then $K_t(x) = O(h^r)$). Let us show that there is a constant $C_0 > 0$ (independent of h) such that $\left| K_t(x) - \hat{K}_t(x) \right| \leq C_0 h^r$ for all $x \in [\tilde{a}, \tilde{b}]$, $t \in [a, b]$. Then, in view of Remark 2, we get $K_t - F_m[\hat{K}_t] = (K_t - F_m[K_t]) + F_m[\hat{K}_t - K_t] = O(h^r)$. Together with (4) this will conclude the proof.

Let $\delta > 0$ and $\delta' = 2k - 1$ be the corresponding parameters of the family $\{\mathfrak{E}_\epsilon\}_\epsilon$. By the definition of $\mathfrak{E}_\epsilon$, we have $K_t(x) - \hat{K}_t(x) = 0$ unless $x < a$ or $x > b$. There are three cases to consider: $t \in [a, a + \delta h)$, $t \in (b - \delta h, b]$, and $t \in [a + \delta h, b - \delta h]$. The last case is trivial (since then $\hat{K}_t \equiv K_t$). In either of the former two cases we have $K_t(x) \leq \delta^r h^r$ when $x \in [a, a + \delta h]$ (in the first case) or when $x \in [b - \delta h, b]$ (in the second case). From the uniform boundedness of $\{\mathfrak{E}_{\epsilon,1}\}_\epsilon$ and $\{\mathfrak{E}_{\epsilon,2}\}_\epsilon$ we have

$$\max_{x \in [a-\delta'h, a]} \left| \hat{K}_t(x) \right| \leq C_\star \delta^r h^r \quad \text{and} \quad \max_{x \in [b, b+\delta'h]} \left| \hat{K}_t(x) \right| \leq C_\star \delta^r h^r.$$

Since $K_t(x) = 0 = O(h^r)$ for $x < a$ and, in the second case, $K_t(x) \le ((\delta' + \delta)h)^r$ for $x \in [b, b + \delta'h]$, it suffices to take $C_0 = (\delta' + \delta)^r + C_\star \delta^r$.

When $r = 0$, the previous arguments fail, since $\mathfrak{E}_h$ was defined for continuous functions. Nevertheless, one can generalize $\mathfrak{E}_h$ to apply them to piecewise continuous functions and similar arguments give the desired $\left\| f - \tilde{F}_m[f] \right\| = O(h)$.

$\square$

For illustrative purposes we consider approximating $f_1(x) = \sin^2(\pi x)$ and $f_2(x) = \sin(\exp(4x))$ over the interval $[0, 1]$ by the modified F^m-transform, for $m \in \{0, 1\}$ and $N \in \{10, 10^2, 10^3, 10^4\}$ (then $h = 1/N$). In these examples the FPBs are given by cubic B-splines (i.e., we consider the $(2, N)$-eFPB of $[0, 1]$).

In each case we compute the absolute error $(f - \tilde{F}_m[f])(t)$ at the nodes t_i, then calculate the maximum among the obtained values. The results for each N, m and $f \in \{f_1, f_2\}$ are depicted in Table 1. The numerical data support the prediction by Theorem 5 that $\tilde{F}_0$-transform approximates f with $O(h^2)$-accuracy, but $\tilde{F}_1$-transform approximates f with $O(h^4)$-accuracy, as both functions are from $C^\infty[a, b]$.

Table 1. Maximum absolute error at nodes

N	10	10^2	10^3	10^4
f_1, $m = 0$	8.33e − 02	6.60e − 04	6.58e − 06	6.58e − 08
f_1, $m = 1$	4.05e − 03	4.33e − 07	4.33e − 11	4.55e − 15
f_2, $m = 0$	1.79e + 00	1.47e + 00	1.52e − 02	1.47e − 04
f_2, $m = 1$	1.93e + 00	4.95e − 01	1.04e − 04	1.09e − 08

In Fig. 3 we also depict the function f_1 and both the usual F_0-transform of f_1 (w.r.t. the FPB $(2, 10)$) and the modified F_0-transform of f_1 (w.r.t. the corresponding $(2, 10)$-eFPB). If $t \in [0.3, 0.7]$, then $F[f_1](t) = \tilde{F}[f_1](t)$ and both functions approximate f_1; however, when t is outside this interval, then $F[f_1]$ approximates f_1 poorly, whereas $\tilde{F}[f_1]$ is still close to the original function f_1.

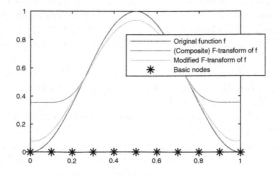

Fig. 3. The function f_1 and its approximation by $F[f_1]$ and $\tilde{F}[f_1]$, for $k = 2$, $N = 10$

5   Conclusion

The proposed modification of the spline-based F-transform technique allows us to eliminate the main obstacle to effective use the advantages of this technique in applications. As it was mentioned before, applications of the F-transforms in numerically solving boundary value problems require high quality of approximation on the whole interval under consideration and especially near the boundary points. Our future research is devoted to such applications of the modified spline-based F-transform.

References

1. Bede, B., Rudas, I.J.: Approximation properties of fuzzy transforms. Fuzzy Sets Syst. **180**(1), 20–40 (2011)
2. Demanet, L., Townsend, A.: Stable extrapolation of analytic functions. arXiv preprint (2016). arXiv:1605.09601
3. Holčapek, M., Tichý, T.: Discrete fuzzy transform of higher degree. In: IEEE International Conference on Fuzzy Systems, FUZZ-IEEE 2014, 6–11 July 2014, Beijing, China, pp. 604–611. IEEE (2014)
4. Kodorane, I., Asmuss, S.: On approximation properties of spline based F-transform with respect to fuzzy m-partition. In: Montero, J., Pasi, G., Ciucci, D. (eds.) Proceedings of EUSFLAT 2013, Advances in Intelligent Systems Research, Milan, Italy, vol. 32, pp. 772–779. Atlantis Press (2013)
5. Kokainis, M., Asmuss, S.: Approximation properties of higher degree F-transforms based on B-splines. In: IEEE International Conference on Fuzzy Systems, FUZZ-IEEE 2015, 2–5 August 2015, Istanbul, Turkey. IEEE (2015)
6. Kokainis, M., Asmuss, S.: Discrete higher degree F-transforms based on bivariate B-splains. In: Proceedings of the 12th International FLINS Conference on Uncertainty Modelling in Knowledge Engineering and Decision Making, 24–26 August 2016, Roubaix, France, pp. 264–269. World Scientific (2016)
7. Kokainis, M., Asmuss, S.: Higher degree F-transforms based on B-splines of two variables. In: Carvalho, J.P., Lesot, M.-J., Kaymak, U., Vieira, S., Bouchon-Meunier, B., Yager, R.R. (eds.) IPMU 2016. CCIS, vol. 610, pp. 648–659. Springer, Cham (2016). https://doi.org/10.1007/978-3-319-40596-4_54
8. Kokainis, M., Asmuss, S.: Approximation by multivariate higher degree F-transform based on B-splines. Soft Comput. **21**(13), 3587–3614 (2017)
9. Kokainis, M., Asmuss, S.: Continuous and discrete higher-degree F-transforms based on B-splines. Soft Comput. **21**(13), 3615–3639 (2017)
10. Perfilieva, I.: Fuzzy transforms: theory and applications. Fuzzy Sets Syst. **157**(8), 993–1023 (2006)
11. Perfilieva, I.: F-Transform. In: Kacprzyk, J., Pedrycz, W. (eds.) Springer Handbook of Computational Intelligence, pp. 113–130. Springer, Heidelberg (2015). https://doi.org/10.1007/978-3-662-43505-2
12. Perfilieva, I., Daňková, M., Bede, B.: Towards a higher degree F-transform. Fuzzy Sets Syst. **180**(1), 3–19 (2011)
13. Perfilieva, I., Haldeeva, E.: Fuzzy transformation and its applications. In: 2001 Proceedings of the 4th Czech-Japan Seminar on Data Analysis and Decision Making under Uncertainty, Czech Republic, pp. 116–124 (2001)
14. Schoenberg, I.J.: Cardinal spline interpolation. In: CBMS, vol. 12. SIAM (1973)

Collocation Method for Linear BVPs via B-spline Based Fuzzy Transform

Martins Kokainis[1(✉)] and Svetlana Asmuss[1,2]

[1] Department of Mathematics, University of Latvia, Zellu 25, Riga 1002, Latvia
{martins.kokainis,svetlana.asmuss}@lu.lv
[2] Institute of Mathematics and Computer Science, University of Latvia,
Raina bulvaris 29, Riga 1459, Latvia

Abstract. The paper is devoted to an application of a modified F-transform technique based on B-splines in solving linear boundary value problems via the collocation method. An approximate solution is sought as a composite F-transform of a discrete function (which allows the solution to be compactly stored as the values of this discrete function). We demonstrate the effectiveness of the described technique with numerical examples, compare it with other methods and propose theoretical results on the order of approximation when the fuzzy partition is based on cubic B-splines.

Keywords: Fuzzy transform · Boundary value problem · Collocation

1 Introduction

In this paper we deal with linear two-point second-order boundary value problems (BVPs for short), subject to linear boundary conditions.

We assume that the reader is familiar with the concept of the ordinary F-transform introduced by I. Perfilieva in 2001 [14] (see also the key paper [11]) and its extension to the higher degree F-transform [13] with degree-m polynomial components (the F^m-transform). We work with fuzzy transforms based on B-splines as basic functions which form a fuzzy partition [7,8] (see also papers [3,6] dealing with fuzzy partitions based on polynomial splines).

The idea to use F-transforms (or their higher degree counterparts) to solve numerically differential equations is not new. For instance, in 2005 the fuzzy transform was applied [19] to solve numerically partial differential equations; more recently, in [5] the authors proposed numerical solutions of Cauchy problems based on fuzzy transform. F-transform based shooting method for solving BVPs (either linear or non-linear) was discussed in [15]. In [16] the authors consider numerically solving second order BVPs with Dirichlet boundary conditions via the F-transform method and the advantages of employing this method.

In contrast to the aforementioned papers, the main idea of our proposal is to ensure that it is the composite F-transform (i.e., the inverse transform of

© Springer International Publishing AG, part of Springer Nature 2018
J. Medina et al. (Eds.): IPMU 2018, CCIS 854, pp. 187–198, 2018.
https://doi.org/10.1007/978-3-319-91476-3_16

the direct F-transform) which solves the BVP. This allows to exploit the good approximation properties provided by the composite transform. The main obstacle to use the spline-based fuzzy partition introduced in [7] (and generalized to for the discrete transform in [8]) is that it does not satisfy the Ruspini condition (where the composite transform approximates the original function) on the whole interval. To deal with this, we introduce a modification of the aforementioned fuzzy partition, as well modify the F-transform itself near the endpoints of the interval.

To actually solve the BVP, we apply the collocation method; the proof techniques are based on the scheme described in [17]. Somewhat unexpectedly, it turns out that the ordinary F-transform already provides high approximation order of the exact solution (and thus it is not necessary to employ higher degree F-transforms). Even though we have considered only linear problems with linear boundary conditions, we expect that these results can be generalized to the non-linear setting as well.

2   Preliminaries

By $[n..m]$ (for integers n, m with $n \leq m$) we denote the set $\{n, n+1, \ldots, m\}$. Let $\mathbb{P}_l$ stand for the space of univariate polynomials of degree at most l.

2.1   Fuzzy Partition Based on B-splines

Central B-splines [18] are even B-splines that have 1-equidistant knots. For a fixed degree the central B-spline is unique (up to a constant factor). The properties of B-splines and construction of a fuzzy partition using central B-spline as the generating function are described in more details in [8].

Definition 1. *The central B-spline of degree $2k - 1$, denoted by ϕ_{2k-1}, is the unique piecewise polynomial function satisfying the following requirements: (1) for each $i \in [-k..k-1]$ the restriction of ϕ_{2k-1} to $[i, i+1]$ is a polynomial of degree at most $2k - 1$; (2) $\phi_{2k-1} \in C^{2k-2}(\mathbb{R})$; (3) $\phi_{2k-1}(t) = 0$ if $t \notin (-k, k)$; (4) $\int_{\mathbb{R}} \phi_{2k-1}(t) \, dt = 1$.*

Fix $N, k \in \mathbb{N}$ such that $N \geq 4k - 1$; let $A = \phi_{2k-1}$. Let an interval $[a, b]$ be fixed; denote $h = (b - a)/N$ and define h-equidistant nodes $t_i = a + hi$, $i \in [0..N]$.

Define $A_i(t) := A\left(\frac{t-t_i}{h}\right)$ (called the basic functions), $i \in [k..N-k]$. Then the basic functions $A_k, \ldots, A_{N-k}$ form a generalized (h, hk)-uniform fuzzy partition (FP) of $[a, b]$, as defined in, e.g., [12]. Let us denote this fuzzy partition by $\mathbb{A}_0$. We refer to $\mathbb{A}_0$ as the FPB (fuzzy partition based on B-splines), or, more specifically, $\text{FPB}(k, N)$.

Recall that a fuzzy partition is said to fulfill the Ruspini condition on some interval $I \subset [a, b]$ if the basic functions sum up to 1 in this interval. In the case of FPB, the Ruspini condition is fulfilled on $[t_{2k-1}, t_{N-2k+1}]$.

Throughout the rest of this section, fix an interval $[a, b] \subset \mathbb{R}$, positive integers k and N, $N \geq 4k - 1$, and $\mathbb{A}_0 = \{A_k, \ldots, A_{N-k}\}$, which is the $\text{FPB}(k, N)$ of $[a, b]$.

2.2   Discrete F-transform

Let $\Delta = \{z_1, \ldots, z_L\} \subset [a, b]$ be a discrete set of the interval $[a, b]$ and $f : \Delta \to \mathbb{R}$. Denote $y_j = f(z_j)$, for each $j \in [1 .. L]$. The set Δ is said to be *sufficiently dense* in the fuzzy partition $\mathbb{A}_0$ if for every basic function $A_i \in \mathbb{A}_0$ there is a discrete point $z_j \in \Delta$, $j \in [1 .. L]$, s.t. it belongs to the fuzzy set A_i with nonzero degree: $A_i(z_j) > 0$.

If the set Δ is sufficiently dense, then the discrete direct F-transform can be defined:

Definition 2 [10]. *The vector $F^{\to}[f]$ is the discrete direct F-transform of function f w.r.t. $\mathbb{A}_0$, if the ith component of this vector is*

$$F_i^{\to}[f] = \frac{\sum_{j=1}^{L} A_i(z_j) y_j}{\sum_{j=1}^{L} A_i(z_j)}, \quad i \in [k .. N - k].$$

One can define the inverse F-transform, which is applied to a vector in $\mathbb{R}^{N-2k+1}$. Composing the direct and inverse F-transforms one obtains the composite F-transform. For the purposes of this paper, we define only the latter.

Definition 3. *Let $f : \Delta \to \mathbb{R}$. Suppose that the direct discrete F-transform of f w.r.t. $\mathbb{A}_0$ is $F^{\to}[f] = (F_k^{\to}[f], \ldots, F_{N-k}^{\to}[f]) \in \mathbb{R}^{N-2k+1}$. Then the function*

$$F[f](t) = \frac{\sum_{i=k}^{N-k} F_i^{\to}[f] A_i(t)}{\sum_{i=k}^{N-k} A_i(t)}, \quad t \in [a, b],$$

is called the (composite) *F-transform of f w.r.t. $\mathbb{A}_0$.*

The main advantage of the FPB comparing to an arbitrary fuzzy partition is that the composite F^m-transform (the generalization of the ordinary F-transform) is exact for polynomials of degree $2m + 1$ (as long as it does not exceed the respective spline degree), as opposed to being exact for polynomials of degree m for a general FP. In the context of the ordinary F-transform, though, this means that the composite transform is exact for the linear polynomials, when the FP is based on B-splines.

Theorem 1. *Let the discrete set Δ consists of the basic nodes t_i, $i \in [0 .. N]$. Let $F[f]$ stand for the discrete F-transform of $f : \Delta \to \mathbb{R}$ w.r.t. $\mathbb{A}_0$. If $f \in \mathbb{P}_1$, then $f(t) = F[f](t)$ for all $t \in [t_{2k-1}, t_{N-2k+1}]$ (i.e., in the interval where the FPB $\mathbb{A}_0$ satisfies the Ruspini condition).*

3   Extended FPB and the Modified F-transform

The aim of this section is to briefly describe how we modify the FPB and fuzzy transform near the endpoints of $[a, b]$ to achieve that the Ruspini condition holds on the whole $[a, b]$ and an analogue of Theorem 1 also applies on the whole $[a, b]$.

The main drawback of the fuzzy partition $FPB(k, N)$ is that it does not fulfill the Ruspini condition on the whole $[a, b]$, which proves to be a forbidding obstacle for many applications. It can be remedied by introducing additional basic functions, generated by the same generating function A, and restricting them to $[a, b]$. Throughout the rest of this paper we introduce the following notation and concepts, in addition to those described in Sect. 2.1.

- $t_i = a + ih$, with $i \in [-2k+1 \ldots -1]$ and $i \in [N+1 .. N+2k-1]$: additional nodes outside the interval $[a, b]$.
- $\tilde{a} := t_{-2k+1} = a - (2k-1)h$ and $\tilde{b} := t_{N+2k-1} = b + (2k-1)h$: the first and the last of the additional nodes.
- $A_i(t) = A((t - t_i)/h)$, $i \in [-k+1 \ldots N+k-1]$: additional basic functions.
- $\mathbb{A}$: the FP of $[a, b]$, formed by the basic functions $A_{-k+1}|_{[a,b]}, \ldots, A_{N+k-1}|_{[a,b]}$.
- $\bar{\mathbb{A}}$: the $FPB(k, N+4k-2)$ of the interval $[\tilde{a}, \tilde{b}]$, formed by the basic functions $A_{-k+1}, \ldots, A_{N+k-1}$.

Fix also the set of basic nodes $\Delta = \{t_0, t_1, \ldots, t_N\} \subset [a, b]$ and the extended set $\tilde{\Delta} = \{t_{-2k+2}, \ldots, t_{N+2k-2}\} \supset \Delta$.

From now on, we will refer to the fuzzy partition $\mathbb{A}$ as the (k, N)-eFPB of $[a, b]$. Notice that $\mathbb{A}$ of $[a, b]$ satisfies the Ruspini condition on the whole $[a, b]$. Furthermore, $\bar{\mathbb{A}}$ (i.e., the FPB of $[\tilde{a}, \tilde{b}]$) satisfies the Ruspini condition on the interval $[a, b]$.

This method, however, does not preserve the aforementioned generalization of Theorem 1, i.e., the composite F^m-transform is not exact for $p \in \mathbb{P}_{2m+1}$ anymore. The solution is to extrapolate the function f to some function $\tilde{f} : \tilde{\Delta} \to \mathbb{R}$. We require that $\tilde{f}$ agrees with f in $[a, b]$ and coincides with f everywhere if $f \in \mathbb{P}_{2k-1}$ (for the sake of generalization of Theorem 1, it suffices to require that for $f \in \mathbb{P}_{2m+1}$; however, this stronger version has its own advantages).

In this paper, we construct $\tilde{f}$ in the simplest way (albeit not the most effective nor stablest way from the viewpoint of numerical methods), by computing the Lagrange polynomial $p \in \mathbb{P}_{2k-1}$ satisfying $p(t_j) = f(t_j)$, $j \in [0 .. 2k-1]$, and letting $\tilde{f}(t_j) = p(t_j)$ for $j < 0$. The values $\tilde{f}(t_j)$ for $j > N$ are defined similarly.

We shall call the direct (composite) F-transformation (w.r.t. $\bar{\mathbb{A}}$ of $[\tilde{a}, \tilde{b}]$) of the extrapolant $\tilde{f}$ a *modified F-transform* of f:

Definition 4. *Let $f : [a, b] \to \mathbb{R}$. Suppose that $\tilde{f}$ is constructed as described previously. Let $F^{\to}[\tilde{f}] = (F^{\to}_{-k+1}[\tilde{f}], \ldots, F^{\to}_{N+k-1}[\tilde{f}])$ be the direct F-transform of $\tilde{f}$ w.r.t. $\bar{\mathbb{A}}$. We call the mapping $f \mapsto F^{\to}[\tilde{f}]$ a modified F-transform or $\tilde{F}$-transform based on the Lagrange extrapolator w.r.t. $\mathbb{A}$, and denote $\tilde{F}^{\to}[f] = F^{\to}[\tilde{f}]$, $\tilde{F}^{\to}_i[f] = F^{\to}_i[\tilde{f}]$. By the term composite $\tilde{F}$-transform we understand the usual inverse F-transform applied to the vector $\tilde{F}^{\to}[f]$.*

This term is justified by the fact that this transform coincides with the usual F-transform for most entries; since we construct $\tilde{f}$ from the values of f in $[a, b]$, one can think of $F^{\to}[\tilde{f}]$ as a single-step transformation of f, instead of two-step procedure.

Now we have a FP which fulfills the Ruspini condition on the whole $[a, b]$, moreover, the $\tilde{F}$-transform (and its higher-degree generalization) has the desired analogue of Theorem 1. But it turns out that there is another, previously unrecognized property of this transform, which turns out to be essential for our applications in this paper:

Proposition 1. *For every $p \in \mathbb{P}_r$, $r \leq 2k - 1$, there is a polynomial $f \in \mathbb{P}_r$ such that $\tilde{F}[f] = p$.*

While its proof is somewhat lengthy, we note that for the $(2, N)$-eFPB we consider in the next section, it suffices to prove this for $k = 2$. This, however, can be checked directly by noting that $\tilde{F}[p] = p$ whenever $\deg p \leq 1$; and $\tilde{F}[t^2 + 2h^2/3] = p$ for $p(t) = t^2$, and $\tilde{F}[t^3 + h^2 t] = p$ for $p(t) = t^3$.

4   Collocation with Composite $\tilde{F}$-transform

Let $e_0, e_1, f \in C[a, b]$. Define a differential operator $\mathfrak{L}u := u'' + e_1 u' + e_0 u$ and boundary conditions (BC) operators $\mathfrak{B}_1, \mathfrak{B}_2$, where $\mathfrak{B}_j u := \alpha_{0,j} u(a) + \alpha_{1,j} u'(a) + \beta_{0,j} u(b) + \beta_{1,j} u'(b)$, $j = 1, 2$, and $\alpha_{i,j}, \beta_{i,j}$ are some reals. Consider the corresponding linear differential equation subject to linear homogeneous boundary conditions (notice that there is no loss of generality in assuming homogeneous boundary conditions):

$$\mathfrak{L}u(s) = f(s), \ a < s < b; \quad \mathfrak{B}_1 u = \mathfrak{B}_2 u = 0. \tag{1}$$

Fix a (k, N)-eFPB of $[a, b]$, denoted by $\mathbb{A}_h$, where $h = (b - a)/N$. Then we can consider the following problem: find a function $g : \Delta \to \mathbb{R}$ such that its $\tilde{F}$-transform w.r.t. $\mathbb{A}_h$ (which we denote by $u_h := \tilde{F}[g]$) satisfies the differential equation $\mathfrak{L}u_h = f$ at the inner basic nodes (i.e., for $s = t_i$, $i \in [1 .. N - 1]$), as well as the boundary conditions $\mathfrak{B}_1 u_h = \mathfrak{B}_2 u_h = 0$.

When the FP is based on cubic B-splines (i.e., $k = 2$) and assuming that the $\tilde{F}$-transform is based on the Lagrange extrapolator, we show that the described collocation problem is uniquely solvable (under standard requirements on the boundary value problem (1)). Moreover, the collocation solution approximates the exact solution u (and its first and second derivative) with order $O(h^2)$. We also conjecture that the statement remains true for higher degree B-splines (i.e., for $k > 2$) and that the order of approximation then is $O(h^{2k-2})$.

Before we proceed, let us introduce some additional notation:

- Spaces $\mathbb{U}_0$ and $\mathbb{U}_2$: let $\mathbb{U}_0$ be the linear space of $\tilde{F}$-transforms of all discrete functions defined on Δ w.r.t. the fuzzy partition $\mathbb{A}_h$, and let $\mathbb{U}_2$ contain the second derivatives of functions in $\mathbb{U}_0$:

$$\mathbb{U}_0 = \left\{ \tilde{F}(g) \,\middle|\, g : \Delta \to \mathbb{R} \right\}, \quad \mathbb{U}_2 = \{ v \mid \exists u \in \mathbb{U}_0 : v \equiv u'' \}.$$

- Operator $\mathfrak{P}_h : C[a, b] \to \mathbb{U}_2$: the linear projector which maps every continuous function f to the unique $u \in \mathbb{U}_2$ satisfying $u(t_i) = f(t_i)$, $i \in [1 .. N - 1]$ (the fact that the operator is well-defined is shown in Proposition 2).

In the following, $\|u\|$, where $u \in C[a, b]$, stands for the usual supremum norm.

Theorem 2. *Let $e_0, e_1, f \in C[a, b]$. Suppose that the BVP (1) has a unique solution and the problem $u'' = 0$ with boundary conditions $\mathfrak{B}_1 u = \mathfrak{B}_2 u = 0$ has only the trivial solution. Let $N \in \mathbb{N}$ be sufficiently large and let $\mathbb{A}_h$ be a $(2, N)$-eFPB of $[a, b]$. Let Δ be as in Sect. 3; let $\tilde{F}$-transform stand for the $\tilde{F}$-transform based on the Lagrange extrapolator w.r.t. $\mathbb{A}_h$.*

Then there is a unique function $g : \Delta \to \mathbb{R}$ such that its $\tilde{F}$-transform, denoted by u_h, satisfies the following constraints:

$$\mathfrak{L}u_h(t_i) = f(t_i), \ i \in [1 .. N - 1]; \quad \mathfrak{B}_1 u_h = \mathfrak{B}_2 u_h = 0. \tag{2}$$

Moreover, for all $j \in \{0, 1, 2\}$ the following estimation holds:

$$\left\| u^{(j)} - u_h^{(j)} \right\| = O(\|u'' - \mathfrak{P}_h(u'')\|) \xrightarrow[h \to 0]{} 0, \tag{3}$$

where u is the unique solution of the BVP (1) and $\mathfrak{P}_h$ is the operator introduced previously. In particular, if $u \in C^4[a, b]$ (which is guaranteed, e.g., when $e_0, e_1, f \in C^2[a, b]$), then

$$\left\| u^{(j)} - u_h^{(j)} \right\| = O(h^2), \quad j \in \{0, 1, 2\}. \tag{4}$$

Proposition 2. *Operator $\mathfrak{P}_h$ is well-defined.*

Proof. We will show that for every $f \in C[a, b]$ there is a unique $g : \Delta \to \mathbb{R}$ such that $\tilde{F}[g](a) = \tilde{F}[g](b) = 0$ and $(\tilde{F}[g])''(t_i) = f(t_i)$, $i \in [1 .. N - 1]$. Then $\mathfrak{P}_h f$ necessarily equals $(\tilde{F}[g])''$.

For simplicity, denote $\tilde{F}_i^{\to}[g]$ by $\tilde{F}_i^{\to}$. Observe that for all $i \in [0 .. N]$

$$\tilde{F}[g](t_i) = \tilde{F}_{i-1}^{\to} A_{i-1}(t_i) + \tilde{F}_i^{\to} A_i(t_i) + \tilde{F}_{i+1}^{\to} A_{i+1}(t_i),$$
$$(\tilde{F}[g])''(t_i) = \tilde{F}_{i-1}^{\to} A_{i-1}''(t_i) + \tilde{F}_i^{\to} A_i''(t_i) + \tilde{F}_{i+1}^{\to} A_{i+1}''(t_i).$$

Specifically, since A is the central cubic B-spline, $A_{i\pm1}(t_i) = 1/6$, $A_i(t_i) = 2/3$ and $A_{i\pm1}''(t_i) = 1/h^2$, $A_i''(t_i) = -2/h^2$. Hence we need to show that for every $f \in C[a, b]$ there is a unique $g : \Delta \to \mathbb{R}$ such that

$$\begin{cases} \tilde{F}_{-1}^{\to} + 4\tilde{F}_0^{\to} + \tilde{F}_1^{\to} = 0 \\ \tilde{F}_{i-1}^{\to} - 2\tilde{F}_i^{\to} + \tilde{F}_{i+1}^{\to} = h^2 f(t_i), \quad i \in [1 .. N - 1] \\ \tilde{F}_{N-1}^{\to} + 4\tilde{F}_N^{\to} + \tilde{F}_{N+1}^{\to} = 0. \end{cases} \tag{5}$$

Denote $y_i = g(t_i)$, $i \in [0 .. N]$. To find $\tilde{F}_0^{\to}$ and $\tilde{F}_{-1}^{\to}$, one must find $y_{-1} := p(a - h)$ and $y_{-2} := p(a - 2h)$, where p is the unique polynomial in $\mathbb{P}_3$ satisfying $p(t_i) = y_i$ for $i \in [0 .. 3]$. Explicit calculation yields $y_{-1} = 4y_0 - 6y_1 + 4y_2 - y_3$ and $y_{-2} = 10y_0 - 20y_1 + 15y_2 - 4y_3$. Similarly one finds also y_{N+1} and y_{N+2}. Then $\tilde{F}_i^{\to} = (y_{i-1} + 4y_i + y_{i+1})/6$ for all i. After simplifying the system (5), we conclude that $\mathfrak{P}_h$ is well-defined iff for every f there is a unique vector $\mathbf{y} = (y_0, \ldots, y_N)^T \in \mathbb{R}^{N+1}$ s.t.

$$\begin{cases} 5y_0 - 5y_1 + 4y_2 - y_3 = 0 \\ y_0 - 2y_1 + y_2 = h^2 f(t_1) \\ y_{i-2} + 2y_{i-1} - 6y_i + 2y_{i+1} + y_{i+2} = 6h^2 f(t_i), \quad i \in [2 .. N - 2] \\ y_{N-2} - 2y_{N-1} + y_N = h^2 f(t_{N-1}) \\ -y_{N-3} + 4y_{N-2} - 5y_{N-1} + 5y_N = 0, \end{cases} \quad (6)$$

which is equivalent to the non-singularity of the matrix $\mathbf{S}$, corresponding to the system (6).

The rows of $\mathbf{S}$, except the first and the last row, are weakly diagonally dominant. The first row $(5\ {-}5\ 4\ {-}1\ 0\ \ldots\ 0)$, however, is not. Let $\mathbf{s}_j$ stand for the jth row of $\mathbf{S}$; replace $\mathbf{s}_1$ with $4\ \mathbf{s}_1 - 10\mathbf{s}_2 + \mathbf{s}_3$ and $\mathbf{s}_{N+1}$ with $4\mathbf{s}_{N+1} - 10\mathbf{s}_N + \mathbf{s}_{N-1}$, obtaining a matrix $\mathbf{S}^\star$. The matrix $\mathbf{S}^\star$ is non-singular, since it is a w.c.d.d. matrix [2, Lemma 3.2]. Hence also $\mathbf{S}$ is non-singular. We conclude that the system (6) is uniquely solvable, thus $\mathfrak{P}_h$ is well-defined. □

Before we can estimate the accuracy of approximation by $\mathfrak{P}_h$, we need the following technical result, whose proof we only briefly sketch.

Proposition 3. *For all $f \in C^2[a, b]$ there is $h_0 > 0$ s.t. $f(a) - S_h(f) = O(h^2)$ for all $h \in (0, h_0)$, where*

$$S_h(f) := (2 + \rho)f(a + h) + 6 \sum_{j=2}^{\lfloor (b-a)/h \rfloor} f(a + hj)(-\rho)^{j-1}, \quad \rho := (2 + \sqrt{3})^{-1}.$$

Proof (sketch of). Let $f_1(x) = f(a) + (x - a)f'(a)$ (i.e., the degree-1 Taylor polynomial of f) and fix any $C > 0.5\,|f''(a)|$. Then there is $h_0 > 0$ such that $|f(x) - f_1(x)| \le C(x - a)^2$ for all $x \in [a, a + h_0]$. Introduce

$$S_{h,1}(f) := (2 + \rho)f(a + h) + 6 \sum_{j=2}^{\lfloor h_0/h \rfloor} f(a + hj)(-\rho)^{j-1},$$

$$S_{h,2}(g) := (2 + \rho)g(a + h) + 6 \sum_{j=2}^{\infty} g(a + hj)(-\rho)^{j-1}, \quad g \in \mathbb{P}_1.$$

It can be verified that $S_{h,2}(g) = g(a)$ for all $g \in \mathbb{P}_1$, thus $|S_h(f) - f(a)|$ is upper bounded by $|S_h(f) - S_{h,1}(f)| + |S_{h,1}(f) - S_{h,1}(f_1)| + |S_{h,1}(f_1) - S_{h,2}(f_1)|$.

To bound the last term, notice $\sum_{j=N}^{\infty} \rho^j = \rho^N/(1 - \rho)$ and $\sum_{j=N}^{\infty}(j+1)\rho^j = N\rho^N/(1 - \rho) + \rho^N/(1 - \rho)^2$. Then the last term is upper bounded by

$$6\left(\frac{N\rho^N}{1 - \rho} + \frac{\rho^N}{(1 - \rho)^2}\right)|f'(a)|\,h + 6\frac{\rho^N}{1 - \rho}\,|f(a)|, \quad N := \lfloor h_0/h \rfloor.$$

Since ρ^N exponentially decreases in N, this sum is of order $O(N^{-2}) = O(h^2)$ (with the big-O constant depending on ρ, h_0 and $f(a)$, $f'(a)$). The first term is upper bounded similarly, by estimating each $f(a+hj)$ with $\|f\|$. For the middle term we use $|f(a+hj) - f_1(a+hj)| \leq Ch^2 j^2$, which allows to bound

$$|S_{h,1}(f) - S_{h,1}(f_1)| \leq Ch^2 \left((2+\rho) + 6\sum_{j=2}^{\infty} j^2 \rho^{j-1} \right) = Ch^2 \left(7 + 5\sqrt{3} \right).$$

Since all three summands are $O(h^2)$, the claim follows.   □

Proposition 4. *For all $v \in C[a,b]$ we have $\mathfrak{P}_h v \xrightarrow[h \to 0]{} v$, i.e., the operators $\mathfrak{P}_h$ strongly converge to the identity operator $\mathfrak{I}$. Furthermore, for all $v \in C^2[a,b]$ we have $\|v - \mathfrak{P}_h v\| = O(h^2)$.*

Proof (sketch of). Let $v_h = \mathfrak{P}_h v$ and $\tilde{v}_h$ be a piecewise linear function, joining the points $(t_i, v(t_i))$, $i \in [0..N]$, with line segments. It is well known [1, Eq. 11.2.4], that $|\tilde{v}_h(t) - v(t)| \leq \omega(v, h)$ for all $t \in [a,b]$, where ω denotes the modulus of continuity, and $\|v - \tilde{v}_h\| = O(h^2)$ when $v \in C^2[a,b]$. Moreover, $v \in C[a,b]$, thus v is also uniformly continuous, hence $\omega(v, h) \to 0$ as $h \to 0$. We shall show that $\|v_h - \tilde{v}_h\| = O(h^2)$, then the claim will follow by the triangle inequality.

To prove this estimate, we recall that the definition of v_h implies $v_h \equiv \tilde{v}_h$ on $[a+h, b-h]$; therefore we must prove the $O(h^2)$ bound only on the rightmost and leftmost intervals, or, equivalently, that $|v(s) - v_h(s)| = O(h^2)$ when $s \in \{a, b\}$ (since both functions are linear in each basic interval). Due to the symmetry, consider only $s = a$.

Technical arguments (which we omit here) gives that $v_h(a) = -3y_0/h^2$, where y_0 is as in (6). The value of y_0 is obtained by multiplying the first row of the inverse matrix of $\mathbf{S}$ with the RHS of (6); carrying out the said calculations yields

$$v_h(a) = \left(2 + \frac{s_{N-3}}{s_{N-2}} \right) v(t_1) + 6\sum_{j=1}^{N-3} v(t_{j+1}) \frac{s_{N-2-j} \cdot (-1)^j}{s_{N-2}} + \frac{2\sqrt{3}\,(-1)^N}{s_{N-2}} v(b-h),$$

where $s_n := (2 + \sqrt{3})^n - (2 - \sqrt{3})^n$. However, asymptotically $s_n \sim (2 + \sqrt{3})^n$ and we obtain $v_h(a) \sim S_h(v)$, in notation from Proposition 3. This proposition also implies $v_h(a) = O(h^2)$, allowing to conclude $\|v_h - \tilde{v}_h\| = O(h^2)$.   □

Proof (of Theorem 2). The hypothesis of Theorem 2 imply existence of the Green's function $G(s,t)$ for the problem $u'' = 0$ with boundary conditions $\mathfrak{B}_1 u = \mathfrak{B}_2 u = 0$. Let u be the exact solution of the BVP (1) and $v := u''$. If we are given the second derivative $v_h := u_h''$ of the approximate solution u_h, then u_h is uniquely determined via the Green's function:

$$u_h(s) = \int_a^b G(s,t) v_h(t)\, dt, \quad u_h'(s) = \int_a^b \frac{\partial G(s,t)}{\partial s} v_h(t)\, dt. \tag{7}$$

Define an integral operator $\mathfrak{K} : C[a,b] \to C[a,b]$ by

$$\mathfrak{K}w(s) := \int_a^b \left(e_1(s) \frac{\partial G(s,t)}{\partial s} + e_0(s) G(s,t) \right) w(t)\, dt, \quad w \in C[a,b].$$

This operator is compact, which together with the strong convergence of the operators $\mathfrak{P}_h$ (Proposition 4) implies [1, Lemma 11.1.4] that the operators $\mathfrak{P}_h\mathfrak{K}$ converge in norm to $\mathfrak{K}$. Therefore for sufficiently small h the inverse operators $(\mathfrak{I}+\mathfrak{P}_h\mathfrak{K})^{-1}$ exist (where $\mathfrak{I}$ stands for the identity operator) [1, Theorem 11.1.2] and are uniformly bounded:

$$\left\|(\mathfrak{I}+\mathfrak{P}_h\mathfrak{K})^{-1}\right\| \le C_*, \quad h \le h_0. \tag{8}$$

The BVP (1) is equivalent to the equation $(\mathfrak{I}+\mathfrak{K})v = f$. Similarly, the problem (2) is equivalent to the equation $\mathfrak{P}_h(\mathfrak{I}+\mathfrak{K})v_h = \mathfrak{P}_h f$. Since $\mathfrak{P}_h v_h = v_h$, this can be simplified to

$$(\mathfrak{I}+\mathfrak{P}_h\mathfrak{K})v_h = \mathfrak{P}_h f. \tag{9}$$

From (8) it follows that for $h \le h_0$ there exists a unique $v_h \in \mathbb{U}_2$ satisfying (9) and the function $u_h \in \mathbb{U}_0$ (which is uniquely determined by (7)) satisfies the collocation problem.

To estimate the rate of convergence, notice $(\mathfrak{I}+\mathfrak{P}_h\mathfrak{K})v = \mathfrak{P}_h f + (v - \mathfrak{P}_h v)$. Subtracting (9) from this equality, we have $(\mathfrak{I}+\mathfrak{P}_h\mathfrak{K})(v-v_h) = (v-\mathfrak{P}_h v)$, hence $\|v - v_h\| \le \left\|(\mathfrak{I}+\mathfrak{P}_h\mathfrak{K})^{-1}\right\| \cdot \|v - \mathfrak{P}_h v\|$. Now from (8) we conclude (3).

Finally, suppose $u \in C^4[a,b]$. From Proposition 4, $\|v - v_h\| = O(h^2)$. Let $C_j = \max_{s\in[a,b]} \int_a^b \left|\frac{\partial^j G(s,t)}{\partial s^j}\right| dt$, then (7) implies $\left\|u^{(j)} - u_h^{(j)}\right\| \le C_j \|v - v_h\|$ and the estimate (4) follows. $\qquad\square$

5 Numerical Examples

Now we demonstrate the proposed collocation method with some examples. Even though we considered only $k = 2$ in Theorem 2, we conjecture that the problem (2) is uniquely solvable also when $\mathbb{A}_h$ is a (k, N)-eFPB for any $k > 2$ (under certain assumptions on the BVP and its exact solution); moreover, the estimate (4) then should be $O(h^{2k-2})$. To support this hypothesis, in the examples below we present also numerical tests with $k > 2$. Moreover, some examples hint at other ways to generalize Theorem 2.

Example 1. Consider the problem from [9, Example 1]:

$$u''(s) + (s^2 - 6s - 1)u'(s) + (5s - s^2 + 6)u(s) = e^s - s^2 + 5s + 6 \quad s \in (0,1),$$

with BC $u(0) + u'(0) = 2$, $2u(1) - u'(1) = 2$. The unique solution of this BVP is $u(s) = se^s + 1$. The maximum error at the nodes t_i is reported in Table 1 for various k, N. The third row is cited from [9, Table 2] (the authors there apply an $O(h^4)$-accurate method, which would correspond to $k = 3$ in our proposal).

For $k = 2$, we expect approximation error $O(h^2)$ due to Theorem 2. This is consistent with the numerical data in the first row of Table 1. However, increasing k, i.e., solving the respective system when collocating w.r.t. (k, N)-eFPB with $k \in \{3, 4\}$, indicates approximation of order $O(h^4)$ and $O(h^6)$, respectively.

Table 1. Maximum error at nodes, Example 1

	$N = 10$	$N = 20$	$N = 40$	$N = 80$	$N = 160$
$k = 2$	4.11e − 03	6.95e − 04	1.42e − 04	3.22e − 05	7.69e − 06
$k = 3$	2.10e − 05	4.71e − 07	1.08e − 08	2.61e − 10	6.27e − 12
[9]	4.37e − 07	2.67e − 08	1.66e − 09	1.03e − 10	—
$k = 4$	1.72e − 07	1.31e − 09	8.02e − 12	5.39e − 14	3.94e − 16

Example 2. We consider a well-known singular BVP, namely, the Bessel's equation of order 0:

$$u''(s) + \frac{u'(s)}{s} + u(s) = 0, \quad s \in (0,1),$$

with BC $u'(0) = 0$, $u(1) = 1$, whose solution is $u(s) = J_0(s)/J_0(1)$ (where J_0 is the Bessel function of the first kind). This example fails to satisfy Theorem 2; nevertheless, in numerical tests our method performs well, which suggests that Theorem 2 could be generalized to cover certain singular BVPs. In [4, Problem 1] this BVP was numerically solved with an $O(h^5)$-accurate method for steps with size $h \in \{0.1, 0.05, 0.02, 0.01\}$. We solve this problem with our proposed method with $k = 3$ (which we conjecture to be $O(h^4)$-accurate), with the same values of h (which correspond in our case to $N \in \{10, 20, 50, 100\}$). The maximal error at the nodes t_i are displayed in Table 2; in its first row we cite the respective errors from [4].

Table 2. Maximum error at nodes, Example 2

	$N = 10$	$N = 20$	$N = 50$	$N = 100$
[4]	1.67e − 06	2.04e − 07	1.42e − 08	1.44e − 09
Proposed method	9.54e − 08	2.04e − 09	3.16e − 11	1.23e − 12

Example 3. Finally, an example which shows that any analogue of Theorem 2 for $k > 2$ will require stricter constraints on the smoothness of u to have better approximation than $O(h^2)$. Consider the equation

$$u''(s) + su'(s) - u(s) = se^s - |s| \left(6 - 12s + 2s^2 - 3s^3\right), \quad s \in (-1,1),$$

Table 3. Maximum error at nodes, Example 3

k	N				
	25	50	100	500	1000
2	3.16e − 03	3.87e − 04	9.76e − 05	3.92e − 06	9.79e − 07
3	1.05e − 03	5.40e − 04	1.36e − 04	5.46e − 06	1.37e − 06

with BC $u(-1) = e^{-1} - 2$, $u(1) = e$ (the example is from [17, Problem 6]). The unique solution of this BVP is $u(s) = e^s - |s|^3 (1 - s)$, which is in $C^2[a, b]$, but its second derivative is not differentiable.

Theorem 2 applies, but not the estimate (4). Nevertheless, the numerical results (Table 3) are consistent with the estimate $O(h^2)$ (which hints at possible generalizations of the obtained bounds). However, when applying the proposed method with $k = 3$, the error rate remains $O(h^2)$, even though generally $O(h^4)$ is expected for sufficiently smooth u (in fact, errors for $k = 3$ are even larger than those for $k = 2$, which could be explained by a larger constant hidden in the big-O notation).

6 Conclusion

The proposed method allows to solve linear BVPs, obtaining a function whose F-transform approximates the exact solution. Employing the discrete F-transform allows to conveniently and compactly store the obtained solution and obtain a continuous function from it with the help of inverse F-transform. Stability of the proposed method must be further investigated. Numerical tests indicate that the 2-norm condition number of the matrix of the linear system (6) is of order $O(h^{-2})$, but this estimate is yet to be carried out theoretically. Furthermore, the computational complexity of the proposed method must be estimated (numerical tests suggest that the time required to solve a BVP via the proposed method is of order $O(h^{-1})$, assuming a fixed k).

The numerical results also suggest that fuzzy partitions based on B-splines of degree $2k - 1$ could generalize Theorem 2 with the error estimate $O(h^{2k-2})$ (in the case of a sufficiently smooth solution of BVP). Generalizing our proposal to non-linear problems is also under consideration. Furthermore, this method might be applicable to certain singular problems. All these generalizations, however, are likely to require substantially different proof techniques than those employed here.

Currently the theoretical proofs heavily rely on the fact that $k = 2$ and require a lot of technical explicit calculations. While it shows that the proposed method works in the case of cubic B-splines, these methods provide little insight in other cases. Our future work will focus on developing theoretical analysis of the proposed method that would allow to apply it in the more general setting.

References

1. Atkinson, K., Han, W.: Theoretical Numerical Analysis: A Functional Analysis Framework. Springer, New York (2001). https://doi.org/10.1007/978-1-4419-0458-4
2. Azimzadeh, P., Forsyth, P.A.: Weakly chained matrices, policy iteration, and impulse control. SIAM J. Numer. Anal. **54**(3), 1341–1364 (2016). arXiv:1510.03928
3. Bede, B., Rudas, I.J.: Approximation properties of fuzzy transforms. Fuzzy Sets Syst. **180**(1), 20–40 (2011)
4. Goh, J., Majid, A.A., Ismail, A.I.M.: A quartic B-spline for second-order singular boundary value problems. Comput. Math. Appl. **64**(2), 115–120 (2012)

5. Khastan, A., Perfilieva, I., Alijani, Z.: A new fuzzy approximation method to Cauchy problems by fuzzy transform. Fuzzy Sets Syst. **288**, 75–95 (2016)
6. Kodorane, I., Asmuss, S.: On approximation properties of spline based F-transform with respect to fuzzy m-partition. In: Montero, J., Pasi, G., Ciucci, D. (eds.) Proceedings of EUSFLAT 2013, Advances in Intelligent Systems Research, Milan, Italy, vol. 32, pp. 772–779. Atlantis Press (2013)
7. Kokainis, M., Asmuss, S.: Approximation properties of higher degree F-transforms based on B-splines. In: IEEE International Conference on Fuzzy Systems, FUZZ-IEEE 2015, 2–5 August 2015, Istanbul, Turkey. IEEE (2015)
8. Kokainis, M., Asmuss, S.: Continuous and discrete higher-degree F-transforms based on B-splines. Soft Comput. **21**(13), 3615–3639 (2017)
9. Lang, F.G., Xu, X.P.: Quintic B-spline collocation method for second order mixed boundary value problem. Comput. Phys. Commun. **183**(4), 913–921 (2012)
10. Novák, V., Perfilieva, I., Jarushkina, N.G.: A general methodology for managerial decision making using intelligent techniques. In: Rakus-Andersson, E., Yager, R.R., Ichalkaranje, N., Jain, L.C. (eds.) Recent Advances in Decision Making, pp. 103–120. Springer, Heidelberg (2009). https://doi.org/10.1007/978-3-642-02187-9_7
11. Perfilieva, I.: Fuzzy transforms: theory and applications. Fuzzy Sets Syst. **157**(8), 993–1023 (2006)
12. Perfilieva, I.: F-transform. In: Kacprzyk, J., Pedrycz, W. (eds.) Springer Handbook of Computational Intelligence, pp. 113–130. Springer, Heidelberg (2015)
13. Perfilieva, I., Daňková, M., Bede, B.: Towards a higher degree F-transform. Fuzzy Sets Syst. **180**(1), 3–19 (2011)
14. Perfilieva, I., Haldeeva, E.: Fuzzy transformation and its applications. In: 2001 Proceedings of the 4th Czech-Japan Seminar on Data Analysis and Decision Making under Uncertainty, Czech Republic. pp. 116–124 (2001)
15. Perfilieva, I., Števuliáková, P., Valášek, R.: F-transform-based shooting method for nonlinear boundary value problems. Soft Comput. **21**(13), 3493–3502 (2017)
16. Perfilieva, I., Števuliáková, P., Valášek, R.: F-transform for numerical solution of two-point boundary value problem. Iran. J. Fuzzy Syst. **14**(6), 1–13 (2017)
17. Russell, R.D., Shampine, L.F.: A collocation method for boundary value problems. Numer. Math. **19**(1), 1–28 (1972)
18. Schoenberg, I.J.: Cardinal spline interpolation. In: CBMS, vol. 12. SIAM (1973)
19. Štěpnička, M., Valášek, R.: Numerical solution of partial differential equations with help of fuzzy transform. In: Proceedings of the 14th IEEE International Conference on Fuzzy Systems, June 2005, Reno, USA, pp. 1104–1109 (2005)

Imprecise Probabilities: Foundations and Applications

Natural Extension of Choice Functions

Arthur Van Camp[1](✉)(iD), Enrique Miranda[2](iD), and Gert de Cooman[1](iD)

[1] IDLab, Ghent University, Zwijnaarde, Belgium
{Arthur.VanCamp,Gert.deCooman}@UGent.be
[2] Department of Statistics and Operations Research, University of Oviedo,
Oviedo, Spain
mirandaenrique@uniovi.es

Abstract. We extend the notion of natural extension, that gives the least committal extension of a given assessment, from the theory of sets of desirable gambles to that of choice functions. We give an expression of this natural extension and characterise its existence by means of a property called avoiding complete rejection. We prove that our notion reduces indeed to the standard one in the case of choice functions determined by binary comparisons, and that these are not general enough to determine all coherent choice function. Finally, we investigate the compatibility of the notion of natural extension with the structural assessment of indifference between a set of options.

Keywords: Choice functions · Coherence · Natural extension
Sets of desirable gambles · Structural assessments

1 Introduction

Since the publication of the seminal works in [1,2], coherent choice functions have been used widely as a model of the rational behaviour of an individual or a group. In particular, [3] established an axiomatisation of coherent choice functions, generalising the axioms in [4] to allow for incomparability.

In previous works [5,6], we have investigated some of the properties of coherent choice functions, their connection with the models considered earlier by Seidenfeld et al. [3] and also those particular coherent choice functions that are related to the optimality criteria of maximality and E-admissibility. In all those cases we took for granted that the choice function is given on the full class of option sets, and that it is coherent. However, it is somewhat unrealistic to assume that the subject always specifies an entire choice function: this means that he would have to specify for *every* option set which are the options he chooses, and this in a manner that is coherent in the sense that we shall discuss later on. Rather, a subject will typically specify a choice function only partially, by specifying the rejection of *some* options from *some* option sets. We call this partial specification of a choice function his *assessment*. Such an assessment can consist of an *arbitrary* amount of rejection statements; we do not rule out here

© Springer International Publishing AG, part of Springer Nature 2018
J. Medina et al. (Eds.): IPMU 2018, CCIS 854, pp. 201–213, 2018.
https://doi.org/10.1007/978-3-319-91476-3_17

the possibility that the subject's assessment consists of an uncountable collection of rejection statements.

The question we shall tackle in this paper is the following: given such an assessment, what is the implied choice between other option sets, using *only* the consequences of coherence?

To answer this question, after giving some preliminary notions in Sect. 2, we shall define in Sect. 3 the natural extension, when it exists, as the least committal coherent choice function that 'extends' a given assessment. In Sect. 4 we shall show that our notion is compatible with the eponymous notion established in the theory of sets of desirable gambles, that correspond to choice functions determined by binary comparisons. Then in Sect. 5 we use our work to show (i) that a coherent choice function may not be determined as the infima of a family of binary choice functions; and (ii) that the notion of natural extension can also be made compatible with a structural assessment of indifference. Finally, some additional comments are given in Sect. 6. Due to the space constraints, proofs have been omitted.

2 Preliminary Concepts

Consider a real vector space V provided with the vector addition $+$ and scalar multiplication. We denote its additive identity by 0. Elements of V are intended as abstract representation of options between which a subject can express his preferences, by specifying choice functions. We therefore call V also the *option space*. We denote by $Q(V)$ the set of all non-empty *finite* subsets of V, a strict subset of the power set $P(V)$ of V. Elements A of $Q(V)$ are the option sets amongst which a subject can choose his preferred options. When it is clear what option space V we are considering, we will also use the simpler notation Q, and use Q_0 to denote those option sets that include 0. We will assume throughout that V is ordered by a vector ordering $\preceq$. We will associate with it the strict partial order $\prec$, as follows: $u \prec v \Leftrightarrow (u \preceq v \text{ and } u \neq v)$, for all u and v in V. For notational convenience, we let $V_{\succ 0} := \{u \in V : 0 \prec u\}$, $V_{\prec 0} := \{u \in V : u \prec 0\}$, and $V_{\preceq 0} := \{u \in V : u \preceq 0\}$.

Definition 1. *A* choice function *C on an option space V is a map*

$$C \colon Q \to Q \cup \{\emptyset\} \colon A \mapsto C(A) \text{ such that } C(A) \subseteq A.$$

The idea underlying this simple definition is that a choice function C selects the set $C(A)$ of 'best' options in the *option set* A. Our definition resembles the one commonly used in the literature [3,7,8], except perhaps for an also not entirely unusual restriction to *finite* option sets [9–11].

Equivalently to a choice function C, we may consider its associated *rejection function* R, defined by $R(A) := A \setminus C(A)$ for all A in Q. It returns the options $R(A)$ that are rejected—not selected—by C. We collect all the rejection functions in the set $\mathbf{R}$. For technical reasons, we shall focus on rejection functions in this paper. Moreover, we shall restrict our attention to those rejection functions that satisfy a number of rationality requirements; they are called *coherent*. For brevity, we will commonly refer to choice functions and rejection functions as *choice models*, in order to distinguish them from models of *desirability* (see Sect. 4).

Definition 2 (Coherent rejection function). *We call a rejection function R on $\mathcal{V}$ coherent if for all A, A_1 and A_2 in $\mathcal{Q}$, all u and v in $\mathcal{V}$, and all λ in $\mathbb{R}_{>0}$:*

R1. $R(A) \neq A$;
R2. *if* $u \prec v$ *then* $u \in R(\{u, v\})$;
R3. a. *if* $A_1 \subseteq R(A_2)$ *and* $A_2 \subseteq A$ *then* $A_1 \subseteq R(A)$;
    b. *if* $A_1 \subseteq R(A_2)$ *and* $A \subseteq A_1$ *then* $A_1 \setminus A \subseteq R(A_2 \setminus A)$;
R4. a. *if* $A_1 \subseteq R(A_2)$ *then* $\lambda A_1 \subseteq R(\lambda A_2)$;
    b. *if* $A_1 \subseteq R(A_2)$ *then* $A_1 + \{u\} \subseteq R(A_2 + \{u\})$.

We collect all coherent rejection functions on $\mathcal{V}$ in the set $\overline{\mathbf{R}}(\mathcal{V})$, often simply denoted as $\overline{\mathbf{R}}$ when it is clear from the context which vector space we are using.

These axioms constitute a subset of the ones introduced by Seidenfeld et al. [3], duly translated from horse lotteries to our abstract options, which are more general as shown in earlier work of ours [5, Sect. 3]. In this respect, our notion of coherence is less restrictive than theirs. On the other hand, our Axiom R2 is more restrictive than the corresponding one in [3]. This is necessary in order to link coherent choice functions and coherent sets of desirable gambles (see [5, Sect. 4]).

In order to be able to use choice models for conservative reasoning, as we will do, we provide them with a partial order $\sqsubseteq$ having the interpretation of 'being at most as informative as'. For any R_1 and R_2 in $\mathbf{R}$, we let $R_1 \sqsubseteq R_2 \Leftrightarrow (\forall A \in \mathcal{Q})(R_1(A) \subseteq R_2(A))$. For any collection $\mathcal{R} \subseteq \mathbf{R}$ of rejection functions, the infimum $\inf \mathcal{R}$ is the rejection function given by $(\inf \mathcal{R})(A) := \bigcap_{R \in \mathcal{R}} R(A)$ for every A in $\mathcal{Q}$.

3   Natural Extension of Rejection Functions

We consider now a rejection function that is defined on some *subset of* the class $\mathcal{Q}$ of all option sets, and investigate under which conditions it is possible to extend it to a rejection function on $\mathcal{Q}$ that satisfies the coherence axioms. Taking into account Axiom R4b, we can assume without loss of generality that our assessment is made in terms of option sets that reject the option 0.

To be more specific, we assume that an assessment $\mathcal{B}$ is a subset of $\mathcal{Q}_0$. It consists of an arbitrary collection of option sets that include 0. Its interpretation is that 0 should be rejected from every option set B in $\mathcal{B}$. We are looking for the least informative coherent rejection function R that *extends* the assessment $\mathcal{B}$, by which we mean that $0 \in R(B)$ for all B in $\mathcal{B}$.[1]

---

[1] This is not an extension of a rejection function defined on a smaller domain $\mathcal{B}$ to a bigger domain $\mathcal{Q}_0$. Rather, it is the extension of an *assessment*, where we do not necessarily know all the rejected options in every option set B in $\mathcal{B}$ (except for 0).

Definition 3 (Natural extension). *Given any assessment $\mathcal{B} \subseteq \mathcal{Q}_0$, the natural extension of $\mathcal{B}$ is the rejection function*

$$\mathcal{E}(\mathcal{B}) := \inf\{R \in \overline{\mathbf{R}} : (\forall B \in \mathcal{B})0 \in R(B)\} = \inf\{R \in \overline{\mathbf{R}} : R \text{ extends } \mathcal{B}\},$$

where we let $\inf \emptyset$ be equal to $\mathrm{id}_\mathcal{Q}$, the identity rejection function that maps every option set to itself.

We can equivalently define the natural extension as a *choice function* instead of a rejection function, but that turns out to be notationally more involved, which is why we have decided to use rejection functions in this paper.

The above definition is not very useful for practical inference purposes: it does not provide an explicit expression for $\mathcal{E}(\mathcal{B})$. To try and remedy this, consider the special rejection function $R_\mathcal{B}$ based on the assessment $\mathcal{B}$, defined as:

$$R_\mathcal{B}(A) := \Big\{u \in A : (\exists A' \in \mathcal{Q})\Big(A' \supseteq A \text{ and } (\forall v \in \{u\} \cup (A' \setminus A))$$

$$\big((A' - \{v\}) \cap \mathcal{V}_{\succ 0} \neq \emptyset \text{ or } (\exists B \in \mathcal{B}, \exists \mu \in \mathbb{R}_{>0})\{v\} + \mu B \preccurlyeq A'\big)\Big)\Big\} \quad (1.1)$$

for all A in $\mathcal{Q}$. From here on, we let $\preccurlyeq$ be the ordering on $\mathcal{Q}$ defined by $A_1 \preccurlyeq A_2 \Leftrightarrow (\forall u_1 \in A_1)(\exists u_2 \in A_2)u_1 \preceq u_2$.

Proposition 1. *Consider $\mathcal{B} \subseteq \mathcal{Q}_0$. Then $R_\mathcal{B}$ is the least informative rejection function that satisfies Axioms R2–R4 and extends $\mathcal{B}$.*

After inspection of the rationality Axioms R1–R4, we see that all axioms but the first are *productive*, in the sense that application of these axioms allow us to identify new rejected options within, possibly, new option sets. Axiom R1 however is a *destructive* one: it indicates how far our rejections can go, and where the inferences should stop. Indeed, it requires that, within a given option set A, not every element of A should be rejected. In other words, it requires that, for any given option set, we should choose at least one of its elements. Therefore we need to be careful and avoid assessments that lead to a violation of Axiom R1, or to a complete rejection of some option set.

Definition 4 (Avoiding complete rejection). *Given any assessment $\mathcal{B} \subseteq \mathcal{Q}_0$, we say that $\mathcal{B}$ avoids complete rejection when $R_\mathcal{B}$ satisfies Axiom R1.*

To see that this notion is not trivial, consider the following example:

Example 1. As an example of an assessment that does not avoid complete rejection, consider $\mathcal{B} := \{\{0, u\}, \{0, -u\}\} \subseteq \mathcal{Q}_0$ for an arbitrary u in $\mathcal{V}$. By Proposition 1, $R_\mathcal{B}$ extends $\mathcal{B}$ (so $0 \in R_\mathcal{B}(\{0, u\})$ and $0 \in R_\mathcal{B}(\{0, -u\})$) and satisfies Axioms R2–R4. By Axiom R4b, from $0 \in R_\mathcal{B}(\{0, -u\})$ we infer that $u \in R_\mathcal{B}(\{0, u\})$. Using that $0 \in R_\mathcal{B}(\{0, u\})$, we infer that $\{0, u\} = R_\mathcal{B}(\{0, u\})$, contradicting Axiom R1. Therefore $\mathcal{B}$ does not avoid complete rejection.   ◊

Theorem 1. *Consider any assessment $\mathcal{B} \subseteq \mathcal{Q}_0$. Then the following statements are equivalent:*

(i) $\mathcal{B}$ avoids complete rejection;
(ii) There is a coherent extension of $\mathcal{B}$: $(\exists R \in \overline{\mathbf{R}})(\forall B \in \mathcal{B})0 \in R(B)$;
(iii) $\mathcal{E}(\mathcal{B}) \neq \mathrm{id}_{\mathcal{Q}}$;
(iv) $\mathcal{E}(\mathcal{B}) \in \overline{\mathbf{R}}$;
(v) $\mathcal{E}(\mathcal{B})$ is the least informative rejection function that is coherent and extends $\mathcal{B}$.

When any of these equivalent statements hold, then $\mathcal{E}(\mathcal{B}) = R_{\mathcal{B}}$.

4   Connection with Desirability

Let us compare our discussion of natural extension with the case of binary preferences and desirability. A *desirability assessment* $B \subseteq \mathcal{V}$ is usually (see for instance Sect. 1.2 of Ref. [12], and also Ref. [13]) a set of options that the agent finds desirable—strictly prefers to the zero option. As we did for choice functions, we pay special attention to *coherent* sets of desirable options. The following is an immediate generalisation of existing coherence definitions [12, 13] from gambles to abstract options.

Definition 5 (Coherent set of desirable options). *We call a set of desirable options $D \subseteq \mathcal{V}$ coherent if for all u and v in $\mathcal{V}$ and λ in $\mathbb{R}_{>0}$:*

D1. $0 \notin D$;
D2. if $0 \prec u$ then $u \in D$;
D3. if $u \in D$ then $\lambda u \in D$;
D4. if $u, v \in D$ then $u + v \in D$.

We collect all coherent sets of desirable options in the set $\overline{\mathbf{D}}(\mathcal{V})$, often simply denoted as $\overline{\mathbf{D}}$ when it is clear from the context which vector space we are using.

Any coherent set of desirable options D gives rise to a coherent rejection function R_D given by $R_D(A) = \{u \in A : (\forall v \in A)v - u \notin D\}$ for all A in $\mathcal{Q}$.

Of course, any desirability assessment $B \subseteq \mathcal{V}$ can be transformed into an assessment for rejection functions: we simply assess that 0 is rejected in the binary choice between 0 and u, for every option u in B. The assessment based on B is therefore given by $\mathcal{B}_B := \{\{0, u\} : u \in B\}$; clearly B and $\mathcal{B}_B$ are in a one-to-one correspondence: given an assessment $\mathcal{B}_B$ that consists of an arbitrary family of binary option sets, we retrieve B as $B = \bigcup(\mathcal{B}_B \setminus \{0\}) = (\bigcup \mathcal{B}_B) \setminus \{0\}$.

Given any desirability assessment $B \subseteq \mathcal{V}$ and any set of desirable options $D \subseteq \mathcal{V}$, we say that D extends B if $B \subseteq D$. Our next proposition expresses this in terms of rejection functions.

Proposition 2. *Consider any desirability assessment $B \subseteq \mathcal{V}$ and any set of desirable options $D \subseteq \mathcal{V}$. Then D extends B if and only if R_D extends $\mathcal{B}_B$.*

For desirability, Axioms D2–D4 are the productive ones, while the only destructive axiom is Axiom D1. The property for desirability that corresponds to avoiding complete rejection for choice models is *avoiding non-positivity*, commonly formulated as (see for instance Ref. [13, Definition 1])

$$\mathrm{posi}(B) \cap \mathcal{V}_{\preceq 0} = \emptyset \tag{1.2}$$

for the desirability assessment $B \subseteq \mathcal{V}$. Here, posi stands for 'positive hull', and is defined by

$$\mathrm{posi}(B) := \left\{ \sum_{k=1}^{n} \lambda_k u_k : n \in \mathbb{N}, \lambda_k \in \mathbb{R}_{>0}, u_k \in B \right\}$$

$$\subseteq \mathrm{span}(B) := \left\{ \sum_{k=1}^{n} \lambda_k u_k : n \in \mathbb{N}, \lambda_k \in \mathbb{R}, u_k \in B \right\} \subseteq \mathcal{V}.$$

Theorem 1 is the equivalent for choice models of the natural extension theorem for desirability. Let us state this natural extension theorem for desirability.

Theorem 2 *[13, Theorem 1]. Consider any desirability assessment $B \subseteq \mathcal{V}$, and define its natural extension as*

$$\mathcal{E}^{\mathbf{D}}(B) := \inf\{D \in \overline{\mathbf{D}} : B \subseteq D\}, \tag{1.3}$$

where we let $\inf \emptyset = \mathcal{V}$. Then the following statements are equivalent:

(i) *B avoids non-positivity;*
(ii) *B is included in some coherent set of desirable options;*
(iii) *$\mathcal{E}^{\mathbf{D}}(B) \neq \mathcal{V}$;*
(iv) *$\mathcal{E}^{\mathbf{D}}(B) \in \overline{\mathbf{D}}$;*
(v) *$\mathcal{E}^{\mathbf{D}}(B)$ is the least informative set of desirable options that is coherent and includes B.*

When any of these equivalent statements hold, $\mathcal{E}^{\mathbf{D}}(B) = \mathrm{posi}(\mathcal{V}_{\succ 0} \cup B)$.

Our next result tells us that the procedure of natural extension we have established for rejection functions is an extension of the procedure of natural extension for coherent sets of desirable gambles considered above.

Theorem 3. *Consider any desirability assessment $B \subseteq \mathcal{V}$. Then B avoids non-positivity if and only if $\mathcal{B}_B$ avoids complete rejection, and if this is the case, then $\mathcal{E}(\mathcal{B}_B) = R_{\mathcal{E}^{\mathbf{D}}(B)}.$*

To summarise these statements, consider the commuting diagram in Fig. 1, where we have used the maps

$$\mathcal{E}^{\mathbf{D}} : \mathcal{P}(\mathcal{V}) \to \mathbf{D} : B \mapsto \mathcal{E}^{\mathbf{D}}(B)$$

$$\mathcal{B}_\bullet : \mathcal{P}(\mathcal{V}) \to \mathcal{Q}_0 : B \mapsto \mathcal{B}_B := \{\{0, u\} : u \in B\}$$

$$\mathcal{E} : \mathcal{P}(\mathcal{Q}_0) \to \mathbf{R} : \mathcal{B} \mapsto \mathcal{E}(\mathcal{B})$$

$$D_\bullet : \mathbf{R} \to \mathbf{D} : R \mapsto D_R := \{u \in \mathcal{V} : 0 \in R(\{0, u\})\}$$

$$R_\bullet : \mathbf{D} \to \mathbf{R} : D \mapsto R_D$$

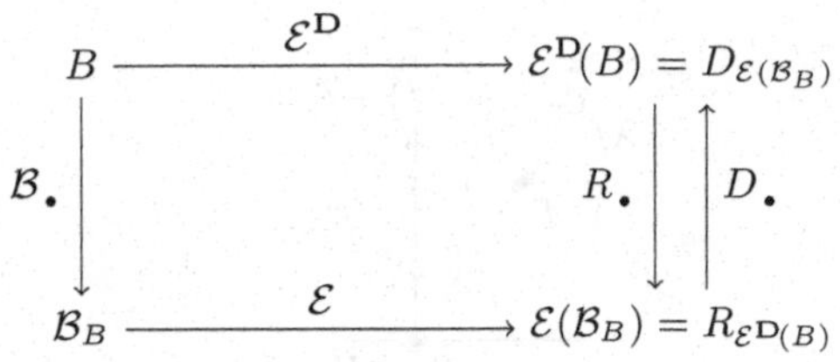

$$B \xrightarrow{\quad \mathcal{E}^{\mathbf{D}} \quad} \mathcal{E}^{\mathbf{D}}(B) = D_{\mathcal{E}(\mathcal{B}_B)}$$

$$\mathcal{B}_{\bullet} \downarrow \qquad\qquad R_{\bullet} \downarrow\uparrow D_{\bullet}$$

$$\mathcal{B}_B \xrightarrow{\quad \mathcal{E} \quad} \mathcal{E}(\mathcal{B}_B) = R_{\mathcal{E}^{\mathbf{D}}(B)}$$

Fig. 1. Commuting diagram for the case of binary assessments

Start with a desirability assessment $B \subseteq \mathcal{V}$ that avoids non-positivity. Taking the natural extension for desirability commutes with taking the corresponding assessment (for choice models), then the natural extension, and eventually going back to the set of desirable options corresponding to this natural extension. Furthermore, taking the natural extension of the corresponding assessment (for choice models) commutes with taking the natural extension for desirability, and then going to the corresponding rejection function.

5 Examples

5.1 Choice Functions That Are No Infima of Binary Choice Functions

Many important choice functions are infima of purely binary choice models: consider, for instance, the E-admissible or M-admissible choice functions [6]. It is an important question whether *all* coherent choice functions are infima of purely binary choice functions; if this question answered positively, this would immediately imply a representation theorem. If this question is answered in the negative, choice functions would constitute a theory that is more general than sets of desirable gambles in two ways: not only because it allows for more than binary choice, also because it is capable of expressing preferences that can never be retrieved as an infimum of purely binary preferences.

Below we will answer this question in the negative: we will define a special rejection function $R_\mathcal{B}$, based some particular assessment $\mathcal{B} \subseteq \mathcal{Q}_0$, and prove that it is no infimum of purely binary rejection functions.

Example 2. We will work with the special vector space of gambles $\mathcal{V} = \mathcal{L}$ on a binary possibility space $\mathcal{X} = \{\mathrm{H}, \mathrm{T}\}$, ordered by the standard point-wise ordering $\leq$: for any f, g in $\mathcal{L}$, we let $f \leq g \Leftrightarrow (\forall x \in \mathcal{X}) f(x) \leq g(x)$.

We consider a *single* assessment $\mathcal{B} := \{B\}$, where B consists of a gamble and one scaled variant of it, together with 0: the assessment we consider is $B := \{0, f, \lambda f\}$ with f a gamble and λ an element of $\mathbb{R}_{>0}$ and different from 1. We assume that $f(\mathrm{H}) < 0 < f(\mathrm{T})$, and that $\lambda > 1$. The idea is that B consists of 0 and two gambles that lie on the same line through 0, and on the same side of that line; see Fig. 2 for an illustration of the assessment.

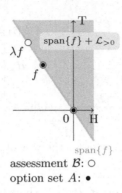

assessment $\mathcal{B}$: ○
option set A: ●

Fig. 2. Illustration of the assessment

Note that this assessment indeed avoids complete rejection: for instance, the coherent set of desirable options $D := \mathrm{posi}(\mathcal{V}_{\succ 0} \cup \{f\})$ satisfies $D \cap B = \{f, \lambda f\} \neq \emptyset$. Therefore, $R_{\mathcal{B}}$ is a coherent rejection function. To prove that $R_{\mathcal{B}}$ is no infimum of purely binary rejection functions, we first show the intermediate result that $0 \notin R_{\mathcal{B}}(A)$, where $A := \{0, f\}$. To prove this, assume *ex absurdo* that $0 \in R_{\mathcal{B}}(A)$, and infer using Eq. (1.1) that then there would be some $A' \supseteq A$ in $\mathcal{Q}$ such that

$$(\forall h \in \{0\} \cup (A' \setminus A))\big((A' - \{h\}) \cap \mathcal{L}_{>0} \neq \emptyset \text{ or } (\exists \mu \in \mathbb{R}_{>0})\{h\} + \mu B \preccurlyeq A'\big). \quad (1.4)$$

At this point, remark already that $A' \neq A$: indeed, if *ex absurdo* $A' = A$, then $\{0\} \cup (A' \setminus A) = \{0\}$, so we need only consider $h = 0$. Infer that $A' \cap \mathcal{L}_{>0} = \emptyset$ and $(\forall \mu \in \mathbb{R}_{>0})\{0, \mu f, \mu \lambda f\} \not\preccurlyeq \{0, f\}$, leading to a contradiction. Therefore, $A' \supset A$.

Without loss of generality, we let $A' := \{0, f, h_1, \ldots, h_n\} \supset A$ where n belongs to $\mathbb{N}$ and $h_1, \ldots, h_n$ to $\mathcal{L}$, so $\{0\} \cup (A' \setminus A) = \{0, h_1, \ldots, h_n\}$.

It then follows that $(\max A') \cap \{0, h_1, \ldots, h_n\} \neq \emptyset$.

Let us prove as an intermediate result that $(\max A') \cap (\mathrm{span}\{f\} + \mathcal{L}_{>0}) = \emptyset$. To see this, since $\{0, f\} \cap (\mathrm{span}\{f\} + \mathcal{L}_{>0}) = \emptyset$, infer that $(\max A') \cap (\mathrm{span}\{f\} + \mathcal{L}_{>0}) \subseteq \{h_1, \ldots, h_n\}$, and assume *ex absurdo* that $(\max A') \cap (\mathrm{span}\{f\} + \mathcal{L}_{>0}) \neq \emptyset$. Let h be an element of $\arg\max\{g(\mathrm{T}) : g \in (\max A') \cap (\mathrm{span}\{f\} + \mathcal{L}_{>0})\}$, then $h(\mathrm{T}) + \mu \lambda f(\mathrm{T}) > h(\mathrm{T})$, so $h + \mu \lambda f \in \{h\} + \mu B$ is undominated in $(\max A') \cap (\mathrm{span}\{f\} + \mathcal{L}_{>0})$ whence $\{h\} + \mu B \not\preccurlyeq (\max A') \cap (\mathrm{span}\{f\} + \mathcal{L}_{>0})$ for all μ in $\mathbb{R}_{>0}$. Note that, since h belongs to $\mathrm{span}\{f\} + \mathcal{L}_{>0}$, also $h + \mu \lambda f$ belongs to $\mathrm{span}\{f\} + \mathcal{L}_{>0}$ for every μ in $\mathbb{R}_{>0}$. Therefore, since an element of $\mathrm{span}\{f\} + \mathcal{L}_{>0}$ can never be dominated by an element of $(\mathrm{span}\{f\} + \mathcal{L}_{>0})^c = \mathrm{span}\{f\} + \mathcal{L}_{\leq 0}$, also $\{h\} + \mu B \not\preccurlyeq \max A'$ for all μ in $\mathbb{R}_{>0}$. We deduce that also $\{h\} + \mu B \not\preccurlyeq A'$ for all μ in $\mathbb{R}_{>0}$. Since h belongs to $\max A'$, also $A' - \{h\} \cap \mathcal{L}_{>0} = \emptyset$, a contradiction. So we have that $(\max A') \cap (\mathrm{span}\{f\} + \mathcal{L}_{>0}) = \emptyset$, and therefore, again because an element of $\mathrm{span}\{f\} + \mathcal{L}_{>0}$ can never be dominated by an element of $\mathrm{span}\{f\} + \mathcal{L}_{\leq 0}$, also $A' \cap (\mathrm{span}\{f\} + \mathcal{L}_{>0}) = \emptyset$.

Now we go back to Eq. (1.4), and consider first $h = 0$. Then $A' \cap \mathcal{L}_{>0} \neq \emptyset$ or $(\exists \mu \in \mathbb{R}_{>0})\mu B \preccurlyeq A'$. Since $A' \cap (\mathrm{span}\{f\} + \mathcal{L}_{>0}) = \emptyset$, in particular $A' \cap \mathcal{L}_{>0} = \emptyset$, so the only possibility left is $(\exists \mu \in \mathbb{R}_{>0})\mu B \preccurlyeq A'$, or, in other words,

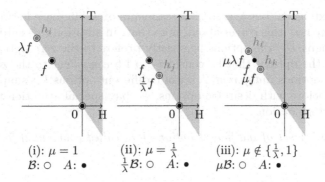

(i): $\mu = 1$
$\mathcal{B}$: ○ A: ●

(ii): $\mu = \frac{1}{\lambda}$
$\frac{1}{\lambda}\mathcal{B}$: ○ A: ●

(iii): $\mu \notin \{\frac{1}{\lambda}, 1\}$
$\mu\mathcal{B}$: ○ A: ●

Fig. 3. Illustration of the three different cases mentioned

$\{0, \mu f, \mu\lambda f\} \prec \{0, f, h_1, \ldots, h_n\}$ for some μ in $\mathbb{R}_{>0}$. There are three possibilities: if (i) $\mu = 1$, then $h_i \geq \lambda f$—and therefore, since $A' \cap (\mathrm{span}\{f\} + \mathcal{L}_{>0}) = \emptyset$, necessarily $h_i = \lambda f$—for some i in $\{1, \ldots, n\}$; if (ii) $\mu = \frac{1}{\lambda}$ then $h_j \geq \frac{1}{\lambda} f$—and therefore, since $A' \cap (\mathrm{span}\{f\} + \mathcal{L}_{>0}) = \emptyset$, necessarily $h_j = \frac{1}{\lambda} f$—for some j in $\{1, \ldots, n\}$; and finally, if (iii) $\mu \notin \{\frac{1}{\lambda}, 1\}$, then $h_k \geq \mu f$ and $h_\ell \geq \mu\lambda f$—and therefore, since $A' \cap (\mathrm{span}\{f\} + \mathcal{L}_{>0}) = \emptyset$, necessarily $h_k = \mu f$ and $h_\ell = \mu\lambda f$—for some k and ℓ in $\{1, \ldots, n\}$. These are illustrated in Fig. 3.

In any case, we find that $\{h_1, \ldots, h_n\} \cap \mathrm{posi}\{f\} \neq \emptyset$. Without loss of generality, let h_1 be the unique gamble in $\{h_1, \ldots, h_n\} \cap \mathrm{posi}\{f\}$ with highest value in T: $\{h_1\} = \arg\max\{g(\mathrm{T}) : g \in \{h_1, \ldots, h_n\} \cap \mathrm{posi}\{f\}\}$. Then, since $h_1 \in \{0\} \cup (A' \setminus A)$, by Eq. (1.4) we have that $(A' - \{h_1\}) \cap \mathcal{L}_{>0} \neq \emptyset$ or $(\exists \mu \in \mathbb{R}_{>0})\{h_1\} + \mu B \prec A'$. Since $A' \cap (\mathrm{span}\{f\} + \mathcal{L}_{>0}) = \emptyset$ and $h_1 \in \mathrm{posi}\{f\}$, we find in particular $A' \cap (\{h_1\} + \mathcal{L}_{>0}) = \emptyset$, whence $(A' - \{h_1\}) \cap \mathcal{L}_{>0} = \emptyset$. Therefore necessarily $\{h_1, h_1 + \mu f, h_1 + \mu\lambda f\} = \{h_1\} + \mu B \prec A'$ for some μ in $\mathbb{R}_{>0}$. Note that both $h_1 + \mu f$ and $h_1 + \mu\lambda f$ belong to $\mathrm{posi}\{f\}$, and have a value in T that is strictly higher than $h_1(\mathrm{T})$. But at least one of $h_1 + \mu f$ or $h_1 + \mu\lambda f$ is not equal to f, and therefore an element of $\{h_1, \ldots, h_n\} \cap \mathrm{posi}\{f\}$, a contradiction with the fact that $h_1 \in \arg\max\{g(\mathrm{T}) : g \in \{h_1, \ldots, h_n\} \cap \mathrm{posi}\{f\}\}$. Therefore indeed $0 \notin R_\mathcal{B}(A)$.

So we have found a rejection function $R_\mathcal{B}$ such that $0 \in R_\mathcal{B}(\{0, f, \lambda f\})$ but $0 \notin R_\mathcal{B}(\{0, f\})$. However, any rejection function R_D that is defined by means of a coherent set of desirable options D satisfies that

$$0 \in R_D(\{0, f, \lambda f\}) \Leftrightarrow 0 \in R_D(\{0, f\}), \tag{1.5}$$

and Eq. (1.5) is preserved when taking infima of rejection functions. As a consequence, $R_\mathcal{B}$ is no infimum of purely binary rejection functions.   ◇

5.2   Natural Extension and Indifference

Next we investigate if it is possible to obtain an extension of a given assessment that takes into account not only the implications of coherence, as we did with the natural extension, but also some assessments of indifference between a set of options. To see how this comes about, note that, in addition to a subject's set of desirable options D—the options he strictly prefers to the zero option—we can also consider the options that he considers to be *equivalent* to the zero option. We call these options *indifferent*. A set of indifferent options I is simply a subset of $\mathcal{V}$, but as before with desirable options, we pay special attention to *coherent* sets of indifferent options.

Definition 6. *A set of indifferent options I is called coherent if for all u, v in $\mathcal{V}$ and λ in $\mathbb{R}$:*

I_1. $0 \in I$;
I_2. *if $u \in \mathcal{V}_{\succ 0} \cup \mathcal{V}_{\prec 0}$ then $u \notin I$;*
I_3. *if $u \in I$ then $\lambda u \in I$;*
I_4. *if $u, v \in I$ then $u + v \in I$.*

Taken together, Axioms I_3 and I_4 are equivalent to $\mathrm{span}(I) = I$, and due to Axiom I_1, I is non-empty and therefore a linear subspace of $\mathcal{V}$.

The interaction between indifferent and desirable options is subject to rationality criteria as well: they should be compatible with one another.

Definition 7. *Given a set of desirable options D and a coherent set of indifferent options I, we call D compatible with I if $D + I \subseteq D$.*

We collect all options that are indifferent to an option $u \in \mathcal{V}$ into the *equivalence class* $[u] := \{v \in \mathcal{V} : v - u \in I\} = \{u\} + I$. We also denote $[u]$ as u/I. Of course, $[0] = \{0\} + I = I$ is a linear subspace, and the classes $[u] = \{u\} + I$ are affine subspaces of $\mathcal{V}$. The set of all these equivalence classes is the *quotient space* $\mathcal{V}/I := \{[u] : u \in \mathcal{V}\} = \{\{u\} + I : u \in \mathcal{V}\} = \{u/I : u \in \mathcal{V}\}$. This quotient space is a vector space under the vector addition and the scalar multiplication. $[0] = I$ is the additive identity of $\mathcal{V}/I$.

Definition 8. *We call a rejection function R on $\mathcal{Q}(\mathcal{V})$ compatible with a coherent set of indifferent options I if there is some representing rejection function R' on $\mathcal{Q}(\mathcal{V}/I)$ such that $R(A) = \{u \in A : [u] \in C'(A/I)\}$ for all A in $\mathcal{Q}(\mathcal{V})$.*

We refer to an earlier paper [6] of ours for a study of the compatibility of the structural assessment of coherence with the theory of coherent rejection functions, and to [14, 15] for other works on this topic.

The natural extension under indifference, if it is coherent, is the least informative coherent rejection function that extends the assessment $\mathcal{B} \subseteq \mathcal{Q}_0(\mathcal{V})$ and is compatible with the set of indifferent options I.

Definition 9. *Given any assessment $\mathcal{B} \subseteq \mathcal{Q}_0(\mathcal{V})$ and any coherent set of indifferent options I, the natural extension of $\mathcal{B}$ under I is the rejection function*

$$\mathcal{E}_I(\mathcal{B}) := \inf\{R \in \overline{\mathbf{R}}(\mathcal{V}) : R \text{ extends } \mathcal{B} \text{ and is compatible with } I\},$$

where, as usual, we let $\inf \emptyset = \mathrm{id}_{\mathcal{Q}(\mathcal{V})}$, the identity rejection function that maps every option set to itself.

To help link this definition with a more constructive and explicit expression, consider the special rejection function $R_{\mathcal{B},I}$, defined by:

$$R_{\mathcal{B},I}(A) := \{u \in A : [u] \in R_{\mathcal{B}/I}(A/I)\} \text{ for all } A \text{ in } \mathcal{Q}(\mathcal{V}), \qquad (1.6)$$

where we let $\mathcal{B}/I := \{B/I : B \in \mathcal{B}\} \subseteq \mathcal{Q}_{[0]}(\mathcal{V}/I)$, being—loosely speaking—the assessment $\mathcal{B}$ expressed in the quotient space $\mathcal{V}/I$. Recall that $R_{\mathcal{B}}$, as defined in Eq. (1.1), is relative to a given but otherwise arbitrary vector space $\mathcal{V}$. Our special rejection function $R_{\mathcal{B},I}$ uses the version $R_{\mathcal{B}/I}$ on $\mathcal{V}/I$ instead of $\mathcal{V}$.

The following is the counterpart of Proposition 1 under indifference:

Proposition 3. *Consider any assessment $\mathcal{B} \subseteq \mathcal{Q}_0(\mathcal{V})$ and any coherent set of indifferent options $I \subseteq \mathcal{V}$. Then $R_{\mathcal{B},I}$ is the least informative rejection function that satisfies Axioms R2–R4, extends $\mathcal{B}$, and is compatible with I.*

Recall from our results on the (normal) natural extension from Sect. 3 that not every assessment is extendible to a coherent rejection function: this is only the case if the assessment avoids complete rejection. Here too, when we deal with the natural extension under indifference, something similar occurs.

Definition 10 (Avoiding complete rejection under indifference). *Given any assessment $\mathcal{B} \subseteq \mathcal{Q}_0(\mathcal{V})$ and any coherent set of indifferent options $I \subseteq \mathcal{V}$, we say that $\mathcal{B}$ avoids complete rejection under I when $R_{\mathcal{B},I}$ satisfies Axiom R1.*

However, and perhaps surprisingly, avoiding complete rejection under indifference is sufficient for avoiding complete rejection:

Proposition 4. *Consider any assessment $\mathcal{B} \subseteq \mathcal{Q}_0(\mathcal{V})$ and any coherent set if indifferent options $I \subseteq \mathcal{V}$. Then $\mathcal{B}$ avoids complete rejection under I if and only if $\mathcal{B}/I$ avoids complete rejection, and both those equivalent conditions imply that $\mathcal{B}$ avoids complete rejection.*

This allows us to formulate a counterpart to Theorem 1 for natural extension under indifference:

Theorem 4. *Consider any assessment $\mathcal{B} \subseteq \mathcal{Q}_0$ and any coherent set of indifferent options $I \subseteq \mathcal{V}$. Then the following statements are equivalent:*

(i) *$\mathcal{B}$ avoids complete rejection under I;*
(ii) *There is some R in $\overline{\mathbf{R}}(\mathcal{V})$ that extends $\mathcal{B}$ that is compatible with I, meaning that $(\forall B \in \mathcal{B})0 \in R(B)$ and*

$$(\forall A \in \mathcal{Q}(\mathcal{V}))R(A) = \{u \in A : [u] \in R(A)/I\};$$

(iii) *$\mathcal{E}_I(\mathcal{B}) \neq \mathrm{id}_{\mathcal{Q}(\mathcal{V})}$;*
(iv) *$\mathcal{E}_I(\mathcal{B}) \in \overline{\mathbf{R}}(\mathcal{V})$;*
(v) *$\mathcal{E}_I(\mathcal{B})$ is the least informative rejection function that is coherent, extends $\mathcal{B}$, and is compatible with I.*

When any of these equivalent statements hold, then $\mathcal{E}_I(\mathcal{B}) = R_{\mathcal{B},I}$.

6 Conclusions

In this paper, we have investigated the natural extension of choice functions, found an expression for it, and characterised the assessments that have coherent extensions. We made the connection with binary choice, and showed how the well-known natural extension for desirability follows from our natural extension.

As future lines of research, we would like to study the compatibility of the notion of natural extension with other structural assessments; in this respect, we have already investigated the compatibility with a notion of *irrelevance* when modelling multivariate choice functions. It is an open problem to study whether something similar can be made with respect to the *exchangeable* choice functions we have considered in [16].

Acknowledgements. The research in this paper has been supported by project TIN2014-59543-P.

References

1. Arrow, K.: Social Choice and Individual Values. Yale University Press (1951)
2. Uzawa, H.: Note on preference and axioms of choice. Ann. Inst. Stat. Math. **8**, 35–40 (1956)
3. Seidenfeld, T., Schervish, M.J., Kadane, J.B.: Coherent choice functions under uncertainty. Synthese **172**(1), 157–176 (2010)
4. Rubin, H.: A weak system of axioms for "rational" behavior and the nonseparability of utility from prior. Stat. Risk Model. **5**(1–2), 47–58 (1987)
5. Van Camp, A., Miranda, E., de Cooman, G.: Lexicographic choice functions. Int. J. Approx. Reason. **92**, 97–119 (2018)
6. Van Camp, A., de Cooman, G., Miranda, E., Quaeghebeur, E.: Coherent choice functions, desirability and indifference. Fuzzy Sets Syst. **341**(C), 1–36 (2018). https://doi.org/10.1016/j.fss.2017.05.019
7. Aizerman, M.A.: New problems in the general choice theory. Soc. Choice Welfare **2**, 235–282 (1985)
8. Sen, A.: Social choice theory: a re-examination. Econometrica **45**, 53–89 (1977)
9. He, J.: A generalized unification theorem for choice theoretic foundations: avoiding the necessity of pairs and triplets. Economics Discussion Paper 2012–23, Kiel Institute for the World Economy (2012)
10. Schwartz, T.: Rationality and the myth of the maximum. Noûs **6**(2), 97–117 (1972)
11. Sen, A.: Choice functions and revealed preference. Rev. Econ. Stud. **38**(3), 307–317 (1971)
12. Quaeghebeur, E.: Desirability. In: Augustin, T., Coolen, F.P.A., de Cooman, G., Troffaes, M.C.M. (eds.) Introduction to Imprecise Probabilities, pp. 1–27. Wiley, Hoboken (2014)
13. de Cooman, G., Quaeghebeur, E.: Exchangeability and sets of desirable gambles. Int. J. Approx. Reason. **53**(3), 363–395 (2012). Precisely imprecise: a collection of papers dedicated to Henry E. Kyburg, Jr
14. Bradley, S.: How to choose among choice functions. In: Augustin, T., Doria, S., Miranda, E., Quaeghebeur, E. (eds.) Proceedings of ISIPTA 2015, pp. 57–66. Aracne (2015)

15. Seidenfeld, T.: Decision without independence and without ordering: what is the difference? Econ. Philos. **4**, 267–290 (1988)
16. Van Camp, A., de Cooman, G.: Exchangeable choice functions. In: Antonucci, A., Corani, G., Couso, I., Destercke, S. (eds.) Proceedings of ISIPTA 2017. Proceedings of Machine Learning Research, vol. 62, pp. 346–357 (2017)

Approximations of Coherent Lower Probabilities by 2-monotone Capacities

Ignacio Montes[1] , Enrique Miranda[1](✉) , and Paolo Vicig[2]

[1] Department of Statistics and Operations Research,
University of Oviedo, Oviedo, Spain
{imontes,mirandaenrique}@uniovi.es
[2] DEAMS, University of Trieste, Trieste, Italy
paolo.vicig@deams.units.it

Abstract. We investigate the problem of approximating a coherent lower probability on a finite space by a 2-monotone capacity that is at the same time as close as possible while not including additional information. We show that this can be tackled by means of a linear programming problem, and investigate the features of the set of undominated solutions. While our approach is based on a distance proposed by Baroni and Vicig, we also discuss a number of alternatives. Finally, we show that our work applies to the more general problem of approximating coherent lower previsions.

Keywords: Coherent lower probabilities · 2-monotonicity
Coherent lower previsions · Distortion models · Total variation distance

1 Introduction

Among the many models of imprecise probabilities [1], one of the most general is that of *coherent lower previsions* [2], that can be regarded as sets of expectations with respect to a convex family of finitely additive probability measures. In addition to its generality, it also has a clear behavioural interpretation in terms of acceptable betting rates, as well as the epistemic interpretation in terms of sets of probability measures. Nevertheless, coherent lower previsions (or their restrictions to events, called *coherent lower probabilities*) also have a number of drawbacks that hinder their use in practice: for instance, they have no easy representation in terms of their extreme points in general, and they lack some attractive mathematical properties possessed by more specific models.

One alternative that somewhat solves these issues is to work with 2-monotone capacities, which can be easily determined by means of a finite number of extreme points [3] and that still include as particular cases many of the imprecise probability models from the literature, such as probability intervals [4], belief functions [5] or possibility measures [6]. It is therefore interesting to determine if we can approximate a coherent lower probability by a 2-monotone one with a minimal loss of information. This is the problem we are tackling in this paper.

ⓒ Springer International Publishing AG, part of Springer Nature 2018
J. Medina et al. (Eds.): IPMU 2018, CCIS 854, pp. 214–225, 2018.
https://doi.org/10.1007/978-3-319-91476-3_18

After giving some preliminary concepts in Sect. 2, in Sect. 3 we study the problem of finding undominated outer approximations that minimize the distance to the original model, in the sense proposed by Baroni and Vicig [7]. In Sect. 4, we focus on outer approximations by means of some particular subfamilies of 2-monotone capacities and prove that this problem has a unique solution. A comparison with some alternative approches is given in Sect. 5. Finally, in Sect. 6 we show that our results allow us to solve the problem of outer approximating coherent lower previsions. Some additional comments are provided in Sect. 7. Due to the space limitations, proofs of the results have been omitted.

2   Preliminary Concepts

Let $\mathcal{X}$ be a finite space with cardinality n, and consider a lower probability $\underline{P} : \mathcal{P}(\mathcal{X}) \to [0,1]$. Its associated *credal set* is given by:

$$\mathcal{M}(\underline{P}) = \{P \text{ probability} \mid P(A) \geq \underline{P}(A) \quad \forall A \subseteq \mathcal{X}\},$$

Under an epistemic interpretation of uncertainty, we may regard $\underline{P}$ as a model for the imprecise knowledge of a probability measure P, and then $\mathcal{M}(\underline{P})$ would be the set of candidates for this unknown probability measure. The notion of coherence means that the bounds $\underline{P}$ gives for the probabilities of the different events are tight:

Definition 1 [2]. *A lower probability $\underline{P} : \mathcal{P}(\mathcal{X}) \to [0,1]$ is called coherent when $\mathcal{M}(\underline{P}) \neq \emptyset$ and $\underline{P}(A) = \min\{P(A) : P \in \mathcal{M}(\underline{P})\}$ for every $A \subseteq \mathcal{X}$.*

The conjugate of a coherent lower probability, given by $\overline{P}(A) = 1 - \underline{P}(A^c)$ for every $A \subseteq \mathcal{X}$, is called coherent *upper* probability.

Coherent lower probabilities include as particular cases most of the models of non-additive measures in the literature; they correspond moreover to *balanced* games within game theory [8]. One particular case of coherent lower probabilities are the 2-monotone capacities.

Definition 2 [9]. *A coherent lower probability $\underline{P} : \mathcal{P}(\mathcal{X}) \to [0,1]$ is called 2-monotone if for every $A, B \subseteq \mathcal{X}$ it satisfies:*

$$\underline{P}(A \cup B) + \underline{P}(A \cap B) \geq \underline{P}(A) + \underline{P}(B). \tag{1}$$

2-monotone capacities are sometimes called *convex* in the literature. They possess a number of interesting properties that are not always shared with coherent lower probabilities: the extreme points of their credal set can be easily determined using the permutations of the possibility space [3]; moreover, they have a unique extension as an expectation operator that preserves 2-monotonicity: their Choquet integral [10].

For all these reasons, it becomes interesting in practice to approximate a coherent lower probability $\underline{P}$ by a 2-monotone capacity Q that at the same time

(a) does not introduce new information; and (b) is as close as possible to the original model.

The first constraint is modelled by requiring that the credal set determined by $\underline{Q}$ includes that of $\underline{P}$, or in other words, that $\underline{Q}(E) \leq \underline{P}(E)$ for every $E \subseteq \mathcal{X}$. In that case, we shall say that $\underline{Q}$ is an *outer approximation* of $\underline{P}$.

With respect to the second, one preliminary idea would be to use the partial order associated with the credal set inclusion and to require $\underline{Q}$ to be *undominated*, in the sense that there is no other 2-monotone capacity $\underline{Q}'$ such that $\mathcal{M}(\underline{P}) \subseteq \mathcal{M}(\underline{Q}') \subsetneq \mathcal{M}(\underline{Q})$. However, this requirement alone does not determine a unique solution, nor does it provide us with a tool to determine the 2-monotone outer approximations, either.

3   Approximations by Linear Programming

In order to overcome the above issues, in this paper we shall consider the outer approximations $\underline{Q}$ of the coherent lower probability $\underline{P}$ that minimize the distance proposed by Baroni and Vicig [7], given by

$$d(\underline{P}, \underline{Q}) := \sum_{E \subseteq \mathcal{X}} (\underline{P}(E) - \underline{Q}(E)). \tag{2}$$

If we interpret $\underline{P}(E) - \underline{Q}(E)$ as the additional imprecision introduced on E when replacing $\underline{P}(E)$ with $\underline{Q}(E)$, then $d(\underline{P}, \underline{Q})$ can be understood as the total imprecision added by the outer approximation $\underline{Q}$.

To solve the minimization problem, we determine $\underline{Q}$ through its *Möbius inverse* $m_{\underline{Q}}$ by means of the formula $\underline{Q}(E) = \sum_{B \subseteq E} m_{\underline{Q}}(B)$ for every $E \subseteq \mathcal{X}$ and consider thus the following linear programming problem:

$$\min d(\underline{P}, \underline{Q}) \tag{LP-2monot}$$

subject to:

$$\sum_{E \subseteq \mathcal{X}} m_{\underline{Q}}(E) = 1, \qquad m_{\underline{Q}}(\emptyset) = 0. \tag{LP-2monot.1}$$

$$\sum_{\{x_i, x_j\} \subseteq A \subseteq E} m_{\underline{Q}}(A) \geq 0, \quad \forall E \subseteq \mathcal{X}, \forall x_i, x_j \in E, \ x_i \neq x_j. \tag{LP-2monot.2}$$

$$m_{\underline{Q}}(\{x_i\}) \geq 0, \quad \forall x_i \in \mathcal{X}. \tag{LP-2monot.3}$$

$$\sum_{A \subseteq E} m_{\underline{Q}}(A) \leq \underline{P}(E) \quad \forall E \neq \emptyset, \mathcal{X}. \tag{LP-2monot.4}$$

In fact, (LP-2monot.2) characterizes 2-monotonicity of $\underline{Q}$ via its Möbius inverse $m_{\underline{Q}}$ [11], while (LP-2monot.1) and (LP-2monot.3) ensure that $\underline{Q}$ is also a coherent lower probability. Finally, (LP-2monot.4) means that $\underline{Q}$ outer approximates $\underline{P}$. It is not difficult to check that the number of constraints in this linear programming problem is $2^n + n + 2^{n-2}\binom{n}{2}$.

The feasible region of this linear programming problem is non-empty: it suffices to take into account that the *vacuous* lower probability, given by $\underline{Q}_v(E) = 0$ for every E that is not equal to $\mathcal{X}$ and $\underline{Q}_v(\mathcal{X}) = 1$, is a 2-monotone outer approximation of any coherent lower probability.

Moreover, the linear programming problem above has an optimal solution by means of Weierstrass' theorem [12]. To see this, note that (i) $d(\underline{P}, \underline{Q}) = \sum_{E \subseteq \mathcal{X}} (\underline{P}(E) - \sum_{B \subseteq E} m_{\underline{Q}}(B))$ is continuous on the variables $m_{\underline{Q}}(B)$; (ii) the feasible region is bounded, since by [13, Theorem 1] the values of $m_{\underline{Q}}$ are bounded when $\underline{Q}$ is 2-monotone; and (iii) it is closed, being a polyhedral set in $\mathbb{R}^{2^n}$.

Given this, our first result tells us that any solution of the linear programming problem is undominated:

Proposition 1. *Let $\underline{P}$ be a coherent lower probability, and let $\underline{Q}$ be an optimal solution of the linear programming problem (LP-2monot). Then, $\underline{Q}$ is an undominated outer approximation of $\underline{P}$.*

Not surprisingly, (LP-2monot) may not have a unique solution:

Example 1. Consider $\mathcal{X} = \{x_1, x_2, x_3, x_4\}$ and let $\underline{P}$ be the coherent lower probability that is the lower envelope of the probability mass functions $P_1 = (0.5, 0.5, 0, 0)$, $P_2 = (0, 0, 0.5, 0.5)$. It is given by:

$$\underline{P}(A) = \begin{cases} 0 & \text{if } |A| = 1 \text{ or } A = \{x_1, x_2\}, \{x_3, x_4\} \\ 1 & \text{if } A = \mathcal{X} \\ 0.5 & \text{otherwise.} \end{cases}$$

To see that $\underline{P}$ is not 2-monotone, note that, given $A = \{x_1, x_3\}$ and $B = \{x_2, x_3\}$,

$$\underline{P}(A \cup B) + \underline{P}(A \cap B) = 0.5 < 1 = \underline{P}(A) + \underline{P}(B).$$

To see that (LP-2monot) may have more than one solution, note that, if $\underline{Q}$ is a 2-monotone outer approximation of $\underline{P}$, it must satisfy $\underline{Q}(\{x_1, x_3\}) + \underline{Q}(\{x_2, x_3\}) \leq \underline{Q}(\{x_1, x_2, x_3\}) + \underline{Q}(\{x_3\}) \leq 0.5$, whence $\overline{P}(\{x_1, x_3\}) + \overline{P}(\{x_2, x_3\}) - \underline{Q}(\{x_1, x_3\}) - \underline{Q}(\{x_2, x_3\}) \geq 0.5$; similarly, we obtain that $\underline{P}(\{x_1, x_4\}) + \overline{P}(\{x_2, x_4\}) - \underline{Q}(\{x_1, x_4\}) - \underline{Q}(\{x_2, x_4\}) \geq 0.5$, and therefore $d(\underline{P}, \underline{Q}) \geq 1$ for any 2-monotone outer approximation of $\underline{P}$. This distance is attained by the 2-monotone capacities $\underline{Q}_1, \underline{Q}_2$ given by

$$\underline{Q}_1(A) = \begin{cases} 0 & \text{if } |A| = 1 \text{ or } A = \{x_1, x_2\}, \{x_3, x_4\} \\ 0.5 & \text{if } |A| = 3 \\ 1 & \text{if } A = \mathcal{X} \\ 0.25 & \text{otherwise.} \end{cases}$$

and

$$\underline{Q}_2(A) = \begin{cases} 0 & \text{if } |A| = 1 \text{ or } A = \{x_1, x_2\}, \{x_3, x_4\} \\ 0.5 & \text{if } |A| = 3 \\ 1 & \text{if } A = \mathcal{X} \\ 0.2 & \text{if } A = \{x_1, x_4\}, \{x_2, x_3\} \\ 0.3 & \text{otherwise.} \end{cases}$$

Their 2-monotonicity can easily be verified by means of Eq. (1).   ♦

Obviously, if our initial model $\underline{P}$ is not 2-monotone, any undominated 2-monotone capacity that outer approximates $\underline{P}$ will not agree with $\underline{Q}$ on some event A. Interestingly, it can be checked that both models always agree on singletons:

Proposition 2. *Let $\underline{P}$ be a coherent lower probability. If $\underline{Q}$ is an undominated 2-monotone capacity that outer approximates $\underline{P}$, then $\underline{Q}(\{x\}) = \underline{P}(\{x\})$ for every $x \in \mathcal{X}$.*

As a consequence, both of them induce the same order on $\mathcal{X}$. It can be checked that this property does not extend to some particular subfamilies of 2-monotone capacities, such as belief functions.

4   Particular Cases

In this section, we investigate the outer approximations of a coherent lower probability in some subfamilies of 2-monotone capacities associated with distortion models. With the term *distortion model* we refer to a model where an initial probability measure P_0 is modified in some sense.

4.1   Pari-Mutuel Models

We begin by considering the *Pari Mutuel Model* [2, 14, 15] (PMM, for short). This is a betting scheme originated in horse racing. It is determined by two elements: a probability measure P_0 and a distortion factor $\delta > 0$. For every event A of $\mathcal{P}(\mathcal{X})$, $P_0(A)$ is interpreted as a fair price for a bet on A, and $\delta > 0$ denotes the loading of the house. They determine a coherent lower probability by:

$$\underline{P}(A) = \max\{0, (1 + \delta)P_0(A) - \delta\} \ \forall A \subseteq \mathcal{X}. \tag{3}$$

The lower probability associated with a PMM is 2-monotone, as shown for instance in [15, Sect. 2]. Moreover, in [14] it is proven that PMMs correspond to particular instances of *probability intervals* [4].

Our next result gives the unique undominated outer approximation of a coherent lower probability in terms of pari mutuel models.

Proposition 3. *Let $\underline{P}$ be a coherent lower probability with conjugate upper probability $\overline{P}$. Define the constant value $\delta > 0$ and the probability P_0 by:*

$$\delta = \sum_{i=1}^{n} \overline{P}(\{x_i\}) - 1, \qquad P_0(\{x_i\}) = \frac{\overline{P}(\{x_i\})}{1 + \delta} \ \ \forall i = 1, \ldots, n.$$

Denote by $\underline{Q}$ the coherent lower probability associated with the PMM (P_0, δ) by means of Eq. (3). Then, $\underline{Q}$ is the unique undominated pari mutuel model that outer approximates $\underline{P}$.

4.2 ε-contamination Models

Another distortion model is the *ε-contamination model*, also called *linear-vacuous mixture* in [2]. Given a probability measure P_0 and $\varepsilon \in (0,1)$, they determine the coherent lower probability

$$\underline{P}(A) = \begin{cases} (1-\varepsilon)P_0(A) & \text{if } A \neq \mathcal{X}. \\ 1 & \text{if } A = \mathcal{X}. \end{cases} \tag{4}$$

Equivalently, $\underline{P} = (1-\varepsilon)P_0 + \varepsilon\underline{Q}_v$. The lower probability induced by such a model is 2-monotone. This follows from the fact that it satisfies an even stronger property: complete monotonicity, as can be deduced for instance from [10, Theorems 5 and 11].

As with the PMM, we prove that there is only one undominated outer approximation for a coherent lower probability in terms of ε-contamination models.

Proposition 4. *Let $\underline{P}$ be a coherent lower probability satisfying the condition $\sum_{j=1}^{n} \underline{P}(\{x_j\}) > 0$. Define $\varepsilon \in (0,1)$ and the probability P_0 by:*

$$\varepsilon = 1 - \sum_{j=1}^{n} \underline{P}(\{x_j\}), \quad P_0(\{x_i\}) = \frac{\underline{P}(\{x_i\})}{\sum_{j=1}^{n} \underline{P}(\{x_j\})} \quad \forall i = 1, \ldots, n.$$

Denote by $\underline{P}_\varepsilon$ the ε-contamination model they determine by means of Eq. (4). Then, $\underline{P}_\varepsilon$ is the unique undominated ε-contamination model that outer approximates $\underline{P}$.

Note that the assumption $\sum_{j=1}^{n} \underline{P}(\{x_j\}) > 0$ in this proposition is necessary for the existence of some outer approximation: if $\underline{P}(\{x_j\}) = 0$ for every $x_j \in \mathcal{X}$, any ε-contamination model that outer approximates $\underline{P}_\varepsilon$ should also satisfy $\underline{P}_\varepsilon(\{x_j\}) = 0$ for every $x_j \in \mathcal{X}$, whence

$$\underline{P}_\varepsilon(\{x_j\}) = (1-\varepsilon)P_0(\{x_j\}) = 0 \quad \forall x_j \in \mathcal{X},$$

where P_0 is the precise probability in the ε-contamination model. However, since $\varepsilon \in (0,1)$, it follows that $P_0(\{x_j\}) = 0$ for every $x_j \in \mathcal{X}$ and $P_0(\mathcal{X}) = \sum_{j=1}^{n} P_0(\{x_j\}) = 0$, a contradiction.

5 Comparison with Other Approaches

In this section, we briefly explore other alternatives to the linear programming approach we have considered so far, in order to justify better our choice.

5.1 Quadratic Problems

As Example 1 shows, the linear programming problem (LP-2monot) may not have a unique solution. One way to overcome this issue is to consider, instead of the distance given by Eq. (2), the quadratic distance given by:

$$\tilde{d}(\underline{P}, Q) := \sum_{E \subseteq \mathcal{X}} (\underline{P}(E) - Q(E))^2.$$

It is not difficult to prove that, for any coherent lower probability $\underline{P} : \mathcal{P}(\mathcal{X}) \to [0, 1]$, there is a unique 2-monotone capacity $Q \leq \underline{P}$ that minimizes $\tilde{d}(\underline{P}, Q)$. From this it follows that Q is undominated in the family of outer approximations of $\underline{P}$ by 2-monotone capacities. Note this outer approximation need not be one of the solutions of the linear programming problem (LP-2monot).

In spite of this positive result, while in our view the distance of Baroni and Vicig may be interpreted as the additional imprecision introduced by the outer approximation, a similar interpretation of the quadratic distance is not immediate; further, summing squares of differences in $[0, 1]$ the solution of the quadratic problem may seem closer to the original model than it actually is.

5.2 The Total Variation Distance

Another possibility would be to consider an extension of the *total variation distance* [16, Chap. 4.1] to the imprecise case. Recall that given two probability measures P_1 and P_2, their *total variation* is defined as

$$||P_1 - P_2|| = \max_{E \subseteq \mathcal{X}} |P_1(E) - P_2(E)|.$$

This definition can be equivalently expressed as:

$$||P_1 - P_2|| = \frac{1}{2} \sum_{x \in \mathcal{X}} |P_1(\{x\}) - P_2(\{x\})|.$$

In an imprecise framework, given two coherent lower probabilities $\underline{P}_1, \underline{P}_2$, we can extend the definition above in a number of (not necessarily equivalent) ways:

$$d_1(\underline{P}_1, \underline{P}_2) = \max_{E \subseteq \mathcal{X}} |\underline{P}_1(E) - \underline{P}_2(E)|,$$

$$d_2(\underline{P}_1, \underline{P}_2) = \frac{1}{2} \sum_{x \in \mathcal{X}} |\underline{P}_1(\{x\}) - \underline{P}_2(\{x\})|,$$

$$d_3(\underline{P}_1, \underline{P}_2) = \sup_{P_1 \in \mathcal{M}(\underline{P}_1), P_2 \in \mathcal{M}(\underline{P}_2)} ||P_1 - P_2||,$$

and we refer to [1, Sect. 11.4] for some comments on d_1 in the context of imprecise Markov chains.

However, all these extensions may lead to outer approximations that are dominated, and therefore cannot be considered adequate for our problem, as the next examples show.

Example 2. Consider a four element space and the lower probability $\underline{P}$ defined in the following table:

A	$\underline{P}(A)$	$\underline{Q}'_1(A)$	$\underline{Q}'_2(A)$
$\{x_1\}$	0.1	0.1	0.1
$\{x_2\}$	0	0	0
$\{x_3\}$	0	0	0
$\{x_4\}$	0.1	0.1	0.1
$\{x_1, x_2\}$	0.4	0.3	0.3
$\{x_1, x_3\}$	0.4	0.3	0.3
$\{x_1, x_4\}$	0.4	0.4	0.3
$\{x_2, x_3\}$	0.2	0.2	0.2
$\{x_2, x_4\}$	0.4	0.3	0.3
$\{x_3, x_4\}$	0.4	0.3	0.3
$\{x_1, x_2, x_3\}$	0.5	0.5	0.5
$\{x_1, x_2, x_4\}$	0.6	0.6	0.6
$\{x_1, x_3, x_4\}$	0.6	0.6	0.6
$\{x_2, x_3, x_4\}$	0.5	0.5	0.5
$\mathcal{X}$	1	1	1

Note that $\underline{P}$ is a coherent lower probability because it is the lower envelope of the probability measures with mass functions

$$(0.4, 0, 0.2, 0.4), \ (0.3, 0.1, 0.1, 0.5), \ (0.3, 0.3, 0.3, 0.1), \ (0.1, 0.3, 0.3, 0.3)$$
$$(0.4, 0.2, 0, 0.4), \ (0.2, 0.2, 0.4, 0.2), \ (0.2, 0.4, 0.2, 0.2), \ (0.5, 0.1, 0.1, 0.3).$$

To see that it is not 2-monotone, note that, taking $A = \{x_1, x_2\}$ and $B = \{x_1, x_3\}$ it holds that:

$$\underline{P}(\{x_1, x_2, x_3\}) + \underline{P}(\{x_1\}) = 0.6 < 0.8 = \underline{P}(\{x_1, x_2\}) + \underline{P}(\{x_1, x_3\}).$$

Therefore, any outer approximation $\underline{Q}$ in the class of 2-monotone lower probabilities must satisfy:

$$\underline{Q}(\{x_1, x_2\}) + \underline{Q}(\{x_1, x_3\}) \leq \underline{P}(\{x_1, x_2\}) + \underline{P}(\{x_1, x_3\}) - 0.2.$$

Hence, $d_1(\underline{P}, \underline{Q}) \geq 0.1$. Also, the previous inequality is indeed an equality, which is attained, for example, by the 2-monotone capacities $\underline{Q}'_1, \underline{Q}'_2$ in the table above. Thus, both $\underline{Q}'_1, \underline{Q}'_2$ are optimal outer approximations with respect to the distance d_1, even if $\underline{Q}'_2$ is dominated by $\underline{Q}'_1$. ◆

Example 3. Consider again the coherent lower probability from Example 1. Any 2-monotone outer approximation $\underline{Q}$ of $\underline{P}$, undominated or not, shall satisfy $\underline{Q}(\{x_j\}) = 0$ for every j, and as a consequence $d_2(\underline{P}, \underline{Q}) = 0$. As for d_3, since $\|P_1 - P_2\| = 1$ for the probability measures $P_1 = (0.5, 0.5, 0, 0)$ and $P_2 = (0, 0, 0.5, 0.5)$ from $\mathcal{M}(\underline{P})$, and by definition this is the maximum value of the total variation, we deduce that $d_3(P_1, P_2) = 1$. Because $\mathcal{M}(\underline{P}) \subset \mathcal{M}(\underline{Q})$, also $d_3(\underline{P}, \underline{Q}) = 1$ for any 2-monotone outer approximation $\underline{Q}$ of $\underline{P}$, even for the 'most dominated' vacuous lower probability $\underline{Q}_v$. Thus, d_2, d_3 do not rule out the undominated solutions, either. ◆

5.3   The Weber Set

We have already mentioned that one of the advantages of 2-monotone capacities is the existence of a simple procedure to obtain the extreme points of the associated credal set. Let $\underline{P}$ be a 2-monotone capacity, and for any permutation σ of $\{1, \ldots, n\}$, define the precise probability P_σ by means of the constraints

$$P_\sigma(\{x_{\sigma(1)}, \ldots, x_{\sigma(k)}\}) = \underline{P}(\{x_{\sigma(1)}, \ldots, x_{\sigma(k)}\}) \tag{5}$$

for $k = 1, \ldots, n$. It was first proven by Shapley [3] that, if S_n denotes the set of permutations of $\{1, \ldots, n\}$, it holds that $ext(\mathcal{M}(\underline{P})) = \{P_\sigma \mid \sigma \in S_n\}$. In general, even when $\underline{P}$ is not 2-monotone but only coherent, we can define the set of probabilities:

$$W(\underline{P}) = \{P_\sigma \mid \sigma \in S_n\}, \tag{6}$$

where P_σ is defined as in Eq. (5). This set is called the *Weber set* of $\underline{P}$, and it holds that [17] $\underline{P}$ is 2-monotone if and only if $ext(\mathcal{M}(\underline{P})) = W(\underline{P})$. Otherwise, $\mathcal{M}(\underline{P})$ is a proper subset of $conv(W(\underline{P}))$. This implies that the lower envelope of $conv(W(\underline{P}))$ is a coherent lower probability that outer approximates $\underline{P}$.

In fact, in the case of cardinality four, the lower envelope of $conv(W(\underline{P}))$ is indeed 2-monotone:

Proposition 5. *Let $\underline{P} : \mathcal{P}(\mathcal{X}) \to [0, 1]$ be a coherent lower probability, where $|\mathcal{X}| \leq 4$, and denote by Q the coherent lower probability defined by $Q(E) = \min\{P(E) \mid P \in conv(\overline{W}(\underline{P}))\}$ for every $E \subseteq \mathcal{X}$, where $W(\underline{P})$ is given by Eq. (6). Then, Q is a 2-monotone outer approximation of $\underline{P}$.*

It can be checked that the lower envelope of the Weber set is not necessarily 2-monotone for cardinalities greater than four. Moreover, even in the case of cardinality four the lower envelope of the Weber set is not in general an undominated outer approximation:

Example 4. Consider a four-element space $\mathcal{X} = \{x_1, x_2, x_3, x_4\}$, and the lower probability $\underline{P}$ given in the following table:

A	$\underline{P}(A)$	$Q(A)$	$Q'(A)$
$\{x_1\}$	0.1	0.1	0.1
$\{x_2\}$	0	0	0
$\{x_3\}$	0	0	0
$\{x_4\}$	0.3	0.3	0.3
$\{x_1, x_2\}$	0.1	0.1	0.1
$\{x_1, x_3\}$	0.3	0.2	0.3
$\{x_1, x_4\}$	0.6	0.5	0.5
$\{x_2, x_3\}$	0.3	0.2	0.2
$\{x_2, x_4\}$	0.4	0.3	0.4
$\{x_3, x_4\}$	0.4	0.3	0.4
$\{x_1, x_2, x_3\}$	0.5	0.5	0.5
$\{x_1, x_2, x_4\}$	0.6	0.6	0.6
$\{x_1, x_3, x_4\}$	0.7	0.7	0.7
$\{x_2, x_3, x_4\}$	0.6	0.6	0.6
$\mathcal{X}$	1	1	1

If we compute $Q = \min\{P \mid P \in conv(W(\underline{P}))\}$, we obtain the values depicted in the table above. However, this 2-monotone capacity is dominated by the 2-monotone outer approximation Q' given in the same table.   ♦

6   Approximations of Coherent Lower Previsions

The problem considered in this paper could be generalized from coherent lower probabilities to the richer framework of coherent *lower previsions* [2]: if we denote by $\mathcal{L}(\mathcal{X})$ the set of bounded real-valued functions on $\mathcal{X}$, a coherent lower prevision is a function $\underline{P} : \mathcal{L}(\mathcal{X}) \to \mathbb{R}$ that satisfies

- $\underline{P}(f) \geq \inf f$
- $\underline{P}(\lambda f) = \lambda \underline{P}(f)$
- $\underline{P}(f + g) \geq \underline{P}(f) + \underline{P}(g)$

for every $f, g \in \mathcal{L}(\mathcal{X})$ and every $\lambda > 0$. Equivalently, $\underline{P}$ is coherent when it is the lower envelope of a set of expectation operators with respect to a family of probability measures on $\mathcal{X}$.

The notion of 2-monotonicity has also been extended to lower previsions: $\underline{P}$ is a 2-monotone lower prevision if and only if

$$\underline{P}(f \wedge g) + \underline{P}(f \vee g) \geq \underline{P}(f) + \underline{P}(g) \ \forall f, g \in \mathcal{L}(\mathcal{X}),$$

where $\wedge$ and $\vee$ denote the pointwise minimum and maximum. In general, a coherent lower probability $\underline{P}$ on $\mathcal{P}(\mathcal{X})$ may have more than one extension as a coherent lower prevision on $\mathcal{L}(\mathcal{X})$; however, if $\underline{P}$ is 2-monotone, then it has a unique extension to $\mathcal{L}(\mathcal{X})$ as a 2-monotone lower prevision: its Choquet integral [10], that is also its least-committal or *natural extension* [2].

Similarly to what we have done in the rest of the paper, we could study the problem of outer approximating a coherent lower prevision by a 2-monotone one. Interestingly, this problem turns out to be equivalent to the one we are considering in this paper, as our next result shows:

Theorem 1. *Let $\underline{P} : \mathcal{L}(\mathcal{X}) \to \mathbb{R}$ be a coherent lower prevision, and let $\underline{P}'$ be its restriction to events. Then, there is a one-to-one correspondence between the sets*

$$\{\underline{Q} : \mathcal{L}(\mathcal{X}) \to \mathbb{R} \ 2\text{-monotone undominated outer approximation of } \underline{P}\}$$

and

$$\{\underline{Q}' : \mathcal{P}(\mathcal{X}) \to [0, 1] \ 2\text{-monotone undominated outer approximation of } \underline{P}'\}.$$

The key in this result is that if we want to outer approximate a coherent lower prevision, we can simply consider its restriction to events, outer approximate it and then apply the procedure of natural extension in [2]. Figure 1 illustrates the procedure.

Therefore, it suffices to focus on outer approximations of coherent lower probabilities instead of lower previsions.

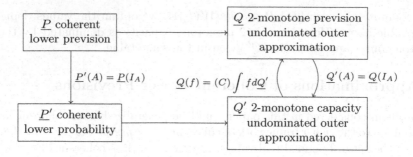

Fig. 1. Correspondence between the 2-monotone outer approximations.

7   Conclusions

Our results allow us to conclude that we can find undominated outer approximations of a coherent lower probability that are at the same time as close as possible, in the sense of Baroni and Vicig, by means of a suitable linear programming problem. Although the approximation is not unique in general, it is so if we focus on some particular subfamilies of 2-monotone capacities, such as those associated with certain distortion models. Moreover, the problem can be immediately applied to the approximation of coherent lower previsions by 2-monotone ones.

While in our view the distance we have considered is the most meaningful for the problem at hand and the results in Sect. 5 support this somewhat, we should also make a more thorough comparison with other distances from the literature, and also with the related study made in [18] about outer approximations with possibility measures.

As other future lines of research, we would like to study in more detail the loss of information entailed by the outer approximations, as well as the elicitation among them when there is more than one solution. In addition, we would also like to investigate more deeply the features of the solutions obtained by means of the quadratic approach. Finally, it may be interesting to consider the problem of the inner approximations of a coherent lower probability, even if they entail removing, perhaps unjustifiedly, some imprecision from our model.

Acknowledgements. The research in this paper has been supported by project TIN2014-59543-P. We would also like to thank Sébastien Destercke for some helpful suggestions.

References

1. Augustin, T., Coolen, F., de Cooman, G., Troffaes, M. (eds.): Introduction to Imprecise Probabilities. Wiley Series in Probability and Statistics. Wiley, Hoboken (2014)
2. Walley, P.: Statistical Reasoning with Imprecise Probabilities. Chapman and Hall, London (1991)

3. Shapley, L.S.: Cores of convex games. Int. J. Game Theor. **1**, 11–26 (1971)
4. de Campos, L.M., Huete, J.F., Moral, S.: Probability intervals: a tool for uncertain reasoning. Int. J. Uncertain. Fuzziness Knowl. Based Syst. **2**, 167–196 (1994)
5. Shafer, G.: A Mathematical Theory of Evidence. Princeton University Press, Princeton (1976)
6. Dubois, D., Prade, H.: Possibility theory: qualitative and quantitative aspects. In: Smets, P. (ed.) Quantified Representation of Uncertainty and Imprecision. Handbook on Defeasible Reasoning and Uncertainty Management Systems, vol. 1, pp. 169–226. Kluwer Academic Publishers, Dordrecht (1998). https://doi.org/10.1007/978-94-017-1735-9_6
7. Baroni, P., Vicig, P.: An uncertainty interchange format with imprecise probabilities. Int. J. Approx. Reason. **40**, 147–180 (2005)
8. Grabisch, M.: Set Functions, Games and Capacities in Decision Making. Springer, Cham (2016). https://doi.org/10.1007/978-3-319-30690-2
9. Choquet, G.: Theory of capacities. Annales de l'Institut Fourier **5**, 131–295 (1953–1954)
10. de Cooman, G., Troffaes, M.C.M., Miranda, E.: n-monotone exact functionals. J. Math. Anal. Appl. **347**, 143–156 (2008)
11. Chateauneuf, A., Jaffray, J.Y.: Some characterizations of lower probabilities and other monotone capacities through the use of Möbius inversion. Math. Soc. Sci. **17**(3), 263–283 (1989)
12. Rudin, W.: Principles of Mathematical Analysis. McGraw Hill, New York (1976)
13. Grabisch, M., Miranda, P.: Exact bounds of the Möbius inverse of monotone set functions. Discret. Appl. Math. **186**, 7–12 (2015)
14. Montes, I., Miranda, E., Destercke, S.: A study of the pari-mutuel model from the point of view of imprecise probabilities. In: Proceedings of the Tenth International Symposium on Imprecise Probability: Theories and Applications (ISIPTA 2017) (2017)
15. Pelessoni, R., Vicig, P., Zaffalon, M.: Inference and risk measurement with the pari-mutuel model. Int. J. Approx. Reason. **51**, 1145–1158 (2010)
16. Levin, D.A., Peres, Y., Wilmer, E.: Markov Chains and Mixing Times. American Mathematical Society, Providence (2009)
17. Ichiishi, T.: Supermodularity: applications to convex games and to the greedy algorithm for LP. J. Econ. Theory **25**, 283–286 (1981)
18. Dubois, D., Prade, H.: Consonant approximations of belief functions. Int. J. Approx. Reason. **4**(5–6), 419–449 (1990)

Web Apps and Imprecise Probabilitites

Jorge Castro, Joaquim Gabarro[✉], and Maria Serna

ALBCOM, CS Department, Universitat Politècnica de Catalunya, Barcelona, Spain
{castro,gabarro,mjserna}@cs.upc.edu

Abstract. We propose a model for the behaviour of Web apps in the unreliable WWW. Web apps are described by orchestrations. An orchestration mimics the personal use of the Web by defining the way in which Web services are invoked. The WWW is unreliable as poorly maintained Web sites are prone to fail. We model this source of unreliability trough a probabilistic approach. We assume that each site has a probability to fail. Another source of uncertainty is the traffic congestion. This can be observed as a non-deterministic behaviour induced by the variability in the response times. We model non-determinism by imprecise probabilities. We develop here an ex-ante normal to characterize the behaviour of finite orchestrations in the unreliable Web. We show the existence of a normal form under such semantics for orchestrations using asymmetric parallelism.

Keywords: Web apps · Orchestrations · Orc
Imprecise probabilities · Normal forms

1 Introduction

The appearance of the World Wide Web [4] deeply changed our every day life and in particular the way to interact with the world. In this paper we address the following problem: How to give an ex-ante (before execution) meaning of our interaction trough the Web. We model such interactions by means of *orchestrations*. An orchestration is the sequence and conditions in which one Web service invokes other Web services in order to realize some useful task [1]. An orchestration defines the flow control from a one-party perspective (in this case us) [17]. In general, before executing a Web app (for instance to search for flight info and get hotel reservations), we have an idea of the possible outcomes of the execution. Orchestrations are designed to address two main issues:

J. Castro was partially supported by the Spanish Ministry for Economy and Competitiveness (MINECO) under grant (TIN2017-89244-R) and the recognition 2017 SGR-856 (MACDA) from AGAUR (Generalitat de Catalunya). J. Gabarro and M. Serna were partially supported by MINECO and FEDER funds under grant GRAMM (TIN2017-86727-C2-1-R) and Generalitat de Catalunya, Agéncia de Gestió d'Ajuts Universitaris i de Recerca, under project 2017 SGR 786 (ALBCOM). M. Serna was also supported by MINECO under grant BGSMath (MDM-2014-044).

© Springer International Publishing AG, part of Springer Nature 2018
J. Medina et al. (Eds.): IPMU 2018, CCIS 854, pp. 226–238, 2018.
https://doi.org/10.1007/978-3-319-91476-3_19

- Interacting trough the Web using big doses of *parallelism*. We can have several browsers (like Mozilla, Chrome or Explorer) an try to get different pieces of information at the same time.
- Web is *unreliable*. Sometimes the invoked service responds but others does not. Perhaps those Web services are no longer maintained or simply they are not available at this moment. A natural way to overcome unreliability is the use of *redundancy*.

The causes of uncertainty depend deeply on the universe we are dealing with. For instance, the causes of uncertainty in economy [9,12] appear to be quite different from those on the Web. When a basic service is invoked, a site call is executed, the site can provide an answer returning some information or it can fail to (broken link). Moreover, this situation is far from being stable. Usually, based on our knowledge on the site behaviour or on external information, it is feasible to assume a *priori probability* for the broken (or silent) site event. In order to minimize the risk of calling a silent site, it is usual to issue several calls to sites providing similar information. In such a case the answer that arrives first is chosen. However, there is no a priori knowledge on which site will respond first [5] because in many cases becomes too hard to get sufficient data on the environment in order to provide precise probabilistic predictions. This lack of precise probabilistic knowledge appears when considering an indeterministic behaviour. Following [3],we propose to model non-determinism in terms of *imprecise probabilities*.

The ex-ante characterization of an orchestration, although formulated in terms of imprecise probabilities, has an obvious practical relevance. Let us consider an orchestration P that guarantees the result *great success* with an imprecise probability greater than $1/3$ and obtains the *satisfactory* result with probability greater than $3/4$. Let Q be another orchestration that guarantees these same results with imprecise probabilities that are greater than, respectively, $1/4$ and $4/5$. Depending on our particular circumstances we can choose in a reasoned way which of the two processes, P or Q, is more convenient for our interests.

Besides proposing the uncertainty model we extend the bag semantics for orchestrations proposed in [7] to deal with daemonic indeterminism through imprecise probabilities. This allows us to generalize the previous theorem on the existence of normal forms to a more general setting. Technically, Theorem 2 shows a normal form characterization for probabilistic orchestrations and Theorem 3 extends this result to the non-deterministic case. We complement this theoretical result by developing a complete example of uncertainty analysis in our proposed model. We also sketch other possible applications of imprecise probabilities to orchestrations. In order to make the paper self-contained we start introducing the Orc language [14]. We also recall the bag semantics [7] providing meaning to Orc expressions.

2   Orchestrations in Reliable Environments

An orchestration is a user-defined program that utilizes services on the Web. In Orc [14] services are modelled by sites which have some predefined semantics.

Typical examples of services are: an eigensolver, a search engine or a database. A *site* accepts an argument and *publishes* a result value[1]. For example, a call to a search engine, *Find(s)*, may publish the set of sites which *currently* offer service *s*. A site is *silent* if it does not publish a result. A site call can publish *at most one response*. Although a site call may have a well-defined result it may be the case that a call to the site, in an untrusted environment, fails (silence). Orc contains a number of inbuilt sites: 0 is always silent while $1(x)$ always publishes x. An orchestration which composes a number of service calls into a complex computation can be represented by an Orc expression. An orchestrator may utilize any service that is available on the grid. In this paper we deal only with *finite orchestrations* where finite means: excluding iteration and recursion. Two Orc expressions P and Q can be combined using the following operators [11,14]:

- Sequence $P > x > Q(x)$: P is evaluated and, for each value v published by P, an instance $Q(v)$ is executed. If P publishes the stream, $v_1, v_2, \ldots v_n$, then $P > x > Q(x)$ publishes some permuted stream of the outputs of the calls $Q(v_1), Q(v_2), \ldots, Q(v_n)$. When the value of x is not needed we write $P \gg Q$.
- Symmetric Parallelism $P \mid Q$: P and Q are evaluated in parallel. $P \mid Q$ publishes *some* interleaving of the streams published by P and Q.
- Asymmetric Parallelism $P(x) < x < Q$: P and Q are evaluated in parallel. Some sub-expressions in P may become blocked by a dependency on x. The first result published by Q is bound to x, the remainder of Q's evaluation is terminated and evaluation of the blocked residue of P is resumed.

Usually orchestrations assume some degree of redundancy. Following an example.

Example 1. Suppose that you need to send news to a group. Usually you prefer the *BBC* but it is uncertain to get a result because there is a call for strike. In such a case, you also try to get news from the *CNN*, *News* = (*BBC* | *CNN*). To inform the group, you send the news to Alice, but at this moment you are uncertain about her capacity to get the email, therefore you send also the news to Bob, *Emails(x)* = (*Alice(x)* | *Bob(x)*). Consider the orchestration *eNews* = *Emails(x)* < *x* < *News*.

    Let us describe the behaviour of *eNews*. A call to *eNews* spans into simultaneous (parallel) calls (or threads) to *News* and *Emails(x)*. The call to *News* span into parallel calls to *BBC* and *CNN*. The call to *Emails(x)* span into parallel calls to *Alice(x)* and *Bob(x)* At this moment the call to *eNews* has evolved into four simultaneous threads (the programs executing the calls) corresponding to *BBC*, *CNN*, *Alice(x)* and *Bob(x)*.

    The thread corresponding to *Alice(x)* will remain blocked until variable x takes a value. The same will happen with *Bob(x)*. Eventually (at some future time) the calls to *BBC* and *CNN* return. Assume that *BBC* returns first, this value will be assigned to x. Once x has a value, threads corresponding to *Alice(x)*

---

[1] The words "publishes","returns" and "outputs" are used interchangeably. The terms "site" and "service" are also used interchangeably.

and $Bob(x)$ proceed and Alice and Bob will receive an email with the BBC news. Another result is possible if CNN returns first. In this case Alice and Bob get the CNN info. Note that $eNews$ has no control about which result will appear, therefore is a non-deterministic program.   □

To reason about a program, we need a semantics *to assign meaning* [6]. Bag semantics was introduced in [7] to give a precise description of the approach taken in Example 1. In such approach we abstract from return time. First, let us start from the operational semantics introduced in [14]. In such a model any variable x contains all the possible values before being used. Therefore, variables keep a stream of values. When an orchestration E publishes a stream $v_1, v_2, \ldots, v_n$, *the relative ordering of the values depends on the relative response time of the sites* appearing in E. However, when we are uncertain about return times, is a strongly desirable *abstract from time*. In such a case we forget about orderings in the streams describing them as a multi-set or bag $\lfloor\!\lfloor v_1, v_2, \ldots, v_n \rfloor\!\rfloor$ (notation $\lfloor\!\lfloor \cdot \rfloor\!\rfloor$ is taken from [13]). In such a case, the "meaning" of E, denoted by $[\![E]\!]$, is the bag $\lfloor\!\lfloor v_1, v_2, \ldots, v_n \rfloor\!\rfloor$ and we write $[\![E]\!] = \lfloor\!\lfloor v_1, v_2, \ldots, v_n \rfloor\!\rfloor$. The fact that site 0 never returns is formalized as site 0 returns nothing, that is $[\![0]\!] = \lfloor\!\lfloor \,\, \rfloor\!\rfloor = \emptyset$. As the pruning operator (or parallel asymmetric composition) can give rise to a non deterministic behaviour, we consider also the "daemonic choice" operator $\sqcap$ [13, p. 4] to denote non deterministic choice. Toni Hoare in [10], considers the non-deterministic choice $P \sqcap Q$ between processes P and Q. In such a case $P \sqcap Q$ denotes a process which behaves like P or Q, where the selection is done without knowledge or control of the external environment. Such a choice is called *daemonic choice*. A semantic characterization of $P \sqcap Q$ in terms of *refusal sets* can be found in [10]. In a reliable environment, a call to a site S always returns a value and we write $[\![S]\!] = \lfloor\!\lfloor s \rfloor\!\rfloor$.

In the following examples we justify the use of bags and how they can be obtained from the simple bags corresponding to site calls.

Example 2. $[\![BBC]\!] = \lfloor\!\lfloor bbc \rfloor\!\rfloor$ and $[\![CNN]\!] = \lfloor\!\lfloor cnn \rfloor\!\rfloor$. A call to *News* in the preceding Example 1 returns a bag containing two items, $[\![News]\!] = [\![BBC \mid CNN]\!] = \lfloor\!\lfloor bbc, cnn \rfloor\!\rfloor$. This result is consistent with the idea that in *News* the BBC and the CNN are called in parallel an they return at different moments. The orchestration has no control on which one will return first. The bag $\lfloor\!\lfloor bbc, cnn \rfloor\!\rfloor$ mimics the idea that eventually we will get both results but we forget temporal information.   □

Sometimes we want to introduce redundancy as, for example in $TwiceBBC = (BBC \mid BBC)$. Observe that this orchestration returns $\lfloor\!\lfloor bbc, bbc \rfloor\!\rfloor$ mimicking the idea of getting twice the same result. Showing the need of using *bags* (or *multisets*). Sometimes we get expressions that depend on the values that a variable gets during the execution. In such a case, the bag semantics provides a meaning for the variable that is used to derive the meaning of the expression. Besides in a reliable environment, asymmetric parallelism introduces indeterminism. The following example illustrates those traits.

Example 3. Now we face the meaning of x appearing in $eNews = Emails(x) < x < News$. According to Example 2 we have $[\![News]\!] = \lfloor\!\lfloor \text{bbc}, \text{cnn} \rfloor\!\rfloor$. As we do not control explicitly the return times, under some (external and uncontrolled) circumstances, a call $News$ returns bbc but in some other cases it returns cnn. So, x can hold either of both values, i.e., $[\![x]\!] = \lfloor\!\lfloor \text{bbc} \rfloor\!\rfloor \sqcap \lfloor\!\lfloor \text{cnn} \rfloor\!\rfloor$. Assuming $[\![Emails(\text{x})]\!] = \lfloor\!\lfloor \text{alice\_x}, \text{bob\_x} \rfloor\!\rfloor$,

$$[\![eNews]\!] = [\![Emails(\text{bbc})]\!] \sqcap [\![Emails(\text{cnn})]\!]$$
$$= \lfloor\!\lfloor \text{alice\_bbc}, \text{bob\_bbc} \rfloor\!\rfloor \sqcap \lfloor\!\lfloor \text{alice\_cnn}, \text{bob\_cnn} \rfloor\!\rfloor$$

This result translates the idea that, depending on external circumstances, two possible output streams are possible: Alice and Bob get the BBC or Alice and Bob get the CNN. The orchestrator has no control on which one will occur.   □

Working in a similar way with the different operations the existence of a normal form can be shown.

Theorem 1 [7]. *Given an Orc expression E it holds that either $[\![E]\!] = \lfloor\!\lfloor\ \rfloor\!\rfloor$ or there is a unique non-deterministic finite decomposition in multi-sets $[\![E]\!] = \sqcap_i M_i$, where elements in M_i corresponds to the possible values returned by site calls.*

3   Orchestrations and Probabilistic Information

Until now, we have considered reliable orchestrations as we were certain about returns. In this section, we consider unreliable settings modelled with probabilities. Let $\Delta_n = \{(p_1, \ldots, p_n) \mid p_i \geq 0, 1 \leq i \leq n, \sum_{i=1}^n p_i = 1\}$. We adopt from [13] the notation $(prog_1 @ p_1 \parallel prog_2 @ p_2 \parallel \cdots \parallel prog_n @ p_n)$ where $(p_1, \ldots, p_n) \in \Delta_n$ and $(prog_1, \ldots, prog_n)$ are sequential programs to represent a probabilistic program that behaves like $prog_i$ with probability p_i.

The probabilistic choice follows two natural laws [13,15]. When the same program $prog_1$ appears twice, we should add the probabilities:

$$(prog_1 @ p_1 \parallel prog_1 @ p_2 \parallel prog_3 @ p_3 \parallel \cdots \parallel prog_n @ p_n)$$
$$= (prog_1 @ (p_1 + p_2) \parallel prog_3 @ p_3 \parallel \cdots \parallel prog_n @ p_n).$$

The second rule assumes distributivity in respect to the daemonic choice operator.

$$((prog_1 \sqcap prog_1') @ p_1 \parallel prog_2 @ p_2 \parallel \cdots \parallel prog_n @ p_n)$$
$$= (prog_1 @ p_1 \parallel \cdots \parallel prog_n @ p_n) \sqcap (prog_1' @ p_1 \parallel prog_2 @ p_2 \parallel \cdots \parallel prog_n @ p_n).$$

Sometimes we can model a faulty uncertain behaviour by a probability distribution on the involved processes, but this is not always possible. We have two semantic models for faulty behaviour.

- *Probabilistic information.* In this case we model a faulty site as a site S returning s with a probability p, and failing to return (behaves like site 0) with probability $(1-p)$. The faulty version of S, denoted as $S_{\mathcal{F}}$ is $S_{\mathcal{F}} = (S@p \parallel 0@(1-p)))$. Moreover $[\![S_{\mathcal{F}}]\!] = ([\![S]\!]@p \parallel \lfloor\!\lfloor \; \rfloor\!\rfloor@(1-p)) = (\lfloor\!\lfloor s\rfloor\!\rfloor@p \parallel \lfloor\!\lfloor \; \rfloor\!\rfloor@(1-p))$. We identify $(\lfloor\!\lfloor s\rfloor\!\rfloor@1 \parallel \lfloor\!\lfloor \; \rfloor\!\rfloor@0) = \lfloor\!\lfloor s\rfloor\!\rfloor$ and $(\lfloor\!\lfloor s\rfloor\!\rfloor@0 \parallel \lfloor\!\lfloor \; \rfloor\!\rfloor@1) = \lfloor\!\lfloor \; \rfloor\!\rfloor$. We assume probabilistic independence on the behaviour of the sites. Two consecutive calls to a given site are considered independent in relation to its probabilistic behaviour.
- *No probabilistic information.* In such a case, we assume indeterminism, i.e., $[\![S_{\mathcal{F}}]\!] = [\![S]\!] \sqcap \lfloor\!\lfloor \; \rfloor\!\rfloor$.

When it is clear from the context that S is faulty, we denote $S_{\mathcal{F}}$ shortly as S.

Example 4. Suppose that, from a user point of view, sites *CNN* and *BBC* are unreliable, $[\![BBC]\!] = (\lfloor\!\lfloor bbc\rfloor\!\rfloor@2/3 \parallel \lfloor\!\lfloor \; \rfloor\!\rfloor@1/3)$ and $[\![CNN]\!] = (\lfloor\!\lfloor cnn\rfloor\!\rfloor@1/2 \parallel \lfloor\!\lfloor \; \rfloor\!\rfloor@1/2)$. A precise semantics for $[\![News]\!]$ comes from the way in which probabilities interact with parallel composition [7]. Assuming independence among executions, the probabilistic behaviour is given in the following table:

	$\lfloor\!\lfloor bbc\rfloor\!\rfloor@2/3$	$\lfloor\!\lfloor \; \rfloor\!\rfloor@1/3$
$\lfloor\!\lfloor cnn\rfloor\!\rfloor@1/2$ $\lfloor\!\lfloor \; \rfloor\!\rfloor@1/2$	$\lfloor\!\lfloor cnn, bbc\rfloor\!\rfloor@(1/2 \times 2/3)$ $\lfloor\!\lfloor bbc\rfloor\!\rfloor@(1/2 \times 2/3)$	$\lfloor\!\lfloor cnn\rfloor\!\rfloor@(1/2 \times 1/3)$ $\lfloor\!\lfloor \; \rfloor\!\rfloor@(1/2 \times 1/3)$

Therefore $[\![News]\!] = (\lfloor\!\lfloor cnn, bbc\rfloor\!\rfloor@1/3 \parallel \lfloor\!\lfloor cnn\rfloor\!\rfloor@1/6 \parallel \lfloor\!\lfloor bbc\rfloor\!\rfloor@1/3 \parallel \lfloor\!\lfloor \; \rfloor\!\rfloor@1/6)$. Different bags can represent the orchestration result. The empty bag $\lfloor\!\lfloor \; \rfloor\!\rfloor$ appears when both sites fail. This result is different form the one in Example 2 where only the bag $\lfloor\!\lfloor cnn, bbc\rfloor\!\rfloor$ appears. □

Probabilistic distributions are parametrized when sites are parametrized $S(x_1, \ldots x_n)$:

$$[\![S(x_1, \ldots x_n)]\!] = \begin{cases} \lfloor\!\lfloor s(v_1, \ldots, v_n)\rfloor\!\rfloor & \text{if } (x_1, \ldots x_n) = (v_1, \ldots v_n) \\ \lfloor\!\lfloor \; \rfloor\!\rfloor & \text{if } \exists i : 1 \leq i \leq n : x_i \text{ undefined} \end{cases}$$

Example 5. Suppose that *Alice* succeeds (or returns) with probability 4/5 and *Bob* returns with probability 5/7. In the case of *Alice* we have:

$$[\![Alice(x)]\!] = \begin{cases} (\lfloor\!\lfloor alice\_v\rfloor\!\rfloor@4/5 \parallel \lfloor\!\lfloor \; \rfloor\!\rfloor@1/5) & \text{if } x = v \\ \lfloor\!\lfloor \; \rfloor\!\rfloor & \text{if } x \text{ is undefined} \end{cases}$$

When $x = \lfloor\!\lfloor cnn\rfloor\!\rfloor$ it holds $[\![Alice(cnn)]\!] = (\lfloor\!\lfloor alice\_cnn\rfloor\!\rfloor@4/5 \parallel \lfloor\!\lfloor \; \rfloor\!\rfloor@1/5)$. When x is undefined, $x = \lfloor\!\lfloor \; \rfloor\!\rfloor$ and $[\![Alice(\lfloor\!\lfloor \; \rfloor\!\rfloor)]\!] = \lfloor\!\lfloor \; \rfloor\!\rfloor$. When it is clear from the context, we write $[\![Alice(x)]\!] = (\lfloor\!\lfloor alice\_x\rfloor\!\rfloor@4/5 \parallel \lfloor\!\lfloor \; \rfloor\!\rfloor@1/5)$ assuming implicitly that, when x is undefined $[\![Alice(x)]\!] = \lfloor\!\lfloor \; \rfloor\!\rfloor$. Let $Emails(x)$ be $(Alice(x) \mid Bob(x))$. The semantics is

$$(\lfloor\!\lfloor alice\_x, bob\_x\rfloor\!\rfloor@4/7 \parallel \lfloor\!\lfloor alice\_x\rfloor\!\rfloor@8/35 \parallel \lfloor\!\lfloor bob\_x\rfloor\!\rfloor@1/7 \parallel \lfloor\!\lfloor \; \rfloor\!\rfloor@2/35).$$

When $x = \lfloor\!\lfloor\mathtt{cnn}\rfloor\!\rfloor$, the semantics of $Emails(\mathtt{cnn})$ is

$$(\lfloor\!\lfloor\mathtt{alice\_cnn}, \mathtt{bob\_cnn}\rfloor\!\rfloor@4/7 \mathbin{\|} \lfloor\!\lfloor\mathtt{alice\_cnn}\rfloor\!\rfloor@8/35 \mathbin{\|} \lfloor\!\lfloor\mathtt{bob\_cnn}\rfloor\!\rfloor@1/7 \mathbin{\|} \lfloor\!\lfloor\ \rfloor\!\rfloor@2/35),$$

and, when x is undefined, $\llbracket Emails(\mathrm{x}) \rrbracket = \lfloor\!\lfloor\ \rfloor\!\rfloor$.   □

The tools described in Examples 4 and 5 can be generalized. We can show that when there is no asymmetric parallelism (no indeterminism) probabilistic information can be carried out through constructors. Based on the approach given in [7], we can shown the existence of a normal form.

Theorem 2. *Let E be a finite orchestration, defined trough sequencing and parallel composition over n different faulty sites. Assume site i succeeds and returns a value with probability p_i. Under the bag semantics, there is a probabilistic finite choice decomposition in multisets $\llbracket E \rrbracket = \mathbin{\|}_j M_j@F_j(p)$, elements in M_j corresponds to the possible values returned by site calls, parameter p is the success probability vector $(p_1, \ldots, p_n)$, and F_j is an arithmetic expression defined on p.*

4   Daemonic Choice and Imprecise Probabilities

We consider in this section the more general case of a non-deterministic orchestration defined on a faulty environment. In order to provide a semantic characterization, we keep probabilistic information as much as possible and encode non-deterministic choices as imprecise probabilities. In this way, the behaviour of a non-deterministic choice on n processes $P_1 \sqcap P_2 \sqcap \cdots \sqcap P_n$ corresponds to any of the possible behaviours defined by an imprecise probability choice:

$$\{(P_1@p_1 \mathbin{\|} P_2@p_2 \mathbin{\|} \cdots \mathbin{\|} P_n@p_n) \mid (p_1, p_2, \ldots p_n) \in \Delta_n)\}.$$

Consequently, we identify the meaning $\llbracket P_1 \sqcap P_2 \sqcap \cdots \sqcap P_n \rrbracket$ with

$$\{(\llbracket P_1 \rrbracket@p_1 \mathbin{\|} \llbracket P_2 \rrbracket@p_2 \mathbin{\|} \cdots \mathbin{\|} \llbracket P_n \rrbracket@p_n) \mid (p_1, p_2, \ldots p_n) \in \Delta_n)\}.$$

Let us observe that in the asymmetric parallelism operation the non deterministic choices are restricted to the selection of an element from a multiset. The following example attempts to grasp the relation between our approach and imprecise probabilities.

Example 6. To assign a meaning to $eNews$, recall from Example 4 that

$$\llbracket News \rrbracket = (\lfloor\!\lfloor\mathtt{cnn}, \mathtt{bbc}\rfloor\!\rfloor@1/3 \mathbin{\|} \lfloor\!\lfloor\mathtt{cnn}\rfloor\!\rfloor@1/6 \mathbin{\|} \lfloor\!\lfloor\mathtt{bbc}\rfloor\!\rfloor@1/3 \mathbin{\|} \lfloor\!\lfloor\ \rfloor\!\rfloor@1/6)$$

We like to keep this probabilistic information as much as possible in $\llbracket x \rrbracket$. As $\llbracket x \rrbracket$ has to be some possible multisets with at most one element we translate $\lfloor\!\lfloor\mathtt{cnn}, \mathtt{bbc}\rfloor\!\rfloor$ into $\lfloor\!\lfloor\mathtt{cnn}\rfloor\!\rfloor \sqcap \lfloor\!\lfloor\mathtt{bbc}\rfloor\!\rfloor$. In a first approach $\llbracket x \rrbracket$ should be the process

$$\left((\lfloor\!\lfloor\mathtt{cnn}\rfloor\!\rfloor \sqcap \lfloor\!\lfloor\mathtt{bbc}\rfloor\!\rfloor)@1/3 \mathbin{\|} \lfloor\!\lfloor\mathtt{cnn}\rfloor\!\rfloor@1/6 \mathbin{\|} \lfloor\!\lfloor\mathtt{bbc}\rfloor\!\rfloor@1/3 \mathbin{\|} \lfloor\!\lfloor\ \rfloor\!\rfloor@1/6\right).$$

Modelling non-determinism by imprecise probabilities

$$[\![(\lfloor\![\mathtt{cnn}]\!\rfloor \sqcap \lfloor\![\mathtt{bbc}]\!\rfloor)]\!] = \{\lfloor\![\mathtt{cnn}]\!\rfloor@p_1 \parallel \lfloor\![\mathtt{bbc}]\!\rfloor@p_2 \mid (p_1,p_2) \in \Delta_2\}$$

Substituting in the previous expression, we get

$$[\![x]\!] = \Big\{(\lfloor\![\mathtt{cnn}]\!\rfloor@\tfrac{1}{2}p_1 \parallel \lfloor\![\mathtt{bbc}]\!\rfloor@\tfrac{1}{3}p_1 \parallel \lfloor\!\lfloor\_\rfloor\!\rfloor@\tfrac{1}{6}p_1$$
$$\parallel \lfloor\![\mathtt{cnn}]\!\rfloor@\tfrac{1}{6}p_2 \parallel \lfloor\![\mathtt{bbc}]\!\rfloor@\tfrac{2}{3}p_2 \parallel \lfloor\!\lfloor\_\rfloor\!\rfloor@\tfrac{1}{6}p_2) \mid (p_1,p_2) \in \Delta_2\Big\}.$$

Regrouping terms we get for $[\![x]\!]$

$$\Big\{\Big(\lfloor\![\mathtt{cnn}]\!\rfloor@\big(\tfrac{1}{2}p_1 + \tfrac{1}{6}p_2\big) \parallel \lfloor\![\mathtt{bbc}]\!\rfloor@\big(\tfrac{1}{3}p_1 + \tfrac{2}{3}p_2\big) \parallel \lfloor\!\lfloor\_\rfloor\!\rfloor@\tfrac{1}{6}\Big) \mid (p_1,p_2) \in \Delta_2\Big\}$$

Observe that, although the characterization of $[\![x]\!]$ is imprecise, we can assure that the probability of having $\mathtt{cnn}$ as a result lies in the interval $[1/6, 1/2]$ and that the probability of $\mathtt{bbc}$ lies in $[1/3, 2/3]$. We rewrite $[\![x]\!]$ as:

$$\{(\lfloor\![\mathtt{cnn}]\!\rfloor@p \parallel \lfloor\![\mathtt{bbc}]\!\rfloor@q \parallel \lfloor\!\lfloor\_\rfloor\!\rfloor@1/6) \mid p \in [1/6, 1/2], q \in [1/3, 2/3], p+q = 5/6\}$$

Assuming the semantics of $Emails(x)$ in Example 5, then $[\![eNews]\!]$ is

$$\Big\{(\parallel \lfloor\![\mathtt{alice\_cnn}, \mathtt{bob\_cnn}]\!\rfloor@\tfrac{4}{7}p \parallel \lfloor\![\mathtt{alice\_cnn}]\!\rfloor@\tfrac{8}{35}p \parallel \lfloor\![\mathtt{bob\_cnn}]\!\rfloor@\tfrac{1}{7}p$$
$$\parallel \lfloor\![\mathtt{alice\_bbc}, \mathtt{bob\_bbc}]\!\rfloor@\tfrac{4}{7}q \parallel \lfloor\![\mathtt{alice\_bbc}]\!\rfloor@\tfrac{8}{35}q \parallel \lfloor\![\mathtt{bob\_bbc}]\!\rfloor@\tfrac{1}{7}q$$
$$\parallel \lfloor\!\lfloor\_\rfloor\!\rfloor@\tfrac{3}{14}) \mid p \in [1/6, 1/2], q \in [1/3, 2/3], p+q = 5/6\Big\}.$$

The meaning $[\![eNews]\!]$ provides, for each possible output stream, a probability interval. This quantitative information may be relevant in any discussion about the appropriateness of this orchestration.   □

Our next result provides a generalization of a similar result in [7]. We are able to include asymmetric parallelism in the bag semantics and devise a normal form. The proof is by induction on the structure of the orchestration and uses formalizations of the preceding ideas.

Theorem 3. *Let E be a finite faulty orchestration, defined trough sequencing, parallel composition and asymmetric parallelism over n different faulty sites. Assume that site i succeeds an returns a value with probability p_i and let $p = (p_1, \ldots, p_n)$. Under the bag semantics, encoding the daemonic choice due to asymmetric parallelism into imprecise probabilities, there are multisets $M_1, \ldots M_\ell$ and a Cartesian product of probability spaces $\Delta_{m_1} \times \cdots \times \Delta_{m_k}$ such that*

$$[\![E]\!] = \big\{(M_1@F_1(p,\delta) \parallel \cdots \parallel M_l@F_l(p,\delta)) \mid \delta \in \Delta_{n_1} \times \cdots \times \Delta_{n_k}\big\}.$$

Multiset's elements correspond to possible values returned by site calls and formulas F_j are arithmetic expressions defined on the success probability vector p and a tuple of distributions δ.

5   An Example of Application

In order to clarify the measurable probabilities $p = (p_1, \ldots, p_n)$ and the source for imprecise probabilities $\delta \in \Delta_{n_1} \times \cdots \times \Delta_{n_k}$ appearing in Theorem 3 we analyse the meaning on a longer orchestration.

Consider the orchestration *EmailFlightHotel* sending to *Alice* information about flights and hotels. Three hotels are asked to give information: $Hotels = (H_1 \mid H_2 \mid H_3)$ and two flight companies are contacted: $Flights = (F_1 \mid F_2)$. Thus,

$$EmailFlightHotel = Alice(f, h) < f < Flights < h < Hotels$$

The set of sites is $\{H_1, H_2, H_3, F_1, F_2, Alice(f, h)\}$. The success probabilities are respectively $\{1/2, 1/3, 1/4, 1/5, 1/6, 2/3\}$. Observe that

$$[\![Hotels]\!] = (\lfloor\!\lfloor h_1, h_2, h_3 \rfloor\!\rfloor @1/24$$
$$\mid \lfloor\!\lfloor h_1, h_2 \rfloor\!\rfloor @3/24 \mid \lfloor\!\lfloor h_1, h_3 \rfloor\!\rfloor @2/24 \mid \lfloor\!\lfloor h_2, h_3 \rfloor\!\rfloor @1/24 \mid$$
$$\mid \lfloor\!\lfloor h_1 \rfloor\!\rfloor @6/24 \mid \lfloor\!\lfloor h_2 \rfloor\!\rfloor @3/24 \mid \lfloor\!\lfloor h_3 \rfloor\!\rfloor @2/24$$
$$\mid \lfloor\!\lfloor \ \rfloor\!\rfloor @6/24)$$

Note that any probability appearing in $[\![Hotels]\!]$ is a funcion of the success probability of H_1, H_2, H_3 given by $p = (1/2, 1/3, 1/4)$. For instance the $6/24$ appearing in $\lfloor\!\lfloor h_1 \rfloor\!\rfloor @6/24$ is computed as $6/24 = 1/2(1-1/3)(1-1/4)$. To assign a meaning to h, each multiset gets an imprecise probability on its elements. That is

$$\lfloor\!\lfloor h_1, h_2, h_3 \rfloor\!\rfloor = \{(\lfloor\!\lfloor h_1 \rfloor\!\rfloor @p_{1,1} \mid \lfloor\!\lfloor h_2 \rfloor\!\rfloor @p_{1,2} \mid \lfloor\!\lfloor h_3 \rfloor\!\rfloor @p_{1,3}) \mid (p_{1,1}, p_{1,2}, p_{1,3}) \in \Delta_3\}$$
$$\lfloor\!\lfloor h_1, h_2 \rfloor\!\rfloor = \{(\lfloor\!\lfloor h_1 \rfloor\!\rfloor @p_{2,1} \mid \lfloor\!\lfloor h_2 \rfloor\!\rfloor @p_{2,2} \mid) \mid (p_{2,1}, p_{2,2}) \in \Delta_2\}$$
$$\lfloor\!\lfloor h_1, h_3 \rfloor\!\rfloor = \{(\lfloor\!\lfloor h_1 \rfloor\!\rfloor @p_{3,1} \mid \lfloor\!\lfloor h_3 \rfloor\!\rfloor @p_{3,3}) \mid (p_{3,1}, p_{3,3}) \in \Delta_2\}$$
$$\lfloor\!\lfloor h_2, h_3 \rfloor\!\rfloor = \{(\lfloor\!\lfloor h_2 \rfloor\!\rfloor @p_{4,2} \mid \lfloor\!\lfloor h_3 \rfloor\!\rfloor @p_{4,3}) \mid (p_{4,2}, p_{4,3}) \in \Delta_2\}$$

Define $\delta = (p_{1,1}, p_{1,2}, p_{1,3}, p_{2,1}, p_{2,2}, p_{3,1}, p_{3,3}, p_{4,2}, p_{4,3}) \in \Delta_3 \times \Delta_2 \times \Delta_2 \times \Delta_2$, then: $[\![h]\!] = \{(\lfloor\!\lfloor h_1 \rfloor\!\rfloor @P_{h_1} \mid \lfloor\!\lfloor h_2 \rfloor\!\rfloor @P_{h_2} \mid \lfloor\!\lfloor h_3 \rfloor\!\rfloor @P_{h_3} \mid \lfloor\!\lfloor \ \rfloor\!\rfloor @P_{h_\emptyset} \mid)\}$ where:

$$P_{h_1} = \frac{1}{24}(p_{1,1} + 3p_{2,1} + 2p_{3,1} + 6) \quad P_{h_2} = \frac{1}{24}(p_{1,2} + 3p_{2,2} + p_{4,2} + 3)$$

$$P_{h_3} = \frac{1}{24}(p_{1,3} + 2p_{3,3} + p_{4,3} + 2) \quad P_{h_\emptyset} = \frac{6}{24}$$

Note that P_{h_i} is a function of p and δ previously defined, that is $P_{h_i} = F_{h_i}(p, \delta)$, similarly for $P_{h_\emptyset}$.

Working in a similar way.

$$[\![Flights]\!] = (\lfloor\!\lfloor f_1, f_2 \rfloor\!\rfloor @1/30 \mid \lfloor\!\lfloor f_1 \rfloor\!\rfloor @5/30 \mid \lfloor\!\lfloor f_2 \rfloor\!\rfloor @4/30 \mid \lfloor\!\lfloor \ \rfloor\!\rfloor @20/30)$$

In this case the probabilities appearing in $[\![Flights]\!]$ are function of $p' = (1/5, 1/6)$. There is just one bag with more than one element, then

$$\lfloor\!\lfloor f_1, f_2 \rfloor\!\rfloor = \{(\lfloor\!\lfloor f_1 \rfloor\!\rfloor @q_{1,1} \mid \lfloor\!\lfloor f_2 \rfloor\!\rfloor @q_{1,2} \mid (q_{1,1}, q_{1,2}) \in \Delta_2)\}$$

Defining $\delta' = (q_{1,1}, q_{1,2}) \in \Delta_2$. So, $[\![f]\!] = \{(\lfloor\!\lfloor f_1 \rfloor\!\rfloor @ Q_{f_1} \parallel \lfloor\!\lfloor f_2 \rfloor\!\rfloor @ Q_{f_2} \parallel \lfloor\!\lfloor \quad \rfloor\!\rfloor$
$@ Q_{f_0})\}$ where

$$Q_{f_1} = \frac{1}{30}(q_{1,1} + 5), \; Q_{f_2} = \frac{1}{30}(q_{1,2} + 4), \; Q_{f_0} = \frac{20}{30}.$$

Finally,

$[\![EmailFightHotel]\!]$

$= \{(\lfloor\!\lfloor \mathtt{alice\_f_1\_h_1} \rfloor\!\rfloor @ P_1 \parallel \cdots \parallel \lfloor\!\lfloor \quad \rfloor\!\rfloor @ P_0) \mid \cdots\}$

$= \{(\lfloor\!\lfloor \mathtt{alice\_f_1\_h_1} \rfloor\!\rfloor @ \frac{2}{3} Q_{f_1} P_{h_1} \parallel \cdots \parallel \lfloor\!\lfloor \quad \rfloor\!\rfloor @ \frac{1}{3} P_{h_0} P_{f_0}) \mid \cdots\}$

$= \{(\lfloor\!\lfloor \mathtt{alice\_f_1\_h_1} \rfloor\!\rfloor @ \frac{2}{3}(\frac{1}{30} q_{1,1} + \frac{5}{30})(\frac{1}{24} p_{1,1} + \frac{3}{24} p_{2,1} + \frac{2}{24} p_{3,1} + \frac{6}{24}) \parallel \cdots$

$\parallel \lfloor\!\lfloor \quad \rfloor\!\rfloor @ \frac{1}{3} \cdot \frac{6}{24} \cdot \frac{20}{30}) \mid \cdots\}$

Define p'' as the array associated to the *Alice* probability of success, $p'' = (2/3)$.
Defining (with a small abuse of notation):

$$\delta = (\delta, \delta') = (p_{1,1}, p_{1,2}, p_{1,3}, p_{2,1}, p_{2,2}, p_{3,1}, p_{3,3}, p_{4,2}, p_{4,3}, q_{1,1}, q_{1,2})$$
$$p = (p, p', p'') = (1/2, 1/3, 1/4, 1/5, 1/6, 2/3)$$

We have got P_1 as an arithmetic expression on (p, δ) like the probability expressions in Theorem 3. Other cases are similar.

6 Other Applications

We consider briefly two settings to which we can extend the preceding approach. We started from the case of fully reliable sites to include probabilistic (but reliable) sites. We can consider the case where sites are fully reliable (response is granted) with uncertain response time. As before we encode demonic choice as an imprecise probability. This case, although a special case of the preceding one, merits special attention because the empty bag cannot appear. Even if the result is uncertain it is less uncertain than in the faulty case. Let us develop those ideas through an example.

Example 7. Let us consider *eNews* introduced in Example 1. According to Example 3 we have $[\![x]\!] = \lfloor\!\lfloor \mathtt{bbc} \rfloor\!\rfloor \sqcap \lfloor\!\lfloor \mathtt{cnn} \rfloor\!\rfloor$. As sites always return, the only probabilities are due to the indeterministic choice and therefore $[\![x]\!]$ is less ambiguous that in Example 6,

$$[\![x]\!] = \{(\lfloor\!\lfloor \mathtt{bbc} \rfloor\!\rfloor @ p_1 \parallel \lfloor\!\lfloor \mathtt{cnn} \rfloor\!\rfloor @ p_2) \mid (p_1, p_2) \in \Delta_2)\}$$

Then $[\![eNews]\!] = \{([\![Emails(\mathtt{bbc})]\!] @ p_1 \parallel [\![Emails(\mathtt{cnn})]\!] @ p_2) \mid (p_1, p_2) \in \Delta_2\}$. The main difference between this example and Example 6 is that both, Alice and Bob, will receive one newspaper for sure but is not known which one. Note that in Example 6 both (Alice and Bob) just one (Alice or Bob) or no-one (neither Alice nor Bob) get a newspaper. $\square$

We have used probabilistic information only on site failures. However, our approach can be extended to orchestrations having probabilistic behaviour. The following examples provides the main ideas in this setting.

Example 8. We assume that all the sites have a reliable behaviour. Consider a probabilistic site (modelled by an orchestration) $infoNews = (BBC@3/4 \parallel CNN@1/4)$ returning news form the BBC with probability 3/4 or CNN with probability 1/4. Note that $infoNews$ returns a result with probability 1, that is $[\![infoNews]\!] \neq \lfloor\!\lfloor \ \rfloor\!\rfloor$. Based on the previous site define $otherNews = (infoNews \mid DISNEY)$ and finally:

$$pr\_toAlice = Alice(x) < x < otherNews.$$

Clearly $[\![otherNews]\!] = (\lfloor\!\lfloor\texttt{bbc},\texttt{disney}\rfloor\!\rfloor@3/4 \parallel \lfloor\!\lfloor\texttt{cnn},\texttt{disney}\rfloor\!\rfloor@1/4)$. Using indeterminism to split bags into individual responses we get

$$[\![x]\!] = ((\lfloor\!\lfloor\texttt{bbc}\rfloor\!\rfloor \sqcap \lfloor\!\lfloor\texttt{disney}\rfloor\!\rfloor)@3/4 \parallel (\lfloor\!\lfloor\texttt{cnn}\rfloor\!\rfloor \sqcap \lfloor\!\lfloor\texttt{disney}\rfloor\!\rfloor)@1/4)$$
$$= (\lfloor\!\lfloor\texttt{bbc}\rfloor\!\rfloor@3/4 \parallel \lfloor\!\lfloor\texttt{cnn}\rfloor\!\rfloor@1/4) \sqcap (\lfloor\!\lfloor\texttt{bbc}\rfloor\!\rfloor@3/4 \parallel \lfloor\!\lfloor\texttt{disney}\rfloor\!\rfloor@1/4)$$
$$\sqcap (\lfloor\!\lfloor\texttt{cnn}\rfloor\!\rfloor@1/4 \parallel \lfloor\!\lfloor\texttt{disney}\rfloor\!\rfloor)@3/4) \sqcap \lfloor\!\lfloor\texttt{disney}\rfloor\!\rfloor.$$

Translating indeterminism into imprecise probabilities, we get

$$[\![x]\!] = \Big\{ \Big(\lfloor\!\lfloor\texttt{bbc}\rfloor\!\rfloor@\frac{3}{4}(p_1 + p_2) \parallel \lfloor\!\lfloor\texttt{cnn}\rfloor\!\rfloor@\frac{1}{4}(p_1 + p_3)$$
$$\parallel \lfloor\!\lfloor\texttt{disney}\rfloor\!\rfloor@(p_4 + \frac{1}{4}p_2 + \frac{3}{4}p_3) \mid (p_1, p_2, p_3, p_4) \in \Delta_4\}.$$

Finally,

$$[\![pr\_toAlice]\!] = \Big\{ \Big(\lfloor\!\lfloor\texttt{alice\_bbc}\rfloor\!\rfloor@\frac{3}{4}(p_1 + p_2) \parallel \lfloor\!\lfloor\texttt{alice\_cnn}\rfloor\!\rfloor@\frac{1}{4}(p_1 + p_3)$$
$$\parallel \lfloor\!\lfloor\texttt{alice\_disney}\rfloor\!\rfloor@(p_4 + \frac{1}{4}p_2 + \frac{3}{4}p_3) \mid (p_1, p_2, p_3, p_4) \in \Delta_4\}.$$

Thus, it is granted that Alice gets a result but the type of the result is uncertain. □

7   Conclusion and Open Problems

The economist Frank Knight has made a distinction between risk and uncertainty [12] as illustrated by the following quotation taken from [2, Chap. 11]:

> Risk refers to something that can be measured by mathematical probabilities. In contrast, uncertainty refers to something that cannot be measured (using probabilities) because there are no objective standards to express these probabilities.

In this paper we model the uncertainty issued by the daemonic choice, represented by $P \sqcap Q$ by the set of imprecise probabilities $\{(P@p_1 \parallel Q@p_2) \mid p_1 + p_2 = 1\}$. Imprecise probabilities overcome the Knightian problem of the existence of a unique probability. In particular we apply this approach to model the uncertain Web. Nevertheless, in asymmetric parallelism, non determinism is limited to the selection of an element from a multiset. It will be of interest to analyse, in the general context of processes' algebra, the existence of normal forms by modelling non determinism by imprecise probabilities.

In this paper we have assumed sites with well defined return probabilities. It could be also possible to consider sites with imprecise return probabilities. For instance, let CNN a site with uncertain return probability in between $[1/6, 1/2]$, then

$$[\![CNN]\!] = \{\lfloor\lfloor \mathrm{cnn}\rfloor\rfloor@p \parallel \lfloor\!\lfloor\ \rfloor\!\rfloor@(1-p) \mid p \in [1/6, 1/2]\}$$

It seems possible to extend the normal forms to this case.

In [8] another approach was undertaken to model Web uncertainty. It is assumed that sites can fail but the number of failures is bounded. As the failing sites are not known, some working hypothesis should be done between the best and the worst scenarios. It is assumed that some sites will fail trying to damage the orchestrations as much as possible (daemons $\eth$) but others will fail trying to minimize damage (angels $\mathfrak{a}$). This approach give rise to a strategic situation analysed trough a zero-sum game (called the $\mathfrak{a}$-$\eth$ game) [16]. It is an open topic if there is any relation between $\mathfrak{a}$-$\eth$ approach and imprecise probabilities approach.

A fundamental question in program design is: when a program is better than another? Partial orders have been considered to tackle this question. Expression $P \sqsubset Q$ points out that program Q is better than program P [10,13]. On highly unreliable environments this question is even more crucial. Although there exists a general approach [13], the application to the Web environment remains open.

Finally, in Theorem 3 a Cartesian product of probability spaces is considered. However there are situations where a richer correlation structure is suitable, or where additional information could be incorporated (i.e. more complex constraints on the probabilities directly). For instance, consider the case of locally congested network evolving along the time. These cases seems hard to study is this framework.

References

1. W3C, Web Services Glossary. http://www.w3.org/TR/ws-gloss/
2. Akerlof, G., Schiller, R.: Animal Spirits. Princeton University Press, Princeton and Oxford (2009)
3. Augustin, T., Coolen, F., Cooman, G., Troffaes, M.: Introduction to Imprecise Probabilities. Wiley (2014). https://doi.org/10.1002/9781118763117
4. Berners-Lee, T., Cailliau, R., Luotonen, A., Nielsen, H.F., Secret, A.: The worldwide web. Commun. ACM **37**(8), 76–82 (1994). https://doi.org/10.1145/179606.179671
5. Dean, J., Barroso, L.: The tail at scale. Commun. ACM **56**(2), 74–80 (2013). https://doi.org/10.1145/2408776.2408794

6. Floyd, R.W.: Assigning meanings to programs. In: Schwartz, J.T. (ed.) Proceedings of Symposium on Applied Mathematical Aspects of Computer Science, pp. 19–32. American Mathematical Society (1967)
7. Gabarro, J., Leon-Gaixas, S., Serna, M.: The computational complexity of QoS measures for orchestrations. J. Comb. Optim. **34**(4), 1265–1301 (2017). https://doi.org/10.1007/s10878-017-0146-9
8. Gabarro, J., Serna, M., Stewart, A.: Analysing web-orchestrations under stress using uncertainty profiles. Comput. J. **57**(11), 1591–1615 (2014). https://doi.org/10.1093/comjnl/bxt063
9. Galbraith, J.K.: The Age of Uncertainty. Houghhton Miffin Company, Boston (1977)
10. Hoare, C.: Communicating Sequential Processes. Prentice-Hall, London (1985)
11. Kitchin, D., Quark, A., Cook, W., Misra, J.: The orc programming language. In: Lee, D., Lopes, A., Poetzsch-Heffter, A. (eds.) FMOODS/FORTE -2009. LNCS, vol. 5522, pp. 1–25. Springer, Heidelberg (2009). https://doi.org/10.1007/978-3-642-02138-1_1
12. Knight, F.: Risk, Uncertainty and Profit. Houghton Mifflin, Boston and New York (1921). http://www.econlib.org/library/Knight/knRUP.html
13. McIver, A., Morgan, C.: Abstraction, Refinement and Proof for Probabilistic Systems. Springer, New York (2005). https://doi.org/10.1007/b138392
14. Misra, J., Cook, W.: Computation orchestration: a basis for wide-area computing. Softw. Syst. Model. **6**(1), 83–110 (2007). https://doi.org/10.1007/s10270-006-0012-1
15. Morgan, C., McIver, A., Sanders, J.W.: Probably Hoare? Hoare probably!. In: Davies, J.W., Roscoe, B., Woodcock, J. (eds.) Millennial Perspectives in Computer Science, pp. 271–282. Palgrave, Basingstoke (2000)
16. von Neumann, J., Morgenstern, O.: Theory of Games and Economic Behavior, 60th Anniversary Commemorative Edition. Princeton University Press, Princeton and Oxford (1953)
17. Peltz, C.: Web services orchestration and choreography. IEEE Comput. **36**(10), 46–52 (2003). https://doi.org/10.1109/MC.2003.1236471

Conditional Submodular Coherent Risk Measures

Giulianella Coletti[1], Davide Petturiti[2(✉)], and Barbara Vantaggi[3]

[1] Dip. Matematica e Informatica, University of Perugia, Perugia, Italy
giulianella.coletti@unipg.it
[2] Dip. Economia, University of Perugia, Perugia, Italy
davide.petturiti@unipg.it
[3] Dip. S.B.A.I., "La Sapienza" University of Rome, Rome, Italy
barbara.vantaggi@sbai.uniroma1.it

Abstract. A family of conditional risk measures is introduced by considering a single period financial market, relying on a notion of conditioning for submodular capacities, which generalizes that introduced by Dempster. The resulting measures are expressed as discounted conditional Choquet expected values, take into account ambiguity towards uncertainty and allow for conditioning to "null" events. We also provide a characterisation of consistence of a partial assessment with a conditional submodular coherent risk measure. The latter amounts to test the solvability of a suitable sequence of linear systems.

Keywords: Coherent risk measure
Conditional Choquet expected value
Conditional submodular capacity

1 Introduction

An important issue in financial risk measurement is to express how risky a given portfolio is and, so, risk measures are quantitative tools developed to determine the capital that the owner of the portfolio should allocate to face possible losses and to ensure their financial stability. Among the different measures we mention the well-known Value At Risk and Expected Shortfall, which are very popular among practitioners.

Recall that the Expected Shortfall is a coherent risk measure (see [2] for an axiomatization and characterization of coherent risk measures), while the Value At Risk is generally not coherent due to lack of subadditivity.

In this paper the interest is directed to those situations where the information is partial and, so, beliefs cannot be encoded in a single probability measure, but they are expressed through a set of probability measures. In such cases we need to deal with ambiguity towards uncertainty. Starting from the given set of probability measures we need to consider its envelopes that turn out to be normalized capacities, often being submodular [supermodular].

© Springer International Publishing AG, part of Springer Nature 2018
J. Medina et al. (Eds.): IPMU 2018, CCIS 854, pp. 239–250, 2018.
https://doi.org/10.1007/978-3-319-91476-3_20

The starting point of this paper is the class of coherent risk measures studied in [11] and defined through the Choquet expected value with respect to a submodular [supermodular] capacity. One of the most appealing properties of these risk measures is that they can be represented as a suitably modified worst or best expected loss over a whole class of probabilistic models. The connection between risk measures and imprecise probabilities has been studied, e.g., in [20,23,29].

In financial applications, as in every decision problem under uncertainty, the arrival of new information has an impact on risk measurement, for this a notion of conditioning for risk measures should be introduced (see, e.g., [8,24]). As is well-known, the problem of defining a suitable notion of conditioning is crucial also for capacities, for which a large debate is still present in the literature [6,9,10,12,13,18,19,22,27,30].

Here, we study conditional risk measures relying on an axiomatic definition of conditional submodular capacities (originally given for plausibility functions in [10]) which generalizes the one introduced in [12], allowing for conditioning to "null" events. Our notion of conditioning differs from that used in [23], the latter being based on the Walley's generalized Bayesian rule. The introduced conditional risk measures are obtained as a discounted conditional Choquet expected value computed with respect to a conditional submodular [supermodular] capacity.

In real risk management problems the risk manager is only able to assess the value of a conditional risk measure "directly" on a finite set of conditional risks (i.e., conditional random quantities). For this we investigate the consistence of a partial assessment with a conditional submodular coherent risk measure, providing a characterization in terms of solvability of a suitable sequence of linear systems.

2   Conditional Submodular Functionals

Let $\Omega = \{\omega_1, \ldots, \omega_n\}$ be a finite set of states of the world and denote by $\wp(\Omega)$ the power set of Ω, whose elements are interpreted as the events of interest. Let $\wp(\Omega)^0 = \wp(\Omega) \setminus \{\emptyset\} = \{A_1, \ldots, A_{2^n-1}\}$ be the set of non-impossible events, whose enumeration is assumed to be fixed.

We recall that a *(normalized) capacity* (see [14,17]) on $\wp(\Omega)$ is a function $\psi : \wp(\Omega) \to [0,1]$ such that $\psi(\emptyset) = 0$, $\psi(\Omega) = 1$ and $\psi(A) \leq \psi(B)$ whenever $A \subseteq B$, for $A, B \in \wp(\Omega)$. A *submodular (or 2-alternating) capacity* ψ on $\wp(\Omega)$ further satisfies, for every $A_1, A_2 \in \wp(\Omega)$,

$$\psi(A_1 \cup A_2) + \psi(A_1 \cap A_2) \leq \psi(A_1) + \psi(A_2).$$

The *dual* function φ defined, for every $A \in \wp(\Omega)$, as $\varphi(A) = 1 - \psi(A^c)$, is a *supermodular (or 2-monotone) capacity* and satisfies the above inequality in the opposite direction.

The functions ψ and φ on $\wp(\Omega)$ are completely singled out by the *Möbius inverse* of φ [7,17], defined for every $A \in \wp(\Omega)$ as

$$m(A) = \sum_{B \subseteq A} (-1)^{|A \setminus B|} \varphi(B).$$

The function $m : \wp(\Omega) \to \mathbb{R}$ is such that $m(\emptyset) = 0$, $\sum_{A \in \wp(\Omega)} m(A) = 1$, and, for every $A \in \wp(\Omega)$,

$$\varphi(A) = \sum_{B \subseteq A} m(B) \quad \text{and} \quad \psi(A) = \sum_{B \cap A \neq \emptyset} m(B).$$

In particular, if m is non-negative then ψ and φ are a *plausibility* and a *belief function* [7,12,17,27]. Recall that submodular and supermodular capacities are distinguished subclasses of *(coherent) upper* and *lower probabilities* [28,30].

A *risk* $X : \Omega \to \mathbb{R}$ is a state-contingent (possibly positive or negative) money payoff. The set of all risks on Ω is denoted as $\mathbb{R}^{\Omega}$, which is easily seen to be a linear lattice, i.e., it is closed under the pointwise operations of linear combination, minimum and maximum, the latter denoted as $\wedge$ and $\vee$.

For a risk $X \in \mathbb{R}^{\Omega}$, if ν is a capacity on $\wp(\Omega)$ and σ is a permutation of $\{1, \dots, n\}$ such that $X(\omega_{\sigma(1)}) \leq \dots \leq X(\omega_{\sigma(n)})$ (see [14]), the *Choquet integral* of X with respect to ν is defined, denoting $E_i^{\sigma} = \{\omega_{\sigma(i)}, \dots, \omega_{\sigma(n)}\}$ for $i = 1, \dots, n$ and $E_{n+1}^{\sigma} = \emptyset$, as

$$\oint X \, d\nu = \sum_{i=1}^{n} X(\omega_{\sigma(i)})(\nu(E_i^{\sigma}) - \nu(E_{i+1}^{\sigma})).$$

In particular, when ν reduces, respectively, to a submodular capacity ψ or to a supermodular capacity φ, the Choquet integral is the maximum/minimum of the prevision (expected value) functionals determined by the *core* of φ (see, e.g., [14,26]), that is the set of probability measures

$$\mathcal{P}_{\varphi} = \{P \,:\, P \text{ is a probability measure}, \varphi \leq P \leq \psi\},$$

for which it holds $\varphi = \min \mathcal{P}_{\varphi}$ and $\psi = \max \mathcal{P}_{\varphi}$.

Given a risk $X \in \mathbb{R}^{\Omega}$ and an event $H \in \wp(\Omega)^0$, a *conditional risk* is a pair (X, H), usually denoted as $X|H$, which consists in regarding X under the hypothesis H. In particular, a conditional event $E|H \in \wp(\Omega) \times \wp(\Omega)^0$ is identified with the conditional risk $\mathbf{1}_E|H$, where $\mathbf{1}_E$ denotes the indicator of event E, and an unconditional risk $X \in \mathbb{R}^{\Omega}$ is identified with $X|\Omega$.

Definition 1. *Let $\mathcal{H} \subseteq \wp(\Omega) \setminus \{\emptyset\}$ be an additive class (i.e., a set of events closed under finite unions). A function $\psi : \wp(\Omega) \times \mathcal{H} \to [0, 1]$ is a* **conditional submodular capacity** *if it satisfies the following conditions:*

(i) $\psi(E|H) = \psi(E \cap H|H)$, for every $E \in \wp(\Omega)$ and $H \in \mathcal{H}$;
(ii) $\psi(\cdot|H)$ is a submodular capacity on $\wp(\Omega)$, for every $H \in \mathcal{H}$;
(iii) $\psi(E \cap F|H) = \psi(E|H) \cdot \psi(F|E \cap H)$, for every $E \cap H, H \in \mathcal{H}$ and $E, F \in \wp(\Omega)$.

Moreover, given a conditional submodular capacity $\psi(\cdot|\cdot)$, the dual **conditional supermodular capacity** *$\varphi(\cdot|\cdot)$ is defined for every event $E|H \in \wp(\Omega) \times \mathcal{H}$ as*

$$\varphi(E|H) = 1 - \psi(E^c|H).$$

Every conditional submodular capacity $\psi(\cdot|\cdot)$ on $\wp(\Omega) \times \mathcal{H}$ is in bijection with a linearly ordered class $\{\psi_0, \ldots, \psi_k\}$ of (unconditional) submodular capacities on $\wp(\Omega)$, called *minimal agreeing class*, such that

- $\psi_0(\cdot) = \psi(\cdot|H_0^0)$ with $H_0^0 = \bigcup_{H \in \mathcal{H}} H$;
- for $\alpha > 0$, $\psi_\alpha(\cdot) = \psi(\cdot|H_0^\alpha)$ with $H_0^\alpha = \bigcup \{H \in \mathcal{H} : \psi_\beta(H) = 0, \beta = 0, \ldots, \alpha - 1\} \neq \emptyset$.

The class $\{\psi_0, \ldots, \psi_k\}$ is such that for every $H \in \mathcal{H}$ there is $\alpha \in \{0, \ldots, k\}$ such that $\psi_\alpha(H) > 0$. Moreover, for every $E|H \in \wp(\Omega) \times \mathcal{H}$, denoting with α_H the minimum index in $\{0, \ldots, k\}$ such that $\psi_{\alpha_H}(H) > 0$, it holds that

$$\psi(E|H) = \frac{\psi_{\alpha_H}(E \cap H)}{\psi_{\alpha_H}(H)}.$$

If condition *(ii)* in Definition 1 is reinforced by requiring that $\psi(\cdot|H)$ is a plausibility function on $\wp(\Omega)$, for every $H \in \mathcal{H}$, the resulting conditional measure is a *conditional plausibility function* according to [10], for which a minimal agreeing class representation has been given in [5].

The conditional Choquet expected values induced by the dual conditional measures $\psi(\cdot|\cdot)$ and $\varphi(\cdot|\cdot)$ on $\wp(\Omega) \times \mathcal{H}$ are conditional functionals defined, for every $X|H \in \mathbb{R}^\Omega \times \mathcal{H}$, as

$$\Psi(X|H) = \oint X \, d\psi(\cdot|H) \quad \text{and} \quad \Phi(X|H) = \oint X \, d\varphi(\cdot|H).$$

The conditional functionals $\Psi(\cdot|\cdot)$ and $\Phi(\cdot|\cdot)$ are, respectively, submodular and supermodular [17, 28], i.e., they satisfy, for every $X_1, X_2 \in \mathbb{R}^\Omega$ and every $H \in \mathcal{H}$,

$$\Psi(X_1 \vee X_2|H) + \Psi(X_1 \wedge X_2|H) \leq \Psi(X_1|H) + \Psi(X_2|H),$$
$$\Phi(X_1 \vee X_2|H) + \Phi(X_1 \wedge X_2|H) \geq \Phi(X_1|H) + \Phi(X_2|H),$$

and are *dual* since, for every $X|H \in \mathbb{R}^\Omega \times \mathcal{H}$, it holds $\Psi(-X|H) = -\Phi(X|H)$.

The following definition allows to provide a linear expression of conditional functionals $\Psi(\cdot|\cdot)$ and $\Phi(\cdot|\cdot)$.

Definition 2. *For every $H \in \wp(\Omega)^0$, the H-cut generalized upper and lower risks corresponding to a risk $X : \Omega \to \mathbb{R}$ are the functions $X^{\mathbf{U},H}, X^{\mathbf{L},H} : \wp(\Omega)^0 \to \mathbb{R}$ defined, for $i = 1, \ldots, 2^n - 1$, as*

$$X^{\mathbf{U},H}(A_i) = \max_{\omega \in A_i \cap H} X(\omega) \quad \text{and} \quad X^{\mathbf{L},H}(A_i) = \min_{\omega \in A_i \cap H} X(\omega),$$

with $X^{\mathbf{U},H}(A_i) = X^{\mathbf{L},H}(A_i) = 0$ if $A_i \cap H = \emptyset$, also denoted as the row vectors

$$X^{\mathbf{U},H} = (X^{\mathbf{U},H}(A_1), \ldots, X^{\mathbf{U},H}(A_{2^n-1})),$$
$$X^{\mathbf{L},H} = (X^{\mathbf{L},H}(A_1), \ldots, X^{\mathbf{L},H}(A_{2^n-1})).$$

In what follows, operations between H-cut generalized upper and lower risks, such as sum, multiplication and multiplication for a constant, are always assumed pointwise on the elements of $\wp(\Omega)^0$.

Proposition 1. *Let* $\psi : \wp(\Omega) \times \mathcal{H} \to [0,1]$ *be a conditional submodular capacity generated by the minimal agreeing class of submodular capacities* $\{\psi_0, \ldots, \psi_k\}$ *on* $\wp(\Omega)$ *with duals* $\{\varphi_0, \ldots, \varphi_k\}$ *determining the Möbius inverses* $\{m_0, \ldots, m_k\}$, *and* $\varphi(\cdot|\cdot)$ *the dual conditional supermodular capacity. For every risk* $X \in \mathbb{R}^\Omega$ *and every* $H \in \mathcal{H}$ *with* $\alpha_H \in \{0, \ldots, k\}$ *the minimum index such that* $\psi_{\alpha_H}(H) > 0$, *it holds*

$$\Psi(X|H) = \frac{1}{\psi_{\alpha_H}(H)} \sum_{h=1}^{2^n-1} X^{\mathbf{U},H}(A_h) m_{\alpha_H}(A_h),$$

$$\Phi(X|H) = \frac{1}{\psi_{\alpha_H}(H)} \sum_{h=1}^{2^n-1} X^{\mathbf{L},H}(A_h) m_{\alpha_H}(A_h).$$

Proof. By the representation of $\psi(\cdot|H)$ through $\{\psi_0, \ldots, \psi_k\}$, we have that

$$\Psi(X|H) = \oint X \, d\psi(\cdot|H)$$

$$= \frac{1}{\psi_{\alpha_H}(H)} \oint X \, d\psi_{\alpha_H}(\cdot \cap H)$$

$$= \frac{1}{\psi_{\alpha_H}(H)} \sum_{i=1}^{n} X(\omega_{\sigma(i)})(\psi_{\alpha_H}(E_i^\sigma \cap H) - \psi_{\alpha_H}(E_{i+1}^\sigma \cap H))$$

$$= \frac{1}{\psi_{\alpha_H}(H)} \sum_{h=1}^{2^n-1} X^{\mathbf{U},H}(A_h) m_{\alpha_H}(A_h),$$

where the last equality follows since $\psi_{\alpha_H}(E_i^\sigma \cap H) - \psi_{\alpha_H}(E_{i+1}^\sigma \cap H)$ is equal to $\sum_{\substack{B \cap E_i^\sigma \cap H \neq \emptyset \\ B \cap E_{i+1}^\sigma \cap H = \emptyset}} m_{\alpha_H}(B)$ and for all B in the sum it holds $X^{\mathbf{U},H}(B) = X(\omega_{\sigma(i)})$.

Analogously, we have

$$\Phi(X|H) = \oint X \, d\varphi(\cdot|H)$$

$$= \frac{1}{\psi_{\alpha_H}(H)} \oint X \, d[\psi_{\alpha_H}(H) - \psi_{\alpha_H}((\cdot)^c \cap H)]$$

$$= \frac{1}{\psi_{\alpha_H}(H)} \sum_{i=1}^{n} X(\omega_{\sigma(i)})(\psi_{\alpha_H}((E_{i+1}^\sigma)^c \cap H) - \psi_{\alpha_H}((E_i^\sigma)^c \cap H))$$

$$= \frac{1}{\psi_{\alpha_H}(H)} \sum_{h=1}^{2^n-1} X^{\mathbf{L},H}(A_h) m_{\alpha_H}(A_h),$$

where the last equality follows since $\psi_{\alpha_H}((E_{i+1}^\sigma)^c \cap H) - \psi_{\alpha_H}((E_i^\sigma)^c \cap H)$ is equal to $\sum_{\substack{B \cap (E_{i+1}^\sigma)^c \cap H \neq \emptyset \\ B \cap (E_i^\sigma)^c \cap H = \emptyset}} m_{\alpha_H}(B)$ and for all B in the sum it holds $X^{\mathbf{L},H}(B) = X(\omega_{\sigma(i)})$.

Conditional submodular and supermodular capacities are generally not pointwise envelopes of a class of conditional probabilities in the sense of [15]. Nevertheless, for each fixed conditioning event $H \in \mathcal{H}$, they are pointwise envelopes of a class of unconditional probabilities, which coincides with the core

$$\mathcal{P}_{\varphi(\cdot|H)} = \{P \ : \ P \text{ is a probability measure on } \wp(\Omega), \varphi(\cdot|H) \leq P \leq \psi(\cdot|H)\},$$

i.e., $\varphi(\cdot|H) = \min \mathcal{P}_{\varphi(\cdot|H)}$ and $\psi(\cdot|H) = \max \mathcal{P}_{\psi(\cdot|H)}$. In particular, for every $X \in \mathbb{R}^{\Omega}$ and every $H \in \mathcal{H}$, it holds

$$\Psi(X|H) = \max \left\{ \int X \, dP \ : \ P \in \mathcal{P}_{\varphi(\cdot|H)} \right\},$$

$$\Phi(X|H) = \min \left\{ \int X \, dP \ : \ P \in \mathcal{P}_{\varphi(\cdot|H)} \right\},$$

so, locally on every $H \in \mathcal{H}$, an upper/lower prevision (expected value) interpretation can be given. This implies that, for every $H \in \mathcal{H}$, $\Psi(\cdot|H)$ and $\Phi(\cdot|H)$ are, respectively, subadditive and superadditive [14].

Finally, for every non-negative $Y \in \mathbb{R}^{\Omega}$, $E \in \wp(\Omega)$ and $E \cap H, H \in \mathcal{H}$, by Definition 1 and the properties of the Choquet integral [14], the following product rule (that generalizes condition *(iii)* of Definition 1) holds

$$\Psi(Y \mathbf{1}_E|H) = \Psi(\mathbf{1}_E|H) \cdot \Psi(Y|E \cap H).$$

3   Conditional Submodular Coherent Risk Measures

Let us consider a single period financial market related to the time period $[0, T]$, where among the traded assets there are (infinitely divisible) default-free zero coupon bonds with maturity T and interest rate $i(0, T)$. Denote with $r = 1 + i(0, T)$ and r^{-1} the compounding and discount factors determined by the default-free zero coupon bonds, i.e., r is the value at time T of 1 money unit at time 0, while r^{-1} is the value at time 0 of 1 money unit at time T. Assume $r > 0$ and Ω is finite.

In the seminal paper [2], a *coherent risk measure* is defined as a mapping $\rho : \mathbb{R}^{\Omega} \to \mathbb{R}$ satisfying:

(i) $\rho(X + \alpha r) = \rho(X) - \alpha$, for every $X \in \mathbb{R}^{\Omega}$ and $\alpha \in \mathbb{R}$;
(ii) $\rho(\lambda X) = \lambda \rho(X)$, for every $X \in \mathbb{R}^{\Omega}$ and $\lambda \geq 0$;
(iii) if $X \leq Y$ then $\rho(X) \geq \rho(Y)$, for every $X, Y \in \mathbb{R}^{\Omega}$;
(iv) $\rho(X + Y) \leq \rho(X) + \rho(Y)$, for every $X, Y \in \mathbb{R}^{\Omega}$.

The number $\rho(X)$ attached to X is interpreted as the least amount that an agent would ask at present time to bear the risk X at time T. It turns out that an agent acts like a risk measure minimizer, since the more risky is X the higher is the value $\rho(X)$. In particular, a position X is desirable when $\rho(X) \leq 0$, while is not desirable otherwise.

As largely acknowledged in the literature, there is an intimate connection between the theory of risk measures and the theory of imprecise previsions [3,20,29]. In particular, in [11] the author introduces a class of risk measures determined by a submodular [supermodular] capacity $\psi(\cdot)$ [$\varphi(\cdot)$] on $\wp(\Omega)$ through the corresponding Choquet expected value $\Psi(\cdot)$ [$\Phi(\cdot)$] on $\mathbb{R}^{\Omega}$ setting, for every $X \in \mathbb{R}^{\Omega}$,

$$\rho(X) = \Psi(-r^{-1}X) = -\Phi(r^{-1}X).$$

As in every decision problem under uncertainty, the arrival of new information has a deep impact on risk measurement, for this a notion of conditioning for risk measures should be introduced. Recurring to the Walley notion of conditioning for imprecise previsions [30], a definition of conditional risk measure has been proposed in [23, 29]. In what follows we give a definition of conditional risk measure which departs from previous proposals and allows for "null" conditioning events.

Denote with $\Psi(\cdot|\cdot)$ and $\Phi(\cdot|\cdot)$ the conditional Choquet expected values on $\mathbb{R}^{\Omega} \times \mathcal{H}$, determined by the conditional submodular and supermodular capacities $\psi(\cdot|\cdot)$ and $\varphi(\cdot|\cdot)$ on $\wp(\Omega) \times \mathcal{H}$.

Definition 3. *Given $r > 0$, a function $\rho : \mathbb{R}^{\Omega} \times \mathcal{H} \to \mathbb{R}$ is a **conditional submodular coherent risk measure** if*

$$\rho(X|H) = \Psi(-r^{-1}X|H) = -\Phi(r^{-1}X|H).$$

The following proposition investigates the properties of the conditional risk measure defined above. In particular, recall that $X_1, X_2 \in \mathbb{R}^{\Omega}$ are said *comonotonic* if, for every $\omega, \omega' \in \Omega$, it holds

$$(X_1(\omega) - X_1(\omega'))(X_2(\omega) - X_2(\omega')) \geq 0.$$

Proposition 2. *For every $H \in \mathcal{H}$, it holds:*

(i) $\rho(\cdot|H)$ is a coherent risk measure;
(ii) $\rho(X_1 \vee X_2|H) + \rho(X_1 \wedge X_2|H) \leq \rho(X_1|H) + \rho(X_2|H)$, for every $X_1, X_2 \in \mathbb{R}^{\Omega}$;
(iii) $\rho(X_1 + X_2|H) = \rho(X_1|H) + \rho(X_2|H)$, for every comonotonic $X_1, X_2 \in \mathbb{R}^{\Omega}$;
(iv) $\rho(X\mathbf{1}_E|H) = (1 + r\rho(\mathbf{1}_{E^c}|H)) \cdot \rho(X|E \cap H)$, for every non-positive $X \in \mathbb{R}^{\Omega}$, $E \in \wp(\Omega)$ and $E \cap H, H \in \mathcal{H}$.

Proof. The proof of *(i)–(iii)* follows directly by Definition 3 and the properties of the Choquet integral [14, 26]. Statement *(iv)* follows by the product rule for $\Psi(\cdot|\cdot)$ since $Y = -r^{-1}X$ is non-negative and $\Psi(\mathbf{1}_E|H) = 1 + r\rho(\mathbf{1}_{E^c}|H)$. $\qed$

The following example describes a situation where conditional risk measures of the type discussed above naturally arise.

Example 1. Consider an Italian insurance company interested in evaluating the risk of premium increment in pension insurance contracts linked to the Italian GDP, in a future time $T = 1$ year. Assume $i(0, T) = 2\%$, which implies $r = 1.02$.

On November 19th, 2017, the ex Prime Minister of Italy belonging to the right party gave a campaign speech in which he mentioned a reform of the Italian public pension system, if his coalition won the next elections. Hence, in December 2017, the risk of premium increment in pension insurance contracts is very influenced by uncertainty on the winning coalition of the next Italian political elections, expected in March 2018.

Consider the following events, which are uncertain before elections take place:

- $G_1 =$ "Italian GDP decreases in 1 year",
- $G_2 =$ "Italian GDP has an increment ranging in 0%–0.5% in 1 year",
- $G_3 =$ "Italian GDP increases more than 0.5% in 1 year",
- $B =$ "The center-right coalition wins 2018 Italian political elections and a reform of the Italian pension system is approved in 1 year".

The above events determine the set of states of the world $\Omega = \{\omega_1, \ldots, \omega_6\}$, with $G_i = \{\omega_i, \omega_{i+3}\}$, for $i = 1, 2, 3$, and $B = \{\omega_4, \omega_5, \omega_6\}$. Notice that, events G_1, G_2, G_3 form a partition of Ω. In particular, the insurance company is interested in evaluating the effect of event B, thus $\mathcal{H} = \{B, B^c, \Omega\}$ is considered as set of hypotheses.

The insurance company asks the opinion of three political observers, that elicit a probability distribution on the set Ω. All the three experts believe that the event B has zero probability, while they assess

	$G_1 \cap B^c$	$G_2 \cap B^c$	$G_3 \cap B^c$
P^1	0.3	0.5	0.2
P^2	0.3	0.3	0.4
P^3	0.1	0.5	0.4

The class $\mathcal{P} = \{P^1, P^2, P^3\}$ of probability measures on $\wp(\Omega)$ determines the submodular and supermodular capacities $\psi_0 = \max \mathcal{P}$ and $\varphi_0 = \min \mathcal{P}$, whose corresponding Möbius inverse is $m_0(\{\omega_3\}) = m_0(\{\omega_i, \omega_j\}) = 0.2$, for $i \neq j$ in $\{1, 2, 3\}$, $m_0(\{\omega_1\}) = 0.1$, $m_0(\{\omega_2\}) = 0.3$, $m_0(B^c) = -0.2$ and 0 otherwise. Notice that $\psi_0(B) = \varphi_0(B) = 0$.

The complete ignorance on what would happen if the event B were true is modelled by choosing as ψ_1 and φ_1 the pair of dual submodular and supermodular capacities vacuous at B, whose corresponding Möbius inverse is such that $m_1(B) = 1$ and 0 otherwise.

Let us consider two contracts whose annual premium increment gives rise to the risks

$$X(\omega) = \begin{cases} €500 & \text{if } \omega \in G_1, \\ -€300 & \text{if } \omega \in G_2, \\ -€600 & \text{if } \omega \in G_3, \end{cases} \quad \text{and} \quad Y(\omega) = \begin{cases} €200 & \text{if } \omega \in G_1, \\ €200 & \text{if } \omega \in G_2, \\ -€800 & \text{if } \omega \in G_3. \end{cases}$$

After simple computations we have that

$$\rho(X|B^c) = €333.33, \rho(Y|B^c) = €196.08, \rho(X|B) = €588.24, \rho(Y|B) = €784.31.$$

This shows that, even though under both hypotheses B^c and B the two positions are not desirable, under hypothesis B^c position Y is less risky, while under hypothesis B position X is less risky.

4   Consistence of a Conditional Risk Measure Assessment

The previous section has shown that, once a conditional submodular [supermodular] capacity $\psi(\cdot|\cdot)$ $[\varphi(\cdot|\cdot)]$ on $\wp(\Omega) \times \mathcal{H}$ is fixed, the corresponding conditional Choquet expected value $\Psi(\cdot|\cdot)$ $[\Phi(\cdot|\cdot)]$ on $\mathbb{R}^{\Omega} \times \mathcal{H}$ is completely determined and, so, the corresponding conditional submodular coherent risk measure $\rho(\cdot|\cdot)$ on $\mathbb{R}^{\Omega} \times \mathcal{H}$, up to the choice of the compounding factor r, which is an exogenous market parameter.

Nevertheless, in many real situations, the risk manager is only able to assess the value of a conditional risk measure "directly" on a finite set of conditional risks $\mathcal{C} = \{X_i|H_i\}_{i \in I} \subseteq \mathbb{R}^{\Omega} \times \wp(\Omega)$.

Since a conditional risk measure is intended as a normative model regulating the activity of a risk manager, given $\rho : \mathcal{C} \to \mathbb{R}$, the primary issue is to determine its consistence with a conditional risk measure of reference.

Definition 4. *Given $r > 0$, an assessment $\rho : \mathcal{C} \to \mathbb{R}$ is consistent with a conditional submodular coherent risk measure, if there is a conditional submodular capacity $\psi(\cdot|\cdot)$ on $\wp(\Omega) \times \mathcal{H}$, determining the conditional functionals $\Psi(\cdot|\cdot)$ and $\Phi(\cdot|\cdot)$ on $\wp(\Omega) \times \mathcal{H}$, such that, for every $i \in I$,*

$$\rho(X_i|H_i) = \Psi(-r^{-1}X_i|H_i) = -\Phi(r^{-1}X_i|H_i),$$

where $\mathcal{H}$ is the additive set obtained closing $\{H_i\}_{i \in I}$ under finite unions.

The following theorem provides a characterization of consistence.

Theorem 1. *Given $r > 0$, the following statements are equivalent:*

(i) the assessment $\rho : \mathcal{C} \to \mathbb{R}$ is consistent with a conditional submodular coherent risk measure;

(ii) there exists a minimal agreeing class $\{\psi_0, \ldots, \psi_k\}$ of submodular capacities on $\wp(\Omega)$ whose Möbius inverses $\{m_0, \ldots, m_k\}$ solve the sequence of linear systems $\mathcal{S}_0, \ldots, \mathcal{S}_k$ with

$$\mathcal{S}_\alpha : \begin{cases} \sum_{h=1}^{2^n-1} [(-r^{-1}X_i)^{\mathbf{U},H_i} - \rho(X_i|H_i) \cdot (\mathbf{1}_{H_i})^{\mathbf{U},H_0^\alpha}](A_h)m_\alpha(A_h) = 0, \, \forall i \in I_\alpha, \\ m_\alpha(\{\omega\}) \geq 0, \, \forall \omega \in \Omega, \\ \sum_{\{\omega,\omega'\} \subseteq B \subseteq A} m_\alpha(B) \geq 0, \forall A \cap H_0^\alpha \neq \emptyset \text{ and} \forall \omega, \omega' \in A, \omega \neq \omega', \\ \sum_{h=1}^{2^n-1} m_\alpha(A_h) = 1, \\ \sum_{B \cap A \neq \emptyset} m_\alpha(B) = 0, \, \forall A \cap H_0^\alpha = \emptyset, \end{cases}$$

with $I_0 = I$, $I_\alpha = \{i \in I : \psi_\beta(H_i) = 0, \beta = 0, \ldots, \alpha - 1\}$, for $\alpha > 0$, and $H_0^\alpha = \bigcup_{h \in I_\alpha} H_h$.

Proof. Let $\mathcal{H}$ be the additive set obtained closing $\{H_i\}_{i \in I}$ with respect to finite unions. By the bijection between conditional submodular capacities $\psi(\cdot|\cdot)$ and minimal agreeing classes $\{\psi_0, \ldots, \psi_k\}$, the assessment $\rho(\cdot|\cdot)$ on $\mathcal{C}$ is consistent with a conditional submodular coherent risk measure if and only if we can solve the following sequence of systems $\mathcal{S}_0^*, \ldots, \mathcal{S}_k^*$ with

$$\mathcal{S}_\alpha^* : \begin{cases} \oint(-rX_i)\,d\psi_\alpha(\cdot \cap H_i) - \rho(X_i|H_i) \cdot \psi_\alpha(H_i) = 0, \forall i \in I_\alpha, \\ \psi_\alpha \text{ is a submodular capacity on } \wp(\Omega), \\ \psi_\alpha(A) = 0, \ \forall A \cap H_0^\alpha = \emptyset, \end{cases}$$

where I_α and H_0^α are defined as in statement *(ii)*. Finally, by our Proposition 1, and Proposition 1 and Corollary 2 in [7], every system $\mathcal{S}_\alpha^*$ has solution if and only if the corresponding system $\mathcal{S}_\alpha$ has solution.

The previous theorem offers an operative tool to build a minimal agreeing class by solving the sequence of systems $\mathcal{S}_0, \ldots, \mathcal{S}_k$, as shown in the following example.

Example 2. Let $\Omega = \{\omega_1, \omega_2, \omega_3\}$, $H = \{\omega_1, \omega_2\}$, $K = \{\omega_1, \omega_3\}$, and $r = 1.1$. Let $X \in \mathbb{R}^\Omega$ be defined as $X(\omega_1) = €2200$, $X(\omega_2) = -€2200$ and $X(\omega_3) = €1100$, and consider the assessment $\rho(X|H) = €2000$ and $\rho(X|K) = -€1400$. To avoid cumbersome notation, let us denote $x_i^\alpha = m_\alpha(\{\omega_i\})$, $x_{ij}^\alpha = m_\alpha(\{\omega_i, \omega_j\})$, and $x_{123}^\alpha = m_\alpha(\Omega)$.

It holds that $\mathcal{H} = \{H, K, \Omega\}$ and $H_0^0 = \Omega$, so, the first system in the sequence is

$$\mathcal{S}_0 : \begin{cases} -4000x_1^0 - 4000x_{13}^0 = 0, \\ -600x_1^0 + 400x_3^0 - 600x_{12}^0 + 400x_{13}^0 + 400x_{23}^0 + 400x_{123}^0 = 0, \\ x_i^0 \geq 0, \text{ for } i = 1, 2, 3, \\ x_{ij}^0 \geq 0, \text{ for } i, j = 1, 2, 3, i \neq j, \\ x_{123}^0 + x_{ij}^0 \geq 0, \text{ for } i, j = 1, 2, 3, i \neq j, \\ x_1^0 + x_2^0 + x_3^0 + x_{12}^0 + x_{13}^0 + x_{23}^0 + x_{123}^0 = 1, \end{cases}$$

for which a solution is $x_2^0 = 1$ and 0 otherwise, determining the Möbius inverse m_0 and the corresponding submodular capacity ψ_0.

Then, we have that $H_0^1 = K$, so, the second system in the sequence is

$$\mathcal{S}_1 : \begin{cases} -600x_1^1 + 400x_3^1 - 600x_{12}^1 + 400x_{13}^1 + 400x_{23}^1 + 400x_{123}^1 = 0, \\ x_i^1 \geq 0, \text{ for } i = 1, 2, 3, \\ x_{ij}^1 \geq 0, \text{ for } i, j = 1, 2, 3, i \neq j, \\ x_{123}^1 + x_{ij}^1 \geq 0, \text{ for } i, j = 1, 2, 3, i \neq j, \\ x_1^1 + x_2^1 + x_3^1 + x_{12}^1 + x_{13}^1 + x_{23}^1 + x_{123}^1 = 1, \\ x_2^1 + x_{12}^1 + x_{23}^1 + x_{123}^1 = 0, \end{cases}$$

for which a solution is $x_1^1 = \frac{2}{5}$, $x_{13}^1 = \frac{3}{5}$ and 0 otherwise, determining the Möbius inverse m_1 and the corresponding submodular capacity ψ_1.

A simple verification shows that $\{\psi_0, \psi_1\}$ is a minimal agreeing class on $\wp(\Omega)$, determining a conditional submodular capacity $\psi(\cdot|\cdot)$ on $\wp(\Omega) \times \mathcal{H}$ such that the corresponding conditional Choquet expected value $\Psi(\cdot|\cdot)$ on $\mathbb{R}^\Omega \times \mathcal{H}$ satisfies

$$\rho(X|H) = \Psi(-r^{-1}X|H) \quad \text{and} \quad \rho(X|K) = \Psi(-r^{-1}X|K).$$

5   Conclusions

By considering a single period financial market we introduce a specific family of conditional coherent risk measures, obtained through the Choquet integral with respect to a conditional submodular capacity. We further deal with the problem of consistence with a conditional submodular coherent risk measure by providing a characterization in terms of solvability of a suitable sequence of linear systems.

The introduced conditional risk measures are based on an axiomatic definition of conditioning for submodular capacities, however, since different notions of conditioning can be chosen, a future aim is to make a comparison of the resulting properties of the conditional risk measures by varying the adopted notion of conditioning (also referring to results in [20,21,23,29]).

Another aspect to address is the extension of the present model in the multiperiod setting that should also take into account the uncertainty on the time value of money. This requires to consider entire cash flow processes rather than total amounts at terminal dates as risky objects. In this extension the properties of time consistence should be considered and compared with the other existing in literature (see, e.g., [1,4,16,25]).

References

1. Artzner, P., Delbaen, F., Eber, J.-M., Heath, D., Ku, H.: Coherent multiperiod risk adjusted values and Bellman's principle. Ann. Oper. Res. **152**, 5–22 (2007). https://doi.org/10.1007/s10479-006-0132-6
2. Artzner, P., Delbaen, P., Eber, J.-M., Heath, D.: Coherent measures of risk. Math. Finan. **9**(3), 203–228 (1999). https://doi.org/10.1111/1467-9965.00068
3. Baroni, P., Pelessoni, R., Vicig, P.: Generalizing Dutch risk measures through imprecise previsions. Int. J. Uncertain. Fuzziness Knowl.-Based Syst. **17**(2), 153–177 (2009). https://doi.org/10.1142/S0218488509005796
4. Bion-Nadal, J.: Dynamic risk measures: time consistency and risk measures from BMO martingales. Finan. Stochast. **12**, 219–244 (2008). https://doi.org/10.1007/s00780-007-0057-1
5. Capotorti, A., Coletti, G., Vantaggi, B.: Standard and nonstandard representability of positive uncertainty orderings. Kybernetika **50**(2), 189–215 (2014). https://doi.org/10.14736/kyb-2014-2-0189
6. Chateauneuf, A., Kast, R., Lapied, A.: Conditioning capacities and Choquet integrals: the role of comonotony. Theor. Dec. **51**, 367–386 (2001). https://doi.org/10.1023/A:1015567329595
7. Chateauneuf, A., Jaffray, J.-Y.: Some characterizations of lower probabilities and other monotone capacities through the use of Möbous inversion. Math. Soc. Sci. **17**, 263–283 (1989). https://doi.org/10.1016/0165-4896(89)90056-5
8. Cheridito, P., Delbaen, F., Kupper, M.: Dynamic monetary risk measures for bounded discrete-time processes. Electron. J. Probab. **11**, 57–106 (2006). https://doi.org/10.1214/EJP.v11-302
9. Coletti, G., Petturiti, D., Vantaggi, B.: Conditional belief functions as lower envelopes of conditional probabilities in a finite setting. Inf. Sci. **339**, 64–84 (2016). https://doi.org/10.1016/j.ins.2015.12.020

10. Coletti, G., Vantaggi, B.: A view on conditional measures through local representability of binary relations. Int. J. Approx. Reason. **47**, 268–283 (2008). https://doi.org/10.1016/j.ijar.2007.05.007
11. Delbaen, F.: Coherent risk measures on general probability spaces. In: Sandman, K., Schönbucher, P.J. (eds.) Advances in Finance Stochastics, pp. 1–37. Springer, Heidelberg (2002). https://doi.org/10.1007/978-3-662-04790-3_1
12. Dempster, A.P.: Upper and lower probabilities induced by a multivalued mapping. Ann. Math. Stat. **2**, 325–339 (1967). https://doi.org/10.1214/aoms/1177698950
13. Denneberg, D.: Conditioning (updating) non-additive measures. Ann. Oper. Res. **52**(1), 21–42 (1994). https://doi.org/10.1007/BF02032159
14. Denneberg, D.: Non-additive Measure and Integral. Kluwer Academic Publishers, Dordrecht (1994). https://doi.org/10.1007/978-94-017-2434-0
15. Dubins, L.E.: Finitely additive conditional probabilities, conglomerability and disintegrations. Ann. Probab. **3**, 89–99 (1975). https://doi.org/10.1214/aop/1176996451
16. Föllmer, H., Penner, I.: Convex risk measures and the dynamics of their penalty functions. Stat. Dec. **24**, 61–96 (2006). https://doi.org/10.1524/stnd.2006.24.1.61
17. Grabisch, M.: Set Functions, Games and Capacities in Decision Making. Springer, Cham (2016). https://doi.org/10.1007/978-3-319-30690-2
18. Halpern, J.H.: Reasoning About Uncertainty. The MIT Press, Cambrige (2005)
19. Jaffray, J.Y.: Bayesian updating and belief functions. IEEE Trans. Man Cybern. **22**, 1144–1152 (1992). https://doi.org/10.1109/21.179852
20. Pelessoni, R., Vicig, P.: Imprecise previsions for risk measurement. Int. J. Uncertain. Fuzziness Knowl.-Based Syst. **11**(4), 393–412 (2003). https://doi.org/10.1142/S0218488503002156
21. Pelessoni, R., Vicig, P.: Uncertainty modelling and conditioning with convex imprecise previsions. Int. J. Approx. Reason. **39**, 297–319 (2005). https://doi.org/10.1016/j.ijar.2004.10.007
22. Pelessoni, R., Vicig, P.: 2-coherent and 2-convex conditional lower previsions. Int. J. Approx. Reason. **77**, 66–86 (2016). https://doi.org/10.1016/j.ijar.2016.06.003
23. Pelessoni, R., Vicig, P., Zaffalon, M.: Inference and risk measurement with the parimutuel model. Int. J. Approx. Reason. **51**, 1145–1158 (2010). https://doi.org/10.1016/j.ijar.2010.08.005
24. Riedel, F.: Dynamic coherent risk measures. Stochast. Process. Appl. **112**(2), 185–200 (2004). https://doi.org/10.1016/j.spa.2004.03.004
25. Roorda, B., Schumacher, J.M., Engwerda, J.: Coherent acceptability measures in multiperiod models. Math. Finan. **15**(4), 589–612 (2005). https://doi.org/10.1111/j.1467-9965.2005.00252.x
26. Schmeidler, D.: Integral representation without additivity. Proc. Am. Math. Soc. **97**(2), 255–261 (1986). https://doi.org/10.1090/S0002-9939-1986-0835875-8
27. Shafer, G.: A Mathematical Theory of Evidence. Princeton University Press, Princeton (1976)
28. Troffaes, M.C.M., de Cooman, G.: Lower Previsions. Wiley, Hoboken (2014). https://doi.org/10.1002/9781118762622
29. Vicig, P.: Financial risk measurement with imprecise probabilities. Int. J. Approx. Reason. **49**, 159–174 (2008). https://doi.org/10.1016/j.ijar.2007.06.009
30. Walley, P.: Statistical Reasoning with Imprecise Probabilities. Chapman and Hall, New York (1991)

Mathematical Fuzzy Logic and Mathematical Morphology

On the Structure of Group-Like FL_e-chains

Sándor Jenei$^{(\boxtimes)}$

Institute of Mathematics and Informatics,
University of Pécs, Ifjúság u. 6., Pécs 7624, Hungary
jenei@ttk.pte.hu

Abstract. Hahn's celebrated embedding theorem asserts that linearly ordered Abelian groups embed in the *lexicographic product* of real groups [13]. In this paper the *partial*-lexicographic product construction is introduced, a class of residuated monoids, namely, group-like FL_e-chains which possess finitely many idempotents are *embedded* into finite partial-lexicographic products of linearly ordered Abelian groups, that is, Hahn's theorem is extended to this residuated monoid class. As a side-result, the finite strong standard completeness of the logic $\mathbf{IUL}^{fp}$ is announced.

Keywords: Residuated monoids · Representation · Hahn embedding
Linearly ordered Abelian groups · Partial lexicographic products

1 Introduction and Preliminaries

Hahn's structure theorem states that *linearly ordered Abelian groups* can be embedded in the *lexicographic product* of *real groups* [13]. Conrad, Harvey, and Holland generalized Hahn's theorem for *lattice-ordered Abelian groups* [10]. We extend Hahn's theorem to a class of linearly ordered, commutative *residuated monoids*. The price for not having inverses will be that the embedding is done into *partial*-lexicographic products, to be introduced in this paper, and not into lexicographic ones. Very surprisingly, even in our residuated monoid setting, real groups[1] occasionally equipped with a top element, or both with a top and a bottom element serve as basic building blocks.

Ward and Dilworth introduced *residuated lattices* in the 30's of the last century to investigate ideal theory of commutative rings with unit [28]. Examples of residuated lattices include Boolean algebras, Heyting algebras [24], complemented semigroups [6], bricks [4], residuation groupoids [7], semiclans [5], Bezout monoids [3], BL-algebras [14], MV-algebras [8], lattice-ordered groups; several other algebraic structures can be rendered as residuated lattices.

S. Jenei—The present scientific contribution is dedicated to the 650th anniversary of the foundation of the University of Pécs, Hungary, and was supported by the GINOP 2.3.2-15-2016-00022 grant.

[1] Real groups are very specific in the class of residuated lattices.

© Springer International Publishing AG, part of Springer Nature 2018
J. Medina et al. (Eds.): IPMU 2018, CCIS 854, pp. 253–264, 2018.
https://doi.org/10.1007/978-3-319-91476-3_21

Residuated lattices[2] are algebraic models of substructural logics [12]. Such logics are meant to be logics, which, when formulated as Gentzen-style systems, lack some of the traditional trio of structural rules, namely: contraction, weakening, or exchange. In residuated lattices the monoidal operation is not necessarily commutative, and the maximum (if exists) may differ from the unit of the monoid. Residuated lattices with an additional constant in their signature are called FL-algebras[3]. FL-algebras are algebraic models of the Full Lambek calculus **FL**, hence their name. By adding the commutativity of the monoid, we obtain the algebraic models of **FL$_e$** (**F**ull **L**ambek calculus with **e**xchange), that is, the exchange rule corresponds to commutativity. For example, the integrality of the monoid[4] corresponds to the weakening rule; thus, integral commutative residuated lattices[5] are the algebraic models of **FL$_{ew}$** (**F**ull **L**ambek calculus with **e**xchange and **w**eakening).

The notion of residuated lattices is of huge generality, and establishing a *structure theorem* or a *classification* requires further assumptions. Hölder's theorem, which states the embeddability of Archimedean, naturally and linearly ordered semigroups into the additive semigroup of the real numbers, is a precursor in this topic [16]. Aczél *did not assume isotonicity* of the semigroup operation but assumed the universe to be an interval of real numbers and also found in [1, page 256] that the cancellative property[6] is sufficient and necessary for the existence of an order-isomorphism to a subsemigroup of the additive semigroup of the real numbers [1, page 268]. Clifford's contribution [9] asserts that Archimedean, naturally and linearly ordered semigroups in which the *cancellation law does not hold* embedded into either the real numbers in the interval $[0,1]$ with the usual ordering and $ab = \max(a+b, 1)$ or the real numbers in the interval $[0,1]$ and the symbol ∞ with the usual ordering and $ab = a+b$ if $a+b \leq 1$ and $ab = \infty$ if $a+b > 1$ (see also [11, Theorem 2 in Sect. 2 of Chapter XI]). Clifford also showed that *by dropping the Archimedean property*, every naturally and linearly ordered, commutative semigroup is uniquely expressible by his ordinal sum construction as a linearly ordered set of ordinally irreducible semigroups of this kind. Mostert and Shields have dropped the *linear order assumption* and gave a complete description of topological semigroups over compact manifolds with connected, regular boundary in [26] by using a subclass of compact connected Lie groups and via classifying semigroups on arcs such that one endpoint is an identity for the semigroup, and the other is a zero. They classified such semigroups as ordinal sums[7] of three basic multiplications which an arc may possess. The word 'topological', refers to the continuity of the semigroup operation with respect to the topology. In the next related classification result, the property of *topological*

---

[2] In the more general modern terminology.

[3] FL-algebras are also called pointed residuated lattices or pointed residuated lattice-ordered monoids.

[4] Integrality means that the unit element of the multiplication is the greatest element of the underlying universe.

[5] Also called FL$_{ew}$-algebras.

[6] He called it reducible.

[7] In the sense of Clifford.

connectedness of the underlying universe was dropped whereas the continuity condition was somewhat strengthened: Under the assumption of divisibility[8], residuated chains were classified as ordinal sums[9] of linearly ordered Wajsberg hoops [2]. Postulating the divisibility condition proved to be sufficient for the classification of commutative, integral, prelinear residuated monoids over *arbitrary lattices*, see [23], where the authors introduced the notion of poset sum[10] of hoops and proved that commutative[11] GBL-algebras embed into the poset sum of a family of MV-chains, thus extending the Conrad-Harvey-Holland theorem to commutative GBL-algebras. A structural representation of absorbent-continuous[12], group-like FL$_e$-algebras over complete and weakly real chains[13] has been given in [21, 22].

The class of FL$_{ew}$-chains is of huge generality. For example, every commutative integral monoid on a finite chain belongs to this class, and commutative integral monoids on finite chains seem to be very far from being described structurally in a satisfactory manner. It has been shown in [20, Theorem 3.1] that any FL$_{ew}$-chain embeds into a densely-ordered FL$_{ew}$-chain[14]. Further, by [18, Theorem 3], any densely-ordered FL$_{ew}$-chain embeds into a densely-ordered, involutive FL$_{ew}$-chain. These suggest that the classification of densely-ordered, involutive FL$_{ew}$-chains is of the same difficulty as the classification of commutative integral monoids on finite chains; a hopeless task at present. Therefore, it is very surprising that when considering the even more general class of involutive FL$_e$-chains (without w and the densely-ordered assumption), the additionally postulated $t = f$ condition with the assumption on the number of idempotent elements results in such a strong structural description, which uses only linearly ordered Abelian groups (Theorem 3).

The naturally ordered[15] condition has been assumed in all previous results. Non-integral residuated structures, and consequently, substructural logics without the weakening rule, are far less understood at present. Our study concerns group-like FL$_e$-chains; a contribution to *non-integral* residuated structures *without* postulating *the naturally ordered condition*.

Group-like FL$_e$-chains are linearly ordered, involutive, commutative, residuated lattices such that the unit of the monoidal operation coincides with the constant which defines the involution (see Definition 1). The latest postulate forces the structure to resemble linearly ordered Abelian groups in several ways:

---

[8] Divisibility is the dual notion of being naturally ordered. For residuated integral monoids, divisibility is equivalent to the continuity of the semigroup operation in the order topology provided that the underlying chain is densely-ordered.

[9] In the sense of Aglianò-Montagna.

[10] A common generalization of ordinal sums and direct products.

[11] It was proved for a wider class of algebras.

[12] Absorbent continuity is a weakened version of the naturally ordered property.

[13] Weakly real chains are densely-ordered chains with two additional properties.

[14] A stronger statement was proved there for *at most countable* algebras. However, the part of the proof, which is about the densification step does not use this assumption.

[15] Or its dual notion, divisibility.

Firstly, like lattice-ordered Abelian groups, if an involutive FL_e-algebra is conic[16] then the monoidal operation of it can be recovered from its restrictions to the positive and negative cones of the algebra, and more importantly, for complete, densely-ordered, group-like FL_e-chains the monoidal operation can be recovered from its restriction solely to its positive cone (see [19, Theorem 1]).

Secondly, as we shall claim in Theorem 1, group-like FL_e-chains can be characterized as generalizations of linearly ordered Abelian groups, roughly, by just weakening their cancellative property.

Thirdly, in quest for establishing a structural description for group-like FL_e-chains, similar to that of Hahn's, two variants of the so-called partial-lexicographic product construction (Definition 4, Theorem 2) along with two related decompositions (Lemma 2) will be introduced. Using them, the structure of group-like FL_e-chains with finitely many idempotents can be described by iterating finitely many times the type I and type II variants of the partial-lexicographic product construction: Each group-like FL_e-chain which possesses finitely many idempotents can be embedded into such a finite partial-lexicographic product of linearly ordered Abelian groups (Theorem 3). A corollary of this representation theorem is the extension of the embedding theorem of linearly ordered Abelian groups by Hahn to group-like FL_e-chains with finitely many idempotents, see Corollary 1. Another side-result is the finite strong standard completeness of the logic $\mathbf{IUL}^{fp}$, see Corollary 3.

Definition 1. An *FL_e-algebra* is a structure $(X, \wedge, \vee, *, \rightarrow_*, t, f)$ such that $(X, \wedge, \vee)$ is a lattice[17], $(X, \leq, *, t)$ is a commutative, residuated[18] monoid[19], and f is an arbitrary constant. One defines $x' = x \rightarrow_* f$ and calls an FL_e-algebra *involutive* if $(x')' = x$ holds. We call an FL_e-algebra *group-like* if it is involutive and $t = f$.[20] Denote its positive cone by X^+.

2   Results

In our investigations a crucial role will be played by the τ function.

Definition 2. (τ) For an involutive FL_e-algebra $(X, \wedge, \vee, *, \rightarrow_*, t, f)$, for $x \in X$, define $\tau(x)$ to be the greatest element of $Stab_x = \{u \in X \mid u * x = x\}$. Since

---

[16] That is, all elements are comparable with the unit element of the monoidal operation.

[17] Sometimes the lattice operators are replaced by their induced ordering $\leq$ in the signature, in particular, if an FL_e-*chain* is considered, that is, if the ordering is linear.

[18] That is, there exists a binary operation $\rightarrow_*$ such that $x * y \leq z$ if and only if $x \rightarrow_* z \geq y$; this equivalence is called residuation condition or adjointness condition, $(*, \rightarrow_*)$ is called an adjoint pair. Equivalently, for any x, z, the set $\{v \mid x * v \leq z\}$ has its greatest element, and $x \rightarrow_* z$ is defined as this element: $x \rightarrow_* z := \max\{v \mid x * v \leq z\}$.

[19] We use the word monoid to mean semigroup with unit element.

[20] Lattice-ordered Abelian groups equipped with $x \rightarrow_* y := y * x^{-1}$ and $f := t$ are group-like FL_e-algebras.

t is the unit element of $\circledast$, $Stab_x$ is nonempty. Since $\circledast$ is residuated, the greatest element of $Stab_x$ exists, and it holds true that

$$\tau(x) \circledast x = x \quad and \quad \tau(x) = x \rightarrow_\circledast x \geq t. \tag{1}$$

Lemma 1. (τ-lemma). *Let $(X, \wedge, \vee, \circledast, \rightarrow_\circledast, t, f)$ be a group-like FL$_e$-chain.*

1. *The τ value of any expression equals the maximum of the τ-values of its variables[21].*

2.1 Group-Like FL$_e$-algebras Vs. Partially Ordered Abelian Groups

The notion of group-like FL$_e$-algebras is defined with respect to the very general notion of residuated lattices by adding further postulates (such as commutativity, an extra constant f, involutivity, and the $t = f$ property). The following theorem relates group-like FL$_e$-algebras to (in the setting of residuated lattices, very specific) lattice-ordered Abelian groups, thus picturing their precise interrelation. In addition, Theorem 1 is also used in the basic step of the induction in the proof of Theorem 3.

Theorem 1. *For a group-like FL$_e$-algebra $(X, \wedge, \vee, \circledast, \rightarrow_\circledast, t, f)$ the following statements are equivalent:*

1. *Each element of X has inverse given by $x^{-1} = x'$, and hence $(X, \wedge, \vee, \circledast, t)$ is a lattice-ordered Abelian group,*
2. *$\circledast$ is cancellative,*
3. *$\tau(x) = t$ for all $x \in X$.*
4. *The only idempotent element in the positive cone of X is t.*

2.2 Construction of Involutive FL$_e$-algebras

We start with a notation.

Definition 3. Let $(X, \leq)$ be a chain (a linearly ordered set). For $x \in X$ define

$$x_\downarrow = \begin{cases} z \text{ if there exists } X \ni z < x \text{ such that there is no element in } X \\ \quad \text{between } z \text{ and } x, \\ x \text{ if for any } X \ni z < x \text{ there exists } v \in X \text{ such that } z < v < x \\ \quad \text{holds.} \end{cases}$$

We define $x_\uparrow$ dually.

We introduce a construction, called *partial-lexicographic product* with three slightly different variations in Definition 4. Roughly, only a subalgebra is used as a first component of a lexicographic product and the rest of the algebra is left unchanged (hence the adjective 'partial'). This results in an involutive FL$_e$-algebra, which is group-like provided that the second component of the lexicographic product is so, see Theorem 2. The type 0 construction is of technical nature, type I and type II constructions will play a key role in Theorem 3. Let $\boldsymbol{\Gamma}$ denote the lexicographic product.

---

[21] We set $\max(\emptyset) = t$.

Definition 4. Let $\mathbf{X} = (X, \wedge_X, \vee_X, *, \rightarrow_*, t_X, f_X)$ be a group-like FL_e-algebra and $\mathbf{Y} = (Y, \wedge_Y, \vee_Y, \star, \rightarrow_\star, t_Y, f_Y)$ be an involutive FL_e-algebra, with residual complement $'$ and $'$, respectively.

(0) Add a new element $\perp$ to Y as a bottom element, and extend $\star$ by $\perp \star y = y \star \perp = \perp$ for $y \in Y \cup \{\perp\}$.
Let $\mathbf{X}_1 = (X_1, \wedge_X, \vee_X, *, \rightarrow_*, t_X, f_X)$ be a prime[22], cancellative subalgebra of $\mathbf{X}$ (in the sense of Theorem 1, it is a linearly ordered group)[23]. We define

$$\mathbf{X}_{\Gamma(\mathbf{X_1}, \mathbf{Y})} = \left(X_{\Gamma(X_1, Y)}, \leq, \circledast, \rightarrow_\circledast, (t_X, t_Y), (f_X, f_Y) \right),$$

where

$$X_{\Gamma(X_1, Y)} = (X_1 \times Y) \cup ((X \setminus X_1) \times \{\perp\}),$$

$\leq$ is the restriction of the lexicographical order of $\leq_X$ and $\leq_{Y \cup \{\perp\}}$ to $X_{\Gamma(X_1, Y)}$, $\circledast$ is defined coordinatewise, and the operation $\rightarrow_\circledast$ is given by

$$(x_1, y_1) \rightarrow_\circledast (x_2, y_2) = ((x_1, y_1) \circledast (x_2, y_2)')',$$

where

$$(x, y)' = \begin{cases} (x^{'}, y^{'}) & \text{if } x \in X_1 \\ (x^{'}, \perp) & \text{if } x \notin X_1 \end{cases}.$$

Call $\mathbf{X}_{\Gamma(\mathbf{X_1}, \mathbf{Y})}$ the *(type 0) partial-lexicographic product* of X, X_1, and Y.

1. Add a new element $\top$ to Y as a top element, and extend $\star$ by $\top \star y = y \star \top = \top$ for $y \in Y \cup \{\top\}$, then add a new element $\perp$ to $Y \cup \{\top\}$ as a bottom element, and extend $'$ by $\perp^{'} = \top$, $\top^{'} = \perp$ and $\star$ by $\perp \star y = y \star \perp = \perp$ for $y \in Y \cup \{\top, \perp\}$.
Let $\mathbf{X}_1 = (X_1, \wedge_X, \vee_X, *, \rightarrow_*, t_X, f_X)$ be a cancellative subalgebra of $\mathbf{X}$ (in the sense of Theorem 1, it is a linearly ordered group). We define

$$\mathbf{X}_{\Gamma(\mathbf{X_1}, \mathbf{Y}^{\top\perp})} = \left(X_{\Gamma(X_1, Y^{\top\perp})}, \leq, \circledast, \rightarrow_\circledast, (t_X, t_Y), (f_X, f_Y) \right),$$

where

$$X_{\Gamma(X_1, Y^{\top\perp})} = (X_1 \times (Y \cup \{\top, \perp\})) \cup ((X \setminus X_1) \times \{\perp\}),$$

$\leq$ is the restriction of the lexicographical order of $\leq_X$ and $\leq_{Y \cup \{\top, \perp\}}$ to $X_{\Gamma(X_1, Y^{\top\perp})}$, $\circledast$ is defined coordinatewise, and the operation $\rightarrow_\circledast$ is given by

$$(x_1, y_1) \rightarrow_\circledast (x_2, y_2) = ((x_1, y_1) \circledast (x_2, y_2)')',$$

where

$$(x, y)' = \begin{cases} (x^{'}, y^{'}) & \text{if } x \in X_1 \\ (x^{'}, \perp) & \text{if } x \notin X_1 \end{cases}.$$

Call $\mathbf{X}_{\Gamma(\mathbf{X_1}, \mathbf{Y}^{\top\perp})}$ the *(type I) partial-lexicographic product* of X, X_1, and Y.

---

[22] We mean that $(X \setminus X_1) * (X \setminus X_1) \subseteq X \setminus X_1$ holds.
[23] We remark that the only choice for $\mathbf{X}_1$ is $\mathbf{X}_{\tau=t}$, see Definition 5.

2. Add a new element $\top$ to Y as a top element, and extend $\star$ by $\top \star y = y \star \top = \top$ for $y \in Y \cup \{\top\}$.

Let $\mathbf{X}_1 = (X_1, \wedge, \vee, *, \rightarrow_*, t_X, f_X)$ be a linearly ordered, discretely embedded[24], prime, cancellative subalgebra of $\mathbf{X}$ (in the sense of Theorem 1, it is a discrete linearly ordered group)[25].

We define

$$\mathbf{X}_{\Gamma(\mathbf{X_1}, \mathbf{Y}^\top)} = \left(X_{\Gamma(X_1, Y^\top)}, \leq, \circledast, \rightarrow_\circledast, (t_X, t_Y), (f_X, f_Y) \right),$$

where

$$X_{\Gamma(X_1, Y^\top)} = (X_1 \times (Y \cup \{\top\})) \cup ((X \setminus X_1) \times \{\top\}),$$

$\leq$ is the restriction of the lexicographical order of $\leq_X$ and $\leq_{Y \cup \{\top\}}$ to $X_{\Gamma(X_1, Y^\top)}$, $\circledast$ is defined coordinatewise, and the operation $\rightarrow_\circledast$ is given by

$$(x_1, y_1) \rightarrow_\circledast (x_2, y_2) = ((x_1, y_1) \circledast (x_2, y_2)')',$$

where

$$(x, y)' = \begin{cases} ((x^{f}), \top) & \text{if } x \notin X_1 \text{ and } y = \top \\ (x^{f}, y^{f}) & \text{if } x \in X_1 \text{ and } y \in Y \\ ((x^{f})_\downarrow, \top) & \text{if } x \in X_1 \text{ and } y = \top \end{cases} \tag{2}$$

Call $\mathbf{X}_{\Gamma(\mathbf{X_1}, \mathbf{Y}^\top)}$ the *(type II) partial-lexicographic product* of X, X_1, and Y.

Theorem 2. $\mathbf{X}_{\Gamma(\mathbf{X_1}, \mathbf{Y})}$, $\mathbf{X}_{\Gamma(\mathbf{X_1}, \mathbf{Y}^{\top\perp})}$, *and* $\mathbf{X}_{\Gamma(\mathbf{X_1}, \mathbf{Y}^\top)}$ *are involutive* FL$_e$-*algebras with the same rank*[26] *as that of* $\mathbf{Y}$. *In particular, if* $\mathbf{Y}$ *is group-like then so are* $\mathbf{X}_{\Gamma(\mathbf{X_1}, \mathbf{Y})}$, $\mathbf{X}_{\Gamma(\mathbf{X_1}, \mathbf{Y}^{\top\perp})}$, *and* $\mathbf{X}_{\Gamma(\mathbf{X_1}, \mathbf{Y}^\top)}$.

2.3 Decomposition of Group-Like FL$_e$-algebras

We introduce two *decompositions* of group-like FL$_e$-chains in Lemma 2 under the assumption that there exists the second smallest idempotent of the positive cone of the algebra. The decompositions are the 'inverse operations' of the type I and type II constructions of the previous section. We 'isolate' a homomorphic image of a subalgebra in the decompositions, which is always a linearly ordered Abelian group. When we 'factorize' only this subalgebra and leave the rest of the original algebra unchanged, for the 'remaining' algebra it holds true that its set of positive idempotents is order-isomorphic to the set of positive idempotents of the original algebra deprived of its least element. In addition, the original algebra can be reconstructed as a partial lexicographic product using these components.

---

[24] We mean that for $x \in X_1$, it holds true that $x \notin \{x_\uparrow, x_\downarrow\} \subset X_1$ ($\downarrow$ and $\uparrow$ are computed in X).

[25] Just like at item (0), the only choice for $\mathbf{X}_1$ is $\mathbf{X}_{\tau=t}$, provided that it is discrete, see Definition 5.

[26] The rank of an involutive FL$_e$-algebra is positive if $t > f$, negative if $t < f$, and 0 if $t = f$.

Definition 5. For a group-like FL_e-chain $(X, \wedge, \vee, *, \to_*, t, f)$, for $u \geq t$ and $\circ \in \{<, =, \geq\}$ denote

$$X_{\tau \circ u} = \{x \in X : \tau(x) \circ u\}.$$

Proposition 1. *Let* $\mathbf{X} = (X, \leq, *, \to_*, t, f)$ *be a group-like* FL_e-*chain. Let* $u \geq t$ *be idempotent. Then* $X_{\tau < u} \cup \{t\}$, $X_{\tau = u} \cup \{t\}$, $X_{\tau \geq u} \cup \{t\}$ *are nonempty subuniverses.*

The respective subalgebras of $\mathbf{X}$ will be denoted by $\mathbf{X}_{\tau < u}$, $\mathbf{X}_{\tau = u}$, and $\mathbf{X}_{\tau \geq u}$.

Definition 6. Let $\mathbf{X} = (X, \leq, *, \to_*, t, f)$ be a group-like FL_e-chain such that there exists u, the smallest idempotent strictly above t. For $x \in X_{\tau < u}$ let

$$\top_{[x]} = \begin{cases} \bigvee_{z \in [x]} z & \text{in case } u \neq t_\uparrow \\ x_\uparrow & \text{in case } u = t_\uparrow \end{cases} \quad \text{and} \quad \bot_{[x]} = \begin{cases} \bigwedge_{z \in [x]} z & \text{in case } u \neq t_\uparrow \\ x_\downarrow & \text{in case } u = t_\uparrow \end{cases}.$$

Denote

$$X^T_{\tau \geq u} = \{\top_{[v]} \mid v \in X_{\tau < u}\},$$
$$X^{Gap_1}_{\tau \geq u} = \{x \in X_{\tau \geq u} \mid x_\downarrow < x, x * u' = x_\downarrow\}.$$

– Let

$$X^G_{[\tau \geq u]} = \{\{x, x_\downarrow\} \mid x \in X^{Gap_1}_{\tau \geq u}\},$$
$$X^E_{[\tau \geq u]} = \{\{\top_{[v]}, \bot_{[v]}\} \mid v \in X_{\tau < u}\},$$
$$X^{EG}_{[\tau \geq u]} = X^E_{[\tau \geq u]} \cup X^G_{[\tau \geq u]}.$$

For $x \in X_{\tau \geq u}$, let

$$[x] = \begin{cases} p & \text{if } x \in p \in X^{EG}_{[\tau \geq u]} \\ \{x\} & \text{otherwise} \end{cases},$$
$$X_{[\tau \geq u]} = \{[x] \mid x \in X_{\tau \geq u}\}.$$

Letting $x \in p$ and $y \in q$, over $X_{[\tau \geq u]}$ define

$$p \leq_1 q \text{ iff } x \leq y,$$

$$p *_1 q = [x * y], \tag{3}$$

$$p \to_{*_1} q = [x \to_* y] \tag{4}$$

$$p^{*_1} = [x'], \tag{5}$$

– Let

$$X^{TG}_{\tau \geq u} = X^T_{\tau \geq u} \cup X^{Gap_1}_{\tau \geq u}$$

and over $X_{\tau \geq u}^{TG}$ define

$$^{\wp} : z \mapsto (z_\perp)',$$

where for $x \in X_{\tau \geq u}$,

$$x_\perp = \begin{cases} z \text{ if there exists } X_{\tau \geq u} \ni z < x \text{ such that there is} \\ \quad \text{no element in } X_{\tau \geq u} \text{ between } z \text{ and } x, \\ x \text{ if for any } X_{\tau \geq u} \ni z < x \text{ there exists } v \in X_{\tau \geq u} \\ \quad \text{such that } z < v < x \text{ holds.} \end{cases}$$

Clearly,

$$x_\perp = \begin{cases} x_\downarrow & \text{if } x \in X_{\tau \geq u} \setminus X_{\tau \geq u}^T \\ \perp_{[v]} & \text{if } x = \top_{[v]} \in X_{\tau \geq u}^T \end{cases}.$$

Define $x_\top$ dually.

Lemma 2 constitutes the tool to manage the induction step in Theorem 3, the main theorem of the paper.

Lemma 2. (Decompositions). *Let* $\mathbf{X} = (X, \leq, \circledast, \rightarrow_\circledast, t, f)$ *be a group-like* *FL$_e$-chain such that there exists* u, *the smallest idempotent above* t *and* $u \neq t_\uparrow$.

1. *Assume that* u' *is idempotent.*
   (a) $\mathbf{X}_{[\tau \geq u]}^{EG} = (X_{[\tau \geq u]}^{EG}, \leq_1, \circledast_1, [u])$ *is a linearly ordered Abelian group with inverse operation* $^{\wp}$, *and*
   $\mathbf{X}_{[\tau \geq u]}^{E} = (X_{[\tau \geq u]}^{E}, \leq_1, \circledast_1, [u])$ *is its subgroup.*
   (b) $\mathbf{X}_{[\tau \geq u]} = (X_{[\tau \geq u]}, \leq_{\circledast_1}, \circledast_1, \rightarrow_{\circledast_1}, [u], [u])$ *is a group-like FL$_e$-chain with involution* $^{\wp}$ *and*
   $\mathbf{X}_{[\tau \geq u]}^{EG}$ *(qua group-like FL$_e$-chain) is a cancellative subalgebra of it. The set of positive idempotents of* $\mathbf{X}_{[\tau \geq u]}$ *is order-isomorphic to the set of positive idempotents of* $\mathbf{X}$ *deprived of* u.
   (c) $\mathbf{X}$ *embeds into* $(\mathbf{X}_{[\tau \geq u]})_{\Gamma(\mathbf{X}_{[\tau \geq u]}^{EG}, [\mathbf{t}]_{[\tau < u]}^{\top \perp})}$.
2. *Assume that* u' *is not idempotent.*
   (a) $\mathbf{X}_{\tau \geq u}^{TG} = (X_{\tau \geq u}^{TG}, \leq, \circledast, u)$ *is a linearly and discretely embedded Abelian group with inverse operation* $^{\wp}$, *and* $\mathbf{X}_{\tau \geq u}^{T} = (X_{\tau \geq u}^{T}, \leq, \circledast, u)$ *is its subgroup.*
   (b) $\mathbf{X}_{\tau \geq u}^{TG}$ *(qua group-like FL$_e$-chain) is a cancellative, discrete, prime subalgebra of the group-like FL$_e$-chain* $\mathbf{X}_{\tau \geq u}$. *The set of positive idempotents of* $\mathbf{X}_{\tau \geq u}$ *is order-isomorphic to the set of positive idempotents of* $\mathbf{X}$ *deprived of* t.
   (c) $\mathbf{X}$ *embeds into* $(\mathbf{X}_{\tau \geq u})_{\Gamma(\mathbf{X}_{\tau \geq u}^{TG}, [\mathbf{t}]_{[\tau < u]}^{\top})}$.

The following theorem, the main theorem of the paper, asserts that any group-like FL$_e$-chain which has finitely many idempotents is a subalgebra of a group-like FL$_e$-chain, which is built by iterating finitely many times the type

I and the type II partial-lexicographic product constructions of Definition 4 using only linearly ordered Abelian groups, as building blocks. Hahn's embedding theorem states that linearly ordered Abelian groups can be embedded in the *lexicographic product* of *real groups*. We assert in Corollary 1 that group-like FL_e-chains with a finite number of idempotents can be embedded in the *partial*-lexicographic product of linearly ordered groups.

Theorem 3. *If* X *is a group-like* FL_e-*chain, which has only* $n \in N$ *idempotents in its positive cone then there exist linearly ordered Abelian groups* G_i ($i \in \{1, 2, \ldots, n\}$), $H_1 \leq G_1$, $H_i \leq \Gamma(H_{i-1}, G_i)$ ($i \in \{2, \ldots, n-1\}$), *and a binary sequence* $\iota \in \{\top\bot, \top\}^{\{2,\ldots,n\}}$ *such that* X *embeds into* X_n, *where* $X_1 := G_1$ *and* $X_i := X_{i-1}\Gamma(H_{i-1}, G_i^{\iota_i})$ ($i \in \{2, \ldots, n\}$).[27]

Definition 7. We say that a group-like FL_e-chain X is represented as a *finite partial-lexicographic product* (of linearly ordered Abelian groups $G_1 \ldots, G_n$), if X arises via finitely many iterations of the type I and type II constructions using the linearly ordered Abelian groups $G_1 \ldots, G_n$ in the way it is described in Theorem 3. Note that linearly ordered Abelian groups are exactly the indecomposable algebras with respect to the type I and type II partial-lexicographic product constructions.

Using the terminology above, we may rephrase Theorem 3: Every group-like FL_e-chain, which has only finitely many idempotents in its positive cone embeds into a finite partial-lexicographic product of linearly ordered Abelian groups. By Hahn's embedding theorem, every linearly ordered Abelian group can be embedded in the lexicographic product of real groups. Therefore, it follows that

Corollary 1 (Hahn-type embedding). *Group-like* FL_e-*chains with a finite number of idempotents embed in the finite partial-lexicographic product of lexicographic products of real groups.*

We remark that lexicographic products can be used instead of partial-lexicographic products if the less ambitious goal of embedding only the monoidal reduct is aimed at. By observing that as a monoid, $X_{\Gamma(X_1, Y^\top)}$ embeds into $X_{\Gamma(X_1, Y^{\top\bot})}$, which, in turn, embeds into $\Gamma(X^{\top\bot}, Y^{\top\bot})$, one obtains the following.

Corollary 2 (Lexicographical embedding of the monoid reduct). *The monoid reduct of any group-like* FL_e-*chain with a finite number of idempotents embeds in the* lexicographic *product of the 'extended' additive group of the reals*[28].

An immediate consequence of Lemma 1 is.

Proposition 2. *Any finitely generated group-like* FL_e-*chain has only a finite number of idempotents.*

---

[27] In the spirit of Theorem 1 we identify linearly ordered Abelian groups by cancellative, group-like FL_e-chains here.

[28] R extended by $\top$ and $\bot$, just like Y in item (1) of Definition 4.

The logic **IUL** was introduced in [25]. Its standard completeness has been left as an open problem. Based on Theorem 3 and Proposition 2 we are able to prove the strong standard completeness of a somewhat simpler logic **IUL**^{fp}, which is defined as the logic **IUL** extended by the axiom **t** $\Leftrightarrow$ **f**.

By relying on Proposition 2 and Theorem 3, we can prove.

Corollary 3. *The logic* **IUL**^{fp} *is finitely strongly standard complete.*

References

1. Aczél, J.: Lectures on Functional Equations and Their Applications. Academic Press, New York/London (1966)
2. Aglianò, P., Montagna, F.: Varieties of BL-algebras I: general properties. J. Pure Appl. Algebra **181**(2–3), 105–129 (2003)
3. Ánh, P.N., Márki, L., Vámos, P.: Divisibility theory in commutative rings: Bezout monoids. Trans. Am. Math. Soc. **364**, 3967–3992 (2012)
4. Bosbach, B.: Concerning bricks. Acta Math. Acad. Sci. Hung. **38**, 89–104 (1981)
5. Bosbach, B.: Concerning semiclans. Arch. Math. **37**, 316–324 (1981)
6. Bosbach, B.: Komplementäre halbgruppen. axiomatik und arithmetik. Fund. Math. **64**, 257–287 (1969)
7. Bosbach, B.: Residuation groupoids and lattices. Stud. Sci. Math. Hung. **13**, 433–451 (1978)
8. Cignoli, R., D'Ottaviano, I.M.L., Mundici, D.: Algebraic Foundations of Many-Valued Reasoning. Kluwer, Dordrecht (2000)
9. Clifford, A.H.: Naturally linearly ordered commutative semigroups. Am. J. Math. **76**(3), 631–646 (1954)
10. Conrad, P.F., Harvey, J., Holland, W.C.: The Hahn embedding theorem for lattice-ordered groups. Trans. Am. Math. Soc. **108**, 143–169 (1963)
11. Fuchs, L.: Partially Ordered Algebraic Systems. Pergamon Press, Oxford/London/New York/Paris (1963)
12. Galatos, N., Jipsen, P., Kowalski, T., Ono, H.: Residuated Lattices: An Algebraic Glimpse at Substructural Logics. Studies in Logic and the Foundations of Mathematics, vol. 151. Elsevier, New York City (2007)
13. Hahn, H.: Über die nichtarchimedischen Grössensysteme. S.-B. Akad. Wiss. Wien. IIa **116**, 601–655 (1907)
14. Hájek, P.: Metamathematics of Fuzzy Logic. Kluwer Academic Publishers, Dordrecht (1998)
15. Horžík, R.: Algebraic semantics: semilinear FL-algebras. In: Cintula, P., Hájek, P., Noguera, C. (eds.) Handbook of Mathematical Fuzzy Logic, pp. 283–353. College Publications, London (2011)
16. Hölder, O.: Die Axiome der Quantität und die Lehre vom Mass. In: Berichte über die Verhandlungen der Königlich Sachsischen Gesellschaft der Wissenschaften zu Leipzig, Mathematisch-Physische Classe, vol. 53, pp. 1–64 (1901)
17. Jenei, S.: On the geometry of associativity. Semigroup Forum **74**(3), 439–466 (2007)
18. Jenei, S.: On the structure of rotation-invariant semigroups. Arch. Math. Log. **42**, 489–514 (2003)
19. Jenei, S.: Structural description of a class of involutive uninorms via skew symmetrization. J. Log. Comput. **21**(5), 729–737 (2011)

20. Jenei, S., Montagna, F.: A proof of standard completeness of Esteva and Godo's monoidal logic MTL. Stud. Log. **70**(2), 184–192 (2002)
21. Jenei, S., Montagna, F.: A classification of certain group-like FL_e-chains. Synthese **192**(7), 2095–2121 (2015)
22. Jenei, S., Montagna, F.: Erratum to "a classification of certain group-like FL_e-chains". Synthese **193**(1), 313–313 (2016)
23. Jipsen, P., Montagna, F.: Embedding theorems for classes of GBL-algebras. J. Pure Appl. Algebra **214**(9), 1559–1575 (2010)
24. Johnstone, P.T.: Stone Spaces. Cambridge University Press, Cambridge (1982)
25. Metcalfe, G., Montagna, F.: Substructural fuzzy logics. J. Symb. Log. **72**(3), 834–864 (2007)
26. Mostert, P.S., Shields, A.L.: On the structure of semigroups on a compact manifold with boundary. Ann. Math. **65**, 117–143 (1957)
27. Tsinakis, C., Wille, A.M.: Minimal varieties of involutive residuated lattices. Stud. Log. **83**(1–3), 407–423 (2006)
28. Ward, M., Dilworth, R.P.: Residuated lattices. Trans. Am. Math. Soc. **45**, 335–354 (1939)

Logics for Strict Coherence and Carnap-Regular Probability Functions

Tommaso Flaminio[✉] [iD]

IIIA - CSIC, Campus de la Universidad Autònoma de Barcelona s/n,
08193 Bellaterra, Spain
tommaso@iiia.csic.es

Abstract. In this paper we provide a characterization of strict coherence in terms of the logical consistency of suitably defined formulas in fuzzy-modal logics for probabilistic reasoning. As a direct consequence of our characterization, we also show the decidability for the problem of checking the strict coherence of rational-valued books on classical events. Further, we introduce a fuzzy modal logic that captures Carnap-regular probability functions, that is normalized and finitely additive measures which maps to 0 only the impossible event.

Keywords: Modal logics · Probabilistic logics · Fuzzy logics
Strict coherence · Decidability · Carnap-regular measures

1 Introduction and Motivation

Modal expansions of Łukasiewicz propositional fuzzy logics for probabilistic reasoning have been firstly introduced in 1995 by Hájek et al. [11] and then further generalized and extended (see [7] for an overview). The very basic idea which lies at the ground of a fuzzy-modal approach to uncertainty (and probability in particular) is to consider formulas of the form $U(\varphi)$ to be read "the formula φ is uncertain" and provide axioms for U in such a way that the truth-degree of the modal formula $U(\varphi)$ becomes the uncertain degree of φ. Specific axioms for the modality U can be provided so as to cope with the peculiar measure we are interested in. Thus, for instance, U can be axiomatized to behave as a probability function by imposing normalization and finite additivity [11,12], but alternative axiomatizations capturing more general uncertainty measures are also feasible [7]. We shall recall the basic notions and results about fuzzy-modal logics for probability in the following Sect. 2.

Besides their intrinsic theoretical interest, fuzzy probabilistic modal logics have been successfully employed in the last years to provide a purely logical characterization of de Finetti's foundation of subjective probability theory (see for instance [5,9]). Let us recall that given a finite set $\mathcal{E} = \{\varphi_1, \ldots, \varphi_n\}$ of events (i.e., formulas of classical propositional logic) a book β is a map from

© Springer International Publishing AG, part of Springer Nature 2018
J. Medina et al. (Eds.): IPMU 2018, CCIS 854, pp. 265–274, 2018.
https://doi.org/10.1007/978-3-319-91476-3_22

$\mathcal{E}$ to the real unit interval $[0, 1]$. The value that a book assigns to any event, can hence be regarded as the selling price that a bookmaker decides and that a gambler accepts to pay in order to participate in a precisely defined game (see [3,6] for further details) where the events φ_i's are uncertain and in which, when the uncertainty about the events is resolved by a possible world w, the gambler wins, on each event φ_i, 1 provided that φ_i holds in w, or 0 otherwise. Thus, a book is said to be *coherent* if there is no strategy which the gambler can perform in order to make the bookmaker incur in a sure loss, i.e., a loss in every possible world.

Formally, let $\mathcal{E} = \{\varphi_1, \ldots, \varphi_n\}$ be a finite set of events (i.e., formulas of classical propositional logic) and let $\beta : \mathcal{E} \to [0, 1]$ be a book. Then β is coherent if for every choice of positive or negative *stakes* $\sigma_1, \ldots, \sigma_n \in \mathbb{R}$, there exists a classical valuation $w : \mathcal{E} \to \{0, 1\}$ such that the balance of the game in w is not negative, i.e.,

$$\sum_{i=1}^n \sigma_i(\beta_i - w(\varphi_i)) \geq 0.$$

The celebrated de Finetti theorem states that a book is coherent if and only if there exists a probability function on the algebra $\mathbf{B}_{\mathcal{E}}$ spanned by the events in $\mathcal{E}$ which extends the book β [3].

De Finetti's coherence guards bookmakers against the possibility of sure loss by simultaneously barring them from what they should reasonably aim to, namely making profit. The condition of strict coherence has been put forward as a natural reaction to this rather odd feature [8,15,17]. Indeed, a book β is *strictly coherent* if every possibility of loss is paired with a possibility of gain. The following example aims at clarifying the difference between coherence and strict coherence.

Example 1. Consider the usual coin tossing game in which events are h, "the coin lands head" and t, "the coin lands tail", together with the following books β_1 and β_2:

1. $\beta_1(h) = 1$, $\beta_1(t) = 0$,
2. $\beta_2(h) = 1/3$, $\beta_2(t) = 2/3$.

It is easy to check that both β_1 and β_2 are coherent. Indeed, modulo de Finetti's theorem, one can easily find two probability measures μ_1 and μ_2 on $\mathbf{B}_{\{h,t\}}$ such that μ_i extends β_i. The notable difference between β_1 and β_2 can be observed by computing the balance for the corresponding games. Imagine in fact that gambler's stakes for h and t are, respectively, σ_h and σ_t for both β_1 and β_2. Thus, given a valuation $v : \{h, t\} \to \{0, 1\}$, the total balance for β_1 is[1]

$$B_1 = \sigma_h(1 - v(h)) + \sigma_t(0 - v(t)) = v(t)(\sigma_h - \sigma_t),$$

while for β_2, the balance equals

---

[1] In the computation we used the fact that $h = \neg t$, whence $1 - v(h) = v(\neg h) = v(t)$.

$$B_2 = \sigma_h(1/3 - v(h)) + \sigma_t(2/3 - v(t)).$$

Now, although the bookmaker will never incur in a sure loss, the gambler has a betting strategy on β_1 (for instance $\sigma_h = 0$ and $\sigma_t = -1$) which makes $B_1 < 0$ if $v(t) = 1$ and $B_1 = 0$ if $v(t) = 0$ (and hence $v(h) = 1$). Vice versa, for every betting strategy σ_h, σ_t on β_2 if the valuation v mapping $h \mapsto 0$ and $t \mapsto 1$, $B_2 < 0$, then the other one $v' = 1 - v$ makes $B_2 > 0$. Therefore, β_2 is strictly coherent.

The following result strengths de Finetti's theorem to the specific case of strictly coherent books in terms of *Carnap-regular* probability functions, i.e., those normalized and finitely additive maps μ from Boolean algebras to the real unit interval $[0,1]$ which further satisfy the following quasi-equation: if $\mu(\varphi) = 0$, then $\varphi = \bot$. In other words, the only event which a Carnap-regular probability function maps to 0 is the impossible event.

Theorem 1 ([8]). *Let $\mathcal{E} = \{\varphi_1, \ldots, \varphi_n\}$ be a finite set of classical formulas and let $\beta : \mathcal{E} \to [0,1]$ be a book. Then β is strictly coherent iff there exists a Carnap-regular probability function $\mu : \mathbf{B}_{\mathcal{E}} \to [0,1]$ such that $\mu(\varphi_i) = \beta(\varphi_i)$.*

Recall that if $\mathbf{B}$ is a finite Boolean algebra with atoms $\alpha_1, \ldots, \alpha_m$, then a probability function μ is Carnap-regular iff $\mu(\alpha_i) > 0$ for all $j = 1, \ldots, m$.

In this paper we provide a logical characterization of rational-valued and strictly coherent books in terms of the satisfiability of a modal formula of a fuzzy modal probabilistic logic. Further, we introduce a modal fuzzy logic that captures, on the semantics side, all Carnap-regular probability measures on finite Boolean algebras. As a consequence of the logical characterization of strict coherence, we obtain the decidability for the problem of checking the strict coherence for a rational-valued book on classical events.

This paper is organized as follows: In the next section we will recall the basics of fuzzy modal logics for probabilistic reasoning and we will show that the main formalisms we shall use in this paper does not enjoy single model completeness property, that is, the is not a single model with respect to which our modal fuzzy logic is complete. In Sect. 3 we prove a first logical characterization for strict coherence. In the same section we will also show that it can be decided if a rational-valued book is strictly coherent or not. Section 4 is devoted to axiomatize a fuzzy modal logic for Carnap-regular probability functions and we will further exhibit a second characterization for strictly coherent books. We end this short paper with Sect. 5 where we will discuss our future work on the present subject.

2   Preliminaries

Let us start this section by recalling some logical preliminaries about the logic $FP(C, \mathbf{L})$ [11] and some of its expansions. We assume the reader to be familiar with propositional Łukasiewicz logic, Ł. Otherwise, we suggest to consult [2,4, 12,16].

Syntax. The syntax of $FP(C, Ł)$ comprises a countable (finite or infinite) set of propositional variables $Var = \{p_1, p_2, \ldots\}$, connectives $\wedge, \vee, \neg, \perp$ of classical logic C and $\oplus, \neg, \perp$ of Łukasiewicz language, plus the unary modality P. Formulas belong to two classes:

(EF): The class of formulas from C. They are inductively defined as usual, and will be used to denote *events*. The class of those formulas will be denoted by E (for *events*).

(MF): The class of modal formulas is defined inductively: for every formula $\varphi \in E$, $P\varphi$ is an atomic modal formula, $\perp$ is also an atomic modal formula. Compound formulas are defined from the atomic ones and using the connectives of Ł. We will denote by MF the class of modal formulas. Note that connectives appearing in the scope of the modal operator P are from classical logic language, while those outside are from propositional Łukasiewicz logic Ł.

Remark 1. As argued in [7] Łukasiewicz logic (Ł in symbols) can be regarded as the minimal propositional framework which allows to properly axiomatize the modality P as a probability function. This fact is witnessed by the presence, in Ł, of the binary connective $\oplus$ which behaves, on $[0, 1]$, as a truncated sum: $x \oplus y = \min\{1, x + y\}$. Thus, by $\oplus$ we can write an axiom (axiom (P3) below) which ensures P to be additive.

Semantics. The semantics for $FP(C, Ł)$ is constituted by a set of *measured Kripke frames*. Those are triples of the form $M = (W, e, \mu)$ where W is a nonempty set of possible worlds, $e : Var \times W \to \{0, 1\}$ is such that, for every $w \in W$, $e(\cdot, w) \to \{0, 1\}$ is a classical valuation, and μ is a finitely additive probability function defined on the Boolean algebra $\mathbf{B}_W$ of μ-measurable sets of the form $\{f_\varphi : w \in W \mapsto e(\varphi, w) \in \{0, 1\} \mid \varphi \in E\}$.

Given a formula ϕ of $FP(C, Ł)$, a measured Kripke frame $M = (W, e, \mu)$ and a world $w \in W$, the *truth-value* of ϕ in M at w (denoted $\|\phi\|_{M,w}$) is inductively defined as follows:

- If $\phi \in E$, $\|\phi\|_{M,e} = e(w, \phi)$;
- If $\phi = P\varphi$, then $\|P\varphi\|_{M,w} = \mu(f_\varphi)$;
- If ϕ is a compound modal formula, $\|\phi\|_{M,w}$ is computed by truth-functionality from its atomic modal subformulas and using the truth-functions of Łukasiewicz connectives.

Notice that, if Φ is modal, its truth-value in a measured Kripke frame M is independent of the chosen world w. Hence we will denote it by $\|\Phi\|_M$ without danger of confusion.

We shall henceforth say that a measured Kripke frame M *models* a modal formula Φ (and we write $M \models \Phi$), if $\|\Phi\|_M = 1$. If Γ is a set of modal formulas, $M \models \Gamma$ means that $M \models \gamma$ for all $\gamma \in \Gamma$.

A measured Kripke frame $M = (W, e, \mu)$ is said to be *rational-valued* (or $\mathbb{Q}$-*valued* for short) if μ takes value in the rational unit interval $[0, 1] \cap \mathbb{Q}$.

2.1   A Complete Axiomatization for Fuzzy Probabilistic Logics

In this section we will recall basic notions and results about the fuzzy modal logic $FP(C, Ł)$, its semantics and expansions.

Definition 1. *The logic $FP(C, Ł)$ is axiomatized by the following axioms and rules:*

(C) All axioms and rules of classical propositional logic restricted to formulas in E.

(Ł) All axioms and rules of Łukasiewicz propositional logic restricted to formulas in MF.

(P) The following axioms for the modality P:
  (P1) ¬P⊥,
  (P2) $P(\neg\varphi) \leftrightarrow \neg P\varphi$,
  (P3) $P(\varphi \vee \psi) \leftrightarrow [(P(\varphi) \to P(\varphi \wedge \psi)) \to P(\psi)]$.

(N) The necessitation rule for P: $\varphi \vdash P(\varphi)$.

The notion of *proof* in $FP(C, Ł)$ is defined as usual. Given a set of modal formulas Γ and a modal formula Φ, we shall henceforth write $\Gamma \vdash \Phi$ to denote that Φ is provable from Γ in $FP(C, Ł)$.

Theorem 2 ([11]). *The logic $FP(C, Ł)$ is sound and complete with respect to the class of measured Kripke frames. In particular, for every finite set of modal formulas $\Gamma \cup \{\Phi\}$, $\Gamma \vdash \Phi$ if and only if for every measured Kripke frame M, if $M \models \Gamma$, then $M \models \Phi$.*

Recalling what we stated in Remark 1 above, any expansion $\mathcal{L}$ of Ł is suitable for axiomatizing P as a (finitely additive) probability function. We will be henceforth concerned with the logics $RŁ$ [10] and $RŁ_\Delta$ [5] (as outer logic for probability). These logics respectively are the expansion of Ł obtained by expanding its language with countably many unary connectives δ_n (for $n \in \mathbb{N} \setminus \{0\}$) ($RŁ$) and the Baaz-Monteiro unary connective Δ [1] (for $RŁ_\Delta$). Taking into account the following formulas,

$$(D1): n.\delta_n\varphi \leftrightarrow \varphi \qquad\qquad (D2): \neg\delta_n\varphi \oplus (n-1).\neg\delta_n\varphi$$
$$(\Delta 1): \Delta(\varphi \to \psi) \to (\Delta\varphi \to \Delta\psi) \quad (\Delta 2): \Delta\varphi \vee \neg\Delta\psi$$
$$(\Delta 3): \Delta\varphi \to \varphi \qquad\qquad (\Delta 4): \Delta\varphi \to \Delta\Delta\varphi$$
$$(\Delta 5): \Delta(\varphi \vee \psi) \leftrightarrow (\Delta\varphi \vee \Delta\psi) \quad (\Delta N): \varphi \vdash \Delta\varphi,$$

the logic $RŁ$ is $Ł + \{(D1), (D2)\}$ and $RŁ_\Delta$ is $RŁ + \{(\Delta 1) - (\Delta 6), (\Delta N)\}$.

Remark 2.(1) It was observed in [10] that in $RŁ$ one can define rational truth-constants by means of the following stipulation: for every pair of mutually prime positive integers n, m with $n < m$, let $\overline{n/m}$ be a shorthand for the $RŁ$ formula $n.\delta_m(\top)$. Then, for every $[0,1]$-valued evaluation e of $RŁ$, one can easily prove that $e(\overline{n/m}) = n/m$.

(2) The standard semantics of Δ provided by any $[0,1]$-valued evaluation e of $R\text{Ł}_\Delta$ is as follows: for every formula φ, $e(\Delta\varphi) = 1$ iff $e(\varphi) = 1$ and $e(\Delta\varphi) = 0$ otherwise. Therefore, one can define in $R\text{Ł}_\Delta$ a further operator (usually denoted by ∇) which is dual to Δ: for every formula φ, $\nabla\varphi = \neg\Delta\neg\varphi$. Thus, every $[0,1]$-valued evaluation e maps $\nabla\varphi$ into 1 iff $e(\varphi) > 0$, while $e(\nabla\varphi) = 0$ if $e(\varphi) = 0$ (see for instance [5] for further details).

Notation 1. *In what follows, we will denote by $FP(C_k, L)$ the logic axiomatized as in Definition 1 where the classical logic language contains only k propositional variables. Furthermore, for any expansion $\mathcal{L}$ of L, $FP(C_k, \mathcal{L})$ will denote the logic obtained from $FP(C_k, L)$ by replacing, in Definition 1, the axioms scheme (Ł) by axioms and rules from $\mathcal{L}$.*

Notice that for any schematic extension $\mathcal{L}$ of L, the notion of measured Kripke frame for $FP(C, \mathcal{L})$ (or $FP(C_k, \mathcal{L})$) remains the same with only necessary modification regarding the evaluation of compound modal formulas. Furthermore the following holds.

Theorem 3. *For any schematic extension $\mathcal{L}$ of L, $FP(C, \mathcal{L})$ and $FP(C_k, \mathcal{L})$ are sound and complete with respect to the class of measured Kripke frames.*

A measured Kripke frame $M = (W, e, \mu)$ is called a *normal frame* if W is a nonempty set of classical evaluations and for each $\varphi \in E$ and $w \in W$, $e(\varphi, w) = w(\varphi)$ [13]. The notion of *normal $\mathbb{Q}$-valued* Kripke frame is defined accordingly. If (W, e, μ) is normal, we shall usually omit e from the signature. Furthermore it is easy to see that for every modal formula Φ and for every measured Kripke frame M there exists a normal frame M' such that $\|\Phi\|_M = \|\Phi\|_{M'}$. Thus, in particular, Theorems 2 and 3 hold for the class of normal frames.

In the rest of this section we will always consider a finite language for events and therefore we will be mainly concerned with a modal logic of the form $FP(C_k, \mathcal{L})$ for $k > 0$.

Let us end this section proving the following result.

Proposition 1. *There is not a unique $\mathbb{Q}$-valued measured Kripke frame with respect to which $FP(C_k, R\text{Ł})$ is complete.*

Proof. Assume, by way of contradiction, that there is a unique $\mathbb{Q}$-valued Kripke frame M_0 such that, for every finite set of modal formulas $\Gamma \cup \{\Phi\}$ such that $\Gamma \nvdash \Phi$, then $M_0 \models \Gamma$, implies $M_0 \nvDash \Phi$. Modulo Theorem 3, the absurd hypothesis can be equivalently stated in the following way: there is a $\mathbb{Q}$-valued measured Kripke frame M_0 such that, for every modal formula Φ, $M_0 \models \Phi$ if and only if $M \models \Phi$ for every $\mathbb{Q}$-valued measured Kripke frame.

Without loss of generality we can assume M_0 to be a $\mathbb{Q}$-valued normal model (W_0, μ_0). Further, we can assume W_0 to be the finite set of all valuations on the k variable of events of $FP(C_k, R\text{Ł})$, whence μ_0 is defined on free k-generated Boolean algebra $\mathbf{B}_{W_0}$ (which coincides with the Boolean algebra of all subsets of W_0). For each $i = 1, \ldots, 2^k$, let

$$\alpha_i = \bigwedge_{j=1}^{k} p_j^* \tag{1}$$

(for $p_j^* \in \{p_j, \neg p_j\}$) and consider the formula

$$\Psi = \bigwedge_{i=1}^{2^k} (P(\alpha_i) \leftrightarrow \overline{\mu_0(\alpha_i)}).$$

The formula Φ is clearly true in M_0, while any other probability function $\mu \neq \mu_0$ defined on $\mathbf{B}_{W_0}$ determines a normal $\mathbb{Q}$-valued frame $M = (W_0, \mu)$ in which Ψ is clearly false. This settles the claim. $\square$

3   A Logical Characterization of Strict Coherence

In this section we are going to characterize the strict coherence of a $\mathbb{Q}$-valued book in terms of the logical consistency of a modal formula of $FP(C_k, R\L_\Delta)$. Further, we will show that the problem of establishing the strict coherence of a $\mathbb{Q}$-valued book is decidable.

Theorem 4. *Let $\mathcal{E} = \{\varphi_1, \ldots, \varphi_n\}$ be a finite set of classical formulas and let $\beta : \mathcal{E} \to [0, 1]$ be a $\mathbb{Q}$-valued book. Then β is strictly coherent if and only if there exists a $k \in \mathbb{N}$ and a modal formula T_β of $FP(C_k, R\L_\Delta)$ such that T_β is consistent.*

Proof. Let Var be the set of variables occurring in the formulas $\varphi_1, \ldots, \varphi_n$ of $\mathcal{E}$, let $k = |Var|$ and let, for all $i = 1, \ldots, 2^k$, α_i be defined as in (1) in the proof of Proposition 1. Consider the formula:

$$T_\beta = \left(\bigwedge_{i=1}^n P(\varphi_i) \leftrightarrow \overline{\beta(\varphi_i)} \right) \wedge \left(\bigwedge_{j=1}^{2^k} \nabla P(\alpha_j) \right).$$

Now, we prove that β si strictly coherent iff T_β is consistent, i.e., it has a model. In particular we are going to show the following claim:

Claim 1. *Let $M = (W, \mu)$ be a measured Kripke frame. Then $M \models T_\beta$ iff $\mu(f_{\varphi_i}) = \beta(\varphi_i)$ and μ is Carnap-regular.*

Proof (of Claim 1). Assume $M \models T_\beta$. Then, in particular $M \models \bigwedge_{i=1}^n P(\varphi_i) \leftrightarrow \overline{\beta(\varphi_i)}$ and this is true iff, for all $i = 1, \ldots, n$, $\|P(\varphi_i)\|_M = \beta(\varphi_i)$, i.e., $\mu(f_{\varphi_i}) = \beta(\varphi_i)$. Furthermore, $M \models \bigwedge_{j=1}^{2^k} \nabla P(\alpha_j)$ iff (recalling Remark 2(2)) $\mu(f_{\alpha_j}) > 0$ iff μ is Carnap-regular. $\square$

Therefore the theorem holds due to Claim 1 and Theorem 1 plus observing that $\mathbf{B}_W = \mathbf{B}_\mathcal{E}$. $\square$

An immediate consequence of the above result and [5, Theorem 7] which shows that the SAT problem for $FP(C_k, R\L_\Delta)$ is NP-complete, is the decidability for strictly coherent books.

In what follows we denote by $\mathscr{S}_\mathcal{E}$ the set of $\mathbb{Q}$-valued strictly coherent books on the finite set of events $\mathcal{E}$.

Corollary 1. *For every finite set $\mathcal{E}$ of events, $\mathscr{S}_{\mathcal{E}}$ is decidable.*

Proof. Theorem 4 characterizes the strict coherence of a $\mathbb{Q}$-valued book $\beta : \mathcal{E} \to [0,1]$ as the satisfiability, in $FP(C_k, \mathrm{R\mathbb{L}}_\Delta)$, of the formula T_β. Thus $\mathscr{S}_{\mathcal{E}}$ is decidable since the SAT problem for $FP(C_k, \mathrm{R\mathbb{L}}_\Delta)$ is decidable as well. □

Remark 3. Notice that the formula T_β whose satisfiability in $FP(C_k, \mathrm{R\mathbb{L}}_\Delta)$ is equivalent to the strict coherence of β has exponential length in β. Thus, although the satisfiability problem for the modal logic $FP(C_k, \mathrm{R\mathbb{L}}_\Delta)$ is NP-complete, the upper bound for strict coherence provided by the above procedure is set, up to now, at EXPTIME.

4   A Modal Logic for Carnap-Regular Probabilities

In this section we are going to strengthen the logic $FP(C_k, \mathrm{R\mathbb{L}}_\Delta)$ in order to capture exactly those probability functions which are Carnap-regular. Precisely, and since Carnap regular measures are not ensured to exist for every Boolean algebra (see for instance [14]), we need first to fix a finite language (i.e., a finite set of propositional variables) for events. Although this constraint might seem to be restrictive, it does not prevent us to prove the main result of this section, namely Theorem 5 below.

In the following definition we adopt a notation which has been already used in the previous sections, namely, letting $p_1, \ldots, p_k$ the variables of C_k we put, for all $j = 1, \ldots, 2^k$

$$\alpha_j = \bigwedge_{i=1}^{k} p_i^*$$

where, for each $i = 1, \ldots, k$, p_i^* stands for either p_i or $\neg p_i$.

Definition 2. *For each $k > 0$, the logic $FR(C_k, \mathrm{R\mathbb{L}}_\Delta)$ is the axiomatic extension of $FP(C_k, \mathrm{R\mathbb{L}}_\Delta)$ given by the following modal axioms for P:*

(R) $\bigwedge_{j=1}^{2^k} \nabla P(\alpha_j)$.

A *Carnap Kripke frame* is a measured Kripke frame (W, e, μ) in which μ is a Carnap-regular measure.

Theorem 5. *The logic $FR(C_k, \mathrm{R\mathbb{L}}_\Delta)$ is sound and complete with respect to the class of Carnap Kripke frames.*

Proof. In the light of Theorem 3 we only need to prove that the added regularity axiom (R) selects, among all measured Kripke frames, only those with a Carnap-regular measure μ. Indeed, as we already proved in Claim 1, a measured Kripke frame $M = (W, e, \mu)$ satisfies (R) iff μ is Carnap-regular. This settles the claim. □

Finally, the strict coherence of a $\mathbb{Q}$-valued book $\beta : \mathcal{E} \to [0,1]$ can be characterized in $FR(C_k, R\text{Ł}_\Delta)$ by a slightly less complex formula than T_β (recall Theorem 4). Let

$$T_\beta^- = \bigwedge_{i=1}^{n} P(\varphi_i) \leftrightarrow \overline{\beta(\varphi_i)}.$$

Then the following easily holds taking into account the proof of Theorem 4 and the axiomatization of $FR(C_k, R\text{Ł}_\Delta)$.

Corollary 2. Let $\mathcal{E} = \{\varphi_1, \ldots, \varphi_n\}$ a set of formulas on k propositional variables. Then a $\mathbb{Q}$-valued book $\beta : \mathcal{E} \to [0,1]$ is strictly coherent if and only if T_β^- is consistent in $FR(C_k, R\text{Ł}_\Delta)$.

Remark 4. In Remark 3, due to the exponential length of T_β, we set the upper bound for the complexity of $\mathscr{S}_\mathcal{E}$ at EXPTIME. Vice versa, Corollary 2 above, provides a characterization of $\mathscr{S}_\mathcal{E}$ in terms of a formula, T_β^-, which is polynomial in β. However, since the axiomatization of $FR(C_k, R\text{Ł}_\Delta)$ includes an axiom (that we denoted by (R)) whose length is exponential in k, the procedure that checks $\mathscr{S}_\mathcal{E}$ via Corollary 2 does not lower the upper bound for the problem of checking strict coherence from EXPTIME.

5   Conclusions and Future Work

Fuzzy modal logics for uncertain reasoning have been quite intensively studied and in the last years have been also successfully applied to characterize de Finetti's foundations of subjective probability theory (see for instance [5, 9] and [7, Sect. 6]). This paper contributes to deepen the investigation on the aforementioned modal logics as suitable formalizations for the foundations of probabilistic reasoning. The side effect of the present investigation, is also to show that these logics are modular and manageable formalisms.

Several interesting problems regarding fuzzy modal logics remain open. First of all, although Proposition 1 shows that $FP(C_k, R\text{Ł})$ does not enjoy the single model completeness property, this is unknown for $FP(C_k, \text{Ł})$. Second, and in our opinion more interesting, is the open problem of providing an NP-algorithm that checks strict coherence. Indeed, the solution of this second problem would immediately show that $\mathscr{S}_\mathcal{E}$ is NP-complete (see [8, Sect. 7]). However, providing an NP-algorithm for that problem seems far from being trivial (recall Remarks 3 and 4).

In our future work we will mainly focus on the latter problem which apparently needs deep mathematical, and in particular geometric techniques to be solved.

Acknowledgments. The author acknowledges partial support by the Spanish Ramon y Cajal research program RYC-2016-19799; the Spanish FEDER/MINECO project TIN2015-71799-C2-1-P and the SYSMICS project (EU H2020-MSCA-RISE-2015 Project 689176).

References

1. Baaz, M.: Infinite-valued Gödel logics with $0-1$-projections and relativizations. In: Hájek, P. (ed.) GÖDEL 1996. LNL, vol. 6, pp. 23–33. Springer, Heidelberg (1996)
2. Cignoli, R., D'Ottaviano, I.M.L., Mundici, D.: Algebraic Foundations of Many-valued Reasoning. Kluwer, Dordrecht (2000). https://doi.org/10.1007/978-94-015-9480-6
3. de Finetti, B.: Theory of Probability, vol. 1. Wiley, Hoboken (1974)
4. Di Nola, A., Leuştean, I.: Łukasiewicz logic and MV-algebras. In: Cintula, P., Hájek, P., Noguera, C. (eds.) Handbook of Mathematical Fuzzy Logic-Volume 2. Studies in Logic, Mathematical Logic and Foundations, Studies in Logic, vol. 38, pp. 469–584. College Publications, London (2011). Chap. VI
5. Flaminio, T.: NP-containment for the coherence test of assessments of conditional probability: a fuzzy logical approach. Arch. Math. Logic **46**, 301–319 (2007)
6. Flaminio, T., Godo, L., Hosni, H.: On the logical structure of de Finettis notion of event. J. Appl. Logic **12**, 279–301 (2014)
7. Flaminio, T., Godo, L., Marchioni, E.: Reasoning about uncertainty of fuzzy events: an overview. In: Cintula, P., et al. (eds.) Understanding Vagueness - Logical, Philosophical, and Linguistic Perspectives, pp. 367–400. College Publications (2011)
8. Flaminio, T., Hosni, H., Montagna, F.: Strict coherence on many-valued events. J. Symbolic Logic (2018). https://doi.org/10.1017/jsl.2017.34
9. Flaminio, T., Montagna, F.: A logical and algebraic approach to conditional probability. Arch. Math. Logic **44**(4), 245–262 (2005)
10. Gerla, B.: Rational Łukasiewicz logic and divisible MV-algebras. Neural Netw. Worlds **25**, 1–13 (2001)
11. Hájek, P., Godo, L., Esteva, F.: Probability and fuzzy logic. In: Besnard, P., Hanks, S. (eds.) Proceedings of Uncertainty in Artificial Intelligence, UAI 1995, pp. 237–244. Morgan Kaufmann, San Francisco (1995)
12. Hájek, P.: Metamathematics of Fuzzy Logics. Kluwer Academic Publishers, Dordrecht (1998)
13. Hájek, P.: Complexity of fuzzy probability logics II. Fuzzy Sets Syst. **158**(23), 2605–2611 (2007)
14. Kelley, J.L.: Measures on Boolean algebras. Pac. J. Math. **9**(4), 1165–1177 (1959)
15. Kemeny, J.G.: Fair bets and inductive probabilities. J. Symb. Logic **20**(3), 263–273 (1955)
16. Mundici, D.: Advanced Łukasiewicz calculus and MV-algebras. Trends in Logic, vol. 35. Springer, Dordrecht (2011). https://doi.org/10.1007/978-94-007-0840-2
17. Shimony, A.: Coherence and the axioms of confirmation. J. Symb. Logic **20**(1), 1–28 (1955)

Connecting Systems of Mathematical Fuzzy Logic with Fuzzy Concept Lattices

Pietro Codara[1], Francesc Esteva[1], Lluis Godo[1]([⊠]), and Diego Valota[2]

[1] Artificial Intelligence Research Institute (IIIA-CSIC),
Campus de la UAB, 08193 Bellaterra, Barcelona, Spain
{codara,esteva,godo}@iiia.csic.es
[2] Dipartimento di Informatica, Università degli Studi di Milano,
Via Comelico 39, 20135 Milano, Italy
valota@di.unimi.it

Abstract. In this paper our aim is to explore a new look at formal systems of fuzzy logics using the framework of (fuzzy) formal concept analysis (FCA). Let L be an extension of MTL complete with respect to a given L-chain. We investigate two possible approaches. The first one is to consider fuzzy formal contexts arising from L where attributes are identified with L-formulas and objects with L-evaluations: every L-evaluation (object) satisfies a formula (attribute) to a given degree, and vice-versa. The corresponding fuzzy concept lattices are shown to be isomorphic to quotients of the Lindenbaum algebra of L. The second one, following an idea in a previous paper by two of the authors for the particular case of Gödel fuzzy logic, is to use a result by Ganter and Wille in order to interpret the (lattice reduct of the) Lindenbaum algebra of L-formulas as a (classical) concept lattice of a given context.

Keywords: Mathematical fuzzy logics · MTL · Concept lattices
FCA · Łukasiewicz logic

1 Introduction

In this paper our aim is to explore a new look at formal systems of fuzzy logics using the framework of (fuzzy) formal concept analysis (FCA).

The possibility of connecting descriptions of real-world contexts with powerful formal instruments is what makes (fuzzy) FCA a promising framework, merging the intuitions of intended semantics with the advantages of formal semantics. In the case of classical logic, a first attempt has been done in [8].

To build a bridge between systems of fuzzy logic and FCA, we explore two possible approaches. In the first one, given a fuzzy logic L we consider fuzzy FCA tables where attributes are described by formulas of the logic L, while L-evaluations play the role of objects: every object (L-evaluation) satisfies attributes (formulas) to a given degree, and vice-versa, every attribute (formula) is satisfied to a given extent by objects (evaluations).

© Springer International Publishing AG, part of Springer Nature 2018
J. Medina et al. (Eds.): IPMU 2018, CCIS 854, pp. 275–286, 2018.
https://doi.org/10.1007/978-3-319-91476-3_23

The second one is, following the idea in [5] for the particular case of Gödel fuzzy logic [12], is to use Ganter and Wille's result [10, Theorem 3] in order to interpret the lattice reduct of the Lindenbaum algebra of L-formulas as a lattice of the set of formal concepts of a given context. Then, in order to endow the lattice of concepts with a structure of L-algebra, suitable operations on formal concepts have to be defined.

The paper is structured as follows. After this brief introduction, we recall some background notions in Sect. 2, in Sect. 3 we introduce concept lattices of formulas and evaluations, and in Sect. 4 we recall the construction of [5]. Both approaches will be used to obtain formal concepts for formulas of the 3-valued Łukasiewicz logic.

2 Preliminaries

2.1 Basic Notions on Formal Concept Analysis

We recollect the basic definitions and facts about formal concept analysis needed in this work. For further details on this topics we refer the reader to [10].

Recall that an element j of a distributive lattice H is called a *join-irreducible* if j is not the bottom of H and if whenever $j = a \sqcup b$, then $j = a$ or $j = b$, for $a, b \in L$. Meet-irreducible elements are defined dually. Given a lattice $\mathbf{H} = (H, \sqcap, \sqcup, 1)$, we denote by $\mathfrak{J}(H)$ the set of its join-irreducible elements, and by $\mathfrak{M}(H)$ the set of its meet-irreducible elements.

Let G and M be arbitrary sets of *objects* and *attributes*, respectively, and let $I \subseteq G \times M$ be an arbitrary binary relation. Then, the triple $\mathbb{K} = (G, M, I)$ is called a *formal context*. For $g \in G$ and $m \in M$, we interpret $(g, m) \in I$ as "the object g has attribute m". For $A \subseteq G$ and $B \subseteq M$, a Galois connection between the powersets of G and M is defined through the following operators:

$$A^* = \{m \in M \mid \forall g \in A : gIm\} \qquad B^\circ = \{g \in G \mid \forall m \in B : gIm\}$$

Every pair (A, B) such that $A^* = B$ and $B^\circ = A$ is called a *formal concept*. A and B are the *extent* and the *intent* of the concept, respectively. Given a context $\mathbb{K}$, the set $\mathfrak{B}(\mathbb{K})$ of all formal concepts of $\mathbb{K}$ is partially ordered by $(A_1, B_1) \leq (A_2, B_2)$ if and only if $A_1 \subseteq A_2$ (or, equivalently, $B_2 \subseteq B_1$). The *basic theorem on concept lattices* [10, Theorem 3] states that the set of formal concepts of the context $\mathbb{K}$ is a complete lattice $(\mathfrak{B}(\mathbb{K}), \sqcap, \sqcup)$, called *concept lattice*, where meet and join are defined by:

$$\prod_{j \in J} (A_j, B_j) = \left(\bigcap_{j \in J} A_j, \left(\bigcup_{j \in J} B_j \right)^{\circ *} \right),$$

$$\bigsqcup_{j \in J} (A_j, B_j) = \left(\left(\bigcup_{j \in J} A_j \right)^{* \circ}, \bigcap_{j \in J} B_j \right),$$

(1)

for a set J of indexes. The following proposition is fundamental for our purposes.

Proposition 1 ([10, **Proposition 12**]). *For every finite lattice H there is (up to isomorphisms) a unique context $\mathbb{K}_H$, with $L \cong \mathfrak{B}(\mathbb{K}_H)$:*

$$\mathbb{K}_H := (\mathfrak{J}(H), \mathfrak{M}(H), \leq).$$

The context $\mathbb{K}_H$ is called the *standard context* of the lattice H.

Since H is finite, $\mathfrak{J}(H)$ is finite as well. Hence, the concept $(\mathfrak{J}(H), \emptyset)$ is the top element of $\mathfrak{B}(\mathbb{K}_H)$. We denote it by $\top_G$, emphasizing the fact that the join-irreducible elements of L are the objects of our context. Analogously, the concept $(\emptyset, \mathfrak{M}(H))$ is the bottom element of $\mathfrak{B}(\mathbb{K}_H)$, and we denote it by $\bot_M$.

2.2 On t-Norm Based Fuzzy Logics

In this paper we investigate logical systems based on left continuous *t-norms*, that are binary, commutative, associative and monotonically non-decreasing operations over $[0,1]$ that have 1 as unit element. A t-norm operator $\odot$ is used to interpret a conjunction connective, while its corresponding implication connective $\rightarrow$ is modelled by the *residuum* of $\odot$, that is, defined by $x \rightarrow y = max\{z \mid x \odot z \leq y\}$ for all $x, y, z \in [0,1]$. It has been shown that the necessary and sufficient condition for a t-norm $\odot$ to have a residuum (i.e. satisfying the condition $x \odot y \leq z$ iff $x \leq y \rightarrow z$ for all $x, y, z \in [0,1]$) is the left-continuity $\odot$.

In [7] the authors introduce MTL, the logic of all left-continuos t-norms and their residua [13]. MTL encompasses the Basic fuzzy Logic BL of Hájek [12], which is the logic of continuous t-norms and their residua. For axiomatisations of MTL and BL, we refer the reader to [7] and [12] respectively.

Other relevant t-norm based fuzzy logics can be obtained as schematic extensions of MTL or BL. Gödel logic G is the schematic extension of BL obtained by adding the *idempotency* axiom, $\varphi \rightarrow (\varphi \odot \varphi)$. Łukasiewicz logic Ł is the schematic extension of BL obtained by adding the *double negation* axiom $\neg\neg\varphi \rightarrow \varphi$. Adding $\varphi \odot \varphi \leftrightarrow \varphi \odot \varphi \odot \varphi$ to Ł we obtain the 3-valued Łukasiewicz logic Ł3.

Our interest in Ł3 is given by the recent paper [6], where authors characterize this logic as the logic of prototypes and counterexamples. Gödel logic will be used as a stepping stone for developing the methodology to be applied to the case of Ł3.

Each schematic extension L of MTL determines a subvariety $\mathbb{V}(L)$ of the variety of MTL algebras $\mathbb{MTL}$, that is the class of algebras $\mathbf{A} = (A, \wedge, \vee, \odot, \rightarrow, \bot, \top)$ of type $(2, 2, 2, 2, 0, 0)$ such that $(A, \wedge, \vee, \bot, \top)$ is a bounded lattice, with top $\top$ and bottom $\bot$, $(A, \odot, \top)$ is a commutative monoid, satisfying the *residuation* equivalence, $x \odot y \leq z$ if and only if $x \leq y \rightarrow z$, and the *prelinearity* equation $(x \rightarrow y) \vee (y \rightarrow x) = \top^1$. Negation is usually defined as $\neg x = x \rightarrow \bot$.

The notion of logical consequence for a logic L relative to a class $\mathcal{A} \subseteq \mathbb{V}(L)$ is defined as follows: for any set of formulas $T \cup \{\varphi\}$, φ is a logical consequence

---

[1] MTL algebras are commutative integral bounded residuated lattices satisfying prelinearity [9].

of T, written $T \models_{\mathcal{A}} \varphi$, whenever for all algebra $\mathbf{A} \in \mathcal{A}$ and each evaluation e of formulas on $\mathbf{A}$, if $e(\psi) = 1$ for all $\psi \in T$, then $e(\varphi) = 1$ as well.

Given a logic L, two formulas φ and ψ are logically equivalent, in symbols $\varphi \equiv_{L} \psi$, if and only if $\varphi \leftrightarrow \psi$ is a L-tautology, that is, if $\models_{\mathbb{V}(L)} \varphi \leftrightarrow \psi$. The *Lindenbaum Algebra* **Lind**(L) of L is the algebra whose elements are the equivalence classes of formulas of L, with respect to $\equiv_{L}$. The free k-generated algebra $\mathbf{F}_k(\mathbb{V}(L))$ in $\mathbb{V}(L)$ is the subalgebra of the Lindenbaum algebra **Lind**(L) of the formulas over the first k variables. Combinatorial representations of $\mathbf{F}_k(\mathbb{G})$ and $\mathbf{F}_k(\mathbb{MV}_3)$, where $\mathbb{G} = \mathbb{V}(G)$ and $\mathbb{MV}_3 = \mathbb{V}(L_3)$, can be found in [1].

3   The Concept Lattice of Formulas and Evaluations

Suppose L is an axiomatic extension of MTL that is complete with respect to a given L-chain M, that is, $\models_{\mathbb{V}(L)} = \models_M$. In what follow, we will denote by $\mathcal{L}$ the set of propositional L-formulas built from a *finite* set of propositional variables V, and by Ω the set of truth-evaluations of propositinal variables into the L-chain M, that is, $\Omega = \{e : V \to M\}$. Of course, every evaluation of variables uniquely extends to an evaluation of any propositional formula using the truth-functions interpreting the connectives.

In our FCA-based analysis of the notion of consequence in the logic L, we will consider attributes described as propositional formulas from $\mathcal{L}$, and objects as evaluations from Ω. In this setting, a formal context will be specified by a triple

$$K = (\Omega_0, \mathcal{L}_0, R),$$

where $\Omega_0 \subseteq \Omega$ and $\mathcal{L}_0 \subseteq \mathcal{L}$ are finite sets, and $R : \Omega_0 \times \mathcal{L}_0 \to M$ is a M-valued fuzzy relation defined as $R(e, \varphi) = e(\varphi)$.

In this way, each attribute or formula $\varphi \in \mathcal{L}_0$ determines a fuzzy set of objects $\varphi^* : \Omega_0 \to M$, with $\varphi^*(e) = R(e, \varphi)$, for all $e \in \Omega_0$, and vice-versa, each object or evaluation $e \in \Omega_0$ determines a fuzzy set of attributes $e^\circ : \mathcal{L}_0 \to M$, with $e^\circ(\varphi) = R(e, \varphi)$, for all $\varphi \in \mathcal{L}_0$. More than that, following Pollandt [14] and Bělohlávek's [2] models of FCA, this correspondence is extended to a Galois connection between fuzzy sets of formulas and fuzzy sets of evaluations as follows.

Definition 1. *Let $F \in \mathcal{F}(\mathcal{L}_0)$ be a fuzzy subset of formulas (fuzzy theory) and let $E \in \mathcal{F}(\Omega_0)$ be a fuzzy set of evaluations. Define:*

- *F^* is the fuzzy subset of Ω_0 defined as $F^*(e) = \inf_{\varphi \in \mathcal{L}_0} F(\varphi) \to R(e, \varphi)$, for all $e \in \Omega_0$,*
- *E° is the fuzzy subset of $\mathcal{L}_0$ defined as $E^\circ(\varphi) = \inf_{e \in \Omega_0} E(e) \to R(e, \varphi)$, for all $\varphi \in \mathcal{L}_0$.*

A pair (E, F) is a logic fuzzy concept *if $F^* = E$ and $E^\circ = F$.*

In other words, F^* is the fuzzy set of models of the fuzzy theory F, and E° is the fuzzy set of formulas satisfied by the fuzzy set of evaluations E. Moreover, as it is known, $*\circ$ is a closure operation on the set $\mathcal{F}(\mathcal{L}_0)$ of M-valued fuzzy sets of formulas, hence $F \leq F^{*\circ}$. Actually the mapping $*\circ : \mathcal{F}(\mathcal{L}_0) \to \mathcal{F}(\mathcal{L}_0)$, defined by

$$F^{*\circ}(\varphi) = \inf_{e \in \Omega_0} [\inf_{\psi \in \mathcal{L}_0} F(\psi) \to e(\psi)] \to e(\varphi)$$

can be considered as a *graded* logical consequence relation, that it is even a bit more general than the one central to the so-called *graded approach to fuzzy logic*, developed by authors like J. A. Goguen, J. Pavelka, V. Nóvak and G. Gerla, as discussed e.g. in [11].

In what follows, we will denote by $\mathbf{C}(K) = (C(K), \preceq)$ the lattice of fuzzy concepts induced by a context K, where the ordering $\preceq$ is defined as

$$(E, F) \preceq (E', F') \text{ iff } E \leq E' \text{ and } F \geq F',$$

and the meet and join operations are defined as:

$$(E, F) \sqcap (E', F') = (E \cap E', (F \cup F')^{*\circ}), \quad (E, F) \sqcup (E', F') = ((E \cup E')^{\circ *}, F \cap F'),$$

where $\cap$ and $\cup$ denote intersection and union of fuzzy sets, defined with the min and max operations respectively.

This lattice is bounded and the bottom element is the concept $\perp_K = (\emptyset, \mathcal{L}_0)$, while the top element is $\top_K = (\Omega_0, T_{\Omega_0})$, where T_{Ω_0} if the fuzzy set of formulas defined by $T_{\Omega_0}(\psi) = \inf_{e \in \Omega_0} e(\psi)$.

Let us see how it looks like the fuzzy concept in $\mathbf{C}(K)$ induced by (the crisp set of) a single formula $\varphi \in \mathcal{L}_0$, i.e. the pair $(\varphi^*, \varphi^{*\circ})$, where for the sake of a simpler notation we have used φ^* for $\{\varphi\}^*$ and $\varphi^{*\circ}$ for $(\{\varphi\}^*)^\circ$. An easy computation shows that:

- $\varphi^*(e) = R(e, \varphi) = e(\varphi)$, for all $e \in \Omega_0$;
- $\varphi^{*\circ}(\psi) = \inf_{e \in \Omega_0} R(e, \varphi) \to R(e, \psi) = \inf_{e \in \Omega_0} e(\varphi \to \psi)$, for all $\psi \in \mathcal{L}_0$.

Further, if we consider a finite set of formulas or theory T, using the same notation convention as above, the corresponding concept $(T^*, T^{*\circ})$ is as follows, where $\bigwedge T$ denotes the $\wedge$-conjunction of all the formulas in T, i.e. $\bigwedge T = \bigwedge_{\varphi \in T} \varphi$:

- $T^*(e) = \inf_{\varphi \in T} R(e, \varphi) = \inf_{\varphi \in T} e(\varphi) = e(\bigwedge T)$, for all $e \in \Omega_0$;
- $T^{*\circ}(\psi) = \inf_{e \in \Omega_0} T^*(e) \to R(e, \psi) = \inf_{e \in \Omega_0} e(\bigwedge T \to \psi)$, for all $\psi \in \mathcal{L}_0$.

Note that, as discussed above, $T^{*\circ}$ accounts for a certain notion of graded consequence from T, in the sense that $T^{*\circ}(\psi)$ provides the degree in which ψ is implied by T, relative to the set of interpretations Ω_0. It is a graded consequence that resembles Pavelka's notion of *truth degree of a formula in a theory* (see e.g. [11,12]), although they do not coincide. It is also related to the so-called degree preserving logic $\models_L^{\leq}$ companion of L, see e.g. [4]. Indeed, it is easy to check the following lemma.

Lemma 1. *For any $\psi \in \mathcal{L}_0$, $T^{*\circ}(\psi) = 1$ iff $e(\bigwedge T \to \psi) = 1$ for all $e \in \Omega_0$.*

Therefore, when $\Omega_0 = \Omega$, $T^{*\circ}(\psi) = 1$ holds if, and only if, $\inf_{\varphi \in T} e(\varphi) \leq e(\psi)$, i.e. iff $T \models_L^{\leq} \psi$. That is, the core of $T^{*\circ}$ is nothing but the set of consequences of T (restricted to $\mathcal{L}_0$) under the degree preserving logic companion of L.

Lemma 2. $T_1^{*\circ} = T_2^{*\circ}$ iff $\bigwedge T_1$ and $\bigwedge T_2$ are logically equivalent relative to Ω_0, i.e. $e(\bigwedge T_1) = e(\bigwedge T_2)$ for every evaluation $e \in \Omega_0$.

Proof. The direction right-to-left is trivial. As for the converse, if $T_1^{*\circ} = T_2^{*\circ}$, then in particular, for all χ, $T_1^{*\circ}(\chi) = 1$ iff $T_2^{*\circ}(\chi) = 1$. Take $\chi = \bigwedge T_1$. Since $T_1^{*\circ}(\bigwedge T_1) = 1$, then $T_2^{*\circ}(\bigwedge T_1) = 1$ as well, and by Lemma 1 this happens iff for all $e \in \Omega_0$, $e(\bigwedge T_2) \leq e(\bigwedge T_1)$. Analogously, if we take $\chi = \bigwedge T_2$, we would get that, for all $e \in \Omega_0$, $e(\bigwedge T_1) \leq e(\bigwedge T_2)$. □

Notice again that in case $\Omega_0 = \Omega$, then $T_1^{*\circ} = T_2^{*\circ}$ iff $\bigwedge T_1$ and $\bigwedge T_2$ are logically equivalent in the usual sense.

The set $C^{cg}(K)$ of concepts of the form $(T^*, T^{*\circ})$, with $T \subseteq \mathcal{L}_0$ a finite (crisp) set of formulas, is in fact what is known as the set of *crisply generated concepts* in the fuzzy concept lattice $\mathbf{C}(K)$ [3]. As already mentioned, for the purpose of building concepts, we can always replace a finite theory T by the $\wedge$-conjunction of its formulas $\bigwedge T$. Indeed, for every concept of the form $(T^*, T^{*\circ})$ with T a finite set of formulas, there is always a formula φ (e.g. $\bigwedge T$) such that $(T^*, T^{*\circ}) = (\varphi^*, \varphi^{*\circ})$. Thus $C^{cg}(K) = \{(\varphi^*, \varphi^{*\circ}) \mid \varphi \in \mathcal{L}_0\}$ and we can safely restrict ourselves to deal with concepts induced by a *single* formula.

The lattice operations in $\mathbf{C}(K)$ over concepts from $C^{cg}(K)$ take the following form.

Lemma 3. For any $\varphi, \psi \in \mathcal{L}$,

$$(\varphi^*, \varphi^{*\circ}) \sqcap (\psi^*, \psi^{*\circ}) = ((\varphi \wedge \psi)^*, (\varphi \wedge \psi)^{*\circ}), \tag{2}$$

$$(\varphi^*, \varphi^{*\circ}) \sqcup (\psi^*, \psi^{*\circ}) = ((\varphi \vee \psi)^*, (\varphi \vee \psi)^{*\circ}). \tag{3}$$

Proof. By definition, $(\varphi^*, \varphi^{*\circ}) \sqcap (\psi^*, \psi^{*\circ}) = (\varphi^* \cap \psi^*, (\varphi^* \cap \psi^*)^\circ)$, but since $(\varphi^* \cap \psi^*)(e) = \min(\varphi^*(e), \psi^*(e)) = \min(e(\varphi), e(\psi)) = e(\varphi \wedge \psi) = (\varphi \wedge \psi)^*(e)$, we have $(\varphi^* \cap \psi^*, (\varphi^* \cap \psi^*)^\circ) = ((\varphi \wedge \psi)^*, (\varphi \wedge \psi)^{*\circ})$.

Analogously, by definition $(\varphi^*, \varphi^{*\circ}) \sqcup (\psi^*, \psi^{*\circ}) = ((\varphi^* \cup \psi^*)^{\circ*}, \varphi^{*\circ} \cap \psi^{*\circ})$, but $(\varphi^{*\circ} \cap \psi^{*\circ})(\chi) = \min(\inf_e e(\varphi \to \chi), \inf_e e(\psi \to \chi)) = \inf_e \min(e(\varphi \to \chi), e(\psi \to \chi)) = \inf_e e(\varphi \vee \psi \to \chi) = (\varphi \vee \psi)^{*\circ}(\chi)$. Therefore $((\varphi^* \cup \psi^*)^{\circ*}, \varphi^{*\circ} \cap \psi^{*\circ}) = ((\varphi^* \cup \psi^*)^{\circ*}, (\varphi \vee \psi)^{*\circ}) = ((\varphi \vee \psi)^*, (\varphi \vee \psi)^{*\circ})$. □

As it proven in [3], $C^{cg}(K)$ is indeed a $\sqcap$-subsemilattice of $\mathbf{C}(K)$ in the general case. Indeed, notice that the $\sqcap$ operation is closed in $C^{cg}(K)$, since the concept induced by the conjunction $\bigwedge T$ of a set of formulas $T \subseteq \mathcal{L}_0$, even if $\bigwedge T$ does not belong to $\mathcal{L}_0$, is the same concept induced by the crisp set of formulas T, and hence it belongs to $C^{cg}(K)$. However, this is not the case for a disjunction of a set of formulas. However, if we can guarantee that the concept induced by a disjunction also belongs to $C^{cg}(K)$, then $C^{cg}(K)$ is actually a sublattice of $\mathbf{C}(K)$.

Lemma 4. *If $\mathcal{L}_0$ is closed by $\vee$ (modulo logical equivalence) then $\sqcup$ is closed in $C^{cg}(K)$, and $\mathbf{C}^{cg}(K) = (C^{cg}(K), \sqcap, \sqcup, \top_K, \perp_k)$ is a sublattice of $\mathbf{C}(K)$.*

In the following we will assume $\mathcal{L}_0 = \mathcal{L}$ to avoid any problem. In such a case, we can also enrich the lattice $\mathbf{C}^{cg}(K)$ with some further operations in a natural way so to come up with a residuated lattice structure, inherited from the L-algebras.

Definition 2. *We define the following two operations on fuzzy concepts from $C^{cg}(K)$. For any $\varphi, \psi \in \mathcal{L}$, let us define:*

$$(\varphi^*, \varphi^{*\circ}) \boxtimes (\psi^*, \psi^{*\circ}) = ((\varphi \odot \psi)^*, (\varphi \odot \psi)^{*\circ}), \tag{4}$$

$$(\varphi^*, \varphi^{*\circ}) \Rightarrow (\psi^*, \psi^{*\circ}) = ((\varphi \to \psi)^*, (\varphi \to \psi)^{*\circ}). \tag{5}$$

It is easy to check that $\boxtimes$ and $\Rightarrow$ endow the lattice $C^{cg}(K)$ with a structure of residuated lattice, in particular with the structure of a L-algebra.

Proposition 2. $\mathbf{C^{cg}(K)} = (C^{cg}(K), \sqcap, \sqcup, \boxtimes, \Rightarrow, \top_K, \perp_K)$ *is an L-algebra that is isomorphic to the quotient algebra $\mathcal{L}/\equiv_{\Omega_0}$, where $\varphi \equiv_{\Omega_0} \psi$ iff $e(\varphi) = e(\psi)$ for all $e \in \Omega_0$.*

Proof. Elements of $\mathcal{L}/\equiv_{\Omega_0}$ are equivalence classes of formulas from $\mathcal{L}$, according to the congruence relation $\equiv_{\Omega_0}$. Given a formula $\varphi \in \mathcal{L}$, let us denote by $[\varphi]$ its equivalence class. Since the class of L-algebras is a variety, it is closed under quotients, hence $\mathcal{L}/\equiv_{\Omega_0}$ is an L-algebra as well. Now consider the mapping $\lambda : \mathcal{L}/\equiv_{\Omega_0} \to C^{cg}(K)$ defined by $\lambda([\varphi]) = (\varphi^*, \varphi^{*\circ})$. It is easy to check that this mapping is one-to-one thanks to Lemma 2, and moreover it is an algebraic homomorphism with respect to the operations involved: $\lambda([\varphi] \wedge [\psi])) = \lambda([\varphi]) \sqcap \lambda([\psi])$, etc. Therefore, $\mathbf{C^{cg}(K)}$ is an L-algebra as well, isomorphic to $\mathcal{L}/\equiv_{\Omega_0}$. $\square$

Corollary 1. *If $\Omega_0 = \Omega$, then $\mathbf{C^{cg}(K)}$ is isomorphic to the Lindenbaum algebra $\mathbf{Lind}(L) = \mathcal{L}/\equiv_L$.*

3.1   An Example: The Case of Ł3

In this section, we provide an example of the construction of the concept lattice of formulas and evaluations for the Łukasiewicz 3-valued logic Ł3.

Let $\mathcal{L}_0 = \{\varphi_1, \varphi_2, \ldots, \varphi_{12}\}$ be the set of all Ł3-formulas (up to logical equivalence) on one variable x, where[2]

$$\varphi_1 = x^2 \wedge (\neg x)^2 = \perp, \qquad \varphi_2 = (\neg x)^2, \qquad \varphi_3 = x \wedge \neg x,$$
$$\varphi_4 = x^2, \qquad \varphi_5 = \neg x, \qquad \varphi_6 = (x \vee \neg x)^2,$$
$$\varphi_7 = \neg x^2 \wedge \neg(\neg x)^2, \qquad \varphi_8 = x, \qquad \varphi_9 = \neg x^2,$$
$$\varphi_{10} = x \vee \neg x, \qquad \varphi_{11} = \neg(\neg x)^2, \qquad \varphi_{12} = \neg x^2 \vee \neg(\neg x)^2 = \top.$$

Further, let us consider all possible 3-valued evaluations on the variable x as the set of objects: $\Omega_0 = \{e_0, e_1, e_2\}$, where $e_0(x) = 0, e_1(x) = \frac{1}{2}, e_2(x) = 1$. The following table shows the values of each formula of $\mathcal{L}_0$ under each evaluation.

---

[2] We use φ^2 as a shorcut for $\varphi \odot \varphi$.

	φ_1	φ_2	φ_3	φ_4	φ_5	φ_6	φ_7	φ_8	φ_9	φ_{10}	φ_{11}	φ_{12}
$e_0(\cdot)$	0	1	0	0	1	1	0	0	1	1	0	1
$e_1(\cdot)$	0	0	1/2	0	1/2	0	1	1/2	1	1/2	1	1
$e_2(\cdot)$	0	0	0	1	0	1	0	0	0	1	1	1

As described before in this section, the triple $K = \{\Omega_0, \mathcal{L}_0, R\}$, where $R : \Omega_0 \times \mathcal{L}_0 \rightarrow \{0, \frac{1}{2}, 1\}$ is a 3-valued fuzzy relation defined as $R(e, \varphi) = e(\varphi)$, identifies a formal context.

First of all, we aim at obtaining all the concepts induced by a single formula. For instance, consider the formula $\varphi_8 = x$. Then, $\varphi_8^*(e_0) = e_0(\varphi_8) = 0$, $\varphi_8^*(e_1) = \frac{1}{2}$, and $\varphi_8^*(e_2) = 1$. We denote the fuzzy set of objects (evaluations) φ_8^* by the tuple $(0, \frac{1}{2}, 1)$. Let us compute the fuzzy set of attributes (formulas) $\varphi_8^{*\circ}$:

$$\varphi_8^{*\circ}(\varphi_1) = \inf_{e \in \Omega_0} e(\varphi_8 \rightarrow \varphi_1) = 0, \qquad \varphi_8^{*\circ}(\varphi_2) = \inf_{e \in \Omega_0} e(\varphi_8 \rightarrow \varphi_2) = 0,$$

$$\varphi_8^{*\circ}(\varphi_3) = \inf_{e \in \Omega_0} e(\varphi_8 \rightarrow \varphi_3) = 0, \qquad \varphi_8^{*\circ}(\varphi_4) = \inf_{e \in \Omega_0} e(\varphi_8 \rightarrow \varphi_4) = 1/2,$$

$$\varphi_8^{*\circ}(\varphi_5) = \inf_{e \in \Omega_0} e(\varphi_8 \rightarrow \varphi_5) = 0, \qquad \varphi_8^{*\circ}(\varphi_6) = \inf_{e \in \Omega_0} e(\varphi_8 \rightarrow \varphi_6) = 1/2,$$

$$\varphi_8^{*\circ}(\varphi_7) = \inf_{e \in \Omega_0} e(\varphi_8 \rightarrow \varphi_7) = 0, \qquad \varphi_8^{*\circ}(\varphi_8) = \inf_{e \in \Omega_0} e(\varphi_8 \rightarrow \varphi_8) = 1,$$

$$\varphi_8^{*\circ}(\varphi_9) = \inf_{e \in \Omega_0} e(\varphi_8 \rightarrow \varphi_9) = 0, \qquad \varphi_8^{*\circ}(\varphi_{10}) = \inf_{e \in \Omega_0} e(\varphi_8 \rightarrow \varphi_{10}) = 1,$$

$$\varphi_8^{*\circ}(\varphi_{11}) = \inf_{e \in \Omega_0} e(\varphi_8 \rightarrow \varphi_{11}) = 1, \qquad \varphi_8^{*\circ}(\varphi_{12}) = \inf_{e \in \Omega_0} e(\varphi_8 \rightarrow \varphi_{12}) = 1.$$

We indicate the value of $\varphi_8^{*\circ}$ by the tuple $(0, 0, 0, \frac{1}{2}, 0, \frac{1}{2}, 0, 1, 0, 1, 1, 1)$. The pair $(\varphi_8^*, \varphi_8^{*\circ})$ is the formal concept induced by the furmula φ_8. In the same way, we can compute all the formal concepts induced by single formulas of $\mathcal{L}_0$, obtaining:

$$(\varphi_1^*, \varphi_1^{*\circ}) = ((0, 0, 0), (1, 1, 1, 1, 1, 1, 1, 1, 1, 1, 1, 1)),$$
$$(\varphi_2^*, \varphi_2^{*\circ}) = ((1, 0, 0), (0, 1, 0, 0, 1, 1, 0, 0, 1, 1, 0, 1)),$$
$$(\varphi_3^*, \varphi_3^{*\circ}) = ((0, 1/2, 0), (1/2, 1/2, 1, 1/2, 1, 1/2, 1, 1, 1, 1, 1, 1)),$$
$$(\varphi_4^*, \varphi_4^{*\circ}) = ((0, 0, 1), (0, 0, 0, 1, 0, 1, 0, 1, 0, 1, 1, 1)),$$
$$(\varphi_5^*, \varphi_5^{*\circ}) = ((1, 1/2, 0), (0, 1/2, 0, 0, 1, 1/2, 0, 0, 1, 1, 0, 1)),$$
$$(\varphi_6^*, \varphi_6^{*\circ}) = ((1, 0, 1), (0, 0, 0, 0, 0, 1, 0, 0, 0, 1, 0, 1)),$$
$$(\varphi_7^*, \varphi_7^{*\circ}) = ((0, 1, 0), (0, 0, 1/2, 0, 1/2, 0, 1, 1/2, 1, 1/2, 1, 1)),$$
$$(\varphi_8^*, \varphi_8^{*\circ}) = ((0, 1/2, 1), (0, 0, 0, 1/2, 0, 1/2, 0, 1, 0, 1, 1, 1)),$$
$$(\varphi_9^*, \varphi_9^{*\circ}) = ((1, 1, 0), (0, 0, 0, 0, 1/2, 0, 0, 0, 1, 1/2, 0, 1)),$$
$$(\varphi_{10}^*, \varphi_{10}^{*\circ}) = ((1, 1/2, 1), (0, 0, 0, 0, 0, 1/2, 0, 0, 0, 1, 0, 1)),$$
$$(\varphi_{11}^*, \varphi_{11}^{*\circ}) = ((0, 1, 1), (0, 0, 0, 0, 0, 0, 0, 0, 1/2, 0, 1/2, 1, 1)),$$
$$(\varphi_{12}^*, \varphi_{12}^{*\circ}) = ((1, 1, 1), (0, 0, 0, 0, 0, 0, 0, 0, 0, 1/2, 0, 1)).$$

Note that all the formal concepts above are precisely all the crisply generated concepts. Indeed, the concept generated by $\{\psi_1, \ldots, \psi_k\} \subseteq \mathcal{L}_0$ coincides with the concept generated by the single formula $\bigwedge_{i=1,\ldots,k} \psi_i$, which is logically equivalent to a formula of $\mathcal{L}_0$. We also observe that $\varphi_{12}^{*\circ} = (\inf_{e \in \Omega_0} e(\varphi_1), \ldots, \inf_{e \in \Omega_0} e(\varphi_{12})) = T_{\Omega_0} \neq \emptyset$.

As described in the previous part of the section, we can endow the set $\mathcal{C}^{cg}(K)$ with the operations defined in (2)–(5). We obtain in this way an algebra $\mathbf{C^{cg}(K)}$ of crisply generated concepts of $\mathcal{L}_0$ which is isomorphic to the free 1-generated Ł$_3$ algebra, depicted in Fig. 1, via the isomorphism λ that associates each formula $\varphi \in \mathcal{L}_0$ with the concept $(\varphi^*, \varphi^{*\circ})$.

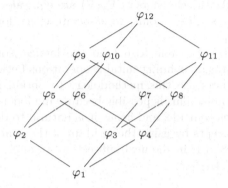

Fig. 1. The Lindenbaum-Tarski algebra of Ł$_3$ over one generator.

Consider now the set of objects (evaluations) $\Omega_B = \{e_0, e_2\} \subseteq \Omega_0$. Again, the triple $K_B = \{\Omega_B, \mathcal{L}_0, R_B\}$, where $R_B : \Omega_B \times \mathcal{L}_0 \to \{0, \frac{1}{2}, 1\}$ is a 3-valued fuzzy relation defined as $R(e, \varphi) = e(\varphi)$, identifies a formal context. Actually, the fuzzy relation R_B is in fact a crisp relation, since the evaluation e_0 and e_2 only evaluate x to either 0 or 1. In this new setting, we can compute all the formal concepts induced by single formulas of $\mathcal{L}_0$, obtaining:

$$(\varphi_1^*, \varphi_1^{*\circ}) = (\varphi_3^*, \varphi_3^{*\circ}) = (\varphi_7^*, \varphi_7^{*\circ}) = ((0,0), (1,1,1,1,1,1,1,1,1,1,1,1)) ,$$
$$(\varphi_4^*, \varphi_4^{*\circ}) = (\varphi_8^*, \varphi_8^{*\circ}) = (\varphi_{11}^*, \varphi_{11}^{*\circ}) = ((0,1), (0,0,0,1,0,1,0,1,0,1,1,1)) ,$$
$$(\varphi_2^*, \varphi_2^{*\circ}) = (\varphi_5^*, \varphi_5^{*\circ}) = (\varphi_9^*, \varphi_9^{*\circ}) = ((1,0), (0,1,0,0,1,1,0,0,1,1,0,1)) ,$$
$$(\varphi_6^*, \varphi_6^{*\circ}) = (\varphi_{10}^*, \varphi_{10}^{*\circ}) = (\varphi_{12}^*, \varphi_{12}^{*\circ}) = ((1,1), (0,0,0,0,0,1,0,0,0,1,0,1)) ,$$

which, in fact, they turn out to be classical concepts. Not surprisingly, endowing this set of concepts $\mathcal{C}^{cg}(K_B)$ with the operations defined in (2)–(5) we obtain an algebra of concepts which is isomorphic to the free 1-generated Boolean algebra. Such algebra is obtained as a quotient of $\mathbf{C^{cg}(K)}$. As it is easily seen using Proposition 2, this holds in general, that is, an algebra of concepts $\mathbf{C^{cg}(K')}$, with $K' = \{\Omega_0', \mathcal{L}_0, R\}$ and $\Omega_0' \subseteq \Omega_0$ is a quotient of the algebra $\mathbf{C^{cg}(K)}$.

4 The Natural Concept Lattice of a Logic

In this section we recall the construction of concept lattices applied in [5] to characterize formal concept lattices associated to Gödel algebras.

Proposition 1 states that for every finite lattice H there is always a canonical way to build the *standard context* $\mathbb{K}_H$, whose concept lattice $\mathfrak{B}(\mathbb{K}_H)$ is isomorphic to H. Let $\mathbf{A} = (A, \wedge, \vee, \rightarrow, \top, \bot)$ be a finite algebra in a variety $\mathbb{V} \subseteq \text{MTL}$, and let $C_{\mathbf{A}} = \mathfrak{B}((\mathfrak{J}(\mathbf{A}), \mathfrak{M}(\mathbf{A}), \leq))$ be the concept lattice of its standard context. Then, the lattice $\mathbf{C_A} = (C_{\mathbf{A}}, \sqcap, \sqcup, \top_G, \bot_M)$, is isomorphic to the lattice reduct of $\mathbf{A}$.

Pushing further this approach, when $\mathbb{V}$ is a locally finite variety the k-generated free algebras $\mathbf{F}_k(\mathbb{V})$ are finite, and hence we can apply to them the above construction. As the elements of $\mathbf{F}_k(\mathbb{V})$ are equivalence classes of logical formulas in k variables, this amount to associate every logical formula to its natural formal concept.

For some cases it is possible to extend the lattice isomorphism to a full isomorphism of algebras by defining suitable operations between the formal concepts. In [5] the authors use this methodology to obtain formal concepts for every Gödel logic formula. This is possible because in Gödel algebras lattice and monoidal conjunctions coincide, and hence it is natural to define an implication operator between concepts by using the residum of the concepts meet.

Comparing to Sect. 3.1, in the next subsection we apply the above sketched construction to $\mathbf{F}_1(\mathbb{V}(\text{Ł}_3))$.

4.1   Constructing the Concept Lattice of the Logic Ł$_3$

Consider the set $\mathcal{L}_0 = \{\varphi_1, \varphi_2, \ldots, \varphi_{12}\}$ of all Ł$_3$-fomulas (up to logical equivalence) on one variable x. The formulas of $\mathcal{L}_0$ are exhibited in Fig. 1. Let $H = (\mathcal{L}_0, \leq)$ be the lattice reduct of the free 1-generated Ł$_3$ algebra $\mathbf{F}_1$ depicted in Fig. 1. The sets of join irreducible elements and meet irreducible elements of L are $\mathfrak{J}(H) = \{\varphi_2, \varphi_3, \varphi_4, \varphi_7\}$, and $\mathfrak{M}(H) = \{\varphi_6, \varphi_9, \varphi_{10}, \varphi_{11}\}$, respectively. By Proposition 1, we can identify $\mathfrak{J}(H)$ and $\mathfrak{M}(H)$ with the set of objects and attributes, respectively, of a standard context $\mathbb{K}_H = (\mathfrak{J}(H), \mathfrak{M}(H), \leq)$. The following table shows the relation $\leq$

$\leq$	φ_6	φ_9	φ_{10}	φ_{11}
φ_2	×	×	×	
φ_3		×	×	×
φ_4	×		×	×
φ_7		×		×

The corresponding standard context lattice is depicted in Fig. 2. Clearly, by Proposition 1, it is isomophic to the lattice reduct of the free 1-generated Ł$_3$ algebra of Fig. 1, via a lattice isomorphim f defined as follows. For each $\varphi \in \mathcal{L}_0$, let J_φ be the maximal subset of $\mathfrak{J}(H)$ such that $\varphi = \bigvee J_\varphi$, and M_φ be the maximal subset of $\mathfrak{M}(H)$ such that $\varphi = \bigwedge M_\varphi$. Then the map f associates each $\varphi \in \mathcal{L}_0$ with the formal concept $(J_\varphi, M_\varphi) \in \mathbb{K}_H$.

To extend the above defined lattice isomorphism to an algebraic isomorphism between $\mathbf{F}_1(\text{Ł}_3)$ and the concept lattice of the standard context $\mathbb{K}_H$, it is necessary to define a *proper* monoidal conjunction between concepts of $\mathbb{K}_H$. Of course,

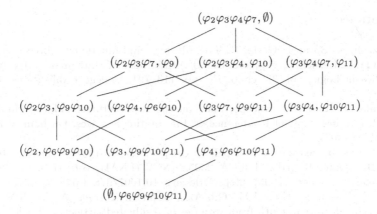

Fig. 2. The concept lattice associated with the lattice reduct of $\mathbf{F}_1$

an obvious way to define such an operation is through the isomorphism f, that is, for each pair of concepts $(E, F), (E', F') \in \mathbb{K}_H$, to define $(E, F) \otimes (E', F') = (J_{\varphi \odot \psi}, M_{\varphi \odot \psi})$, where $f^{-1}((E, F)) = \varphi$ and $f^{-1}((E', F')) = \psi$. However, this does not shed any light on how the operation works on the elements of the concepts. To have a much better insight in the operation seems not to be an easy task, even in the case of locally finite subvarieties of MTL (such as Ł3), and it will be faced in some future paper.

5 Conclusions and Further Developments

To obtain a direct relation between a formal concept and a fuzzy logic formula, in this work we have explored two ways to obtain concept lattices isomorphic to Lindenbaum algebras of many-valued logics. The first approach naturally gives the desired isomorphism between the concept lattice and the algebra of formulas, while to complete the second approach additional research has to be done.

To depict the two constructions we have chosen the logic Ł3. In [6], Ł3 has been characterized as a logic of prototypes and counterexamples. The construction of *possible worlds* in [6] gives a lattice of functions $\Omega_0^{\Omega_0^n}$ very similar to the concept lattice of our first approach in Sect. 3.1. Hence, putting together the characterization of [6] with the constructions presented here, it will be ideally possible to build a formal concept semantics of prototypes and counterxamples for the logic Ł3.

Acknowledgments. The authors are thankful to the anonymous reviewers for their helpful comments. Pietro Codara is supported by the INdAM-Marie Curie Cofund project *LaVague* (FP7-PEOPLE-2012-COFUND 600198). Francesc Esteva and Lluis Godo acknowledge partial support by the FEDER/MINECO project TIN2015-71799-C2-1-P.

References

1. Aguzzoli, S., Bova, S., Gerla, B.: Free algebras and functional representation. In: Cintula, P., Háajek, P., Noguera, C. (eds.) Handbook of Mathematical Fuzzy Logic. Studies in Logic, vol. 38, chap. IX, pp. 713–791. College Publications, London (2011)

2. Bělohlávek, R.: Fuzzy concepts and conceptual structures: induced similarities. In: Proceedings of Joint Conference on Information Sciences, Durham, vol. 1, pp. 179–182 (1998)

3. Bělohlávek, R., Sklenář, V., Zacpal, J.: Crisply generated fuzzy concepts. In: Ganter, B., Godin, R. (eds.) ICFCA 2005. LNCS (LNAI), vol. 3403, pp. 269–284. Springer, Heidelberg (2005). https://doi.org/10.1007/978-3-540-32262-7_19

4. Bou, F., Esteva, F., Font, J.M., Gil, À.J., Godo, L., Torrens, A., Verd, V.: Logics preserving degrees of truth from varieties of residuated lattices. J. Logic Comput. **19**(6), 1031–1069 (2009)

5. Codara, P., Valota, D.: On Gödel algebras of concepts. In: Hansen, H.H., Murray, S.E., Sadrzadeh, M., Zeevat, H. (eds.) TbiLLC 2015. LNCS, vol. 10148, pp. 251–262. Springer, Heidelberg (2017). https://doi.org/10.1007/978-3-662-54332-0_14

6. Dutta, S., Esteva, F., Godo, L.: On a three-valued logic to reason with prototypes and counterexamples and a similarity-based generalization. In: Luaces, O., Gámez, J.A., Barrenechea, E., Troncoso, A., Galar, M., Quintián, H., Corchado, E. (eds.) CAEPIA 2016. LNCS (LNAI), vol. 9868, pp. 498–508. Springer, Cham (2016). https://doi.org/10.1007/978-3-319-44636-3_47

7. Esteva, F., Godo, L.: Monoidal t-norm based logic: towards a logic for left-continuous t-norms. Fuzzy Sets Syst. **124**(3), 271–288 (2001)

8. Ferré, S., Ridoux, O.: A logical generalizationof formal concept analysis. In: Ganter, B., Mineau, G.W. (eds.) ICCS-ConceptStruct 2000. LNCS (LNAI), vol. 1867, pp. 371–384. Springer, Heidelberg (2000). https://doi.org/10.1007/10722280_26

9. Galatos, N., Jipsen, P., Kowalski, T., Ono, H.: Residuated Lattices: An Algebraic Glimpse at Substructural Logics. Elsevier, New York (2007)

10. Ganter, B., Wille, R.: Formal Concept Analysis: Mathematical Foundations, 1st edn. Springer, New York (1997). https://doi.org/10.1007/978-3-642-59830-2

11. Gerla, G.: Mathematical Tools for Approximate Reasoning. Kluwer Academic Press, Dordrecht (2001). https://doi.org/10.1007/978-94-015-9660-2

12. Hájek, P.: Metamathematics of Fuzzy Logic. Kluwer, Dordrecht (1998). https://doi.org/10.1007/978-94-011-5300-3

13. Jenei, S., Montagna, F.: A proof of standard completeness for Esteva and Godo's logic MTL. Stud. Logica. **70**(2), 183–192 (2002)

14. Pollandt, S.: Fuzzy-Begriffe. Formale Begriffsanalyse unscharfer Daten. Springer, Heidelberg (1997). https://doi.org/10.1007/978-3-642-60460-7

Spatio-Temporal Drought Identification Through Mathematical Morphology

Hilde Vernieuwe[1]($\boxtimes$) ⓘ, Bernard De Baets[1] ⓘ, and Niko E. C. Verhoest[2] ⓘ

[1] KERMIT, Department of Data Analysis and Mathematical Modelling,
Ghent University, Coupure links 653, 9000 Ghent, Belgium
hilde.vernieuwe@ugent.be
[2] Laboratory of Hydrology and Water Management,
Ghent University, Coupure links 653, 9000 Ghent, Belgium

Abstract. Droughts are initiated by a lack of precipitation over a large area and a long period of time. In order to be able to estimate the possible impacts of droughts, it is important to identify and characterise these events. Describing a drought is, however, not such an easy task as it represents a spatio-temporal phenomenon, with no clear start and ending, trailing from one place to another. This study tries to objectively identify droughts in space and time by applying operators from mathematical morphology. On the basis of the identified droughts, OWA operators are employed to characterise the events in order to aid farmers, water managers, *etc.* in coping with these events.

Keywords: Mathematical morphology · Droughts · Spatio-temporal OWA operators

1 Introduction

Droughts can entail severe socio-economic damage such as crop failure, shortage in water and energy supply. Eventually, famines and severe conflicts can be caused by major drought events. Although it is relatively easy in the field to tell whether or not one is experiencing a drought, identifying a drought event is a quite difficult task. Droughts can last for several months or even years, they have no clear starting and ending point, they can start at one place and move to another place, and their intensity can change with place and time.

Commonly, four types of droughts are distinguished [1]. A deficiency in precipitation over a large area and a prolonged time period is the primary cause of a drought [2] and is regarded as a *meteorological drought*. The combination with high evaporation rates can result in a large period of low soil moisture and, hence, lead to an *agricultural drought* as crops become affected. In a later stage, the recharge to aquifers and rivers may be reduced and a *hydrological drought* develops. When water demands cannot be met by the water supply systems and economic activities and ecosystems seriously suffer, a *socio-economic drought* is experienced.

© Springer International Publishing AG, part of Springer Nature 2018
J. Medina et al. (Eds.): IPMU 2018, CCIS 854, pp. 287–298, 2018.
https://doi.org/10.1007/978-3-319-91476-3_24

Although scientific literature w.r.t. drought characterisation shows that efforts have already been made to try to quantitatively characterise droughts, capturing all aspects of a drought still remains a difficult task. Andreadis et al. [3] were among the first to introduce a spatial identification procedure in which the fact that droughts can merge or break up at subsequent time steps is acknowledged. Lloyd-Hughes [4] continued on this identification procedure and extended it to the space-time domain.

Furthermore, in order to determine whether a location is dry, one generally relies upon a drought index on the basis of which a threshold is set. A value below the threshold then indicates that drought conditions are met. The standardised precipitation index [5] or the Palmer drought severity index [6] are examples of two commonly used drought indices. However, all drought indices have their own advantages and shortcomings [7,8] and applications of these indices will on their turn also suffer from these shortcomings. In this respect and to overcome inconsistencies between different drought indices, Sheffield and Wood [8,9] suggested to use percentile values of a drought variable as a basis for drought characterisation. By using percentile values, a fair comparison between values at different locations is possible.

Besides, soil moisture appears to be a fine candidate to be used as drought variable [7,9]. Soil moisture is a key variable in the hydrological cycle as it controls the majority of processes in the hydrological cycle, e.g. evaporation, runoff, infiltration and drainage. It furthermore reflects the impact of meteorological variables such as temperature and radiation. Soil moisture values in the top layer of the soil are also related to short-term precipitation, soil moisture values in the root zone indicate the amount of water that is available for plant growth and soil moisture values in the deep soil layers reflect the amount of water that is available for rivers and aquifers.

In this study, we attempt to identify and characterise droughts by taking into account their spatio-temporal nature, such that the phenomenon can be better understood. In order to objectively identify droughts in space and time, mathematical morphological operators [10] will be applied to a 30-year daily time series of spatially distributed soil moisture data over Australia. Applications and extensions of mathematical morphology w.r.t. image filtering, image segmentation, etc. have already been reported in the processing of remote sensing data [11] and medical image analysis [12]. Some three-dimensional applications of mathematical morphology have also been described in literature. Peters II and Nichols [13] and Paris and Sillon [14] applied operators from mathematical morphology to image sequences. Pierre et al. [15] tried to unravel the three-dimensional complexity of the soil structure and Mao et al. [16] employed morphological filtering to extract information regarding geologic bodies.

Section 2 first elaborates on the data and the study region chosen for this study. Section 3 then explains the data preprocessing, the selection of the threshold and the application of mathematical morphology for drought identification. Section 4 further elaborates on the determined drought characteristics. Section 5 formulates the conclusions and perspectives for future research.

2   Data and Study Region

In order to be able to identify drought events in space and time, a long time series of historical data is required. Ideally, such time series should be available at large scales in order to also capture the spatial characteristics of the events. With the emergence of satellite remote sensing data in the late seventies, obtaining information at a high temporal and spatial resolution has become easier. The Global Land Evaporation Amsterdam Model (GLEAM) [17] benefits from the use of satellite-derived observations to estimate terrestrial evaporation and soil moisture. The data set used in this study (the GLEAM v3.0a data set [18]) spans a period of 35 years (from 1/1/1980 till 31/12/2014) of global daily root-zone soil moisture values at a 0.25° resolution. The soil consists of soil particles and voids in between them. As the soil gets wet by infiltration of precipitation, the voids get filled with water. The soil moisture value θ of a soil expresses the amount of water in the soil relative to the total soil volume:

$$\theta = \frac{\text{volume of water}}{\text{total soil volume}}. \tag{1}$$

The amount of voids in the soil is expressed by the porosity η:

$$\eta = \frac{\text{volume of voids}}{\text{total soil volume}}. \tag{2}$$

Further, we will make use of relative soil moisture values, which are defined as $\frac{\theta}{\eta}$, and reflects the fraction of pores that are filled with water. A relative soil moisture value of 1 hence indicates that all the voids of the soil are filled with water.

From this GLEAM data set, daily data covering Australia were selected. It is well known that Australia contains large areas of arid and semi-arid land and is vulnerable to the effects of climate change, in particular to the expected drying trend for the next 50–100 years [19]. Already now, substantial agricultural areas are affected by periodic droughts.

Unlike in many other drought characterization studies, it was decided not to convert the daily values to monthly values as these daily values will allow for the detection of drought periods lasting less than one month. As pointed out by Byun and Wilhite [20], an affected drought region can return to normal conditions with only one day's rainfall. In order to be able to calculate spatial characteristics on the data set, the selected GLEAM data set was reprojected to the Lambert Azimuthal Equal Area coordinate system, resulting in a pixel resolution of 27.442 km × 29.079 km.

3   Data Pre-processing and Drought Identification

3.1   Selection of the Threshold for Defining a Drought

Identifying droughts directly on the basis of the soil moisture values is not appropriate, as the same soil moisture value can indicate a rather dry soil for one soil type whereas for another soil type it can indicate a relatively moist soil. Therefore, one has to resort to a variable with an unambiguous interpretation. Sheffield and Wood [8,9] introduced the idea of using a percentile level of a drought variable as threshold for defining a drought. This idea is also used in this study. A percentile value of 10%, reflecting that one is facing drought conditions in 10% of the time, is chosen for each pixel. This value can also be regarded as the value that separates moderate from severe and more extreme droughts [3]. Furthermore, as to allow the soil moisture values of the neighbouring locations to take part in the threshold for the pixel at hand, and to establish a smoother transition between the relative soil moisture values at the thresholds for neighbouring pixels, a 3×3-neighbourhood was identified around the pixel at hand. The relative soil moisture value at the threshold for this pixel is then determined on the basis of the empirical cumulative distribution function of all soil moisture values observed at all time steps within the neighbourhood. This idea can be extended easily when one aims at taking into account larger neighbourhoods such as regions with the same land cover type.

3.2   Delineation of Droughts Through Mathematical Morphology

After selecting only those pixels with a value below the 10th percentile value, a time series of binary maps indicating which locations possibly belong to a drought are obtained (see Fig. 1 for an example of such a map). It is clear that applying the threshold results in a scattered pattern, showing isolated single pixels indicated as dry and larger dry regions containing pixels that are denoted as not dry. This also occurs in the time dimension. Pixels can be indicated as dry at one time step, whereas at subsequent time steps, they are denoted as not dry, followed by being dry in the time steps thereafter. Hence, a processing procedure is required in order to smooth these irregularities away. A method well suited for this purpose is mathematical morphology [10], which aims at simplifying images by retaining the essential shape characteristics and removing irrelevancies [21].

By using the basic operators from mathematical morphology, i.e. erosion and dilation, the salt and pepper noise, i.e. the holes within the larger events and the smaller droughts, can be filled or removed, respectively. To apply these operators, a structuring element is first determined whose size influences the size of the droughts that will be removed and the holes that will be filled. As a drought has a spatio-temporal character, it is chosen to employ a structuring element that has space-time dimensions. To this end, the thresholded maps of the time series are placed one after the other, and a three-dimensional structuring element is applied to this series. Different sizes of structuring elements were employed of which the smallest a $3 \times 3 \times 3$- and the largest a $7 \times 7 \times 7$-box, of which the first two dimensions indicate the spatial size of the structuring element and the third dimension indicates the number of time steps that is taken into account. It should be noted that other choices could be made one could for instance opt to take into account more time steps resulting in a structuring element of size $3 \times 3 \times 7$. Future research will encompass drought identification with a more diverse set of structuring elements.

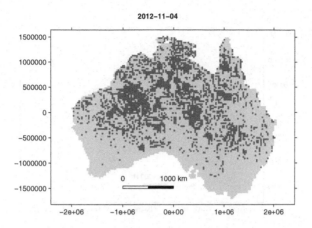

Fig. 1. Possible drought locations after thresholding a percentile map

Consider the space E of all pixels at all time steps, of which the spatio-temporal series of dry-denoted pixels A is a subset. With each point x of the space E, a structuring element $B(x)$, e.g. a box, is associated. The erosion ϵ and dilation δ of A are then given by:

$$\epsilon(A) = \{x \mid B(x) \subseteq A\} \tag{3}$$

$$\delta(A) = \{x \mid B(x) \cap A \neq \emptyset\}. \tag{4}$$

The compositions $\gamma = \delta \circ \epsilon$ and $\phi = \epsilon \circ \delta$ are called the morphological opening and closing respectively. By first applying a morphological opening followed by a morphological closing, an open-close filter is created and the salt and pepper noise is removed.

Figure 2 illustrates a resulting map after applying an open-close filter to the thresholded time series (see Fig. 1 for an example of a map from this time series) and with the $3 \times 3 \times 3$-box structuring element. Different colors are used to illustrate the different droughts. This figure clearly shows that at the given time step, the green-coloured drought is not spatially contiguous. However, as in former or later time steps, the currently isolated green parts merge, these parts belong to the same event. Events smaller than the size of the chosen structuring element will not be identified. By using the dilation operator it is possible that pixels that were originally denoted as not dry, *i.e.* their value is higher than the imposed threshold, will become part of the identified drought. By enlarging the size of the structuring element to e.g. a $5 \times 5 \times 5$-box, fewer drought events will be identified of which none will be smaller than the size of this enlarged structuring element.

When applying a $7 \times 7 \times 7$-structuring element to the thresholded time series, it was noted that for the date given in Fig. 1, no single drought event was obtained (data not shown), not even a part of the larger green drought of Fig. 2. The reason

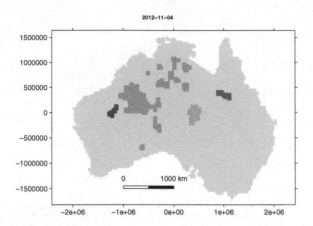

Fig. 2. Resulting droughts after application of an open-close filter to the scattered map of Fig. 1 (Color figure online)

herefore is that the larger $7 \times 7 \times 7$-structuring element does not fit between the salt noise. After applying the erosion operator in the first step of the open-close filter, none of the pixels is still indicated as dry. This result is not desirable as from Fig. 1 it is clear that a quite large contiguous area, apart from some isolated pixels, is dry. This area is larger than the $7 \times 7 \times 7$-structuring element and should hence be identified. This can be dealt with by gradually eliminating the noise components by using an open-close filter iteratively, starting from a small-sized structuring element in the first iteration and by gradually enlarging the structuring element to the desired size in the different iterations. Such a filter is called an alternating sequential filter (ASF) [22] and has already been applied to filter images for which the elimination of large noise components with an open-close filter destroys the original image too much [23,24]:

$$\text{ASF}_i = (\phi_i \circ \gamma_i) \circ (\phi_{i-1} \circ \gamma_{i-1}) \circ \ldots \circ (\phi_1 \circ \gamma_1), \tag{5}$$

with i representing the i-th iteration, γ_1 and ϕ_1 representing the opening, respectively closing operator with the smallest structuring element. ASF_1 hence corresponds to the open-close filter applied in the previous paragraph. In this study larger ASF filters were also applied in order to compare the results of directly applying a larger-sized structuring element to the results of gradually enlarging the structuring element. ASF_2, with a $3 \times 3 \times 3$- and a $5 \times 5 \times 5$-structuring element, and ASF_3 with a $3 \times 3 \times 3$-, a $5 \times 5 \times 5$- and a $7 \times 7 \times 7$-structuring element were applied to the time series. Figure 3 illustrates the obtained results after applying ASF_2 and ASF_3. It can be seen that the larger drought (green-coloured in Fig. 2) also appears in the result after applying ASF_3. This favours the application of an ASF-filter instead of directly applying an open-close filter with a larger structuring element. The splitted drought, purple-coloured in the top panel of Fig. 3, is identified by ASF_2, however, its parts are too small to be identified by ASF$_3$. Furthermore, it is noticed that these parts turn out to belong to an event different from the larger drought (green-coloured in Figs. 2 and 3), whereas after application of the open-close filter, they result to be part of this larger drought (green-coloured in Fig. 2). Table 1 lists the number of completed drought events obtained for different sizes of the structuring element. As is expected, the larger the structuring element, the smaller the number of droughts identified.

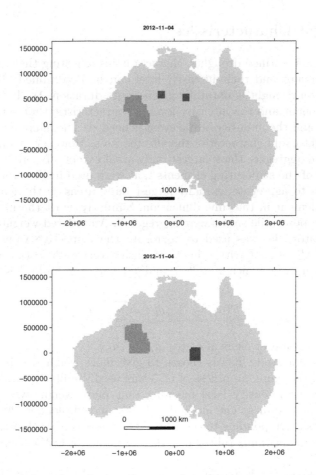

Fig. 3. Delineated droughts after application of an alternating sequential filter with a $3 \times 3 \times 3$- and $5 \times 5 \times 5$-structuring element (top panel) and a $3 \times 3 \times 3$-, a $5 \times 5 \times 5$- and a $7 \times 7 \times 7$-structuring element. (Color figure online)

Table 1. Number of completed drought events for different sizes of neighbourhoods and the structuring element

Size of the structuring element	Number of droughts
$3 \times 3 \times 3$	1866
ASF_2	348
ASF_3	118

4 Drought Characteristics

For each of the identified droughts, characteristics reflecting their temporal and spatial component and their intensity level can be determined. The temporal component of a drought is naturally given by its duration. With respect to the spatial component and the intensity level, a characteristic that summarizes the affected area and the intensity of the event is needed. The maximal area covered by the drought could characterise the affected area. However, it might be more informative to aggregate the τ largest daily areal extents. In order to be in line with the size of the structuring elements that were used in the previous section, it was chosen to aggregate as many largest daily areas as the length τ of the structuring element in the time dimension. Similarly, w.r.t the intensity, the τ smallest daily percentile values were aggregated. An ordered-weighted-averaging (OWA) operator [25] was used to aggregate the respective values. An OWA operator $F : \mathbb{R}^n \to \mathbb{R}$ of arity n has a weighting vector $\mathbf{W} = (w_1, w_2, \ldots, w_n)^T$ associated with it, for which $w_i \in [0,1]$ and $\sum_{i=1}^{n} w_i = 1$, and takes the following form:

$$F(a_1, a_2, \ldots, a_n) = \sum_{i=1}^{n} w_i b_i, \tag{6}$$

with b_j the j-th largest element of $\{a_1, a_2, \ldots, a_n\}$.

In the current study, it was chosen to give more weight to the larger daily areal extents, therefore, an orness of 0.75 was used. Similarly, an orness of 0.25 was used for the intensity level meaning that more weight was given to the smaller percentile values. The method of Fullér and Majlender [26] was used to obtain weights corresponding to these orness-values ensuring that the dispersion of the weights is maximal. Table 2 lists these weights for OWA operators of arity 3, 5 and 7.

Figure 4 shows the histograms of the obtained duration, affected area and intensity, for the different droughts obtained after applying the open-close filter with a $3 \times 3 \times 3$-structuring element, ASF_2 and ASF_3. It can be seen that the longest droughts become shorter when larger structuring elements are used. Similarly, the largest affected area becomes smaller when larger structuring elements are used. This is due to the fact that in order to retain a drought event with a large structuring element, the structuring element has to fit entirely, in space and time, within the thresholded time series. This also holds for the tails of the droughts, resulting in longer and larger major droughts when smaller

Table 2. Weights used by the OWA operators of different arities corresponding to the length τ of the structuring element $B(x)$ in the time dimension.

	orness = 0.75				orness = 0.25		
Length τ of $B(x)$	3	5	7		3	5	7
w_1	0.6162	0.4594	0.3637	w_1	0.1162	0.0477	0.0279
w_2	0.2676	0.2608	0.2390	w_2	0.2676	0.0840	0.0429
w_3	0.1162	0.1480	0.1556	w_3	0.6162	0.1480	0.0659
w_4	-	0.0840	0.1012	w_4	-	0.2608	0.1012
w_5	-	0.0477	0.0659	w_5	-	0.4594	0.1556
w_6	-	-	0.0429	w_6	-	-	0.2390
w_7	-	-	0.0279	w_7	-	-	0.3637

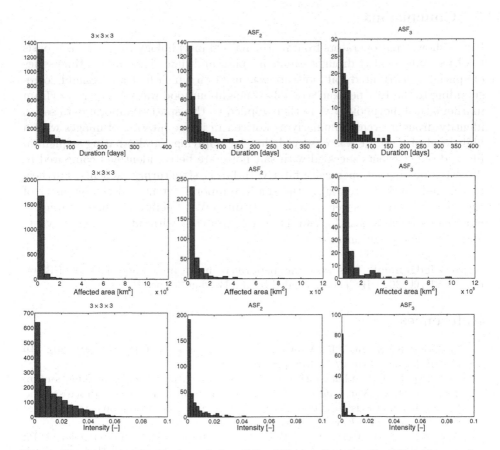

Fig. 4. Histograms of the characteristics of the different droughts identified after applying an open-close filter (left panels), ASF$_2$ (middle panels) or ASF$_3$ (right panels) to the time series of thresholded images. Neighbourhoods used for thresholding the images are 3×3. Please note the different scaling of the axes.

structuring elements are used. Regarding the identified drought intensities, it can be seen that the highest intensity values become smaller, *i.e.* only the more intense droughts are obtained, for larger structuring elements. Depending on the goal of the end-user, e.g. water managers, insurance companies, governments, farmers, *etc.* one might opt to use other sizes of structuring elements to identify droughts. Furthermore, the spatial resolution of the data set might also influence the size of the structuring element. On the basis of the identified and characterised droughts, all recorded historical droughts can then be queried such that end-users can relate a possible current drought to historical ones in order to better cope with the event and take appropriate measures. Furthermore, the characteristics can serve as a basis for a probabilistic model such that probabilities of occurrence of droughts can be calculated, as will be explored in future research.

5   Conclusions

It was shown that operators from mathematical morphology can serve as a basis to objectively identify drought events in space and time. Therefore, a time series of spatially distributed soil moisture was used on which first a threshold corresponding to the 10% percentile of a 3×3-neighbourhood was set. Operators from mathematical morphology were then applied to this spatio-temporal data set to identify drought events. Similarly as noticed in image filtering of images with a high level of noise, it was observed that by applying an alternating sequential filter, drought events speckled with salt noise are better identified compared to directly applying an open-close filter with larger structuring elements. Furthermore, characteristics reflecting the spatio-temporal extent and the intensity of the identified droughts were determined using OWA operators. These characteristics can serve as a basis for end-users to better cope with future drought events and take appropriate measures.

Acknowledgement. This work was performed in the framework of the STEREO-project SR/00/302 ('Hydras+'), funded by the Belgian Science Policy.

References

1. Mishra, A.K., Singh, V.P.: A review of drought concepts. J. Hydrol. **391**, 202–216 (2010). https://doi.org/10.1016/j.jhydrol.2010.07.012
2. Tallaksen, L.M., Van Lanen, H.A.J.: Drought as a natural hazard: introduction. In: Tallaksen, L.M., Van Lanen, H.A.M. (eds.) Hydrological Drought Processes and Estimation Methods for Streamflow and Groundwater. Developments in Water Science, pp. 3–17. Elsevier, Amsterdam (2004)
3. Andreadis, K.M., Clark, E.A., Wood, A.W., Hamlet, A.F., Lettenmaier, D.P.: Twentieth-century drought in the conterminous United States. J. Hydrometeorol. **6**(6), 985–1001 (2005). https://doi.org/10.1175/JHM450.1
4. Lloyd-Hughes, B.: A spatio-temporal structure-based approach to drought characterisation. Int. J. Climatol. **32**(3), 406–418 (2011). https://doi.org/10.1002/joc.2280

5. McKee, T.B., Doesken, N.J., Kleist, J.: The relationship of drought frequency and duration to time scales. In: Proceedings of the Eight Conference on Applied Climatology, pp. 179–184. American Meteorological Society, Anaheim (1993)
6. Palmer, W.C.: Meteorologic drought. Research Paper 45, US Department of Commerce, Weather Bureau (1965)
7. Sheffield, J., Goteti, G., Wen, F., Wood, E.F.: A simulated soil moisture based drought analysis for the United States. J. Geophys. Res. **109**, D24108 (2004). https://doi.org/10.1029/2004JD005182
8. Sheffield, J., Wood, E.F.: Drought Past Problems and Future Scenarios. Earthscan, London (2011)
9. Sheffield, J., Wood, E.F.: Characteristics of global and regional drought, 1950–2000: an analysis of soil moisture data from off-line simulation of the terrestrial hydrological cycle. J. Geophys. Res. **112**, D17115 (2007). https://doi.org/10.1029/2006JD008288
10. Serra, J.: Introduction to mathematical morphology. Comput. Vis. Graph. Image Process. **35**(3), 283–305 (1986)
11. Soille, P., Pesaresi, M.: Advances in mathematical morphology applied to geoscience and remote sensing. IEEE Trans. Geosci. Remote Sens. **40**(9), 2042–2055 (2002). https://doi.org/10.1109/TGRS.2002.804618
12. Dufour, A., Tankyevych, O., Naegel, B., Talbot, H., Ronse, C., Baruthio, J., Dokládal, P., Passat, N.: Filtering and segmentation of 3D angiographic data: advances based on mathematical morphology. Med. Image Anal. **17**, 147–164 (2013). https://doi.org/10.1016/j.media.2012.08.004
13. Peters II, R.A., Nichols, J.A.: Rocket plume image sequence enhancement using 3D operators. IEEE Trans. Aerospace Electron. Syst. **33**(2), 485–498 (1997)
14. Paris, S., Sillon, F.: Robust acquisition of 3D informations from short image sequences. Graph. Models **65**, 222–238 (2003)
15. Pierret, A., Capowiez, Y., Belzunces, L., Moran, C.J.: 3D reconstruction and quantification of macropores using X-ray computed tomography and image analysis. Geoderma **106**, 247–271 (2002). https://doi.org/10.1016/S0016-7061(01)00127-6
16. Mao, X., Zhang, B., Deng, H., Zou, Y., Chen, J.: Three-dimensional morphological analysis method for geologic bodies and its parallel implementation. Comput. Geosci. **96**, 11–22 (2016). https://doi.org/10.1016/j.cageo.2016.07.004
17. Miralles, D.G., Holmes, T.R.H., De Jeu, R.A.M., Gash, J.H., Meesters, A.G.C.A., Dolman, A.J.: Global land-surface evaporation estimated from satellite-based observations. Hydrol. Earth Syst. Sci. **15**, 453–469 (2011). https://doi.org/10.5194/hess-15-453-2011
18. Martens, B., Miralles, D.G., Lievens, H., van der Schalie, R., de Jeu, R.A.M., Fernández-Prieto, D., Beck, H.E., Dorigo, W.A., Verhoest, N.E.C.: GLEAM v3: satellite-based land evaporation and root-zone soil moisture. Geosci. Model Dev. **10**, 1903–1925 (2017). https://doi.org/10.5194/gmd-10-1903-2017
19. McCarthy, J.J., Canziani, O.F., Leary, N.A., Dokken, D.J., White, K.S.: Working group II: impacts, adaptation, and vulnerability. Technical report, International Panel on Climate Change IPCC: Third Assessment Report (2001)
20. Byun, H.R., Wilhite, D.A.: Objective quantification of drought severity and duration. J. Clim. **12**(9), 2747–2756 (1999)
21. Haralick, R.M., Sternberg, S.R., Zhuang, X.: Image analysis using mathematical morphology. IEEE Trans. Pattern Anal. Mach. Intell. **9**(4), 532–550 (1987)
22. Serra, J., Vincent, L.: An overview of morphological filtering. Circ. Syst. Sig. Process. **11**(1), 47–108 (1992)

23. Dougherty, E.R., Lotufo, R.A.: Hands-on morphological image processing. In: Tutorial Texts in Optical Engineering, p. 272. SPIE Publications (2003)
24. Soille, P.: Morphological Image Analysis: Principles and Applications. Springer, Heidelberg (2004). https://doi.org/10.1007/978-3-662-05088-0
25. Yager, R.R.: Ordered weighted averaging aggregation operators in multi-criteria decision making. IEEE Trans. Syst. Man Cybern. **18**, 183–190 (1988)
26. Fullér, R., Majlender, P.: An analytic approach for obtaining maximal entropy OWA operator weights. Fuzzy Sets Syst. **124**, 53–57 (2001). https://doi.org/10.1016/S0165-0114(01)00007-0

Measures of Comparison and Entropies
for Fuzzy Sets and Their Extensions

On Dissimilarity Measures at the Fuzzy Partition Level

Grégory Smits[(⊠)], Olivier Pivert, and Toan Ngoc Duong

IRISA - University of Rennes, UMR 6074, Lannion, France
{gregory.smits,olivier.pivert,ngoc-toan.duong}@irisa.fr

Abstract. On the one hand, a user vocabulary is often used by soft-computing-based approaches to generate a linguistic and subjective description of numerical and categorical data. On the other hand, knowledge extraction strategies (as e.g. association rules discovery or clustering) may be applied to help the user understand the inner structure of the data. To apply knowledge extraction techniques on subjective and linguistic rewritings of the data, one first has to address the question of defining a dedicated distance metric. Many knowledge extraction techniques indeed rely on the use of a distance metric, whose properties have a strong impact on the relevance of the extracted knowledge. In this paper, we propose a measure that computes the dissimilarity between two items rewritten according to a user vocabulary.

Keywords: Fuzzy partition · Data personalization
Dissimilarity measure

1 Introduction

Helping users extract and understand the content of a raw data set is a crucial task in data mining. Most of the datasets contain the description of items on attributes that are generally of a numerical or a categorical nature. It is cognitively difficult for an end-user to browse and analyze a large collection of numerical and categorical data, and it is moreover, technically speaking, almost impossible to generate an interpretable graphical view of a set of data described on more than 3 dimensions. To overcome these difficulties, soft-computing-based approaches of data management leverage a user vocabulary to turn numerical and categorical variables (i.e. attributes) into linguistic variables. Once rewritten according to the user vocabulary, concise and easily interpretable views of the data may be generated to give the user an insight into the content of the dataset [1]. In addition, data mining techniques, as clustering algorithms for instance, may be used to discover the inner structure of the data, whose description also constitutes valuable knowledge [2]. Many data mining techniques rely on a distance measure to determine the similarity of two items. In this work, we address the question of computing the distance between two items rewritten according to a user vocabulary formalized by means of strong fuzzy partitions.

© Springer International Publishing AG, part of Springer Nature 2018
J. Medina et al. (Eds.): IPMU 2018, CCIS 854, pp. 301–312, 2018.
https://doi.org/10.1007/978-3-319-91476-3_25

This question of a distance measure at the partition level has been notably studied by Guillaume et al. [3], but the measure they proposed sometimes leads to questionable results as we will see in Sect. 2.4.

We propose in this paper a new dissimilarity measure at the partition level that somehow reconsiders the indistinguishability relation introduced by the use of a fuzzy vocabulary for the sake of a better interpretability of the generated results. The final objective is to use the proposed dissimilarity measure to build clusters of data rewritten according to a user vocabulary instead of considering their numerical and categorical values. Motivation for that are manifold. First, the indistinguishability area defined by the cores of the fuzzy sets will reduce the number of distinct rewritings to consider, thus making it possible to handle larger datasets. Second, translating numerical and categorical values into linguistic terms allows for the conception of graphical views representing the obtained clusters on many dimensions at the same time [1], which cannot be envisaged on numerical/categorical data. And third, starting with a rewriting step of the data is a way to personalize the data-to-knowledge translation process and to make it more easily interpretable for end-users. But the relevance of the structure built by a clustering highly depends on the properties of its underlying distance measure.

In this paper, we focus on the definition of dissimilarity measures at the fuzzy partition level and the study of their properties. Their use by a clustering process will be the next step. The rest of the document is structured as follows. In Sect. 2, preliminary notions regarding fuzzy-set-based vocabularies and dissimilarity measures at the partition level are recalled. Sections 3.1 and 3.2 detail our proposed dissimilarity measures, respectively for numerical and categorical domains.

Motivating Example

To illustrate the motivation for a new dissimilarity measure, let us consider the vocabulary, i.e. fuzzy partition, illustrated in Fig. 1 that turns the mileage of a car into a linguistic variable that may take the values {veryLow, low, medium, high, veryHigh}. In the situation illustrated by Fig. 1, a dissimilarity measure at the partition level has to be able to capture the fact that t_1 is closer to t_3 than to t_6 because the linguistic value that describes t_1, namely *low mileage*, is closer to *medium mileage* than to *veryHigh mileage*, this case being well covered by the measure defined in [3]. However, contrary to [3], we argue that the indistinguishability relation should be limited to the core of the fuzzy sets (as e.g. between t_3 and t_4), and that it appears more natural and interpretable to consider t_3 as closer to t_2 than to t_5 even if these last two points satisfy the linguistic value *medium mileage* at the same degree (in this case $\mu_{medium}(t_2) = \mu_{medium}(t_5) = 0.7$). This expected behavior is all the more important if the considered task is to build groups of items having close linguistic rewritings. Using a dissimilarity measure that is more appropriate to compare rewritten data, we expect that more meaningful groups of items will be obtained especially by avoiding grouping tuples that are significantly different.

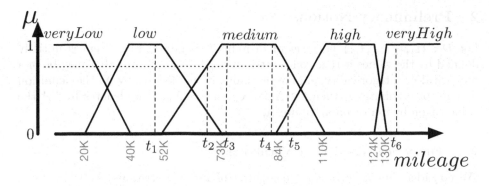

Fig. 1. Distance computation at the partition level between numerical values

Even if meaningful fuzzy partitions may be built on categorical attributes, using a dedicated graphical interface as ReqFlex for instance [4] (Fig. 2), distance measures between categories or discrete fuzzy sets are generally reduced to a Boolean test of equality. A second contribution of this paper is to propose a measure to compute the dissimilarity between two categorical values that takes into account the structure of its underlying user vocabulary. The idea is to consider that discrete fuzzy sets sharing some categories should be considered as somewhat semantically related. By doing so, one may infer a weak partial order on discrete fuzzy sets defined on top of a categorical attribute. Thus, categorical values taken from these two sets should be considered closer to each other than categorical values taken from two sets having an empty intersection. To illustrate this proposal, let us consider a possible fuzzy vocabulary describing different car brands according to their relative reliability reputation (Fig. 2). Then, we argue that the brand *Chrysler*, characterized as a fully *moderatelyReliable brand*, should be considered as closer to *VW*, a *reliable brand*, than *Daewoo*, that belongs to the set of *poorlyReliable brands*, because *moderatelyReliable brands* and *reliable brands* have in this case much more in common than with *poorlyReliable brands*. Obviously, the relevance of this interpretation of semantic closeness is context-dependent, and most of all depends on the point-of-view expressed by the user through the definition of his/her vocabulary.

reliable		moderatelyReliable	poorlyReliable		*toAvoid*		*others*	
1	*VW, Mercedes*	1 *Chrysler*	1 *Daewoo*		1 *Lada*		1 *Jaguar, Porsche, ...*	
0.8	*AUDI, Toyota*	0.8 *KIA, Nissan, Suzuki*	0.8		0.8		0.8	
0.6	*Ford*	0.6 *Peugeot, Renault*	0.6 *Chevrolet*		0.6		0.6	
0.4	*Peugeot*	0.4 *Ford*	0.4 *Renault*		0.4 *Chevrolet*		0.4	
0.2		0.2 *AUDI, Toyota*	0.2 *KIA, Nissan, Suzuki*		0.2		0.2	

brand

Fig. 2. Example of a subjective vocabulary on a categorical attribute

2   Preliminary Notions

Let $\mathcal{D} : \{t_1, t_2, \ldots, t_m\}$ be a set of m items to analyze. Each item is initially defined by the values it takes on n attributes $\{A_1, A_2, \ldots, A_n\}$ that may be of a numerical or categorical type. More formally, if one denotes by X_i the definition domain of attribute A_i then $t \in X_1 \times X_2 \times \ldots \times X_n$. One denotes by $t.A$ the value taken by item t on attribute A.

2.1   Fuzzy-Set-Based User Vocabulary

We consider that a vocabulary composed of Fuzzy Partitions (FP) is defined on the attributes $\{A_1, A_2, \ldots, A_n\}$. Such a vocabulary, denoted by $\mathcal{V} = \{V_1, \ldots, V_n\}$, formally consists of a set of linguistic variables, associated with each attribute: V_j is a triple $\langle A_i, \{v_{i,1}, \ldots, v_{i,q_i}\}, \{l_{i,1}, \ldots, l_{i,q_i}\}\rangle$ where q_i denotes the number of modalities associated with attribute A_i, the v_i's denote their respective membership functions defined on domain X_i and the l_i's their respective linguistic labels, generally adjectives of the natural language. For instance, an attribute A_i describing prices may be associated with $q_i = 3$ modalities, in turn associated with the labels $l_{i,1} = $ 'cheap', $l_{i,2} = $ 'reasonable' and $l_{i,3} = $ 'expensive'.

It is assumed that for all attributes, each value may be completely rewritten in terms of $V : \forall y \in D_j, \sum_{s=1}^{q_j} v_{js}(y) = 1$. Moreover, it is assumed that the partitions defined on numerical attributes form a strong FP [5], which leads to the constraint that y can partially satisfy up to two adjacent modalities. Figures 1 and 2 are examples of such partitions defined on a numerical and a categorical attribute respectively.

2.2   Item Rewriting Vector

Initially defined in a numerical and categorical space, an item may be rewritten using the linguistic terms from the user vocabulary. The result of such a rewriting step is called an item rewriting vector.

Definition 1. *One denotes by R_t the rewriting vector of an item t wrt. a user vocabulary $\mathcal{V}$, this vector being the concatenation of the satisfaction degrees obtained by t on the different terms that compose $\mathcal{V}$. Such a vector is represented in the following way:*

$$R_t = \langle \mu_{v_{1,1}}(t), \mu_{v_{1,2}}(t), \ldots, \mu_{v_{1,q_1}}(t), \ldots, \mu_{v_{n,1}}(t), \mu_{v_{1,n}}(t), \ldots, \mu_{v_{1,q_n}}(t)\rangle.$$

We also denote by $R_t^{A_i}$ the part of the whole rewriting vector R_t that concerns the attribute A_i, $R_t^{A_i} = \langle \mu_{v_{i,1}}(t.A_i), \mu_{v_{i,2}}(t.A_i), \ldots, \mu_{v_{i,q_i}}(t.A_i)\rangle$.

Example 1. Table 1 shows the data (attribute values and rewriting vectors from Fig. 1) that have to be considered when computing a dissimilarity at the FP level.

Table 1. Items from Fig. 1 and their rewriting vector

t	$t.mileage$	$R_t^{mileage}$	t	$t.mileage$	$R_t^{mileage}$
t_1	$50K$	$\langle 0, 1, 0, 0, 0 \rangle$	t_2	$70K$	$\langle 0, 0.3, 0.7, 0, 0 \rangle$
t_3	$74K$	$\langle 0, 0, 1, 0, 0 \rangle$	t_4	$80K$	$\langle 0, 0, 1, 0, 0 \rangle$
t_5	$90K$	$\langle 0, 0, 0.7, 0.3, 0 \rangle$	t_6	$134K$	$\langle 0, 0, 0, 0, 1 \rangle$

2.3   Properties of a Dissimilarity Measure at the Partition Level

When it comes to defining a dissimilarity that takes into account fuzzy sets, then three types of comparison may be envisaged [6]: (1) between two points that belong to a same fuzzy set, (2) between a point and a fuzzy set and, (3) between two fuzzy sets [7]. As shown in [3], (that is, to the best of our knowledge, the only existing approach addressing the question of a distance calculation at the fuzzy partition level) the measure we have to define has, in some sense, to combine these three types of fuzzy distances.

In fine, we aim at computing the dissimilarity between two items wrt. the considered vocabulary $\mathcal{V}$. This measure obviously relies on the aggregation of dissimilarities computed on each considered dimension. On a given dimension A_i, the dissimilarity at the partition level of two items, say t and t', has to combine the dissimilarity between the two numerical/categorical values ($t.A_i$ and $t'.A_i$) and between their rewriting wrt. $\mathcal{V}$: $R_t^{A_i}$ and $R_{t'}^{A_i}$. The expected behavior of the function to build is that the farther $t.A_i$ and $t'.A_i$, the higher the returned dissimilarity value. But, this function also has to take into account the indistinguishability relation embedded in the definition of a fuzzy subset, which means that the dissimilarity between $R_t^{A_i}$ and $R_{t'}^{A_i}$ should be 0 if $t.A_i$ and $t'.A_i$ fall in the core of a same partition element.

On any dimension involved in a rewriting vector, the function to define has to fulfil the following properties to constitute a *dissimilarity*:

- positiveness: $d(t, t') \geq 0$,
- identity of indiscernibles: a property that is generally defined in the following way $d(t, t') = 0 \Leftrightarrow t = t'$ but extended as follows in our particular context $d(t, t') = 0 \Leftrightarrow R_t = R'_t$ to capture the indistinguishability relation embedded in the FP,
- symmetry: $d(t, t') = d(t', t)$.

A dissimilarity that also satisfies the triangle inequality: $d(t, t') \leq d(t, t'') + d(t'', t')$, is called a semi-distance.

2.4   Behavior of Existing Approaches

In this subsection, we show that the existing approaches (a dedicated one [3] and a naive one) to the computation of a dissimilarity degree at the FP level lead, in some particular cases, to results difficult to understand and interpret.

A Generic Dissimilarity Measure. Whatever the type of the attribute A_i concerned, numerical or categorical, a way to compute the dissimilarity of two items t and t', or more precisely their rewriting vectors $R_t^{A_i}$ and $R_{t'}^{A_i}$, is to simply compare one-by-one the respective membership degrees of t and t' on the different terms of the vocabulary. Such a dissimilarity measure, denoted by $d_1^i(t, t')$ may be formalized as follows:

$$d_1^i(t, t') = \frac{1}{q_i} \sum_{j=1}^{q_i} |\mu_{v_{i,j}}(t) - \mu_{v_{i,j}}(t')|.$$

The main advantage of this basic strategy is that it can be applied to both numerical and categorical attributes. However it suffers from the fact that it does not take into account the structure of the concerned FP. It indeed considers at the same distance of 1 any pair of values falling in the core of two distinct partition elements, whatever the position of these elements in the partition. In the example illustrated in Fig. 1, $d_1^{mile.}(t_1, t_3) = d_1^{mile.}(t_3, t_6) = d_1^{mile.}(t_1, t_6) = 1$.

A Pseudo-Metric at the FP-Level. In [3], the authors address the question of distance calculation at the FP level, but for numerical attributes only. They especially define a pseudo-metric for the case of strong FP. This metric relies on a strict discretization of the universe of the concerned attribute as shown in Fig. 3 that form crisp areas denoted $\{I_1, I_2, \ldots, I_{q_i}\}$. Then, to compute the distance between two points on a given attribute A_i, their position within this discretization is first computed using the following function:

$$P(t) = I(t) - \mu_{v_i, I(t)}(t),$$

where $I(t)$ is the index of the area ($I(t) \in \{I_1, I_2, \ldots, I_{q_i}\}$) in which t is located. Then, the dissimilarity is quantified by the function $d_2^i(t, t')$:

$$d_2^i(t, t') = \frac{|P(t) - P(t')|}{q_i - 1}.$$

Example 2. To illustre how dissimilarity degrees are computed using the measure d_2, let us consider the points t_2, t_4 and t_5 from Fig. 1. Then, these points are assigned to the following areas: $I(t_2) = I(t_4) = 3$ and $I(t_5) = 4$. Considering that $\mu_{v_{medium}}(t_2) = 0.7$ and $\mu_{v_{high}}(t_5) = 0.3$, we thus obtain the following distance degrees:

- $d_2^{mile.}(t_2, t_4) = \frac{|2.3-2|}{4} = 0.075$,
- $d_2^{mile.}(t_4, t_5) = \frac{|2-3.7|}{4} = 0.425$,
- $d_2^{mile.}(t_2, t_5) = \frac{|2.3-3.7|}{4} = 0.35$.

The metric d_2 handles well the distance between the partition elements to which the two points belong. If one goes back to the situation illustrated in Fig. 1, then $d_2(t_1, t_3) < d_2(t_1, t_6)$. However, despite the fact that the core of a

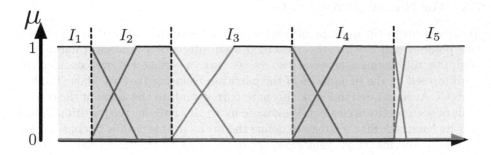

Fig. 3. Discretization of a numerical domain used by the metric d_2

partition element introduces an area of indistinguishability, it appears desirable to take into account the position of the points within the indistinguishability area when computing a distance with points outside this area. For the sake of understandability and interpretability, but also to improve the relevance of the data mining task that relies on a distance calculation, it indeed appears relevant and desirable to consider t_4 (Fig. 1) closer to t_5 than to t_2. However, in this particular case, as t_2 and t_4 fall in the same area according to the crisp discretization suggested in [3] ($I(t_2) = I(t_4) = 3$), then $d_2(t_2, t_4) < d_2(t_4, t_5)$, which, we think, is highly questionable.

3   A Dissimilarity at the FP Level

In this section, we propose a measure to compute the dissimilarity between two points that combines their respective position at the partition level as well as their value dissimilarity. In this sense, the proposed measure is inspired from works done in the context of distance calculation in image processing, and especially the fuzzy geodesic distance suggested in [8].

Definition 2. *We denote by $d_*(t, t')$ the global dissimilarity to determine between t and t' taking into account the structure of the FPs that form the vocabulary $\mathcal{V}$. $d_*(t, t')$ relies on the aggregation of dissimilarity degrees on the different considered dimensions, we thus denote by $d_*^i(t, t')$ the dissimilarity between t and t' on attribute A_i:*

$$d_*(t, t') = \frac{1}{n} \sum_{i=1}^{n} d_*^i(t, t'). \tag{1}$$

The functions d_*^i's are defined in such a way that they return a dissimilarity degree in the unit interval, hence the co-domain of $d_*(t, t')$ is also $[0, 1]$. In the rest of this section, we provide definitions of $d_*^i(t, t')$, first when the concerned attribute is of a numerical type associated with strong FPs (Sect. 3.1), then when it concerns a categorical attribute associated with a discrete fuzzy partition (Sect. 3.2).

3.1   For Numerical Attributes

We first address the question of computing the distance at the FP level between two points t and t' when the concerned attribute is of a numerical nature. To compute the distance between two values wrt. a strong FP, we consider the path formed by the boundaries of the partition elements that are above the line $y = 0.5$. As illustrated in Fig. 4, this path corresponds to the union of the convex hulls of each partition element. We denote by $\mathcal{L}_i$ this path for the partition V_i and $|\mathcal{L}_i|$ its length. A first strategy to define the limits of this path is to consider the minimum and maximum values present in the data on the concerned attribute. This strategy being very sensitive to extremum values, we propose a second one leveraging the fact that all the values inside the core of a partition element are indistinguishable. We thus consider that all the values fully satisfying the first (resp. last) element of the partition are at the same distance wrt. a point taken outside the core of this element. This allows us to consider that the path $\mathcal{L}_i$ starts with the right bound of the core of the first partition element and ends with the left bound of the core of the last element (See. Fig. 4). So every value inside the core of the first (resp. last) element of the partition is treated as the right (resp. left) bound of the core of the element in the dissimilarity calculation.

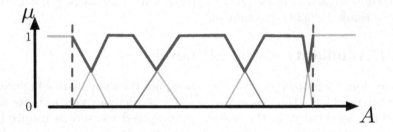

Fig. 4. Hull of a strong FP

To compute the dissimilarity between two points wrt. a strong FP, we then distinguish between two cases. When the two values to compare fall in the core of a same modality, then we assume their distance to be equal to 0 so as to satisfy the indistinguishability relation introduced by the different fuzzy sets. In all other cases, the distance between two values corresponds to the length of the path following $\mathcal{L}_i$ between these two values. Such a path between two values, say t and t'^1, is denoted by $\mathcal{L}_i(t, t')$ as illustrated in Fig. 5.

Definition 3. *Let A_i be a numerical attribute, V_i its FP and $\mathcal{L}_i$ its upper delimiting path. Then $d^i_{*_1}(t, t')$ is defined as follows:*

---

[1] For the sake of simplicity, t and t' are used instead of $t.A_i$ and $t'.A_i$ respectively to lighten the notation.

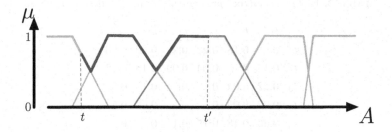

Fig. 5. Path between two values t and t'

$$d^i_{*_1}(t, t') = \begin{cases} 0 & if\ \exists v \in V_i,\ st.\ \mu_v(t) = \mu_v(t') = 1, \\ \frac{|\mathcal{L}_i(t,t')|}{|\mathcal{L}_i|} & otherwise. \end{cases} \tag{2}$$

Proposition 1. *The proposed definition of d^i_* when A_i is numerical is a dissimilarity.*

Proof. The dissimilarity between two values t and t' wrt. a strong FP being computed as the ratio between two path lengths, then the obtained dissimilarity degree is obviously positive and symmetrical. About the identity of indiscernibles, that should be interpreted in our case as the identity of indistinguishables, the conditional definition of $d^i_*(t, t')$ is used to guarantee such an indistinguishability relation between values inside the core of a fuzzy set. If $\mu_v(t) = \mu_v(t') = 1$ and due to the structural properties of the strong FP used on numerical attributes then $d^i_*(t, t') = 0 \Leftrightarrow R^{A_i}_t = R^{A_i}_{t'}$.

Remark 1. The satisfaction of the identity of indistinguishables is in opposition with the triangle inequality. Indeed, considering a partition element v and three points t, t' and t'' such that $\mu_v(t) = \mu_v(t') = 1$, $\mu_v(t'') < 1$ and $t \leq t' < t''$ (resp. $t'' < t \leq t'$), then $d^i_*(t, t') = 0$ and $d^i_*(t, t'') \geq d^i_*(t', t'')$ (resp. $d^i_*(t, t'') \leq d^i_*(t', t'')$). Thus, one observes that $d^i_*(t, t'') > d^i_*(t, t') + d^i_*(t', t'')$ which violates the triangle inequality property. The triangle inequality is however satisfied if there is no situation of indistinguishability between the three values considered.

Example 3. If one goes back to the situation depicted in Fig. 1 and Table 1, then the proposed definition of $d^i_*(t, t')$ leads to the expected behavior as shown by the dissimilarity matrix Table 2.

3.2   For Categorical Attributes

Contrary to numerical attributes, categorical ones are generally defined on non-ordered domains. Hence, no explicit distance can be directly defined for a categorical attribute.

Table 2. Distance matrix between the items detailed in Table 1

	t_1	t_2	t_3	t_4	t_5	t_6
t_1	0	0.18	0.22	0.27	0.36	0.76
t_2	0.18	0	0.04	0.09	0.18	0.58
t_3	0.22	0.04	0	0	0.15	0.55
t_4	0.27	0.09	0	0	0.09	0.5
t_5	0.36	0.18	0.15	0.1	0	0.4
t_6	0.76	0.58	0.55	0.49	0.4	0

The question of computing a distance between categorical values has already largely been addressed, especially by the data mining community [9,10]. Most of the proposed measures rely on contextual information, structural properties (considering clusters of data for instance) or correlations with other dimensions than the concerned categorical one [11,12]. The seldom measures that only make use of the concerned categorical attribute deduce links between categorical values if their frequency of appearing is close [13,14]. So, to the best of our knowledge, the dissimilarity defined in this section is the first one that addresses the question of comparing categorical values according to a discrete FP.

By defining a fuzzy-set-based vocabulary on a categorical attribute, the user expresses a subjective point-of-view about the way the categories have to be interpreted. A discrete fuzzy set gathers categories that define, combined all together, a "semantic concept". Categories regrouped in a same fuzzy partition element can be discriminated according to their respective membership degree within the set. When the user gradually assigns a categorical value to two different partition elements, we consider that he/she creates a semantic link between the two fuzzy-sets concerned. The idea behind the dissimilarity we propose for categorical attributes is to deduce, not an order, but semantical links between partition elements based on their intersections. These links are used in the proposed dissimilarity measure to compute a distance between two categorical values that belong to two different partition elements. The relevance of the links deduced between fuzzy sets based on their intersections obviously depends on the concerned applicative context and the semantics of point-of-view expressed by the user in his/her vocabulary.

Principles and Properties of the Proposed Dissimilarity. Let t and t' be two categorical values satisfying the fuzzy terms v and v' respectively (v and v' may be identical) from an FP V_j. The principle of the dissimilarity measure is to combine the membership of t and t' to their respective partition elements (i.e. v and v') and the semantic closeness of v and v'. In other words, the more t and t' belong to a "semantically" close partition elements, the closer they are.

This semantic closeness between two elements from an FP is denoted by $C_J(v, v')$ and may be defined by means of the Jaccard index that quantifies the proportion of elements v and v' share.

$$C_J(v, v') = \frac{\sum_{x \in \mathcal{D}} \min(\mu_v(x), \mu_{v'}(x))}{\sum_{x \in \mathcal{D}} \max(\mu_v(x), \mu_{v'}(x))}.$$

Definition 4. *For a categorical attribute A_i, the measure d_*^i is defined as follows:*

$$d_*^i(t, t') = 1 - \max_{j,k=1..q_i} \top(\mu_{v_{i,j}}(t), \mu_{v_{i,k}}(t'), C_J(v_{i,j}, v_{i,k})), \tag{3}$$

where the product t-norm $\top$ is used in our case for aggregating $\mu_{v_{i,j}}(t)$, $\mu_{v_{i,k}}(t')$ and $C_J(v_{i,j}, v_{i,k})$ to introduce compensation between the aggregated criteria.

Proposition 2. d_*^i *as defined in Eq. 3 is a dissimilarity.*

Proof. The definition of d_*^i when the concerned attribute A_i is categorical is obviously positive as both $\mu_{v_{i,j}}(t)$, $\mu_{v_{i,k}}(t')$ and $C_J(v, v')$ are defined in the unit interval. The Jaccard index and the product t-norm being symmetric, their combination in d_*^i is so as well. $d_*^i(t, t') = 0$ iff. $\mu_{v_{i,j}}(t) = 1$, $\mu_{v_{i,k}}(t') = 1$ and $C_J(v, v') = 1$. Due to the constraints imposed on the FP (Sect. 2.1) and especially the fact that each item is completely rewritten by $\mathcal{V}$ then $\mu_{v_{i,j}}(t) = 1$ (resp. $\mu_{v_{i,k}}(t') = 1$) and $C_J(v, v') = 1$ implies $v = v'$ and $R_t^i = R_{t'}^i$.

Remark 2. We consider that it would be artificial and senseless to introduce a notion of transitivity in the definition of d_*^i. It would indeed be debatable to consider that a value belonging to a partition element v_i is somewhat similar to a value belonging to an element v_j because v_i has a non-empty intersection with v_k that itself has a non-empty intersection with v_j, especially if v_i and v_j have an empty intersection.

Example 4. Table 3 gives some dissimilarities computed between different car brands wrt. the FP illustrated in Fig. 2.

Table 3. Dissimilarity matrix for some car brands according to the FP Fig. 2

	VW	Mercedes	AUDI	Ford	Peugeot	Daewoo
VW	0	0	0.2	0.4	0.8	1
Mercedes	0	0	0.2	0.4	0.8	1
AUDI	0.2	0.2	0	0.4	0.8	1
Ford	0.4	0.4	0.4	0	0.8	1
Peugeot	0.8	0.8	0.8	0.8	0	0.75
Daewoo	1	1	1	1	0.75	0

4 Conclusion and Perspectives

The rewriting of data according to a fuzzy user vocabulary makes it possible to personalize a data-to-knowledge process. In order to be able to apply data mining tools on linguistic and subjective representations of the data, it is first necessary to address the question of quantifying the dissimilarity between two such representations. We thus provide in this paper a dissimilarity measure that takes into account the structure of the fuzzy partitions that form the user vocabulary. We show on some examples that the proposed dissimilarities return relevant results and better discriminate the compared values without sacrificing the indistinguishability relation introduced by the use of fuzzy partition elements.

The next step is obviously to show that the use of this dissimilarity by a clustering algorithm leads to more meaningful and relevant results thanks to a better discrimination of the compared items.

References

1. Smits, G., Pivert, O., Yager, R.R.: A soft computing approach to agile business intelligence. In: 2016 IEEE International Conference on Fuzzy Systems (FUZZ-IEEE), pp. 1850–1857. IEEE (2016)
2. Smits, G., Pivert, O.: Linguistic and graphical explanation of a cluster-based data structure. In: Beierle, C., Dekhtyar, A. (eds.) SUM 2015. LNCS (LNAI), vol. 9310, pp. 186–200. Springer, Cham (2015). https://doi.org/10.1007/978-3-319-23540-0_13
3. Guillaume, S., Charnomordic, B., Loisel, P.: Fuzzy partitions: a way to integrate expert knowledge into distance calculations. Inf. Sci. **245**, 76–95 (2013)
4. Smits, G., Pivert, O., Girault, T.: Reqflex: fuzzy queries for everyone. Proc. VLDB Endow. **6**(12), 1206–1209 (2013)
5. Ruspini, E.H.: A new approach to clustering. Inf. Control **15**(1), 22–32 (1969)
6. Bloch, I.: On fuzzy distances and their use in image processing under imprecision. Pattern Recogn. **32**(11), 1873–1895 (1999)
7. Montes, S., Couso, I., Gil, P., Bertoluzza, C.: Divergence measure between fuzzy sets. Int. J. Approximate Reason. **30**(2), 91–105 (2002)
8. Bloch, I.: Fuzzy geodesic distance in images. In: Martin, T.P., Ralescu, A.L. (eds.) FLAI 1995. LNCS, vol. 1188, pp. 153–166. Springer, Heidelberg (1997). https://doi.org/10.1007/3-540-62474-0_12
9. Boriah, S., Chandola, V., Kumar, V.: Similarity measures for categorical data: a comparative evaluation. In: Proceedings of the 2008 SIAM International Conference on Data Mining, SIAM, pp. 243–254 (2008)
10. Alamuri, M., Surampudi, B.R., Negi, A.: A survey of distance/similarity measures for categorical data. In: 2014 International Joint Conference on Neural Networks (IJCNN), pp. 1907–1914. IEEE (2014)
11. Gibson, D., Kleinberg, J., Raghavan, P.: Clustering categorical data: an approach based on dynamical systems. Databases **1**, 75 (1998)
12. Guha, S., Rastogi, R., Shim, K.: Rock: a robust clustering algorithm for categorical attributes. Inf. Syst. **25**(5), 345–366 (2000)
13. Goodall, D.W.: A new similarity index based on probability. Biometrics 882–907 (1966)
14. Lin, D., et al.: An information-theoretic definition of similarity. In: ICML, vol. 98, pp. 296–304 (1998)

Monotonicity of a Profile of Rankings with Ties

Raúl Pérez-Fernández[1], Irene Díaz[2(✉)],
Susana Montes[3], and Bernard De Baets[1]

[1] KERMIT, Department of Data Analysis and Mathematical Modelling,
Ghent University, Coupure links 653, 9000 Gent, Belgium
{raul.perezfernandez,bernard.debaets}@ugent.be
[2] Department of Computer Science, University of Oviedo, Oviedo, Spain
sirene@uniovi.es
[3] Department of Statistics O.R. and Mathematics Didactics,
University of Oviedo, Oviedo, Spain
montes@uniovi.es

Abstract. A common problem in social choice theory concerns the aggregation of the rankings expressed by several voters. Two different settings are often discussed depending on whether the aggregate is assumed to be a latent true ranking that voters try to identify or a compromise ranking that (partially) satisfies most of the voters. In a previous work, we introduced the notion of monotonicity of a profile of rankings and used it for statistically testing the existence of this latent true ranking. In this paper, we consider different extensions of this property to the case in which voters provide rankings with ties.

Keywords: Social choice · Monotonicity · Ranking · Signature

1 Introduction

The aggregation of the rankings expressed by several voters is a classical problem in social choice theory that can be traced back to the 18-th century. Arrow [1] pointed out that "each individual has two orderings, one which governs him in his everyday actions, and one which would be relevant under some ideal conditions and which is in some sense truer than the first ordering. It is the latter which is considered relevant to social choice, and it is assumed that there is complete unanimity with regard to the truer individual ordering". From this reflection, one could conclude that there are two different settings for the aggregation of rankings: there exists a latent true ranking that voters try to identify, the goal of the aggregation being to identify said true ranking, or, contrarily, voters have conflicting opinions, the goal of the aggregation being to agree on a compromise ranking. In [2], we described a statistical test for testing the existence of a latent true ranking based on the notion of monotonicity of a profile of rankings.

Unfortunately, since the monotonicity of a profile of rankings with ties has not been defined, the aforementioned statistical test cannot be used in case the

© Springer International Publishing AG, part of Springer Nature 2018
J. Medina et al. (Eds.): IPMU 2018, CCIS 854, pp. 313–322, 2018.
https://doi.org/10.1007/978-3-319-91476-3_26

rankings provided by the voters contain ties. This is a typical problem in real-life problems where voters might consider that two or more candidates are equally suitable [3]. Some existing methods for the aggregation of rankings, such as the method of Kemeny [4] and the method of Schulze [5], are explicitly defined to aggregate rankings with ties. Others, such as the Borda count [6], need to be adapted [7]. In the case of our statistical test, we will see that there is an immediate extension of the property of monotonicity, but we will also propose other extensions based on the notions of signature and ordered signature that might play an interesting role when using real-life data.

The remainder of the paper is structured as follows. In Sect. 2, we recall the notion of monotonicity of a profile of rankings. The natural generalization of this notion to rankings with ties is provided in Sect. 3. In Sect. 4, we introduce the notions of signature and ordered signature and discuss their relation with the property of monotonicity of a profile of rankings with ties. We end with some conclusions and open problems in Sect. 5.

2   Monotonicity of a Profile of Rankings Without Ties

We consider the problem where several voters express their preferences on a set $\mathscr{C}$ of k candidates. In particular, each of the r voters expresses a ranking $\succ_j$ on $\mathscr{C}$, i.e., the asymmetric part of a total order relation $\succeq_j$ on $\mathscr{C}$. The set of all possible rankings on $\mathscr{C}$ is denoted by $\mathcal{L}(\mathscr{C})$.

Each ranking $\succ$ on $\mathscr{C}$ defines an order relation $\trianglerighteq_\succ$ on $\mathcal{L}(\mathscr{C})$ according to how far two rankings in $\mathcal{L}(\mathscr{C})$ are from $\succ$ in terms of reversals[1]. For any $\succ_i, \succ_j \in \mathcal{L}(\mathscr{C})$, the fact that $(\succ_i, \succ_j) \in \trianglerighteq_\succ$ is denoted by $\succ_i \trianglerighteq_\succ \succ_j$.

Definition 1. *Let $\mathscr{C}$ be a set of k candidates and $\succ$ be a ranking on $\mathscr{C}$. The order relation $\trianglerighteq_\succ$ on $\mathcal{L}(\mathscr{C})$ is defined as*

$$\trianglerighteq_\succ = \left\{ (\succ_i, \succ_j) \in \mathcal{L}(\mathscr{C})^2 \,\middle|\, \begin{array}{c} (\forall (a_{i_1}, a_{i_2}) \in \mathscr{C}^2) \\ ((a_{i_1} \succ a_{i_2} \wedge a_{i_1} \succ_j a_{i_2}) \Rightarrow a_{i_1} \succ_i a_{i_2}) \end{array} \right\}.$$

Figure 1 displays the Hasse diagram of the order relation $\trianglerighteq_\succ$ on $\mathcal{L}(\mathscr{C})$ for the set of candidates $\mathscr{C} = \{a, b, c\}$ and the ranking $a \succ b \succ c$.

In [2], we described a statistical test for testing the existence of a latent true ranking. This test requires the given profile of rankings to be (close to being) monotone, i.e., it requires the frequencies with which each ranking is expressed by the voters to be (close to being) decreasing on the Hasse diagram of the order relation $\trianglerighteq_{\succ_0}$, for some $\succ_0 \in \mathcal{L}(\mathscr{C})$. This 'closeness to being monotone' was used for determining whether the hypothesis of existence of a latent true ranking should or should not be rejected. For more details, we refer to [2].

---

[1] A reversal is a switch of consecutive elements in a ranking. The minimum number of reversals needed for changing a given ranking into another one is measured by the Kendall distance function [8].

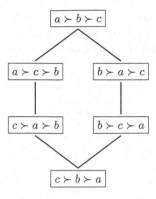

Fig. 1. Hasse diagram of the order relation $\trianglerighteq_{\succ}$ on $\mathcal{L}(\mathscr{C})$ for the ranking $a \succ b \succ c$.

3   Monotonicity of a Profile of Rankings with Ties

In real-life problems, voters might consider that two or more candidates are equally suitable, and, thus, express a tie between these candidates. In this case, every voter should be allowed to provide a weak order relation $\succsim_j$ on $\mathscr{C}$, i.e., a complete and transitive relation on $\mathscr{C}$ that might not be antisymmetric. Any weak order relation $\succsim$ can be written as the union of two relations $\succ$ and $\sim$, where $\succ$ (called a ranking with ties) represents the antisymmetric part of $\succsim$ and $\sim$ represents the symmetric part of $\succsim$. Note that $\sim$ is an equivalence relation, and, thus, partitions $\mathscr{C}$ into equivalence classes. The set of all rankings with ties on $\mathscr{C}$ is denoted by $\mathcal{L}^*(\mathscr{C})$. As a ranking is a particular case of a ranking with ties, it obviously holds that $\mathcal{L}(\mathscr{C}) \subseteq \mathcal{L}^*(\mathscr{C})$.

In case the voters express rankings with ties instead of rankings, the relation $\trianglerighteq_{\succ}$ needs to be extended to $\mathcal{L}^*(\mathscr{C})$. Note that, for a ranking with ties $\succ_i$, the conditions $a_{i_1} \succ_i a_{i_2}$ and $a_{i_2} \not\succ_i a_{i_1}$ are no longer equivalent. Therefore, the former unique condition for rankings (without ties) now needs to be divided in two parts.

Proposition 1. *Let $\mathscr{C}$ be a set of k candidates and $\succ$ be a ranking on $\mathscr{C}$. The relation $\trianglerighteq_{\succ}$ defined as*[2]

$$
\trianglerighteq_{\succ} = \left\{ (\succ_i, \succ_j) \in \mathcal{L}^*(\mathscr{C})^2 \;\middle|\; \begin{array}{c} \left(\forall (a_{i_1}, a_{i_2}) \in \mathscr{C}^2\right) \\ \left(\begin{array}{c} (a_{i_1} \succ a_{i_2} \wedge a_{i_1} \succ_j a_{i_2}) \Rightarrow (a_{i_1} \succ_i a_{i_2}) \\ \wedge \\ (a_{i_1} \succ a_{i_2} \wedge a_{i_2} \succ_i a_{i_1}) \Rightarrow (a_{i_2} \succ_j a_{i_1}) \end{array} \right) \end{array} \right\}
$$

is an order relation on $\mathcal{L}^*(\mathscr{C})$.

---

[2] For any ranking $\succ$ on $\mathscr{C}$, the restriction of the relation $\trianglerighteq_{\succ}$ on $\mathcal{L}^*(\mathscr{C})$ to $\mathcal{L}(\mathscr{C})$ coincides with the relation $\trianglerighteq_{\succ}$ on $\mathcal{L}(\mathscr{C})$. Therefore, the use of the same notation is justified.

Figure 2 displays the Hasse diagram of the order relation $\trianglerighteq_\succ$ on $\mathcal{L}^*(\mathscr{C})$ for the set of candidates $\mathscr{C} = \{a, b, c\}$ and the ranking $a \succ b \succ c$. Note that this Hasse diagram coincides with the one used by Kemeny [4] for defining a distance function on $\mathcal{L}^*(\mathscr{C})$.

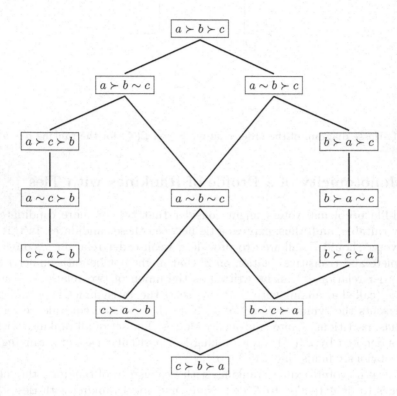

Fig. 2. Hasse diagram of the order relation $\trianglerighteq_\succ$ on $\mathcal{L}^*(\mathscr{C})$ for the ranking $a \succ b \succ c$.

The statistical test introduced in [2] might also be extended by requiring the frequencies with which each ranking with ties is expressed by the voters to be (close to being) decreasing on the Hasse diagram of the order relation $\trianglerighteq_{\succ_0}$, for some $\succ_0 \in \mathcal{L}(\mathscr{C})$.

Unfortunately, the inclination of voters towards expressing ties between candidates might prevent the frequencies from being decreasing on the Hasse diagram of the order relation $\trianglerighteq_{\succ_0}$ (for any $\succ_0 \in \mathcal{L}(\mathscr{C})$). For instance, the fact that the ranking with ties $a \sim b \sim c$ in Fig. 2 would appear with the highest frequency in a given profile of rankings with ties will make the frequencies with which each ranking with ties is expressed to be far from being decreasing on the Hasse diagram of the order relation $\trianglerighteq_{\succ_0}$, for any $\succ_0 \in \mathcal{L}(\mathscr{C})$. Nevertheless, this might not be due to the absence of a true latent ranking, but due to the indecision of some of the voters. In particular, the frequencies with which the ranking with ties $a \sim b \sim c$ is expressed might be higher than that of $a \succ b \succ c$, even though the latter is the true latent ranking.

4   Signatures and Ordered Signatures

In this section, we study the notions of signature and ordered signature that will help to deal with the aforementioned problem.

Definition 2. *Let $\mathscr{C}$ be a set of k candidates.*

(i) *The signature $\mathscr{S}$ of a ranking with ties $\succ$ on $\mathscr{C}$, denoted by $\mathscr{S}(\succ)$, is a vector where the i-th component equals the size of the i-th equivalence class in $\succ$.*

(ii) *The ordered signature $\mathscr{O}$ of a ranking with ties $\succ$ on $\mathscr{C}$, denoted by $\mathscr{O}(\succ)$, is a vector where the i-th component equals the size of the i-th largest equivalence class in $\succ$.*

The set of all the signatures on $\mathscr{C}$ is denoted by $\mathbb{S}(\mathscr{C})$ and the set of all the ordered signatures on $\mathscr{C}$ is denoted by $\mathbb{O}(\mathscr{C})$.

Remark 1. Each signature $\mathscr{S} \in \mathbb{S}(\mathscr{C})$ leads to a unique ordered signature $\mathscr{O} \in \mathbb{O}(\mathscr{C})$ by ordering the numbers in $\mathscr{S}$ in a decreasing manner. Note that the lengths of $\mathscr{S}$ and $\mathscr{O}$ coincide. The fact that a signature $\mathscr{S}$ leads to an ordered signature $\mathscr{O}$ is denoted by $\mathscr{S} \rightsquigarrow \mathscr{O}$.

Example 1. Consider the set of candidates $\mathscr{C} = \{a, b, c\}$. The signature of the ranking with ties $a \succ b \succ c$ is the vector $(1, 1, 1)$ and its ordered signature is $(1, 1, 1)$. Therefore, it holds that

$$\mathscr{S}(a \succ b \succ c) = \mathscr{O}(a \succ b \succ c) = (1, 1, 1).$$

Analogously, the signature of the ranking with ties $a \succ b \sim c$ is the vector $(1, 2)$ and its ordered signature is $(2, 1)$. Therefore, it holds that

$$\mathscr{S}(a \succ b \sim c) = (1, 2) \quad \text{and} \quad \mathscr{O}(a \succ b \sim c) = (2, 1).$$

In general, the set of all signatures on $\mathscr{C}$ is given by:

$$\mathbb{S}(\mathscr{C}) = \{(1, 1, 1), (1, 2), (2, 1), (3)\}.$$

Analogously, the set of all ordered signatures on $\mathscr{C}$ is given by:

$$\mathbb{O}(\mathscr{C}) = \{(1, 1, 1), (2, 1), (3)\}.$$

These (ordered) signatures can be used for defining two natural order relations on $\mathcal{L}^*(\mathscr{C})$. In the first order relation, only couples of rankings with ties belonging to $\trianglerighteq_{\succ}$ and that have the same signature are considered to be comparable, while, in the second order relation, only couples of rankings with ties belonging to $\trianglerighteq_{\succ}$ and that have the same ordered signature are considered to be comparable.

Proposition 2. *Let $\mathscr{C}$ be a set of k candidates and $\succ$ be a ranking on $\mathscr{C}$. The relation $\trianglerighteq_\succ^S$ defined as*

$$\trianglerighteq_\succ^S = \trianglerighteq_\succ \cap \{(\succ_i, \succ_j) \in \mathcal{L}^*(\mathscr{C})^2 \mid \mathscr{S}(\succ_i) = \mathscr{S}(\succ_j)\},$$

is an order relation on $\mathcal{L}^*(\mathscr{C})$.

Proposition 3. *Let $\mathscr{C}$ be a set of k candidates and $\succ$ be a ranking on $\mathscr{C}$. The relation $\trianglerighteq_\succ^O$ defined as*

$$\trianglerighteq_\succ^O = \trianglerighteq_\succ \cap \{(\succ_i, \succ_j) \in \mathcal{L}^*(\mathscr{C})^2 \mid \mathscr{O}(\succ_i) = \mathscr{O}(\succ_j)\},$$

is an order relation on $\mathcal{L}^*(\mathscr{C})$.

Figures 3 and 4 display the Hasse diagram of the order relations $\trianglerighteq_\succ^S$ and $\trianglerighteq_\succ^O$ on $\mathcal{L}^*(\mathscr{C})$ for the set of candidates $\mathscr{C} = \{a, b, c\}$ and the ranking $a \succ b \succ c$. Note that we use dashed lines for separating sets of incomparable rankings with ties.

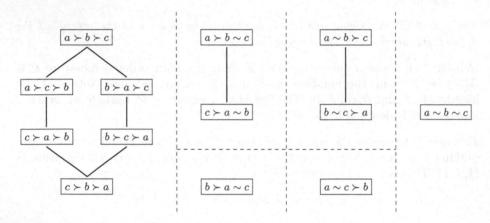

Fig. 3. Hasse diagram of the order relation $\trianglerighteq_\succ^S$ on $\mathcal{L}^*(\mathscr{C})$ for the ranking $a \succ b \succ c$.

Obviously, there exists an immediate connection between the three relations.

Proposition 4. *Let $\mathscr{C}$ be a set of k candidates and $\succ$ be a ranking on $\mathscr{C}$. The following statement holds:*

$$\trianglerighteq_\succ^S \subseteq \trianglerighteq_\succ^O \subseteq \trianglerighteq_\succ.$$

Consider the relation $\gg^S$ on $\mathbb{S}(\mathscr{C})$, where '$\mathscr{S}_1 \gg^S \mathscr{S}_2$' represents that the length of the signature $\mathscr{S}_1$ equals the length of the signature $\mathscr{S}_2$ plus one and, at the same time, the signature $\mathscr{S}_2$ can be obtained by merging two consecutive components of $\mathscr{S}_1$. For instance, the signature $(1, 2)$ is obtained by merging the last two components of the signature $(1, 1, 1)$, therefore $(1, 1, 1) \gg^S (1, 2)$. We consider its pre-order closure[3] for defining a natural order relation on $\mathbb{S}(\mathscr{C})$.

---

[3] The pre-order closure of a relation R is the smallest reflexive and transitive relation containing R [9].

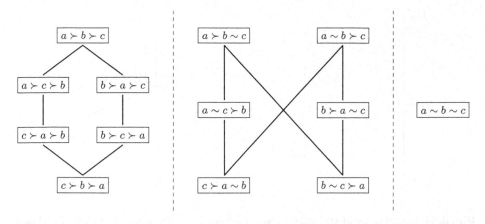

Fig. 4. Hasse diagram of the order relation $\unrhd^0_{\succ}$ on $\mathcal{L}^*(\mathscr{C})$ for the ranking $a \succ b \succ c$.

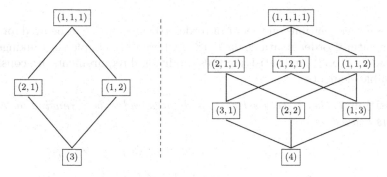

Fig. 5. Hasse diagram of the order relation $\geq^{\mathbb{S}}$ on $\mathbb{S}(\mathscr{C})$ for a set $\mathscr{C}$ of three (left) and of four (right) candidates.

Proposition 5. *Let $\mathscr{C}$ be a set of k candidates. The relation $\geq^{\mathbb{S}}$, defined as the pre-order closure of $\gg^{\mathbb{S}}$, is an order relation on $\mathbb{S}(\mathscr{C})$.*

Analogously, a natural order relation can be defined for ordered signatures. Consider the relation $\gg^{\mathbb{O}}$ on $\mathbb{O}(\mathscr{C})$, where '$\mathscr{O}_1 \gg^{\mathbb{O}} \mathscr{O}_2$' represents that there exist two signatures $\mathscr{S}_1, \mathscr{S}_2 \in \mathbb{S}(\mathscr{C})$ such that $\mathscr{S}_1 \rightsquigarrow \mathscr{O}_1$, $\mathscr{S}_2 \rightsquigarrow \mathscr{O}_2$ and $\mathscr{S}_1 \gg^{\mathbb{S}} \mathscr{S}_2$. For instance, for the ordered signatures $(2,1)$ and $(1,1,1)$, it holds that $(1,2) \rightsquigarrow (2,1)$, $(1,1,1) \rightsquigarrow (1,1,1)$ and $(1,1,1) \gg^{\mathbb{S}} (1,2)$. Therefore, it holds that $(1,1,1) \gg^{\mathbb{O}} (2,1)$. This relation is used for defining a natural order relation on $\mathbb{O}(\mathscr{C})$.

Proposition 6. *Let $\mathscr{C}$ be a set of k candidates. The relation $\geq^{\mathbb{O}}$, defined as the pre-order closure of $\gg^{\mathbb{O}}$, is an order relation on $\mathbb{O}(\mathscr{C})$.*

Figures 5 and 6 display the Hasse diagram of the order relations $\geq^{\mathbb{S}}$ and $\geq^{\mathbb{O}}$ on $\mathbb{S}(\mathscr{C})$ for a set $\mathscr{C}$ of three and of four candidates.

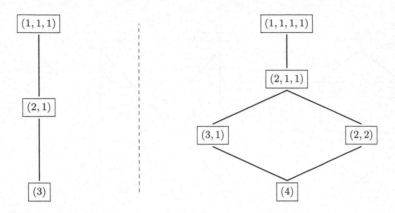

Fig. 6. Hasse diagram of the order relation $\geq^O$ on $\mathbb{O}(\mathscr{C})$ for a set $\mathscr{C}$ of three (left) and of four (right) candidates.

These order relations on the set of (ordered) signatures can be used for defining four natural order relations on $\mathcal{L}^*(\mathscr{C})$, where only couples of rankings with ties belonging to $\trianglerighteq_{\succ}$ and satisfying these additional requirements are considered comparable elements.

Proposition 7. *Let $\mathscr{C}$ be a set of k candidates and $\succ$ be a ranking on $\mathscr{C}$. The relations defined as*

$$\trianglerighteq_{\succ}^{S\downarrow} = \trianglerighteq_{\succ} \cap \{(\succ_i, \succ_j) \in \mathcal{L}^*(\mathscr{C})^2 \mid \mathscr{S}(\succ_i) \geq^S \mathscr{S}(\succ_j)\},$$

$$\trianglerighteq_{\succ}^{S\uparrow} = \trianglerighteq_{\succ} \cap \{(\succ_i, \succ_j) \in \mathcal{L}^*(\mathscr{C})^2 \mid \mathscr{S}(\succ_j) \geq^S \mathscr{S}(\succ_i)\},$$

$$\trianglerighteq_{\succ}^{O\downarrow} = \trianglerighteq_{\succ} \cap \{(\succ_i, \succ_j) \in \mathcal{L}^*(\mathscr{C})^2 \mid \mathscr{O}(\succ_i) \geq^O \mathscr{O}(\succ_j)\},$$

$$\trianglerighteq_{\succ}^{O\uparrow} = \trianglerighteq_{\succ} \cap \{(\succ_i, \succ_j) \in \mathcal{L}^*(\mathscr{C})^2 \mid \mathscr{O}(\succ_j) \geq^O \mathscr{O}(\succ_i)\},$$

are four order relations on $\mathcal{L}^*(\mathscr{C})$.

For instance, we illustrate in Fig. 7 the Hasse diagram of the order relation $\trianglerighteq_{\succ}^{O\downarrow}$ on $\mathcal{L}^*(\mathscr{C})$ for the set of candidates $\mathscr{C} = \{a, b, c\}$ and the ranking $a \succ b \succ c$. The fact that the frequencies of the rankings with ties are decreasing on the Hasse diagram of $\trianglerighteq_{\succ}^{O\downarrow}$ (and not on that of $\trianglerighteq_{\succ}$) would imply that voters do not express ties between candidates as often as they express strict preferences.

All different relations between the seven order relations are described in Fig. 8. One can see that $\trianglerighteq_{\succ}$ contains the other six order relations, whereas $\trianglerighteq_{\succ}^{S}$ is contained in the other six order relations.

The statistical test introduced in [2] might now be extended by requiring the frequencies with which each ranking is expressed to be (close to being) decreasing on the Hasse diagram of one of the seven different order relations. Obviously, since $\trianglerighteq_{\succ}^{S}$ is contained in the other six order relations, decreasingness w.r.t. $\trianglerighteq_{\succ}^{S}$ will be the closest to being satisfied, and, thus, the most unlikely of leading to a rejection of the hypothesis of existence of a latent true ranking. Different

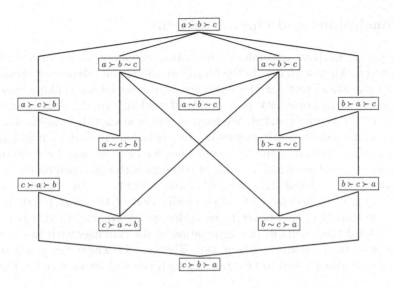

Fig. 7. Hasse diagram of the order relation $\geqq_{\succ}^{0\downarrow}$ on $\mathcal{L}^*(\mathscr{C})$ for the ranking $a \succ b \succ c$.

conclusions concerning the tendency of voters towards expressing ties can be drawn from the results of the test by selecting different order relations among the seven ones defined in this manuscript.

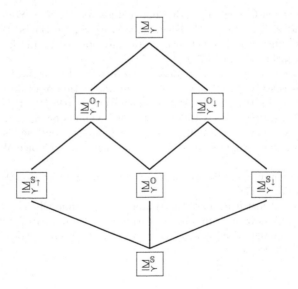

Fig. 8. Hasse diagram of the order relation $\subseteq$ on the set of all order relations on $\mathcal{L}^*(\mathscr{C})$ defined in this section.

5   Conclusions and Open Problems

In this paper, we have generalized the notion of monotonicity of a profile of rankings to rankings with ties. In the future, we will aim at adapting a previously-proposed statistical test for the existence of a true latent ranking based on the property of monotonicity of a profile of rankings to the setting in which rankings with ties are provided. We have conjectured that, in real-life data, the inclination of voters towards expressing ties between candidates might play a big role in the rejection of the test, and, thus, we have proposed six alternative definitions of monotonicity of a profile of rankings with ties based on the notions of signature and ordered signature. A thorough study on the influence of the chosen notion of monotonicity (and especially that of the least restrictive one $\geqq_\succ^S$) in the statistical test is yet to be addressed. Moreover, in this paper, we have restricted the results of the aggregation of the rankings with ties given by the voters to be a ranking without ties. The case in which the result of this aggregation is also allowed to be a ranking with ties will be analysed in the near future.

Acknowledgments. This research has been partially supported by the Research Foundation of Flanders (FWO17/PDO/160) and by the Ministry of Economy, Industry and Competitiveness (TIN2017-87600-P).

References

1. Arrow, K.J.: Social Choice and Individual Values. Wiley, New York (1963)
2. Pérez-Fernández, R., Alonso, P., Díaz, I., Montes, S., De Baets, B.: Monotonicity as a tool for differentiating between truth and optimality in the aggregation of rankings. J. Math. Psychol. **77**, 1–9 (2017)
3. Pérez-Fernández, R., Alonso, P., Díaz, I., Montes, S., De Baets, B.: Monotonicity-based consensus states for the monometric rationalisation of ranking rules and how they are affected by ties. Int. J. Approximate Reason. **91**, 131–151 (2017)
4. Kemeny, J.G.: Mathematics without numbers. Daedalus **88**(4), 577–591 (1959)
5. Schulze, M.: A new monotonic, clone-independent, reversal symmetric, and condorcet-consistent single-winner election method. Soc. Choice Welfare **36**, 267–303 (2011)
6. Borda, J.C.: Mémoire sur les Élections au Scrutin. Histoire de l'Académie Royale des Sciences, Paris (1781)
7. Black, D.: Partial justification of the Borda count. Public Choice **28**, 1–15 (1976)
8. Kendall, M.G.: A new measure of rank correlation. Biometrika **30**, 81–93 (1938)
9. Foldes, S.: Fundamental Structures of Algebra and Discrete Mathematics. Wiley, New York (1994)

Consistency Properties for Fuzzy Choice Functions: An Analysis with the Łukasiewicz t-norm

Susana Díaz[1]([✉]), José Carlos R. Alcantud[2], and Susana Montes[1]

[1] Department of Statistics and O. R., UNIMODE Research Unit,
University of Oviedo, Oviedo, Spain
{diazsusana,montes}@uniovi.es

[2] BORDA Research Unit and Multidisciplinary Institute of Enterprise (IME),
Campus Miguel de Unamuno, University of Salamanca, Salamanca, Spain
jcr@usal.es

Abstract. In continuation of the research in Alcantud and Díaz [1], we investigate the relationships between consistency axioms in the framework of fuzzy choice functions. In order to help disclose the role of a t-norm in such analyses, we start to study the situation that arises when we use other t-norms instead. We conclude that unless we impose further structure on the domain of application for the choices, the use of the Łukasiewicz t-norm as a replacement for the minimum t-norm does not guarantee a better performance.

Keywords: Choice function · Fuzzy choice function
Fuzzy arrow axiom · Consistency · Triangular norm

1 Introduction

The notion of fuzzy choice function has been thoroughly studied from many perspectives since their introduction in economics by Dasgupta and Deb [5]. Georgescu [7] is a monograph on the topic which refers to a successful notion that extends Banerjee's [4] previous concept by the fuzzification of the available domain. It has also been studied by Alcantud and Díaz [1], Georgescu [8] (rationality indicators of a fuzzy choice function), Martinetti et al. [9,10], Wang [11] (congruence conditions of fuzzy choice functions), Wang et al. [12], Wu and Zhao [13], et cetera.

By reference to the standard crisp analysis, the choice functions that can be derived from fuzzy relations by some reasonable procedure are called rational fuzzy choice functions. Alternatively, their consistency can be verified in terms of rationality axioms. And more generally, rationality indicators can be used as a proxy of their closeness to being rational.

It is therefore important to have a precise knowledge about various forms that rationality axioms can adopt. In particular, their relationships with different

© Springer International Publishing AG, part of Springer Nature 2018
J. Medina et al. (Eds.): IPMU 2018, CCIS 854, pp. 323–331, 2018.
https://doi.org/10.1007/978-3-319-91476-3_27

specifications of the underlying fuzzy concepts like implications, is a natural field for investigation.

Alcantud and Díaz [1] introduce some consistency axioms and then investigate some relationships between these and other known axioms. All these axiomatic conditions depend upon the choice of a t-norm that defines the logical implications. As a first analysis of this issue as well as a new research programme for sequential application of fuzzy choice functions, [1] concentrates on the case of the minimum t-norm. The structure of the feasible domain of choice situations is another important ingredient in the design and consequences of the consistency axioms. However the most prominent formulations are the subject of a complete analysis in [1].

The current state of matters does not reveal the role of the t-norm in the aforementioned analyses. We start to study the situation that arises when we use other t-norms instead. In this regard we perform a partial investigation of the relevant case of the Łukasiewicz t-norm. We conclude that unless we impose further structure on the domain of application for the choices, the application of the Łukasiewicz t-norm to define implications is of no avail.

This paper is organized as follows. Section 2 gives basic concepts and defines the rationality axioms that we need to analyze. Section 3 contains our (negative) results. Section 4 gives a preliminary research summary of a more specialized analysis where the structural restriction on the domain of choice is tightened. We conclude in Sect. 5.

2   Background and Definitions

A basic notion that we need in the fuzzy analysis of choices is the following (cf., Georgescu [7, Definition 5.13]):

Definition 1. *Let X be a non-empty set. Let $\mathcal{B}$ denote a non empty set of non-zero fuzzy subsets of X. A fuzzy choice function on $(X, \mathcal{B})$ is a function $C : \mathcal{B} \to \mathcal{F}(X)$ verifying the following two conditions:*

- *$C(S) \subseteq S$,*
- *$C(S)$ is non-zero (that is to say, $C(S)(x) \leq S(x)$ for all $x \in X$).*

for each $S \in \mathcal{B}$.

For the purpose of studying the salient characteristics of these ideas we usually impose properties on the structure of the domain of choices $\mathcal{B}$. As explained in [1], two basic options come to mind. The first one is Definition 2 below, proposed by Georgescu [7, Subsect. 5.2]. The second one is Definition 3 below, proposed by Martinetti et al. [9]. Since we wanted to consider the most general possible situation, in this contribution we work under Definition 3.

Definition 2. *We say that a fuzzy choice function C satisfies*

- *(H1) if all S and $C(S)$ are normal fuzzy sets, i.e. if $\exists x \in X$ such that $C(S)(x) = 1$.*

- *(H2) if $\mathcal{B}$ contains all the non-empty crisp finite subsets of X, i.e. if $\mathcal{B}$ contains $f[X] = \{[x_1, \ldots, x_n] : n \geq 1, x_1, \ldots, x_n \in X\}$.*

Definition 3. *We say that a fuzzy choice function C satisfies*

- *(WH1) if for all $S \in \mathcal{B}$, there exists an element $x \in X$ such that $S(x) > 0$ and $C(S)(x) = S(x)$.*
- *(WH2) if $\mathcal{B}$ contains $\{x\}, \{x, y\}$ and $\{x, y, z\}$ for all $x, y, z \in X$.*

Implications are used in many consistency properties and therefore, their definition in the fuzzy choice context are a recurrent issue. They are functions $I : [0, 1] \times [0, 1] \to [0, 1]$ that must extend the classical definition of implication: $I(0, 0) = I(0, 1) = I(1, 1) = 0$ and $I(1, 0) = 0$. There is an extensive literature on the study of other properties that implications can satisfy and the characterization of the families of operators obtained in each case (see for example, [2, 3]). Many of the definitions of fuzzy implication are based on triangular norms.

Triangular norms (t-norms for short) play a key role in fuzzy logic. They are used to model conjunction.
A t-norm is a binary operation $*$ on $[0, 1]$ that is commutative, associative, monotone and has 1 as neutral element. The most popular t-norm is the minimum. It is also the greatest t-norm. Another two important t-norms are the product, denoted $a *_P b$ and the Łukasiewicz operator, $a *_L b = \max(a + b - 1, 0)$.

Given a t-norm $*$, the implication $I : [0, 1] \times [0, 1] \to [0, 1]$ based on $*$, is defined as follows:

$$I(a, b) = \sup\{z \in [0, 1] \mid a * z \leq b\}, \text{ for any } a, b \in [0, 1].$$

And the equality of two values a and b in $[0, 1]$, denoted $E(a, b)$ is defined as follows $E(a, b) = \min(I(a, b), I(b, a))$.

Both definitions are extended to the fuzzy sets context using the minimum: given $S, T \subseteq X$,

$$I(S, T) = \min_{x \in X} I(S(x), T(x)) \qquad E(S, T) = \min(I(S, T), I(T, S)).$$

Observe that since we only consider finite sets, the previous minimums are well defined.

Triangular conorms (t-conorms) are the dual operators of the t-norms and they are used to define disjunction in the fuzzy context. A t-conorm is a binary operation on $[0, 1]$ that is commutative, associative, monotone and has 0 as neutral element. The most popular t-conorm is the maximum.

Let us now recap the definitions of some consistency properties of choice. We address the reader to Alcantud and Díaz [1] for motivation and interpretations.

Definition 4. *A fuzzy choice function C satisfies the Fuzzy Arrow Axiom **FAA**, if for any $S, T \in \mathcal{B}$ and $x \in X$ we have*

$$I(S, T) * S(x) * C(T)(x) \leq E(S \cap C(T), C(S)).$$

Definition 5. *A fuzzy choice function* C *satisfies the Fuzzy Chernoff condition* **FCH**, *if for any* $S, T \in \mathcal{B}$ *and* $x \in X$ *we have*

$$I(S,T) \leq I(S \cap C(T), C(S)).$$

Definition 6. *A fuzzy choice function* C *satisfies the Fuzzy Binariness property* **FB**, *if for any* $S \in \mathcal{B}$ *we have*

$$S(x) * \bigwedge_{y \in X} (S(y) \to_L C(\{x,y\})(x)) \leq C(S)(x), \text{ for all } x \in X.$$

Definition 7. *A fuzzy choice function* C *satisfies the Fuzzy Concordance property* **FC**, *if for any* $S_1, S_2, T \in \mathcal{B}$, *the following holds true:*

$$E(S_1 \cup S_2, T) \leq I(C(S_1) \cap C(S_2), C(T)).$$

Definition 8. *A fuzzy choice function* C *satisfies the Fuzzy Superset property* **FSUP**, *if for all* $S, T \in \mathcal{B}$ *the following holds true:*

$$I(S,T) * I(C(T), C(S)) \leq E(C(T), C(S)).$$

3 Connection Between Consistency Properties

In this contribution we extend the analysis in Alcantud and Díaz [1]. We assume the classical Zadeh's definitions of the union and intersection of fuzzy sets, *i.e.* for any S and T fuzzy sets defined on the universe X,

$$(S \cap T)(x) = \min(S(x), T(x)) \text{ for all } x \in X$$

$$(S \cup T)(x) = \max(S(x), T(x)) \text{ for all } x \in X.$$

Moreover, we define implication using the Lukasiewicz t-norm, *i.e.*

$$x \to_L y = \min(1, 1 - x + y).$$

Thus for any S and T fuzzy sets defined on the universe X, the degree of inclusion corresponding to the Lukasiewicz t-norm is

$$I(S,T) = \bigcap_{x \in X} S(x) \to_L T(x) = \min(\min_{x \in X}(1 - S(x) + T(x)), 1).$$

It follows from this expression that for two fuzzy sets S and T defined on the same universe X their degree of equality using the Lukasiewicz t-norm is

$$E(S,T) = \bigcap_{x \in X} (1 - |S(x) - T(x)|).$$

In [1] the following relationships among consistency properties were established, when both the minimum t-norm (in order to define implications) and conditions WH1 and WH2 hold true:

$$\text{FSUP}$$
$$\Uparrow \hspace{-0.5em}/$$
$$\text{FAA} \implies \begin{cases} \text{FCH} \\ \\ \text{FB} \end{cases}$$
$$\Downarrow \hspace{-0.5em}/$$
$$\text{FC}$$

Our aim is to study if these relationships still hold when fuzzy implication and equality are defined through the Łukasiewicz operator. Unfortunately we are led to conclude that not only the false implications remain false (cf., Counterexamples 3 and 4 below), but also the implications that hold true under the definitions with the minimum t-norm now become false too (cf., Counterexamples 1 and 2 below).

Under the assumptions of [1], the Fuzzy Arrow Axiom implies both the fuzzy Chernoff condition and the fuzzy Binariness property. However under the current assumptions where the Łukasiewicz t-norm replaces the minimum for the purpose of defining implications, these consequences no longer holds:

Fact 1. The Fuzzy Arrow Axiom does not imply the fuzzy Chernoff condition.

Counterexample 1. Let $X = \{x, y, z\}$ and $\mathcal{B} = \{X, T, S, \{x, y\}, \{x, z\}, \{y, z\}, \{x\}, \{y\}, \{z\}\}$, where T and S are defined in Table 1. Consider the fuzzy choice function C also defined in this table. Such choice function satisfies the Fuzzy Arrow Axiom. Conditions WH1 and WH2 are also satisfied, but C does not satisfy the fuzzy Chernoff condition because $I(S, T) = 0.7 \not< 0.6 = I(C(T) \cap S, C(S))$.

Table 1. Feasible subsets and choice function corresponding to Counterexample 1.

	S	T	$C(S)$	$C(T)$	$C([x, y])$	$C([x, z])$	$C([y, z])$	$C(X)$
x	0.9	0.6	0.9	0.6	1	1	0	1
y	0.9	0.9	0.5	0.9	0.5	0	1	0.5
z	0.5	0.6	0.5	0.5	0	0.5	0.5	0.5

Fact 2. The Fuzzy Arrow Axiom does not imply fuzzy Binariness condition.

Counterexample 2. Consider $X = \{x, y\}$ and $\mathcal{B} = \{X, S, \{x\}, \{y\}\}$, where S is defined in Table 2. This choice function satisfies the Fuzzy Arrow Axiom. Conditions WH1 and WH2 are also satisfied, but C does not satisfy the fuzzy Binariness condition:

$$S(x) *_L \bigwedge_{y \in X} (S(y) \to_L C(\{x, y\})(x)) = 0.9 *_L ((0.9 \to_L 1) \wedge (0.7 \to_L 0.7))$$
$$= 0.9 *_L 1 = 0.9 \not< 0.8 = C(S)(x).$$

Table 2. Feasible subsets and choice function corresponding to Counterexample 2.

	S	$C(S)$	$C(X)$
x	0.9	0.8	1
y	0.7	0.7	0.7

Fact 3. Condition **FAA** does not imply **FSUP** under hypothesis WH1 and WH2 when the Łukasiewicz t-norm is considered.

Counterexample 3. *In order to check that* **FAA** *does not imply* **FSUP** *it suffices to check the counterexample provided in [1] for the minimum t-norm. We reproduce it here for the sake of completeness (cf., Table 3). In that example the set X contains three elements, $X = \{x, y, z\}$. Using the Łukasiewicz t-norm, the set $\mathcal{B} = \{X, T, S, \{x, y\}, \{x, z\}, \{y, z\}, \{x\}, \{y\}, \{z\}\}$ satisfies the fuzzy Arrow Axiom but it does not satisfy the superset property: $I(S, T) = 1$, $I(C(T), C(S)) = 1$ and $E(C(S), C(T)) = 1 - |0.9 - 0.7| = 0.8$, then*

$$I(S, T) *_L I(C(T), C(S)) = 1 \nleq 0.8 = E(C(S), C(T)).$$

Table 3. Feasible subsets and choice function corresponding to Counterexample 3.

	T	S	$C(T)$	$C(S)$	$C([x, y])$	$C([x, z])$	$C([y, z])$	$C(X)$
x	0.9	0.9	0.7	0.9	1	0.7	0	0.7
y	0.7	0.6	0.6	0.6	0.6	0	0.6	0.6
z	0.7	0.7	0.7	0.7	0	1	1	1

In order to prove that the fuzzy Arrow Axiom does not imply the fuzzy Concordance property in the current setting, we can also benefit from the corresponding counterexample given in [1]:

Fact 4. **FAA** does not imply **FC** under conditions WH1 and WH2 when the Łukasiewicz t-norm is considered to define implication and equality.

Counterexample 4. *We reproduce the choice setting provided in [1] for the minimum t-norm in Table 4, for the sake of completeness. In that example the set X contains three elements, $X = \{x, y, z\}$.*
Using the Łukasiewicz t-norm,
$\mathcal{B} = \{X, T, S_1, S_2, \{x, y\}, \{x, z\}, \{y, z\}, \{x\}, \{y\}, \{z\}\}$ *satisfies the fuzzy Arrow Axiom but it does not satisfy the fuzzy Concordance property: $I(S, T) = 1$, $I(C(T), C(S)) = 1$ and $E(C(S), C(T)) = 1 - |0.9 - 0.7| = 0.8$, thus*

$$I(S, T) *_L I(C(T), C(S)) = 1 \nleq 0.8 = E(C(S), C(T)).$$

Table 4. Feasible subsets and choice function corresponding to Counterexample 4.

T	S_1	S_2	$C(X)$	$C(T)$	$C(S_1)$	$C(S_2)$	$C([x,y])$	$C([x,z])$	$C([y,z])$
x	0.9	0.9	0.6	0.5	0.5	0.9	0.6	0.6	0.5	0
y	0.9	0.6	0.9	0.5	0.5	0.6	0.9	1	0	0.5
z	0.5	0.5	0.5	1	0.5	0.5	0.5	0	1	1

4 Open Problems

When the minimum t-norm is used to define implication and equality between two fuzzy sets, some properties that are not true under conditions WH1 and WH2, do hold under the more stringent conditions H1 and H2. For example, when both the minimum t-norm (in order to define implications) and conditions H1 and H2 hold true, Alcantud and Díaz [1] establish the following relationships:

$$\textbf{FAA} \Rightarrow \begin{cases} \textbf{FCH} \\ \textbf{FSUP} \\ F\gamma \\ \textbf{FB} \end{cases}$$

We observe that we can prove more implications among consistency properties under conditions H1 and H2 than under the milder conditions WH1 and WH2.

In the light of the results obtained in the previous section, conditions WH1 and WH2 are clearly insufficient to provide relationships when we consider the implication defined by the Łukasiewicz t-norm. The question that naturally arises is if we can obtain positive results under conditions H1 and H2.

The first implication we have studied under H1 and H2 is whether Fuzzy Arrow Axiom guarantees Fuzzy Chernoff Condition in the context of the Łukasiewicz implicator. Unfortunately, the result we have obtained is negative.

Fact 5. FAA does not imply **FC** even under conditions H1 and H2 when the definitions of implication and equality are based on the Łukasiewicz t-norm.

Counterexample 5. *As a counterexample, we consider the set* $X = \{x, y, z\}$ *and the choice set* $\mathcal{B} = \{X, S, T, \{x, y\}, \{x, z\}, \{y, z\}\}$ *where S and T are defined in Table 5. Using the Łukasiewicz t-norm, one can check that the set $\mathcal{B}$ satisfies the fuzzy Arrow Axiom but it does not satisfy the fuzzy Chernoff condition:* $I(S, T) = 0.8$ *and* $I(S \cap C(T), C(S)) = 0.75$, *thus*

$$I(S, T) = 0.8 \nleq 0.75 = I(S \cap C(T), C(S)).$$

Table 5. Feasible subsets and choice function corresponding to Counterexample 5.

	T	S	$C(T)$	$C(S)$	$C([x,y])$	$C([x,z])$	$C([y,z])$	$C(X)$
x	1	0.7	1	0.7	1	1	0	1
y	0.8	1	0.75	1	0.75	0	1	0.75
z	0.75	0.75	0.75	0.5	0	0.75	0.5	0.75

5   Conclusions

Alcantud and Díaz [1] established a number of relationships among some consistency properties of fuzzy choice functions in a concrete setting. However the results in this paper lead us to conjecture that when we use the Łukasiewicz t-norms to define logical implications, those axioms are independent from each other. Therefore the hypothesis that we can advance is that the role of the t-norm is crucial in the analysis of consistency axioms.

In the future we intend to complete this inspection with a full investigation of the case in Sect. 4. By doing so we will check whether our conjecture is true. In order to fully reveal the role of the t-norm in the study of the rationality axioms of fuzzy choice functions, other cases should be looked into carefully. The ultimate goal is to answer the general question, which exact families of t-norms ensure the relationships among consistency properties in Sect. 3?

We shall also investigate the framework of the sequential application of fuzzy choice functions with the Łukasiewicz and other t-norms. The inspiration for this approach lies in the research programme posed by [1] and previous research in the crisp case like García-Sanz and Alcantud [6] and the references therein.

Acknowledgment. The authors are grateful to Humberto Bustince for suggesting this line of research, and to the four anonymous referees for their detailed coments. Susana Díaz and Susana Montes acknowledge financial support by the Spanish Ministerio de Economía y Competitividad under Project TIN2014-59543-P.

References

1. Alcantud, J.C.R., Díaz, S.: Rational fuzzy and sequential fuzzy choice. Fuzzy Sets Syst. **315**, 76–98 (2017)
2. Baczyński, M., Jayaram, B.: Fuzzy Implications. Studies in Fuzziness and Soft Computing, vol. 231. Springer, Heidelberg (2008). https://doi.org/10.1007/978-3-540-69082-5
3. Baczyński, M., Beliakov, G., Bustince, H., Pradera, A.: Advances in Fuzzy Implication Functions. Springer, Heidelberg (2013). https://doi.org/10.1007/978-3-642-35677-3
4. Banerjee, A.: Fuzzy choice functions, revealed preference and rationality. Fuzzy Sets Syst. **70**, 31–43 (1995)
5. Dasgupta, M., Deb, R.: Fuzzy choice functions. Soc. Choice Welfare **8**, 171–182 (1991)

6. García-Sanz, M.D., Alcantud, J.C.R.: Sequential rationalization of multivalued choice. Math. Soc. Sci. **74**, 29–33 (2015)
7. Georgescu, I.: Fuzzy Choice Functions: A Revealed Preference Approach, vol. 214. Springer, Heidelberg (2007). https://doi.org/10.1007/978-3-540-68998-0
8. Georgescu, I.: Acyclic rationality indicators of fuzzy choice functions. Fuzzy Sets Syst. **160**, 2673–2685 (2009)
9. Martinetti, D., Montes, S., Díaz, S., De Baets, B.: On Arrow-Sen style equivalences between rationality conditions for fuzzy choice functions. Fuzzy Optim. Decis. Making **13**, 369–396 (2014)
10. Martinetti, D., De Baets, B., Díaz, S., Montes, S.: On the role of acyclicity in the study of rationality of fuzzy choice functions. Fuzzy Sets Syst. **239**, 35–50 (2014)
11. Wang, X.: A note on congruence conditions of fuzzy choice functions. Fuzzy Sets Syst. **145**, 355–358 (2004)
12. Wang, X., Wu, C., Wu, X.: Choice functions in fuzzy environment: an overview. In: Cornelis, C., Deschrijver, G., Nachtegael, M., Schockaert, S., Shi, Y. (eds.) 35 Years of Fuzzy Set Theory. Studies in Fuzziness and Soft Computing, vol. 261, pp. 149–169. Springer, Heidelberg (2010). https://doi.org/10.1007/978-3-642-16629-7_8
13. Wu, X., Zhao, Y.: Research on bounded rationality of fuzzy choice functions. Sci. World J. **2014**, 928279 (2014)

Entropy and Monotonicity

Bernadette Bouchon-Meunier and Christophe Marsala[✉]

Sorbonne Université, CNRS, Laboratoire d'Informatique de Paris 6, LIP6,
75005 Paris, France
{Bernadette.Bouchon-Meunier,Christophe.Marsala}@lip6.fr

Abstract. Measuring the information provided by the observation of
events has been a challenge for seventy years, since the simultaneous
inception of entropy by Claude Shannon and Norbert Wiener in 1948.
Various definitions have been proposed, depending on the context, the
point of view and the chosen knowledge representation. We show here
that one of the most important common feature in the choice of an
entropy is its behavior with regard to the refinement of information and
we analyse various definitions of monotonicity.

Keywords: Entropy · Monotonicity · Refinement of information
Measure of fuzziness · Intuitionistic entropy measure

1 Introduction

Measuring the information provided by the observation of events has been a challenge for seventy years, since the simultaneous inception of entropy by Shannon [16] and Wiener [18] in 1948. Various definitions have been proposed since then, depending on the context, the point of view and the chosen knowledge representation. They have been called entropies or measures of information in the original probabilistic framework. Their extension to other frameworks such as fuzzy knowledge representation or its generalizations have given rise to the study of other aspects of information, for instance fuzziness or specificity. They have often been constructed by analogy with the probabilistic case, which may look artificial, but their properties go far beyond a simple analogy.

Continuing the analysis proposed in [5], we will show in the sequel that they have in common a property of monotonicity, based on an order supporting the differences of context, point of view or knowledge representation and expressing a refinement of the tool to perform observations.

Additivity and recursivity were among the main axioms underlying the definition of entropy by Shannon [16], and they imply an increase of the entropy resulting from the refinement of information acquired on an event through observations. Later on, Renyi [15] introduced the first of a long list of generalizations

---

A homage to Claude Shannon and Norbert Wiener for the 70th anniversary of their
inception of entropy.

J. Medina et al. (Eds.): IPMU 2018, CCIS 854, pp. 332–343, 2018.
https://doi.org/10.1007/978-3-319-91476-3_28

of Shannon entropy, still satisfying a property of additivity. It is worth mentioning Mugur-Schächter's work [14] on the general relativity of descriptions. She considers that any process of knowledge extraction is associated with epistemic operators called a delimitator and a view, representing the influence of the context and the observation tool on the considered event. A refinement of information results from a change in the observation tool. In his generalized information theory, Kampé de Fériet [10,11] takes into account observers and also requests a monotonicity of information with respect to an order on events.

In Sect. 2, we introduce entropy measures and their monotonicity according to three different forms associated with different visions of the refinement of information. In Sect. 3, we illustrate these visions on a sample of classic entropy measures, namely Shannon and weighted entropies, measures of fuzziness, similarity relation-based entropy measures and we develop the same visions on several entropies proposed in the framework of Atanassov intuitionistic fuzzy sets. We present conclusions and perspectives in Sect. 4.

Given the number of entropy measures introduced in the literature, we choose to study in more details a class of measures in the framework of Atanassov intuitionistic fuzzy sets. Future developments will review other classes.

2   Monotonicity of Entropy Measures

The evaluation of information is a complex problem, dealing with the form and the content of a piece of information. Information theory do not pretend to evaluate all aspects, and it provides an evaluation of the decrease of uncertainty after an observation of events by means of entropies.

Given the amount of data available in the numerical world, which is covering all aspects of modern life, evaluating information is a major issue. All tools enabling the user to compare two pieces of information, to evaluate the information available in a given environment, to make diagnosis or prediction on the basis of information provided by observations or data, to aggregate chunks of information, are useful. Unfortunately, there are many such tools and it is difficult to see their common features. This is why we propose to analyse measures of information and to revisit classic approaches of information evaluation in order to focus on monotonicity which we consider the most natural and relevant property requested from such a tool.

Let us consider a set of objects or events that represent the real world. In this paper, for the sake of simplicity, we only consider finite sets, but this work could be generalised to non-countable sets. We consider a σ-algebra B defined on a finite universe U. We use the notation proposed in the seminal paper by Aczél and Daróczy on the so-called inset entropy [2] to formalize the available information on the set of objects or events and taking into account the context, the point of view and the chosen knowledge representation.

For any integer n, we note: $X_n = \{(x_1, \ldots, x_n) \mid x_i \in B, \forall i = 1, \ldots, n\}$, $P_n = \{(p_1, \ldots, p_n) \mid p_i \in [0,1]\}$, p_i being associated with x_i through a function $p : B \to [0,1]$, a particular case being a probability distribution defined on

(U,B), $W_n = \{(w_{x_1}, \ldots, w_{x_n}) \mid w_{x_i} \in R^+, \forall i = 1, \ldots, n\}$, a family of n-tuples of weights[1] associated with n-tuples of B through a function $f : B \to R^{+n}$, such that $f(x_1, \ldots, x_n) = (w_{x_1}, \ldots, w_{x_n})$.

Similarly to the definition of inset entropy [2], we introduce an **entropy measure** as a sequence of mappings $E_n : X_n \times P_n \times W_n \to R^+$ satisfying several properties among a long list for instance available in [1] or in [12].

We claim that the most significant properties to characterize an entropy measure are relative to a form of monotonicity with respect to a refinement of information which can take various forms, depending on a chosen order. Such a monotonicity corresponds to the natural idea that the more details, precision, certainty we obtain from the observation of objects or events, or equivalently the more refined information we have on them, the bigger the amount of information we have on these objects or events.

To use a metaphor, we can consider that we are facing a picture of an object providing some amount of information on it. We can first improve the light on the object before taking another picture in order to decrease the fuzziness of the details, or take another picture with a higher resolution, which gives more information on the object according to an increase of the precision, both cases corresponding to a monotonicity described in Sect. 2.1. We can also select a part of the object and make several pictures of this part, in a form of weak recursivity described in Sect. 2.2. We can finally partition the object into different parts and, for each of them, make more pictures providing a bigger amount of information on the object, in a form of weak additivity presented in Sect. 2.3.

We present these three forms of monotonicity which can be adapted to the knowledge representation we choose, as highlighted in the next sections.

In the sequel, for the sake of simplicity, we choose the notation

$$
\begin{pmatrix}
x_1, & \ldots, & x_n \\
p_1, & \ldots, & p_n \\
w_{x_1}, & \ldots, & w_{x_n}
\end{pmatrix}
$$

rather than $((x_1, \ldots, x_n), (p_1, \ldots, p_n), (w_{x_1}, \ldots, w_{x_n}))$ to represent an element of $X_n \times P_n \times W_n$, according to the notation used in [2].

2.1   Partial Order

The first form of monotonicity, noted **O-monotonicity**, is based on some **partial order** $\prec$ on $\bigcup_n X_n \times P_n \times W_n$, describing a more precise, detailed or reliable observation of the objects or events and it can be written as follows:

$$
E_n \begin{pmatrix}
x_1, & \ldots, & x_n \\
p_1, & \ldots, & p_n \\
w_1, & \ldots, & w_n
\end{pmatrix}
\leq E_n \begin{pmatrix}
x_1', & \ldots, & x_n' \\
p_1', & \ldots, & p_n' \\
w_1', & \ldots, & w_n'
\end{pmatrix}
$$

---

[1] In the following, for the sake of simplicity, w_{x_i} will be denoted w_i when the meaning of i is clear.

if

$$\begin{pmatrix} x_1, \ldots, x_n \\ p_1, \ldots, p_n \\ w_1, \ldots, w_n \end{pmatrix} \prec \begin{pmatrix} x'_1, \ldots, x'_n \\ p'_1, \ldots, p'_n \\ w'_1, \ldots, w'_n \end{pmatrix}$$

Examples of monotonicity can be based on the following partial orders:

$$\begin{pmatrix} x_1, \ldots, x_n \\ p_1, \ldots, p_n \\ w_1, \ldots, w_n \end{pmatrix} \prec \begin{pmatrix} x_1, \ldots, x_n \\ p_1, \ldots, p_n \\ w'_1, \ldots, w'_n \end{pmatrix}$$

if and only if, $\forall i$ one of the following conditions $(O1)$, called **sharpness** in [7], or $(O2)$ is satisfied:

$(O1)$ $- w'_i \leq w_i \Leftrightarrow w'_i \geq \frac{1}{2}$;
          $- w'_i \geq w_i \Leftrightarrow w'_i < \frac{1}{2}$.
$(O2)$ $w'_i \geq w_i$.

Other examples will be studied later.

2.2   Weak Recursivity

The second form of monotonicity, noted **R-monotonicity**, is based on a decrease of the coarseness of a partition of the universe of discourse, corresponding to a property of **weak recursivity** defined as follows:

$$E_n \begin{pmatrix} x_1, & x_2, & \ldots, & x_n \\ p_1, & p_2, & \ldots, & p_n \\ w_{x_1}, & w_{x_2}, & \ldots, & w_{x_n} \end{pmatrix} \geq E_{n-1} \begin{pmatrix} x_1 \cup x_2, & x_3, & \ldots, & x_n \\ p_1 + p_2, & p_3, & \ldots, & p_n \\ w_{x_1 \cup x_2} & w_{x_3}, & \ldots, & w_{x_n} \end{pmatrix}$$

A particular case of weak recursivity is what we call the ψ-recursivity, defined as:

$$E_n \begin{pmatrix} x_1, & x_2, & \ldots, & x_n \\ p_1, & p_2, & \ldots, & p_n \\ w_{x_1}, & w_{x_2} & \ldots, & w_{x_n} \end{pmatrix} =$$

$$E_{n-1} \begin{pmatrix} x_1 \cup x_2, & x_3 & \ldots, & x_n \\ p_1 + p_2, & p_3 & \ldots, & p_n \\ w_{x_1 \cup x_2}, & w_{x_3} & \ldots, & w_{x_n} \end{pmatrix} + \psi \begin{pmatrix} x_1, & x_2 \\ p_1, & p_2 \\ w_{x_1}, & w_{x_2} \end{pmatrix} E_2 \begin{pmatrix} x_1, & x_2 \\ \frac{p_1}{p_1+p_2}, & \frac{p_2}{p_1+p_2} \\ w_{x_1}, & w_{x_2} \end{pmatrix}$$

for a function $\psi : X_2 \times P_2 \times W_2 \rightarrow R^+$.

The classic property of **recursivity** corresponds to:

$$\psi_0 \begin{pmatrix} x_1, & x_2 \\ p_1, & p_2 \\ w_{x_1}, & w_{x_2} \end{pmatrix} = p_1 + p_2,$$

where the weights are not taken into account.

2.3   Weak Additivity

The third form of monotonicity, noted **A-monotonicity**, is based on the consideration of a secondary finite universe U' and a σ-algebra B' on U' providing more details on the observed phenomenon or object, through additional observations.

Similarly to the situation on U, we consider for any integer m

- $X'_m = \{(x'_1, \ldots, x'_m) \mid x'_i \in B', \forall i\}$;
- $P'_m = \{(p'_1, \ldots, p'_m) \mid p'_i \in [0,1]\}$, p'_i being associated with x'_i through a function $p' : B' \to [0,1]$;
- $W'_m = \{(w'_1, \ldots, w'_m) \mid w'_i \in R^+; \forall i\}$, a family of m-tuples of weights associated with m-tuples of elements of B' through a function $f' : B' \to R^+$, such that $f'(x'_i) = w'_i$.

We further suppose that there exist two combination operators $\star$ and $\circ$ enabling us to equip the cartesian product of $U \times U'$ with similar distributions:

- $P_n \star P'_m = \{(p_1 \star p'_1, \ldots, p_i \star p'_j, \ldots) \mid p_i \star p'_j \in [0,1]\}$, $p_i \star p'_j$ being associated with (x_i, x'_j) for any i and j through a function $p \star p'$,
- $W_n \circ W'_m = \{(w_{1,1}, \ldots, w_{i,j}, \ldots) \mid w_{i,j} \in R^+, \; \forall i, j\}$, is defined through a function $f \circ f' : B \times B' \to R^+$, such that: $f \circ f'(x_i, x'_j) = w_{i,j}$ for all $i = 1, \ldots, n$ and $j = 1, \ldots, m$.

Such a refinement leads to a property of **weak additivity** stating the following:

$$
E_{n \times m}
\begin{pmatrix}
(x_1, x'_1), & \cdots, & (x_n, x'_m) \\
p_1 \star p'_1, & \cdots, & p_n \star p'_m \\
w_1 \circ w'_1, w_1 \circ w'_2, & \cdots, & w_n \circ w'_m
\end{pmatrix}
\geq
$$
$$
\max \left[E_n
\begin{pmatrix}
x_1, \ldots, x_n \\
p_1, \ldots, p_n \\
w_1, \ldots, w_n
\end{pmatrix},
E_m
\begin{pmatrix}
x_1, \ldots, x_m \\
p_1, \ldots, p_m \\
w'_1, \ldots, w'_m
\end{pmatrix}
\right]
$$

The classic **additivity** property stands in the case where U and U' are independent universes, p and p' being probability distributions on (U, B) and (U', B'), weights generally not being taken into account. It yields:

$$
E_{n \times m}
\begin{pmatrix}
(x_1, x'_1), (x_1, x'_2), \ldots, (x_i, x'_j), \ldots, (x_n, x'_m) \\
p_1 \star p'_1, p_1 \star p'_2, \ldots, p_i \star p'_j, \ldots, p_n \star p'_m \\
w_{x_1, x'_1}, w_{x_1, x'_2}, \ldots, w_{x_i, x'_j}, \ldots, w_{x_n, x'_m}
\end{pmatrix}
=
$$
$$
E_n
\begin{pmatrix}
x_1, \ldots, x_n \\
p_1, \ldots, p_n \\
w_{x_1}, \ldots, w_{x_n}
\end{pmatrix}
+ E_m
\begin{pmatrix}
x'_1, \ldots, x'_m \\
p'_1, \ldots, p'_m \\
w_{x'_1}, \ldots, w_{x'_m}
\end{pmatrix}
$$

3 Diverse Entropy Measures

3.1 Shannon and Weighted Entropies

It is well-known that the classic **Shannon entropy** defined as:

$$E_n^S(p) = -\sum_{i=1}^{n} p_i \log p_i$$

is additive and recursive and then R-monotonous and A-monotonous. Its generalization to the case where weights are associated with events to represent a cost or an importance is a **weighted entropy** defined as follows [8]:

$$E_n^w \begin{pmatrix} x_1, \ x_2, \ \ldots, \ x_n \\ p_1, \ p_2, \ \ldots, \ p_n \\ w_1, \ w_2, \ \ldots, \ w_n \end{pmatrix} = -\sum_i w_i\, p_i \log p_i.$$

It is O-monotonous with regard to the partial order (O2).

The weighted entropy is also recursive, and then R-monotonous when considering

$$w_{x_1 \cup x_2} = \frac{(p_1 w_1 + p_2 w_2)}{(p_1 + p_2)}.$$

3.2 Measure of Fuzziness

Shortly after the weighted entropy, another entropy measure was introduced by De Luca and Termini in a non-probabilistic framework [7] by analogy with the Shannon entropy. It is a measure of fuzziness, in the case where f is the membership function of a fuzzy set on U:

$$E_n^{DLT} \begin{pmatrix} x_1, \ x_2, \ \ldots, \ x_n \\ p_1, \ p_2, \ \ldots, \ p_n \\ w_1, \ w_2, \ \ldots, \ w_n \end{pmatrix} = -\sum_i w_i \log w_i - \sum_i (1 - w_i) \log(1 - w_i).$$

A major property of this quantity is its O-monotonicity with respect to the above mentioned partial order *(O1)* defining the sharpness.

It can further be observed that, in the case where the weights are possibility degrees, *ie.* $\max(w_1, \ldots, w_n) = 1$, this measure of fuzziness is also weakly recursive and then R-monotonous:

$$E_n^{DLT} \begin{pmatrix} x_1, \ x_2, \ \ldots, \ x_n \\ p_1, \ p_2, \ \ldots, \ p_n \\ w_1, \ w_2, \ \ldots, \ w_n \end{pmatrix} \geq E_{n-1}^{DLT} \begin{pmatrix} x_1 \cup x_2, & x_3, \ \ldots, \ x_n \\ p_1 + p_2, & p_3, \ \ldots, \ p_n \\ \max(w_1, w_2), & w_3, \ \ldots, \ w_n \end{pmatrix}$$

3.3 Entropy Measures Under Similarity Relations

We consider a similarity relation S on $U = \{x_1, \ldots, x_n\}$, reflexive, symmetric and min-transitive. Yager [19] defines the following entropy measure:

$$E_n \begin{pmatrix} x_1, \ x_2, \ \ldots, \ x_n \\ p_1, \ p_2, \ \ldots, \ p_n \\ \overline{S}_1, \ \overline{S}_2, \ \ldots, \ \overline{S}_n \end{pmatrix} = - \sum_{x_i \in U} p_i \log \overline{S}_i$$

with $\overline{S}_i = \sum\limits_{x_j \in U} p_j S(x_i, x_j)$ for all $i = 1, \ldots n$.

The similarity reflects a point of view on the n events, expliciting to which extent they are similar with regard to a given criterion. If we consider two different points of view, symbolised by two similarity relations S and S'. We can show that this entropy measure is O-monotonous with respect to the order (O3):

$$\begin{pmatrix} x_1, \ \ldots, \ x_n \\ p_1, \ \ldots, \ p_n \\ \overline{S}_1, \ \ldots, \ \overline{S}_n \end{pmatrix} \leq \begin{pmatrix} x_1, \ \ldots, \ x_n \\ p_1, \ \ldots, \ p_n \\ \overline{S}'_1, \ \ldots, \ \overline{S}'_n \end{pmatrix}$$

if and only if similarities S and S' satisfy:

(O3) $S \prec S' \Leftrightarrow S(x_i, x_j) \leq S'(x_i, x_j) \ \forall i, j$.

This entropy measure is also A-monotonous, if we define a joint similarity relation $S \times S'$ on the cartesian product $U \times U'$ as follows, for two similarity relations S defined on U and S' defined on U':

$$S \times S'((x_i, y_j), (x_k, y_l)) = \min \left(S(x_i, x_k), S'(y_j, y_l) \right)$$

for any x_i and x_k in U, any y_j and y_l in U'.

3.4 Intuitionistic Entropy Measures

In this section, we consider the setting of the Atanassov intuitionistic fuzzy sets (AIFS) where several entropy measures have been introduced [9,17]. First of all, some basics of AIFS are recalled.

Let X be a universe, an *Atanassov intuitionistic fuzzy set* (AIFS) A of X is defined [3] by:

$$A = \{(x, \mu_A(x), \nu_A(x)) | x \in X\}$$

with $\mu : X \to [0,1]$, $\nu : X \to [0,1]$ and $0 \leq \mu_A(x) + \nu_A(x) \leq 1$, $\forall x \in X$. Here, $\mu_A(x)$ and $\nu_A(x)$ represent respectively the membership degree and the non-membership degree of x in A. Given an intuitionistic fuzzy set A of X, the hesitancy lying on the membership of x to A is the *intuitionistic index of x to A* defined for all $x \in X$ as $\pi_A(x) = 1 - (\mu_A(x) + \nu_A(x))$.

The inclusion of AIFS is defined as: $A \subseteq B$ if and only if $\mu_A(x) \leq \mu_B(x)$ and $\nu_A(x) \geq \nu_B(x)$, $\forall x \in X$.

The union of two AIFS A and B is defined as the AIFS $A \cup B$ such that $\mu_{A \cup B}(x) = \max(\mu_A(x), \mu_B(x))$ and $\nu_{A \cup B}(x) = \min(\nu_A(x), \nu_B(x))$, $\forall x \in X$. The intersection of two AIFS A and B is defined as the AIFS $A \cap B$ such that $\mu_{A \cap B}(x) = \min(\mu_A(x), \mu_B(x))$ and $\nu_{A \cap B}(x) = \max(\nu_A(x), \nu_B(x))$, $\forall x \in X$.

A classical representation of an AIFS uses a 3D-representation as for instance, in [17]. It leads us to propose the representation of an AIFS as a complex number from $\mathbb{C}$: for all $x \in X$, $w_A(x) = \mu_A(x) + i\nu_A(x)$. Here, we have $\mu_A(x) = \mathrm{Re}(w_A(x))$, $\nu_A(x) = \mathrm{Im}(w_A(x))$, $\mu_A(x)$ is the real part of w_A and $\nu_A(x)$ is its imaginary part.

An AIFS is a point in the region under the line $y = 1 - x$ such that $x \in [0, 1]$ and $y \in [0, 1]$.

Definitions of Entropy Measures for AIFS. Several works in AIFS theory have proposed the definition for an *entropy of an intuitionistic fuzzy set A*.

These forms can be represented in the model introduced in Sect. 2 if we extend the definition of W_n to complex numbers:

$$W_n = \{(w_{x_1}, \ldots, w_{x_n}) \mid w_{x_i} \in \mathbb{C}, \forall i = 1, \ldots, n\}.$$

There exists various definitions of entropy measures in the AIFS settings [9]. In our setting, entropy measures of an Atanassov intuitionistic fuzzy set are summarised as:

$$E_n^{IFS} \begin{pmatrix} x_1, & \cdots, & x_n \\ p_1, & \cdots, & p_n \\ w_A(x_1) & \cdots, & w_A(x_n). \end{pmatrix}.$$

For instance, the entropy measure given in [17] is defined as:

$$E_n^{S} \begin{pmatrix} x_1, & \cdots, & x_n \\ p_1, & \cdots, & p_n \\ w_A(x_1) & \cdots, & w_A(x_n) \end{pmatrix} = 1 - \frac{1}{2n} \sum_{i=1}^{n} |\mu_A(x_i) - \nu_A(x_i)|$$

with $\mu_A(x) = \mathrm{Re}(w_A(x))$, and $\nu_A(x) = \mathrm{Im}(w_A(x))$.

In [9], the following entropy measure is also introduced:

$$E_n^{G} \begin{pmatrix} x_1, & \cdots, & x_n \\ p_1, & \cdots, & p_n \\ w_A(x_1) & \cdots, & w_A(x_n) \end{pmatrix} = \frac{1}{2n} \sum_{i=1}^{n} \left(1 - |\mu_A(x_i) - \nu_A(x_i)|\right)(1 + \pi_A(x_i))$$

with $\mu_A(x) = \mathrm{Re}(w_A(x))$, $\nu_A(x) = \mathrm{Im}(w_A(x))$ and $\pi_A(x) = 1 - \mathrm{Re}(w_A(x)) - \mathrm{Im}(w_A(x))$.

Another way to define an entropy measure is presented in [6] where the definition is based on extensions of the Hamming distance and the Euclidian distance to AIFS. For instance, the following entropy measure is proposed:

$$E_n^{B} \begin{pmatrix} x_1, & \cdots, & x_n \\ p_1, & \cdots, & p_n \\ w_A(x_1) & \cdots, & w_A(x_n) \end{pmatrix} = \sum_{i=1}^{n} \pi_A(x_i)$$

with $\pi_A(x) = 1 - \mathrm{Re}(w_A(x)) - \mathrm{Im}(w_A(x))$.

Entropy Measures for AIFS and Monotonicity. In [9], it is recalled that, in the AIFS setting, a monotonicity property for an entropy measure could be ensured by definition. The authors present several definitions that lie on the definition of a partial order on W_n and the concept of *less fuzzy than*. The following three definitions of partial order could be used.

$(O4)$ A is *less fuzzy* than B if
$\mu_A(x) \leq \mu_B(x)$ and $\nu_A(x) \geq \nu_B(x)$ for $\mu_B(x) \leq \nu_B(x), \forall x \in X$,
or $\mu_A(x) \geq \mu_B(x)$ and $\nu_A(x) \leq \nu_B(x)$ for $\mu_B(x) \geq \nu_B(x), \forall x \in X$.

$(O5)$ A is *less fuzzy* than B if
$A \subseteq B$ for $\mu_B(x) \leq \nu_B(x), \forall x \in X$,
or $B \subseteq A$ for $\mu_B(x) \geq \nu_B(x), \forall x \in X$;

and

$(O6)$ A is *less fuzzy* than B if
$\mu_A(x) \leq \mu_B(x)$ and $\nu_A(x) \leq \nu_B(x), \forall x \in X$,

O-monotonicity. It is easy to see that $(O4)$, $(O5)$, and $(O6)$ enable the definition of O-monotonicities. In [6], it is stated that these definitions of monotonicity produces particular forms of E^{IFS}:

- E_n^S satisfies the monotonicity based on $(O4)$;
- E_n^G satisfies the monotonicity based on $(O5)$;
- E_n^B satisfies the monotonicity based on $(O6)$.

R-monotonicity. If we consider E_n^S, this measure satisfies the R-monotonicity if we have

$$E_n^S \begin{pmatrix} x_1, & \cdots, & x_n \\ p_1, & \cdots, & p_n \\ w_A(x_1) \ldots, w_A(x_n) \end{pmatrix} \geq E_{n-1}^S \begin{pmatrix} x_1 \cup x_2, & x_3, & \cdots, & x_n \\ p_1 + p_2, & p_3, & \cdots, & p_n \\ w_A(x_1) \cup w_A(x_2), w_A(x_3), & \ldots, & w_A(x_n) \end{pmatrix}$$

with $w_A(x_1) \cup w_A(x_2)$ the union of the two AIFS $w_A(x_1)$ and $w_A(x_2)$.

Hereafter, to simplify the notations, we denote these two measures E_n^S and E_{n-1}^S respectively.

It can be shown that we have:

$$E_n^S - E_{n-1}^S = \frac{1}{2(n-1)} \left(\frac{1}{n} \sum_{i=1}^{n} |\mu_A(x_i) - \nu_A(x_i)| \right.$$
$$+ |\max(\mu_A(x_1), \mu_A(x_2)) - \min(\nu_A(x_1), \nu_A(x_2))|$$
$$\left. - |\mu_A(x_1) - \nu_A(x_1)| - |\mu_A(x_2) - \nu_A(x_2)| \right).$$

and thus E_n^S satisfies the R-monotonicity if

$$|\max(\mu_A(x_1), \mu_A(x_2)) - \min(\nu_A(x_1), \nu_A(x_2))| \geq$$
$$|\mu_A(x_1) - \nu_A(x_1)| + |\mu_A(x_2) - \nu_A(x_2)|$$

This inequality is not satisfied for any AIFS. For instance, the inequality is not satisfied with $w_A(x_1) = 0.1 + 0.9i$ and $w_A(x_2) = 0.2 + 0.7i$.

It is different if we consider E_n^B. This measure satisfies the R-monotonicity if

$$E_n^B \begin{pmatrix} x_1, & \ldots, & x_n \\ p_1, & \ldots, & p_n \\ w_A(x_1), & \ldots, & w_A(x_n) \end{pmatrix} \geq E_{n-1}^B \begin{pmatrix} x_1 \cup x_2, & x_3, & \ldots, & x_n \\ p_1 + p_2, & p_3, & \ldots, & p_n \\ w_A(x_1) \cup w_A(x_2), & w_A(x_3), & \ldots, & w_A(x_n) \end{pmatrix}.$$

Hereafter, for the sake of simplicity, we denote these two measures E_n^B and E_{n-1}^B respectively.

We have

$$\begin{aligned}
E_n^B - E_{n-1}^B &= 1 - \mu_A(x_1) - \nu_A(x_1) + 1 - \mu_A(x_2) - \nu_A(x_2) \\
&\quad -1 + \max(\mu_A(x_1), \mu_A(x_2) + \min(\nu_A(x_1), \nu_A(x_2)) \\
&= 1 + (\max(\mu_A(x_1), \mu_A(x_2) - \mu_A(x_1) - \mu_A(x_2)) \\
&\quad (\min(\nu_A(x_1), \nu_A(x_2)) - \nu_A(x_1) - \nu_A(x_2))
\end{aligned}$$

and thus

$$\begin{aligned}
E_n^B - E_{n-1}^B &= 1 - \min(\mu_A(x_1), \mu_A(x_2) - \max(\nu_A(x_1), \nu_A(x_2)) \\
&= \pi_A(x_1 \cap x_2).
\end{aligned}$$

As a consequence, we have $E_n^B - E_{n-1}^B \geq 0$ and E_n^B satisfies the R-monotonicity.

4 Conclusion

Entropy and measures of information have been extensively studied for 70 years. Extensions to fuzzy sets, intuitionistic fuzzy sets and other representation models of uncertainty and imprecision have been proposed in many papers. These extensions are often only based on a formal analogy between the introduced quantities and classic entropies, in spite of the fact that their purpose is different, entropies measuring the decrease of uncertainty resulting from the occurrence of an event, while fuzzy set related measures evaluate the imprecision of events.

General approaches have been proposed to compare and organize all these quantities, pointing out diverse so-called fundamental properties for instance in [1,12] and showing that a particular quantity satisfies or does not satisty them. Attempts have also been done to exhibit classes of quantities with a similar behavior with regard to sets of properties [4].

In this paper, we highlight the common property of monotonicity of entropy measures with regard to a refinement of information, showing that the main differences between these quantities come from the diversity of orders defining such a refinement. This paper is not intended to provide a review of all entropy measures existing in the literature, but to clarify the concept of refinement of information and the underlying monotonicity, and to illustrate this paradigm by classic examples in a sample of knowledge representation environments, namely the classic probabilistic one, the fuzzy one, the similarity-based one and the intuitionistic fuzzy framework.

In the future, we propose to extend this study to so-called relative entropy measures enabling to compare two sets of observations, for instance divergences or relative entropies as the most popular [13].

We will also analyse the use of such measures to evaluate the quality of information obtained from numerical media, selecting the most appropriate knowledge representation and the relevant definition of refinement of information for a given context.

References

1. Aczél, J., Daróczy, Z.: On Measures of Information and their Characterizations, Mathematics in Science and Engineering, vol. 115. Academic Press, New York (1975)
2. Aczél, J., Daróczy, Z.: A mixed theory of information. I: symmetric, recursive and measurable entropies of randomized systems of events. R.A.I.R.O. Informatique théorique/Theoret. Comput. Sci. **12**(2), 149–155 (1978)
3. Atanassov, K.T.: Intuitionistic fuzzy sets. Fuzzy Sets Syst. **20**, 87–96 (1986)
4. Bouchon, B.: Entropic models. Cybern. Syst.: Int. J. **18**(1), 1–13 (1987)
5. Bouchon-Meunier, B., Marsala, C.: Entropy measures and views of information. In: Kacprzyk, J., Filev, D., Beliakov, G. (eds.) Granular, Soft and Fuzzy Approaches for Intelligent Systems. SFSC, vol. 344, pp. 47–63. Springer, Cham (2017). https://doi.org/10.1007/978-3-319-40314-4_3
6. Burillo, P., Bustince, H.: Entropy on intuitionistic fuzzy sets and on interval-valued fuzzy sets. Fuzzy Sets Syst. **78**, 305–316 (1996)
7. de Luca, A., Termini, S.: A definition of a nonprobabilistic entropy in the setting of fuzzy sets theory. Inf. Control **20**, 301–312 (1972)
8. Guiaşu, S.: Weighted entropy. Report Math. Phys. **2**(3), 165–179 (1971)
9. Guo, K., Song, Q.: On the entropy for Atanassov's intuitionistic fuzzy sets: an interpretation from the perspective of amount of knowledge. Appl. Soft Comput. **24**, 328–340 (2014)
10. Kampé de Fériet, J.: Mesures de l'information par un ensemble d'observateurs. In: Gauthier-Villars (ed.) Comptes Rendus des Scéances de l'Académie des Sciences, série A, Paris, vol. 269, pp. 1081–1085, Décembre 1969
11. Kampé de Fériet, J.: Mesure de l'information fournie par un événement (1971). séminaire sur les questionnaires
12. Klir, G., Wierman, M.J.: Uncertainty-Based Information. Elements of Generalized Information Theory. Studies in Fuzziness and Soft Computing. Springer, Heidelberg (1998). https://doi.org/10.1007/978-3-7908-1869-7
13. Montes, I., Montes, S., Pal, N.: On the use of divergences for defining entropies for atanassov intuitionistic fuzzy sets. In: Kacprzyk, J., Szmidt, E., Zadrożny, S., Atanassov, K.T., Krawczak, M. (eds.) IWIFSGN/EUSFLAT -2017. AISC, vol. 642, pp. 554–565. Springer, Cham (2018). https://doi.org/10.1007/978-3-319-66824-6_49
14. Mugur-Schächter, M.: The general relativity of descriptions. Analyse de Systèmes **11**(4), 40–82 (1985)
15. Rényi, A.: On measures of entropy and information. In: Proceedings of 4th Berkeley Symposium on Mathematical Statistics and Probability, vol. 1, pp. 547–561 (1960)

16. Shannon, C.E.: The Mathematical Theory of Communication. University of Illinois Press, Urbana (1948). Ed. by C.E. Shannon, W. Weaver
17. Szmidt, E., Kacprzyk, J.: New measures of entropy for intuitionistic fuzzy sets. In: Proceedings of the Ninth International Conference on Intuitionistic Fuzzy Sets (NIFS), Sofia, Bulgaria, vol. 11, pp. 12–20, May 2005
18. Wiener, N.: Cybernetics, Or Control and Communication in the Animal and the Machine. Hermann & Cie, Cambridge. (MIT Press), Paris, 2nd revised edn. 1961 (1948)
19. Yager, R.R.: Entropy measures under similarity relations. Int. J. Gen. Syst. **20**(4), 341–358 (1992)

On the Problem of Comparing Ordered Ordinary Fuzzy Multisets

Ángel Riesgo[1], Pedro Alonso[1], Irene Díaz[1], Vladimír Janiš[2], Vladimír Kobza[2], and Susana Montes[1(✉)]

[1] UNIMODE Research Unit, University of Oviedo, Oviedo, Spain
ariesgo@yahoo.com, {palonso,sirene,montes}@uniovi.es
[2] Matej Bel University, Banská Bystrica, Slovakia
{vladimir.janis,vladimir.kobza}@umb.sk

Abstract. In this work we deal with a particular type of hesitant fuzzy set, in the case where membership values can appear multiple times and are ordered. They are called ordered ordinary fuzzy multisets. Some operations between them are introduced by means of an extension principle. In particular, the divergence measures between two of these multisets are defined and we have studied in detail the local family of divergences. Finally, these measures are related to the ones given for ordinary fuzzy sets.

Keywords: Divergence measure · Hesitant set
Ordered ordinary fuzzy multiset · Aggregation function

1 Introduction

The comparison of two sets is a very important tool in many areas. This can be done using different approaches. In some cases we want to measure the equality degree and in others the difference degree. A study of measures of comparison was given by Bouchon-Meunier et al. [2]. Since then more measures for comparing fuzzy sets have been introduced (see [1,10,17,18,21,22]). Recently, a review of the measures based on the differences was proposed by Couso et al. [4]. The more usual measures of comparison are dissimilarities [11]. But distances are also considered in some papers. In both cases, the main difference is the axiom related to the proximity; that is, some kind of triangular inequality axiom. A good alternative, in some cases where they are not appropriate, are the divergence measures. They were introduced in [14] as a way to compare two fuzzy sets. A deep study about these measures in the most general case can be found in [9]. Because of their utility, they were also studied as a tool to compare two intuitionistic fuzzy sets [12,13].

Another way to deal with imprecision is by means of hesitant fuzzy sets, first introduced by Torra [16], which expand on an idea first advanced by Grattan-Guinness more than two decades earlier [6], where an element can have multiple

J. Medina et al. (Eds.): IPMU 2018, CCIS 854, pp. 344–355, 2018.
https://doi.org/10.1007/978-3-319-91476-3_29

membership values. In a common intuitive interpretation of the hesitant fuzzy sets, each membership value is regarded as an independent verdict on membership that one particular "expert" or "decision maker" (DM) has produced. A realistic scenario occurs when the number of experts is fixed and it may be possible to link the membership values to the expert that has produced it. But in such a situation, the hesitant fuzzy sets cannot be a good model, as it should be possible for a membership value to appear more than once. This need for repeated membership values was already acknowledged by Torra in his original description of hesitant fuzzy sets and is referred to as the **hesitant fuzzy multisets** (HFM's). On the other hand, if we want to treat the experts as being distinguishable, which seems sensible for comparison purposes, then such n-tuples should be ordered. In this framework, the **ordered ordinary fuzzy multisets** appear from the hesitant fuzzy sets. Here, we have tried to introduce the divergence measures between them as a way to compare two sets. The definition and study of this concept is the main goal of this work.

This paper is organized as follows. Section 2 gives the basic concepts. Section 3 studies the ordered ordinary fuzzy multisets. Section 4 is devoted to divergence measures between two of these multisets and the case of local divergence measures is characterized. We conclude in Sect. 5.

2   Basic Concepts

In the definitions that follow, we assume that there is always a finite axiomatic reference set or **universe** (also called the **universe of discourse**), which we will denote by X, and that fuzzy sets are defined by associating each element in the universe with a membership function whose precise definition depends on the particular flavor of fuzzy sets we are dealing with.

Definition 1 *([20]). Let X be the universe. An **ordinary fuzzy set** is a function $\mu\colon X \to [0, 1]$.*

Mainly for notational convenience, it will also be useful to give a name to the family of all the ordinary fuzzy sets over a universe. Thus, given a universe X, the family of all the ordinary fuzzy sets over X is called the **ordinary fuzzy power set** over X and it is denoted by $\mathscr{F}(X)$.

In this set we can consider the two classical operations of union and intersection as follows:

- The union of A and B is the ordinary fuzzy set on X with the following membership function: $\mu_{A \cup B}(x) = s(\mu_A(x), \mu_B(x))$, for any $x \in X$, where s is any t-conorm (see [7])
- The intersection of A and B is the ordinary fuzzy set on X with the following membership function: $\mu_{A \cap B}(x) = t(\mu_A(x), \mu_B(x))$, for any $x \in X$, where t is any t-norm (see [7])

for any $A, B \in \mathscr{F}(X)$. A general study about fuzzy sets can be found in [5,8].

In many applications of fuzzy sets, the need arises to measure how similar or different two fuzzy sets are. Ideally, such measures must take into account the level of fuzziness and there are now many proposals in the literature that tackle this problem. In his recent book about hesitant fuzzy sets, Xu mentions many of the existing approaches [19]. In this work, we shall focus on the **divergence measure** originally proposed by Montes et al. [14] for the ordinary fuzzy sets, which was later extended to the intuitionistic case [12,13]. We will start by recalling the definition of divergence for the ordinary fuzzy sets.

Definition 2. *Let X be the universe. A map $D\colon \mathscr{F}(X) \times \mathscr{F}(X) \to \mathbb{R}$ is a* **divergence measure** *if for all $A, B \in \mathscr{F}(X)$ the following three conditions are met:*

1. $D(A, B) = D(B, A)$
2. $D(A, A) = 0$
3. $\max\{D(A \cup C, B \cup C), D(A \cap C, B \cap C)\} \le D(A, B), \ \forall C \in \mathscr{F}(X)$

The first two conditions are completely straightforward. A measure of the difference between two sets should obviously be symmetric and should be zero when the two sets being compared are the same. The third one is a bit less intuitive; basically, the more similar two sets become, the smaller the measure of their difference should be and this can be formalized through the use of a third non-empty set that dilutes the difference between the original sets. Note that we also expect such a measure to be non-negative, but this can be deduced from the third and second conditions (by making C the empty set), and consequently it can be left out of the definition. It should also be noted that the third condition in the previous definition is actually a compact form of what can be (and often is) expressed as two separate conditions: the divergence between A and B being an upper bound for the divergence of both the unions and the intersections with any third set C. It is also important to note that the third condition depends on how the intersection and union operations are defined, and these operations can be based on a pair of a t-norm and a t-conorm different from the standard ones, so the definition above is actually a family of definitions in the more general case (see [9]).

Among all the possible divergence measures that adhere to the definition above, it is often useful to single out those that exhibit the additional property that a change in the membership values for one element in the universe results in a change in the measure value that will only depend on the membership values for that element. Divergence measures that behave in this way are referred to as being "local" and the formal definition is as follows:

Definition 3. *Let X be the universe. A divergence measure $D\colon \mathscr{F}(X) \times \mathscr{F}(X) \to \mathbb{R}$ is a* **local divergence measure** *if $\forall A, B \in \mathscr{F}(X)$ and $\forall x \in X$ there is a function $h_x\colon [0, 1] \times [0, 1] \to \mathbb{R}$ such that:*

$$D(A, B) - D(A \cup \{x\}, B \cup \{x\}) = h_x(\mu_A(x), \mu_B(x))$$

where $\{x\}$ is the ordinary fuzzy set such that $\mu_{\{x\}}(x) = 1$ and $\mu_{\{x\}}(x') = 0$ for any other $x' \ne x$.

In general, the function h_x can be different for each element x in the universe X [9], although the assumption that there is one common function h, independent of x, is usually made in most of the existing literature [14]. In that case, we refer to such a function as the **characteristic function** of the local divergence measure. Intuitively, h_x uses the two membership values to measure the contribution to the divergence of a particular element of the universe.

A powerful characterization of the local divergence measures is given by the following theorem:

Theorem 1 (Representation Theorem). *Let X be the universe. A divergence measure $D\colon \mathscr{F}(X) \times \mathscr{F}(X) \to \mathbb{R}$ is a **local divergence measure** if and only if $\forall A, B \in \mathscr{F}(X)$ and $\forall x \in X$ there is a function $h_x\colon [0,1] \times [0,1] \to \mathbb{R}$ such that:*

$$D(A,B) = \sum_{x \in X} h_x(\mu_A(x), \mu_B(x))$$

And h_x fulfills the conditions:

(i) $h_x(u,v) = h_x(v,u)$, $\forall u, v \in [0,1]$;
(ii) $h_x(u,u) = 0$, $\forall u \in [0,1]$;
(iii) $h_x(u,w) \geq h_x(u,v)$, $\forall u, v, w \in [0,1]$ with $u \leq v \leq w$;
(iv) $h_x(u,w) \geq h_x(v,w)$, $\forall u, v, w \in [0,1]$ with $u \leq v \leq w$.

After stating these definitions, the problem that we need to tackle is how to build specific examples of divergence measures. A first idea consists in generalizing similar concepts of distance and dissimilarity from classical set theory. For example, the **Hamming distance** between two crisp sets A and B over a finite universe X can be defined as the count of the different members between A and B (basically, the cardinality of their union minus that of their intersection). This can be generalized to two ordinary fuzzy sets as $D_{\text{Hamming}}(A, B) = \sum_{x_i \in X} |\mu_A(x_i) - \mu_B(x_i)|$. It is obvious that the Hamming distance thus defined is a local divergence for ordinary fuzzy sets with the characteristic function $h_{Hamming}(x,y) = |x-y|$. However, we can find in [9] examples of divergence measures which are neither dissimilarities nor distances.

An extension of the fuzzy sets was proposed by Torra in 2010 under the name **hesitant fuzzy set** theory [16]. The hesitant theoretical framework broadens the range of the membership functions to encompass any subset of the $[0,1]$ interval. This leads to the following formal definition:

Definition 4. *Let X be the universe. A **hesitant fuzzy set** A on X is defined by a membership function $\mu_A\colon X \to \mathcal{P}([0,1])$, where $\mathcal{P}([0,1])$ is the family of all the subsets of the real closed interval $[0,1]$.*

In the next section we will explain how they are useful and what sort of real-life problems they attempt to model. Before that, we only have to introduce the idea of aggregation function. It can be proved [8] that the min and max functions are the only t-norm and t-conorm that fulfill the more restrictive property of idempotency $(t(a,a) = a)$. But if we do not impose associativity, we can define a more general form of operation on the real unit interval $[0,1]$ called an **aggregation operation**. The aggregation operations on $[0,1]$ can be further extended to any real interval $[a,b]$ [3] and we will refer to them as **aggregation functions** in the general case.

Definition 5. *Given a closed real interval* $\mathbb{I} = [a,b]$ *$(a,b \in \mathbb{R}, a < b)$, a function* $f\colon \mathbb{I}^n \to \mathbb{I}$ *is an* **aggregation function** *if it satisfies the following three conditions [3]:*

1. $f(a,\ldots,a) = a$ *(lower boundary condition)*
2. $f(b,\ldots,b) = b$ *(upper boundary condition)*
3. $x_i \le y_i (i = 1,\ldots,n) \implies f(x_1,\ldots,x_n) \le f(y_1,\ldots,y_n)$ *(monotonicity)*

Aggregation functions such as the arithmetic mean can usually be extended to the whole real line, in the form of $\mathbb{R}^n \to \mathbb{R}$ functions, so we may also use the term "aggregation function" for such extensions, under the assumption that they fulfill the boundary conditions when their domain is restricted to a closed real interval.

3   Ordered Ordinary Fuzzy Multisets

In the case of the hesitant fuzzy sets, we have to be particularly careful about their interpretation. In a typical hesitant element like $\{0.2, 0.3\}$, we might be misled into thinking of the hesitancy as a lack of precision about the actual value, but that sort of situation would be better modeled through an interval-valued fuzzy set with a membership of $[0.2, 0.3]$ or the equivalent intuitionistic fuzzy set with membership $(0.2, 0.7)$. The important thing in a typical hesitant fuzzy set is that the different membership values do not represent an interval, but distinct possible values, even widely differing ones [15]. In a common intuitive interpretation of the hesitant fuzzy sets, each membership value is regarded as an independent verdict on membership that one particular "expert" or "decision maker" (DM) has produced. In this view, a hesitant membership value of $\{0.2, 0.3\}$ would be regarded as the result of an expert considering that the set has a membership value of 0.2 and a second expert considering that the set has a membership value of 0.3. These "experts" can be human individuals or, more often, simply different criteria or methodologies that produce fuzzy membership values for a common phenomenon. Note that if the experts evaluate different phenomena, then it would make more sense to use separate fuzzy sets.

A problem with the experts' interpretation when we consider the hesitant membership value as simply a subset of $[0,1]$ is that these experts should be indistinguishable and even the number of experts involved is subject to variation. A more realistic scenario occurs when the number of experts is fixed and it may be possible to link the membership values to the expert that has produced it. In such a situation, the typical hesitant fuzzy sets cannot be a good model. On the one hand, it should be possible for a membership value to appear more than once. If both the first and the second expert produce a membership value of 0.2, then the hesitant membership value should be something like $(0.2, 0.2)$, which is a pair (an n-tuple for the general case with n experts) rather than a subset. As we commented in the Introduction, this need for repeated membership values was already considered by Torra in [16]. On the other hand, if we want to treat the experts as being distinguishable, which seems sensible for comparison purposes, then such n-tuples should be ordered, so that $(0.1, 0.2)$ is different from $(0.2, 0.1)$ as a membership value. We will refer to these ordered vector-like membership values as **ordered fuzzy multisets**.

From this discussion, we can see that there are four distinct cases in the experts' model.

1. If the experts produce unrelated values for different phenomena, then the values for each expert should be treated as **independent fuzzy sets**, as there is no link between them.
2. If the experts produce related values that account for one phenomenon and there is a fixed number of them producing a membership value for each element in the universe, then we can model the problem with **ordered fuzzy multisets**.
3. If the experts produce related values that account for one phenomenon and there is a fixed number of them producing a membership value for each element in the universe and they are indistinguishable (as if in a secret vote; we cannot know which expert produced which value for one element, but we can know that n experts chose the same value) then the problem should be modeled using **hesitant fuzzy multisets**.
4. If the experts produce related values that account for one phenomenon and there is a variable or unknown number of them producing membership values for each element in the universe and they are indistinguishable then the problem should be modeled using **hesitant fuzzy sets**.

There has been little or no research so far concerning the ordered fuzzy multisets; they are pretty straightforward if we consider them as a Cartesian product, which may explain the scant interest such mathematical constructions have elicited. Now, we will present some formal definitions.

Thus, we need to define a special type of hesitant fuzzy set where membership values can appear multiple times and are labeled with a coordinate index. We can do this easily by replicating the notion of a fuzzy set over the various dimensions of the Cartesian product of real intervals $[0, 1]^n$:

Definition 6. *Let X be the universe. An n-**dimensional ordered ordinary fuzzy multiset (OOFM)** A on X is a function $\mu_A \colon X \to [0, 1]^n$.*

With this definition, each restriction of μ_A to the i-th coordinate in the image set is an ordinary fuzzy set. These n ordinary fuzzy sets, denoted by $A_1, \ldots, A_n$, can be referred to as the **fuzzy coordinates** of A.

As in the ordinary fuzzy sets, we will usually differentiate between the OOFM $A = (A_1, \ldots, A_n)$ and its membership function $\mu_A = (\mu_{A_1}, \ldots, \mu_{A_n})$, although there is no formal difference between the "sets" and the "membership functions".

The family of all the possible n-dimensional ordered ordinary fuzzy multisets on X is called the n-**dimensional ordered ordinary fuzzy power multiset** of X, which we will denote by $\mathscr{F}^n(X)$.

The usual fuzzy set operations such as complement, union and intersection can be carried over to the $[0,1]^n$ space coordinatewise in a straightforward way. We can express this as a relatively trivial extension principle:

Definition 7 (OOFM Extension Principle). *Given a reference universe X and an additional reference set Y and a function $f: \mathscr{F}(X) \to Y$, for a natural number $n > 1$ we can define a function $\tilde{f}: \mathscr{F}^n(X) \to Y^n$ by:*

$$\tilde{f}(A_1, \ldots, A_n) = (f(A_1), \ldots, f(A_n))$$

If the image set Y is also $\mathscr{F}(X)$, then the Cartesian product $[\mathscr{F}(X)]^n$ can be identified with $\mathscr{F}^n(X)$ in an obvious way.

A more general form of the principle can be stated for multivariate functions, so that a function $f: [\mathscr{F}(X)]^m \to Y$, with $m > 1$ being natural number, can be similarly extended to a function $\tilde{f}: [\mathscr{F}^n(X)]^m \to Y^n$. In particular, we will often be using the two-dimensional ($m = 2$) OOFM Extension Principle.

The complement, intersection and union can then be defined by means of this OOFM Extension Principle (in its two-dimensional form, for the union and intersection). For the sake of completeness, we are going to define them explicitly:

Definition 8. *Let n be a natural number and let $\tilde{A} \in \mathscr{F}^n(X)$ be an n-dimensional OOFM. Let $c: [0,1] \to [0,1]$ be a function that defines an ordinary fuzzy complement ($c(x) := 1 - x$ in the standard case) on the family of ordinary fuzzy sets $\mathscr{F}(X)$ such that given $A \in \mathscr{F}(X)$, its complement A^c is defined by the relation $\mu_{A^c}(x) = c(\mu_A(x))$. Then the c-**based complement** of $\tilde{A}$ is the n-dimensional OOFM $\tilde{A}^c$ defined by a membership function $\mu_{\tilde{A}^c}: X \to [0,1]^n$ such that $\mu_{\tilde{A}_i^c}(x) = c(\mu_{\tilde{A}_i}(x))$.*

If $c(x) = 1 - x$, then the c-based complement is referred to as the standard complement of $\mathscr{F}^n(X)$.

Definition 9. *Let n be a natural number and let $\tilde{A}, \tilde{B} \in \mathscr{F}^n(X)$ be two n-dimensional OOFM's. Let $t: [0,1] \to [0,1]$ be a t-norm that defines an ordinary fuzzy intersection on the family of ordinary fuzzy sets $\mathscr{F}(X)$ such that given $A, B \in \mathscr{F}(X)$, their intersection $A \cap B$ is defined by the relation $\mu_{A \cap B}(x) = t(\mu_A(x), \mu_B(x))$. Then the t-**based intersection** of $\tilde{A}$ and $\tilde{B}$ is the n-dimensional OOFM $\tilde{A} \cap \tilde{B}$ defined by a membership function $\mu_{\tilde{A} \cap \tilde{B}}: X \to [0,1]^n$ such that $\mu_{(\tilde{A} \cap \tilde{B})_i}(x) = t(\mu_{\tilde{A}_i}(x), \mu_{\tilde{B}_i}(x))$.*

If $t(x,y) = \min\{x, y\}$, then the t-based intersection is referred to as the standard intersection of $\mathscr{F}^n(X)$.

The definition for the union is completely analogous:

Definition 10. *Let n be a natural number and let $\tilde{A}, \tilde{B} \in \mathscr{F}^n(X)$ be two n-dimensional OOFM's. Let $s \colon [0,1] \to [0,1]$ be a t-conorm that defines an ordinary fuzzy union on the family of ordinary fuzzy sets $\mathscr{F}(X)$ such that given $A, B \in \mathscr{F}(X)$, their union $A \cap B$ is defined by the relation $\mu_{A \cup B}(x) = s(\mu_A(x), \mu_B(x))$. Then the s-**based union** of $\tilde{A}$ and $\tilde{B}$ is the n-dimensional OOFM $\tilde{A} \cup \tilde{B}$ defined by a membership function $\mu_{\tilde{A} \cup \tilde{B}} \colon X \to [0,1]^n$ such that $\mu_{(\tilde{A} \cup \tilde{B})_i}(x) = s(\mu_{\tilde{A}_i}(x), \mu_{\tilde{B}_i}(x))$.*

If $s(x,y) = \max\{x, y\}$, then the s-based union is referred to as the standard union of $\mathscr{F}^n(X)$.

As these operations have been defined in terms of the ones for the ordinary fuzzy sets, the properties of the latter are replicated in an obvious way. In particular, the intersection and union are commutative and associative and the identity element is the OOFM $I \colon X \to [0,1]^n$ defined by the unity membership function $\mu_I(x) = (1, \ldots, 1) \ \forall x \in X$ for the intersection and the OOFM $\emptyset \colon X \to [0,1]^n$ defined by the null membership function $\mu_\emptyset(x) = (0, \ldots, 0) \ \forall x \in X$ for the union. As in the case of the ordinary fuzzy sets, the union and intersection are idempotent and distributive if and only if the t-norm and t-conorm considered are the standard ones.

4 Measures of Divergence Between OOFMs

For the OOFM's, we can define a vector form of divergence by applying the two-dimensional OOFM Extension Principle. This allows extending the existing notion of divergence to a multidimensional arrangement in a straightforward way. We will use the term "multidivergence" for this concept as it is a vector, rather than a real number like the divergence measures.

Definition 11. *Let X be the universe. A map $D^n \colon \mathscr{F}^n(X) \times \mathscr{F}^n(X) \to \mathbb{R}^n$ is an n-**dimensional ordered ordinary fuzzy multidivergence measure** if for all $A, B \in \mathscr{F}^n(X)$ the following three conditions are met:*

1. $D^n(A, B) = D^n(B, A)$
2. $D^n(A, A) = (0, \ldots, 0)$
3. $\max\{D_i^n(A \cup C, B \cup C), D_i^n(A \cap C, B \cap C)\} \leq D_i^n(A, B), \ \forall C \in \mathscr{F}^n(X)$ and $\forall i \in [1, \ldots, n]$, where D_i^n is the i-th component function of D^n.

The simplest cases of ordered ordinary multidivergence measures will be those where the image by the i-th component function of the multidivergence depends upon the i-th component of the OOFM's A and B only. We will refer to such multidivergence measures as "pure" or "Cartesian", as opposed to the general case where the various coordinates in the domain get mixed up in the resulting vector, which we will call "mixed multidivergences".

Definition 12. *Let X be the universe. An n-dimensional ordered ordinary multidivergence measure $D^n \colon \mathscr{F}^n(X) \times \mathscr{F}^n(X) \to \mathbb{R}^n$ is called a **pure** ordered ordinary multidivergence measure if each component function $D_i^n \colon \mathscr{F}^n(X) \times \mathscr{F}^n(X) \to \mathbb{R}$ is a function of the i-th coordinate fuzzy sets only. Formally, for any $A, B \in \mathscr{F}^n(X)$ there exist n functions $\phi_i \colon \mathscr{F}(X) \times \mathscr{F}(X) \to \mathbb{R}$ such that $D_i^n(A, B) = \phi_i(A_i, B_i)$.*

In the case of pure ordered ordinary multidivergence measures, Definition 11 means that each dimension can be treated as an ordinary divergence measure. We can sum this up in a proposition:

Proposition 1. *Given a universe X and an n-dimensional pure ordered ordinary fuzzy multidivergence measure D^n defined over $\mathscr{F}^n(X)$ such that for any $x \in X$ and any pair $A, B \in \mathscr{F}^n(X)$, D^n is decomposed as $D^n(A,B)(x) = (D_1^n(A_1, B_1)(x), \ldots, D_n^n(A_n, B_n)(x))$, each component function D_i^n is a divergence measure.*

Such pure ordered ordinary fuzzy multidivergence measures can be constructed by simply combining several ordinary divergences through a Cartesian product. So, for example, we can take a Hamming-style divergence or a Euclidean-style divergence, both defined on $\mathscr{F}(X)$, and construct with either of them a pure ordered fuzzy multidivergence on $\mathscr{F}^2(X)$.

Example 1. Let $X = \{a, b\}$ be a universe made up of two elements. If we take two of its 2-dimensional ordered ordinary fuzzy multisets A and B defined by $\mu_A(a) = (0.2, 0.3)$, $\mu_A(b) = (0.7, 0.9)$, $\mu_B(a) = (0.5, 0.4)$ and $\mu_B(b) = (0.1, 0.2)$, then the Hamming multidivergence D_{Hamming}^2 would be given by:

$$
D_{\text{Hamming}}^2(A, B) = \begin{bmatrix} D_{\text{Hamming}}(A_1, B_1) \\ D_{\text{Hamming}}(A_2, B_2) \end{bmatrix}
$$
$$
= \begin{bmatrix} D_{\text{Hamming}}((0.2, 0.7), (0.5, 0.1)) \\ D_{\text{Hamming}}((0.3, 0.9), (0.4, 0.2)) \end{bmatrix} = \begin{bmatrix} 0.9 \\ 0.8 \end{bmatrix}
$$

Just as it was done with the ordinary fuzzy divergence measures, we can refer to an ordered ordinary multidivergence measure as being **local** when the variation in the divergence between two pairs of fuzzy sets that is triggered by varying the membership value for one element of the universe can be expressed through a characteristic function:

Definition 13. *Let X be the universe. An n-dimensional ordered ordinary multidivergence measure $D^n \colon \mathscr{F}^n(X) \times \mathscr{F}^n(X) \to \mathbb{R}^n$ is said to be **local** if for every $x \in X$ there is a function $h_x^n \colon [0,1]^n \times [0,1]^n \to \mathbb{R}^n$ such that:*

$$
D_i^n(A, B) - D_i^n(A \cup \{x\}, B \cup \{x\}) = h_{x_i}^n(\mu_A(x), \mu_B(x)) \quad (i = 1, \ldots, n)
$$

where $\{x\}$ is the n-dimensional ordered ordinary fuzzy set such that $\mu_{\{x\}}(x) = \{1, \ldots, 1\}$ and $\mu_{\{x\}}(y) = \{0, \ldots, 0\}$ for any other $y \neq x$.

Local multidivergences are very closely related to the local fuzzy divergence measures, as we can see:

Proposition 2. *Let X be the universe and let $\mathscr{F}^n(X)$ be its n-dimensional ordered ordinary fuzzy power multiset. If an n-dimensional ordered ordinary multidivergence measure $D^n \colon \mathscr{F}^n(X) \times \mathscr{F}^n(X) \to \mathbb{R}^n$ is both pure and local, then each one of the n component functions D_i^n is a local fuzzy divergence measure.*

When the multidivergence is pure, as each coordinate is a local divergence itself, we can apply the representation Theorem 1, leading to a multi-dimensional version:

Theorem 2 (Local Multidivergence Representation Theorem). *Let X be the universe. An n-dimensional pure ordered ordinary multidivergence measure $D^n \colon \mathscr{F}^n(X) \times \mathscr{F}^n(X) \to \mathbb{R}^n$ is a **local multidivergence measure** if and only if $\forall A, B \in \mathscr{F}^n(X)$ and for every $x \in X$ there are n functions $h_{x_i}^n \colon [0,1] \times [0,1] \to \mathbb{R}$ (with $i = 1, \ldots, n$) such that:*

$$D_i^n(A, B) = \sum_{x \in X} h_{x_i}^n(\mu_{A_i}(x), \mu_{B_i}(x))$$

And all the h_{x_i} fulfill the conditions:

(i) $h_{x_i}(u, v) = h_{x_i}(v, u)$, $\forall u, v \in [0, 1]$;
(ii) $h_{x_i}(u, u) = 0$, $\forall u \in [0, 1]$;
(iii) $h_{x_i}(u, w) \geq h_{x_i}(u, v)$, $\forall u, v, w \in [0, 1]$ with $u \leq v \leq w$;
(iv) $h_{x_i}(u, w) \geq h_{x_i}(v, w)$, $\forall u, v, w \in [0, 1]$ with $u \leq v \leq w$.

Up to this point, we have simply combined ordinary fuzzy sets into a multidimensional vector arrangement. The multidivergence measures as vector quantities are not really useful when we want to quantify or compare how close or distant different ordered fuzzy multisets are from one another. Even if we can rely on the product order or a lexicographic order in some simple cases, ideally we would rather arrive at a single real number as the useful divergence measure. This can be done by simply applying an aggregation function to the multidivergence. This idea leads to the definition below:

Definition 14. *Given a universe X, an n-dimensional ordered ordinary fuzzy multidivergence measure D^n defined over $\mathscr{F}^n(X)$ and an aggregation function $f \colon \mathbb{R}^n \to \mathbb{R}$, the **$f$-aggregated ordered ordinary fuzzy divergence measure** is the function $D_f^{*n} \colon \mathscr{F}^n(X) \times \mathscr{F}^n(X) \to \mathbb{R}$ defined by the f aggregation of the components of D^n:*

$$D_f^{*n}(A, B) := f(D_1^n(A, B), \ldots, D_n^n(A, B))$$

Example 2. In our previous example, we arrived at a Hamming-style multidivergence made up of two values. If we use the arithmetic mean as our aggregation function, we can turn the Hamming 2-multidivergence (0.9, 0.8) into the Hamming arithmetic-mean-aggregated ordered ordinary fuzzy divergence value of 0.85.

5   Conclusions

In this work we have considered the classical hesitant fuzzy sets to define the ordered ordinary fuzzy multisets, which can be more useful in some particular cases in decision making. An interesting tool in this area, a measure for comparing two multisets, was also defined and studied in detail, in particular the case of local multidivergences.

In the future we would like to be able to relate the OOFM's with the hesitant fuzzy multisets and thereby extend the idea of divergence measure to these multisets.

Acknowledgment. Authors acknowledge financial support by the Spanish Ministry under Projects TIN2014-59543-P and TIN2017-87600-P.

References

1. Anthony, M., Hammer, P.L.: A Boolean measure of similarity. Discrete Appl. Math. **154**(16), 2242–2246 (2006)
2. Bouchon-Meunier, B., Rifqi, M., Bothorel, S.: Towards general measures of comparison of objects. Fuzzy Sets Syst. **84**, 143–153 (1996)
3. Beliakov, G., Bustince, H., Calvo, T.: A Practical Guide to Averaging Functions. Studies in Fuzziness and Soft Computing, vol. 329. Springer, Heidelberg (2016). https://doi.org/10.1007/978-3-319-24753-3. ISBN 978-3-319-24751-9
4. Couso, I., Garrido, L., Sánchez, L.: Similarity and dissimilarity measures between fuzzy sets: a formal relational study. Inf. Sci. **229**, 122–141 (2013)
5. Dubois, D., Prade, H.: Fundamentals of Fuzzy Sets. Kluwer Academic Publishers, Massachusetts (2000)
6. Grattan-Guinness, I.: Fuzzy membership mapped onto intervals and many-valued quantities. Math. Logic Q. **22–1**, 149–160 (1976)
7. Klement, P., Mesiar, R., Pap, E.: Triangular Norms. Trends in Logic, vol. 8. Springer, Heidelberg (2000). https://doi.org/10.1007/978-94-015-9540-7
8. Klir, G.J., Folger, T.A.: Fuzzy Sets Uncertainty and Information. Prentice-Hall, Englewood Cliffs (1988)
9. Kobza, V., Janiš, V., Montes, S.: Generalizated local divergence measures. J. Intell. Fuzzy Syst. **33**, 337–350 (2017)
10. Li, Y., Qin, K., He, X.: Some new approaches to constructing similarity measures. Fuzzy Sets Syst. **234**(1), 46–60 (2014)
11. Lui, X.: Entropy, distance measure and similarity measure of fuzzy sets and their relations. Fuzzy Sets Syst. **52**, 305–318 (1992)
12. Montes, I., Pal, N.R., Janiš, V., Montes, S.: Divergence measures for intuitionistic fuzzy sets. IEEE Trans. Fuzzy Syst. **23**(2), 444–456 (2015)
13. Montes, I., Janiš, V., Pal, N.R., Montes, S.: Local divergences for Atanassov intuitionistic fuzzy sets. IEEE Trans. Fuzzy Syst. (in press). https://doi.org/10.1109/TFUZZ.2015.2457447
14. Montes, S., Couso, I., Gil, P., Bertoluzza, C.: Divergence measure between fuzzy sets. Int. J. Approx. Reason. **30**, 91–105 (2002)

15. Rodríguez, R.M., Bedregal, B., Bustince, H., Dong, Y.C., Farhadinia, B., Kahraman, C., Martínez, L., Torra, V., Xu, Y.J., Xu, Z.S., Herrera, F.: A position and perspective analysis of hesitant fuzzy sets on information fusion in decision making. Towards high quality progress. Inf. Fusion **29**, 89–97 (2016)
16. Torra, V.: Hesitant fuzzy sets. Int. J. Intell. Syst. **25**–**6**, 529–539 (2010)
17. Valverde, L., Ovchinnikov, S.: Representations of T-similarity relations. Fuzzy Sets Syst. **159**(17), 2211–2220 (2008)
18. Wilbik, A., Keller, J.M.: A fuzzy measure similarity between sets of linguistic summaries. IEEE Trans. Fuzzy Syst. **21**(1), 183–189 (2013)
19. Xu, Z.S.: Hesitant Fuzzy Sets Theory. Studies in Fuzziness and Soft Computing, vol. 314. Springer, Heidelberg (2014). https://doi.org/10.1007/978-3-319-04711-9
20. Zadeh, L.A.: Fuzzy sets. Inf. Control **8**, 338–353 (1965)
21. Zadeh, L.A.: A note on similarity-based definitions of possibility and probability. Inf. Sci. **267**, 334–336 (2014)
22. Zhang, C., Fu, H.: Similarity measures on three kinds of fuzzy sets. Pattern Recogn. Lett. **27**(12), 1307–1317 (2006)

New Trends in Data Aggregation

The Median Procedure as an Example of Penalty-Based Aggregation of Binary Relations

Raúl Pérez-Fernández[(✉)] and Bernard De Baets

KERMIT, Department of Data Analysis and Mathematical Modelling,
Ghent University, Coupure links 653, 9000 Ghent, Belgium
{raul.perezfernandez,bernard.debaets}@ugent.be

Abstract. The aggregation of binary relations is a common topic in many fields of application such as social choice and cluster analysis. In this paper, we discuss how the median procedure – probably the most common method for aggregating binary relations – fits in the framework of penalty-based data aggregation.

Keywords: Aggregation · Penalty function · Binary relation · Median

1 Introduction

The use of penalty functions has been a common approach in data aggregation since Yager introduced the concept in 1993 [1]. Intuitively, a penalty function measures the disagreement of a list of objects with a consensus object. The result of the aggregation is thus considered to be the consensus object(s) that minimizes the penalty w.r.t. the list of objects to be aggregated. Unfortunately, in contrast with the initial ideas proposed by Yager, the use of penalty functions in data aggregation is nowadays mostly confined to the aggregation of real numbers [2,3]. In a recent paper [4], we pointed out that the aggregation on other structures different than the set of real numbers also obeys some similar laws, and proposed a more general definition of a penalty function based on the compatibility with a betweenness relation.

For instance, the aggregation of binary relations is a long-standing problem in many fields of application. The aggregation of linear order relations (and rankings) [5], weak order relations [6] and tournament relations [7] is a common topic in social choice theory, and the aggregation of equivalence relations [8] is of interest to the field of cluster analysis. One common approach for the aggregation of all these types of binary relations is the median procedure [9], in which the result of the aggregation is the binary relation that minimizes the symmetric difference distance to the list of binary relations to be aggregated. In this paper, we will demonstrate that the median procedure is indeed a case of penalty-based aggregation of binary relations.

© Springer International Publishing AG, part of Springer Nature 2018
J. Medina et al. (Eds.): IPMU 2018, CCIS 854, pp. 359–366, 2018.
https://doi.org/10.1007/978-3-319-91476-3_30

The remainder of the paper is structured as follows. First, we recall some preliminary notions on betweenness relations and penalty functions in Sect. 2. In Sect. 3, we discuss some structural properties of the set of binary relations on a finite set. In Sect. 4, the median procedure for aggregating binary relations is discussed, whereas we prove that this median procedure is a prominent example of penalty-based aggregation of binary relations in Sect. 5. We round up with some conclusions in Sect. 6.

2   Preliminaries

The notion of an element lying in between two other elements is a common topic in mathematics dating back to the foundations of geometry. In the following, we provide a definition requiring a minimal set of axioms [10]. Further additional axioms have been proposed concerning different types of transitivity [11–13].

Definition 1. *A ternary relation B on a set X is called a betweenness relation if it satisfies the following three properties:*

(i) Symmetry in the end points: for any $x, y, z \in X$, it holds that

$$(x, y, z) \in B \Leftrightarrow (z, y, x) \in B.$$

(ii) Closure: for any $x, y, z \in X$, it holds that

$$((x, y, z) \in B \wedge (x, z, y) \in B) \Leftrightarrow y = z.$$

(iii) End-point transitivity: for any $o, x, y, z \in X$, it holds that

$$((o, x, y) \in B \wedge (o, y, z) \in B) \Rightarrow (o, x, z) \in B.$$

In a partially ordered set or in a metric space, we naturally have an associated betweenness relation [4].

Definition 2. *Consider an order relation $\leq$ on a set X. The betweenness relation induced by $\leq$ is the ternary relation $B_{\leq}$ on X defined as*

$$B_{\leq} = \left\{ (x, y, z) \in X^3 \mid (x = y) \vee (y = z) \vee (x \leq y \leq z) \vee (z \leq y \leq x) \right\}.$$

Definition 3. *Consider a distance function d on a set X. The betweenness relation induced by d is the ternary relation B_d on X defined as*

$$B_d = \left\{ (x, y, z) \in X^3 \mid d(x, z) = d(x, y) + d(y, z) \right\}.$$

Any betweenness relation on a set X is easily extended to X^n by the so-called product betweenness relation.

Definition 4. *Consider $n \in \mathbb{N}$ and a betweenness relation B on a set X. The product betweenness relation is the ternary relation $B^{(n)}$ on X^n defined as*

$$B^{(n)} = \left\{ (\mathbf{x}, \mathbf{y}, \mathbf{z}) \in (X^n)^3 \mid (\forall i \in \{1, \ldots, n\})((x_i, y_i, z_i) \in B) \right\}.$$

Since Yager proposed for the first time the use of penalty functions in the field of data aggregation [1], the definition of a penalty function has suffered many changes (see, for instance, [3,14] or the recent survey [2]). Nevertheless, these penalty functions have mainly been used for aggregating real numbers. In the following, we consider the definition given in [4] that allows to deal with aggregation on structures equipped with a betweenness relation (e.g. partially ordered sets and metric spaces).

Definition 5. *Consider $n \in \mathbb{N}$, a set X and a betweenness relation B on X^n. A function $P : X \times X^n \to \mathbb{R}^+$ is called a penalty function (compatible with B) if the following four properties hold:*

(P1) $P(y; \mathbf{x}) \geq 0$, for any $y \in X$ and any $\mathbf{x} \in X^n$;
(P2) $P(y; \mathbf{x}) = 0$ if and only if $\mathbf{x} = (y, \ldots, y)$;
(P3) The set of minimizers of $P(\cdot; \mathbf{x})$ is non-empty, for any $\mathbf{x} \in X^n$.
(P4) $P(y; \mathbf{x}) \leq P(y; \mathbf{x}')$, for any $y \in X$ and any $\mathbf{x}, \mathbf{x}' \in X^n$ such that $((y, \ldots, y), \mathbf{x}, \mathbf{x}') \in B$.

Remark 1. Two additional desirable properties for a penalty function are:

(P5) $P(y; \mathbf{x}) \leq P(y'; \mathbf{x})$, for any $y, y' \in X$, any $\mathbf{x} \in X^n$ and any minimizer $z \in X$ of $P(\cdot; \mathbf{x})$ such that $((z, \ldots, z), (y, \ldots, y), (y', \ldots, y')) \in B$;
(P6) $P(y; \mathbf{x}) = P(z; \mathbf{x})$, for any $y \in X$, any $\mathbf{x} \in X^n$ and any two minimizers $z, z' \in X$ of $P(\cdot; \mathbf{x})$ such that $((z, \ldots, z), (y, \ldots, y), (z', \ldots, z')) \in B$.

A penalty function is then used for determining the result of an aggregation process: we select as the aggregate the value(s) that minimizes the penalty given the list of objects to be aggregated.

Definition 6. *Consider $n \in \mathbb{N}$, a set X, a betweenness relation B on X^n and a penalty function $P : X \times X^n \to \mathbb{R}^+$ compatible with B. The function $f : X^n \to \mathcal{P}(X)$ such that $f(\mathbf{x})$ equals the set of minimizers of $P(\cdot; \mathbf{x})$, for any $\mathbf{x} \in X^n$, is called the penalty-based function associated with P.*

Note that penalty-based functions are not aggregation functions in the most classical sense [15] since they do not need to fulfill the monotonicity property (an interesting discussion on this topic is addressed in [3]). Actually, one should note that when we no longer deal with the aggregation of elements in a partially ordered set, the property of monotonicity might not even be definable.

3   Binary Relations

Throughout the rest of this paper, we consider a fixed finite set X. A binary relation (on X) is a set of couples $(x, y) \in X^2$ or, equivalently, a subset of X^2. We denote the set of all binary relations by $\mathcal{B}$. For a binary relation R it is common to write xRy instead of $(x, y) \in R$. A binary relation R is said to be included in another binary relation S, denoted by $R \subseteq S$, if, for any $x, y \in X$, xRy implies that xSy. The union of two binary relations R and S, denoted by $R \cup S$, is the binary relation defined as $R \cup S = \{(x, y) \in X^2 \mid xRy \vee xSy\}$. Similarly, the intersection of two binary relations R and S, denoted by $R \cap S$, is the binary relation defined as $R \cap S = \{(x, y) \in X^2 \mid xRy \wedge xSy\}$. The set difference of two binary relations R and S, denoted by $R\backslash S$, is the binary relation defined as $R\backslash S = \{(x, y) \in X^2 \mid xRy \wedge xS^c y\}$. Finally, the symmetric difference of two binary relations R and S, denoted by $R\Delta S$, is the binary relation defined as $R\Delta S = (R\backslash S) \cup (S\backslash R)$, or, equivalently, as $R\Delta S = (R \cup S)\backslash(R \cap S)$.

The symmetric difference is commonly used for defining a natural distance function on the set of binary relations. For any two binary relations R and S, the symmetric difference distance between R and S is given by:

$$d_\Delta(R, S) = |R\Delta S| = |R \cup S| - |R \cap S|,$$

where $|T|$ denotes the number of couples in a given binary relation T.

Note that the set of binary relations is then equipped with a natural distance function d_Δ and with a natural order $\subseteq$. Thus, two natural betweenness relations arise:

$$B_\subseteq = \left\{(R, R', R'') \in \mathcal{B}^3 \mid R = R' \vee R' = R'' \vee R \subseteq R' \subseteq R'' \vee R'' \subseteq R' \subseteq R\right\},$$
$$B_{d_\Delta} = \left\{(R, R', R'') \in \mathcal{B}^3 \mid d_\Delta(R, R'') = d_\Delta(R, R') + d_\Delta(R', R'')\right\}.$$

It is easy to verify that $B_\subseteq \subset B_{d_\Delta}$.

Remark 2. An interesting observation is that, if $R \subseteq R''$ (or $R'' \subseteq R$), then $(R, R', R'') \in B_\subseteq$ holds if and only if $(R, R', R'') \in B_{d_\Delta}$ holds (for any $R' \in \mathcal{B}$).

Several properties of binary relations are interesting to be studied. A binary relation R is called:

(i) *reflexive*, if, for any $x \in X$, it holds that xRx;
(ii) *symmetric*, if, for any $x, y \in X$, it holds that xRy implies that yRx;
(iii) *antisymmetric*, if, for any $x, y \in X$, it holds that xRy and yRx imply that $x = y$;
(iv) *transitive*, if, for any $x, y, z \in X$, it holds that xRy and yRz imply that xRz;
(v) *complete*, if, for any $x, y \in X$, either xRy or yRx holds.

Some specific types of binary relations have attracted the attention of the scientific community. In the following, we highlight:

(i) A *linear order relation* is an antisymmetric, transitive and complete (thus reflexive)[1] binary relation. The set of all linear order relations is denoted by $\mathcal{L}$.

---

[1] The irreflexive part of a linear order relation is often called a ranking. Since the (ir)reflexivity of a relation does not usually play a role in the aggregation process, both linear order relations and rankings are often used interchangeably.

(ii) A *weak order relation* is a transitive and complete (thus reflexive) binary relation. The set of all weak order relations is denoted by $\mathcal{W}$.
(iii) A *tournament relation* is an antisymmetric and complete (thus reflexive)[2] binary relation. The set of all tournament relations is denoted by $\mathcal{T}$.
(iv) An *equivalence relation* is a reflexive, symmetric and transitive binary relation. The set of all equivalence relations is denoted by $\mathcal{E}$.

4   The Median Procedure

The median procedure [9] is a common technique for aggregating binary relations. In general, a binary relation R is said to be a median of a list $(R_i)_{i=1}^n$ of n binary relations if

$$\sum_{i=1}^n d_\Delta(R, R_i) = \min_{R' \in \mathcal{B}} \sum_{i=1}^n d_\Delta(R', R_i). \tag{1}$$

It is known that if n is an odd number the median R of any list $(R_i)_{i=1}^n$ of n binary relations is unique and given by:

$$R = \left\{ (x, y) \in X^2 \mid \tfrac{n+1}{2} \leq |\{i \in \{1, \ldots, n\} \mid xR_i y\}| \right\}. \tag{2}$$

If n is an even number, the median does not need to be unique. However, it is known that the medians of any list $(R_i)_{i=1}^n$ of n binary relations are all those binary relations R that satisfy that $(\underline{R}, R, \overline{R}) \in B_\subseteq$ where:

$$\underline{R} = \left\{ (x, y) \in X^2 \mid \tfrac{n}{2} < |\{i \in \{1, \ldots, n\} \mid xR_i y\}| \right\},$$
$$\overline{R} = \left\{ (x, y) \in X^2 \mid \tfrac{n}{2} \leq |\{i \in \{1, \ldots, n\} \mid xR_i y\}| \right\}. \tag{3}$$

Interestingly, the median procedure is also used in the aggregation of special types of binary relations. Consider a subset $\mathcal{S}$ of the set of all binary relations $\mathcal{B}$. A binary relation $R \in \mathcal{S}$ is said to be an $\mathcal{S}$-median of the list $(R_i)_{i=1}^n$ of n binary relations in $\mathcal{S}$ if

$$\sum_{i=1}^n d_\Delta(R, R_i) = \min_{R' \in \mathcal{S}} \sum_{i=1}^n d_\Delta(R', R_i).$$

Unfortunately, unlike the computation of the median[3], the computation of other $\mathcal{S}$-medians is not trivial. Many studies have analysed the complexity of computing different types of $\mathcal{S}$-medians (mainly concerning subsets $\mathcal{S}$ defined by properties of binary relations such as reflexivity, symmetry, antisymmetry, transitivity and completeness) [16], probably the cases in which $\mathcal{S} \in \{\mathcal{L}, \mathcal{W}, \mathcal{T}, \mathcal{E}\}$ being the most prominent examples [8, 17].

---

[2] The term tournament relation is sometimes used for referring to the irreflexive part of what we call a tournament relation in this paper.

[3] Note that the notions of $\mathcal{B}$-median and median are equivalent.

The method of Kemeny [5], probably one of the most common methods for aggregating linear order relations, selects as the aggregate of a given list of linear order relations the linear order relation(s) that minimizes the sum of Kendall distances [18]. Since the Kendall distance between two linear order relations is half of their symmetric difference distance, it is immediate to see that the method of Kemeny actually amounts to identifying the $\mathcal{L}$-medians. The method of Kemeny is also considered for aggregating weak order relations by selecting as the aggregate of a given list of weak order relations the weak order relation(s) that minimizes the sum of Kemeny distances [5]. Since the Kemeny distance between two weak order relations coincides with the symmetric difference distance between the strict parts of these weak order relations, it is immediate to see the link between $\mathcal{W}$-medians and the method of Kemeny for weak order relations.

5 The Median Procedure as an Example of Penalty-Based Aggregation of Binary Relations

Consider a subset $\mathcal{S}$ of the set of all binary relations $\mathcal{B}$, and consider the problem of aggregating a list $(R_i)_{i=1}^n$ of n binary relations in $\mathcal{S}$. The function $P : \mathcal{S} \times \mathcal{S}^n \to \mathbb{R}^+$, defined as

$$P(R, (R_1, \ldots, R_n)) = \sum_{i=1}^n d_\Delta(R, R_i)$$

$$= \sum_{(x,y) \in X^2} |\{i \in \{1, \ldots, n\} \mid (xRy \wedge xR_i^c y) \vee (xR^c y \wedge xR_i y)\}| .$$

(4)

trivially satisfies conditions (P1) and (P2) of a penalty function.

For any list $(R_i)_{i=1}^n$ of n binary relations in $\mathcal{S}$, the set of minimizers of $P(\cdot; (R_1, \ldots, R_n))$ obviously is non-empty (since $P(R; (R_1, \ldots, R_n))$ is an integer number for all $R \in \mathcal{S}$ and, moreover, $\mathcal{S}$ is a finite set). Thus, condition (P3) of a penalty function is satisfied.

Consider now any $R \in \mathcal{S}$ and any $(R_i)_{i=1}^n, (R_i')_{i=1}^n \in \mathcal{S}^n$ such that $((R, \ldots, R), (R_i)_{i=1}^n, (R_i')_{i=1}^n) \in B_{d_\Delta}^{(n)}$. By definition of the product between-ness relation, it follows that $(R, R_i, R_i') \in B_{d_\Delta}$ for any $i \in \{1, \ldots, n\}$. There-fore, it holds that $d_\Delta(R, R_i) = d_\Delta(R, R_i') - d_\Delta(R_i, R_i') \leq d_\Delta(R, R_i')$ (for any $i \in \{1, \ldots, n\}$), and, thus,

$$P(R, (R_1, \ldots, R_n)) = \sum_{i=1}^n d_\Delta(R, R_i) \leq \sum_{i=1}^n d_\Delta(R, R_i') = P(R, (R_1', \ldots, R_n')) .$$

We conclude that P satisfies condition (P4) of a penalty function, and, thus, is a penalty function compatible with $B_{d_\Delta}^{(n)}$. Note that, since $B_\subseteq \subseteq B_{d_\Delta}$, it also holds that P is a penalty function compatible with $B_\subseteq^{(n)}$.

Unfortunately, the desirable conditions (P5) and (P6) might not be satisfied in general for any S (for instance, we proved in [4] these conditions to fail for the case in which S is the set of all rankings).

Fortunately, in the general case of binary relations (with no extra properties required) both conditions (P5) and (P6) are fulfilled. Consider any $R, R' \in \mathcal{B}$, any $(R_i)_{i=1}^n \in \mathcal{B}^n$ and any minimizer $S \in \mathcal{B}$ of $P(\cdot; (R_i)_{i=1}^n)$ such that $((S, \ldots, S), (R, \ldots, R), (R', \ldots, R')) \in B_{d_\Delta}^{(n)}$. First, since S is a minimizer of $P(\cdot; (R_i)_{i=1}^n)$, it trivially follows from Eq. (4) that

$$
|\{i \in \{1, \ldots, n\} \mid (xSy \wedge xR_i^c y) \vee (xS^c y \wedge xR_i y)\}|
$$
$$
\leq |\{i \in \{1, \ldots, n\} \mid (xS'y \wedge xR_i^c y) \vee (xS'^c y \wedge xR_i y)\}|,
$$

for any $(x, y) \in X^2$ and any $S' \in \mathcal{B}$. Second, since $((S, \ldots, S), (R, \ldots, R), (R', \ldots, R')) \in B_{d_\Delta}^{(n)}$, it holds that any couple (x, y) satisfying both that $(x, y) \in S$ and $(x, y) \in R'$ also needs to satisfy that $(x, y) \in R$.

Combining the last two statements, we conclude that $P(R, (R_1, \ldots, R_n)) \leq P(R', (R_1, \ldots, R_n))$, i.e., condition (P5) is satisfied. Finally, condition (P6) follows straightforwardly after considering the characterization of the minimizers of $P(\cdot; (R_i)_{i=1}^n)$ given by Eq. (2) (if n is an odd number) or Eq. (3) (if n is an even number). Again, since $B_\subseteq \subseteq B_{d_\Delta}$, conditions (P5) and (P6) are also satisfied for the betweenness relation $B_\subseteq^{(n)}$.

6   Conclusions

In this paper, we have discussed some prominent examples of aggregation of (families of) binary relations, and proved that the commonly-used median procedure turns out to be an example of penalty-based data aggregation. Moreover, two desirable properties of a penalty function are proved to hold in case the aggregation of binary relations (and not a particular family of binary relations) is considered. Unfortunately, these two properties are known to fail for the aggregation of some families of binary relations, e.g., for the aggregation of rankings. Future research concerns the study of the impact of replacing d_Δ by another distance function in Eq. (1). Certainly, this has been a popular study subject in the field of social choice theory for the particular setting of rankings [19–22]. To the best of our knowledge, the literature is sparser for the more general setting of binary relations.

Acknowledgments. Raúl Pérez-Fernández is supported as a postdoc by the Research Foundation of Flanders (FWO17/PDO/160).

References

1. Yager, R.R.: Toward a general theory of information aggregation. Inf. Sci. **68**, 191–206 (1993)
2. Bustince, H., Beliakov, G., Dimuro, G.P., Bedregal, B., Mesiar, R.: On the definition of penalty functions in data aggregation. Fuzzy Sets Syst. **323**, 1–18 (2017)
3. Calvo, T., Beliakov, G.: Aggregation functions based on penalties. Fuzzy Sets Syst. **161**, 1420–1436 (2010)
4. Pérez-Fernández, R., De Baets, B.: On the role of monometrics in penalty-based data aggregation (submitted)
5. Kemeny, J.G.: Mathematics without numbers. Daedalus **88**(4), 577–591 (1959)
6. Black, D.: Partial justification of the Borda count. Public Choice **28**, 1–15 (1976)
7. Monjardet, B.: An axiomatic theory of tournament aggregation. Math. Oper. Res. **3**(4), 334–351 (1978)
8. Grötschel, M., Wakabayashi, Y.: A cutting plane algorithm for a clustering problem. Math. Program. **45**, 59–96 (1989)
9. Barthelemy, J.P., Monjardet, B.: The median procedure in cluster analysis and social choice theory. Math. Soc. Sci. **1**, 235–267 (1981)
10. Pitcher, E., Smiley, M.F.: Transitivities of betweenness. Trans. Am. Math. Soc. **52**(1), 95–114 (1942)
11. Fishburn, P.C.: Betweenness, orders and interval graphs. J. Pure Appl. Algebra **1**(2), 159–178 (1971)
12. Huntington, E.V., Kline, J.R.: Sets of independent postulates for betweenness. Trans. Am. Math. Soc. **18**(3), 301–325 (1917)
13. Pasch, M.: Vorlesungen über neuere Geometrie, vol. 23. Teubner, Leipzig (1882)
14. Calvo, T., Mesiar, R., Yager, R.R.: Quantitative weights and aggregation. IEEE Trans. Fuzzy Syst. **12**, 62–69 (2004)
15. Grabisch, M., Marichal, J.L., Mesiar, R., Pap, E.: Aggregation Functions. Cambridge University Press, Cambridge (2009)
16. Wakabayashi, Y.: The complexity of computing medians of relations. Resenhas **3**(3), 323–349 (1998)
17. Brandt, F., Conitzer, V., Endriss, U., Lang, J., Procaccia, A.D. (eds.): Handb. Comput. Soc. Choice. Cambridge University Press, New York (2016)
18. Kendall, M.G.: A new measure of rank correlation. Biometrika **30**, 81–93 (1938)
19. Lerer, E., Nitzan, S.: Some general results on the metric rationalization for social decision rules. J. Econ. Theory **37**, 191–201 (1985)
20. Andjiga, N.G., Mekuko, A.Y., Moyouwou, I.: Metric rationalization of social welfare functions. Math. Soc. Sci. **72**, 14–23 (2014)
21. Elkind, E., Faliszewski, P., Slinko, A.: Distance rationalization of voting rules. Soc. Choice Welf. **45**(2), 345–377 (2015)
22. Viappiani, P.: Characterization of scoring rules with distances: application to the clustering of rankings. In: Proceedings of the 24th International Joint Conference on Artificial Intelligence, Buenos Aires (2015)

Least Median of Squares (LMS)
and Least Trimmed Squares (LTS) Fitting
for the Weighted Arithmetic Mean

Gleb Beliakov[1], Marek Gagolewski[2,3], and Simon James[1]([✉]) [iD]

[1] School of Information Technology, Deakin University, Burwood, Victoria, Australia
{gleb,sjames}@deakin.edu.au
[2] Systems Research Institute, Polish Academy of Sciences,
ul. Newelska 6, 01-447 Warsaw, Poland
gagolews@ibspan.waw.pl
[3] Faculty of Mathematics and Information Science,
Warsaw University of Technology, ul. Koszykowa 75, 00-662 Warsaw, Poland

Abstract. We look at different approaches to learning the weights of the weighted arithmetic mean such that the median residual or sum of the smallest half of squared residuals is minimized. The more general problem of multivariate regression has been well studied in statistical literature, however in the case of aggregation functions we have the restriction on the weights and the domain is also usually restricted so that 'outliers' may not be arbitrarily large. A number of algorithms are compared in terms of accuracy and speed. Our results can be extended to other aggregation functions.

Keywords: Aggregation functions · Robust statistics
Least median of squares fitting · Least trimmed squares fitting

1 Introduction

In the application of aggregation functions, a key problem is how to determine the weights or function parameters that give the best fit with respect to some penalty or objective to an observed dataset. The learned parameters can then be used either for data analysis or in the prediction of new values.

A standard approach is to use programming methods such that the sum of residuals is optimized [1–4], e.g. for a weighted arithmetic mean with respect to an unknown n-dimensional vector of weights $\mathbf{w}$, and an observed dataset consisting of m input-output pairs $(\mathbf{x}_i, y_i), \mathbf{x}_i \in [0,1]^n, y_i \in [0,1]$, we have

© Springer International Publishing AG, part of Springer Nature 2018
J. Medina et al. (Eds.): IPMU 2018, CCIS 854, pp. 367–378, 2018.
https://doi.org/10.1007/978-3-319-91476-3_31

$$\text{Minimize}_{\mathbf{w}} \quad \sum_{i=1}^{m} |r_i|^p \tag{1}$$

$$\text{s.t.} \quad r_i = \left(\sum_{j=1}^{n} w_j x_{ij} \right) - y_i, \qquad i = 1, \ldots, m,$$

$$\sum_{j=1}^{n} w_j = 1, \qquad w_j \geq 0, \qquad j = 1, \ldots, n.$$

For $p = 2$ we are minimizing the sum of squared residuals or least squares (LS), which can be solved as a quadratic programming problem, while for $p = 1$, we have the least absolute deviation (LAD), which can be solved using linear programming methods by introducing two decision variables for each observed instance and setting $r_i = r_i^+ - r_i^-$ and $r_i^+, r_i^- \geq 0$.

The LAD approach should be less susceptible to outliers, however as has been well observed in statistical literature [5], leverage points can still exert influence if the residual associated with outliers is significantly larger than residuals associated with other points.

For example, consider the 2-variate data depicted in blue in Fig. 1(a).

When fitting using Eq. (1) and $p = 1$ (LAD[1]), we obtain the weighting vector $\mathbf{w} = (0.300, 0.700)$ with a total fitting error of 4.9×10^{-4}. Using $p = 2$ (LS), we also obtain a good result with the same weighting vector (to 3 d.p.) and error 2.8×10^{-8}.

However suppose we introduce outlying points at $\mathbf{x} = (0, 1)$, $y = 0$. These are indicated by the red point depicted in Fig. 1(a). With the introduction of a single outlier, the LS results in $\mathbf{w} = (0.740, 0.260)$, the penalty or objective value increasing to 1.3×10^{-1} (note that these weights reverse the importance allocated to each variable). The outlier effect on the LS method is illustrated visually in Fig. 1(b)–(c). With the single outlier, the weights determined by LAD are almost unchanged (when evaluated to 3 d.p.), the error increasing to 7.0×10^{-1}. However when we introduce 2 outliers (at the same point), the LS continues to allocate more weight to the first variable, $\mathbf{w} = (0.841, 0.159)$ with overall penalty 2.2×10^{-1} and, at this point, the LAD fitting results in the vector $\mathbf{w} = (1, 0)$, i.e. interpolating the two outliers, because the sum of the residuals when fitting to these points is 1.4, which is less than the error that would result if the original model's weighting vector $\mathbf{w} = (0.7, 0.3)$ were used.

In the 80s, this problem for standard linear regression prompted Rousseeuw and others [5,7,8] to consider optimizing with respect to the median residual (least median of squares or LMS) or the sum of the smallest 50% of residuals (least trimmed squares or LTS) instead, i.e. minimizing $|r_{(k)}|^p$ or $\sum_{i=1}^{k} |r_{(k)}|^p$ where $k = \lceil m/2 \rceil$ and $|r_{(i)}|$ indicates the i-th smallest residual.

---

[1] All fitting performed in R [6] with details available at http://aggregationfunctions. wordpress.com.

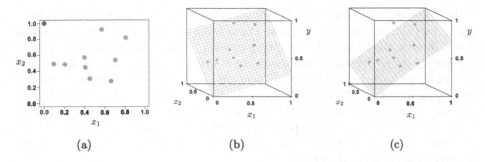

(a)   (b)   (c)

Fig. 1. (a) Randomly generated data (uniformly over $[0, 1]^2$) in blue and an outlying point in red shown as projection onto the 2-dimensional plane. (b)–(c) Data from (a) with a well fitting weighted mean (b) and a weighted mean affected by the outlier in red (c), both determined using least squares fitting. In the latter case a single outlier pulls the function towards the outlying point and in 3-dimensional space. (Color figure online)

In [5], Rousseeuw notes that the breakdown point, i.e. the percentage of data that can be arbitrarily large before a reliable result is obtained, is $((m/2) - n + 2)/m$.

Rousseeuw's method involves sampling n points (or $n + 1$ in the case of standard regression requiring an intercept), solving the exact interpolation problem, then checking the residuals. After multiple iterations, the weighting vector that minimizes the objective function of the residuals is taken as the approximate solution. The number of samples can be chosen such that the probability of a 'good' solution appearing in one of the samples is high. An underlying assumption then is that there exists a sample of n points that are representative enough of the non-outlier dataset. Rousseeuw also has investigated the reliability in terms of estimating accuracy assuming normally distributed error.

In the case of weighted means, solving the interpolation problem for n points could result in negative weights if the data contains noise, and depending on the granularity at which data is collected, real data is likely to include subsets of observed points resulting in singular matrices and hence be unsolvable. It is noted in [9] that minimizing the median residual has a relationship to the infinity norm (L_∞), i.e. the problem can be expressed as a mixed integer program where the maximum residual is minimized for a subset consisting of half of the data (which theoretically could be implemented using binary variables indicating whether a datum is included or not). Of course, for any reasonable sized dataset this quickly becomes infeasible, however we can still look to use the minimization of the maximum residual as the basis of a number of approximation algorithms. Furthermore, it should be noted that with computing power and the availability of general-purpose solvers, many real applications would have the luxury of being able to spend a little extra computing time if high accuracy is needed, so a range of approaches are practically feasible.

In this contribution, we introduce and investigate a number of algorithms that aim to find the best approximation to the weights of a weighted arithmetic mean that minimize the LMS and LTS fitting criteria. We test the algorithms against synthetic data to determine whether their respective performance is dependent on factors such as the number of outliers, the structure of the outliers, and the variable parameters of each algorithm. In Sect. 2 we give a brief overview of aggregation functions (of which the weighted arithmetic mean is an archetypical example) and the data-fitting problem. In Sect. 3 a number of algorithms are presented and compared with numerical experiments. Some concluding remarks are provided in the final section.

2   Preliminiaries

We are concerned with the modelling of data with aggregation functions [1,3,4,10,11], a class of multi-variate functions $A : [0,1]^n \rightarrow [0,1]$ satisfying monotonicity in each argument and boundary conditions $A(0,\ldots,0) = 0$ and $A(1,\ldots,1) = 1$.

Although a broad definition, in the context of machine learning, the monotonicity of aggregation functions ensures a degree of robustness and conceptual reliability in the obtained model (provided monotonicity makes sense in the application), while the boundary conditions to some extent ensure that the scale of the output can be interpreted over a similar scale to the inputs. In particular, we will focus on use of the weighted arithmetic mean, $\text{WAM}(\mathbf{x}) = \sum_{j=1}^{n} w_i x_i$, with $\mathbf{w} = (w_1, w_2, \ldots, w_n)$ an n-dimensional weighting vector that satisfies $\sum_{i=1}^{n} w_i = 1$ and $w_i \geq 0, \forall i$.

The weighted arithmetic mean is said to be averaging, i.e. for all $\mathbf{x} \in [0,1]^n$, $\min(\mathbf{x}) \leq \text{WAM}(\mathbf{x}) \leq \max(\mathbf{x})$.

There are countless families of aggregation functions with various interesting properties, including those that are averaging and defined with respect to weighting vectors. While we focus on the simplest family, most of our results would be easily extended to the cases of OWA operators, weighted quasi-arithmetic means and the Choquet integral to name a few. We note too that other intervals can be considered, however we will contain ourselves to $[0,1]$ here.

How to fit weighted arithmetic means to data based on least absolute deviation has been addressed in [2,12–14]. We recall that Eq. (1) can be used as the basis for finding the best fitting aggregation function, while further requirements on the weights may also be desired (see, e.g. the summaries and references in [1]).

More complicated aggregation functions can be fit to data using more or less the same approach. Ordered weighted averaging (OWA) functions merely require each of the input vectors to be sorted, while the fitting can be performed on weighted quasi-arithmetic means by transforming the inputs and outputs (although this can result in residuals being over- or under-estimated, see [15]).

3 Least Median of Squares (LMS) and Least Trimmed Squares (LTS) Fitting for the Weighted Arithmetic Mean

The difficult aspect of solving Eq. (1) with respect to the LMS or LTS is determining the subset $S \subset \{1, \ldots, m\}$ such that $|S| = \lceil m/2 \rceil$ and there exists an observation k with $|r_k| \geq |r_i|, \forall\, i \in S$ and $|r_k| \leq |r_i|, \forall i \notin S$.

Once we have S, the LMS problem could be solved by finding the maximum error $z = |r_k|$ using the following linear program

$$\underset{\mathbf{w}}{\text{Minimize}} \quad z \tag{2}$$

$$\text{s.t.} \quad z \geq \left(\sum_{j=1}^{n} w_j x_{ij} \right) - y_i, \quad z \geq y_i - \left(\sum_{j=1}^{n} w_j x_{ij} \right), \quad i = 1, \ldots, m,$$

$$\sum_{j=1}^{n} w_j = 1, \quad w_j \geq 0, \quad j = 1, \ldots, n,$$

$$z \geq 0.$$

This requires only $n + 1$ decision variables and $2m + 1$ linear constraints if all decision variables are assumed to be positive. The LTS is solved merely by solving Eq. (1) on the given subset. We first describe our experimental setup before testing multiple approaches.

3.1 Random Test Data

We considered two simple data creation methods, differing in the outliers generated in order to detect whether certain LMS or LTS approaches are more susceptible to their structure and distribution throughout the data.

We generated random 5-dimensional vectors such that one or two of the variables held most of the importance (to ensure the potential for high residuals). A random integer q between 400 and 600 was selected for each test, then each w_j calculated as $w_j = j^{q/100} - (j-1)^{q/100}$ before being normalized so that the vectors added to 1. The non-outlier data were generated with $\mathbf{x}_i$ drawn randomly from the unit hypercube (with uniform probability) and y-values calculated using $y_i = \sum_{j=1}^{n} w_j x_j$. Guassian noise was then added with standard deviation $\sigma = 0.05$.

Outliers were generated according to two methods. The first method assumes these values are just extra noisy values that follow the same model. Increasing the number of outliers present would not be expected to have a drastic impact on the fitted weighting vectors. The second method strategically positions the values so that the importance of the highest weight should be brought down and the fitted weighting vector would not represent the non-outlier data very well (See Fig. 2).

rand.data.1. x and y values were determined in the same way as for the non-outlier data, however with $\sigma = 0.1$ and an extra 0.3 added to the y value in the same direction as the noise, i.e. these data points are at least 6 standard deviations (with respect to the noise of non-outlying values) away from values calculated using the model **w**. Values outside the unit interval were discarded and redrawn.

rand.data.2. x values are centered according to the generating weighting vector **w** with the weights squared and divided by the maximum w_j^2 before Gaussian noise is added with $\sigma = 0.005$. The y values are set to 0 with Guassian noise added $\sigma = 0.1$ and negative values made positive.

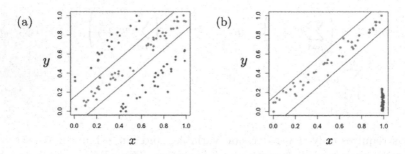

Fig. 2. Structure of random data generated for experiments using (a) rand.data.1 - data randomly distributed at least 6 standard deviations away from the generating function points, and (b) rand.data.2 - data distributed near $y = 0$, close to the corner of the hypercube corresponding to the dimension allocated the highest importance. This data is for the special case of 1 dimension - in our experiments we used 5-dimensional **x** vectors. Lines indicate 3 standard deviations (with respect to Gaussian noise of non-outlier data) either side of the generating function.

3.2   Algorithms Based on Random Sampling

We first tested 4 approaches based on Rousseeuw's approach [5,7,8] where we randomly sample sets of n inputs and use them to estimate the weights. In each case, we assume an input dataset consisting of m observed n-dimensional **x** inputs and the corresponding y values.

LMS1/LTS1. The weighting vector is initialized at $(1/n, 1/n, \ldots, 1/n)$ and objective value at m. For each of Q iterations, n observed instances are sampled and the corresponding matrix is solved[2] to give the hyperplane through those sampled points. *If* all weights are positive, the squared residual values between this hyperplane and all m points is calculated and the objective value determined (median residual for LMS, mean of smallest 50% of residuals for LTS). If this objective is lower than the current best, the weighting vector

---

[2] Achieved in R using solve(), provided the matrix is non-singular. In the event of singular matrices, the particular iteration contributed nothing to the output.

and best objective value are updated. The iteration is skipped if any of the weights are negative. After Q iterations, the current best weighting vector (not necessarily normalized) and square root of the objective are given as output.

LMS2/LTS2. Same setup as for the LMS1/LTS1 approach, however residuals are still calculated for weighting vectors that include negative values. After Q iterations, the residuals are re-calculated and all inputs with values lower than or equal to the median residual are allocated to the inclusion set S. For LMS, The fitting method of Eq. (2) is then used to find the final weighting vector and the corresponding median residual is then calculated. For LTS, the least squares fitting approach is used on S and then the corresponding root mean squared error of the smallest 50% of residuals according to the resulting weighting vector is calculated. In other words, the method of sampling and solving the system of n points is used to make a best guess at S and then exact fitting approaches are used on S.

LMS3/LTS3. As with previous approaches, subsets of n observations are randomly sampled in each iteration, however rather than solving the linear system, a weighting vector is found by optimizing with respect to the n points (which will always result in appropriate weighting vectors). LMS3a, LTS3a optimize with respect to the maximum error, LMS3b, LTS3b optimize with respect to the least squares criterion and LMS3c and LTS3c optimize with respect to the least absolute deviation. For each iteration, the weighting vector that minimizes the LMS or LTS objective is checked and stored if better than the current best. As with LMS2/LTS2, the best performing vector with respect to the objective is then used to establish the subset S and either the maximum error or least squares are minimized for S.

LMS4/LTS4. Same as LMS3a and LTS3b, however drawing $2n$ random observations.

In each of these methods, thousands of iterations can be used to sample the observations and find the best performing weight vector.

Experiments. To observe the effect of increasing iterations and to compare the approaches, these methods were tested for varying values of Q, and varying number of outliers. In each experiment, 100 non-outlier instances were generated along with 99 outliers, then each method was tested for each setting of $Q = \{5, 10, 15, 20, 25, 30, 35, 40, 45, 50, 100, 200, 500, 1000, 2000\}$, incorporating progressively more of the outliers in 10s, i.e. $10, 20, 30, \ldots, 90$ and 99. There were 100 random datasets generated using each of rand.data.1 and rand.data.2.

Influence of Outliers. Firstly, we are interested in the performance for the highest number of iterations in terms of whether the outliers influenced the weighting vector obtained. In each experiment, the outliers were deemed to have affected the output if the maximum error of the non-outlier data calculated was greater than the minimum error for the outliers.

In the case of data generated by rand.data.1, with the exception of one instance for LMS4, only LMS1 and LTS1 resulted in weighting vectors that were influenced by outliers with increasing frequency as more outliers were included. Table 1 shows the proportion of the 100 experiments where this occurred for each setting of the number of outliers.

Table 1. Prop. of fitted **w** influenced by outliers for LMS1 and LTS1 using rand.data.1

Outliers	10	20	30	40	50	60	70	80	90	99
LMS1	0.01	0	0.01	0.02	0.01	0.02	0.08	0.18	0.24	0.31
LTS1	0	0	0.01	0.01	0.02	0.05	0.07	0.16	0.18	0.32

The main reason these methods may be more susceptible to bad fitting is because an iteration is essentially wasted if the sample generates any negative weights.

For the data generated by rand.data.2, the proportion of tests where the methods resulted in weighting vectors affected by outliers was much higher. We show only LMS1/LTS1, LMS2/LTS2 and LMS3a/LTS3b to give an indication of the performance (all LMS3/LTS3 and LMS4/LTS4 results were similar) (Table 2).

Table 2. Prop. of fitted **w** influenced by outliers for LMS1 and LTS1 with rand.data.2

Outliers	10	20	30	40	50	60	70	80	90	99
LMS1	0	0	0	0.01	0.09	0.43	0.82	0.99	1	1
LTS1	0	0	0	0	0.10	0.31	0.76	0.96	0.99	1
LMS2	0	0	1	0	1	0.17	0.81	1	1	1
LTS2	0	0	0	0	0	0.19	0.67	0.96	1	1
LMS3a	0	0	0.02	0.12	0.28	0.58	0.88	0.99	1	1
LTS3b	0.01	0.01	0.05	0.14	0.31	0.60	0.84	0.98	1	1

For these results, it is not easy to determine whether the outliers exert an influence due to a 'bad fit' or because the objective is actually minimized by using the outliers. For example, where there were 50 outliers present, LMS1 had 9 instances where the resulting model was influenced by the outliers, however in 4 of those cases an unaffected model with a better objective was achieved using LMS3a. Conversely, of the 28 affected models using LMS3a, for 4 of these, there was an unaffected model with better error using LMS1. We can infer that the error rate in the presence of this many outliers with this structure in the data can be similar for affected and unaffected models.

Influence of Iterations. Our next question is how many iterations are required to achieve a good level of accuracy. For these particular data generation methods, all methods except for LMS1/LTS1 actually achieved a reasonable accuracy once the number of iterations was above 15, with only marginal improvements after 100. This is not overly surprising, since with these datasets and approaches, if the 5 sampled points happen to be non-outliers then the sampling should identify the plane closest to the non-outlying set and the final step should obtain the optimal error measure. The best methods overall for varying number of outliers were those that used LAD on the random subsets. Figure 3 shows the improvement from 5 to 100 iterations for all methods except for LTS1/LMS1 on both datasets with 50 outliers. LTS1 and LMS1 were not comparable to the remaining methods until the number of iterations was above 500 and at 2000 performed worst overall.

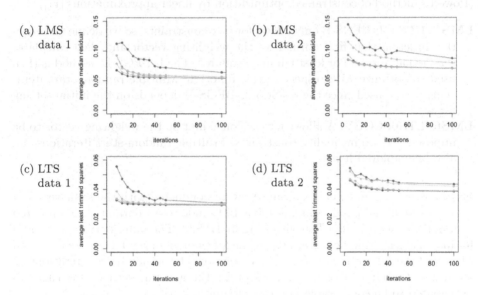

Fig. 3. Average error measures performance over 100 tests with 50 outliers present using rand.data.1 (data 1) and rand.data 2 (data 2). Red = LMS2/LTS2, Blue = LMS3a/LTS3a, Green = LMS3b/LTS3b, Yellow = LMS3c/LTS3b and Grey = LMS4/LTS4. (Color figure online)

Running Time. Lastly we can comment on the time taken to execute the algorithms, which increased close to linearly with the number of iterations. Table 3 shows average times for each of the methods. The LMS3b/LTS3b methods were the slowest, since implementation of LAD requires two extra decision variables for each observation.

3.3   General-Purpose Optimization

We can also look to whether general-purpose solvers can achieve a better trade-off between accuracy and time. We consider two multivariate optimization methods:

Table 3. Time taken on average (in seconds) to run each method with 100 non-outlier and 50 outlier data.

Iterations	LMS						LTS					
	1	2	3a	3b	3c	4	1	2	3a	3b	3c	4
100	0.003	0.047	0.081	0.103	0.065	0.088	0.003	0.049	0.065	0.102	0.081	0.075
500	0.013	0.228	0.392	0.461	0.307	0.419	0.014	0.242	0.308	0.471	0.401	0.317
1000	0.028	0.449	0.765	0.917	0.596	0.821	0.029	0.453	0.602	0.929	0.774	0.617
2000	0.056	0.882	1.523	1.820	1.173	1.625	0.059	0.888	1.198	1.839	1.539	1.231

the derivative based L-BFGS-B method (Broyden-Fletcher-Goldfarb-Shanno with lower and upper box constraints [16]) and the derivative-free COBYLA method (Powell's method of constrained optimization by linear approximations [17]).

LMS5/LTS5. L-BFGS-B only allows for box constraints, so we define an objective function that first normalizes the weighting vector and then calculates the median residual or least trimmed squares. The L-BFGS-B method is then used to optimize with respect to this function. Multiple random-start iterations can be used since the result of L-BFGS-B depends on the initial setting for **w**.

LMS6/LTS6. COBYLA allows for the constraint on the weighting vector to be imposed via two inequality constraints. Multiple random-start iterations can also be employed here.

Experiments. Numerical experiments were conducted with the same setup as for the random sampling techniques. For 100 random test datasets, we compared LMS5/LTS5 and LMS6/LTS6 with LMS3b/LTS3b. The same settings were used for increasing the number of outliers, while for number of random starts we tested $\{1, 3, 5, 10, 20, 50, 100\}$. For the comparison we set the number of iterations to 100 times the number of random starts for the general methods (as this was anticipated to be comparable in terms of time taken).

Influence of Outliers. For rand.data.1, for the highest number of random starts and iterations tested, LMS5 and LMS6 were influenced by outliers only for high number of outliers present. For 90 outliers, 2 and 1 instance respectively were influenced, while for 99 outliers, this rose to 33 and 8 respectively. In fact, even for 20 random starts, it was still only these two methods that were susceptible. For rand.data.2, all methods were similarly susceptible to outliers. For 50 outliers, the LMS methods had between 24 and 26 tests affected by outliers, while for LTS this was 30–31. Where there was 30 outliers, each method only had one instance where the outliers affected the result.

Influence of Number of Random Starts. As the methods were similarly affected by outliers, we can consider the accuracy in terms of median residual and least trimmed squares values obtained with respect to increases in the number

of random starts (or iterations for LMS3b/LTS3b). The general optimization
methods were more inaccurate where the number of random starts was below
50, however beyond this the methods were comparable.

Running Time. The time taken to run LMS3b/LTS3b was comparable to
LMS6 and LTS5, i.e. LMS used with the derivative-free COBYLA method and
LTS with L-BFGS-B, however LMS with L-BFGS-B and LTS with COBYLA
took considerably longer. This makes some sense given that once the non-outlier
data are found, the LTS problem is essentially a smooth quadratic problem while
for LMS there are points of discontinuity in the optimization function. Table 4
shows the average results, showing LMS6 and LTS3b to be slightly faster overall
over these tests although not significantly.

Table 4. Time taken on average to run general-purpose solvers (LMS5/LTS5,
LMS6/LTS6) compared with LMS3b/LTS3b for data generated by rand.data.1 with
10 outliers. Iterations* indicates number of random starts for general-purpose solvers
and number of iterations divided by 100 for LMS3b/LTS3b, i.e. 20 represents 2000.

Iterations*	LMS			LTS		
	3b	5	6	3b	5	6
1	0.087	0.428	0.073	0.084	0.093	0.256
3	0.250	1.399	0.229	0.252	0.266	0.774
5	0.416	2.289	0.380	0.418	0.460	1.371
10	0.825	4.542	0.771	0.838	0.912	2.630
20	1.651	9.145	1.547	1.666	1.808	5.249
50	4.124	22.436	3.805	4.170	4.563	13.408
100	8.245	45.162	7.649	8.350	9.049	26.902

4   Conclusions and Future Work

We have tested various approaches to LMS and LTS fitting of the weighted
arithmetic mean. Overall we found that random sampling techniques were fairly
competitive with general purpose solvers, however in the latter case there could
be improvements made by fine-tuning some of the parameters or altering the
objective functions slightly to make them smoother. We did investigate peeling
methods, i.e. removing outer points based on the initial optimization, however
these were not competitive for the techniques we have shown results for.

We can recommend the LMS3b or LMS6 approaches for fitting to the median
residual and LTS3b or LTS5 for fitting with respect to least trimmed squares,
although there is much more to investigate.

The techniques could be extended to other aggregation functions with some
additional problems arising in some cases, e.g., in the case of the Choquet integral, a random sample of observations, even if 2^n are taken, would not necessarily

cover all orderings and hence all simplexes over which the Choquet integral needs to be defined. It also has many additional constraints.

References

1. Beliakov, G., Pradera, A., Calvo, T.: Aggregation Functions: A Guide for Practitioners. Springer, Heidelberg (2007). https://doi.org/10.1007/978-3-540-73721-6
2. Beliakov, G.: Construction of aggregation functions from data using linear programming. Fuzzy Sets Syst. **160**, 65–75 (2009)
3. Beliakov, G., Bustince, H., Calvo, T.: A Practical Guide to Averaging Functions. Springer, Cham (2016). https://doi.org/10.1007/978-3-319-24753-3
4. Gagolewski, M.: Data Fusion: Theory, Methods and Applications. Institute of Computer Science, Polish Academy of Sciences, Warsaw (2015)
5. Rousseeuw, P.J.: Least median of squares regression. J. Am. Stat. Assoc. **79**(388), 871–880 (1984)
6. R Core Team: R: A Language and Environment for Statistical Computing. R Foundation for Statistical Computing, Vienna, Austria (2017)
7. Dallal, G.E., Rousseeuw, P.J.: LMSMVE: a program for least median of squares regression and robust distances. Comput. Biomed. Res. **25**, 384–391 (1992)
8. Rousseeuw, P.J., Hubert, M.: Recent developments in PROGRESS. In: L_1-Statistical Procedures and Related Topics. IMS Lecture Notes - Monograph Series, vol. 31, pp. 201–214 (1997)
9. Farebrother, R.W.: The least median of squared residuals procedure. In: Farebrother, R.W. (ed.) L_1-Norm and L_∞-Norm Estimation. BRIEFSSTATIST, pp. 37–41. Springer, Heidelberg (2013). https://doi.org/10.1007/978-3-642-36300-9_6
10. Grabisch, M., Marichal, J.L., Mesiar, R., Pap, E.: Aggregation Functions. Cambridge University Press, Cambridge (2009)
11. Torra, V., Narukawa, Y.: Modeling Decisions: Information Fusion and Aggregation Operators. Springer, Heidelberg (2007). https://doi.org/10.1007/978-3-540-68791-7
12. Beliakov, G., James, S.: Citation-based journal ranks: the use of fuzzy measures. Fuzzy Sets Syst. **167**, 101–119 (2011)
13. Beliakov, G., James, S.: Using linear programming for weights identification of generalized Bonferroni means in R. In: Torra, V., Narukawa, Y., López, B., Villaret, M. (eds.) MDAI 2012. LNCS (LNAI), vol. 7647, pp. 35–44. Springer, Heidelberg (2012). https://doi.org/10.1007/978-3-642-34620-0_5
14. Bloomfield, P., Steiger, W.: Least Absolute Deviations: Theory Applications and Algorithms. Birkhauser, Basel (1983)
15. Bartoszuk, M., Beliakov, G., Gagolewski, M., James, S.: Fitting aggregation functions to data: part I - linearization and regularization. In: Carvalho, J.P., Lesot, M.-J., Kaymak, U., Vieira, S., Bouchon-Meunier, B., Yager, R.R. (eds.) IPMU 2016. CCIS, vol. 611, pp. 767–779. Springer, Cham (2016). https://doi.org/10.1007/978-3-319-40581-0_62
16. Byrd, R.H., Lu, P., Nocedal, J., Zhu, C.: A limited memory algorithm for bound constrained optimization. SIAM J. Sci. Comput. **16**, 1190–1208 (1995)
17. Powell, M.J.D.: A direct search optimization method that models the objective and constraint functions by linear interpolation. In: Gomez, S., Hennart, J.P. (eds.) Advances in Optimization and Numerical Analysis. MAIA, vol. 275, pp. 51–67. Kluwer Academic, Dordrecht (1994). https://doi.org/10.1007/978-94-015-8330-5_4

Combining Absolute and Relative Information in Studies on Food Quality

Marc Sader$^{(\boxtimes)}$, Raúl Pérez-Fernández, and Bernard De Baets

KERMIT, Department of Data Analysis and Mathematical Modelling,
Ghent University, Coupure links 653, 9000 Ghent, Belgium
{marc.sader,raul.perezfernandez,bernard.debaets}@ugent.be

Abstract. A common problem in food science concerns the assessment of the quality of food samples. Typically, a group of panellists is trained exhaustively on how to identify different quality indicators in order to provide absolute information, in the form of scores, for each given food sample. Unfortunately, this training is expensive and time-consuming. For this very reason, it is quite common to search for additional information provided by untrained panellists. However, untrained panellists usually provide relative information, in the form of rankings, for the food samples. In this paper, we discuss how both scores and rankings can be combined in order to improve the quality of the assessment.

Keywords: Consensus evaluation · Absolute information
Relative information

1 Introduction

We consider the problem in which several panellists are asked to score a food sample on a given ordinal scale, the goal being to reach a consensus evaluation of the sample. This problem commonly appears in food science, for instance, when identifying the degree of spoilage [1,2] or when evaluating the appearence [3,4] of a given sample. Unfortunately, training and (subsequently) collecting information from panellists usually carries big expenses. For this reason, there usually is a limited amount of data available to reach a consensus evaluation. It is thus quite common to invoke untrained panellists and to gather some additional information [5]. However, untrained panellists are obviously not as skilled as trained panellists, and might be unable to accurately evaluate a given sample. Since it is a conceptually easier task, untrained panellists are then just asked to rank different samples according to their personal appreciation. In this paper, we propose to combine both types of information in order to improve the quality of the assessment. Moreover, we illustrate our proposal by discussing an experiment concerning the freshness of raw Atlantic salmon (*Salmo salar*) [6].

The remainder of the paper is organized as follows. In Sect. 2, we recall the well-known notions of median and Kemeny median. Section 3 is devoted to the introduction of a method for reaching a consensus evaluation of the given samples while combining both scoring and ranking information. We end with some conclusions in Sect. 4.

2   Preliminaries

2.1   Obtaining a Consensus Vector of Scores

We consider the setting where n_T trained panellists are asked to assign a score on a k-point scale to each (food) sample in a set $X = \{x_1, \ldots, x_n\}$ of n (food) samples. The goal is to agree on the consensus score that should be assigned to each of the samples based on the scores provided by the trained panellists. For any $i \in \{1, \ldots, n_T\}$, we denote by $\mathbf{s}_i$ the vector of scores assigned by the i-th panellist. The scale we use throughout this paper is shown in Fig. 1.

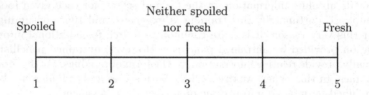

Fig. 1. Example of a 5-point scale, where the extreme scores of "1" and "5" represent spoiled and fresh, respectively, and the intermediate score of "3" represents a neutral response of neither spoiled nor fresh.

A common method for determining the consensus vector of scores is based on the minimization of a distance, i.e., the consensus vector of scores $\mathbf{s}^*$ should satisfy

$$\mathbf{s}^* = \operatorname*{arg\,min}_{\mathbf{s} \in \{1,\ldots,k\}^n} \sum_{i=1}^{n_T} d(\mathbf{s}, \mathbf{s}_i),$$

where d is a fixed distance function on the set of vectors of scores. Note that there can be multiple minimizers $\mathbf{s}^*$.

One could note that several examples of this procedure are commonly used in practice. For instance, when we consider the sum of zero-one distances[1] over all components, the preceding method amounts to identifying the mode(s). Similarly, when we consider the sum of ℓ_1-distances[2], the preceding method amounts to identifying the median(s), and, when we consider the sum of ℓ_2-distances[3], it amounts to identifying the mean(s). One could note that the latter method presumes the existence of a certain notion of distance between labels, something that is not advisable in case the considered scale is defined by abstract words [7].

---

[1] The zero-one distance function is defined as $d_0(s, s') = 0$ if $s = s'$ and $d_0(s, s') = 1$ otherwise.
[2] The ℓ_1-distance function is defined as $d_1(s, s') = |s - s'|$.
[3] The ℓ_2-distance function is defined as $d_2(s, s') = (s - s')^2$.

Example 1. Consider the set of $n = 4$ samples $X = \{x_1, x_2, x_3, x_4\}$ and the vectors of scores on the fixed 5-point scale provided by $n_T = 10$ trained panellists shown in Table 1. Note that these data come from a real-life dataset concerning an experiment on raw Atlantic salmon (*Salmo salar*) [6].

Table 1. The scores assigned to samples x_1, x_2, x_3 and x_4 by the trained panellists.

	s_1	s_2	s_3	s_4	s_5	s_6	s_7	s_8	s_9	s_{10}
x_1	5	4	5	2	3	3	5	5	2	4
x_2	5	2	1	2	4	2	5	4	3	3
x_3	2	1	5	2	2	4	3	2	2	2
x_4	3	1	2	1	2	2	2	3	3	1

For each of the 625 possible vectors of scores, we compute the sum of ℓ_1-distances to the vectors of scores provided by the trained panellists. We conclude that the vector of scores that minimizes this value is $s^* = (4, 3, 2, 2)$. As expected, this vector coincides with the one obtained by identifying the median for each of the samples. ◁

2.2 Obtaining a Consensus Ranking

We consider the setting where n_U untrained panellists are asked to rank all the samples on the considered set $X = \{x_1, \ldots, x_n\}$ of n samples. Untrained panellists are asked to provide a complete ranking of the samples, however, they are allowed to express ties in case they consider two or more samples to be equally suitable. The goal is to agree on the consensus ranking that should be assigned to each of the samples based on the scores provided by the trained panellists. For any $i \in \{1, \ldots, n_U\}$, we denote by $\precsim_i$ the ranking assigned by the i-th panellist, which can be split into the usual symmetric $\sim_i$ and antisymmetric $\prec_i$ parts. We denote by $\mathcal{W}$ the set of all rankings (with ties) on X.

A common method for determining the consensus ranking is due to Kemeny [8] in which a consensus ranking $\precsim^*$ is one that satisfies

$$\precsim^* = \arg\min_{\precsim \in \mathcal{W}} \sum_{i=1}^{n_U} K(\precsim, \precsim_i),$$

where $K(\precsim^1, \precsim^2)$ denotes the Kemeny distance between two rankings $\precsim^1$ and $\precsim^2$. We recall that the Kemeny distance[4] between two rankings is computed as follows. For each pair of samples $\{x_u, x_v\}$, if both rankings agree on the order of the samples, we write down 0; if, in one ranking, x_u is ranked above x_v (or x_v is ranked above x_u) and, in the other ranking, x_u and x_v are tied, we write

---

[4] When the rankings contain no ties, the Kemeny distance is equal to the double of the Kendall distance [9].

down 1; and, if, in one ranking, x_u is ranked above x_v and, in the other ranking, x_v is ranked above x_u, we write down 2. After writing down the numbers for all $n(n-1)/2$ possible pairs, the Kemeny distance between the two rankings equals the sum of these numbers.

Example 2. Consider the same set of $n = 4$ samples $X = \{x_1, x_2, x_3, x_4\}$ of Example 1, and the rankings provided by $n_U = 28$ untrained panellists shown in Table 2. Note that these data also come from the experiment on raw Atlantic salmon (*Salmo salar*) in [6].

Table 2. The rankings expressed by the untrained panellists.

i	$\precsim_i$	i	$\precsim_i$
1	$x_2 \sim x_4 \prec x_3 \sim x_2$	15	$x_2 \prec x_1 \prec x_4 \prec x_3$
2	$x_4 \prec x_3 \prec x_2 \prec x_1$	16	$x_4 \prec x_3 \sim x_2 \prec x_1$
3	$x_4 \prec x_2 \prec x_3 \sim x_1$	17	$x_4 \prec x_3 \sim x_2 \prec x_1$
4	$x_1 \prec x_4 \prec x_3 \prec x_2$	18	$x_4 \prec x_3 \sim x_2 \prec x_1$
5	$x_1 \sim x_4 \prec x_3 \prec x_2$	19	$x_2 \prec x_4 \prec x_3 \prec x_1$
6	$x_4 \prec x_3 \sim x_1 \sim x_2$	20	$x_4 \prec x_3 \prec x_2 \prec x_1$
7	$x_4 \sim x_3 \prec x_1 \sim x_2$	21	$x_1 \sim x_4 \prec x_3 \sim x_2$
8	$x_4 \prec x_1 \sim x_2 \prec x_2$	22	$x_1 \sim x_2 \prec x_3 \prec x_4$
9	$x_1 \prec x_3 \sim x_2 \prec x_4$	23	$x_4 \prec x_3 \prec x_2 \prec x_1$
10	$x_2 \prec x_4 \prec x_1 \prec x_3$	24	$x_4 \prec x_1 \prec x_2 \prec x_3$
11	$x_4 \prec x_2 \prec x_3 \prec x_1$	25	$x_4 \prec x_2 \prec x_3 \prec x_1$
12	$x_4 \prec x_1 \prec x_3 \sim x_2$	26	$x_1 \prec x_4 \prec x_3 \prec x_2$
13	$x_2 \sim x_4 \sim x_3 \prec x_1$	27	$x_4 \prec x_2 \prec x_3 \prec x_1$
14	$x_2 \prec x_4 \prec x_1 \prec x_3$	28	$x_4 \prec x_2 \prec x_3 \prec x_1$

For each of the 75 possible rankings, we compute the sum of Kemeny distances to the rankings provided by the untrained panellists. We conclude that the ranking that minimizes this value is $\precsim^* = x_4 \prec x_3 \sim x_2 \prec x_1$. $\triangleleft$

3 Combining Scores and Rankings

We now consider the setting where n_T trained panellists each have assigned a score to each of the n samples in X and n_U untrained panellists each have ranked the n samples in X. The goal is to combine both types of information in order to improve the quality of the assessment of the consensus vector of scores and/or ranking. We propose to consider a combination of the median and the Kemeny median.

3.1 Improving the Quality of the Assessment of a Consensus Vector of Scores

To compute the 'distance'[5] between each possible vector of scores $\mathbf{s}$ and the rankings provided by the untrained panellists, we define the set $\theta_{\mathbf{s}}$ of all possible rankings that do not contradict $\mathbf{s}$, as follows:

$$\theta_{\mathbf{s}} = \left\{ \precsim \in \mathcal{W} \,\middle|\, (\forall i,j \in \{1,\dots,n\})\big(\mathbf{s}(i) < \mathbf{s}(j) \Rightarrow x_i \prec x_j\big) \right\}. \qquad (1)$$

Note that the set $\theta_{\mathbf{s}}$ is always non-empty.

Incorporating the rankings provided by the untrained panellists into the vectors of scores provided by the trained panellists requires defining a cost function. Thus, we define a convex combination of the 'distances' associated with the vectors of scores provided by the trained panellists and the rankings provided by the untrained panellists, as follows:

$$C_\alpha(\mathbf{s}) = \frac{\alpha}{B_T} \sum_{i=1}^{n_T} d_1(\mathbf{s}, \mathbf{s}_i) + \frac{(1-\alpha)}{B_U} \min_{\precsim \in \theta_{\mathbf{s}}} \sum_{i=1}^{n_U} K(\precsim, \precsim_i). \qquad (2)$$

where $B_T = n_T \cdot n \cdot (k-1)$ and $B_U = n_U \cdot n \cdot (n-1)$ are normalizing constants, and $\alpha \in [0,1]$ is a parameter that controls the influence of the scoring and ranking information. In particular, larger values of α give more importance to the trained panellists, whereas smaller values of α give more importance to the untrained panellists.

Finally, we consider the consensus vector(s) of scores to be the minimizer(s) of Eq. (2) for a fixed α, as follows:

$$\mathbf{s}_\alpha^* = \operatorname*{arg\,min}_{\mathbf{s} \in \{1,\dots,k\}^n} C_\alpha(\mathbf{s}). \qquad (3)$$

Note that there can be multiple minimizers $\mathbf{s}_\alpha^*$ for the same α.

Example 3. We continue with the data from Examples 1 and 2. To determine the consensus score that should be assigned to each of these samples, we consider the problem defined by Eq. (3) by computing $C_\alpha(\mathbf{s})$ for each of the 625 vectors of scores. For simplicity, we show one computation for the vector of scores $\mathbf{s} = (4,3,2,2)$, which was determined as the consensus vector of scores in Example 1, with $\sum_{i=1}^{10} d_1(\mathbf{s}, \mathbf{s}_i) = 34$. The distances associated with the vectors of scores are bounded by the upper bound $B_T = 10 \cdot 4 \cdot 4 = 160$, whereas the distances associated with the rankings are bounded by the upper bound $B_U = 28 \cdot 4 \cdot 3 = 336$. Now, we consider the set $\theta_{\mathbf{s}}$ of all possible rankings that do not contradict $\mathbf{s}$. Since the score of x_1 is the largest, x_1 is ranked above the other samples. Similarly, x_2 is ranked above x_3 and x_4. Since the scores of x_3 and x_4 are equal,

---

[5] Note that we write the word 'distance' between quotation marks since we are comparing objects of a different nature, and, thus, we are lacking the semantics associated with the mathematical formalization of a distance (metric).

any of the following cases applies: x_3 is ranked above x_4, x_3 and x_4 are tied, and x_4 is ranked above x_3, as follows:

$$\theta_{(4,3,2,2)} = \left\{ \begin{array}{l} x_4 \prec x_3 \prec x_2 \prec x_1, \\ x_3 \sim x_4 \prec x_2 \prec x_1, \\ x_3 \prec x_4 \prec x_2 \prec x_1 \end{array} \right\}.$$

We compute the sum of the Kemeny distances between each $\precsim \in \theta_{\mathbf{s}}$ and the rankings provided by the untrained panellists. The results are summarized in Table 3.

Table 3. Sum of Kemeny distances between each ranking $\precsim$ that does not contradict $\mathbf{s} = (4, 3, 2, 2)$ and the rankings provided by the untrained panellists.

$\precsim$	$\sum_{i=1}^{28} K(\precsim, \precsim_i)$
$x_4 \prec x_3 \prec x_2 \prec x_1$	109
$x_3 \sim x_4 \prec x_2 \prec x_1$	129
$x_3 \prec x_4 \prec x_2 \prec x_1$	153

Finally, we select the ranking that minimizes the sum of Kemeny distances among those in $\theta_{\mathbf{s}}$ and compute $C_\alpha(\mathbf{s})$ as follows:

$$C_\alpha(\mathbf{s}) = \frac{\alpha}{160} \sum_{i=1}^{10} d_1(\mathbf{s}, \mathbf{s}_i) + \frac{(1-\alpha)}{336} \min \left(\begin{array}{l} \sum_{i=1}^{28} K(x_4 \prec x_3 \prec x_2 \prec x_1, \precsim_i), \\ \sum_{i=1}^{28} K(x_3 \sim x_4 \prec x_2 \prec x_1, \precsim_i), \\ \sum_{i=1}^{28} K(x_3 \prec x_4 \prec x_2 \prec x_1, \precsim_i) \end{array} \right)$$

$$= \frac{109}{336} - \frac{6016}{53760} \alpha.$$

After computing $C_\alpha(\mathbf{s})$ for each of the 625 possible vectors of scores $\mathbf{s}$, we illustrate in Fig. 2 all the $\mathbf{s}_\alpha^*$ that minimize $C_\alpha(\mathbf{s})$ for at least one value of $\alpha \in [0, 1]$. One should note that, for $\alpha = 0$, there will always be multiple minimizers $\mathbf{s}_0^*$ associated with all vectors of scores that are not contradicted by the Kemeny median. Since we know from Example 2 that $\precsim^* = x_4 \prec x_3 \sim x_2 \prec x_1$ is the Kemeny median, we illustrate (in black) all the vectors of scores $\mathbf{s}_0^*$ that are not contradicted by this $\precsim^*$. These vectors of scores form a fan-shaped pattern starting at $\alpha = 0$ since, at the left end, they all result in the same value $C_0(\mathbf{s})$, and, at the right end, they result in (mostly) different values $C_1(\mathbf{s})$.

Since we do not intend to rely only on the rankings provided by the untrained panellists, we ignore the minimizers for $\alpha = 0$. The obtained minimizers $\mathbf{s}_\alpha^*$ are summarized as follows:

$$\mathbf{s}_\alpha^* = \left\{ \begin{array}{ll} \{(4, 2, 2, 2)\} & , \text{ if } 0 < \alpha < \frac{50}{71}, \\ \{(4, 2, 2, 2), (4, 3, 2, 2)\} & , \text{ if } \quad \alpha = \frac{50}{71}, \\ \{(4, 3, 2, 2)\} & , \text{ if } \quad \alpha > \frac{50}{71}. \end{array} \right. \tag{4}$$

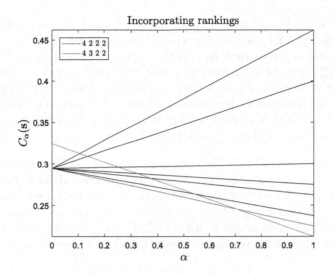

Fig. 2. Illustration of the vectors of scores $\mathbf{s}_\alpha^*$ that minimize $C_\alpha(\mathbf{s})$ for $\alpha \in [0, 1]$.

We deduce based on the vectors of scores provided by the trained panellists that x_2 is ranked at a better position than x_3 and x_4, since the former is assigned a higher score than the former in the consensus vector of scores for large values of α. However, incorporating the rankings provided by the untrained panellists hints that these samples are similar.   ◁

3.2 Improving the Quality of the Assessment of a Consensus Ranking

To compute the 'distance' between each possible ranking $\precsim$ and the vectors of scores provided by the trained panellists, we define the set $\varphi_{\precsim}$ of all possible vectors of scores that do not contradict $\precsim$, as follows:

$$\varphi_{\precsim} = \left\{ \mathbf{s} \in \{1, \ldots, k\}^n \, \middle| \, (\forall i, j \in \{1, \ldots, n\})\big(x_i \precsim x_j \Rightarrow \mathbf{s}(i) \leq \mathbf{s}(j)\big) \right\}. \tag{5}$$

Note that the set $\varphi_{\precsim}$ is always non-empty.

The convex combination of the 'distances' associated with the vectors of scores provided by the trained panellists and the rankings provided by the untrained panellists is now defined as follows:

$$D_\alpha(\precsim) = \frac{\alpha}{B_T} \min_{\mathbf{s} \in \varphi_{\precsim}} \sum_{i=1}^{n_T} d_1(\mathbf{s}, \mathbf{s}_i) + \frac{(1-\alpha)}{B_U} \sum_{i=1}^{n_U} K(\precsim, \precsim_i). \tag{6}$$

Finally, we consider the consensus ranking(s) to be the minimizer(s) of Eq. (6) for a fixed α, as follows:

$$\precsim_\alpha^* = \arg\min_{\precsim \in \mathcal{W}} D_\alpha(\precsim). \tag{7}$$

Note that there can be multiple minimizers $\precsim_\alpha^*$ for the same α.

Since $\alpha \in [0,1]$ can take infinite values, it will be impossible to compute $\mathbf{s}_\alpha^*$ (resp. $\precsim_\alpha^*$) for each α. Therefore, bearing in mind that, for any fixed vector of scores $\mathbf{s}$ (resp. ranking $\precsim$), the corresponding $f(\alpha) := C_\alpha(\mathbf{s})$ (resp. $g(\alpha) := D_\alpha(\precsim)$) can be visualized as a line, we can compare the lines of each possible pair of vectors of scores (resp. rankings). When comparing two lines, we distinguish three cases: there are no points of intersection, there is exactly one point of intersection, or both lines coincide. These facts can then be used to analytically compute $\mathbf{s}_\alpha^*$ and $\precsim_\alpha^*$ as a function of α.

Example 4. We continue with the data from Example 3. To determine the consensus ranking of the samples, we consider the problem defined by Eq. (7) by computing $D_\alpha(\precsim)$ for each of the 75 rankings. For simplicity, we show one computation for the ranking $\precsim = x_4 \prec x_3 \sim x_2 \prec x_1$, which was determined as the consensus ranking in Example 2, with $\sum_{i=1}^{28} K(\precsim, \precsim_i) = 99$. Now, we consider the set $\varphi_{\precsim}$ of all possible vectors of scores that do not contradict $\precsim$.

$$\varphi_{x_4 \prec x_3 \sim x_2 \prec x_1} = \left\{ \mathbf{s} \in \{1, \ldots, 5\}^4 \mid \mathbf{s}(4) \leq \mathbf{s}(3) = \mathbf{s}(2) \leq \mathbf{s}(1) \right\}.$$

We compute the sum of the ℓ_1-distances between each $\mathbf{s} \in \varphi_{\precsim}$ and the vectors of scores provided by the untrained panellists. We note that the vector of scores among those in $\varphi_{\precsim}$ that minimizes the sum of the ℓ_1-distances is $(4, 2, 2, 2)$ with $\sum_{i=1}^{10} d_1(\mathbf{s}, \mathbf{s}_i) = 36$. Finally, we obtain:

$$D_\alpha(\precsim) = \frac{113}{336} - \frac{5984}{53760} \alpha.$$

After computing $D_\alpha(\precsim)$ for each of the 75 possible rankings $\precsim$, we illustrate in Fig. 3 all the $\precsim_\alpha^*$ that minimize $D_\alpha(\precsim)$ for at least one value of $\alpha \in [0,1]$. One should note that, for $\alpha = 1$, there will always be multiple minimizers $\mathbf{s}_1^*$ associated with all rankings that are not contradicted by the median. Since we know from Example 1 that $\mathbf{s}^* = (4, 3, 2, 2)$ is the median, we illustrate all the rankings $\precsim_1^*$ that are not contradicted by this $\mathbf{s}^*$. These rankings form a fan-shaped pattern starting at $\alpha = 1$ since, at the right end, they all result in the same value $D_1(\precsim)$, and, at the left end, they result in (mostly) different values $D_0(\precsim)$.

Since we do not intend to rely only on the scores provided by the trained panellists, we ignore the minimizers for $\alpha = 1$. The obtained minimizers $\precsim_\alpha^*$ are summarized as follows:

$$\mathbf{s}_\alpha^* = \begin{cases} \{x_4 \prec x_3 \sim x_2 \prec x_1\} & , \text{ if } \quad \alpha < \frac{50}{71}, \\ \left.\begin{cases} x_4 \prec x_3 \sim x_2 \prec x_1 \\ x_4 \prec x_3 \prec x_2 \prec x_1 \end{cases}\right\} & , \text{ if } \quad \alpha = \frac{50}{71}, \\ \{x_4 \prec x_3 \prec x_2 \prec x_1\} & , \text{ if } \frac{50}{71} < \alpha < 1. \end{cases} \tag{8}$$

We deduce based on the rankings provided by the untrained panellists that x_2 and x_3 are similar. However, incorporating the vectors of scores provided by the trained panellists hints that sample x_2 might be ranked above sample x_3.   ◁

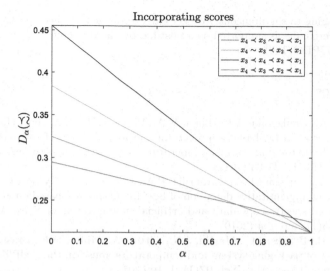

Fig. 3. Illustration of the rankings $\precsim_\alpha^*$ that minimize $D_\alpha(\precsim)$ for $\alpha \in [0, 1]$.

3.3 Discussion

A deeper analysis of the results of the preceding subsections shows that both trained and untrained panellists agree that samples x_1 and x_4 are, respectively, the best and worst samples in $X = \{x_1, x_2, x_3, x_4\}$. However, there is a disagreement with regard to samples x_2 and x_3. While trained panellists considered sample x_2 to be better than sample x_3, untrained panellists did not see significant differences between both samples. Thus, as can be concluded from both Eqs. (4) and (8), samples x_2 and x_3 result to be similar ($\mathbf{s}^*(3) = \mathbf{s}^*(2)$ and $x_3 \sim^* x_2$) for smaller values of α (i.e., in case more importance is given to the untrained panellists), whereas sample x_2 results to be better than sample x_3 ($\mathbf{s}^*(3) < \mathbf{s}^*(2)$ and $x_3 \prec^* x_2$) for larger values of α (i.e., in case more importance is given to the trained panellists).

4   Conclusions

In this paper, we have discussed how to combine absolute and relative information in order to improve the quality of the assessment of food samples. In particular, we have proposed a method based on a convex combination of the distances associated with the median and the Kemeny median. We have illustrated the use of this method using real-life examples, where the freshness of Atlantic salmon was studied, and showed the influence of combining scores and rankings on obtaining the consensus vector of scores and consensus ranking.

Acknowledgments. We gratefully acknowledge Innovation by Science and Technology (IWT) (now known as Flanders Innovation and Entrepreneurship (VLAIO)) for their support of the project CheckPack (IWT-SBO-130036) - Integrated optical sensors in food packaging to simultaneously detect early-spoilage and check package integrity. Raúl Pérez-Fernández is supported as a postdoc by the Research Foundation of Flanders (FWO17/PDO/160).

References

1. Argyri, A.A., Doulgeraki, A.I., Blana, V.A., Panagou, E.Z., Nychas, G.J.E.: Potential of a simple HPLC-based approach for the identification of the spoilage status of minced beef stored at various temperatures and packaging systems. Int. J. Food Microbiol. **150**(1), 25–33 (2011)
2. Argyri, A.A., Panagou, E.Z., Tarantilis, P.A., Polysiou, M., Nychas, G.J.E.: Rapid qualitative and quantitative detection of beef fillets spoilage based on Fourier transform infrared spectroscopy data and artificial neural networks. Sens. Actuators B: Chem. **145**(1), 146–154 (2010)
3. Rogers, H.B., Brooks, J.C., Martin, J.N., Tittor, A., Miller, M.F., Brashears, M.M.: The impact of packaging system and temperature abuse on the shelf life characteristics of ground beef. Meat Sci. **97**(1), 1–10 (2014)
4. Smolander, M., Hurme, E., Latva-Kala, K., Luoma, T., Alakomi, H.L., Ahvenainen, R.: Myoglobin-based indicators for the evaluation of freshness of unmarinated broiler cuts. Innov. Food Sci. Emerg. Technol. **3**(3), 279–288 (2002)
5. Amerine, M.A., Pangborn, R.N., Poessler, E.B.: Principles of Sensory Evaluation of Food. Academic Press, New York (1965)
6. Sader, M., Pérez-fernández, R., Kuuliala, L., Devlieghere, F., De Baets, B.: A combined scoring and ranking approach for determining overall food quality (2018, submitted)
7. Franceschini, F., Galetto, M., Varetto, M.: Qualitative ordinal scales: the concept of ordinal range. Qual. Eng. **16**(4), 515–524 (2004)
8. Kemeny, J.G.: Mathematics without numbers. Daedalus **88**(4), 577–591 (1959)
9. Kendall, M.G.: A new measure of correlation. Biometrika **30**(1–2), 81–93 (1938)

Twofold Binary Image Consensus for Medical Imaging Meta-Analysis

Carlos Lopez-Molina[1,3](✉), Javier Sanchez Ruiz de Gordoa[2],
Victoria Zelaya-Huerta[2], and Bernard De Baets[3]

[1] Departamento de Automatica y Computacion, Universidad Publica de Navarra,
Pamplona, Spain
carlos.lopez@unavarra.es
[2] NavarraBiomed, Servicio Navarro de Salud/Osasunbidea, Pamplona, Spain
[3] KERMIT, Department of Data Analysis and Mathematical Modelling,
Ghent University, Ghent, Belgium

Abstract. In the field of medical imaging, ground truth is often gathered from groups of experts, whose outputs are generally heterogeneous.
This procedure raises questions on how to compare the results obtained
by automatic algorithms to multiple ground truth items. Secondarily, it
raises questions on the meaning of the divergences between experts. In
this work, we focus on the case of immunohistochemistry image segmentation and analysis. We propose measures to quantify the divergence in
groups of ground truth images, and we observe their behaviour. These
measures are based upon fusion techniques for binary images, which is
a common example of non-monotone data fusion process. Our measures
can be used not only in this specific field of medical imagery, but also
in any task related to meta-quality evaluation for image processing, e.g.
ground truth validation or expert rating.

Keywords: Data fusion · Twofold Consensus Ground Truth
Meta-analysis · Medical imagery · Immunohistochemistry (IHC)

1 Introduction

Data fusion pursues rather different goals in very disparate contexts. An common goal is to produce a reduced (compact) representation of a certain amount
of data objects. Whichever specific technique the fusion is based upon, and
whichever data objects are to be fused, reduction is the main goal in most fusion
processes. However, fusion can lead the way to some other subsidiary goals just
as interesting as reduction. For example, the result of a fusion process can be
used as starting point to study the data to be processed, including its individual and group characteristics. Otherwise said, it can be used for data analysis,
specifically to generate metadata (data about data).

The application of data fusion techniques to produce metadata is certainly
not novel; in this regard, a relevant example is the standard deviation. The

© Springer International Publishing AG, part of Springer Nature 2018
J. Medina et al. (Eds.): IPMU 2018, CCIS 854, pp. 389–400, 2018.
https://doi.org/10.1007/978-3-319-91476-3_33

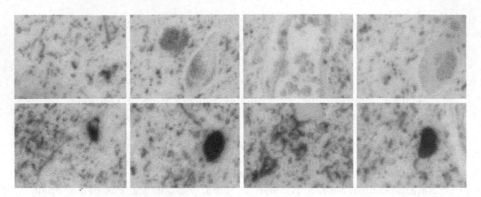

Fig. 1. Subimages extracted from Immunohistochemistry (IHC) images. The upper row displays regions without artifacts affected by tau protein, while the lower row displays artifacts or regions in which the presence of such protein is evident.

arithmetic mean can be seen as procedure to fuse scalar data into a compact representation with minimal loss of information, such loss being measured as the sum of the squared distance to the original values. At the same time, it is also a key to compute the standard deviation, which is a dispersion measurement. Even when dealing with non-Gaussian distribution of values, the standard deviation is used as a feature in data meta-analysis. We believe that principles similar to those by the mean and standard deviation can be ported to scenarios in which monotonicity plays no role. That is, we believe that fusion of non-standard data can also be taken as starting point to produce metadata in non-motonote universes.

In this work we elaborate on images in the context of neurology and neuropathology. This work is part of a research effort on immunohistochemistry (IHC) images for the measurement of deposits of tau protein in patients affected by Progressive Supranuclear Palsy (PSP). In this research effort, expert neuropathologists analyze microscope images of brain tissue and perform manual labelling of the areas affected by tau protein (see Fig. 1). Such binary labelling is further used to perform quantitative measurements with interest for posthumous analysis and disease profiling. Since this process is extremely time-consuming, automatic segmentation methods are being proposed to alleviate the workload of the pathologists. These methods shall be designed to produce results similar to those by expert humans. A key problem found in the evaluation and tuning of such automatic methods is the fact that pathologists often feature severe differences of opinion and/or precision. From a computational point of view, they generate different binary images which shall be taken as ground truth (that is, perfect solutions) for automatic segmentation algorithms. Of course, the multiplicity of ground truth solutions severely hinder the evaluation (and training) of such algorithms. Understanding and evaluating set-based or multivalued ground truth is hence a priority for our applied developments.

We propose to fuse the binary images produced by neuropathologists using the Twofold Consensus Ground Truth (TCGT, [4]). Our approach is rather different to that by other binary image fusion techniques (e.g. [2,9]), in the sense that we avoid the statistical counting of visual items, and rather focus on the spatial interpretation of coincidences and divergences. The TCGT takes as input a set of binary images and yields a set-valued consensus based on the coincidences and divergences in the input images. The resulting set allows for a compact representation of the input set of images, and also for the quantification of some of its characteristics. In this regard, we attempt to quantify two facets of the ground truth images. Firstly, we intend to quantify heterogeneity of the set of ground truth images, since it could be related to the difficulties faced by neuropathologists in the labelling of the original image. Secondly, we aim to evaluate the dissimilarity (degree of coincidence and divergence) of a ground truth image w.r.t. a group og ground truth images. In a sense, the first question relates to the group dispersion or heterogeneity, while the second one relates to the one-to-many dissimilarity of the images. Note that, although initially designed to elaborate on binary edge images, the TCGT can be ported to scenarios in which binary images hold different semantics.

The remainder of this work is organized as follows. In Sect. 2 we introduce the idea of weak and strong consensus, together with the Twofold Consensus Ground Truth. The usefulness of this concept is explained in Sect. 3, in which we develop the application for the meta analysis of IHC ground truth. Finally, Sect. 4 features some conclusions and future lines of research.

2   Twofold Consensus Ground Truth

2.1   Preliminary Notations

In this work we consider images to have some fixed dimensions $\mathcal{M} \times \mathcal{N}$, so that $\Omega = \{1, \ldots, \mathcal{M}\} \times \{1, \ldots, \mathcal{N}\}$ represents the set of positions in an image. The set of all binary images is denoted $\mathbb{B}$, and can be dually seen as the set of mappings $\Omega \mapsto \{0, 1\}$, or as the power set $\wp(\Omega)$. Individual binary images will be referred to with upper case (e.g. E, I), while bold-faced upper case is reserved for sets of images (e.g. $\mathbf{A} = \{A_1, \ldots, A_n\}$).

In this work we consider positive information in binary images to be represented by 1's, while negative information takes 0's. When it comes to the processing of binary images, we can use a dual signal-logical interpretation of this fact. Hence, apart from image-oriented operators, we use the classical set-theoretic operations on binary images, namely intersection ($\wedge$), union ($\vee$), and inclusion ($\subseteq$, $\subset$). The symbols $\cap$ and $\cup$ are reserved for the intersection and union of sets of images, respectively. According to the reference works on binary image morphology [1,8], the dilation of a binary image A by some structuring element K is given by $\mathcal{D}_K(A) = \{c \in \Omega \mid c = a + b \text{ for some } a \in A \text{ and } b \in K\}$.

2.2   Strong and Weak Consensus on Binary Images

Binary images are a very common format to express the output of image processing tasks, despite being barely useful to represent visual information in human terms. This holds, for example, for object recognition or binary segmentation. The nature and shape of the information in a binary image can greatly diverge from task to task, examples being regions (for object recognition or salient region identification), lines (for boundary detection), points (for critical point detection), etc. In many of such cases there is a need to combine different images, either to fuse ground truth images [9] or to fusion different candidate images generated by different algorithms. In [4] we present a technique for binary image fusion, namely the Twofold Consensus Ground Truth (TCGT). Due to the variable understanding of the term consensus, our technique narrows down its goals to three facts, enunciated as follows:

G1. *Preserving discordances:* The consensus should represent non-unanimous features in the images.

G2. *Highlighting agreement:* The consensus must point out those aspects on which the original images agree, either positively (features appearing at all images) or negatively (those appearing at none).

G3. *Keeping original images as perfect:* As long the input images are the only source of ground truth, the result of the fusion must somehow include them. This guarantees that any automatic method performing exactly as a the sources (probably, humans) is evaluated as perfect.

The TCGT is supported by two different consensus operators, namely the *strong* and *weak consensus*.

Definition 1. *The strong consensus image of a set of binary images* $\mathbf{I} = \{I_1, \ldots, I_k\}$ *is the binary image* $s_T(\mathbf{I})$ *defined as*

$$s_T(\mathbf{I}) = \mathcal{D}_T(I_1) \wedge \mathcal{D}_T(I_2) \wedge \ldots \wedge \mathcal{D}_T(I_k) \, , \tag{1}$$

where $\mathcal{D}_T(I_i)$ *denotes the dilation of image* I_i *using the structuring element* T.

Definition 2. *The weak consensus image of a set of binary images* $\mathbf{I} = \{I_1, \ldots, I_k\}$ *is the binary image* $w_T(\mathbf{I})$ *defined as*

$$w_T(\mathbf{I}) = \mathcal{D}_T(I_1) \vee \mathcal{D}_T(I_2) \vee \ldots \vee \mathcal{D}_T(I_k), \tag{2}$$

where $\mathcal{D}_T(I_i)$ *denotes the dilation of image* I_i *using the structuring element* T.

The strong and weak consensus of a set of images materialize as the tightest and loosest agreement that can be reached given a set of binary images $\mathbf{I}$. In this sense, they resemble the upper and lower bounds of interval-valued data, or the boundaries of rough sets [6]. Note that their result is influenced by a structuring element T. This element is used, in the present context, to consider the variable position of the same objects when delineated by different experts. The characteristics of T must fit the conditions of the specific problem. For example, if we consider a spatial tolerance of 7 pixels, T might be a disk with radius 7. If the task allows for no tolerance at all, then a radius 1 disk can be used to perform no dilation in the generation of the strong and weak consensus.

2.3   The Twofold-Consensus Ground Truth

From goals G1-G3, is evident that the consensus must be expressed as a set or multivalued object. Otherwise, it could not allocate the different images we attempt to fuse (as required in G3). The consensus shall not be an image in $\wp(\Omega)$, but a subspace in $\wp(\Omega)$. We seek the set of images which (a) contain all of the positive information in which all ground truth images agree on and (b) does not include positive information not featured by any ground truth image.

Definition 3. *The consensus of a set of binary images* $\mathbf{I}$ *is the set of images* $c_T(\mathbf{I})$ *defined as*

$$c_T(\mathbf{I}) = \{B \in \mathbb{B} \mid B \subseteq w_T(\mathbf{I}) \ and \ s_T(\mathbf{I}) \subseteq \mathcal{D}_T(B)\}. \tag{3}$$

The consensus set satisfies some practical properties, which we review in Sect. 2.4. Also, it has some interesting theoretical properties:

(i)   For any $\mathbf{I} \in \wp(\mathbb{B})$, it holds that $\mathbf{I} \subseteq c_T(\mathbf{I})$. This guarantees goal G3.

(ii) For any $\mathbf{I} \in \wp(\mathbb{B})$, it holds that $c_T(\mathbf{I}) = c_T(\{s_T(\mathbf{I}), w_T(\mathbf{I})\})$.

(iii) For any $\mathbf{I} \in \wp(\mathbb{B})$, it holds that $c_T(\mathbf{I}) = c_T(c_T(\mathbf{I}))$.

(iv) For any $\mathbf{I} \in \wp(\mathbb{B})$ and $B \in \mathbb{B}$, it holds that $B \in c_T(\mathbf{I})$ if and only if $c_T(\mathbf{I}) = c_T(\mathbf{I} \cup \{B\})$. Hence, the information in images within the set does not exceed that in the set itself.

(v)   For any $I \in \wp(\mathbb{B})$, $c_T(\mathbf{I})$ defines a connected subspace of $\mathbb{B}$, *i.e.*, for any $B_1, B_2 \in c_T(\mathbf{I})$, there exists a sequence of images $B_1^*, \ldots, B_r^*$ in $c_T(\mathbf{I})$, so that $B_1^* = B_1$, $B_r^* = B_2$, and two consecutive images B_i^* and B_{i+1}^* only differ in one pixel.

2.4   Visual Properties of the Set Consensus

The set $c_T(\mathbf{I})$, which we refer to as TCGT in the remainder of this work, has interesting visual properties related to the information in the images in $\mathbf{I}$.

The first property is that of *information combination*. This property refers to the ability to combine information from different ground truth images, meaning that the resulting set selectively picks information from each image. An example can be found in Fig. 2. Considering the original image in Fig. 2(a), two humans have created the ground truth images S_1 and S_2 in Figs. 2(b)–(c). The strong and weak consensus of the set of images are included in Figs. 2(d)–(e). The candidate image in Fig. 2(f), which is a selective combination of the images S_1 and S_2, actually belongs to their TCGT (i.e., $D \in c_T(\{S_1, S_2\})$). This illustrates how the TCGT is able to implicitly produce derived information from the combination of divergent solutions. Otherwise said, images which are not in the original set, but similar to (or composed of parts of) them, are included in the TCGT.

Although the example in Fig. 2 is intentionally simplistic, we can observe that, in the definition of the set-valued consensus, we construct something much more powerful than a closed list of images. There is an actual, yet implicit, knowledge construction process.

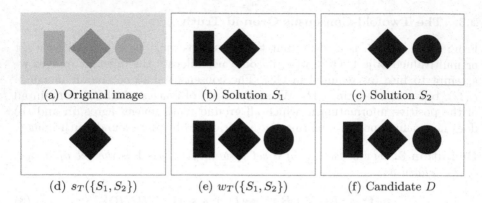

(a) Original image   (b) Solution S_1 (c) Solution S_2

(d) $s_T(\{S_1, S_2\})$ (e) $w_T(\{S_1, S_2\})$ (f) Candidate D

Fig. 2. Example of *information fusion* using the Twofold Consensus Ground Truth. We have (a) an image, (b, c) two hand-made segmentations from it, (d, e) the strong and weak consensus images and (f) a candidate image. The candidate image belongs to $c_T(\{S_1, S_2\})$, although it does not match any of the original images. The structuring element T used for the dilation is a disk of radius 5.

The example in Fig. 2 involves the presence or absence of information in a binary image. However, it is also interesting to analyze the alterations in such information, may they be due to contamination, errors or simple interpretation. Regarding this, an interesting property of the TCGT is the *smart tolerance for spatial displacements*.

The TCGT of a set of images includes images containing objects that do not coincide exactly with those delineated by humans in the generation of the ground truth. Moreover, it implicitly discriminates variations as acceptable/unacceptable not only based upon their magnitude (how different), but also upon their congruence of that variation with the existing variations in the original images in **I**. That is, the acceptance of an object depends upon *the amount of spatial variation*, but also upon *its direction*.

Figure 3 includes a binary image with two ground truth solutions (images S_1 and S_2 in Fig. 3(b)). Note that only the boundaries of the regions are drawn, so that they can be comfortably compared. In order for an image to be part of the TCGT, the object it features must be in between those of S_1 and S_2. Any image featuring a circle-like region will belong to the TCGT of $\{S_1, S_2\}$ as long as its boundaries are confined between those of S_1 and S_2. Hence, it is not only the fact that distorted solutions (in this case, reduced or enlarged circles) do belong to the TCGT. That distortion is not only measured in terms of *distance to the existing solutions*, but also in terms of congruency w.r.t. the divergences already existing in the TCGT. In this case, a solution created as a slight enlargement of the circle in S_2 (as E_{t1}), or a slight decrease of S_1 (as E_{t2}) are not included in the TCGT. However, greater distortions can be considered within the TCGT, as long as still confined in between the limits of S_1 and S_2.

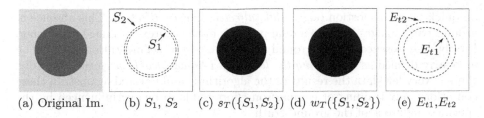

(a) Original Im.   (b) S_1, S_2 (c) $s_T(\{S_1, S_2\})$ (d) $w_T(\{S_1, S_2\})$ (e) E_{t1}, E_{t2}

Fig. 3. Example of information fusion based on strong and weak consensus images. We have (a) an image, (b) two hand-made solutions and (c,d) the strong and weak consensus images, respectively. The candidates in (e) are E_{t1}, a slight shrink of S_1, and E_{t2}, a slight enlargment of S_2. We find $E_{t1} \notin c_T(\{S_1, S_2\})$ and $E_{t2} \notin c_T(\{S_1, S_2\})$. In figures (b) and (e) only the limits of the regions are included, for an easier visual inspection. The structuring element for the dilation is a disk of radius 3.

3   Heterogeneity Measurement in Immunohistochemistry Imagery

3.1   Imaging in Immunochemistry

Immunohistochemistry (IHC) is an imaging method for studying the localization of antigens in tissue sections (e.g., brain tissue) using antibodies. Different antibodies can be used to demonstrate normal anatomy, protein aggregates, or to indicate pathological conditions such as apoptotic cells. In the images in Fig. 1 antibodies are used against Tau, a protein normally localized in the axon of neuron cells that can be pathologically deposited in some neurodegenerative diseases such as Progressive Supranuclear Palsy. The final stage of the tissue is that in which the regions affected by Tau protein take a distinctive color. The measurement and analysis of these images relates, hence, to the localization of pixel clusters with the visual characteristics of the affected regions.

3.2   Heterogeneity Measurement in IHC Imagery

IHC imaging is a costly technique, specially in terms of the time consumed by experts. Depending on the expected output of the IHC image analysis, experts can take hours analyzing and labelling visible artefacts in one image. As an example, the images from which the patches in Fig. 4 are taken contain around 5 megapixels, and often feature hundreds of size-variable tau-affected regions. A detailed analysis of these images cannot be tackled in less than few hours by an expert neuropathologist. Hence, it is very interesting to create automatic procedures that can measure the amount of tau protein visible in IHC images. That is, to create algorithms to replace humans in IHC image analysis.

The first problem encountered to design specific image processing algorithms for IHC is the absence of a large number of reliably-labelled ground truth images. The reason for this absence is the amount of time required to generate them, which forces the neuropathologists to perform semi-quantitative analyses based on quick visual inspection (e.g. *mildly affected* or *very affected*). This absence of ground truth images leads to a dual problem in the context of image processing. Firstly, the absence of the ground truth makes the segmentation task to be as poorly defined as *replicating the labelling a human would perform*. Secondly, there is very few data the results of the algorithm can be tested against. In these conditions, any training or comparison effort tends to be overinfluenced by the specific conditions of the ground truth.

We intend to overcome the lack of ground truth by requiring pathologists to label small, randomly selected subregions within some images. This would cut down the amount of time required from the experts, and would give partial, yet reliable, data about the expected results. Also, this brings a subsidiary problem: different pathologists produce very different label maps for the same image. A significant part of the tau-affected artifacts is homogeneously identified as positive detections. But, there is also a large margin for heterogeneity, especially related to (a) the margins of the artifacts and (b) the interpretation of some unclear regions/artifacts. As the size or number of subregions is increased, a new source of heterogeneous decisions appears: (c) lack of attention or tedium. As a result, we have highly variable results by each expert, which is in fact a typical case of multi-valued ground truth.

Problems with multiple ground truth are not unseen in literature, and solutions range from ground truth fusion [2] to performance measure fusion [5]. For example, for the present problem we can compare the results by an algorithm to each image labelled by pathologists, then fuse those results to get an aggregated or *average* performance of an algorithm. However, our goals in this work are different, and root back to the reasons why heterogeneity appears. Questions we face when divergent solutions are produced are: Should we consider all the images in the dataset as equally important, regardless of how heterogeneous their ground truth images are? What does it mean, having a ground truth set with very high (alternatively, low) heterogeneity? Could we measure how well a ground truth fits in a set of ground truth images? Moreover, could we learn to discard those ground truth solutions that are too different from other ground truth images? We intend to use the TCGT to quantify the heterogeneity of a set-valued ground truth; also, to measure the dissimilarity of an ground truth image w.r.t. a set of ground truth images.

We propose to use the TCGT for the generation of metadata about a IHC imagery dataset. Firstly, we want to measure the heterogeneity of a set of solutions. Normally, these measures are constructed from the analysis of one-to-one distances. However, we can also exploit the fact that the TCGT explicitly materializes the coincidences and divergences in a set of binary images.

Definition 4. *Let* $\mathbf{I} = \{I_i, \ldots, I_n\}$ *be a set of binary images. The* heterogeneity *of* $\mathbf{I}$ *is given by*

$$H_T(\mathbf{I}) = 1 - \frac{|s_T(\mathbf{I})|}{|w_T(\mathbf{I})|}$$

where w_T *and* s_T *are the weak and strong consensus, as in Sect. 2, and* $|\cdot|$ *is the number of featured (1-valued) pixels in an image.*

Definition 4 has one major problem: The use of a quotient makes the measure oblivious of the number of pixels in which divergence of opinion exists. Let an extreme case be that in which $\mathbf{I}$ is a set such that $I_1, \ldots, I_{n-1}$ contain one (same) featured pixels and I_n contains one (extra) featured pixel. We have $H(\mathbf{I}) = 0.5$, despite the very subtle difference between images. This problem is partially due to the orientation of the consensus towards the featured information (assuming it is more important than the non-featured one). In this case, two pixels are more important that all of the remaining ones. Still, it feels confusing that a difference of one pixel in one image can have such great impact in the output yielded by the heterogeneity measure.

We propose an alternative version of the heterogeneity measure that solves the aforementioned problem.

Definition 5. *Let* $\mathbf{I} = \{I_i, \ldots, I_n\}$ *be a set of binary images. The* scaled heterogeneity *of* $\mathbf{I}$ *is given by*

$$H_T^*(\mathbf{I}) = \frac{|w_T(\mathbf{I}) \setminus s_T(\mathbf{I})|}{|\Omega|}$$

where w_T *and* s_T *are the weak and strong consensus, as in Sect. 2.*

There is a list of differences between H and H^*. The most important one is probably the reference for scaling, since they both map to $[0, 1]$ (ignoring the undefined case with $w_T((I)) = \emptyset$). However, they also feature some coincidences. If all images in $\mathbf{I}$ are equal, then $H_T(\mathbf{I}) = H_T^*(\mathbf{I}) = 0$. Also, they both reach maximum values when $s_T(\mathbf{I}) = \emptyset$, although a further analysis of such cases sheds light on a significant difference. In case of H_T, $H_T(\mathbf{I}) = 1$ if and only if $s_T(\mathbf{I}) = w_T(\mathbf{I}) = \emptyset$, except (again) for the undefined case in which all images in $\mathbf{I}$ are empty. However, for H_T^*, the maximum heterogeneity is reached when $s_T(\mathbf{I}) = \emptyset$ and $w_T(\mathbf{I}) = \Omega$.

In our interpretation, the dissimilarity of an image w.r.t. a set of images can be put in terms of the heterogeneity of a set. In fact, to the variation of the heterogeneity when a set is altered.

Definition 6. *Let* $\mathbf{I} = \{I_i, \ldots, I_n\}$ *be a set of binary images, and let* $B \in \mathbb{B}$ *be any binary image. The* dissimilarity *of* B *w.r.t.* $\mathbf{I}$ *is given by*

$$\delta_T(B, \mathbf{I}) = H_T(\{B\} \cup \mathbf{I}) - H_T(\mathbf{I}),$$

where H_T *is a heterogeneity measure, as in Definition 4.*

The dissimilarity measure δ_T is affected by special cases similar to those generating unexpected outputs of H_T. Hence, we also present the scaled dissimilarity δ_T^*.

Definition 7. *Let* $\mathbf{I} = \{I_i, \ldots, I_n\}$ *be a set of binary images, and let* $B \in \mathbb{B}$ *be any binary image. The* scaled dissimilarity *of* B *w.r.t.* $\mathbf{I}$ *is given by*

$$\delta_T^*(B, \mathbf{I}) = H_T^*(\{B\} \cup \mathbf{I}) - H_T^*(\mathbf{I}),$$

where H_T *is a heterogeneity measure, as in Definition 5.*

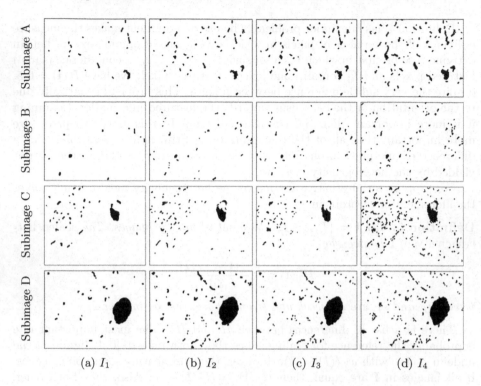

Subimage A   Subimage B   Subimage C   Subimage D

(a) I_1 (b) I_2 (c) I_3 (d) I_4

Fig. 4. Hand-labelled images produced by four neuropathologists on four of the subimages in Fig. 1. Neuropathologists have been told to be mark the tau-affected areas *conservatively* (column (a)), *normally* (columns (b, c)), or *generously* (column (d)).

3.3 Case Study: Measurement of Tau Protein

It is certainly complicated to know whether metadata is faithful to the actual facts or not [3,7]. Given the limited amount of space available in the present work, we intend to do a small experiment to see whether the measures capture pathologists' proneness to label more or less regions. Specifically, we induce a certain bias on neuropathologists, and we check whether our measures are able to detect and quantify it.

In order to complete our experiment we have requested four different neuropathologists to label the four leftmost subimages in Fig. 1. One of the neuropathologists was requested to label the area with tau protein in a conservative manner, two other were requested to act normally, and the fourth was requested to label the featured areas in a generous manner. In this way, we expect to have two extreme ground truth images and two solutions that lie somewhere *in the middle*. Of course, pathologists do not take any kind of suggestion on how to perform their work in a normal situation, neither they have bias on the analysis. However, it is, in our opinion, a legitimate way to produce binary images whose behaviour in terms of heterogeneity and dissimilarity is predictable.

Table 1. Results obtained in the quantification of heterogeneity and dissimilarity of the sets displayed in Fig. 4. For each subimage, $\mathbf{I}$ refers to all of the ground truth solutions for each image, while $\mathbf{I}_{i-j}$ refers to the images in columns from i to j, both included. The structuring element T (which is a circle with radius 5) is ommitted from the formulation in order to ease the interpretation of the table.

(a) Results using heterogeneity and dissimilarity						
Subimage	$H(\mathbf{I})$	$H(\mathbf{I}_{2-4})$	$H(\mathbf{I}_{1-3})$	$H(\mathbf{I}_{2-3})$	$\delta(I_1,\mathbf{I}_{2-3})$	$\delta(I_1,\mathbf{I}_{2-4})$
Subimg. A	.758	.669	.484	.271	.213	.089
Subimg. B	.643	.424	.536	.235	.301	.219
Subimg. C	.747	.485	.636	.258	.377	.262
Subimg. D	.815	.639	.683	.367	.316	.177
Total	.741	.554	.585	.283	.302	.187

(b) Results using scaled heterogeneity and dissimilarity						
Subimage	$H^*(\mathbf{I})$	$H^*(\mathbf{I}_{2-4})$	$H^*(\mathbf{I}_{1-3})$	$H^*(\mathbf{I}_{2-3})$	$\delta^*(I_1,\mathbf{I}_{2-3})$	$\delta^*(I_1,\mathbf{I}_{2-4})$
Subimg. A	.272	.239	.082	.044	.038	.033
Subimg. B	.236	.154	.153	.066	.087	.082
Subimg. C	.249	.162	.147	.060	.087	.087
Subimg. D	.154	.120	.076	.040	.036	.034
Total	.228	.169	.114	.053	.062	.059

The images produced for the experiments are included in Fig. 4. Each row in the figure corresponds to one of the four leftmost images in Fig. 1, while each column corresponds to one of the instructions given to the pathologists. Specifically, the leftmost column is the most conservative inspection of the images, while the rightmost column contains the images in which the neuropathologists was proner to label tau protein.

We have used the measures presented in Sect. 3.2, as recap in Table 1. The standing assumption of our experiment is that images generated under extreme biases should be identified as such by inspecting the values yielded by our measures. Table 1 displays the values gathered in different evaluations for the image sets at each of the rows of Fig. 4.

From the results in Table 1, we can confirm that our measures actually behave according to the semantics of the images. For example, in terms of heterogeneity, the values yielded by H or H^* suffer a severe increase when the set $\mathbf{I}$ includes the images I_1 or I_4, compared to when it does not. For both H and H^* the heterogeneity of $\mathbf{I}_{2-3}$ is significantly increased by adding the images I_1 or I_4, which play the role of extreme cases. This holds for all subimages and heterogeneity measures. In Table 1 we also observe that δ_T and δ_T^* identify the outlying images I_1 and I_4 w.r.t. the *neutral* images I_2 and I_3.

It is relevant to mention that, considering the very small size of the experiment, results shall be put to the test in a more complete scenario. Still, it is rather complicated to find input for metadata evaluation, and typically one must rely on either experiment-driven data (as in this case), or on questionable assumptions on the way in which ground truth data was generated.

4   Conclusions

In this work we have tackled the problem of multiple ground truth in medical imagery, specifically in immunohistochemistry imagery of brain tissue. We have questioned the reasons on the divergences between experts when required to label such images, and proposed four different measures to quantify the heterogeneity in a set of images, as well as the 1-to-n dissimilarity of images. In order to do so, we have applied the notions and developments of the Twofold Consensus Ground Truth (TCGT), a set-valued operator created for binary image fusion. This application intends to illustrate how fusion operators in can be used for purposes other than information aggregation or compression. In our example, the twofold consensus ground truth is used not only to fusion hand-labelled IHC images, but also to analyze the heterogeneity in a set of them, as well as to create one-to-many or many-to-many dissimilarity measures.

Despite the innovative nature of the application, we consider that our work has a solid, context-agnostic mathematical background. However, it requires more comprehensive experimental validation, considering the oriented nature of this research. Hence, the design and analysis of such a experimental setup is a key future line of research.

References

1. Chen, S., Haralick, R.: Recursive erosion, dilation, opening, and closing transforms. IEEE Trans. Image Process. **4**(3), 335–345 (1995)
2. Fernández-García, N., Carmona-Poyato, A., Medina-Carnicer, R., Madrid-Cuevas, F.: Automatic generation of consensus ground truth for the comparison of edge detection techniques. Image Vis. Comput. **26**(4), 496–511 (2008)
3. Lopez-Molina, C., Bustince, H., De Baets, B.: Separability criteria for the evaluation of boundary detection benchmarks. IEEE Trans. Image Process. **25**(3), 1047–1055 (2016)
4. Lopez-Molina, C., De Baets, B., Bustince, H.: Twofold consensus for boundary detection ground truth. Knowl.-Based Syst. **98**, 162–171 (2016)
5. Martin, D., Fowlkes, C., Tal, D., Malik, J.: A database of human segmented natural images and its application to evaluating segmentation algorithms and measuring ecological statistics. In: Proceedingngs of IEEE International Conference on Computer Vision, vol. 2, pp. 416–423 (2001)
6. Pawlak, Z.: Rough sets. Int. J. Comput. Inf. Sci. **11**(5), 341–356 (1982)
7. Pont-Tuset, J., Marques, F.: Measures and meta-measures for the supervised evaluation of image segmentation. In: Proceedings of IEEE Conference on Computer Vision and Pattern Recognition, pp. 2131–2138 (2013)
8. Serra, J.: Image Analysis and Mathematical Morphology. Academic Press Inc., Cambridge (1983)
9. Warfield, S.K., Zou, K.H., Wells, W.M.: Simultaneous truth and performance level estimation (STAPLE): an algorithm for the validation of image segmentation. IEEE Trans. Med. Imaging **23**(7), 903–921 (2004)

Pre-aggregation Functions and
Generalized Forms of Monotonicity

Penalty-Based Functions Defined by Pre-aggregation Functions

Graçaliz Pereira Dimuro[1,2]([✉]) [iD], Radko Mesiar[3], Humberto Bustince[2,4],
Benjamín Bedregal[5], José Antonio Sanz[2,4], and Giancarlo Lucca[4]

[1] Centro Ciências Computacionais,
Universidade Federal do Rio Grande Campus Carreiros, Rio Grande 96201-900, Brazil
gracalizdimuro@furg.br
[2] Institute of Smart Cities, Universidad Publica de Navarra Campus Arrosadía,
31006 Pamploma, Spain
{gracaliz.pereira,bustince,joseantonio.sanz}@unavarra.es
[3] Department of Mathematics, Slovak University of Technology, Radlinského 11,
81005 Bratislava, Slovakia
mesiar@math.sk
[4] Depto. of Automática y Computación,
Universidad Publica de Navarra Campus Arrosadía, 31006 Pamplona, Spain
lucca.112793@e.unavarra.es
[5] Depart. de Informática e Matemática Aplicada,
Universidade Federal do Rio Grande do Norte Campus Universitário,
Natal 59072-970, Brazil
bedregal@dimap.ufrn.br

Abstract. Pre-aggregation function (PAF) is an important concept that has emerged in the context of directional monotonicity functions. Such functions satisfy the same boundary conditions of an aggregation functions, but it is not required the monotone increasingness in all the domain, just in some fixed directions. On the other hand, penalty functions is another important concept for decision making applications, since they can provide a measure of deviation from the consensus value given by averaging aggregation functions, or a penalty for not having such consensus. This paper studies penalty-based functions defined by PAFs. We analyse some properties (e.g.: idempotency, averaging behavior and shift-invariance), providing a characterization of idempotent penalty-based PAFs and a weak characterization of averaging penalty-based PAFs. The use of penalty-based PAFs in spatial/tonal filters is outlined.

Keywords: Pre-aggregation function · Penalty function
Idempotency · Shift-invariance · Average function · Spatial/tonal filters

1 Introduction

Aggregation functions are very important for dealing with some computation problems, such as fuzzy rule based systems and classification systems [20–23]

© Springer International Publishing AG, part of Springer Nature 2018
J. Medina et al. (Eds.): IPMU 2018, CCIS 854, pp. 403–415, 2018.
https://doi.org/10.1007/978-3-319-91476-3_34

and decision making [2,15]. Averaging aggregation functions, for example, are a useful class of aggregation functions, since they provide output values that are bounded by the minimum and the maximum of the inputs, so representing a consensus value of the inputs. Examples of such functions are quasi-arithmetic means, medians, OWA functions, Choquet integral, Sugeno integral, CC-integrals, which appear in the literature in applications such as preference aggregation, aggregation of expert opinions, judgements, fuzzy-rule based classification systems, as properly discussed by Beliakov et al. [5] and Lucca et al. [22].

Observe that aggregation functions are defined considering the monotone increasingness. However, this property may not be required in many applications (see [6,20,26]). Examples of useful functions that are not monotonic are some statistical tools, such as the mode. Wilkin and Beliakov [26] introduced the notion of weak monotonicity, aiming at extending the standard concept of monotonicity, where the monotonicity of functions is required only along the direction of the first quadrant diagonal. Generalizing the concept of weak monotonicity, Bustince et al. [6] introduced the notion of directional monotonicity, allowing the monotonicity property along (some) fixed ray.

Then, Lucca et al. [20] used this concept in order to define pre-aggregation functions (PAFs), which satisfy the same boundary conditions of an aggregation functions, but may be just directional monotonic. Pre-aggregation functions play an important role in the context of fuzzy rule based classification systems, where generalization of the Choquet integral presented excellent results (see, e.g., [20,23]). In particular, some family of non-averaging pre-aggregation functions derived from the Choquet integral are appearing as promising tools proved to be statically equivalent or superior to the best state-of-the-art methods (such as, e.g., FURIA [19]). See also [11], for other properties and construction methods of PAFs.

On the other hand, penalty functions [8] have been discussed in the literature for its application in decision making (see, e.g., [7,15]). They provide a measure of deviation from the consensus value given by averaging aggregation functions, or a penalty for not having a consensus. In fact, there exist some advantages in expressing averaging functions based on penalty functions. They have an intuitive interpretation, namely, the cost related to the disagreement among the output and the input data, and, in some cases, they can be defined in a simple and intuitive way, whereas the associated averaging functions sometimes can not be given in well-defined forms. Examples of functions that minimize some penalty functions, called P-functions, are the weighted arithmetic and geometric means, the median and the mode [26].

This paper enlarges the knowledge about penalty functions in context of PAFs, which is important for many applications, e.g., in image processing, filtering and smoothing [27]. For that, the objectives of the paper are: (i) to analyse the properties of idempotency, the averaging behavior and shift-invariance (in Sect. 3); (ii) to provide a characterization of idempotent penalty-based PAFs and a weak characterization of averaging penalty-based PAFs (in Sect. 4) (iii) to outline the use of penalty-based PAFs in spatial/tonal filters (in Sect. 5). Section 2 presents basic concepts used in the paper.

2 Preliminary Concepts

We denote by $\mathbb{I}$ a closed subinterval of the extended real line, i.e., $\mathbb{I} = [a, b] \subseteq \mathbb{R}$.

Definition 1. *A function $F : \mathbb{I} \to \mathbb{R}$ is said to be quasi-convex if for every $x, y \in \mathbb{I}$ and for every $\lambda \in [0, 1]$ the inequality $F(\lambda x + (1 - \lambda)y) \leq \max\{F(x), F(y)\}$ holds.*

Definition 2. *A function $F : \mathbb{I} \to \mathbb{R}$ is lower semicontinueous at $x_0 \in \mathbb{I}$ if $\lim \inf_{x \to x_0} F(x) \geq F(x_0)$.*

Definition 3 [25]. *A function $A : [0, 1]^n \to [0, 1]$ is an aggregation function whenever the following conditions hold, for all $(x_1, \ldots, x_n) \in [0, 1]^n$:*

(A1) *A is increasing in each argument: for each $i \in \{1, \ldots, n\}$, if $x_i \leq y$, then $A(x_1, \ldots, x_n) \leq A(x_1, \ldots, x_{i-1}, y, x_{i+1}, \ldots, x_n)$;*
(A2) *A satisfies the boundary conditions: $A(0, \ldots, 0) = 0$ and $A(1, \ldots, 1) = 1$.*

Definition 4. *An aggregation function $A : [0, 1]^n \to [0, 1]$ is said to be averaging if it is bounded by the minimum and maximum of its arguments, that is, for all $(x_1, \ldots, x_n) \in [0, 1]^n$, it holds that: $\min\{x_1, \ldots, x_n\} \leq A(x_1, \ldots, x_n) \leq \max\{x_1, \ldots, x_n\}$.*

Due to the monotonicity of aggregation functions, the averaging behavior is equivalent to the idempotency property.

Definition 5 [6]. *Let $r = (r_1, \ldots, r_n)$ be a real n-dimensional vector such that $r \neq 0$. A function $F : [0, 1]^n \to [0, 1]$ is said to be r-increasing if, for all $(x_1, \ldots, x_n) \in [0, 1]^n$ and $c > 0$ such that $(x_1 + cr_1, \ldots, x_n + cr_n) \in [0, 1]^n$, it holds that*

$$F(x_1 + cr_1, \ldots, x_n + cr_n) \geq F(x_1, \ldots, x_n).$$

Similarly, one defines an r-decreasing function.

Pre-aggregation functions, the key concept in this work, were introduced by Lucca et al. [20] aiming at applications on fuzzy-rule based classification systems.

Definition 6 [11,20]. *A function $PA : [0, 1]^n \to [0, 1]$ is said to be an n-ary pre-aggregation function if it satisfies (A2) and:*

(PA) *PA is directionally increasing for some vector $r \in \mathbb{R}^n$, $r \neq 0$, that is, it is r-increasing.*

PA is said an r-pre-aggregation function.

Similarly to Definition 4, one can define averaging pre-aggregation functions. Observe that all pre-aggregation functions that are averaging are also idempotent. However the converse does not hold. See Remark 2 in Sect. 3.

Of particular relevance is the notion of shift invariance [24] (which is also called difference scale invariance [17]). A constant change in every input should result in a corresponding change of the output. In [11], we have modified this concept to consider whenever this change in the output is in the same direction of the related change in the output or in the opposite direction.

Definition 7. *Consider $r = (r_1, \ldots, r_n)$, $r \neq 0$, and $i \in \{1, \ldots, n\}$ such that $r_i \neq 0$. A function $F : [0,1]^n \to [0,1]$ is (r, i)-shift invariant (stable for translations in the direction i of the vector r) if:*

$$\forall c > 0, \boldsymbol{x} \in [0,1]^n : x_1 + cr_1, \ldots, x_n + cr_n \in [0,1] \Rightarrow$$
$$F(x_1 + cr_1, \ldots, x_n + cr_n) = F(x_1, \ldots, x_n) + cr_i.$$

Whenever $r_1 = \ldots = r_n = r \neq 0$ and it holds that:

$$\forall c > 0, \boldsymbol{x} \in [0,1]^n : x_1 + cr, \ldots, x_n + cr \in [0,1] \Rightarrow$$
$$F(x_1 + cr, \ldots, x_n + cr) = F(x_1, \ldots, x_n) + cr,$$

then F is said to be shift invariant (stable for translations in all directions of the vector r).

Note that the concept of shift invariance is equivalent to the notion of shift invariance in the works by Lázaro et al. [24] and Calvo et al. [9].

Proposition 1. *Consider $r = (r_1, \ldots, r_n)$, $r \neq 0$, and $i \in \{1, \ldots, n\}$ such that $r_i \neq 0$. Whenever a function $F : [0,1]^n \to [0,1]$ is (r, i)-shift invariant and $r_i > 0$ ($r_i < 0$) then F is r-increasing (non r-increasing).*

Proof. It is immediate.

Remark 1. Observe that the converse of Proposition 1 does not hold. For example, the $(0, 1)$-increasing function $F : [0,1]^2 \to [0,1]$ defined by $F(x, y) = x - (\max\{0, x - y\})^2$, introduced in [20], is not $((0, 1), 2)$-shift invariant. In fact, for $c = 0.4$, one has that $F(0.6 + 0.4 \cdot 0, 0.5 + 0.4 \cdot 1) = 0.6$, but $F(0.6, 0.5) + 0.4 \cdot 1 = 0.99 \neq 0.6$. Also, the $(0, -1)$-increasing function $F' : [0,1]^2 \to [0,1]$ defined by $F'(x, y) = 1 - F(x, y)$, is not $((0, -1), 2)$-shift invariant. For $c = 0.4$, it holds that: $1 - F(0.6 + 0.4 \cdot 0, 0.5 + 0.4 \cdot 1) = 0.4$, but $(1 - F(0.6, 0.5)) - 0.4 \cdot 1 = 0.01 \neq 0.4$.

3   PAFs Based on Penalty Functions

There exist various slightly different definitions of penalty functions in the literature (see [8] and the references therein). Analysing the discussion presented by Bustince et al. in [8], we decided to adopt the following definition.

Definition 8 [8, Definition 4.1]. *For any closed interval $\mathbb{I} \subseteq \mathbb{R}$, the function $P : \mathbb{I}^{n+1} \to \mathbb{R}^+$ is a penalty function if and only if there exists $c \in \mathbb{R}^+$ such that:*

(P1) $P(\boldsymbol{x}, y) \geq c$, *for all* $\boldsymbol{x} \in \mathbb{I}^n, y \in \mathbb{I}$;
(P2) $P(\boldsymbol{x}, y) = c$ *if and only if* $x_i = y$, *for all* $i = 1 \ldots n$, *and*
(P3) P *is quasi-convex lower semi-continuous in* y, *for each* $\boldsymbol{x} \in \mathbb{I}^n$.

Definition 9 [8, Definition 4.2]. *Let* $P : \mathbb{I}^{n+1} \to \mathbb{R}^+$ *be a penalty function. The function* $F_P : \mathbb{I}^n \to \mathbb{I}$ *is said a P-function, if, for each* $\boldsymbol{x} \in \mathbb{I}^n$:

$$F_P(\boldsymbol{x}) = \frac{a+b}{2}, \tag{1}$$

where $[a, b] = cl(Minz(P(\boldsymbol{x}, \cdot)))$, *and* $Minz(P(\boldsymbol{x}, \cdot))$ *is the set of minimizers of* $P(\boldsymbol{x}, \cdot)$, *that is,*

$$Minz(P(\boldsymbol{x}, \cdot)) = \{y \in \mathbb{I} \mid P(\boldsymbol{x}, y) \le P(\boldsymbol{x}, z), \text{ for each } z \in \mathbb{I}\},$$

and $cl(S)$ *is the closure of* $S \subseteq \mathbb{I}$.

One can think of P as describing the dissimilarity or disagreement between the inputs in $\boldsymbol{x}$ and the value y. It follows that the P-function F_P is a function that minimizes the chosen dissimilarity. Observe that the properties of quasi convexity and lower semicontinuity of a penalty function P imply that the set of minimizers of $P(\boldsymbol{x}, \cdot)$ is either a singleton or an interval (see [8, Remark 4.2]).

In [8], Bustince et al. have proved the following important result, which relates the idempotency property to P-functions:

Theorem 1 [8, Theorem 4.1]. *A function* $F : \mathbb{I}^n \to \mathbb{I}$ *is a P-function, for some penalty function* $P : \mathbb{I}^{n+1} \to \mathbb{R}^+$, *if and only if* F *is idempotent.*

Thus, any idempotent function F may be represented by a P-function $F_P : \mathbb{I}^n \to \mathbb{I}$ (i.e., $\forall \boldsymbol{x} \in \mathbb{I}^n : F(\boldsymbol{x}) = F_P(\boldsymbol{x})$), for a certain penalty function P.

In what follows, we consider penalty functions defined on $\mathbb{I} = [0, 1]$.

Theorem 1 may be adapted for the context of PAFs:

Theorem 2 *[Idempotency & Directional Monotonicity].* *Consider* $\boldsymbol{r} = (r_1, \dots, r_n) \in \mathbb{R}^n$, $\boldsymbol{r} \ne \boldsymbol{0}$. *A function* $F : [0,1]^n \to [0,1]$ *is an* $\boldsymbol{r}$*-pre-aggregation P-function, for some penalty function* $P : [0,1]^{n+1} \to [0,1]$, *if and only if* F *is idempotent and* $\boldsymbol{r}$*-increasing.*

Proof. ($\Rightarrow$) It is immediate, following from Theorem 1 and Definition 6. ($\Leftarrow$) Since F is idempotent, it satisfies **(A2)**, and, by Theorem 1, F is a P-function, for some penalty function P. Thus, by Definition 6, F is an $\boldsymbol{r}$-pre-aggregation P-function, for some penalty function P.

Theorem 3 *[Idempotency & Shift Invariancy].* *Consider* $\boldsymbol{r} = (r_1, \dots, r_n) \in \mathbb{R}^n$, $\boldsymbol{r} \ne \boldsymbol{0}$, *such that there exists* $i \in \{1, \dots, n\}$ *with* $r_i > 0$. *If a function* $F : [0,1]^n \to [0,1]$ *is idempotent and* $(\boldsymbol{r}, i)$*-shift invariant, then* F *is an* $\boldsymbol{r}$*-pre-aggregation P-function, for some penalty function* $P : [0,1]^{n+1} \to [0,1]$.

Proof. By Proposition 1, since F is $(\boldsymbol{r}, i)$-shift invariant then F is $\boldsymbol{r}$-increasing. The result follows from Theorem 2.

Example 1. Consider $r = (r_1, \ldots, r_n) \in \mathbb{R}^n$, $r \neq \mathbf{0}$, an idempotent r-increasing function $F : [0,1]^n \to [0,1]$, and let $P_F : [0,1]^{n+1} \to \mathbb{R}^+$ be a penalty function, defined, for all $\boldsymbol{x} \in [0,1]^n$ and $y \in [0,1]$ by:

$$P_F(x_1, \ldots, x_n, y) = \begin{cases} c & \text{if } x_i = y \text{ for each } i \\ \mid F(x_1, \ldots, x_n) - y \mid + c + \epsilon & \text{otherwise,} \end{cases} \tag{2}$$

where $\epsilon > 0$ and $c \geq 0$, which was introduced in [8, Proof of Theorem 4.1]. Since the set of minimizers of $P_F(\boldsymbol{x}, \cdot)$ is the singleton $\{F(\boldsymbol{x})\}$, one has that

$$F(x_1, \ldots, x_n) = \arg \min_y P_F(x_1, \ldots, x_n, y), \tag{3}$$

that is, F is an r-pre-aggregation P_F-function.

Remark 2. Observe that, unlike aggregation functions, idempotency and averaging are not equivalent properties for PAFs, due to the lack of the monotonicity in some directions of the function's domain. Notice, however, that any averaging PAF is idempotent. For example, consider the mode, which could be defined, for all $\boldsymbol{x} \in [0,1]^n$, as $\mathrm{Mode}(\boldsymbol{x}) = \{x_i \in \boldsymbol{x} \mid x_i$ appears most often in $\boldsymbol{x}$ and $i = 1, \ldots, n\}$, which, however, would not be a well-defined function of signature $[0,1]^n \to [0,1]$. Then, for the purpose of this paper, we consider the definition of the mode by a pair of well-defined functions of signature $[0,1]^n \to [0,1]$, as defined in Remark 2, which can be expressed by the pair of functions $(\mathrm{Mode}_{\min}, \mathrm{Mode}_{\max})$, where $\mathrm{Mode}_{\min} : [0,1]^n \to [0,1]$ and $\mathrm{Mode}_{\max} : [0,1]^n \to [0,1]$ are defined, respectively, by the least and the greatest values that appear most often in the input vector. For $r > 0$, $\mathrm{Mode}_{\min}$ and $\mathrm{Mode}_{\max}$ are obviously averaging and idempotent $(r, \ldots, r)$-PAFs. On the other hand, idempotency does not imply in the averaging behaviour of PAFs. For example, for $r > 0$, consider the (r, r)-PAF $F : [0,1]^2 \to [0,1]$, defined, for all $x, y \in [0,1]$, by:

$$F(x, y) = \begin{cases} x & \text{if } x = y \\ 1 & \text{otherwise.} \end{cases}$$

It is immediate that F is idempotent. However, F is not averaging, since for any $x, y \neq 1$ such that $x \neq y$, it holds that $F(x, y) = 1 \nleq \max\{x, y\}$.

Corollary 1 [*Averaging & Directional Monotonicity*]. *Consider* $r = (r_1, \ldots, r_n) \in \mathbb{R}^n$, $r \neq \mathbf{0}$. *If a function* $A : [0,1]^n \to [0,1]$ *is averaging and r-increasing then A is an r-pre-aggregation P-function, for some penalty function* $P : [0,1]^{n+1} \to [0,1]$.

Proof. Since A is an averaging r-increasing function, then, from Remark 2, it is idempotent. The result follows from Theorem 2.   □

Corollary 2 [*Averaging & Shift Invariancy*]. *Consider* $r = (r_1, \ldots, r_n) \in \mathbb{R}^n$, *with* $r \neq \mathbf{0}$, *such that there exists* $i \in \{1, \ldots, n\}$ *with* $r_i > 0$. *If a function* $A : [0,1]^n \to [0,1]$ *is averaging and (r, i)-shift invariant then A is an r-pre-aggregation P-function, for some penalty function* $P : [0,1]^{n+1} \to [0,1]$.

Proof. It follows from Theorem 3 and Corollary 1.   □

4   Characterization of PAFs Based on Penalty Functions

In the following, we provide characterizations of (averaging/idempotent) PAFs based on penalty functions.

Lemma 1 [8, Lemma 4.1]. *Let $F : [0,1]^n \to [0,1]$ be an idempotent function. Then there is a penalty function $P : [0,1]^{n+1} \to [0,1]$ such that F is a P-function and, for each $\boldsymbol{x} \in [0,1]^n$, P has just one minimizer, that is, $Minz(P(\boldsymbol{x}, \cdot))$ is a degenerate interval.*

Proposition 2. *Consider $\boldsymbol{r} = (r_1, \ldots, r_n) \in \mathbb{R}^n$, $\boldsymbol{r} \neq \boldsymbol{0}$. If a function $A : [0,1]^n \to [0,1]$ is idempotent and $\boldsymbol{r}$-increasing then A is a P-function for some penalty function $P : [0,1]^{n+1} \to [0,1]$ satisfying:*

$$\forall \boldsymbol{x} \in [0,1]^n, c > 0 :$$
$$\boldsymbol{x} + c\boldsymbol{r} \in [0,1]^n \Rightarrow cl(Minz(P(\boldsymbol{x}, \cdot))) \leq_{KM} cl(Minz(P(\boldsymbol{x} + c\boldsymbol{r}, \cdot))), \quad (4)$$

where $\leq_{KM}$ is the Kulisch-Miranker interval order [12] restricted to $\mathbb{III} = \{[x,y] | 0 \leq x \leq y \leq 1\}$, defined, for all $[x,y], [x',y'] \in \mathbb{III}$, by

$$[x,y] \leq_{KM} [x',y'] \Leftrightarrow x \leq x' \text{ and } y \leq y'. \quad (5)$$

Proof. Let $A : [0,1]^n \to [0,1]$ be an idempotent and $\boldsymbol{r}$-increasing function. By Theorem 2, A is a P-function, for some penalty function $P : [0,1]^2 \to [0,1]$. By Lemma 1, the set of minimizers of such penalty function P is a degenerate interval. Now, consider $\boldsymbol{x} \in [0,1]^n$ such that $\boldsymbol{x} + c\boldsymbol{r} \in [0,1]^n$, for all $c > 0$. By Lemma 1, one has that

$$Minz(P(\boldsymbol{x}, \cdot)) = [a, a], Minz(P(\boldsymbol{x} + c\boldsymbol{r}, \cdot)) = [a', a'],$$

for some $a, a' \in [0,1]$. Then, since A is $\boldsymbol{r}$-increasing, one has that, $A(\boldsymbol{x}) \leq A(\boldsymbol{x}+c\boldsymbol{r})$ and, thus, by Eq. (1), one has that $a = \frac{a+a}{2} \leq \frac{a'+a'}{2} = a'$. It follows that $Minz(P(\boldsymbol{x}, \cdot)) \leq_{KM} Minz(P(\boldsymbol{x} + c\boldsymbol{r}, \cdot))$, and, thus, $cl(Minz(P(\boldsymbol{x}, \cdot))) \leq_{KM} cl(Minz(P(\boldsymbol{x} + c\boldsymbol{r}, \cdot)))$. Thus, P satisfies (4).   □

Proposition 3. *Consider $\boldsymbol{r} = (r_1, \ldots, r_n) \in \mathbb{R}^n$, $\boldsymbol{r} \neq 0$. If $F : [0,1]^n \to [0,1]$ is a P-function for some penalty function $P : [0,1]^{n+1} \to [0,1]$ satisfying (4) then F is an idempotent $\boldsymbol{r}$-increasing function.*

Proof. Consider $\boldsymbol{r} = (r_1, \ldots, r_n) \in \mathbb{R}^n$, $\boldsymbol{r} \neq 0$. Denote

$$[a, b] = cl(Minz(P(\boldsymbol{x}, \cdot))), [a', b'] = cl(Minz(P(\boldsymbol{x} + c\boldsymbol{r}, \cdot))),$$

for $\boldsymbol{x}, \boldsymbol{x} + c\boldsymbol{r} \in [0,1]^{n+1}$ and $c > 0$, and suppose that F is a P-function for some penalty function P satisfying (4). For all $\boldsymbol{x} \in [0,1]^n$ and $c > 0$ such that $\boldsymbol{x} + c\boldsymbol{r} \in [0,1]^n$, it holds that $[a, b] \leq_{KM} [a', b']$, i.e., $a \leq a'$ and $b \leq b'$. By Eq. (1), $F(\boldsymbol{x}) = \frac{a+b}{2} \leq \frac{a'+b'}{2} = F(\boldsymbol{x} + c\boldsymbol{r})$. Thus, F is $\boldsymbol{r}$-increasing. It is immediate that F is idempotent.   □

Theorem 4 [Characterization of idempotent pre-aggregation P-functions]. *Consider* $r = (r_1, \ldots, r_n) \in \mathbb{R}^n$, $r \neq 0$. *A function* $F : [0, 1]^n \to [0, 1]$ *is an idempotent* r-*pre-aggregation function if and only if* F *is a P-function for some penalty function* $P : [0, 1]^{n+1} \to [0, 1]$ *satisfying (4).*

Proof. ($\Rightarrow$) It follows from Proposition 2. ($\Leftarrow$) It follows from Proposition 3, since F is idempotent, and, then, it satisfies **(A2)**.   □

Corollary 3. *Consider* $r = (r_1, \ldots, r_n) \in \mathbb{R}^n$, $r \neq 0$. *If* $A : [0, 1]^n \to [0, 1]$ *is an averaging* r-*pre-aggregation function then* A *is a P-function for some penalty function* P *satisfying (4).*

Theorem 5 [Weak characterization of averaging pre-aggregation P-functions]. *Consider* $r = (r_1, \ldots, r_n) \in \mathbb{R}^n$, $r \neq 0$ *and let* $A : [0, 1]^n \to [0, 1]$ *be an averaging function.* A *is* r-*pre-aggregation function if and only if* A *is a P-function for some penalty function* P *satisfying (4).*

Proof. It follows from Theorem 4 and Corollary 3.   □

Example 2. The penalty-based function presented in [8, Example 4.2], when restricted to $[0, 1]$, may be defined by a PAF. Consider a weight vector $W = (w_1, \ldots, w_n)$, where $\sum\limits_{i=1}^{n} w_i = 1$, and the vector $r = (r_1, \ldots, r_n) \in \mathbb{R}^n$, such that $r \neq 0$ and $\sum_{i=1}^{n} r_i \geq 0$. Let $F : [0, 1]^n \to [0, 1]$ be a function such that: (i) $F(x_1, \ldots, x_n) = \sum\limits_{i=1}^{n} w_i x_i \Leftrightarrow x_1 = \ldots = x_n$; (ii) F is r-increasing; (iii) F is averaging. By Corollary 1, F is an r-pre-aggregation P-function, with penalty function $P_F : [0, 1]^{n+1} \to \mathbb{R}^+$, defined, for all $x \in [0, 1]^n$ and $y \in [0, 1]$, by:

$$P_F(x_1, \ldots, x_n, y) = \begin{cases} \mid y - F(x_1, \ldots, x_n) \mid & \text{if } y = \sum\limits_{i=1}^{n} w_i x_i, \\ 1 & \text{otherwise.} \end{cases}$$

See [8, Example 4.2] to see the proof that P_F is a penalty function. The set of minimizers of $P_F(x, \cdot)$ is $\{\sum_{i=1}^{n} w_i x_i\}$, and, then, $\sum_{i=1}^{n} w_i x_i = \arg\min\limits_{y} P_F(x, y)$. Whenever $x_1 = \ldots = x_n$, it holds that $F(x) = \arg\min\limits_{y} P_F(x, y)$.

5   Pre-aggregation Penalty-Based Functions in Image Processing

The following results may be interesting for applying pre-aggregation penalty-based functions in image processing (as suggested in [26, Sect. 4.3]).

Theorem 6. *Consider* $r = (r, \ldots, r) \in \mathbb{R}^n$, $r \neq 0$, *and* $g : [0, 1] \to [0, 1]$. *Let* $f_k : [0, 1]^n \to [0, 1]$ *be* r-*shift invariant functions, for* $k \in \{1, \ldots, n\}$, *and* $F : [0, 1]^n \to [0, 1]$ *be a P-function, for the penalty function* $P : [0, 1]^{n+1} \to \mathbb{R}^+$ *defined on the terms* $g(x_i - f_i(x_1, \ldots, x_n))(x_i - y)^2$, *for all* $x_1, \ldots, x_n, x_i - f_i(x_1, \ldots, x_n) \in [0, 1]$, *and* $i = 1, \ldots, n$. *Then: (i)* F *is* r-*shift invariant; (ii) If* $r > 0$ *then* F *is* r-*increasing.*

Proof. To prove **(i)**, since F is a P-function, define, according to Definition 9, for all $x_1, \ldots, x_n, x_i - f_i(x_1, \ldots, x_n) \in [0, 1]$, and $i = 1, \ldots, n$:

$$F(x_1, \ldots, x_n) = \tag{6}$$
$$\arg\min_y P(g(x_1 - f_1(x_1, \ldots, x_n))(x_1 - y)^2, \ldots,$$
$$g(x_n - f_n(x_1, \ldots, x_n))(x_n - y)^2, y).$$

Since, for all $k \in \{1, \ldots, n\}$, f_k is r-shift invariant, then, for all $x_1 + cr, \ldots, x_n + cr, x_i + cr - f_i(x_1 + cr, \ldots, x_n + cr) \in [0, 1]$, with $c > 0$, it follows that:

$$F(x_1 + cr, \ldots, x_n + cr)$$
$$= \arg\min_y P(g(x_1 + cr - f_1(x_1 + cr, \ldots, x_n + cr))(x_1 + cr - y)^2,$$
$$\ldots, g(x_n + cr - f_n(x_1 + cr, \ldots, x_n + cr))(x_n + cr - y)^2, y) \text{ by Eq. (6)}$$
$$= \arg\min_y P(g(x_1 + cr - f_1(x_1, \ldots, x_n) - cr)(x_1 + cr - y)^2,$$
$$\ldots, g(x_n + cr - f_n(x_1, \ldots, x_n) - cr)(x_n + cr - y)^2, y)$$
$$= \arg\min_y P(g(x_1 - f_1(x_1, \ldots, x_n))(x_1 + cr - y)^2,$$
$$\ldots, g(x_n - f_n(x_1, \ldots, x_n))(x_n + cr - y)^2, y)$$
$$= F(x_1, \ldots, x_n) + cr.$$

and, then, F is r-shift invariant. The proof of **(ii)** follows from Proposition 1. □

Theorem 7. *Consider $r = (r, \ldots, r) \in \mathbb{R}^n$, with $r > 0$, and $g : [0, 1] \to [0, 1]$. Let $f_k : [0, 1]^n \to [0, 1]$ be r-shift invariant functions, for $k \in \{1, \ldots, n\}$. Let $F : [0, 1]^n \to [0, 1]$ be a P-function, for a penalty function $P : [0, 1]^{n+1} \to \mathbb{R}^+$ defined on the terms $g(x_i - f_i(x_1, \ldots, x_n))(x_i - y)^2$, for all $x_1, \ldots, x_n, x_i - f_i(x_1, \ldots, x_n) \in [0, 1]^n$, and $i = 1, \ldots, n$. Then F is an r-PAF.*

Proof. By Theorem 6**(ii)**, F is r-increasing. By Theorem 1, since F is a P-function, then F is idempotent. Thus, conditions **(A2)** hold, and, then, F is an r-PAF. □

Remark 3. In Theorem 7, notice that, for each $k \in \{1, \ldots, n\}$, f_k is an r-PAF. Then:

$$F(0, \ldots, 0)$$
$$= \arg\min_y P(g(0 - f_1(0, \ldots, 0))(0 - y)^2, \ldots, g(0 - f_n(0, \ldots, 0))(0 - y)^2, y)$$
$$= \arg\min_y P(w_1 y^2, \ldots, w_n y^2, y),$$

for $w_i = g(0 - f_i(0, \ldots, 0)) \in [0, 1]$, $i = 1, \ldots, n$. Since P is a penalty function, then, if $y = 0$, there exists $c \in \mathbb{R}^+$ such that $P(0, \ldots, 0, 0) = c \in \mathbb{R}$, with $P(x_1, \ldots, x_n, y) \geq c$, for all $(x_1, \ldots, x_n, y) \in [0, 1]^{n+1}$. Thus:

$$F(0, \ldots, 0) = \arg\min_y P(w_1 y^2, \ldots, w_n y^2, y) = 0.$$

On the other hand, since, for each $k \in \{1, \ldots, n\}$, f_k is an r-PAF, then:

$$F(1, \ldots, 1)$$
$$= \arg\min_y P(g(1 - f_1(1, \ldots, 1))(1 - y)^2, \ldots, g(1 - f_n(1, \ldots, 1))(1 - y)^2, y)$$
$$= \arg\min_y P(v_1(1 - y)^2, \ldots, v_n(1 - y)^2, y)$$
$$\text{for } v_i = g(1 - f_i(1, \ldots, 1)) \in [0, 1], i = 1, \ldots, n$$

Again, since P is a penalty function, then, whenever $y = 1$, there exists $c \in \mathbb{R}^+$ such that $P(0, \ldots, 0, 0) = c \in \mathbb{R}$, with $P(x_1, \ldots, x_n, y) \geq c$, for all $(x_1, \ldots, x_n, y) \in [0, 1]^{n+1}$. Therefore, one concludes that $F(1, \ldots, 1) = \arg\min_y P(v(1 - y)^2, \ldots, v(1 - y)^2, y) = 1$. So, F satisfies the boundary conditions of a PAF.

Corollary 4. *Consider $r = (r, \ldots, r) \in \mathbb{R}^n$, with $r > 0$, and $g : [0, 1] \to [0, 1]$. Let $f_k : [0, 1]^n \to [0, 1]$ be r-increasing functions that may be not r-shift invariant. Let $F : [0, 1]^n \to [0, 1]$ be a P-function, for some penalty function $P : [0, 1]^{n+1} \to \mathbb{R}^+$ defined on the terms $g(x_i - f_i(x_1, \ldots, x_n))(x_i - y)^2$, for all $x_1, \ldots, x_n, x_i - f_i(x_1, \ldots, x_n) \in [0, 1]^n$, and $i = 1, \ldots, n$. Then F is an r-pre-aggregation function.*

5.1 Spatial/Tonal Filters

In this section, we discuss the ideas presented in [5, Sect. 7.8] considering the more generalized point of view of pre-aggregation functions. For that, we generalize the definition of the filter function, in order to be possible to use our theorems.

In image processing, spatial-tonal filters are used to preserve edges within images when performing some tasks, e.g., the filtering, denoising or smoothing processes [18] (see, e.g., the weighted mode filter [16] and the Gauss bilateral filter [28]). Those filters are implemented in discrete form over a finite set of pixels that assume finite values in $[0, 1]$, defined by the family of averaging and idempotent functions $F_{W,g}^k : [0, 1]^{n+1} \to [0, 1]$, given, for all $x \in [0, 1]^n$ and $x_k \in [0, 1]$, with $k = 1, \ldots, n$, by:

$$F_{W,g}^k(x, x_k) = \frac{\sum_{i=1}^n g(|x_i - x_k|)x_i}{\sum_{i=1}^n g(|x_i - x_k|)},$$

where $g : [0, 1] \to [0, 1]$, $W = \{w_1, \ldots, w_n\}$ is a vector of weights (which are obtained by nonlinear and nonconvex functions of the locations of the pixels, which, in general, are constant), x_i is the intensity of the pixel i, and x_k is the intensity of pixel k to be filtered/smoothed/denoised, taking its new value as $\bar{x}_k = F_{W,g}^k(x, x_k)$. In [5], it was shown that $F_{W,g}^k$ are nonlinear and non monotonic. Since they are idempotent, by Theorem 1, they and can be expressed as P-based functions, for the family of penalty functions $P_{F_{W,g}^k} : [0, 1]^{n+1} \to [0, 1]$, defined, for all $x \in [0, 1]^n$ and $y, x_k \in [0, 1]$, by:

$$P_{F_{W,g}^k}(\boldsymbol{x}, y) = \sum_{i=1}^{n} w_i g(|\, x_i - x_k \,|)(x_i - y)^2, \tag{7}$$

with $k = 1, \ldots, n$. In applications of image-filtering, the penalty minimizes the mean squared error between the filtered image and the noisy source image (see, e.g., [14]).

Now, consider $r = (r, \ldots, r)$, with $r > 0$, and the family of projection functions $proj_k : [0,1]^n \rightarrow [0,1]$, defined, for all $\boldsymbol{x} \in [0,1]^n$, by $proj_k(x_1, \ldots, x_k, \ldots, x_n) = x_k$. Then, Eq. (7) can be written as:

$$P_{F_{W,g}^k}(\boldsymbol{x}, y) = \sum_{i=1}^{n} w_i g(|\, x_i - proj_k(x_i, \ldots, x_n) \,|)(x_i - y)^2, \tag{8}$$

with $k = 1, \ldots, n$. It is immediate that $proj_k$ are r-shift-invariant, and, then, by Theorem 7, the filter functions $F_{W,g}^k$ are r-pre-aggregation functions.

Equation (8) is generalized, considering r-shift invariant functions $F_k : [0,1]^n \rightarrow [0,1]$, with $k = 1, \ldots, n$, and replacing $proj_k$ by F_k in the scaling function g, obtaining:

$$P_{F_{W,g}^k}(\boldsymbol{x}, y) = \sum_{i=1}^{n} w_i g(|\, x_i - f_k(x_i, \ldots, x_n) \,|)(x_i - y)^2.$$

Here, again, by Theorem 7, the filter functions $F_{W,g}^k$ are r-pre-aggregation function.

Whenever F_k are all r-increasing, even if they are not r-shift invariant, then, by Corollary 4, the filter functions $F_{W,g}^k$ are r-pre-aggregation function. This generalization means that one may adopt r-pre-aggregation (which may be r-shift invariant or not) of the vector $\boldsymbol{x}$, which allows to deal with the possibility that x_k is itself an outlier within the local region of the image, as suggested in [5]. Thus, one can use, e.g., the median, the mode (defined by either $\mathrm{Mode_{min}}$ or $\mathrm{Mode_{max}}$) or the shorth for the functions F_k.

6 Conclusion

This paper study PAFs in the context of penalty functions, which is important for many applications, e.g., in image processing. We also showed the relation between the idempotency, averaging, shift invariance concepts concerning PAFs, presenting a characterization of idempotent PAFs based on penalty functions and a weak characterization of averaging PAFs based on penalty functions. We also outlined the use of penalty-based PAFs is spatial/tonal filters. Future work is concerned with the study of pre-aggregation P-functions on interval data [1], in a fuzzy interval approach [3,4,10,13].

Acknowledgments. Supported by Caixa and Fundación Caja Navarra of Spain, the Brazilian National Counsel of Technological and Scientific Development CNPq (Proc. 307781/2016-0, 33950/2014-1, 306970/2013-9), the Spanish Ministry of Science and Technology (TIN2016-77356-P) and by grant APVV-14-0013.

References

1. Aguiar, M.S., Dimuro, G.P., Costa, A.C.R.: ICTM: an interval tessellation-based model for reliable topographic segmentation. Numer. Algorithms **37**(1–4), 3–11 (2004)
2. Barrenechea, E., Fernandez, J., Pagola, M., Chiclana, F., Bustince, H.: Construction of interval-valued fuzzy preference relations from ignorance functions and fuzzy preference relations. Appl. Decis. Mak. Knowl.-Based Syst. **58**, 33–44 (2014)
3. Bedregal, B.C., Dimuro, G.P., Reiser, R.H.S.: An approach to interval-valued R-implications and automorphisms. In: Carvalho, J.P., Dubois, D., Kaymak, U., da Costa Sousa, J.M. (eds.) Proceedings of Joint 2009 International Fuzzy Systems Association World Congress and 2009 European Society of Fuzzy Logic and Technology Conference, IFSA/EUSFLAT, pp. 1–6 (2009)
4. Bedregal, B.C., Dimuro, G.P., Santiago, R.H.N., Reiser, R.H.S.: On interval fuzzy S-implications. Inf. Sci. **180**(8), 1373–1389 (2010)
5. Beliakov, G., Bustince, H., Calvo, T.: A Practical Guide to Averaging Functions, vol. 329. Springer, Berlin, New York (2016). https://doi.org/10.1007/978-3-319-24753-3
6. Bustince, H., Fernandez, J., Kolesárová, A., Mesiar, R.: Directional monotonicity of fusion functions. Eur. J. Oper. Res. **244**(1), 300–308 (2015)
7. Bustince, H., Jurio, A., Pradera, A., Mesiar, R., Beliakov, G.: Generalization of the weighted voting method using penalty functions constructed via faithful restricted dissimilarity functions. Eur. J. Oper. Res. **225**(3), 472–478 (2013)
8. Bustince, H., Beliakov, G., Dimuro, G.P., Bedregal, B., Mesiar, R.: On the definition of penalty functions in data aggregation. Fuzzy Sets Syst. **323**, 1–18 (2017)
9. Calvo, T., Kolesárov, A., Komorníková, M., Mesiar, R.: Aggregation operators: properties, classes and construction methods. In: Calvo, T., Mayor, G., Mesiar, R. (eds.) Aggregation Operators. New Trends and Applications, vol. 97, pp. 3–104. Physica-Verlag, Heidelberg (2002). https://doi.org/10.1007/978-3-7908-1787-4_1
10. Dimuro, G.P.: On interval fuzzy numbers. In: 2011 Workshop-School on Theoretical Computer Science, WEIT 2011, pp. 3–8. IEEE, Los Alamitos (2011)
11. Dimuro, G.P., Bedregal, B., Bustince, H., Fernandez, J., Lucca, G., Mesiar, R.: New results on pre-aggregation functions. In: Uncertainty Modelling in Knowledge Engineering and Decision Making, Proceedings of 12th International FLINS Conference (FLINS 2016), World Scientific Proceedings Series on Computer Engineering and Information Science, vol. 10, pp. 213–219. World Scientific, Singapura (2016)
12. Dimuro, G.P., Bedregal, B.C., Santiago, R.H.N., Reiser, R.H.S.: Interval additive generators of interval t-norms and interval t-conorms. Inf. Sci. **181**(18), 3898–3916 (2011)
13. Dimuro, G.P., Bedregal, B.C., Reiser, R.H.S., Santiago, R.H.N.: Interval additive generators of interval T-norms. In: Hodges, W., de Queiroz, R. (eds.) WoLLIC 2008. LNCS (LNAI), vol. 5110, pp. 123–135. Springer, Heidelberg (2008). https://doi.org/10.1007/978-3-540-69937-8_12
14. Elad, M.: On the origin of the bilateral filter and ways to improve it. IEEE Trans. Image Process **11**(10), 1141–1151 (2002)
15. Elkano, M., Galar, M., Sanz, J.A., Schiavo, P.F., Pereira, S., Dimuro, G.P., Borges, E.N., Bustince, H.: Consensus via penalty functions for decision making in ensembles in fuzzy rule-based classification systems. Appl. Soft Comput. (2017). http://www.sciencedirect.com/science/article/pii/S1568494617303150

16. Fu, M., Zhou, W.: Depth map super-resolution via extended weighted mode filtering. In: 2016 Visual Communications and Image Processing (VCIP), pp. 1–4. IEEE, Los Alamitos (2016)
17. Grabisch, M., Marichal, J., Mesiar, R., Pap, E.: Aggregation Functions. Cambridge University Press, Cambridge (2009)
18. Grazzini, J., Soille, P.: Adaptive morphological filtering using similarities based on geodesic time. In: Coeurjolly, D., Sivignon, I., Tougne, L., Dupont, F. (eds.) DGCI 2008. LNCS, vol. 4992, pp. 519–528. Springer, Heidelberg (2008). https://doi.org/10.1007/978-3-540-79126-3_46
19. Hühn, J., Hüllermeier, E.: FURIA: an algorithm for unordered fuzzy rule induction. Data Min. Knowl. Disc. **19**(3), 293–319 (2009)
20. Lucca, G., Sanz, J., Pereira Dimuro, G., Bedregal, B., Mesiar, R., Kolesárová, A., Bustince Sola, H.: Pre-aggregation functions: construction and an application. IEEE Trans. Fuzzy Syst. **24**(2), 260–272 (2016)
21. Lucca, G., Dimuro, G.P., Mattos, V., Bedregal, B., Bustince, H., Sanz, J.A.: A family of Choquet-based non-associative aggregation functions for application in fuzzy rule-based classification systems. In: 2015 IEEE International Conference on Fuzzy Systems (FUZZ-IEEE), pp. 1–8. IEEE, Los Alamitos (2015)
22. Lucca, G., Sanz, J.A., Dimuro, G.P., Bedregal, B., Asiain, M.J., Elkano, M., Bustince, H.: CC-integrals: Choquet-like copula-based aggregation functions and its application in fuzzy rule-based classification systems. Knowl.-Based Syst. **119**, 32–43 (2017)
23. Lucca, G., Sanz, J.A., Dimuro, G.P., Bedregal, B., Bustince, H., Mesiar, R.: C_F-Integrals: a new family of pre-aggregation functions with application to fuzzy rule-based classification systems. Inf. Sci. **435**, 94–110 (2018)
24. Lázaro, J., Rückschlossová, T., Calvo, T.: Shift invariant binary aggregation operators. Fuzzy Sets Syst. **142**(1), 51–62 (2014)
25. Mayor, G., Trillas, E.: On the representation of some aggregation functions. In: Proceedings of IEEE International Symposium on Multiple-Valued Logic, pp. 111–114. IEEE, Los Alamitos (1986)
26. Wilkin, T., Beliakov, G.: Weakly monotonic averaging functions. Int. J. Intell. Syst. **30**(2), 144–169 (2015)
27. Wilkin, T., Beliakov, G.: Robust image denoising and smoothing with generalised spatial-tonal averages. In: 2017 IEEE International Conference on Fuzzy Systems, pp. 1–7. IEEE, Los Alamitos (2017)
28. Yoshizawa, S., Belyaev, A., Yokota, H.: Fast gauss bilateral filtering. Comput. Graph. Forum **29**(1), 60–74 (2010)

Strengthened Ordered Directional and Other Generalizations of Monotonicity for Aggregation Functions

Mikel Sesma-Sara[1(✉)], Laura De Miguel[1], Julio Lafuente[2], Edurne Barrenechea[1], Radko Mesiar[3,4], and Humberto Bustince[1]

[1] Departamento de Automatica y Computacion and Institute of Smart Cities, Universidad Publica de Navarra, Campus Arrosadia s/n, 31006 Pamplona, Spain
{mikel.sesma,laura.demiguel,edurne.barrenechea,bustince}@unavarra.es
[2] Departamento de Matematicas, Universidad Publica de Navarra, Campus Arrosadia s/n, 31006 Pamplona, Spain
lafuente@unavarra.es
[3] Department of Mathematics and Descriptive Geometry, Faculty of Civil Engineering, Slovak University of Technology, Radlinského 11, Bratislava, Slovakia
mesiar@math.sk
[4] Institute for Research and Applications of Fuzzy Modelling, University of Ostrava, 30. dubna 22, Ostrava, Czech Republic

Abstract. A tendency in the theory of aggregation functions is the generalization of the monotonicity condition. In this work, we examine the latest developments in terms of different generalizations. In particular, we discuss strengthened ordered directional monotonicity, its relation to other types of monotonicity, such as directional and ordered directional monotonicity and the main properties of the class of functions that are strengthened ordered directionally monotone. We also study some construction methods for such functions and provide a characterization of usual monotonicity in terms of these notions of monotonicity.

Keywords: Aggregation functions · Directional monotonicity
Generalizations of monotonicity
Strengthened ordered directional monotonicity

1 Introduction

The problem of finding a single representative number for a set of values is common to every field that handles real data. There exist several works in the literature addressing this issue prior to the introduction of the theory of aggregation functions per se. That is the case, for example, of triangular norms [12], copulas [17] and Choquet integrals [7]. According to [13], the inception of the theory of aggregation functions as an independent theory dates back to 1988 [10] and it was not until 2001 that the first monograph on the subject came out [6].

© Springer International Publishing AG, part of Springer Nature 2018
J. Medina et al. (Eds.): IPMU 2018, CCIS 854, pp. 416–426, 2018.
https://doi.org/10.1007/978-3-319-91476-3_35

An aggregation function, in the classical sense, is a function $A : [0,1]^n \to [0,1]$ such that $A(\mathbf{0}) = 0$, $A(\mathbf{1}) = 1$ and it is increasing with respect to every argument (the standard partial order on $[0,1]^n$). Since its appearence, the aggregation theory has been extended to new domains beside real numbers [18] and aggregation functions have been applied in diverse real world problems [8,9,14].

In addition to the extension of aggregation operators to be able to deal with more general scales than numbers, such as lattices, a relevant trend in the theory of aggregation functions is the relaxation of the monotonicity condition. Monotonicity with respect to every argument may lead to exclude from the framework of aggregation functions mappings that are valid to provide a representative value from a set of numerical values. Examples of such functions are the mode operator, or the Lehmer mean [3], among others (see [2]).

On the account of broadening the framework of functions that are sound for fusing data, some generalizations of monotonicity have emerged [1]. One of the most significant forms of monotonicity is that of directional monotonicity, introduced in [5], which, similar to the concept of directional derivatives, deals with monotonicity along a ray in $\mathbb{R}^n$. This type of monotonicity generalizes the formerly presented notion of weak monotonicity [19], which is the particular case of restricting directional monotonicity to the ray $(1, 1, \ldots, 1)$. The fact that the ray of increasingness could be any vector in $\mathbb{R}^n$ permits to select functions that adjust better to particular problems or applications. Nevertheless, that direction is the same for all the points in the domain.

Recently, influenced by the concept of OWA operator [20], the concept of ordered directional (OD) monotonicity has been introduced [4]. The direction of increasingness or decreasingness for ordered directionally monotone function varies depending on the point of the domain that is being considered. Specifically, the ray of increasingness (decreasingness) varies according to the relative size of the inputs, as long as a fixed comonotonicity requirement is satisfied. Directionally and ordered directionally monotone functions have yielded good results in classification problems [11] and in the field of image processing [15].

In this work, we discuss the notion of strengthened ordered directional (SOD) monotonicity [16], a concept based on ordered directional monotonicity, for whose definition no comonotonicity condition is required. This relaxation makes the family of strengthened ordered directionally monotone functions a proper subset of the class of ordered directionally monotone functions, meaning that if a function is strengthened ordered directionally monotone, then it is ordered directionally monotone, but not contrarily. Furthermore, we address some relevant properties of the three forms of monotonicity, i.e., directional, ordered directional and strengthened ordered directional monotonicity, and we point out some links and differences among them. We also expose some construction methods for functions that are monotone in each of the discussed senses and we characterize classical monotonicity with regard to its various generalizations.

This work is organized as follows. We start the next section with some remarks about the notation that is used throughout the paper, as well as recalling some preliminary notions and basic definitions. In Sect. 3 we present the definition of the latest generalization of monotonicity in the literature regarding directional monotonicity; the concept of strengthened ordered directional monotonicity. We also study the class of SOD functions and we expose a scheme of the points that trivially satisfy the conditions for each of the types of monotonicity. In Sect. 4 we discuss some relevant properties of the different notions of monotonicity, as well as the relations that exist among them. In Sect. 5 we present various construction methods for functions that are monotone in each sense. We finish this work with a characterization of the usual condition of monotonicity in terms of the different generalizations, in Sect. 5, followed by some brief concluding remarks.

2   Preliminaries

Let $n \in \mathbb{N}$, with $n \geq 2$. We refer as $\mathbf{x} = (x_1, \ldots, x_n) \in [0,1]^n$ to points in the unit hypercube and as $\overrightarrow{r} = (r_1, \ldots, r_n) \in \mathbb{R}^n$ to vectors connoting a direction in $\mathbb{R}^n$.

The notion of monotonicity is highly related to the concept of order. In this work we consider the usual partial order of $[0,1]^n$, i.e., given $\mathbf{x}, \mathbf{y} \in [0,1]^n$, we say $\mathbf{x} \leq \mathbf{y}$ if $x_i \leq y_i$ for every $i \in \{1, \ldots, n\}$.

Like in the case of OD monotone functions, the points of the domain whose components are decreasingly ordered play an important role in the framework of SOD monotone functions. We use the following notation for the set of these points: Let $H \subset \mathbb{R}^n$, then we set $H_{(\geq)} = \{(h_1, \ldots, h_n) \in H \mid h_1 \geq \cdots \geq h_n\}$ and $H_{(\leq)}, H_{(>)}, H_{(<)}, H_{(=)}$ accordingly.

In order to impose that some points' components are decreasingly ordered, it is common to use permutations. Let $\mathcal{S}_n$ be the set of all permutations of n elements, $\sigma \in \mathcal{S}_n$ and $\mathbf{x} \in [0,1]^n$, we denote by $\mathbf{x}_\sigma$ the tuple $(x_{\sigma(1)}, \ldots, x_{\sigma(n)})$. Note that for $\mathbf{x}, \mathbf{y} \in \mathbb{R}^n$, it holds that $\mathbf{x} \in [0,1]^n$ if and only if $\mathbf{x}_\sigma \in [0,1]^n$. Moreover, for $\mathbf{x}, \mathbf{y} \in [0,1]^n$ and $\sigma \in \mathcal{S}_n$, it holds that $(\mathbf{x} + \mathbf{y})_\sigma = \mathbf{x}_\sigma + \mathbf{y}_\sigma$, and $\mathbf{x} \cdot \mathbf{y} = \mathbf{x}_\sigma \cdot \mathbf{y}_\sigma$, where $\mathbf{x} \cdot \mathbf{y}$ denotes the scalar product given by $\mathbf{x} \cdot \mathbf{y} = \sum_{i=1}^n x_i y_i$.

The notion of directional monotonicity, or monotonicity along a ray $\overrightarrow{r}$, was introduced in [5], generalizing the notion of monotonicity for functions from $[0,1]^n$ to $[0,1]$.

Definition 1. *Let $F : [0,1]^n \rightarrow [0,1]$ and $\overrightarrow{r} \in \mathbb{R}^n$, we say that F is $\overrightarrow{r}$-increasing (decreasing), if for all $c > 0$ and $\mathbf{x} \in [0,1]^n$ such that $\mathbf{x} + c\overrightarrow{r} \in [0,1]^n$, it holds that $F(\mathbf{x} + c\overrightarrow{r}) \geq F(\mathbf{x})$ ($F(\mathbf{x} + c\overrightarrow{r}) \leq F(\mathbf{x})$).*

Directional monotonicity generalizes weak monotonicity, introduced in [19], which is the particular case of considering as direction the vector $\overrightarrow{1} = (1, \ldots, 1)$.

OD monotonicity was presented in [4]. The direction of increasingness of functions that are OD $\overrightarrow{r}$-increasing varies in terms of the relative sizes of the input.

Definition 2. *Let $F : [0,1]^n \rightarrow [0,1]$ and $\overrightarrow{r} \in \mathbb{R}^n$, we say that F is ordered directionally, OD, $\overrightarrow{r}$-increasing (decreasing), if for all $c > 0$, $\sigma \in \mathcal{S}_n$ and $\mathbf{x} \in$*

$[0,1]^n$, *it holds that if* $\mathbf{x}_\sigma$, $\mathbf{x}_\sigma + c\overrightarrow{r} \in [0,1]^n_{(\geq)}$, *then* $F(\mathbf{x} + c\overrightarrow{r}_{\sigma^{-1}}) \geq F(\mathbf{x})$
$(F(\mathbf{x} + c\overrightarrow{r}_{\sigma^{-1}}) \leq F(\mathbf{x}))$.

3 The Class of Strengthened Ordered Directionally Monotone Functions

The concept of SOD monotonicity has been more recently introduced in [16]. It is based on OD monotonicity, in fact the difference between both concepts is that SOD monotone functions are not asked to satisfy the condition of comonotonicity between $\mathbf{x}_\sigma$ and $\mathbf{x}_\sigma + c\overrightarrow{r}$.

Definition 3. *Let* $F \colon [0,1]^n \to [0,1]$ *and* $\overrightarrow{r} \in \mathbb{R}^n$, *we say that* F *is strengthened ordered directionally, SOD,* $\overrightarrow{r}$-*increasing (decreasing), if for all* $c > 0$, $\sigma \in \mathcal{S}_n$ *and* $\mathbf{x} \in [0,1]^n$, *it holds that if* $\mathbf{x}_\sigma \in [0,1]^n_{(\geq)}$ *and* $\mathbf{x}_\sigma + c\overrightarrow{r} \in [0,1]^n$, *then* $F(\mathbf{x} + c\overrightarrow{r}_{\sigma^{-1}}) \geq F(\mathbf{x})$ $(F(\mathbf{x} + c\overrightarrow{r}_{\sigma^{-1}}) \leq F(\mathbf{x}))$.

If a function is simultaneously (OD, SOD) $\overrightarrow{r}$-increasing and (OD, SOD) $\overrightarrow{r}$-decreasing, we say that the function is (OD, SOD) $\overrightarrow{r}$-constant.

Remark 1. The notation (OD, SOD) $\overrightarrow{r}$-increasing refers to directional ($\overrightarrow{r}$-increasing), ordered directional (OD $\overrightarrow{r}$-increasing) and strengthened ordered directional (SOD $\overrightarrow{r}$-increasing) monotonicity.

The case of $\overrightarrow{r} = \overrightarrow{0}$ is trivial for the three notions of monotonicity. In fact, every function is (OD, SOD) $\overrightarrow{0}$-constant.

An OD monotone function F is required to satisfy the inequality $F(\mathbf{x} + c\overrightarrow{r}_{\sigma^{-1}}) \geq F(\mathbf{x})$ for points that satisfy the comonotonicity condition $\mathbf{x}_\sigma$, $\mathbf{x}_\sigma + c\overrightarrow{r} \in [0,1]^n_{(\geq)}$, whereas a SOD monotone function F is required to satisfy the same inequality for points that satisfy that condition and for points that do not. Therefore, SOD $\overrightarrow{r}$-increasingness implies OD $\overrightarrow{r}$-increasingness. However, the converse statement does not hold.

One of the particularities of each notion of monotonicity is the set of points that satisfy the monotonicity conditions trivially. On the one hand, in the case of directional monotonicity, i.e., $\overrightarrow{r}$-increasing functions, the points that trivially satisfy the conditions are those $\mathbf{x} \in [0,1]^n$ such that $\mathbf{x} + c\overrightarrow{r} \notin [0,1]^n$ for all $c > 0$. On the other hand, for a function F that is OD (SOD) $\overrightarrow{r}$-increasing, such set of points is formed by those $\mathbf{x} \in [0,1]^n$ such that if $\sigma \in \mathcal{S}_n$ with $\mathbf{x}_\sigma \in [0,1]^n_{(\geq)}$, then $\mathbf{x}_\sigma + c\overrightarrow{r} \notin [0,1]^n_{(\geq)}$ $(\mathbf{x}_\sigma + c\overrightarrow{r} \notin [0,1]^n)$ for all $c > 0$.

For the case $n = 2$, we show in Table 1 the relation of directions (given in terms of the angle that they form with respect to the non-negative horizontal axis) and points that, for each notion of monotonicity, trivially satisfy the conditions.

Table 1. Directions (in terms of their angle α w.r.t. the non-negative horizontal axis) and points that trivially satisfy the monotonicity conditions for directional, ordered directional and strengthened ordered directional monotonicity

Directions $\vec{r}$ such that	D monotonicity $\mathbf{x} \in [0,1]^2$ such that	OD monotonicity $\mathbf{x} \in [0,1]^2$ such that	SOD monotonicity $\mathbf{x} \in [0,1]^2$ such that
$\alpha = 0$	$x_1 = 1$	$x_1 = 1$ or $x_2 = 1$	$x_1 = 1$ or $x_2 = 1$
$0 < \alpha \leq \frac{\pi}{4}$	$x_1 = 1$ or $x_2 = 1$		
$\frac{\pi}{4} < \alpha < \frac{\pi}{2}$		$x_1 = 1$ or $x_2 = 1$ or $x_1 = x_2$	
$\alpha = \frac{\pi}{2}$	$x_2 = 1$	$x_1 = x_2$	$x_1 = x_2 = 1$
$\frac{\pi}{2} < \alpha < \pi$	$x_1 = 0$ or $x_2 = 1$		$x_1 = x_2 = 1$ or $x_1 = x_2 = 0$
$\alpha = \pi$	$x_1 = 0$		$x_1 = x_2 = 0$
$\pi < \alpha < \frac{5\pi}{4}$	$x_1 = 0$ or $x_2 = 0$	$x_1 = 0$ or $x_2 = 0$ or $x_1 = x_2$	$x_1 = 0$ or $x_2 = 0$
$\frac{5\pi}{4} \leq \alpha < \frac{3\pi}{2}$		$x_1 = 0$ or $x_2 = 0$	
$\alpha = \frac{3\pi}{2}$	$x_2 = 0$		
$\frac{3\pi}{2} < \alpha < 2\pi$	$x_1 = 1$ or $x_2 = 0$	$x_1 = 0$ or $x_1 = 1$ or $x_2 = 0$ or $x_2 = 1$	$x_1 = 0$ or $x_1 = 1$ or $x_2 = 0$ or $x_2 = 1$

4   Properties and Connections of the Different Notions of Monotonicity

We use the following notation to refer to the set of vectors for which a function is increasing (and constant) according to the three different notions of monotonicity with which we deal in this paper. Let $F \colon [0,1]^n \to [0,1]$, thus we set

$$\mathcal{D}^{\uparrow}(F) = \{\, \vec{r} \in \mathbb{R}^n \mid F \text{ is } \vec{r}\text{-increasing}\,\},$$
$$\mathcal{C}(F) = \{\, \vec{r} \in \mathbb{R}^n \mid F \text{ is } \vec{r}\text{-constant}\,\},$$

and the remaining sets of directions $\mathcal{D}^{\uparrow}_{OD}(F)$, $\mathcal{D}^{\uparrow}_{SOD}(F)$, $\mathcal{C}_{OD}(F)$ and $\mathcal{C}_{SOD}(F)$ accordingly.

We derive from the definition the first relation among these sets of directions.

Proposition 1. *Let $F \colon [0,1]^n \to [0,1]$. Then the following items hold:*

(i) $\mathcal{C}(F) \subseteq \mathcal{D}^{\uparrow}(F)$, $\mathcal{C}_{OD}(F) \subseteq \mathcal{D}^{\uparrow}_{OD}(F)$ and $\mathcal{C}_{SOD}(F) \subseteq \mathcal{D}^{\uparrow}_{SOD}(F)$;
(ii) $\mathcal{C}_{SOD}(F) \subseteq \mathcal{C}_{OD}(F)$;
(iii) $\mathcal{D}^{\uparrow}_{SOD}(F) \subseteq \mathcal{D}^{\uparrow}_{OD}(F)$.

The following two results are also obtained from the definition of the different notions of monotonicity.

Proposition 2. *Let $F \colon [0,1]^n \to [0,1]$ and $\vec{r} \in \mathbb{R}^n$ be such that $r_1 \geq \ldots \geq r_n$. Then F is SOD $\vec{r}$-increasing (decreasing) if and only if F is OD $\vec{r}$-increasing (decreasing). Moreover, F is SOD $\vec{r}$-constant if and only if F is OD $\vec{r}$-constant.*

Proposition 3. *Let $F: [0,1]^n \to [0,1]$ be a (OD, SOD) $\overrightarrow{r}$-increasing function and let $\varphi: [0,1] \to [0,1]$ be an increasing (decreasing) function. Then, the composition $\varphi \circ F: [0,1]^n \to [0,1]$ is (OD, SOD) $\overrightarrow{r}$-increasing (decreasing).*

The vectors' magnitude has no influence whatsoever in the qualification of such vectors as directions of (OD, SOD) increasingness. Therefore, it is possible to limit the set of directions to normalized vectors, i.e., vectors of norm 1, as it is shown in the next Proposition.

Proposition 4. *Let $F: [0,1]^n \to [0,1]$ and $k > 0$. Then, F is (OD, SOD) $\overrightarrow{r}$-increasing (decreasing) if and only if F is (OD, SOD) $(k\overrightarrow{r})$-increasing (decreasing).*

The following result reveals a difference between the classes of directionally and ordered directionally monotone functions and the class of strengthened ordered directionally monotone functions.

Proposition 5. *Let $F: [0,1]^n \to [0,1]$. F is (OD) $\overrightarrow{r}$-increasing if and only if (OD) F is $(-\overrightarrow{r})$-decreasing.*

Proof. Case of directional monotonicity:
Let F be $\overrightarrow{r}$-increasing and let $\mathbf{x} \in [0,1]^n$ and $c > 0$ such that $\mathbf{x} - c\overrightarrow{r} \in [0,1]^n$. Set $\mathbf{y} = \mathbf{x} - c\overrightarrow{r}$. Thus,

$$F(\mathbf{x} + c(-\overrightarrow{r})) = F(\mathbf{y}) \leq F(\mathbf{y} + c\overrightarrow{r}) = F(\mathbf{x}),$$

hence F is $(-\overrightarrow{r})$-decreasing. Similarly, one can show the converse statement.
Case of ordered directional monotonicity:
Let F be OD $\overrightarrow{r}$-increasing and let $\mathbf{x} \in [0,1]^n$, $c > 0$ and $\sigma \in \mathcal{S}_n$ such that $\mathbf{x}_\sigma$ and $\mathbf{x}_\sigma + c(-\overrightarrow{r}) \in [0,1]^n_{(\geq)}$. Set $\mathbf{y} = \mathbf{x} + c(-\overrightarrow{r})_{\sigma^{-1}}$. Thus, it holds that $\mathbf{y}_\sigma = \mathbf{x}_\sigma + c(-\overrightarrow{r}) \in [0,1]^n_{(\geq)}$ and $\mathbf{y}_\sigma + c\overrightarrow{r} = \mathbf{x}_\sigma \in [0,1]^n_{(\geq)}$. Now, since F is OD $\overrightarrow{r}$-increasing, it holds that

$$F(\mathbf{x} + c(-\overrightarrow{r})_{\sigma^{-1}}) = F(\mathbf{y}) \leq F(\mathbf{y} + c\overrightarrow{r}_{\sigma^{-1}}) = F(\mathbf{x}),$$

therefore F is OD $(-\overrightarrow{r})$-decreasing. The converse is analogous.

In [16] it is shown that Proposition 5 does not generally hold for SOD monotonicity, which indicates that whereas the results for (ordered) directional increasingness can be readily extended to (ordered) directional decreasingness, it is not generally the case for the results of SOD monotonicity. This fact is patent in the upcoming results on duality (Sect. 5).

The next three theorems concern the directions of (OD, SOD) increasingness for functions with some known directions of (OD, SOD) increasingness. The first one, regarding directional monotonicity can be found in [5]; the second, about ordered directional monotonicity, in [4]; and the third, regarding strengthened ordered directional monotonicity, in [16].

Theorem 1 ([5]). *Let* $\overrightarrow{r}, \overrightarrow{s} \in \mathbb{R}^n$ *and* $a, b \geq 0$, *with* $a + b > 0$. *Let* $\mathbf{x} \in [0,1]^n$, $c > 0$, *and assume that if* $\mathbf{x}$ *and* $\mathbf{x} + c(a\overrightarrow{r} + b\overrightarrow{s}) \in [0,1]^n$, *then either* $\mathbf{x} + ca\overrightarrow{r}$ *or* $\mathbf{x} + cb\overrightarrow{s} \in [0,1]^n$. *Then, if a function* $F : [0,1]^n \to [0,1]$ *is both* $\overrightarrow{r}$-*increasing and* $\overrightarrow{s}$-*increasing, then* F *is also* $(a\overrightarrow{r} + b\overrightarrow{s})$-*increasing.*

Theorem 2 ([4]). *Let* $\overrightarrow{r}, \overrightarrow{s} \in \mathbb{R}^n$ *and* $a, b \geq 0$, *with* $a + b > 0$. *Let* $\mathbf{x} \in [0,1]^n$, $c > 0$, $\sigma \in \mathcal{S}_n$ *and assume that if* $\mathbf{x}_\sigma$ *and* $\mathbf{x}_\sigma + c(a\overrightarrow{r} + b\overrightarrow{s}) \in [0,1]^n_{(\geq)}$, *then either* $\mathbf{x} + ca\overrightarrow{r}$ *or* $\mathbf{x} + cb\overrightarrow{s} \in [0,1]^n_{(\geq)}$. *Then, if a function* $F : [0,1]^n \to [0,1]$ *is both OD* $\overrightarrow{r}$-*increasing and OD* $\overrightarrow{s}$-*increasing, then* F *is also OD* $(a\overrightarrow{r} + b\overrightarrow{s})$-*increasing.*

Theorem 3 ([16]). *Let* $\overrightarrow{r}, \overrightarrow{s} \in \mathbb{R}^n$ *and* $a, b \geq 0$, *with* $a + b > 0$. *Let* $\mathbf{x} \in [0,1]^n$, $c > 0$, $\sigma \in \mathcal{S}_n$ *and assume that if* $\mathbf{x}_\sigma \in [0,1]^n_{(\geq)}$ *and* $\mathbf{x}_\sigma + c(a\overrightarrow{r} + b\overrightarrow{s}) \in [0,1]^n$, *then either* $\mathbf{x} + ca\overrightarrow{r}$ *or* $\mathbf{x} + cb\overrightarrow{s} \in [0,1]^n$. *Then, if a function* $F : [0,1]^n \to [0,1]$ *is both SOD* $\overrightarrow{r}$-*increasing and SOD* $\overrightarrow{s}$-*increasing, then* F *is also SOD* $(a\overrightarrow{r} + b\overrightarrow{s})$-*increasing.*

Therefore if a function F is (OD, SOD) increasing in two directions $\overrightarrow{r}$ and $\overrightarrow{s}$, under the assumptions of the preceding theorems, it is also (OD, SOD) increasing in the direction resulting from a positive linear combination of $\overrightarrow{r}$ and $\overrightarrow{s}$.

5   Construction Methods

In this Section we show how to construct (ordered, strengthened ordered) directionally monotone functions from functions that have the same type of monotonicity. First, we present some results that establish the relation of a function and its dual according to these generalizations of monotonicity.

Proposition 6. *Let* $F : [0,1]^n \to [0,1]$ *and* $F^c : [0,1]^n \to [0,1]$ *be given by* $F^c(\mathbf{x}) = 1 - F(\mathbf{x})$. *Let* $\overrightarrow{r} \in \mathbb{R}^n$. *Then,* F *is (OD, SOD)* $\overrightarrow{r}$-*increasing if and only if* F^c *is (OD, SOD)* $\overrightarrow{r}$-*decreasing.*

Proof. Case of directional monotonicity:
Let $c > 0$ and $\mathbf{x}, \mathbf{x} + c\overrightarrow{r} \in [0,1]^n$, then $F(\mathbf{x}) \leq F(\mathbf{x} + c\overrightarrow{r})$ if and only if $F^c(\mathbf{x}) = 1 - F(\mathbf{x}) \geq 1 - F(\mathbf{x} + c\overrightarrow{r}) = F^c(\mathbf{x} + c\overrightarrow{r})$. The cases of OD and SOD monotonicity are straightforward.

As a consequence, the fact that $(F^c)^c = F$ yields the following result.

Corollary 1. *Let* $F : [0,1]^n \to [0,1]$ *and* $F^c : [0,1]^n \to [0,1]$ *be given by* $F^c(\mathbf{x}) = 1 - F(\mathbf{x})$. *Let* $\overrightarrow{r} \in \mathbb{R}^n$. *Then,* F *is (OD, SOD)* $\overrightarrow{r}$-*constant if and only if both* F *and* F^c *are (OD, SOD)* $\overrightarrow{r}$-*increasing. Additionally, the following equalities hold:*

(i) $\mathcal{C}(F) = \mathcal{C}(F^c)$;
(ii) $\mathcal{C}_{OD}(F) = \mathcal{C}_{OD}(F^c)$;
(iii) $\mathcal{C}_{SOD}(F) = \mathcal{C}_{SOD}(F^c)$.

Under the conditions of Proposition 6, it is clear that F is (OD) $\overrightarrow{r}$-increasing if and only if F^c is (OD) $(-\overrightarrow{r})$-increasing. However, that statement is not generally so for SOD monotonicity.

Proposition 6 and the next result provide the relation of a function and its dual in terms of the directions for which each is (ordered, strengthened ordered) directionally monotone.

Proposition 7. *Let $F : [0,1]^n \to [0,1]$, $G: [0,1]^n \to [0,1]$, defined by $G(\mathbf{x}) = F(\mathbf{1} - \mathbf{x})$, and $\overrightarrow{r} \in \mathbb{R}^n$. Let $\overrightarrow{r}^{\,d} = (r_n, \ldots, r_1)$. Then*

(i) F is $\overrightarrow{r}$-increasing if and only if G is $(-\overrightarrow{r})$-increasing; and
(ii) F is OD (SOD) $\overrightarrow{r}$-increasing if and only if G is OD (SOD) $(-\overrightarrow{r})^{\,d}$-increasing.

Proof. (i) It is straightforward.
(ii) *Case of ordered directional monotonicity:*

Let F be OD $\overrightarrow{r}$-increasing and let $\mathbf{x} \in [0,1]^n$. Let $\sigma \in \mathcal{S}_n$ and $c > 0$ such that $\mathbf{x}_\sigma \in [0,1]^n_{(\geq)}$ and $\mathbf{x}_\sigma + c(-\overrightarrow{r})^d \in [0,1]^n_{(\geq)}$.
Set $\mathbf{y} = \mathbf{1} - \mathbf{x}$ and $\tau \in \mathcal{S}_n$ such that $\tau(i) = \sigma(n-i+1)$ for all $i \in \{1, \ldots, n\}$. Thus, one can verify that $\tau^{-1}(i) = n - \sigma^{-1}(i) + 1$, and that $(\overrightarrow{r}^{\,d})_{\sigma^{-1}} = \overrightarrow{r}_{\tau^{-1}}$. Besides, we have that $\mathbf{y}_\tau \in [0,1]^n_{(\geq)}$ and $\mathbf{y}_\tau + c\overrightarrow{r} \in [0,1]^n_{(\geq)}$. Now, since F is OD $\overrightarrow{r}$-increasing, it holds that

$$G\left(\mathbf{x} + c\left(-\overrightarrow{r}^{\,d}\right)_{\sigma^{-1}}\right) = F\left(\mathbf{y} + c\overrightarrow{r}_{\tau^{-1}}\right) \geq F(\mathbf{y}) = G(\mathbf{x}),$$

therefore G is OD $(-\overrightarrow{r})^{\,d}$-increasing.

The converse is straight since $-\left(-\overrightarrow{r}^{\,d}\right)^d = \overrightarrow{r}$.

Similarly, one can show the case of strengthened ordered directional monotonicity.

Recall that given a function $F: [0,1]^n \to [0,1]$, we can define its dual F^d by $F^d(\mathbf{x}) = 1 - F(\mathbf{1} - \mathbf{x})$.

Corollary 2. *Let $F : [0,1]^n \to [0,1]$ and $F^d: [0,1]^n \to [0,1]$ be its dual function. Then, for $\overrightarrow{r} \in \mathbb{R}^n$, it holds that*

(i) F is $\overrightarrow{r}$-increasing if and only if F^d is $\overrightarrow{r}$-increasing;
(ii) F is OD $\overrightarrow{r}$-increasing if and only if F^d is OD $\overrightarrow{r}^{\,d}$-increasing;
(iii) F is SOD $\overrightarrow{r}$-increasing if and only if F^d is SOD $(-\overrightarrow{r})^{\,d}$-decreasing.

At this point, we expose how a(n) (ordered, strengthened ordered) directionally monotone function can be constructed from a set of n functions with the same type of monotonicity.

Theorem 4. *Let $\overrightarrow{r} \in \mathbb{R}^n$ and $F_i: [0,1]^n \to [0,1]$, $1 \leq i \leq m$, be m (OD, SOD) $\overrightarrow{r}$-increasing functions. Let $H : [0,1]^n \to [0,1]$ be an aggregation function. Then the function $H(F_1, \ldots, F_m): [0,1]^n \to [0,1]$, given by $H(F_1, \ldots, F_m)(\mathbf{x}) = H(F_1(\mathbf{x}), \ldots, F_m(\mathbf{x}))$, is (OD, SOD) $\overrightarrow{r}$-increasing.*

In particular, the convex combination of m (OD, SOD) directionally monotone functions is a function with the same kind of monotonicity.

6 Characterization of Monotonicity in Terms of Its Different Generalizations

The following theorem gives a characterization of the usual condition of increasingness in terms of the different forms of monotonicity discussed in this paper. This result has acquired its present form along with the introduction of the different notions of monotonicity. In [5], one can find the proof of the equivalence of the first two items. the third item was added in [4], and one can find its current presentation in [16].

Theorem 5 ([16]). *Let $F\colon [0,1]^n \to [0,1]$ and $(\overrightarrow{e}_1, \ldots, \overrightarrow{e}_n)$ be the canonical basis of $\mathbb{R}^n$. The following are equivalent:*

(i) *F is increasing;*
(ii) *F is $\overrightarrow{e}_i$-increasing for every $i \in \{1, \ldots, n\}$;*
(iii) *F is OD $\overrightarrow{e}_i$-increasing for every $i \in \{1, \ldots, n\}$;*
(iv) *F is SOD $\overrightarrow{e}_i$-increasing for every $i \in \{1, \ldots, n\}$.*

As a consequence, we have another characterization of usual monotonicity, in terms of vectors for which all the components are positive instead of canonical vectors.

Corollary 3. *Let $F\colon [0,1]^n \to [0,1]$. The following are equivalent:*

(i) *F is increasing;*
(ii) *F is OD $\overrightarrow{r}$ − increasing for every $\overrightarrow{r} \in (\mathbb{R}^+)^n$;*
(iii) *F is SOD $\overrightarrow{r}$ − increasing for every $\overrightarrow{r} \in (\mathbb{R}^+)^n$.*

Another interesting result is the characterization of weak monotonicity in terms of the other generalizations of monotonicity. Weak monotonicity is a particular case of directional monotonicity which only considers the vector $\overrightarrow{1} = (1, \ldots, 1)$ as a direction.

Theorem 6. *Let $F\colon [0,1]^n \to [0,1]$. The following are equivalent:*

(i) *F is weakly increasing;*
(ii) *F is $\overrightarrow{1}$-increasing;*
(iii) *F is OD $\overrightarrow{1}$-increasing;*
(iv) *F is SOD $\overrightarrow{1}$-increasing.*

Proof. (i) and (ii) are equivalent by the definition of weak monotonicity and Proposition 2 yields that (iii) and (iv) are also equivalent.

Let us show that (ii) and (iii) are equivalent. Assume that F is $\overrightarrow{1}$-increasing and let $\mathbf{x} \in [0,1]$, $\sigma \in S_n$ and $c > 0$ such that $\mathbf{x}_\sigma \in [0,1]^n_{(\geq)}$ and $\mathbf{x}_\sigma + c\overrightarrow{1} \in [0,1]^n_{(\geq)}$. Since F is $\overrightarrow{1}$-increasing, it holds that $F(\mathbf{x} + c\overrightarrow{1}_{\sigma-1}) = F(\mathbf{x} + c\overrightarrow{1}) \geq F(\mathbf{x})$.

The converse implication is analogous.

7   Conclusions

We have presented the concept of strengthened ordered directional monotonicity, the latest generalization of monotonicity in the literature. We have also discussed some of the main properties and the links among the different state-of-the-art weaker forms of monotonicity, namely, weak monotonicity, directional monotonicity, ordered directional monotonicity and strengthened ordered directional monotonicity. Moreover, we have highlighted some construction methods for classes of functions that are monotone according to the discussed types of monotonicity and we have provided a characterization of the usual condition of monotonicity in terms of directional, ordered directional and strengthened ordered directional monotonicity.

Acknowledgements. This work is partially supported by the Research Service of Universidad Pública de Navarra and the grants APVV-14-0013 and TIN2016-77356-P (AEI/FEDER, UE).

References

1. Beliakov, G., Calvo, T., Wilkin, T.: Three types of monotonicity of averaging functions. Knowl.-Based Syst. **72**, 114–122 (2014). https://doi.org/10.1016/j.knosys.2014.08.028
2. Beliakov, G., Špirková, J.: Weak monotonicity of Lehmer and Gini means. Fuzzy Sets Syst. **299**, 26–40 (2016). https://doi.org/10.1016/j.fss.2015.11.006
3. Bullen, P.S.: Handbook of Means and Their Inequalities, vol. 560. Springer, Dordrecht (2013). https://doi.org/10.1007/978-94-017-0399-4
4. Bustince, H., Barrenechea, E., Sesma-Sara, M., Lafuente, J., Dimuro, G.P., Mesiar, R., Kolesárová, A.: Ordered directionally monotone functions. Justification and application. IEEE Trans. Fuzzy Syst. (in press). https://doi.org/10.1109/TFUZZ.2017.2769486
5. Bustince, H., Fernandez, J., Kolesárová, A., Mesiar, R.: Directional monotonicity of fusion functions. Eur. J. Oper. Res. **244**(1), 300–308 (2015). https://doi.org/10.1016/j.ejor.2015.01.018
6. Calvo, T., Kolesárová, A., Komornıková, M., Mesiar, R.: A Review of Aggregation Operators. Servicio de Publicaciones de la UAH, Madrid (2001)
7. Choquet, G.: Theory of capacities. Ann. de l'institut Fourier **5**, 131–295 (1954). https://doi.org/10.5802/aif.53
8. De Miguel, L., Sesma-Sara, M., Elkano, M., Asiain, M., Bustince, H.: An algorithm for group decision making using n-dimensional fuzzy sets, admissible orders and OWA operators. Inf. Fusion **37**, 126–131 (2017). https://doi.org/10.1016/j.inffus.2017.01.007
9. García-Lapresta, J., Martínez-Panero, M.: Positional voting rules generated by aggregation functions and the role of duplication. Int. J. Intell. Syst. **32**(9), 926–946 (2017). https://doi.org/10.1002/int.21877
10. Klir, G., Folger, T.: Fuzzy Sets, Uncertainty, and Information. Prentice Hall, Upper Saddle River (1988)

11. Lucca, G., Sanz, J., Dimuro, G., Bedregal, B., Asiain, M.J., Elkano, M., Bustince, H.: CC-integrals: choquet-like copula-based aggregation functions and its application in fuzzy rule-based classification systems. Knowl.-Based Syst. **119**, 32–43 (2017). https://doi.org/10.1016/j.knosys.2016.12.004
12. Menger, K.: Statistical metrics. Proc. Nat. Acad. Sci. **28**(12), 535–537 (1942)
13. Mesiar, R., Kolesárová, A., Stupňanová, A.: Quo vadis aggregation? Int. J. Gen Syst **47**(2), 97–117 (2018). https://doi.org/10.1080/03081079.2017.1402893
14. Paternain, D., Fernandez, J., Bustince, H., Mesiar, R., Beliakov, G.: Construction of image reduction operators using averaging aggregation functions. Fuzzy Sets Syst. **261**, 87–111 (2015). https://doi.org/10.1016/j.fss.2014.03.008
15. Sesma-Sara, M., Bustince, H., Barrenechea, E., Lafuente, J., Kolsesárová, A., Mesiar, R.: Edge detection based on ordered directionally monotone functions. In: Kacprzyk, J., Szmidt, E., Zadrożny, S., Atanassov, K.T., Krawczak, M. (eds.) IWIFSGN/EUSFLAT -2017. AISC, vol. 643, pp. 301–307. Springer, Cham (2018). https://doi.org/10.1007/978-3-319-66827-7_27
16. Sesma-Sara, M., Lafuente, J., Roldán, A., Mesiar, R., Bustince, H.: Strengthened ordered directionally monotone functions. Links between the different notions of monotonicity. Fuzzy Sets and Systems (submitted for publication)
17. Sklar, A.: Fonctions de répartition à n dimensions et leurs marges (1959)
18. Torra, V.: Hesitant fuzzy sets. Int. J. Intell. Syst. **25**(6), 529–539 (2010). https://doi.org/10.1002/int.20418
19. Wilkin, T., Beliakov, G.: Weakly monotonic averaging functions. Int. J. Intell. Syst. **30**(2), 144–169 (2015). https://doi.org/10.1002/int.21692
20. Yager, R.: On ordered weighted averaging aggregation operators in multicriteria decisionmaking. IEEE Trans. Syst. Man Cybern. **18**(1), 183–190 (1988). https://doi.org/10.1109/21.87068

A Study of Different Families of Fusion Functions for Combining Classifiers in the One-vs-One Strategy

Mikel Uriz, Daniel Paternain[✉], Aranzazu Jurio, Humberto Bustince, and Mikel Galar

Dpto. de Automática y Computación, Universidad Publica de Navarra, Campus Arrosadia s/n, 31006 Pamplona, Spain
{mikelxabier.uriz,daniel.paternain,aranzazu.jurio,bustince, mikel.galar}@unavarra.es

Abstract. In this work we study the usage of different families of fusion functions for combining classifiers in a multiple classifier system of *One-vs-One* (OVO) classifiers. OVO is a decomposition strategy used to deal with multi-class classification problems, where the original multi-class problem is divided into as many problems as pair of classes. In a multiple classifier system, classifiers coming from different paradigms such as support vector machines, rule induction algorithms or decision trees are combined. In the literature, several works have addressed the usage of classifier selection methods for these kinds of systems, where the best classifier for each pair of classes is selected. In this work, we look at the problem from a different perspective aiming at analyzing the behavior of different families of fusion functions to combine the classifiers. In fact, a multiple classifier system of OVO classifiers can be seen as a multi-expert decision making problem. In this context, for the fusion functions depending on weights or fuzzy measures, we propose to obtain these parameters from data. Backed-up by a thorough experimental analysis we show that the fusion function to be considered is a key factor in the system. Moreover, those based on weights or fuzzy measures can allow one to better model the aggregation problem.

Keywords: Aggregations · Fusion functions · Classification
One-vs-One · Multiple classifier system

1 Introduction

In Machine Learning, classification consists in learning a classifier from labeled data capable of assigning the correct label to new patterns. Among classification problems, two different scenarios can be considered depending on the number of classes to be distinguished: binary (two-class) and multi-class problems. Multi-class classification is usually more difficult because the establishment of the decision boundaries become more complex. One possible solution to cope with

© Springer International Publishing AG, part of Springer Nature 2018
J. Medina et al. (Eds.): IPMU 2018, CCIS 854, pp. 427–440, 2018.
https://doi.org/10.1007/978-3-319-91476-3_36

this difficulty is the usage of decomposition strategies [20], which divide the original multi-class problem into easier to solve binary problems. Evidently, this simplification in the learning phase come at a cost in the combination phase, where the outputs of all the classifiers that were learned for each new sub-problem needs to be combined.

One of the most commonly employed decomposition strategy is *One-vs-One* (OVO). In OVO, as many new sub-problems as possible pairs of classes are created and each one is addressed by an independent base classifier. New instances are classified by being submitted to all the base classifiers, whose outputs are combined. One important advantage of this technique is that it usually performs better even when the underlying classifier is able to address the multi-class problem directly [12].

In this work, we focus on the OVO strategy, and more specifically on the combination phase of Multiple Classifier Systems (MCSs) formed of OVO classifiers. A MCSs is a set formed of classifiers coming from different learning paradigms [17]. In the case of OVO, the idea is that different classifiers may suit better the classification of each pair of classes. For this reason, several previous works have considered the selection of the best classifier for each pair of classes in the MCS [19,22]. In this work, our aim is to look at this problem as a multi-expert decision making problem, where we have the different experts (types of classifiers) and their preference matrices for the considered alternatives (classes). In this context, we want to study the influence of the fusion function considered to combine the matrices from the different experts into a single one in the classification performance.

In the last decades, the study of aggregation functions has grown significantly, since the necessity of fusing or aggregating quantitative information arises in almost every application [3,4,6,16]. However, in the last years, new extensions of aggregation functions have been proposed, which are able to model the interaction among data in a better way even though classical properties of aggregation functions, such as monotonicity, are not satisfied [21,23]. From a broad point of view, these extensions are called fusion functions [5].

One of the prominent examples of fusion functions that are able to model the importance of the inputs or the interactions among them is the discrete Choquet integral [8] and its extensions (Choquet-like preaggregation functions) [21], which are based on fuzzy measures. In this work, we propose to construct these measures directly from the knowledge that we can extract from the experts (classifiers) using the training data.

In order to perform this study, we use twenty eight datasets from KEEL [2] and we consider the usage of non-parametric statistical tests to analyze the results obtained [14]. Since we are dealing with multiple classes datasets we will not only consider accuracy measure to evaluate the results, but we will also make use of other measures that give more focus to the correct classification of all classes, such as the average accuracy and the geometric mean. We will develop a hierarchical study, where we consider intra- and inter-family comparisons, to analyze usage of different fusion functions.

The structure of the paper is as follows. In Sect. 2, recall the different fusion functions considered in this work. Section 3 contains an introduction to the decomposition of multi-class problems, the OVO strategy and the MCSs formed of OVO classifiers. In Sect. 4, we describe in detail the experimental framework considered for this study, including how to set up the parameters of the parameterizable fusion functions. Section 5 contains the analysis of the results obtained. Finally, in Sect. 6 we draw the conclusions.

2   Fusion Functions

In recent literature, aggregation of quantitative information has been faced by the use of aggregation functions. An aggregation function is defined as a mapping $f : [0,1]^n \rightarrow [0,1]$ (the interval $[0,1]$ can be extended to any other interval) such that $f(0,\ldots,0) = 0$, $f(1,\ldots,1) = 1$ satisfying the monotonicity property, i.e., if $x_i \leq y_i$ for all $i \in \{1,\ldots,n\}$, then $f(x_1,\ldots,x_n) \leq f(y_1,\ldots,y_n)$ [3,4,6,16]. According to [3,4], the main classes of aggregation functions are the following: averaging, conjunctive, disjunctive and mixed. In this work we mainly (but not only) focus on averaging functions, those which are bounded by the minimum and maximum of inputs.

However, in the last two years the monotonicity property of aggregation functions has been dropped or generalized to new types of monotonicity (see for example [5]). From these studies, new concepts such as preaggregation functions [21] or internal fusion functions [23] have been defined. Since in this paper we model data aggregation from a very broad point of view and we use several non-monotone functions, we have used the more general definition of fusion function (see [5]).

In order to classify the big number of fusion functions considered in this work, we have established a classification based on the necessity of defining weights or measures associated to them. Basically we have considered: unweighted fusion functions, weighted fusion functions and measure-based fusion functions.

Unweighted Fusion Functions. In this subsection we consider classical aggregation functions:

- The arithmetic mean $AM(x_1,\ldots,x_n) = \frac{1}{n}(x_1,\ldots,x_n)$;
- The median $MED(x_1,\ldots,x_n) = \begin{cases} \frac{1}{2}\left(x_{(k)} + x_{(k+1)}\right) & \text{if } n = 2k \text{ is even,} \\ x_{(k)} & \text{if } n = 2k-1 \text{ is odd,} \end{cases}$
  where $x_{(k)}$ stands for the k-th largest (smallest) element of $x_1,\ldots,x_n$;
- The geometric mean $GM(x_1,\ldots,x_n) = \left(\prod_{i=1}^n x_i\right)^{\frac{1}{n}}$;
- The harmonic mean $HM(x_1,\ldots,x_n) = n\left(\sum_{i=1}^n \frac{1}{x_i}\right)^{-1}$.

Weighted Fusion Functions. In this subsection we consider fusion functions whose behaviour is modeled by a weighting vector. This means that not every input is equally important for the calculation of the fused value, a fact that clearly allows the incorporation of certain outside information to the fusion process. We will consider a weighting vectors $w = (w_1, \ldots, w_n)$ satisfying $w_i \in [0, 1]$ and $\sum_{i=1}^{n} w_i = 1$ [3,4].

The weighted fusion functions considered, which in fact are weighted aggregation functions, are:

- The weighted arithmetic mean $WAM(x_1, \ldots, x_n) = \sum_{i=1}^{n} w_i x_i$;
- The ordered weighted averaging $OWA(x_1, \ldots, x_n) = \sum_{i=1}^{n} w_i x_{(i)}$, where $(.)$ is a permutation such that $x_{(1)} \geq \cdots \geq x_{(n)}$.

Measure-Based Fusion Functions. In this subsection we consider a set of fusion functions that are based on fuzzy measures. Unlike the case of weighted fusion functions, which allow one to model the importance of each individual input, the use of fuzzy measures allows one to model more general interactions among inputs. In this sense, the importance is given to collections (groups or coalitions) of inputs. Obviously, the construction of the fuzzy measure is the key point for this family of fusion functions.

Definition 1. *Let $\mathcal{N} = \{1, \ldots, n\}$. A discrete fuzzy measure is a set function $m : 2^{\mathcal{N}} \to [0, 1]$ which is monotonic, i.e., $m(S) \leq m(T)$ whenever $S \subseteq T$ and satisfies $m(\emptyset) = 0$ and $m(\mathcal{N}) = 1$.*

We start mentioning the Choquet integral, which is a prominent example of measure-based averaging operator. We start considering a permutation σ such that $x_{\sigma(1)} \leq \cdots \leq x_{\sigma(n)}$ with the convention $x_{\sigma(0)} = 0$:

- The discrete Choquet integral

$$Ch(x_1, \ldots, x_n) = \sum_{i=1}^{n} (x_{\sigma(i)} - x_{\sigma(i-1)}) * m(\{\sigma(i), \ldots, \sigma(n)\})$$

As we have mentioned before, in [21] a new type of operator, called pre-aggregation function, was given. One of the easiest ways to construct pre-aggregation is by changing certain operations in the Choquet integral. We have considered the following pre-aggregation functions:

- The Choquet-based operator based on minimum t-norm

$$Ch_M(x_1, \ldots, x_n) = \sum_{i=1}^{n} \min\{x_{\sigma(i)} - x_{\sigma(i-1)}, m(\{\sigma(i), \ldots, \sigma(n)\})\};$$

- The Choquet-based operator based on Lukasiewicz t-norm

$$Ch_L(x_1, \ldots, x_n) = \sum_{i=1}^{n} \max\{0, x_{\sigma(i)} - x_{\sigma(i-1)} + m(\{\sigma(i), \ldots, \sigma(n)\} - 1)\}.$$

3   One-vs-One Decomposition of Multi-class Problems and Multiple Classifier Systems

In this section we introduce classification problems, and more specifically, the One-vs-One (OVO) strategy to deal with multi-class classification problems and multiple classifier systems aimed at improving classification performance by the combination of several classifiers.

In Machine Learning a classification problem consists in learning a system (classifier) capable of predicting the desired output (label) for each input pattern. Formally, the objective is to find a mapping function $\mathbb{A}^i \to \mathbb{C}$ where $a_1, \ldots, a_i \in \mathbb{A}$ are the i features that characterize each input example $x_1, \ldots, x_n$ and each input example has associated a desired output $y_j \in \mathbb{C} = \{c_1, \ldots, c_m\}$. The classifier is expected to generalize well to examples from the problem that has not been considered in training, that is, it should have a good generalization ability.

A classification problem is said to be a multi-class problem when the number of classes is greater than two ($|\mathbb{C}| > 2$). These problems are considered to be more difficult than binary classification problems since the classification boundaries are usually more complex and there is a greater overlapping among classes. This is why decomposition strategies [20] came up, to deal with multi-class problems by dividing the original problem into easier to solve binary class classification problems. Therefore, a binary classifier is learned for each new problem, known as base learners, and the outputs of these classifiers are combined when classifying a new unlabeled example. These strategies have proved to be not only useful when working with classifiers that only support binary problems (such as Support Vector Machines, SVMs [25]), but also when considering classifiers with inherent multi-class support. In these cases, the final performance of can also be improved if the problem is decomposed [12].

3.1   The One-vs-One Strategy

The OVO strategy is among the most commonly employed decomposition strategies. In this strategy, an m-class problem is divided into as many problems as possible pair of classes, generating $m(m-1)/2$ sub-problems that are faced by independent base classifiers. In each sub-problem, only the examples belonging to a pair of classes are considered, while discarding the rest of them. Then, to classify a new example, it is submitted to all the classifiers whose outputs needs to be combined to decide the final class label. In order to perform the combination, all the outputs are usually stored in a score-matrix (Eq. 1) where each position $r_{ij}, r_{ji} \in [0, 1]$ corresponds to the confidence degree of the classifier distinguishing classes $\{C_i, C_j\}$. Since most of the classifiers provide confidence estimates based on probabilities, usually r_{ji} is computed as $r_{ji} = 1 - r_{ij}$. However, if this is not the case, as it occurs with fuzzy rule-based classification systems [10], the score-matrix should be normalized so that $r_{ij} + r_{ji} = 1$ [10].

$$R = \begin{pmatrix} - & r_{12} & \cdots & r_{1m} \\ r_{21} & - & \cdots & r_{2m} \\ \vdots & & & \vdots \\ r_{m1} & r_{m2} & \cdots & - \end{pmatrix} \qquad (1)$$

Finally, the outputs of the base classifiers are combined for each row (class) and the predicted class label is assigned to the one achieving the greatest total confidence. In the literature, several combination strategies have been developed for this purpose. A thorough review was performed in [12] and several extended combinations have been developed by considering the usage of classifier selection and weighting mechanism [11,13]. In this work, we consider the Weighted Voting (WV) [18] strategy as it has shown to be a robust yet simple method. In this method, each base classifier votes for both classes based on the confidences provided for the pair of classes. Finally, the class having the largest value is given as output.

$$Class = arg \max_{i=1,\ldots,m} \sum_{1 \le j \ne i \le m} r_{ij}. \qquad (2)$$

3.2 Combining Several OVO in a Multiple Classifier Systems

The OVO strategy can be seen as a ensemble model [12]. Ensembles refer to the combination of classifiers aiming at improving the results of single classifiers. This term is usually considered to describe the combination of minor variants of the same classifiers. Otherwise, multiple classifier systems (MCSs) is a broader category also including those combinations considering the hybridization of different classification models [17].

Recently, several works have considered the hybridization of OVO ensembles (where the same base classifier is used for each sub-problem, e.g., SVMs) with MCSs. That is, to construct several OVO ensembles with different classifiers (for example, one using SVMs, another using a rule induction method and the other using Decision Trees) and to combine the outputs of all the OVO ensembles to make the final decision.

In previous works, the authors have focused on dynamically or statically selecting the best classifier for distinguishing each pair of classes [19,22]. However, in this work we aim to look at the problem from a different perspective so as to test the usage of different fusion functions in the combination of the different classifiers.

Once all the OVO classifiers from the MCS have been trained (assuming that we have three different classifiers and a four class problem we would have $3 \cdot 4 \cdot (4-1)/2$ classifiers), a new instance is classified by submitting it to all the classifiers. As a result, instead of obtaining a single score-matrix, we would obtain as many score-matrices as classifiers considered (three in our example). The problem is how to combine these score-matrices into a single one in which we can apply the WV strategy to classify the example. This is why we can understand the problem as a multi-expert decision making problem. Our proposal in this work is to combine the different score-matrices by the usage of fusion function. Our aim is to study how the usage of different fusion functions affects the performance of the MCS. In order to do so, we will consider the different fusion functions reviewed in the previous section and we will propose different mechanism to assign the weights or create the fuzzy measures in the functions requiring these parameters. More details on how these parameters are obtained are given in Sect. 4.2.

4   Experimental Framework

4.1   Datasets, Performance Measures, Statistical Tests and Algorithms

In order to carry out the experimental study, we use twenty-eight numerical datasets selected from the KEEL dataset repository [2], whose main features are introduced in Table 1.

Table 1. Summary of the features of the datasets used in the experimental study.

Dataset	#Ex.	#Atr.	#Clas.	Dataset	#Ex.	#Atr.	#Clas.
autos	159	25	6	nursery	1296	8	5
balance	625	4	3	pageblocks	548	10	5
car	1728	6	4	penbased	1100	16	10
cleveland	297	13	5	satimage	643	36	7
contraceptive	1473	9	3	segment	2310	19	7
dermatology	358	34	6	shuttle	2175	9	7
ecoli	336	7	8	splice	319	60	3
flare	1066	11	6	tae	151	5	3
glass	214	9	7	thyroid	720	21	3
hayes-roth	132	4	3	vehicle	846	18	4
iris	150	4	3	vowel	990	13	11
led7digit	500	7	10	wine	178	13	3
lymphography	148	18	4	yeast	1484	8	10
newthyroid	215	5	3	zoo	101	16	7

The result for each method and dataset is obtained using a 5 fold cross-validation scheme. Moreover, in order to properly analyze the results obtained, we have applied non-parametric statistical tests [14]. More specifically, we use the Wilcoxon test to compare a pair of methods, whereas the Friedman aligned ranks test is considered to compare a group of methods in order to detect whether statistical differences exist. In such a case, the Holm *post-hoc* test is performed to find the algorithms that reject the null hypothesis of equivalence against the selected control method.

Given that we are dealing with multi-class problems, we have considered three different performance measures to analyze the results obtained: Accuracy rate (Acc), that is, the ratio of correctly classified examples; Average Accuracy Rate (AvgAcc), which refers to the average of the ratio of correctly classified examples per class; Geometric Mean (GM), the geometric mean of the ratio of correctly classified examples per class. Hence, Acc gives us a global measure of quality of the algorithm, whereas AvgAcc and GM are more focused on properly measuring whether all the classes of the problem are being properly classified or not (being the GM much more restrictive than AvgAcc).

Regarding the classification algorithms considered to form our MCS of OVO classifiers, we have considered the following ones (which were also considered in our previous works on the topic [11–13]): *Support Vector Machine* (SVM) [25], *C4.5 decision tree* [24], *k−Nearest Neighbors* (*k*NN) [1], *Repeated Incremental Pruning to Produce Error Reduction* (Ripper) [9], *Positive Definite Fuzzy Classifier* (PDFC) [7].

These classifiers were trained using the parameters shown in Table 2. These values are common for all problems, and they were selected according to the recommendation of the corresponding authors, which is also the default setting of the parameters included in KEEL[1] software [2] used to develop our experiments. We treat nominal attributes in SVM and PDFC as scalars to fit the data into the systems using a polynomial kernel.

Table 2. Parameter specification for the base learners employed in the experimentation.

Algorithm	Parameters
SVM_{Poly}	C = 1.0, Tolerance parameter = 0.001, Epsilon = 1.0E-12, Kernel type = Polynomial, Polynomial degree = 1 Fit logistic models = True
SVM_{Puk}	C − 100.0, Tolerance parameter = 0.001, Epsilon = 1.0E-12, Kernel Type = Puk, PukKernel ω = 1.0, PukKernel σ = 1.0 Fit logistic models = True
C4.5	Prune = True, Confidence level = 0.2, Minimum number of item-sets per leaf = 2
3NN	k = 3, Distance metric = HVDM
Ripper	Size of growing subset = 66%, Repetitions of the optimization stage = 2
PDFC	C = 100.0, Tolerance parameter = 0.001, Epsilon = 1.0E-12, Kernel type = Polynomial, Polynomial degree = 1, PDRF type = Gaussian

We should notice that score-matrices should store the confidences obtained from the classifiers. Since not all the classifiers provide confidences straightforwardly, we detail how they have been obtained hereafter.

---

[1] http://www.keel.es.

- **SVM** – Probability estimates from the SVM.
- **C4.5** – Accuracy of the leaf making the prediction (correctly classified train examples divided by the total number of covered train instances).
- **kNN** – Distance-based confidence estimation. $Confidence = \frac{\sum_{l=1}^{k} \frac{e_l}{d_l}}{\sum_{l=1}^{k} \frac{1}{d_l}}$ where d_l is the distance between the input pattern and the lth neighbor and $e_l = 1$ if the neighbor l is from the class and 0 otherwise.
- **Ripper** – Accuracy of the rule used in the prediction (computed as in C4.5 considering rules instead of leafs).
- **PDFC** – The prediction of the classifier, that is, confidence equal to 1 is given for the predicted class.

4.2   Estimation of the Parameters for the Fusion Functions

Hereafter, we present the way in which the parameters required for some of the fusion functions are estimated.

Weight Calculation. For the weighted arithmetic mean we need to set the weights for each input (classifier, e.g., SVM, 3NN, ...). We set each weight as the normalized accuracy of each method in the training dataset, that is, $w_i = \frac{Acc_i}{\sum_{j=1}^{n} Acc_j}$ for all $i \in \{1, \ldots, n\}$.

Moreover, we have used two different versions for weighted fusion functions: a global and a local approach. In the global approach, we set one weight per classifier. However, in the local approach, each classifier gets a weight for each individual problem (accuracy over the pair of classes).

The calculation of the weights for OWA operators is done by means of increasing fuzzy quantifiers (see [26]), which are given by $w_i = Q\left(\frac{i}{n}\right) - Q\left(\frac{i-1}{n}\right)$ for all $i \in \{1, \ldots, n\}$. In this work we have considered 3 different fuzzy quantifiers yielding three OWA operators : 'at least half' (OWA_alh) with $a = 0, b = 0.5$; 'as many as possible' (OWA_amap) with $a = 0.5, b = 1$; and 'most of them' (OWA_mot) with $a = 0.3, b = 0.8$.

Fuzzy Measure Values. For the measure-based fusion functions, we need to build a fuzzy measure $m : 2^{\mathcal{N}} \rightarrow [0,1]$ with $\mathcal{N} = \{1, \ldots, n\}$, being n the number of classifiers considered. We will start by considering the uniform fuzzy measure m_U which is given by $m_U(A) = \frac{|A|}{n}$ for every $A \subseteq \mathcal{N}$. It is clear that the Choquet integral with respect to a uniform measure is nothing but the arithmetic mean.

However, in order to capture the interactions among classifiers by means of the fuzzy measure, we will take the individual accuracy of each classifier as well as the accuracy of each possible combination of classifiers. We will denote these accuracies as Acc_A, for all $A \subseteq \mathcal{N}$. Now, for each level of the fuzzy measure (all the elements of the fuzzy measure with the same cardinality), we calculate the arithmetic mean of accuracies in the corresponding level, namely $MeanAcc_i$ for every $i \in \{1, \ldots, n\}$. Finally, the value of the fuzzy measure for each $A \subseteq \mathcal{N}$ will be given by

$$m(A) = m_U(A)(1 + Acc_A - MeanAcc_{|A|}). \tag{3}$$

Taking this expression into account, the accuracies of classifiers that are better than the average accuracy in the same level will be increased and those that are worse will be decreased with respect to the uniform measure. In a similar way as in the previous calculation of weights, we will consider a global and a local approach for each measure-based fusion functions.

Notice that we cannot guarantee the monotonicity of m for every possible value of Acc and $MeanAcc$. To correct it, and based on the monotonicity verification given in [15], we use a top-down monotonicity correction: we start from the top level of the measure ($m(\mathcal{N})$ and we evaluate the measure values of the level above ($m(A)$ where $|A| = n - 1$). If we find some A such that $m(A) > m(\mathcal{N})$, then we set $m(A) = m(\mathcal{N})$. Once the $n - 1$-th level is verified (w.r.t. the n-th level), we check the $n - 2$-th level w.r.t. the $n - 1$-th level. We repeat the procedure until the whole measure satisfies the monotonicity criterion.

5   Experimental Study

On the one hand, Table 3 shows the accuracy (Acc), the average accuracy per class (AvgAcc) and the geometric mean of each class accuracy (GM) obtained in testing using the different fusion functions to combine the OVO score-matrices in the MCS. The best result in each performance measure is underlined

Table 3. Average test results over all datasets obtained with the different fusion functions for each performance measure

Family	Fusion	Acc	AvgAcc	GM
Unweighted	AM	0.8544	0.7911	0.6240
	MED	0.8580	0.7951	0.6332
	GM	0.8285	0.7535	0.5588
	HM	0.8252	0.7515	0.5610
Weighted	WAM	0.8544	0.7916	0.6308
	WAM_local	0.8481	0.7893	0.6344
	OWA_alh	0.8573	0.7996	0.6448
	OWA_amap	0.8496	0.7815	0.6073
	OWA_mot	0.8554	0.7921	0.6254
Choquet	Ch	0.8552	0.7940	0.6305
	Ch_local	0.8541	0.7924	0.6334
	Ch_L	0.8487	0.7789	0.6087
	Ch_L_local	0.8502	0.7803	0.6088
	Ch_M	0.8548	0.7939	0.6395
	Ch_M_local	0.8556	0.7964	0.6397

On the other hand, Fig. 1 summarizes the statistical study carried out for each performance measure in order to analyze which is the best performer fusion

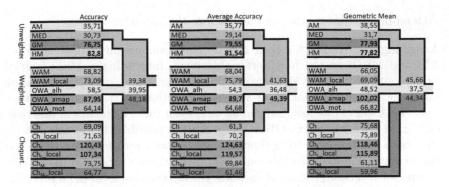

Fig. 1. Hierarchical statistical study comparing the fusion functions in each family and the best performers of each family for each performance measure using Friedman Aligned ranks test.

function in each case. In order to create this figure, for each performance measure, we have confronted the functions in each family following Friedman Aligned ranks test. Then, the best performers of each family are compared in the final stage that gives us the best fusion function. In each comparison, we show the ranks obtained by each method (the lower the better) and we remark in **bold-face** the ranks when the post-hoc test shows that there exist significant differences (with $\alpha = 0.1$) in favor of the winning method.

Finally, we have completed our statistical analysis by comparing the arithmetic mean (AM, which the most commonly considered function) with the winner of each intra-family comparison. These comparisons are presented in Table 4, where the p-values obtained for each comparison between AM and the corresponding fusion function are presented. Statistically significant differences are presented in **bold-face**.

Table 4. Wilcoxon's tests comparing AM vs the best fusion function in each performance measure.

Perf. measure	Unweighted	Weighted	Choquet
Acc	MED	OWA_alh	Ch_M_local
	0.0152	**0.0298**	0.7610
AvgAcc	MED	OWA_alh	Ch
	0.0194	**0.0126**	**0.0994**
GM	MED	OWA_alh	Ch_M_local
	0.0169	**0.0036**	**0.0400**

Attending at these results, we can observe the following facts.

- Analyzing the results for each family, first, among unweighted functions AM and MED are the best performing ones. Interestingly, MED is statistically outperforming AM following the Wilcoxon test in all the three performance measures. Looking at weighted functions it is interesting to note that OWA_alh is the best performing one, even though statistical differences only exist with respect to OWA_amap. This is possibly due to the fact that the corresponding weighting function acts as an average of the three most competitive classifiers. In this case, obtaining the weights from data (WAM and its local version) has result in worse results than establishing a predefined weights. Finally, regarding fuzzy measure-based functions, pre-aggregations considering the minimum are constantly the best in almost all cases, showing its robustness independently of the measure considered (although no statistical differences are found).

  One would expect better performance in the cases where the parameters have been obtained from data, i.e., weighted and measure-based functions. Even though no significant differences are found with respect to WAM and Choquet, in the future our aim is to focus on this functions and try to better model the parameters in order to make the more competitive. In fact, Choquet can recover any OWA operator and hence, intuitively, one should be able to obtain a fuzzy measure leading to at least the same behavior as any OWA (and probably better).

- Finally, looking at Table 4 one can observe that the most commonly considered fusion function in ensembles and MCSs need not be the performing one. AM is statistically outperformed by MED and OWA_alh in all cases and by Choquet in the cases of AvgAcc and GM. Hence, there is margin for improvement by considering different fusion functions.

6   Conclusions

In this work, we have considered an MCSs formed of OVO classifiers and looked at the combination phase as a multi-expert decision making problem. Consequently, we have developed a thorough empirical study in order to analyze the behavior of different families of fusion functions. We have also proposed different ways to obtain the parameters of weighted and fuzzy measure-based fusion functions from data. Even though one could expect better performance from these kind of fusion functions, OWAs with specific weights are the ones with the best results. Since OWAs are a particular case of some fuzzy measure-based functions, this fact encourages us to further study different ways of building the fuzzy measures in order to improve the quality of their results.

Acknowledgments. This work was supported in part by the Spanish Ministry of Science and Technology under Project TIN2016-77356-P (AEI/FEDER, UE).

References

1. Aha, D.W., Kibler, D., Albert, M.K.: Instance-based learning algorithms. Mach. Learn. **6**, 37–66 (1991)
2. Alcalá-Fdez, J., Fernandez, A., Luengo, J., Derrac, J., García, S., Sánchez, L., Herrera, F.: KEEL data-mining software tool: data set repository, integration of algorithms and experimental analysis framework. J. Multiple-Valued Logic Soft Comput. **17**(2–3), 255–287 (2011)
3. Beliakov, G., Bustince, H., Pradera, A.: A Practical Guide to Averaging Functions, 2nd edn. Springer, Cham (2015). https://doi.org/10.1007/978-3-319-24753-3
4. Beliakov, G., Pradera, A., Calvo, T.: Aggregation Functions: A Guide for Practitioners. Springer, Heidelberg (2007). https://doi.org/10.1007/978-3-540-73721-6
5. Bustince, H., Fernandez, J., Kolesárová, A., Mesiar, R.: Directional monotonicity of fusion functions. Eur. J. Oper. Res. **244**, 300–308 (2015)
6. Calvo, T., Mayor, G., Mesiar, R.: Aggregation Operators. New Trends and Applications. Physica-Verlag, Heidelberg (2002). https://doi.org/10.1007/978-3-7908-1787-4
7. Chen, Y., Wang, J.Z.: Support vector learning for fuzzy rule-based classification systems. IEEE Trans. Fuzzy Syst. **11**(6), 716–728 (2003)
8. Choquet, G.: Theory of capacities. Ann. Inst. Fourier **5**, 1953–1954 (1953)
9. Cohen, W.W.: Fast effective rule induction. In: Proceedings of the Twelfth International Conference on Machine Learning, ICML1995, pp. 1–10 (1995)
10. Elkano, M., Galar, M., Sanz, J., Fernandez, A., Barrenechea, E., Herrera, F., Bustince, H.: Enhancing multi-class classification in farc-hd fuzzy classifier: on the synergy between n-dimensional overlap functions and decomposition strategies. IEEE Trans. Fuzzy Syst. **23**(5), 1562–1580 (2015)
11. Galar, M., Fernández, A., Barrenechea, E., Bustince, H., Herrera, F.: Dynamic classifier selection for one-vs-one strategy: avoiding non-competent classifiers. Pattern Recogn. **46**(12), 3412–3424 (2013)
12. Galar, M., Fernández, A., Barrenechea, E., Bustince, H., Herrera, F.: An overview of ensemble methods for binary classifiers in multi-class problems: experimental study on one-vs-one and one-vs-all schemes. Pattern Recogn. **44**(8), 1761–1776 (2011)
13. Galar, M., Fernández, A., Barrenechea, E., Herrera, F.: DRCW-OVO: distance-based relative competence weighting combination for one-vs-one strategy in multiclass problems. Pattern Recogn. **48**(1), 28–42 (2015)
14. García, S., Fernández, A., Luengo, J., Herrera, F.: Advanced nonparametric tests for multiple comparisons in the design of experiments in computational intelligence and data mining: experimental analysis of power. Inf. Sci. **180**, 2044–2064 (2010)
15. Grabisch, M.: A new algorithm for identifying fuzzy measures and its application to pattern recognition. In: International Joint Conference of the 4th IEEE International Conference on Fuzzy Systems and the 2nd International Fuzzy Engineering Symposium, pp. 145–150 (1995)
16. Grabisch, M., Marichal, J.L., Mesiar, R., Pap, E.: Aggregation Functions. Cambridge University Press, Cambridge (2009)
17. Ho, T.K., Hull, J.J., Srihari, S.N.: Decision combination in multiple classifier systems. IEEE Trans. Pattern Anal. Mach. Intell. **16**(1), 66–75 (1994)
18. Hüllermeier, E., Vanderlooy, S.: Combining predictions in pairwise classification: an optimal adaptive voting strategy and its relation to weighted voting. Pattern Recogn. **43**(1), 128–142 (2010)

19. Kang, S., Cho, S., Kang, P.: Multi-class classification via heterogeneous ensemble of one-class classifiers. Eng. Appl. Artif. Intell. **43**, 35–43 (2015)
20. Lorena, A.C., Carvalho, A.C., Gama, J.M.: A review on the combination of binary classifiers in multiclass problems. Artif. Intell. Rev. **30**(1–4), 19–37 (2008)
21. Lucca, G., Sanz, J., Dimuro, G., Bedregal, B., Mesiar, R., Kolesárová, A., Bustince, H.: Preaggregation functions: construction and an application. IEEE Trans. Fuzzy Syst. **24**, 260–272 (2016)
22. Mendialdua, I., Martnez-Otzeta, J.M., Rodriguez-Rodriguez, I., Ruiz-Vazquez, T., Sierra, B.: Dynamic selection of the best base classifier in one versus one. Knowl.-Based Syst. **85**, 298–306 (2015)
23. Paternain, D., Campión, M.J., Bustince, H., Perfilieva, I., Mesiar, R.: Internal fusion functions. IEEE Trans. Fuzzy Syst. **26**, 487–503 (2017)
24. Quinlan, J.R.: C45: Programs for Machine Learning, 1st edn. Morgan Kaufmann Publishers, San Mateo (1993)
25. Vapnik, V.: Statistical Learning Theory. Wiley, New York (1998)
26. Yager, R.: Quantifier guided aggregation using owa operators. Int. J. Intell. Syst. **11**, 49–73 (1998)

Rough and Fuzzy Similarity Modelling Tools

Object [Re]Cognition with Similarity

Łukasz Sosnowski[1(✉)] and Julian Skirzyński[2]

[1] Systems Research Institute, Polish Academy of Sciences,
ul. Newelska 6, 01-447 Warsaw, Poland
sosnowsl@ibspan.waw.pl
[2] School of Computer Science, McGill University,
3480 rue University, Montreal, QC H3A 0E9, Canada
julian.skirzynski@mail.mcgill.ca

Abstract. We discuss the origin of the notion of similarity, basic concepts connected with it and some methods of representing this conception in mathematical setting. We present a framework of recognition that is based on multi-aspects similarity. The framework is implemented in form of a network of comparators, that processes similarity expressed in terms of fuzzy sets. Our approach introduces a new standard to the field of similarity computing and processing.

Keywords: Similarity · Compound objects · Network of comparators
Pattern recognition · Object recognition based on similarity

1 Introduction

The notion of similarity was present in the scientific discourse at least as long as there were the ideas of ancient philosophers. In Plato's *The Republic*, similarity was invoked to advocate arguments on how the State functions and what is its nature. Aristotle put similarity as one of the pillars of his theory of how human behavior is learned, and one of his laws stated that the experience or recall of an object (a situation) will evoke a recall of something similar to that object (situation) [2]. These views gave rise to a theory called associationism which states that people perform complex psychological actions through the act of association between similar mental states they experienced in the past. The main proponents of this stance were members of the school of British Empiricism, so philosophers like David Hume, John Locke or John Stuart Mill. According to Hume, for one example, similarity, besides contiguity of time and space as well as cause and effect, was one of the principles by which ideas are connected. Associationism also affected the first psychologists, like the pioneer of this field of study, William James, who saw similarity at the root of mental associations. On the other hand, there were people who regarded similarity as much of a troublesome idea. Bertrand Russell held that if we accept it, we must also accept the existence of at least one universal – a mind-independent characteristic with which we may describe multiple things and, which he believed, does not exist. Quine went even

© Springer International Publishing AG, part of Springer Nature 2018
J. Medina et al. (Eds.): IPMU 2018, CCIS 854, pp. 443–454, 2018.
https://doi.org/10.1007/978-3-319-91476-3_37

further and called similarity *logically repugnant* as it cannot be explicated in terms of more basic notions. Although the concept served to establish important philosophical dependencies and inflamed disputes across the years, the formal definition was given to it only at the beginning of the 20th century, thanks to thinkers like Rudolf Carnap, Hans Wallach or Roger Shepard. We may divide these definitions into two groups: mathematical and non-mathematical.

Non-mathematical definitions stemmed mainly from psychology. Wertheimer in his classical article from 1923 formed the Factor of Similarity which gave the notion a descriptive specification. His law assumed that objects which are grouped together in the process of cognition are in fact similar. This has been further enhanced by behaviorists like Pavlov who viewed similarity between two stimuli as their relative distance on sensory dimensions. It was until the beginning of the second half of the 20th century, however, that associationism began to be slowly discarded and new ideas came into the scene. Wallach's *On Psychological Similarity* marked a new era in thinking about the titular concept. In his work, the author juxtaposed older views on the topic with his *perceived similarity* conception in which people decide which features to select and which to ignore when judging resemblance of two stimuli. He also showed experiments when such decisions were based on the context in which stimuli were presented, and included external features, independent of the structure of a stimulus (like a potential use) into similarity judgment.

Mathematical definitions for most of the time were based on geometrical understanding of similarity. Carnap set all binary, reflexive and symmetric relations to be equivalent to the notion. This understanding was later adopted by psychologists (see [3] for more references) and similarity was treated as a metric defined in the set of objects being compared. The distance from one point to another defined the level of their difference. Thus, it was possible to quantitatively state that objects a and b resemble each other more than objects c and d or that objects e and f are approximately identical since their dissimilarity does not exceed some threshold t.

Both of the aforementioned accounts, whether strictly mathematical or not, if they drew on geometrical understanding, were later deemed as inappropriate. It was mainly due to Nelson Goodman's criticism who, as Quine, had very little opinion on the concept of similarity and treated it as devoid of any explanatory power. His main argument was that for any three objects it is always possible to state that any two of them are more similar to each other than to the third one. After Wallach, he used this observation to argue that there can be no similarity metric that is context-independent, thus, voicing against models of similarity of his time. This critique was later partially backed by the works of Amos Tversky and most notably, his famous paper *Features of Similarity* which introduced new formal view on similarity and provided psychological data against geometrical stance [16]. Tversky showed how people's judgments often violate each of the metric's axioms with symmetry being almost impossible to keep as corresponding to our behavior. His model, which we shall discuss in detail farther in the text, did not address all the philosophical remarks made by

Goodman or Quine though – the context was still overlooked. Gentner tried to account for the lack of that information and created a conception of relational similarity which he expressed in terms of unary predicates. Nevertheless, in this work we will be considering Tversky's breakthrough formulation as not only is it consistent with psychological evidence but is also very robust in terms of its use in computer science.

The work is constructed as follows. In the second section we lay out basic definitions of concepts that underlie the paper as a whole. Next section describes the similarity concept as well as selected methods of expressing it. Section four contains a detailed description of the recognition framework based on a similarity fuzzy relation and presents its implementation in the form of a network of comparators. The last section provides a summary and some comments about the methodology of the framework.

2   Preliminaries

The basic element that was under the scope of interest of ancient philosophers, just like it is now of modern researchers of artificial intelligence, is a compound object. The structure of a compound object is formulated by utilizing the notion of ontology which comes from philosophy, but now it is also frequently found in the field of artificial intelligence (AI). The formal definition that we will use in here was introduced in 2001 in [15]. It states that ontology is a system marked as $O = \{C, R, H_c, rel, A, L\}$, which specifies the structure of concepts, relationships between them as well as theory defined on a model. C is understood as the set of all concepts of the model and a singular concept is equated with a group of objects with common characteristics. Then, R is a set of named connections between concepts [1], H_c – a collection of taxonomic relationships between concepts, rel – defined, non-taxonomic relationships between concepts, A – a set of axioms, and L – a lexicon defining the meaning of concepts (including relations). L is a set of the form $\{L_c, L_r, F, G\}$, where L_c stands for the lexicon definitions for concepts, L_r – the lexicon of definitions for the set of relationships, F – references to concepts, G – references to relationships.

In the simplest sense, ontology is as a set of concepts connected with one another through named relationships. If we group specific concepts into more general entities, then we can make use of the resulting hierarchies in defining mereologic relations – that is descriptions of dependencies between parts of objects. The literature described many other interesting applications of ontology in computer science, most notably in pattern recognition, image analysis or modeling situational awareness by AI systems. The main problem there, is to understand the structure of an object and, on the basis of the results of perception, discover the similarities. In the literature there are some convergent approaches which treat about interactive granular computing [10]. In the context of this work, ontology is used as a set of concepts describing objects, the structure of this set, and its relations. It is used for designating reducts of features as well as describing features to which they are compared, and hence becomes a necessary tool for recognition and identification processes.

In general, objects can be divided into two groups: compound objects (X_c) and simple objects (X_s). A simple object is any element of the real world that has its representation capable of being expressed by the adopted ontology (O). In addition, the following properties arise from their ontological representation:

1. Objects always belong to a certain class or a fixed number of classes in ontology. A single object may belong to several classes.
2. An object has a property within a class. Features may vary by class.
3. An object may be in relation to other objects in the same ontology.

A compound object is composed of other objects defined by means of ontology (connects them) and creates a new entity. A compound object has its specification, which describes the structure, relations and connections between subobjects. Compound objects satisfy the following additional properties:

1. We can extract from them a minimum of two objects that can be independent entities.
2. Component objects are interrelated with ontology through the formal definition of relationship.

3   Similarity Concept

In some sense, similarity can be seen as a relationship that comes from identity. Identity is an intuitive equality of objects, with the intuition formalized as the equality of attributes of entities that are compared. It is thus the supreme form of similarity. Rules for determining the identity of objects have been already proposed in the 17th century by Gottfried Wilhelm Leibniz who called them 'identity of indiscernibles'. They are as follows:

$$\forall x \forall y [\forall P(Px \leftrightarrow Py) \rightarrow x = y] : x, y \in U \tag{1}$$

and

$$\forall x \forall y [x \neq y \rightarrow \neg \forall P(Px \leftrightarrow Py)] : x, y \in U, \tag{2}$$

where x and y are objects and P is a property. Formula (1) means that for any objects x and y from the universe U, if they have exactly the same values of all properties, these objects are identical in the space in question. Similarly, formula (2) means that for any object x and y, if x is not identical to y, then in the space U there must exist at least one discriminatory characteristic for the two.

Intuitively, similarity is a certain kind of incomplete identity. Two similar objects are those that are primarily comparable and for which a degree of similarity can be obtained. The latter is feasible only if these objects have common or distinguishing features that we understand as descriptive attributes attaining different values. Thus, comparing similar objects' attributes gives the possibility of determining the degree of their similarity. It is commonly understood that the statement *a is similar to b* means that one object resembles the other or is *almost* the same. These statements are, of course, very imprecise, but it is certainly possible to map them using appropriate modelling techniques (e.g. fuzzy sets [4]). By following this intuition, one can determine when two objects fail to fulfill the definition of identity, but if that happens, there is very little left to be fulfilled. The first option then is to use so-called quantitative approach. We are dealing with a set of attributes describing both objects, where most of the attributes of these objects are equal, although there is at least one attribute for which equality does not hold. These objects are almost certainly identical in colloquial speech, but from the strict point of view they are only similar to a certain degree. The second approach is not limited to examining attributes that characterize identities and it focuses on the remaining attributes. These attributes do not meet the condition of identity, but one can try to determine the degree of similarity for them. This is called a qualitative approach. It may involve a situation in which no identities are found on any attribute, and yet these objects are judged similar to a certain degree.

The scale of similarity is most often the interval $[0, 1]$, where 0 means a total lack of similarity and 1 is interpreted as indiscernibility between given attributes, and thus, according to the principle of Leibniz, as an *absolute identity*. Similarity and the very comparison operation are indispensable elements of the world around us, and in many cases, they are necessary to determine the state of an object. In practice, it is weight, size, capacity, duration or other characteristic of objects that is determined. Each of these elements requires knowledge of a certain reference concept, by means of which one can specify a given object's parameter, e.g. a kilogram, a liter, a second, etc. In spite of the introduction of reference values, the feature of the object can be expressed in a countable way. At the same time, objects have common reference points for all.

One can distinguish several types of approaches to defining similarities, and we shall discuss a selected few shortly.

3.1   Selected Methods of Expressing Similarities

In the literature, the problem of similarity is quite widespread, but it is usually not the main research point, but merely a means to achieve other goals. In most cases similarity is equated with the distance in a certain space of features. In this case, the metric is considered in the form:

$$d : X \times X \to [0, +\infty), \tag{3}$$

which satisfies the following properties $\forall x, y, z \in X$:

1. $d(x, y) = 0 \Leftrightarrow x = y$
2. $d(x, y) = d(y, x)$
3. $d(x, y) \leq d(x, z) + d(z, y)$

There are various metrics that suit the type of space and the problem that is to be solved. This solution allows one to convert the problem of determining similarity between objects to the problem of distance measurement in a coordinate system determined by features. This is a relatively common approach, but not always sufficient to solve complex problems. It should be noted that there are very strong constraints associated with the metric. In the case of a generally understood similarity, the condition of symmetry is often not possible to be met, not to mention the condition of transitivity. Therefore, there is a need for other approaches as well. The common element of many solutions is the use of feature vectors. We will try to stress out throughout this paper that the essence of the problem lies in how these vectors are constructed and how they can adapt to new situations.

The next step in evolution related to methods of implementing similarity involves approaches based on ontological relationships between objects and concepts [17]. In this context, individual ontological concepts are treated as features that contribute to comparing objects. The set of these features constitutes an input into the process of determining the minimum set of essential features. This process comes down to the designation of a kind of reduction of features similar to information reducts encountered in data mining [11], i.e. a minimum set of attributes that uniquely identify or classify a given object. There are many reducts that consist of different features and selecting the best reduct is based on domain knowledge about the problem, information about the implementation and many other factors. Ontology and reduct ensure the proper design of a feature vector, however, they do not directly support the method of calculating similarity. Therefore, after the selection of features, we use other methods described earlier, or come up with dedicated methods based on the comparison of ontology. These methods are very complicated and depend on the construction of a particular ontology.

Another approach that replaced distance thinking was the *contrast model* created by Amos Tversky on the basis of study on how people perceive similarity [16]. In this model, not only the common features, but also distinguishing features of objects play an important role. Consequently, the model also examines aspects of reducing similarity between objects and determines their impact on the value of its degree. The common formula of the similarity function in the proposed contrast model is:

$$sim(x,y) = \theta f(X \cap Y) - \alpha f(X - Y) - \beta f(Y - X) : \ \theta, \alpha, \beta \geq 0, \qquad (4)$$

where X and Y are sets of features describing object x and y respectively, $X \cap Y$ determines common features for x and y, $X - Y$ determines feature existing in x and not existing in y, $Y - X$ determines features not existing in x, and existing in y. Function f is a scale factor, while θ, α and β are parameters of the model. It is easy to see that for $\alpha = 0$ and $\beta = 0$ the model is limited to common features of objects. On the other hand for parameters $\theta = 0$ and $\alpha = 1, \beta = 1$ we get:

$$- sim(x,y) = f(X - Y) + f(Y - X), \qquad (5)$$

which is a dissimilarity [7].

From the point of view of modeling similarity, it is important to be able to deal with imprecision of the description and its effect on the result. Another method of representing object similarities involves fuzzy sets [6], as the fuzzy relation is an ideal tool for such purposes. It is defined on the Cartesian product of two crisp sets [4] which in this case include elements for which similarity is determined. There are many similarity measures based on fuzzy sets in the literature. The usual approach is based on the analysis of common features of objects, i.e. those at the intersection of sets $A \cap B$ or complement, in the form:

$$sim(x, y) = 1 - \mu(x, y), \tag{6}$$

where $\mu(x, y)$ is the membership function of a relation designating the degree of difference between two objects. The same approach can be used in building similarity functions, which will be used for the purposes of calculation degrees for individual features or distilling full feature vector. An important aspect of this method is its ability to obtain the results in terms of fuzzy sets.

Slightly different methods can be used when comparing object's structures or their topological relationships. In cases like these, apart from attributes and their values, constraints related to the location of the object in space or the internal structure of the object are imposed. This kind of similarity can also be expressed by means of methods described above, but only on a case-by-case basis. This is why certain standardized methods that deal with such problems have been sought, c.f. rough mereology or near sets [8,9].

The main idea behind rough mereology is to examine an extent to which an object is a part of another object using a properly selected function of rough inclusion. A typical example of the inclusion function, and at the same time an instance of asymmetrical measure of similarity that is based on multiplicity of common components, is the following formula [9]:

$$sim(X, Y) = \mu(X, Y) = \frac{card(X \cap Y)}{card(X)}, \ card(X) \neq 0, \tag{7}$$

where X is a set of sub-objects included in the object x, and Y is a set of ingredients of object y. The rough inclusion function provides a method for comparing parts of objects, their quantities, types or other relationships in the ontological hierarchy. Therefore, it can be interpreted as a measure of similarity that takes into account structural dependencies of objects.

In this paper, structural similarity is calculated on the basis of sum of similarities between sub-components of a fixed structure object. The sub-components are extracted by means of decomposition. We treat their similarity values as additive, and multiply by respective weighting factors. Consequently, arising similarity function is based on the knowledge of composition of a given object and the significance of each component. To define the relationship between an object and its parts, we use functions which state how to construct it from its underlying constituents. Then, these functions and the modeled dependencies are applied to similarities which in consequence allows to interpret the outcome as a similarity value referred to the main object. An example of similarity function of this kind can be as follows:

$$sim(x, y) = \frac{w_1 sim(x_1, y_1) + w_2 sim(x_2, y_2) + ... + w_n sim(x_n, y_n)}{(w_1 + w_2 + ... + w_n)} \tag{8}$$

where x_i are sub-objects of x, and y_i are sub-objects of y for $i = 1, ..., n$.

To summarize, there is a handful of methods of processing and defining similarities. Many of them are related to specific cases of use, where use is subject to special considerations. It is worth pointing out that the methods listed here were chosen from among many other equally useful techniques (e.g. similarity and processing graphs [12]). At the same time, an universal approach that is proposed in this paper, combines the majority of methods described in this section and makes the comparison of similarity results easier. In addition, it considers different possible cases and establishes proper methodologies and facilitations for them.

4   Recognition Framework

There are many ways to implement object recognition solutions. The method considered in this paper is based on multi-similarity calculations, gathering many aspects of similarity between pairs of objects and synthesizing them to get global similarity snapshot. Objects belonging to multi-dimensional space are described by similarity values between input and reference objects measured on a given set of features. The result of a recognition is thus a similarity vector which represents the closeness between input object and reference points in the domain space. Further in the text, units responsible for single-feature calculations will be called *comparators*, and networks allowing processing input objects through the layers of multiple comparators will be called *comparator networks*.

The compound objects comparator (COC) is a construct denoted as com^{ref} and can be expressed in the following form:

$$\mu_{com}^{ref} : X \times 2^{ref} \rightarrow [0,1]^{ref}, \tag{9}$$

where $X \subseteq U$ is the set of input objects to be compared and ref is the set of reference objects that we infer the similarity from. $[0,1]^{ref}$ denotes the space of vectors $\boldsymbol{v}$ of dimension $|ref|$, where each i-th coordinate $v[i] \in [0,1]$ corresponds to an element $y_i \in ref$, $ref = \{y_1, ..., y_{|ref|}\}$. We will further call ref a *reference set*, while each $Y \subseteq ref$ will be referred to as a *reference subset*. Additionally, $a(x)$ will be the function that provides a representation of object $x \in X$ with respect to an attribute a corresponding to some feature. This representation is then used by the comparator while processing x. Similarly, each reference object $y \in Y$ is processed using its representation $a(y)$ for a given attribute a. If we are given an ordering on elements of the reference set ref, i.e. $ref = \{y_1, ..., y_{|ref|}\}$ we can represent the function corresponding to the COC as:

$$\mu_{com}^{ref}(x, Y) = Sh(F(\boldsymbol{v})). \tag{10}$$

We shall now elaborate on subsequent components of this expression. Let's start with $\boldsymbol{v}$ which is the *proximity vector* defined as:

$$\boldsymbol{v}[i] = \begin{cases} 0 & y_i \notin Y \\ sim(x, y_i) & y_i \in Y \end{cases} \tag{11}$$

Note that when Y is a proper subset of ref the positions in v corresponding to $y_i \notin Y$ are filled with zeros. Non-zero elements of v determine the degree of similarity between the object x in question and each element of reference subset Y. In general, the value of similarity $sim(x, y)$ is calculated by the means of fuzzy relation [4] but in reality it is a combination of three mechanisms expressed in the following formula of similarity:

$$sim(x, y_i) = \begin{cases} 0: & Exc^{ref}_{Rules_i}(x) = 1 \vee y_i \notin Y \\ t_h(\mu(x, y_i)): & otherwise \end{cases} \tag{12}$$

This formula also needs explication which is the following: Y is the reference subset; t_h is a threshold function given as

$$t_h(z) = \begin{cases} 0: & z<p \\ z: & z\geq p \end{cases}, p \in [0, 1]. \tag{13}$$

with p corresponding to the lowest similarity acceptable by a single comparator and set independently for each one of them; μ is the basic similarity function defined by the means of traditional fuzzy relation between two objects x and y_i; i is the index of the coordinate of proximity vector for which the similarity is derived; $Exc^{ref}_{Rules_i}$, i.e

$$Exc^{ref}_{Rules_i}(x) = max^{|Rules_i|}_{j=1}\{r_j(x)\}, x \in X \tag{14}$$

is a function associated with exception rules in the form of:

$$r_j : X \to \{0, 1\}, \tag{15}$$

where j is an index of a rule (its id number) in the set $Rules_i$.

The second element of COC, F, is a function responsible for filtering the result before applying the Sh function. Typically, F is based on combination of some standard, idempotent functions such as min, max, top, or simply identity. It introduces competitiveness between reference objects which distinguishes this mechanism from threshold function defined inside $sim(x, y)$.

Finally, Sh, called a *sharpening function* is a mapping that satisfies three basic conditions:

$$\forall i \in \{1, ..., |ref|\} : (v[i] = 0) \Rightarrow (Sh(v)[i] = 0), \tag{16}$$

which ensures keeping the zero values to prevent getting artificially high results;

$$\forall i \in \{1, ..., |ref|\} : (v[i] = max^{|ref|}_{j=1}(v[j])) \Rightarrow (Sh(v)[i] = v[i]), \tag{17}$$

which ensures keeping the maximum value so that the best result retained its original properties;

$$\forall i, j \in \{1, ..., |ref|\} : (v[i] < v[j]) \Rightarrow (Sh(v)[i] < Sh(v)[j]), \tag{18}$$

which ensures strong monotonicity with purpose to increase the difference between the average and the best results.

If we take a wider look at the COC we may notice that if all the notions introduced above are composed, it can be expressed as:

$$\mu_{com}^{ref}(x, Y) = Sh(F(\langle sim(x, y_1), \ldots, sim(x, y_{|ref|})\rangle)) \tag{19}$$

Network of Comparators (NoC) can play different roles depending on their settings. They can serve as multi-stage classifiers whose purpose is to limit the reference set of objects and identify the most probable candidate to be the final result. The scenario of processing in such networks is to compute relatively simple features at the first layers and to filter out the reference objects to only those that are the most promising in the final perspective. Particular comparators can be also specialized in recognition of different features based on the nature of sub-objects. The idea is that the similarity of parts of objects can help in resolving the similarity of the whole objects. From the mathematical perspective a comparator network can be interpreted as a calculation of a function:

$$\mu_{net}^{ref_{out}} : X \rightarrow [0, 1]^{|ref_{out}|}, \tag{20}$$

which takes the input object $x \in X$ as an argument and ref_{out} is a reference set for the network's output layer. The target set of $\mu_{net}^{ref_{out}}$ is the space of proximity vectors. In this way we get the value of the network's function:

$$\mu_{net}^{ref}(x) = \langle SIM(x, y_1), \ldots, SIM(x, y_{|ref|})\rangle, \tag{21}$$

where $SIM(x, y_i)$ is the value of *global similarity* established by the network for input object x and reference object y_i. Global similarity depends on partial (local) similarities calculated by the elements of the network (unit comparators). Through the application of aggregation (in a sense of consensus reaching [5]) and translation procedures at subsequent layers of the network these local similarities ultimately lead to the global one. Particular constituents of the network have been described in detail in previous publications [14]. Figure 1 shows an example of the NoC with all possible elements, interactions between them, and signal granule arising around the input object x.

The models for COC and NoC are functions, and both of them return results which are vectors. Fortunately, there is a simple method for converting these proximity vectors into type I fuzzy sets [4] which allows using fuzzy sets machinery in their further processing and interpretation. Note that individual vector coordinates define similarity of a particular pair of objects (x, y), where $x \in X$ and $y \in Y \subseteq ref$. Since these values reflect the degree of memberships to the fuzzy set, we actually deal with a fuzzy relation, which is also a fuzzy set. Consequently, the result described in functional terms can be converted to a fuzzy set notation in the following way:

$$R(x, y) = \{((x, y_i), v[i]) : i = 1, \ldots, |ref|\}, \tag{22}$$

where $v[i]$ is the i'th coordinate of the proximity vector, which simultaneously fulfills the condition of the fuzzy relation in the form:

$$\mu : X \times ref \rightarrow [0, 1] \tag{23}$$

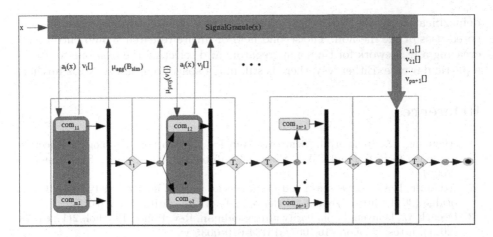

Fig. 1. General scheme of a comparator network in UML-like representation. Notation: com_{ji} – comparators, T_j – translators. Symbols: oval – comparator, thick vertical line – aggregator, rhombus – translator, encircled cross – projection module.

This method is also consistent with the definition of similarity function of the COC and global similarity in the NoC. The form of formula (23) is equivalent to Zadeh's notation:

$$R = \frac{v[1]}{(x, y_1)} + \frac{v[2]}{(x, y_2)} + ... + \frac{v[\|ref\|]}{(x, y_{|ref|})} \tag{24}$$

5   Summary

We analyzed how similarity was perceived and understood over the centuries, and made a brief review of the philosophical currents in search of this notion and its use in formulating concepts. We gathered several approaches to represent similarity and showed methods of processing it. Finally, we described the NoC approach as one which introduces new standards into the field of computing similarity and represents one of the main frameworks for building similarity based recognition systems. This framework provides the ability to build large and complex logical structures that use fuzzy sets as a communication language and express similarity between objects. It is worth noting that even though the method bases on established patterns, it is possible to perform dynamic search of the object space and approximate the optimal solution adequately. By selecting an appropriate defuzzification method, it is also possible to obtain results from the outside of the reference set. Few practical applications have been described in previous publications [13].

Further research should be focused on development of the NoC framework in a defuzzification aspect. Particularly, it is valuable to consider an extension of the catalog of network components with a new element responsible for the defuzzification of the NoC. This could significantly broaden the circle of targeted uses of the method. The second field of future research should encompass creating a framework for tuning aggregators and selecting the best one to use in a particular case. Either way, there is still much space to optimize NoCs further.

References

1. Allemang, D., Hendler, J.: Semantic Web for the Working Ontologist: Effective Modeling in RDFS and OWL. Morgan Kaufmann Publishers Inc., San Francisco (2008)
2. Aristotle, Ross, G.: De sensu and De memoria. Cambridge University Press, Cambridge (2014). https://books.google.co.uk/books?id=7ojRAwAAQBAJ
3. Decock, L., Douven, I.: Similarity after goodman. Rev. Philos. Psychol. **2**(1), 61–75 (2011). https://doi.org/10.1007/s13164-010-0035-y
4. Kacprzyk, J.: Multistage Fuzzy Control: A Model-based Approach to Fuzzy Control and Decision Making. Wiley, Hoboken (2012)
5. Kacprzyk, J., Zadrozny, S.: Soft computing and web intelligence for supporting consensus reaching. Soft. Comput. **14**(8), 833–846 (2010). https://doi.org/10.1007/s00500-009-0475-4
6. Marin, N., Medina, J.M., Pons, O., Sanchez, D., Vila, M.A.: Complex object comparison in a fuzzy context. Inf. Softw. Technol. **45**, 431–444 (2003)
7. Pekalska, E., Duin, R.P.W.: The Dissimilarity Representation for Pattern Recognition: Foundations and Applications (Machine Perception and Artificial Intelligence). World Scientific Publishing Co. Inc., River Edge (2005)
8. Peters, J.F.: Near sets: an introduction. Math. Comput. Sci. **7**(1), 3–9 (2013)
9. Polkowski, L.: Approximate Reasoning by Parts: An Introduction to Rough Mereology. Intelligent Systems Reference Library. Springer, Heidelberg (2011). https://doi.org/10.1007/978-3-642-22279-5
10. Skowron, A., Jankowski, A.: Towards W2T foundations: interactive granular computing and adaptive judgement. In: Zhong, N., Ma, J., Liu, J., Huang, R., Tao, X. (eds.) Wisdom Web of Things. WISEITBS, pp. 47–71. Springer, Cham (2016). https://doi.org/10.1007/978-3-319-44198-6_3
11. Ślęzak, D.: Approximate reducts in decision tables. In: 6th International Conference on Information Processing and Management of Uncertainty in Knowledge-Based Systems, pp. 1159–1164. Universidad de Granada (1996)
12. Sosnowska, J., Skibski, O.: Attachment centrality for weighted graphs. In: Proceedings of the Twenty-Sixth International Joint Conference on Artificial Intelligence, IJCAI 2017, Melbourne, Australia, 19–25 August 2017, pp. 416–422 (2017)
13. Sosnowski, Ł.: Applications of comparators in data processing systems. Technical Transactions, Automatic Control, pp. 81–98 (2013)
14. Sosnowski, Ł.: Framework of compound object comparators. Intell. Decis. Technol. **9**(4), 343–363 (2015)
15. Staab, S., Maedche, A.: Knowledge portals: ontologies at work. AI Mag. **22**(2), 63–75 (2001)
16. Tversky, A., Shafir, E.: Preference, Belief, and Similarity: Selected Writings. MIT Press, Cambridge (2004)
17. Zadeh, P.D.H., Reformat, M.: Feature-based similarity assessment in ontology using fuzzy set theory. In: FUZZ-IEEE, pp. 1–7 (2012)

Attribute Reduction of Set-Valued Decision Information System

Jun Hu$^{(\boxtimes)}$, Siyu Huang, and Rui Shao

Chongqing Key Laboratory of Computational Intelligence,
Chongqing University of Posts and Telecommunications,
Chongqing 400065, China
hujun@cqupt.edu.cn

Abstract. In practice, we may obtain data which is set-valued due to the limitation of acquisition means or the requirement of practical problems. In this paper, we focus on how to reduce set-valued decision information systems under the disjunctive semantics. First, a new relation to measure the degree of similarity between two set-valued objects is defined, which overcomes the limitations of the existing measure methods. Second, an attribute reduction algorithm for set-valued decision information systems is proposed. At last, the experimental results demonstrate that the proposed method can simplify set-valued decision information systems and achieve higher classification accuracy than existing methods.

Keywords: Set-valued decision information system · Rough set
Attribute reduction · Uncertainty

1 Introduction

With the development of information technology, the means of data acquisition becomes more and more diverse. Meanwhile, the cost of data storage is getting lower and lower. These make it possible to acquire and store large amount of data, which stimulate the urgent need for automatic data processing. In many real world problems, data uncertainty is pervasive. In the past few years, it can be found that more attention has been paid to uncertain data. Rough set theory [1] is a powerful tool for dealing with uncertainty. Generally, the data processed by rough set theory is complete, accurate and atomic. However, the data in many real problems may be incomplete, inaccurate or non-atomic due to the limitation of acquisition means or the requirement of practical problems. It has become an important issue how to process incomplete data, interval-valued data and set-valued data.

A set-valued information system is that the value for an object on an attribute is not an exact value, but a set containing all possible values. For the processing of set-valued information system, a number of related researches have been studied. Orlowska and Pawlak [3, 4] investigated set-valued information system considering non-deterministic information and introduced the concept of a non-deterministic information system. Sakai et al. [5, 6] set up the theoretical foundations and algorithmic background for adapting rough set methods for the purposes of the analysis of non-deterministic information systems. Yao [7, 8] proposed a number of set-based computation methods

© Springer International Publishing AG, part of Springer Nature 2018
J. Medina et al. (Eds.): IPMU 2018, CCIS 854, pp. 455–466, 2018.
https://doi.org/10.1007/978-3-319-91476-3_38

based on set-based information system. In addition, some new relations for set-valued information system and their corresponding attribute reduction methods were proposed. And then, Zhang et al. proposed the concept of set-valued information system and a similarity relation [11, 12]. A tolerance relation was defined and the largest tolerance class was used to divide the universe in that paper. Qian et al. [13] proposed a dominant relation for set-valued information system and a corresponding attribute reduction method. Dai and Tian [14] gave a fuzzy relation, that was used to measure the degree of similarity between two set-valued objects. Wang [15] pointed out that the family of reducts defined by Dai need not be a subset of the family of reducts defined within the standard rough set model for set-valued information system. Bao and Yang [16] proposed a δ-dominance relation and a corresponding attribute reduction approach. Two types of fuzzy rough approximations, and two corresponding relative positive domain reductions were proposed by Wei et al. [17]. Moreover, some researchers transformed incomplete information systems into set-valued information systems to achieve reduction. Lipski [9, 10] discussed set-valued information systems from the view of incomplete information systems under the case of missing values. And as a special case of set values, incomplete information is probabilistically dealt with by some authors [18, 19] in the case of missing values.

It is an important issue how to define a binary relation dealing with set-valued information systems by rough set theory. Generally, there are two different semantic interpretations of set-valued data, namely, the conjunctive semantics and the disjunctive semantics [2]. Many different definitions have been proposed for these two semantic interpretations. However, we find that they are not appropriate when dealing with the set-valued data under the disjunctive semantics. For this reason, we developed a new approach based on probability, which can characterize the relation between two set-valued objects more reasonable under the disjunctive semantics. Then, an attribute reduction algorithm based on keeping positive domain for set-valued decision information systems was proposed and some experiments were conducted to prove the effect of the algorithm.

The study is structured as follows: In Sect. 2, some basic concepts of set-valued information system are reviewed. A new approach based on probability is proposed to measure the degree of the similarity between two set-valued objects under the disjunctive semantics in Sect. 3. In Sect. 4, we put forward an attribute reduction algorithm for set-valued decision information system. In Sect. 5, the experimental results and relative analysis are presented. Section 6 concludes the paper.

2 Preliminaries

In this section, some basic concepts about set-valued information system will be reviewed.

Definition 1 [12]. An information system is defined as (U, A, V, F), where U is a non-empty finite set of objects called the universe. A is a non-empty finite set of attributes. V is a union of attribute domains ($V = \cup_{a \in A} V_a$), V_a is a set including all possible values for $a \in A$. $F : U \times A \rightarrow V$ is a function that assigns particular values from attribute domain to objects. For $\forall a \in A$, $x \in U$, $F(a, x) \in V_a$, $F(a, x)$ is the value of a for x. If for any a and x, $F(a, x)$ is a single value, then the information system is called a single-valued information system. Otherwise, it is called a set-valued information system.

There are two different semantic interpretations for set-valued information system [2]. The one is conjunctive semantics, the other is disjunctive semantics. Under the conjunctive semantics, a set value represents all the values of an object to an attribute, and all the values in the set are true. Under the disjunctive semantics, a set value represents all possible values of an object to an attribute, but there is only one true value in the set. For example, for $x \in U$ and $a \in A$, where a means the languages that x can speak. Let $a(x) = \{English, Chinese, French\}$. If $a(x)$ is interpreted conjunctively, it means x can speak English, Chinese, and French. If $a(x)$ is interpreted disjunctively, it means x can speak only one of English, Chinese, and French. We mainly study set-valued information system under the disjunctive semantics in this paper. In the following, a set-valued information system is under the disjunctive semantics if not otherwise specified.

To characterize the relation between two objects in a set-valued information system, many relations have been developed. Here we cite two important definitions of them.

Definition 2 [10]. Let $S = (U, A, V, F)$ be a set-valued information system. For $\forall b \in A$, a tolerance relation can be defined as follows:

$$R_b^{\cap} = \{(x, y) \in U \mid b(x) \cap b(y) \neq \varnothing\}. \tag{1}$$

Then, for $B \subseteq A$, a tolerance relation can be defined as follows:

$$R_B^{\cap} = \{(x, y) \in U \mid \forall b \in B, b(x) \cap b(y) \neq \varnothing\} = \bigcap_{b \in B} R_b^{\cap}. \tag{2}$$

It is obvious that $R_B^{\cap}$ is reflexive and symmetric, but not necessarily transitive.

Dai et al. [14] defined a fuzzy relation, then the fuzzy relation is used to measure the similarity between set-valued objects.

Definition 3. Let $S = (U, A, V, F)$ be a set-valued information system. For $\forall b \in A$, a fuzzy relation $\widetilde{R_b}$ can be defined as follows:

$$\mu_{\widetilde{R_b}}(x, y) = \frac{|b(x) \cap b(y)|}{|b(x) \cup b(y)|}. \tag{3}$$

Thus, for $B \subseteq A$, a fuzzy relation can be defined as follows:

$$\mu_{\widetilde{R_B}}(x, y) = \inf_{b \in B} \mu_{\widetilde{R_b}}(x, y). \tag{4}$$

The above two definitions are not reasonable in some real problems. For example, let $a(x) = \{English, Chinese, French\}$, $a(y) = \{German, Japanese, English,$

Chinese}, and $a(z) = \{$*English, Chinese, French*}. According to Definition 2, $(y, z) \in R_a^\cap$ because $c(y) \cap c(z) = \{$*English, Chinese*} is not empty. That is to say, y and z are indiscernible with respect to the tolerance relation. In other words, y and z speak the same language. However, they may speak different languages. For example, y can only speak *English* and z can only speak *French*. By Definition 3, we know that $\mu_{\widetilde{R_a}}(x, z) = 1$. Thus, x and z are indiscernible with respect to the fuzzy relation. It means that x and z speak the same language. However, they may speak different language either. For example, x can only speak *English* and z can only speak *French*.

From this example, we find that two objects under the disjunctive semantics satisfied the existing relations only denote that they have possibility to be similar. Therefore, a new relation, which can describe the relation between two set-valued objects under the disjunctive semantics, should be defined.

3   A New Similarity Relation Based on Probability

As the analysis in Sect. 2, the existing methods may get unreasonable results under the disjunctive semantics. In order to solve this problem, a similarity relation based on probability is proposed in this section.

Definition 4. Let $S = (U, A, V, F)$ be a set-valued information system. For $\forall b \in A$, the similarity between x and y on b is defined as follows.

$$\mu_{R_b}(x, y) = \begin{cases} \frac{|b(x) \cap b(y)|}{|b(x)| * |b(y)|}, & x \neq y \\ 1, & x = y \end{cases}. \tag{5}$$

From the view of probability, $\mu_{R_b}(x, y)$ is the possibility that x and y take the same value. For $B \subseteq A$, the similarity relation R_B between x and y on B can be defined as follows:

$$\mu_{R_B}(x, y) = \prod_{b \in B} \mu_{R_b}(x, y). \tag{6}$$

There are some important properties of the similarity relation defined above:

(1) R_B is reflective.

Proof. Since $\mu_{R_b}(x, x) = 1$, then $\mu_{R_B}(x, y) = 1$, we know that R_B is reflective.

(2) R_B is symmetric.

Proof. Since $\mu_{R_b}(y, x) = \frac{|b(y) \cap b(x)|}{|b(y)| * |b(x)|} = \frac{|b(x) \cap b(y)|}{|b(x)| * |b(y)|} = \mu_{R_b}(x, y)$, then $\mu_{R_B}(y, x) = \mu_{R_B}(x, y)$, we know that R_B is symmetric.

For the same example used before, according to Definition 4, the similarity between x and y is $\mu_{R_a}(x, y) = \frac{1}{6}$, and the similarity between x and z is $\mu_{R_a}(x, z) = \frac{1}{3}$. That is to say, the probability of x and y speaking the same language is $\frac{1}{6}$, and the probability of x

and z speaking the same language is $\frac{1}{3}$. Comparing to the existing definitions, the definition we proposed gives a more reasonable and meaningful description between two objects with set values.

Definition 5. Let $S = (U, A, V, F)$ be a set-valued information system, for $B \subseteq A$, $x \in U$, the δ similarity class of x with respect to B can be defined as follows:

$$\delta_B(x) = \{y \in U | \mu_{R_B}(x, y) \geq \delta\}(0 \leq \delta \leq 1), \tag{7}$$

where δ is a threshold. We can use δ to control the size of the information granules generated by B. Specifically, the bigger the δ, the smaller size of the information granules generated by B.

Theorem 1. Let $S = (U, A, V, F)$ be a set-valued information system, B and B' be two subsets of A, δ be a threshold. For $x \in U$, if $B \supseteq B'$, then $\delta_B(x) \subseteq \delta_{B'}(x)$.

Proof. Let $B' = \{b_1, b_2, \ldots, b_n\}$, $B = B' \cup \{b_{n+1}, b_{n+2}, \ldots, b_{n+m}\}$. $\forall x \in U, \forall y \in U$. Then:

$$\mu_{R_{B'}}(x, y) = \prod_{b \in B'} \mu_{R_b}(x, y) = \mu_{R_{b_1}}(x, y) * \mu_{R_{b_2}}(x, y) * \ldots * \mu_{R_{b_n}}(x, y)$$

$$\mu_{R_B}(x, y) = \prod_{b \in B} \mu_{R_b}(x, y)$$

$$= \mu_{R_{b_1}}(x, y) * \mu_{R_{b_2}}(x, y) * \ldots * \mu_{R_{b_{n+1}}}(x, y) * \ldots * \mu_{R_{b_{n+m}}}(x, y)$$

$$= \mu_{R_{B'}}(x, y) * \mu_{R_{b_{n+1}}}(x, y) * \ldots * \mu_{R_{b_{n+m}}}(x, y)$$

Because $\forall b \in A$, $\mu_{R_b}(x, y) \in [0, 1]$, there must be $\mu_{R_B}(x, y) \leq \mu_{R_{B'}}(x, y)$.

Suppose $\forall x \in \delta_B(x)$, there exist $\mu_{R_B}(x, y) \geq \delta$, then $\delta \leq \mu_{R_B}(x, y) \leq \mu_{R_{B'}}(x, y)$, that is to say, $x \in \delta_{B'}(x)$. Therefore, $\delta_B(x) \subseteq \delta_{B'}(x)$.

Theorem 1 shows that the number of the elements in the similarity class will be changed with the variation of condition attribute set. The smaller the condition attribute set is, the more elements will be in the similarity class.

Theorem 2. Let $S = (U, A, V, F)$ be a set-valued information system, B be a subset of A, δ_1 and δ_2 be two thresholds. For $x \in U$, if $\delta_1 \leq \delta_2$, then $\delta_{1B}(x) \supseteq \delta_{2B}(x)$.

Proof. Suppose $\forall x \in \delta_{2B}(x)$, that is to say, $\mu_{R_B}(x, y) \geq \delta_2$. Because $\delta_1 \leq \delta_2$, then there must be $\mu_{R_B}(x, y) \geq \delta_1$, it can be inferred that $x \in \delta_{1B}(x)$. Therefore, $\delta_{1B}(x) \supseteq \delta_{2B}(x)$.

Theorem 2 indicates that we can control the elements in the similarity class by the threshold. The smaller the threshold is, the more elements will be in the similarity class.

Definition 6. Let $S = (U, A, V, F)$ be a set-valued information system, B be a subset of A. Given an arbitrary set $X \subseteq U$, the δ-upper approximation $\overline{B_\delta}(X)$ and the δ-lower approximation $\underline{B_\delta}(X)$ of X with respect to B are:

$$\overline{B_\delta}(X) = \{x \in U \mid \delta_B(x) \cap X \neq \emptyset\}, \tag{8}$$

$$\underline{B_\delta}(X) = \{x \in U \mid \delta_B(x) \subseteq X\} \tag{9}$$

where $\underline{B_\delta}(X)$ contains all the objects that can be classified into X definitely, and $\overline{B_\delta}(X)$ contains all the objects that can be classified into X approximately.

Theorem 3. Let $S = (U, A, V, F)$ be a set-valued information system, B and B' be two subsets of A, δ be a threshold. For $X \subseteq U$, if $B \supseteq B'$, then $\overline{B_\delta}(X) \subseteq \overline{B'_\delta}(X)$, and $\underline{B_\delta}(X) \supseteq \underline{B'_\delta}(X)$.

Proof. The proof comes directly from Theorem 1, and hence it is omitted here.

Theorem 3 shows that the number of the elements in the upper and lower approximation sets will be changed with the variation of the condition attribute set. The smaller the condition attribute set is, the more elements will be in the δ-upper approximation set and less in the δ-lower approximation set.

Theorem 4. Let $S = (U, A, V, F)$ be a set-valued information system, B be a subset of A, δ_1 and δ_2 be two thresholds. For $X \subseteq U$, if $\delta_1 \leq \delta_2$, then $\overline{B_{\delta_1}}(X) \supseteq \overline{B_{\delta_2}}(X)$, and $\underline{B_{\delta_1}}(X) \subseteq \underline{B_{\delta_2}}(X)$.

Proof. The proof comes directly from Theorem 2, and hence it is omitted here.

Theorem 4 indicates that we can control the elements in the upper and lower approximation sets by changing the threshold. The smaller the threshold is, the more elements will be in the δ-upper approximation set and less in the δ-lower approximation set.

4 Attribute Reduction of Set-Valued Decision Information System

Generally, most of decision information systems have some redundant attributes. These redundant attributes, on the one hand, waste storage space and reduce the efficiency of data processing. On the other hand, they may be our interference to make correct and concise decisions. Next, we will discuss how to reduce a set-valued decision information system.

Definition 7. Let $S = (U, C \cup D, V, F)$ be a set-valued decision information system, where U is a non-empty finite set of objects called the universe. C is a set of condition attribute, D is a decision attribute, F is a function that assigns particular values from attribute domain to objects. $U/D = \{d_1, d_2, \ldots, d_m\}$ is a division of U. For $B \subseteq C$, the positive domain of D with respect to B is defined as:

$$POS_B(D) = \bigcup_{i=1}^{m} \underline{B_\delta}(d_i), \tag{10}$$

and the negative domain of D with respect to B is:

$$NEG_B(D) = U - POS_B(D), \tag{11}$$

where $POS_B(D)$ is the set of objects in U that can be classified into D definitely. $NEG_B(D)$ is the set of objects in U that can not be classified into D definitely.

Theorem 5. Let $S = (U, C \cup D, V, F)$ be a set-valued decision information system, B and B' be two subsets of C. If $B' \subseteq B$, then $POS_{B'}(D) \subseteq POS_B(D)$.

Proof. If $x \in POS_{B'}(D)$, there exist $d_i \in U/D$ such that $\delta_{B'}(x) \subseteq d_i$. According to Theorem 1, we have $\delta_B(x) \subseteq \delta_{B'}(x)$. So, $\delta_B(x) \subseteq d_i$, that is $x \in POS_B(D)$. Thus, $POS_{B'}(D) \subseteq POS_B(D)$.

Theorem 5 shows that the positive domain of D with respect to an attribute subset, is also a subset of the positive domain of D with respect to all the attributes. That means, if some attributes are deleted from a set-valued decision information system, the positive domain may decrease or remain unchanged. If the positive domain remains unchanged after an attribute is deleted, it means this attribute is redundant for keeping the positive domain. In other words, removing redundant attributes does not affect the correct classification ability of the system.

Definition 8. Let $S = (U, C \cup D, V, F)$ be a set-valued decision information system. For $\forall a \in C$, a is reducible in C with respect to D if $POS_{C-\{a\}}(D) = POS_C(D)$. Otherwise, a is irreducible in C with respect to D.

Definition 9. Let $S = (U, C \cup D, V, F)$ be a set-valued decision information system, and B be a subset of C. B is a reduction of C with respect to D if:

(1) $POS_B(D) = POS_C(D)$, and
(2) $\forall a \in B$, $POS_{B-\{a\}}(D) \neq POS_B(D)$.

According to Definitions 8 and 9, it is obvious that B, the reduction of C, has the same classification ability as C, and deleting any attributes from B will decrease the correct classification ability of the system. To obtain a reduction of a set-valued decision information system, we can develop the following algorithm.

```
Algorithm 1: Attribute Reduction of set-valued decision
  information system
Require: A set-valued decision information system S
Ensure: A reduction B
1: Let  B=C
2: for each condition attribute a∈C do:
3:     if POS_{B-{a}}(D)=POS_C(D) then:
4:         Eliminate a from B
5:     end if
6: end for
```

For a given set-valued decision information system, Algorithm 1 check each attribute by the conditions stated in Definition 9. If the conditions are satisfied, then the attribute will be deleted. Otherwise it will be retained. Finally, we can get a reduction of the set-valued decision information system.

The following example illustrates how to form a reduction using Algorithm 1.

Example 1. For the set-valued decision information system S shown in Table 1, we can compute the reduction of S as follow.

Table 1. A set-valued decision information system

Objects	a_1	a_2	a_3	a_4	a_5	d
x_1	{2}	{1,2}	{1}	{2,3}	{1}	1
x_2	{1,2}	{3}	{1,2}	{2}	{1,2}	3
x_3	{1}	{2,3}	{2}	{1,2}	{1,2}	2
x_4	{1}	{2}	{2}	{2}	{1}	2
x_5	{3}	{2}	{1,2}	{1}	{3}	3
x_6	{1,3}	{2}	{1,2}	{1,2}	{2,3}	3
x_7	{2}	{1,2,3}	{1,2}	{2}	{1,3}	1
x_8	{1}	{3}	{2}	{2}	{2}	3

Let, $\delta = 1/16$, we have:

$$\delta_C(x_1) = \{x_1\}, \ \delta_C(x_2) = \{x_2, x_8\}$$
$$\delta_C(x_3) = \{x_3, x_4, x_8\}, \ \delta_C(x_4) = \{x_3, x_4\}$$
$$\delta_C(x_5) = \{x_5, x_6\}, \ \delta_C(x_6) = \{x_5, x_6\}$$
$$\delta_C(x_7) = \{x_7\}, \ \delta_C(x_8) = \{x_2, x_3, x_8\}$$
$$U/D = \{\{x_1, x_7\}, \{x_3, x_4\}, \{x_2, x_5, x_6, x_8\}\}$$
$$POS_C(D) = \{x_1, x_2, x_3, x_4, x_5, x_6, x_7, x_8\}$$

After a_1 is deleted, we have:

$$\delta_{C-a_1}(x_1) = \{x_1\}, \ \delta_{C-a_1}(x_2) = \{x_2, x_3, x_8\}$$
$$\delta_{C-a_1}(x_3) = \{x_2, x_3, x_4, x_8\}, \ \delta_{C-a_1}(x_4) = \{x_3, x_4, x_7\}$$
$$\delta_{C-a_1}(x_5) = \{x_5, x_6\}, \ \delta_{C-a_1}(x_6) = \{x_5, x_6\}$$
$$\delta_{C-a_1}(x_7) = \{x_4, x_7\}, \ \delta_{C-a_1}(x_8) = \{x_2, x_3, x_8\} a_3$$
$$POS_{C-a_1}(D) = \{x_1, x_5, x_6\}$$

It can be found that the positive domain has changed after a_1 is deleted. So, a_1 is irreducible. Likewise, a_2, a_3, a_4 and a_5 are all irreducible. Then, the reduction is $\{a_1, a_2, a_3, a_4, a_5\}$. Let, it can be found that a_1, and a_5 are irreducible, but and a_4 are reducible. Then, the reduction is $\{a_1, a_2, a_5\}$. According to the above analysis, we can get different reducts by changing the value of δ.

5 Experimental Results

The experiments in this section are to demonstrate the effectiveness of the method proposed in this paper. There are five groups of data sets used in the experiments and the information of all datasets is shown in Table 2.

Table 2. Data description

Source of datasets	Datasets	Number of attributes	Number of samples
weka	vote	17	435
weka	breast-cancer	10	286
UCI	annealing	33	798
UCI	audiology	70	200
UCI	zoo	17	17

In the experiments, we first reduce the five data sets using the tolerance relation, the fuzzy relation and the similarity relation proposed in this paper respectively, then J48 and SMO were used to make comparisons on classification accuracy with the reduction results. The reduction results gotten by different relations are shown in Table 3, and the classification accuracy comparisons are shown as Figs. 1 and 2:

Table 3. Attribute numbers after reduction

Dataset	Tolerance relation	Fuzzy relation	Similarity relation
vote	15	15	16
breast-cancer	9	9	9
annealing	9	10	14
audiology	21	22	63
zoo	13	13	14

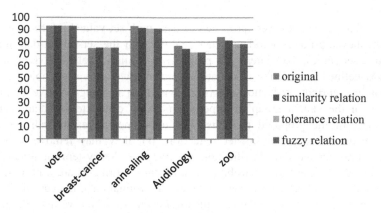

Fig. 1. Comparison of classification accuracy by J48

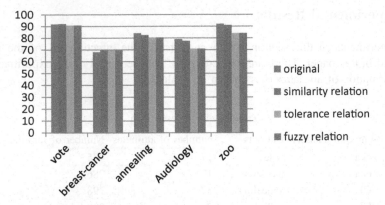

Fig. 2. Comparison of classification accuracy by SMO

It is known that only one or two attributes were removed by attribute reduction in vote and breast-cancer from Table 2, and the classification accuracy gotten by the proposed method and the existing methods has no obvious difference on these two datasets. In other three datasets, the results on attribute reduction are significant. By comparison, the method proposed in this paper retains more attributes than the existing methods. On the other hand, the classification accuracy gotten by the proposed method is almost the same as the classification accuracy of the original data, and is obviously higher than the classification accuracy of the other methods. According to the experimental results, we can draw the conclusion that the method proposed in this paper maintains more attributes than the existing methods, but it ensures that the classification accuracy of the system does not change significantly. Although the existing methods remove more attributes, the classification accuracy of the system is also greatly decreased. That means the existing methods remove some useful attributes, and it is unacceptable in some practical problems.

6   Conclusions

A lot of different methods were proposed to deal with set-valued information system. However, they are not appropriate when dealing with the set-valued data under the disjunctive semantics. To address this problem, a similarity relation based on probability was defined to describe the relation between two set-valued objects. Then, a corresponding attribute reduction algorithm based on keeping positive domain was proposed. In the end, a group of experiments were conducted to demonstrate the effectiveness of the proposed methods. The experimental results indicate that the existing methods can get a smaller reduction, but they may remove some useful attributes, thus reduce the classification accuracy of the system significantly. The proposed method retains more attributes, but all these attributes are useful, and it can always get the classification accuracy close to the original data. Because the threshold has an important influence on the classification accuracy when we used the proposed methods, it will be our future work how to choose an appropriate threshold.

Acknowledgments. This work was supported by the National Natural Science Foundation of China (61472056, 61533020, 61751312, 61379114), the Social Science Foundation of the Chinese Education Commission (15XJA630003), the Scientific and Technological Research Program of Chongqing Municipal Education Commission (KJ1500416), Chongqing Research Program of Basic Research and Frontier Technology (cstc2017jcyjAX0406).

References

1. Pawlak, Z.: Rough set theory and its applications to data analysis. Cybern. Syst. **29**(29), 661–688 (2010)
2. Guan, Y.Y., Wang, H.K.: Set-valued information systems. Inf. Sci.: Int. J. **176**(17), 2507–2525 (2006)
3. Orłowska, E.: Logic of nondeterministic information. Stud. Log. **44**(1), 91–100 (1985)
4. Orłowska, E., Pawlak, Z.: Representation of nondeterministic information. Theoret. Comput. Sci. **29**(1), 27–39 (1984)
5. Sakai, H., Ishibashi, R., Nakata, M.: Lower and upper approximations of rules in non-deterministic information systems. In: Chan, C.-C., Grzymala-Busse, J.W., Ziarko, W. P. (eds.) RSCTC 2008. LNCS (LNAI), vol. 5306, pp. 299–309. Springer, Heidelberg (2008). https://doi.org/10.1007/978-3-540-88425-5_31
6. Sakai, H., Ishibashi, R., Koba, K., Nakata, M.: Rules and apriori algorithm in non-deterministic information systems. In: Peters, J.F., Skowron, A., Rybiński, H. (eds.) Transactions on Rough Sets IX. LNCS, vol. 5390, pp. 328–350. Springer, Heidelberg (2008). https://doi.org/10.1007/978-3-540-89876-4_18
7. Yao, Y.Y.: Information granulation and rough set approximation. Int. J. Intell. Syst. **16**(1), 87–104 (2001)
8. Yao, Y.Y., Noroozi, N.: A unified framework for set-based computations. In: Proceedings of the 3rd International Workshop on Rough Sets and Soft Computing. The Society for Computer Simulation, pp. 252–255(1995)
9. Lipski Jr., W.: On semantic issues connected with incomplete information databases. Trans. Database Syst. **4**(3), 262–296 (1979)
10. Lipski Jr., W.: On databases with incomplete information. J. ACM **28**(1), 41–70 (1981)
11. Zhang, W.X.: Information Systems and Knowledge Discovery. Science Press, Beijing (2003)
12. Zhang, W.X., Mi, J.S.: Incomplete information system and its optimal selections. Comput. Math Appl. **48**(5), 691–698 (2004)
13. Qian, Y., Dang, C., Liang, J., Tang, D.: Set-valued ordered information systems. Inf. Sci. **179**(16), 2809–2832 (2009)
14. Dai, J., Tian, H.: Fuzzy rough set model for set-valued data. Fuzzy Sets Syst. **229**, 54–68 (2013)
15. Wang, C.Y.: A note on a fuzzy rough set model for set-valued data. Fuzzy Sets Syst. **294**, 44–47 (2016)
16. Bao, Z., Yang, S.: Attribute reduction for set-valued ordered fuzzy decision system. In: Sixth International Conference on Intelligent Human-Machine Systems and Cybernetics, pp. 96–99 (2014)

17. Wei, W., Cui, J., Liang, J., Wang, J.: Fuzzy rough approximations for set-valued data. Inf. Sci. **360**, 181–201 (2016)
18. Stefanowski, J., Tsoukiàs, A.: Incomplete information tables and rough classification. Comput. Intell. **17**(3), 545–566 (2001)
19. Nakata, M., Sakai, H.: Rough sets handling missing values probabilistically interpreted. In: Ślęzak, D., Wang, G., Szczuka, M., Düntsch, I., Yao, Y. (eds.) RSFDGrC 2005. LNCS (LNAI), vol. 3641, pp. 325–334. Springer, Heidelberg (2005). https://doi.org/10.1007/11548669_34

Defuzzyfication in Interpretation
of Comparator Networks

Łukasz Sosnowski[1]([✉]) and Marcin Szczuka[2]

[1] Systems Research Institute, Polish Academy of Sciences, Newelska 6, 01-447
Warsaw, Poland
sosnowsl@ibspan.waw.pl
[2] Institute of Informatics, The University of Warsaw, Banacha 2, 02-097 Warsaw,
Poland
szczuka@mimuw.edu.pl

Abstract. We present an extension to the methods and algorithms
for approximation of similarity known as Networks of Comparators. By
interpreting the output of the network in terms of discrete fuzzy set we
make it possible to employ various defuzzyfication techniques for the purpose of establishing a unique value of the output of comparator network.
We illustrate the advantages of this approach using two examples.

Keywords: Comparators · Similarity · Approximation
Defuzzyfication · Decision rules

1 Introduction

There are many ways to implement object recognition solutions. The method
considered in this paper is based on multi-similarity calculations, gathering
many aspects of the similarity between pairs of objects. The objects belong
to multi-dimensional space and are described by the similarity values between
input objects and reference objects, measured on the given set of features. From
this perspective it is a kind of approximation of an input object by the objects
belonging to the reference set. The result of this approximation is in a form
of proximity vector expressed by similarity, which shows the closeness between
input object and reference points in the domain space. The units responsible for
single-feature calculations are called *comparators*. The network-like structure
allowing to process input objects through the layers of multiple comparators
will be called *Network of Comparators* (NoC) or *Comparator Network*.

Comparator networks can play different roles depending on their types. They
can serve as multi-stage classifiers whose purpose is to limit the reference set of
objects and identify the most probable candidate to be a final result. Another
type of NoCs specializes in recognition of different features based on the nature of
sub-objects. The idea is that the similarity of parts of objects can help resolving
the similarity of the whole objects.

© Springer International Publishing AG, part of Springer Nature 2018
J. Medina et al. (Eds.): IPMU 2018, CCIS 854, pp. 467–479, 2018.
https://doi.org/10.1007/978-3-319-91476-3_39

In this paper we move a step further with making the output of a comparator network interpretable. Insofar the proximity vector was interpreted case by case depending on the problem. We want to make the procedure of using NoC's output more regulated by introducing a special overlay for Defuzzyfication of its results. First of all, we make an observation that the collection (vector) of similarity values returned by the network can be converted into of a discrete fuzzy set. By applying Defuzzyfication to this set we may construct the final, single outcome.

The default strategy used this far in NoC-based computations comprises of taking the element with the highest similarity value as final answer can be viewed as an application of the *Maximum defuzzyfication rule*. There is nothing, however, that forbids us from applying another defuzzyfication principle at this stage. In some cases it may be desirable for this outcome to not just be selected among the existing elements of reference set, but constructed (combined) as a new object on the basis of existing ones that are relevant. With more sophisticated defuzzyfication this is doable.

As an important example of aggregation task for a network of comparators we discuss the case where the final reference set consists of prototype decision rules. The overall goal of the network is to provide a decision value that is best suited for the case under consideration and reflects the structure of similarity provided by the network. Decision rules are traditionally regarded as a convenient and capable decision support model. There exist several established approaches to applying decision rules learned from training data to classification of newly provided, previously unseen objects. Various voting and conflict resolution schemes make it possible to improve both applicability and effectiveness of decision rule collections.

With a network of comparators that output a vector of similarity with respect to a set of decision rules we are capable to go even further. Since the network provides us with the vector of similarities between the investigated object and the rules we may treat this as a framework for constructing a rule-based classification (decision-making) ensemble. Each particular rule that has a non-zero similarity to investigated object is a component of this ensemble. Such rule can be viewed as a localized, potentially weak classifier. By making use of similarity values that comparator network provided us with we can now generate the final answer. Thanks to the fact that the process of comparator network construction can be parameterized and tweaked it is possible to factor-in the requirements for the properties of output similarity vector.

The article begins with explanation of the concept and operation of NoC in Sect. 2 followed by explanation of the interpretation of NoC output by means of defuzzyfication in Sect. 3. We follow up with illustrative case study of rule-based classification with use of comparators in Sect. 4. Section 5 concludes the paper and provides some ideas about future research in this area.

2   Networks of Comparators

Networks of Comparators (NoC) were previously described in [7,11]. Generally, it is an approach to reasoning about compound objects based on their multi-aspect similarity. In some sense it models analogy between objects and their relation. It is well suited for dealing with situations that involve information granularity (see [14]). A major advantage of the NoC approach is that the network architecture and settings can be discovered (learned) from data as described in [9,10]. The existing implementations of comparator networks have proven their worth is several real-life applications (see [7,8]).

NoCs may be of different kind depending on their architecture and intended use. They can serve as multi-stage classifiers whose purpose is to limit the reference set of objects and identify the most probable candidate to be a final result. The scenario of processing in such networks is to compute relatively simple features at the first layers and to filter out the reference objects to only those that are the most promising in the final perspective. This kind of network is called homogeneous and it was described in detail in [7].

Second type of network is called heterogeneous and it specializes in recognition of different features based on the nature of sub-objects. The idea is that the similarity of parts of objects can help in resolving the similarity of the whole objects. It is connected with special structures called *composition rules* which are responsible for translating similarity of parts into similarity of whole objects. Sometimes the knowledge about parts only is not enough for strict description of a bigger object, but it brings us closer to the solution in the form of an approximation.

The operation of a NoC can be interpreted as a calculation of a function:

$$\mu_{net}^{ref_{out}} : X \rightarrow [0,1]^{|ref_{out}|}, \tag{1}$$

which takes the input object $x \in X$ as an argument and ref_{out} is a reference set for the network's output layer. The target set (codomain) of $\mu_{net}^{ref_{out}}$ is the space of proximity vectors. As in the previous situation, the proximity vector from the target space will be denoted by v. Such a vector encapsulates information about similarities between a given input object x and objects from the reference set ref. Similarly to the case of a single comparator, by ordering the reference set, i.e. taking $ref = \{y_1, \ldots, y_{|ref|}\}$, we get the value network's function of:

$$\mu_{net}^{ref}(x) = \langle SIM(x, y_1), \ldots, SIM(x, y_{|ref|}) \rangle, \tag{2}$$

where $SIM(x, y_i)$ is the value of *global similarity* established by the network for an input object x and a reference object y_i. Global similarity depends on partial (local) similarities calculated by the elements of the network: layers, comparators, local aggregators, transaltors, projection modules and global aggregators.

Each NoC is composed of three types of layers: input, intermediate (hidden/internal) and output. A given network may have several internal layers. Layer consists of comparators that are grouped together by the common purpose of processing a particular piece of information (attributes) about the object in question. Each layer contains a set of comparators working in parallel and a specific translating/aggregating mechanism. The translating and aggregating mechanisms are necessary to facilitate the flow of information (similarity vectors) between layers. As sets of comparators in a particular layer correspond to a specific combination of attributes, the output of the previous layer has to be aggregated and translated to fit the requirements. This is done by elements called translators and aggregators, respectively. The translator converts comparator outputs to information about reference objects that would be useful for the next layer. The role of the aggregator is to choose the most likely outputs of the translator, in case there was any non-uniqueness in assigning information about input objects to comparators. The operation of a layer in the NoC can be represented as a mapping:

$$\mu_{layer}^{ref} : X \rightarrow [0,1]^{ref_l}, \tag{3}$$

where $x \in X$ is an input object and ref_l is the reference set for the layer.

Within a given layer only the local reference sets associated with comparators in that layer are used to establish (local) similarities. However, through aggregation and translation these local similarities become the material for synthesis of the output similarity and reference set for the layer. This synthesis is based on a translation matrix, as described in [12]. Function (3) is created as a superposition of comparator's function

$$\mu_{com}^{ref} : X \times 2^{ref} \rightarrow [0,1]^{ref}, \tag{4}$$

with local (layer) aggregation function and translation. Local translation operation is responsible for filtering the locally aggregated results.

The input and internal (hidden) layers in the comparator network contain comparators with function (4) together with translators and local aggregators. Local aggregators are a mandatory part of the network responsible for the synthesis of the results obtained by comparators. Aggregators are functions that operate on partial results of comparators. In the simplest case the network only needs a single *global aggregator* in the output layer. However, in the other network architectures it is included in other layers as well, in form of a local aggregator. The local aggregator processes partial results of the network at the level of a given layer. The aggregator's operation depends on the type of reference objects and the output of comparators. It can be represented as:

$$f_{agg}^{ref_l} : [0,1]^{ref_1} \times \ldots \times [0,1]^{ref_k} \rightarrow [0,1]^{ref_l}, \tag{5}$$

where k is the number of comparators in a given layer l, i.e., the number of inputs in the aggregating unit (local aggregator). ref_l is the output (resulting) reference set for layer l composed by means of the *composition rules* from the reference sets ref_i ($i = 1, \ldots, k$) used by comparators in layer l.

The translator is a network component associated with the adaptation of results of one layer to the context of another layer (the one to be fed with). In other words, this element expresses the results of the previous layer (their reference objects) in reference objects of the current one. It uses reference objects of the next layer, taking into account the relationships between the objects of both layers. The translator is defined by means of the translation matrix:

$$M_{ref_l}^{ref_k} = [m_{ij}],\tag{6}$$

where $i \in \{1, \ldots, m\}$, $j \in \{1, \ldots, n\}$ for m and n denoting cardinality of ref_k and ref_l, respectively. The matrix $M_{ref_l}^{ref_k}$ defines the mapping of objects in the set ref_k onto objects in the set ref_l. In practice ref_k is just a union of reference sets for all comparators in a given layer and ref_l is the target reference set. Values in the matrix are within $[0, 1]$.

The *Projection Module* appears in selected layers whenever there is a need for selecting a subset of coordinates (project the vector onto subspace) in proximity vector that will be further used in calculations. The selection of a particular coordinate may be based on its value (above/below threshold) and/or on the limitations regarding the number of coordinates that can be preserved. For the i-th coordinate in the proximity vector the projection can be the following:

$$\mu_{proj}(v[i]) = \begin{cases} \boldsymbol{v}[i] & projection(\boldsymbol{v}[i]) = 1 \\ 0 & projection(\boldsymbol{v}[i]) = 0 \end{cases}\tag{7}$$

where $i \in \{1, \ldots, |ref|\}$ and $projection(a)$ for $a \in [0, 1]$ is a function of the form:

$$projection : [0, 1] \to \{0, 1\},\tag{8}$$

The function *projection* is the actual selecting mechanism. It decides whether a given coordinate is set at 0 or not. This function can be defined as a threshold, maximum, ranking function, etc.

The *global aggregator* is a compulsory element of the output layer. Unlike local aggregators, which process results within a single layer, the global one may process values resulting from all layers at the same time. In the simplified, homogeneous case, when all layers use exactly the same reference set, the global aggregator may be expressed by:

$$\mu_{agg}^{ref_{out}} : \left([0, 1]^{ref}\right)^m \to [0, 1]^{ref_{out}},\tag{9}$$

where m is the number of **all** comparators in the networks, i.e. the number of inputs to the global aggregator.

In the more complicated, heterogeneous case, the sets in subsequent layers and comparators may differ. In this case the aggregator constructs the resulting (global) reference set ref_{out} in such a way that every element $y \in ref_{out}$ is decomposed into y_1 in reference set ref_1, y_2 in reference set ref_2 and so on, up to y_m in reference set ref_m. For a given input object $x \in X$ the value of similarity between x and each element of in $ref_1, \ldots, ref_m$ is known, as this is the output of the corresponding comparator. To obtain the aggregated result we use:

$$\mu_{agg}^{ref_{out}} : [0, 1]^{|ref_1|} \times \ldots \times [0, 1]^{|ref_m|} \to [0, 1]^{ref_{out}}\tag{10}$$

Note, that formula (10) is similar to the one for local aggregator (5). The essential difference is in the fact that the local aggregator is limited to a subset of comparators contained in a given layer, while the global one looks at all comparators in the network.

With all the definitions of units the comparator network can be expressed as a composition of mappings in subsequent layers:

$$\mu_{net}^{ref_{out}}(x) = \mu_{layer-out}^{ref_{out}}(\mu_{layer-int}^{ref_{k-1}} \cdots (\mu_{layer-in}^{ref_1}(x))\ldots), \qquad (11)$$

where ref_i stands for the reference set corresponding to layer i and ref_{out} is the reference set for the network as a whole.

To sum up, the result of the operation of NoC is an intentionally sparse vector of similarity values with respect to the output reference set. In the next section we advocate how this vector can be converted to a final, single answer.

3   Comparator Network Interpretation via Defuzzyfication

Defuzzyfication process in fuzzy sets entails converting a fuzzy grade into a crisp output with some kind of (possibly convoluted) mapping. In general, it can be denoted as $f(A) : A \rightarrow \mathbf{Z}$, where A is a fuzzy set and $\mathbf{Z}$ is a support of set A (elements with non-zero membership value). What is important is that the crisp result $z \in Z$ belongs to the support of A, but does not have to be among elements for which we know membership values.

In both theoretical and practical considerations the mapping f may be very varied and application-specific (see [3,15,16]). Below we list four most typical that are of use in our approach:

1. *Center of Gravity* (CoG) method, also frequently called *Center of Area* or *Centroid* method that can be expressed as $z^* = \frac{\int_Z \mu_A(z)z dz}{\int_Z \mu_A(z)dz}$ for Type II fuzzy sets or $z^* = \frac{\sum_{k=1}^{N} \mu_A(z_k)z_k dz}{\sum_{k=1}^{N} \mu_A(z_k)dz}$ for standard (Type I) fuzzy sets. This is a very popular method although there are some computational problems for more complex membership functions $\mu_A(z)$.
2. *Max-membership* takes all elements in output fuzzy set that have the highest membership value. It can be denoted as $z^* = \arg\max_Z \mu_A(z)$. This method was previously the most commonly used in the NoCs. This method can return more than one result and hence it is often extended to *Smallest Max, Largest Max* or *Mean of Maxima* in order to narrow it down. It is noteworthy that *Mean of Maxima* has a possibility to choose the result from outside of the set of considered elements.
3. *Center of Sums* is a modification of the first one (CoG). It considers the fuzzy set (membership) to be a composition of several fuzzy sets (memberships). It then defuzzyfies each component using CoG and combines them into final result taking an average weighted and normalized using the area of component sets.

4. *Weighted Average* method is quite easy to implement and use but requires the membership function to be symmetric on output set. If symmetry is assured it can be expressed as $z^* = \frac{\sum_z \mu_A(z)z}{\sum_z \mu_A(z)}$.

The model of NoC present in previous section provides its outputs by means of functions, vectors, etc. The final result of processing NoC is a sparse vector, as shown in formula (2), called *proximity vector*. In order to apply defuzzyfication techniques we need first to unambiguously convert the result to a fuzzy set. Taking this step makes it possible to move between models in an easy and precise way. It should be noted that individual coordinates of the proximity vector define the similarity of a particular pair of objects (x, y), where $x \in X$ and $y \in Y \subseteq ref$. These values reflect the degree of memberships of the fuzzy set we are constructing. At the same time each coordinate is parameterized by a pair of objects and so it can be treated as a fuzzy relation, which is itself a fuzzy set (over Cartesian product of $X \times ref$).

The result described proximity vector can be converted to fuzzy set notation in the following way:

$$R(x, y) = \{((x, y_i), \boldsymbol{v}[i]) : i = 1..|ref|\}, \tag{12}$$

where $\boldsymbol{v}[i]$ is i'th coordinate of proximity vector, which simultaneously fulfills the condition of the membership function of the fuzzy relation in the form:

$$\mu : X \times ref \to [0, 1] \tag{13}$$

This is also consistent with the definition of the similarity function of the comparator and NoC.

The formula (12) can be also made compatible with the Zadeh's notation by taking:

$$R = \frac{\boldsymbol{v}[1]}{(x, y_1)} + \frac{\boldsymbol{v}[2]}{(x, y_2)} + \ldots + \frac{\boldsymbol{v}[|ref|]}{(x, y_{|ref|})} \tag{14}$$

In addition, this method is reversible, that means that one can take a fuzzy set membership and reconstruct a proximity vector. It is noteworthy, that we do not need to know the algebraic form of the membership function formula to make conversion, we just need the knowledge about reference objects and their similarity.

The natural next step is to build the membership function as an explicit formula. This allows to use wide spectrum of computational methods developed for fuzzy sets. So, the overall form of the NoC's defuzzyfication layer is provided by the formula:

$$f_{def} : [0, 1]^{|ref_{out}|} \to Z \tag{15}$$

where Z is the support of a fuzzy set approximated by the NoC.
Full NoC model after being extended by the defuzzyfication function is of the form:

$$\mu_{net_{def}}^{ref_{out}} : X \to Z \tag{16}$$

and can be rewritten as:

$$f_{def}(\mu_{net}^{ref_{out}}(x)) = z \tag{17}$$

where $z \in Z$. Note, that the final, crisp result z doesn't have to be an element of the ref_{out}, but $ref_{out}^+ \subseteq Z$, where $ref_{out} \subseteq ref_{out}^+$. ref_{out}^+ consists of only these reference objects for which the similarity value in the proximity vector is positive.

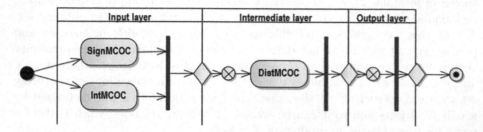

Fig. 1. Scheme of the NoC for recognition real numbers. Notation: SignMCOC – comparator for signs, IntMCOC – comparator for integer parts, DistMCOC - comparator for distance.

To better illustrate the nature of this operation we present a simple example. The task is to identify a real number through comparison to integer numbers that make reference set. Let us assume that the $ref_{out} = \{-100, -99, \ldots - 1, 0, 1, \ldots, 99, 100\}$. Our NoC consists of three layers and three comparators (two in the first layer and one in the second one) denoted as: *SignMCOC, IntMCOC, DistMCOC*. The architecture of the network is shown in Fig. 1. All aggregators use arithmetic average (mean) and translations are implemented with identity and projection module is $TOP\,5$ for the output layer. Input object is $x = 10.34$. The more detailed description of particular comparators is provided in Table 1 together with similarity measures applied inside.

Table 1. Comparators used in the NoC for recognizing numbers

Comparator	Description
SignMCOC	Comparator of signs using nominal scale. Returns 1 only if both signs are equal
IntMCOC	Comparator of integer parts of two numbers. Uses $\mu(x,y) = \frac{|x-y|}{span}$, where span is maximum possible difference
DistMCOC	Comparator using a distance to determine similarity. Uses the same measure like IntMCOC

The result after processing x in our NoC is: $v = \langle 0, \ldots, 0, 0.9883, 0.9933, 0.9983, 0.9967, 0.9917, 0, \ldots, 0 \rangle$, where $0.9883, 0.9933, 0.9983, 0.9967, 0.9917$ correspond to reference objects $8, 9, 10, 11, 12$, respectively. All others were eliminated by the projection module because their similarity did not made the $TOP\,5$.

Next, the resulting proximity vector is converted to Zadeh's notation:

$$R = \frac{0.9883}{(10.34, 8)} + \frac{0.9933}{(10.34, 9)} + \frac{0.9983}{(10.34, 10)} + \frac{0.9967}{(10.34, 11)} + \frac{0.9917}{(10.34, 12)}$$

Next we are looking for a triangular fuzzy sets by designating two straight lines passing through two points each. We have five points returned by NoC: $(8, 0.9883)$, $(9, 0.9933)$, $(10, 0.9983)$, $(11, 0.9967)$, $(12, 0.9917)$. We take first two and input into the general formula for linear function $y = ax + b$ to derive its coefficients. We get a couple of simple equations:

$$\begin{cases} 0.9883 = 8a + b \\ 0.9933 = 9a + b \end{cases} \tag{18}$$

In the same way we take last two (reference) points and derive another formula for a line going through them. Finally we get the following two lines: $y = 0.005x + 0.9483$ and $y = -0.005x + 1.0517$. The intersection of these two corresponds to value 1 on the Y axis. This is a maximum possible value for membership of a fuzzy set and in our case there is only one such point. By using the *Max-membership* defuzzyfication we get a final crisp result as a value on the X axis corresponding to the maximum. By simple substitution we can find that the result is indeed the 10.34 we were looking for.

4   Illustrative Example

Previously we have shown how the output of the NoC can be taken as a special kind of a fuzzy set and then defuzzyfied to get a single, meaningful outcome. The situation becomes more intricate when the elements of ref_{out} are themselves objects with structure and semantics. In our example these are decision rules.

The overall goal is now to construct a decision support system that uses decision rules to establish the decision value and employs the NoC for the purpose of selecting applicable rules and combining their recommendations. Rules themselves are obtained from the outside source, such as algorithms and methods originating in rough sets. There are several data analytics tools that can be used for this purpose, for details consult the overview in [4].

In order to keep the example simple and straightforward we make assumption that the rules we use are in a simplest form. A decision rule is a formula:

$$(a_1(x) \in A_1) \wedge (a_2(x) \in A_2) \wedge \ldots \wedge (a_k(x) \in A_k) \Rightarrow (d = v),$$

where $x \in X$ is an input data object, $a_1, \ldots, a_k$ are attributes, $A_1, \ldots, A_k$ are attribute values (value sets), and d is the decision attribute. Sub-formulæ $a_i(x) \in A_i$ are referred to as descriptors or selectors. In our case we assume that all descriptors are *simple*, i.e., $\forall_i |A_i| = 1$ and the take form $a_i(x) = v_i$.

As mentioned before, we assume that we have acquired a set of decision rules for our data set. These rules are all put into the output reference set, so that $ref_{out} = \{r_1, \ldots, r_m\}$. Now, when we need to establish decision (classify) a previously unseen object from X we usually match it against the rules that we have and then from their output try to figure out the ultimate answer. There are two fundamental problems with such simplistic approach. First of all, the *incomplete coverage problem*, i.e., there may be no rule in our set that exactly matches the object we want to classify. Secondly, we may encounter a *conflict*, i.e., there are two or more rules that match the new object but they point at different decision values. The two issues with rule-based classifiers have been extensively studied and discussed in literature. There exist numerous approaches to both partial rule matching and rule conflict resolution. In the context of rough sets one can consult [1] for approaches to these problems.

In our example both the coverage- (matching) and conflict-related issues can be addressed by making use of the values of similarity (1) outputted by the NoC. The coverage issue is fairly easy to deal with once we have the NoC's similarity vector for our rule set:

$$\mu_{net}^{ref}(x) = \langle SIM(x, r_1), \ldots, SIM(x, r_k)\rangle,$$

where $k = |ref_{out}|$. Depending on the overall strategy we can use a "winner takes all" approach and establish decision value using only the rule r_i that has the highest value of similarity. In the unlikely event that there are several conflicting rules with the same similarity to the investigated object we would apply one of conflict resolution methods discussed later in this section. Another, potentially more flexible and extensible strategy is to select a subset of rules that are sufficiently similar to the investigated new object x. The selection can be based on the cutoff value for similarity. All rules that are sufficiently similar (above threshold) qualify and form a *local ensemble* classification model. The cutoff value can be set arbitrary or learned from the data in parallel to NoC's training. If finding the right threshold proves problematic or if the resulting rule subsets are too large one can resort to simpler schemes. For instance, we can decide to include only the rules with similarity value greater than the average or just a pre-set number of them.

It is worth mentioning that in comparison with "traditional" methods for dealing with partial matching and incomplete coverage in rule-based systems the approach based on NoC has some clear advantages. Typically, when new object has no exactly matching rule the solution is to take a rule or rules that provide the closest match. The closeness is usually established in an arbitrary way. Either the rule that has the highest ratio of exactly matched descriptors is chosen or the rule which descriptors are the closest to the investigated object in terms of distance in the attribute-value space. Both such simple strategies can be implemented with help of NoC, but the NoC has also an added quality. Since we train the NoC, the resulting similarity values it outputs may be much more relevant and better suited for the decision support task at hand.

The second of the rule-related issues, the potential existence of conflicting rules that match the object can as well be addressed using the output of NoC. For starters, as explained in previous paragraphs, the similarities provided by

NoC make it possible to restrict the subset of rules that will be used to determine the decision. Rules that are chosen to take part in classification of the new object x can be viewed as on the one hand an *ensemble classifier* and on the other hand as a *localized decision model*.

With the subset of rules and the associated values of similarity we perform the defuzzyfication of NoC in the manner reflecting the general scheme presented in Sect. 3. In the context of rules this step ca be viewed as a special case of election scenario. The elements (members) of the rule-based ensemble cast votes and the mode of vote tallying is tantamount to processing of the output proximity vector. There exists an extensive literature on application of various election schemes in the context of recognition, classification and decision support (see [2,5,6]). Some of the most popular and/or most useful of them have already been studied in the context of processing the results from NoC, see [8]. These include: *plurality voting, Borda count, Copeland's method, approval voting, weighted voting* and *range voting*.

In the case of a reference set consisting of decision rules with similarities, attached there are many strategies for establishing values that may be applied by the NoC. The already mentioned "winner takes all" approach corresponds to the *plurality voting* strategy and can be trivially implemented using the *Largest Max* defuzzyfication. Implementation of another popular scheme – *weighted voting* – is also straightforward. The final decision is chosen by taking the value that has the highest total of similarities for rules pointing at this particular outcome. Note, that that just as in case of our previous example, the resulting compound classifier may not be equivalent to any rule existing in the output reference set.

5 Summary

In this paper we presented a new step for processing results returned by a Network of Classifiers. We adapted a standardized defuzzyfication procedure thereby introducing a new level of interpretation of results. It makes NoC method even more capable of recognizing compound objects. What is important, we did not change any assumptions about NoC, and the processing flow remains unaltered. All previously reported results concerning NoC applications remain valid.

This article contains two examples. The first, very simple one is meant to give better understanding the idea. The second, based on rules-based classification outlines the possible field of application. In our example in Sect. 4 we have considered only the simplest setting for both rules and NoC. In a real-life scenario we can expect that both the rules and the ways of their application will be much more complex, possibly multi-stage. It is conceivable that for a given decision support task and set of decision rules we may be willing to train several NoCs in such a way that at the end we will have multiple values for each rule in reference set and object, a feature that reflects the multitude of ways we can understand and approximate similarity.

The immediate next step in our work will be focused on preparing the implementation of the defuzzyfication overlay. It will complement the existing *comparators-lib* software library [13]. Once the new methods and algorithms are implemented it will be possible to extend the area of NoC's applicability to more types of compound objects, in particular time series.

References

1. Bazan, J.G.: Hierarchical classifiers for complex spatio-temporal concepts. In: Peters, J.F., Skowron, A., Rybiński, H. (eds.) Transactions on Rough Sets IX. LNCS, vol. 5390, pp. 474–750. Springer, Heidelberg (2008). https://doi.org/10.1007/978-3-540-89876-4_26
2. Faliszewski, P., Hemaspaandra, E., Hemaspaandra, L.A.: Using complexity to protect elections. Commun. ACM **53**(11), 74–82 (2010)
3. Hellendoorn, H., Thomas, C.: On quality defuzzification. In: Bien, Z., Min, K.C. (eds.) Fuzzy Logic and its Applications to Engineering, Information Sciences, and Intelligent Systems. TDLD, vol. 16, pp. 167–176. Springer, Dordrecht (1995). https://doi.org/10.1007/978-94-009-0125-4_16
4. Janusz, A., Stawicki, S., Szczuka, M., Ślęzak, D.: Rough set tools for practical data exploration. In: Ciucci, D., Wang, G., Mitra, S., Wu, W.-Z. (eds.) RSKT 2015. LNCS (LNAI), vol. 9436, pp. 77–86. Springer, Cham (2015). https://doi.org/10.1007/978-3-319-25754-9_7
5. Lin, X., Yacoub, S., Burns, J., Simske, S.: Performance analysis of pattern classifier combination by plurality voting. Pattern Recogn. Lett. **24**(12), 1959–1969 (2003)
6. Pomerol, J., Barba-Romero, S.: Multicriterion Decision in Management: Principles and Practice. International Series in Operations Research and Management Science. Springer, New York (2012). https://doi.org/10.1007/978-1-4615-4459-3
7. Sosnowski, Ł.: Framework of compound object comparators. Intell. Decis. Technol. **9**(4), 343–363 (2015). https://doi.org/10.3233/IDT-140229
8. Sosnowski, Ł., Pietruszka, A., Łazowy, S.: Election algorithms applied to the global aggregation in networks of comparators. In: Proceedings of the 2014 Federated Conference on Computer Science and Information Systems, Warsaw, Poland, 7–10 September 2014, pp. 135–144. IEEE (2014). https://doi.org/10.15439/2014F494
9. Sosnowski, Ł., Ślęzak, D.: Learning in comparator networks. In: Kacprzyk, J., Szmidt, E., Zadrożny, S., Atanassov, K.T., Krawczak, M. (eds.) IWIFSGN/EUSFLAT 2017. AISC, vol. 643, pp. 316–327. Springer, Cham (2018). https://doi.org/10.1007/978-3-319-66827-7_29
10. Sosnowski, Ł., Ślęzak, D.: How to design a network of comparators. In: Imamura, K., Usui, S., Shirao, T., Kasamatsu, T., Schwabe, L., Zhong, N. (eds.) BHI 2013. LNCS (LNAI), vol. 8211, pp. 389–398. Springer, Cham (2013). https://doi.org/10.1007/978-3-319-02753-1_39
11. Sosnowski, Ł., Ślęzak, D.: Networks of compound object comparators. In: FUZZ-IEEE, pp. 1–8 (2013). https://doi.org/10.1109/FUZZ-IEEE.2013.6622547
12. Sosnowski, Ł., Ślęzak, D.: Fuzzy set interpretation of comparator networks. In: Kryszkiewicz, M., Bandyopadhyay, S., Rybinski, H., Pal, S.K. (eds.) PReMI 2015. LNCS, vol. 9124, pp. 345–353. Springer, Cham (2015). https://doi.org/10.1007/978-3-319-19941-2_33

13. Sosnowski, Ł., Szczuka, M.: Recognition of compound objects based on network of comparators. In: Position Papers of the 2016 Federated Conference on Computer Science and Information Systems, FedCSIS 2016, Gdańsk, Poland, 11–14 September 2016, pp. 33–40 (2016). https://doi.org/10.15439/2016F571
14. Sosnowski, Ł., Szczuka, M., Ślęzak, D.: Granular modeling with fuzzy comparators. In: 2015 IEEE International Conference on Big Data, Big Data 2015, Santa Clara, CA, USA, 29 October–1 November 2015, pp. 1550–1555. IEEE (2015). https://doi.org/10.1109/BigData.2015.7363919
15. Sugeno, M.: An introductory survey of fuzzy control. Inf. Sci. $36(1)$, 59–83 (1985). http://www.sciencedirect.com/science/article/pii/002002558590026X
16. Yager, R.R., Filev, D.: Essentials of Fuzzy Modeling and Control. Wiley, Hoboken (1994)

A Comparison of Characteristic Sets and Generalized Maximal Consistent Blocks in Mining Incomplete Data

Patrick G. Clark[1], Cheng Gao[1], Jerzy W. Grzymala-Busse[1,2(✉)], and Teresa Mroczek[2]

[1] Department of Electrical Engineering and Computer Science, University of Kansas, Lawrence, KS 66045, USA
patrick.g.clark@gmail.com, {cheng.gao,jerzy}@ku.edu
[2] Department of Expert Systems and Artificial Intelligence, University of Information Technology and Management, 35-225 Rzeszow, Poland
tmroczek@wsiz.rzeszow.pl

Abstract. We discuss two interpretations of missing attribute values, lost values and "do not care" conditions. Both interpretations may be used for data mining based on characteristic sets. On the other hand, maximal consistent blocks were originally defined for incomplete data sets with "do not care" conditions, using only lower and upper approximations. We extended definitions of maximal consistent blocks to both interpretations while using probabilistic approximations, a generalization of lower and upper approximations. Our main objective is to compare approximations based on characteristic sets with approximations based on maximal consistent blocks in terms of an error rate.

Keywords: Incomplete data mining · Characteristic sets
Maximal consistent blocks · Rough set theory
Probabilistic approximations

1 Introduction

In this paper we discuss incomplete data sets with two interpretations of missing attribute values, lost values and "do not care" conditions, introduced in [5]. Lost values are, e.g., erased or forgotten. During mining data sets with lost values, we try to use only existing attribute values. "Do not care" conditions are usually results of a refusal to answer a question. During mining incomplete data with "do not care" conditions we assume that a missing attribute value may be any value from the attribute domain.

For data mining, we use probabilistic approximations, a generalization of lower and upper approximations known from rough set theory. A probabilistic approximation is defined by using an additional parameter, denoted by α and interpreted as a probability. Lower approximations are probabilistic approximations with $\alpha = 1$, upper approximations are probabilistic approximations with α

© Springer International Publishing AG, part of Springer Nature 2018
J. Medina et al. (Eds.): IPMU 2018, CCIS 854, pp. 480–489, 2018.
https://doi.org/10.1007/978-3-319-91476-3_40

only slightly greater than zero. Such approximations, restricted to complete data sets, i.e., data sets without missing attribute values, were extensively studied in [8,13–20]. Probabilistic approximations were generalized to incomplete data sets in [7].

For mining incomplete data we apply two different ideas, characteristic sets and maximal consistent blocks. Characteristic sets were introduced for data with lost values and "do not care" conditions in [5], while maximal consistent blocks were introduced only to "do not care" conditions in [10]. Additionally, in [10] only lower and upper approximations were considered. Note that some algorithms for determining maximal consistent blocks, restricted to data with only "do not care" conditions, were discussed in [11,12]. We have extended the definition of maximal consistent blocks to incomplete data sets with missing attribute values interpreted as lost values and "do not care" conditions in [1]. In this paper we also applied probabilistic approximations to both characteristic sets and maximal consistent blocks.

Our previous experiments, based on only three types of probabilistic approximations: lower ($\alpha = 1$), middle ($\alpha = 0.5$) and upper ($\alpha = 0.001$), have shown that middle approximations are most promising [1]. Therefore, in our current experiments we used the entire spectrum of probabilistic approximations, with $\alpha = 1$, $\alpha = 0.001$ and any α from 0.1 to 0.9, with an increment equal to 0.1.

Our main objective is to compare approximations based on characteristic sets with approximations based on maximal consistent blocks in terms of an error rate measured by stratified ten-fold cross validation. Our secondary objective is to compare two interpretations of missing attribute values, lost values and "do not care" conditions, again, in terms of an error rate.

2   Incomplete Data Sets

Table 1 presents an example of the incomplete data set. Lost values and "do not care" conditions are denoted by symbols of "?" and "*", respectively. The set of all cases will be denoted by U. In our example, $U = \{1, 2, 3, 4, 5, 6, 7\}$. The set of all cases with the same decision value is called a *concept*. For example, the set $\{1, 2, 3, 4\}$ is a concept. We say that $a(x) = v$ if an attribute a has value v for a case x. For example, $Temperature(1) = high$.

For completely specified data sets, for an attribute-value pair (a, v), a *block* of (a, v), denoted by $[(a, v)]$, is defined as follows

$$[(a,\ v)] = \{x | x \in U, a(x) = v\}.$$

For incomplete decision tables the definition of a block of an attribute-value pair was modified in [5,6] in the following way

– If for an attribute a and a case x we have $a(x) = ?$, the case x should not be included in any blocks $[(a, v)]$ for all values v of attribute a,
– If for an attribute a and a case x we have $a(x) = *$, the case x should be included in blocks $[(a, v)]$ for all specified values v of attribute a.

Table 1. An incomplete data set

Case	Attributes			Decision
	Temperature	Wind	Humidity	Trip
1	High	*	Low	Yes
2	*	Medium	*	Yes
3	Medium	?	*	Yes
4	?	High	*	Yes
5	Medium	Low	High	No
6	High	*	?	No
7	*	High	High	No

For our example from Table 1, all blocks of attribute-value pairs are

$[(Temperature, medium)] = \{2, 3, 5, 7\}$,
$[(Temperature, high)] = \{1, 2, 6, 7\}$,
$[(Wind, low)] = \{1, 5, 6\}$,
$[(Wind, medium)] = \{1, 2, 6\}$,
$[(Wind, high)] = \{1, 4, 6, 7\}$,
$[(Humidity, low)] = \{1, 2, 3, 4\}$,
$[(Humidity, high)] = \{2, 3, 4, 5, 7\}$.

3   Characteristic Sets and Maximal Consistent Blocks

The *characteristic set* $K_B(x)$ is defined as the intersection of the sets $K(x, a)$, for all $a \in B$, where $x \in U$, B is a subset of the set A of all attributes and the set $K(x, a)$ is defined as follows:

- If $a(x)$ is specified, then $K(x, a)$ is the block $[(a, a(x))]$ of attribute a and its value $a(x)$,
- If $a(x) = ?$ or $a(x) = *$, then $K(x, a) = U$.

For the data set from Table 1 and $B = A$, the characteristic sets are
$K_A(1) = \{1, 2\}$,
$K_A(2) = \{1, 2, 6\}$,
$K_A(3) = \{2, 3, 5, 7\}$,
$K_A(4) = \{1, 4, 6, 7\}$,
$K_A(5) = \{5\}$,
$K_A(6) = \{1, 2, 6, 7\}$,
$K_A(7) = \{4, 7\}$.

A binary relation $R(B)$ on U, defined for $x, y \in U$ in the following way

$$(x, y) \in R(B) \; if \; and \; only \; if \; y \in K_B(x)$$

will be called the *characteristic relation*. In our example $R(A) = \{(1, 1), (1, 2),$ $(2, 1), (2, 2), (2, 6), (3, 2), (3, 3), (3, 5), (3, 7), (4, 1), (4, 4), (4, 6), (4, 7), (5,$ $5), (6, 1), (6, 2), (6, 6), (6, 7), (7, 4), (7, 7)\}$.

We quote some definitions from [9]. Let X be a subset of U. The set X is *B-consistent* if $(x, y) \in R(B)$ for any $x, y \in X$. If there does not exist a B-consistent subset Y of U such that X is a proper subset of Y, the set X is called a *maximal B-consistent block*. The set of all maximal B- consistent blocks will be denoted by $\mathscr{C}(B)$. In our example, $\mathscr{C}(A) = \{\{1, 2\}, \{2, 6\}, \{3\}, \{4, 7\}, \{5\}\}$.

Let $B \subseteq A$ and $Y \in \mathscr{C}(B)$. The set of all maximal B-consistent blocks which include an element x of the set U, i.e. the set

$$\{Y | Y \in \mathscr{C}(B), x \in Y\}$$

will be denoted by $\mathscr{C}_x(B)$.

For data sets in which all missing attribute values are "do not care" conditions, an idea of a maximal consistent block of B was defined in [9]. Note that in our definition, the maximal consistent blocks of B are defined for arbitrary interpretations of missing attribute values. For Table 1, the maximal A-consistent blocks $\mathscr{C}_x(A)$ are

$\mathscr{C}_1(A) = \{\{1, 2\}\},$
$\mathscr{C}_2(A) = \{\{1, 2\}, \{2, 6\}\},$
$\mathscr{C}_3(A) = \{\{3\}\},$
$\mathscr{C}_4(A) = \{\{4, 7\}\},$
$\mathscr{C}_5(A) = \{\{5\}\},$
$\mathscr{C}_6(A) = \{\{2, 6\}\},$
$\mathscr{C}_7(A) = \{\{4, 7\}\}.$

4   Probabilistic Approximations

In this section we will discuss two types of probabilistic approximations: based on characteristic sets and on maximal consistent blocks.

4.1   Probabilistic Approximations Based on Characteristic Sets

Three kinds of probabilistic approximations, called singleton, subset and concept, were introduced in [2]. It was proved in [3] that the differences between all three probabilistic approximations, in terms of an error rate computed as a result of ten-fold cross validation, are insignificant. In this paper we restrict our attention only to concept probabilistic approximations, for simplicity calling them probabilistic approximations based on characteristic sets.

A *probabilistic approximation based on characteristic sets* of the set X with the threshold $\alpha, 0 < \alpha \leq 1$, denoted by $appr_\alpha^{CS}(X)$, is defined as follows

$$\cup\{K_A(x) \mid x \in X, \ Pr(X|K_A(x)) \geq \alpha\}.$$

Table 2. Conditional probabilities $Pr([(Trip, yes)]|K_A(x))$

x	1	2	3	4
$K_A(x)$	$\{1, 2\}$	$\{1, 2, 6\}$	$\{2, 3, 5, 7\}$	$\{1, 4, 6, 7\}$
$Pr(\{1, 2, 3, 4\} \mid K_A(x))$	1	0.667	0.5	0.5

Table 3. Conditional probabilities $Pr([(Trip, no)]|K_A(x))$

x	5	6	7
$K_A(x)$	$\{5\}$	$\{1, 2, 6, 7\}$	$\{4, 7\}$
$Pr(\{5, 6, 7\} \mid K_A(x))$	1	0.5	0.5

For Table 1 and both concepts $\{1, 2, 3, 4\}$ and $\{5, 6, 7\}$, all conditional probabilities $Pr(X|K_A(x))$, where X is a concept, are presented in Tables 2 and 3.

All distinct probabilistic approximations based on characteristic sets are

$$appr_{0.5}^{CS}(\{1, 2, 3, 4\}) = U,$$

$$appr_{0.667}^{CS}(\{1, 2, 3, 4\}) = \{1, 2, 6\},$$

$$appr_1^{CS}(\{1, 2, 3, 4\}) = \{1, 2\},$$

$$appr_{0.5}^{CS}(\{5, 6, 7\}) = \{1, 2, 4, 5, 6, 7\},$$

$$appr_1^{CS}(\{5, 6, 7\}) = \{5\}.$$

4.2 Probabilistic Approximations Based on Maximal Consistent Blocks

By analogy with the definition of a probabilistic approximation based on characteristic sets, we may define a probabilistic approximation as follows:

A *probabilistic approximation* based on maximal consistent blocks of the set X with the threshold $\alpha, 0 < \alpha \leq 1$, and denoted by $appr_\alpha^{MCB}(X)$ is defined as follows

$$\cup\{Y \mid Y \in \mathscr{C}_x(A),\ x \in X,\ Pr(X|Y) \geq \alpha\}.$$

All conditional probabilities $Pr(X|Y)$, where X is a concept and $Y \in \mathscr{C}(A)$, are presented in Table 4.

Table 4. Conditional probabilities $Pr(X|Y)$

Y	{1,2}	{2, 6}	{3}	{4, 7}	{5}
$Pr(\{1,2,3,4\} \mid Y)$	1	0.5	1	0.5	0
$Pr(\{5,6,7\} \mid Y)$	0	0.5	0	0.5	1

All distinct probabilistic approximations based on maximal consistent blocks are

$appr_{0.5}^{MCB}(\{1,2,3,4\}) = \{1,2,3,4,6,7\},$

$appr_1^{MCB}(\{1,2,3,4\}) = \{1,2,3\},$

$appr_{0.5}^{MCB}(\{5,6,7\}) = \{2,4,5,6,7\},$

$appr_1^{MCB}(\{5,6,7\}) = \{5\}.$

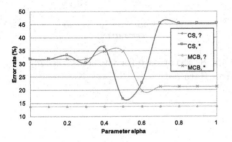

Fig. 1. Error rate for the *bankruptcy* data set

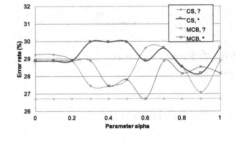

Fig. 2. Error rate for the *breast cancer* data set

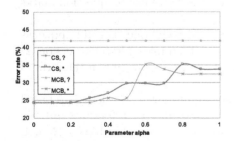

Fig. 3. Error rate for the *echocardiography* data set

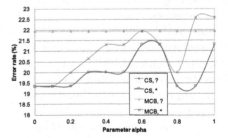

Fig. 4. Error rate for the *hepatitis* data set

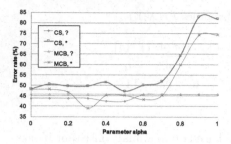

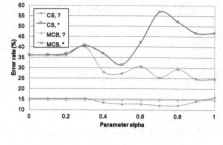

Fig. 5. Error rate for the *image seg-mentation* data set

Fig. 6. Error rate for the *iris* data set

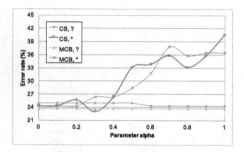

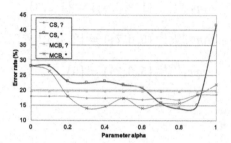

Fig. 7. Error rate for the *lymphography* data set

Fig. 8. Error rate for the *wine recognition* data set

5   Experiments

We conducted our experiments on eight data sets obtained from the *Machine Learning Repository*, University of California at Irvine. Originally, these data sets were completely specified. However, we randomly replaced 35% of the existing, specified attribute values by question marks, indicating lost values. Then we created new data sets with "do not care" conditions by replacing question marks with asterisks.

Our data mining was based on rule induction technique. For this purpose we used the MLEM2 (Modified Learning from Examples Module, version 2) system [4]. The MLEM2 system does not need any pre-processing in the form of discretization or handling missing attribute values.

In our experiments we used four approaches for mining incomplete data based on characteristic sets and maximal consistent blocks and on lost values and "do not care" conditions. Results of our experiments are presented in Figs. 1, 2, 3, 4, 5, 6, 7 and 8. In these figures, "CS" means a characteristic set, "MCB" means a maximal consistent block, "?" means a lost value and "*" means a "do not care" condition.

To compare these four approaches we applied the Friedman rank sum test combined with multiple comparisons, with a 5% level of significance. The Friedman test is nonparametric, i.e., no assumptions about normal distribution etc.

are necessary. This test, conducted in the MS Excel environment, was applied for our eight data sets. In general, for all eight data sets the hypothesis H_0 that all four approaches are equivalent was rejected. However, for one data set (*wine recognition*) the post-hoc test (distribution-free multiple comparisons based on Friedman rank sums) shows that the differences between all four approaches are statistically insignificant. For remaining seven data sets results of statistical analysis are presented in Table 5.

Table 5. Results of statistical analysis

Data set	The best approaches	The worst approaches
Bankruptcy	CS?, MCB?	CS*, MCB*
Breast cancer	MCB?	CS?, CS*, MCB*
Echocardiogram	CS*, MCB*	CS?, MCB?
Hepatitis	CS*	CS?, MCB?
Image recognition	CS?	CS*
Iris	CS?, MCB?	CS*, MCB*
Lymphography	MCB?	CS?, CS*, MCB*

Our results, presented in Figs. 1, 2, 3, 4, 5, 6, 7 and 8, show that for one of the four approaches used for mining incomplete data sets, for probabilistic approximations based on maximal consistent blocks and lost values, the error rate does not depend on the choice of the parameter α. It is caused by the fact that for our data sets, with exception of two data sets: *breast cancer* and *iris*, all maximal consistent blocks are singletons (sets with just one element). For the *breast cancer* data set there is only one maximal consistent block different from a singleton (it contains two cases, 12 and 21, both belong to the same concept). On the other hand, for the *iris* data set there are only two maximal consistent blocks with more cases than one, the first is {6, 16, 45}, all three cases are members of the same concept, the second one is {125, 145}, and again, both cases are members of the same concept. Note, that this approach is quite successful, it is the best or one of the two best approaches when applied to four out of eight data sets used for experiments.

6 Conclusions

In our experiments we compared four approaches to mining incomplete data sets, combining two interpretations of missing attribute values, lost values and "do not care" conditions, with two types of probabilistic approximations, based on characteristic sets and maximal consistent blocks. As follows from our results, presented in Table 5, there are significant differences between all four approaches, depending on a data set. Therefore, for an incomplete data set, the best approach to mining should be chosen by trying all four approaches.

References

1. Clark, P.G., Gao, C., Grzymala-Busse, J.W., Mroczek, T.: Characteristic sets and generalized maximal consistent blocks in mining incomplete data. In: Polkowski, L., Yao, Y., Artiemjew, P., Ciucci, D., Liu, D., Ślęzak, D., Zielosko, B. (eds.) IJCRS 2017. LNCS (LNAI), vol. 10313, pp. 477–486. Springer, Cham (2017). https://doi.org/10.1007/978-3-319-60837-2_39
2. Clark, P.G., Grzymala-Busse, J.W.: Experiments on probabilistic approximations. In: Proceedings of the 2011 IEEE International Conference on Granular Computing, pp. 144–149 (2011)
3. Clark, P.G., Grzymala-Busse, J.W., Rzasa, W.: Mining incomplete data with singleton, subset and concept approximations. Inf. Sci. **280**, 368–384 (2014)
4. Grzymala-Busse, J.W.: MLEM2: a new algorithm for rule induction from imperfect data. In: Proceedings of the 9th International Conference on Information Processing and Management of Uncertainty in Knowledge-Based Systems, pp. 243–250 (2002)
5. Grzymala-Busse, J.W.: Rough set strategies to data with missing attribute values. In: Notes of the Workshop on Foundations and New Directions of Data Mining, in Conjunction with the Third International Conference on Data Mining, pp. 56–63 (2003)
6. Grzymala-Busse, J.W.: Three approaches to missing attribute values–a rough set perspective. In: Proceedings of the Workshop on Foundation of Data Mining, in Conjunction with the Fourth IEEE International Conference on Data Mining, pp. 55–62 (2004)
7. Grzymała-Busse, J.W.: Generalized parameterized approximations. In: Yao, J.T., Ramanna, S., Wang, G., Suraj, Z. (eds.) RSKT 2011. LNCS (LNAI), vol. 6954, pp. 136–145. Springer, Heidelberg (2011). https://doi.org/10.1007/978-3-642-24425-4_20
8. Grzymala-Busse, J.W., Ziarko, W.: Data mining based on rough sets. In: Wang, J. (ed.) Data Mining: Opportunities and Challenges, pp. 142–173. Idea Group Publishing, Hershey (2003)
9. Leung, Y., Li, D.: Maximal consistent block technique for rule acquisition in incomplete information systems. Inf. Sci. **153**, 85–106 (2003)
10. Leung, Y., Wu, W., Zhang, W.: Knowledge acquisition in incomplete information systems: a rough set approach. Eur. J. Oper. Res. **168**, 164–180 (2006)
11. Liang, J.Y., Wang, B.L., Qian, Y.H., Li, D.Y.: An algorithm of constructing maximal consistent blocks in incomplete information systems. Int. J. Comput. Sci. Knowl. Eng. **2**(1), 11–18 (2008)
12. Liu, X., Shao, M.: Approachs to computing maximal consistent block. In: Wang, X., Pedrycz, W., Chan, P., He, Q. (eds.) ICMLC 2014. CCIS, vol. 481, pp. 264–274. Springer, Heidelberg (2014). https://doi.org/10.1007/978-3-662-45652-1_27
13. Pawlak, Z., Skowron, A.: Rough sets: some extensions. Inf. Sci. **177**, 28–40 (2007)
14. Pawlak, Z., Wong, S.K.M., Ziarko, W.: Rough sets: probabilistic versus deterministic approach. Int. J. Man Mach. Stud. **29**, 81–95 (1988)
15. Ślęzak, D., Ziarko, W.: The investigation of the Bayesian rough set model. Int. J. Approx. Reason. **40**, 81–91 (2005)
16. Wong, S.K.M., Ziarko, W.: INFER–an adaptive decision support system based on the probabilistic approximate classification. In: Proceedings of the 6th International Workshop on Expert Systems and their Applications, pp. 713–726 (1986)

17. Yao, Y.Y.: Probabilistic rough set approximations. Int. J. Approx. Reason. **49**, 255–271 (2008)
18. Yao, Y.Y., Wong, S.K.M.: A decision theoretic framework for approximate concepts. Int. J. Man Mach. Stud. **37**, 793–809 (1992)
19. Ziarko, W.: Variable precision rough set model. J. Comput. Syst. Sci. **46**(1), 39–59 (1993)
20. Ziarko, W.: Probabilistic approach to rough sets. Int. J. Approx. Reason. **49**, 272–284 (2008)

Rules Induced from Rough Sets in Information Tables with Continuous Values

Michinori Nakata[1(✉)], Hiroshi Sakai[2], and Keitarou Hara[3]

[1] Faculty of Management and Information Science, Josai International University,
1 Gumyo, Togane, Chiba 283-8555, Japan
nakatam@ieee.org
[2] Department of Mathematics and Computer Aided Sciences, Faculty of Engineering,
Kyushu Institute of Technology, Tobata, Kitakyushu 804-8550, Japan
sakai@mns.kyutech.ac.jp
[3] Department of Informatics, Tokyo University of Information Sciences, 4-1 Onaridai,
Wakaba-ku, Chiba 265-8501, Japan
hara@rsch.tuis.ac.jp

Abstract. Rule induction based on neighborhood rough sets is described in information tables with continuous values. An indiscernible range that a value has in an attribute is determined by a threshold on that attribute. The indiscernibility relation is derived from using the indiscernible range. First, lower and upper approximations are described in complete information tables by directly using the indiscernibility relation. Rules are obtained from the approximations. To improve the applicability of rules, a series of rules is put into one rule expressed with an interval value, which is called a combined rule. Second, these are addressed in incomplete information tables. Incomplete information is expressed by a set of values or an interval value. The indiscernibility relations are constructed from two viewpoints: certainty and possibility. Consequently, we obtain four types of approximations: certain lower, certain upper, possible lower, and possible upper approximations. Using these approximations, rough sets are expressed by interval sets. From these approximations we obtain four types of combined rules: certain and consistent, certain and inconsistent, possible and consistent, and possible and inconsistent ones. These combined rules have greater applicability than single rules that individual objects support.

Keywords: Neighborhood rough sets · Rule induction
Incomplete information · Indiscernibility relation
Lower and upper approximations · Continuous values

1 Introduction

Rough sets, constructed by Pawlak [12], are used as an effective method for data mining. The framework is usually applied to information tables with nominal

© Springer International Publishing AG, part of Springer Nature 2018
J. Medina et al. (Eds.): IPMU 2018, CCIS 854, pp. 490–502, 2018.
https://doi.org/10.1007/978-3-319-91476-3_41

attributes and creates fruitful results in various fields. However, we are frequently faced with attributes taking continuous values, when we describe properties of an object in our daily life. Therefore, we describe rough sets in information tables with continuous values.

Ways how to deal with attributes taking continuous values are broadly classified into two approaches. One is to discretize a continuous domain by dividing it into a collection of disjunctive intervals. Objects included in an interval are regarded as indistinguishable. From this indistinguishability the indiscernibility relation is derived [1]. Results strongly depend on how discretization is made. Especially, objects that are located in the proximity of the boundary of intervals are strongly affected by discretization. This leads to that results abruptly change by a little alteration of discretization. The other is a way using neighborhood [7]. In this approach when the distance of an object to another one on an attribute is less than or equal to a given threshold, two objects are regarded as indistinguishable on the attribute. Results gradually change as the threshold changes. So, we use the latter approach.

Rules are induced from lower and upper approximations. Concretely speaking, when objects o and o' are included in the approximations, let single rules $a_i = 3.60 \rightarrow a_j = v$ and $a_i = 3.73 \rightarrow a_j = v$ be induced, where objects o and o' are characterized by values 3.60 and 3.73 of attribute a_i and the set approximated is specified by value v of attribute a_j. For example, value 3.66 of attribute a_i is not indiscernible with 3.60 and 3.73 under the threshold 0.05. Therefore, we cannot say anything from these rules for a rule supported by an object with value 3.66 of attribute a_i. This means that the rules are short of applicability. To improve such applicability, we consider a combined rule that is derived from a series of single rules supported by individual objects.

In addition, we are frequently confronted with incomplete information in daily life. We cannot sufficiently utilize information obtained from our daily life unless we deal with incomplete information. We express incomplete information in a partial value or an interval value. The interval value contains a missing value that means unknown as a special case. Most of authors fix the indiscernibility of an object with incomplete information with another object [3,14–16], as was done by Kryszkiewicz [4]. However, object o characterized by a value with incomplete information has two possibilities. One possibility is that the object o may have the same value as another one o'; namely, the two objects may be indiscernible. The other possibility is that o may have a different value from o'; namely, the two objects may be discernible. To fix the indiscernibility is to take into account only one of the two possibilities. Therefore, this treatment creates poor results and induces information loss [9,13]. We do not fix the indiscernibility of objects with incomplete information and simultaneously deal with both possibilities. This can be realized by dealing with objects having incomplete information from viewpoints of certainty and possibility [10], as was done by Lipski in the field of incomplete databases [5,6].

We have three approaches from the viewpoints of certainty and possibility. One is based on possible world semantics. This way creates possible tables.

Unfortunately, infinite possible tables can be derived from an information table with continuous values. Another way uses possible classes obtained from the indiscernibility relation [8]. The number of possible classes grows exponentially, as the number of values with incomplete information increases, although this difficulty can be avoided by using minimum and maximum possible classes in the case of nominal attributes [10]. The other directly uses the indiscernibility relation [11]. The way has no problem of computational complexity. Therefore, we use this way.

The paper is organized as follows. In Sect. 2, an approach directly using indiscernibility relations is addressed in complete information tables. In Sect. 3, we develop the approach in incomplete information tables. This is described from two viewpoints of certainty and possibility. In Sect. 4, conclusions are addressed.

2 Rough Sets by Indiscernibility Relations in Complete Information Systems with Continuous Values

A data set is represented as a two-dimensional table, called an information table. In the information table, each row and each column represent an object and an attribute, respectively. A mathematical model of an information table with complete information is called a complete information system. The complete information system is a triplet expressed by $(U, AT, \{D(a_i) \mid a_i \in AT\})$. U is a non-empty finite set of objects, which is called the universe. AT is a non-empty finite set of attributes such that $a_i : U \to D(a_i)$ for every $a_i \in AT$ where $D(a_i)$ is the domain of attribute a_i. Binary relation R_{a_i} expressing indiscernibility of objects on attribute $a_i \in AT$ is called the indiscernibility relation for a_i:

$$R_{a_i} = \{(o, o') \in U \times U \mid |a_i(o) - a_i(o')| \leq \delta_{a_i}\}, \tag{1}$$

where $a_i(o)$ is the value for attribute a_i of object o and δ_{a_i} is a threshold that denotes a range in which $a_i(o)$ is indiscernible with $a_i(o')$. From the indiscernibility relation, indiscernible class $[o]_{a_i}$ for object o is obtained:

$$[o]_{a_i} = \{o' \mid (o, o') \in R_{a_i}\}. \tag{2}$$

Directly using indiscernibility relation R_{a_i}, lower approximation $\underline{apr}_{a_i}(\mathcal{O})$ and upper approximation $\overline{apr}_{a_i}(\mathcal{O})$ for a_i of set $\mathcal{O}$ of objects are:

$$\underline{apr}_{a_i}(\mathcal{O}) = \{o \mid \forall o' \in U \ (o, o') \notin R_{a_i} \vee o' \in \mathcal{O}\}, \tag{3}$$

$$\overline{apr}_{a_i}(\mathcal{O}) = \{o \mid \exists o' \in U \ (o, o') \in R_{a_i} \wedge o' \in \mathcal{O}\}. \tag{4}$$

Proposition 1. If $\delta 1_{a_i} \leq \delta 2_{a_i}$, then $\underline{apr}_{a_i}^{\delta 1}(\mathcal{O}) \supseteq \underline{apr}_{a_i}^{\delta 2}(\mathcal{O})$ and $\overline{apr}_{a_i}^{\delta 1}(\mathcal{O}) \subseteq \overline{apr}_{a_i}^{\delta 2}(\mathcal{O})$, where $\underline{apr}_{a_i}^{\delta 1}(\mathcal{O})$ and $\overline{apr}_{a_i}^{\delta 1}(\mathcal{O})$ are lower and upper approximations under threshold $\delta 1_{a_i}$ and $\underline{apr}_{a_i}^{\delta 2}(\mathcal{O})$ and $\overline{apr}_{a_i}^{\delta 2}(\mathcal{O})$ are lower and upper approximations under threshold $\delta 2_{a_i}$.

For object o in the lower approximation of $\mathcal{O}$, all objects with which o is indiscernible are included in $\mathcal{O}$; namely, $[o]_{a_i} \subseteq \mathcal{O}$. On the other hand, for an object in the upper approximation of $\mathcal{O}$, some objects with which o is indiscernible are in $\mathcal{O}$; namely, $[o]_{a_i} \cap \mathcal{O} \neq \emptyset$. Thus, $\underline{apr}_{a_i}(\mathcal{O}) \subseteq \overline{apr}_{a_i}(\mathcal{O})$.

Rules are induced from lower and upper approximations. When $\mathcal{O}$ is characterized by value x of attribute a_j, objects $o \in \underline{apr}_{a_i}(\mathcal{O})$ and $o \in \overline{apr}_{a_i}(\mathcal{O})$ consistently and inconsistently support a single rule $a_i = a_i(o) \rightarrow a_j = x$, respectively. The degree of consistency, called accuracy, is $|[o]_{a_i} \cap \mathcal{O}|/|\mathcal{O}|$.

Since attribute a_i has a continuous domain, the antecedent part of single rules that individual objects support is usually different. We obtain lots of single rules, but the single rule is short of applicability. For example, let two values $a_i(o)$ and $a_i(o')$ be 3.65 and 3.75 for objects o and o' in $\underline{apr}_{a_i}(\mathcal{O})$. When $\mathcal{O}$ is characterized by value x of attribute a_j, o and o' support single rules $a_i = 3.65 \rightarrow a_j = x$ and $a_i = 3.75 \rightarrow a_j = x$, respectively. By using these rules, we can say that a object having value 3.68 of a_i, indiscernible with 3.65 under $\delta_{a_i} = 0.03$, supports $a_i = 3.68 \rightarrow a_j = x$. However, we cannot at all say for a rule supported by an object with value 3.70 discernible with 3.65 and 3.75. This shows that a single rule is short of applicability.

To improve the applicability of rules, we combine multiple single rules into one rule, which is called a combined rule. Let objects in U be aligned in ascending order of $a_i(o)$ and be attached the serial superscript with 1 to N_U where $|U| = N_U$. $\underline{apr}_{a_i}(\mathcal{O})$ and $\overline{apr}_{a_i}(\mathcal{O})$ consist of collections of objects with serial superscripts. For example, $\underline{apr}_{a_i}(\mathcal{O}) = \{\cdots, o^h, o^{h+1}, \cdots, o^{k-1}, o^k, \cdots\}$ $(h \leq k)$. Let o^l in $\underline{apr}_{a_i}(\mathcal{O})$ support a single rule $a_i = a_i(o^l) \rightarrow a_j = x$. Then, single rules derived from collection $(o^h, o^{h+1}, \cdots, o^{k-1}, o^k)$ can be put into one combined rule $a_i = [a_i(o^h), a_i(o^k)] \rightarrow a_j = x$. Next, when a_j is a numerical attribute, $\mathcal{O}$ is characterized by an interval value. The interval value has the lower bound and the upper bound that are existing values of attribute. Let, the objects be aligned in ascending order of values of a_j and be attached the serial superscript with 1 to N_U. For example, using the ordered objects, $\mathcal{O}$ is specified like $\mathcal{O} = \{o \mid a_j(o) \geq a_j(o^m) \wedge a_j(o) \leq a_j(o^n)\}$ with $m \leq n$. In the case, the combined rlue, derived from collection $(o^h, o^{h+1}, \cdots, o^{k-1}, o^k)$, is expressed with $a_i = [a_i(o^h), a_i(o^k)] \rightarrow a_j = [a_i(o^m), a_i(o^n)]$. The accuracy of the combined rule is $\min_{h \leq s \leq k} |[o^s]_{a_i} \cap \mathcal{O}|/|\mathcal{O}|$.

Proposition 2. Let $\underline{r}$ and $\overline{r}$ be sets of combined rules obtained from $\underline{apr}_{a_i}(\mathcal{O})$ and $\overline{apr}_{a_i}(\mathcal{O})$, respectively. If $(a_i = [l, u] \rightarrow \overline{w}) \in \underline{r}$, then $\exists l' \leq l, \exists u' \geq u$ $(a_i = [l', u'] \rightarrow \overline{w}) \in \overline{r}$, where $\mathcal{O}$ is characterized by $\overline{w}$.

Example 1. Information tables are depicted in Fig. 1. T0 is the original information table. U is $\{o_1, o_2, \cdots, o_{18}, o_{19}\}$. T1, T2, and T3 are derived from T0, where objects are aligned in ascending order of values of attributes a_1, a_2, and a_3, respectively.

U	a_1	a_2	a_3	a_4	U	a_1	a_2	a_3	a_4	U	a_1	a_2	a_3	a_4	U	a_1	a_2	a_3	a_4
	T0					**T1**					**T2**					**T3**			
1	3.11	2.98	3.02	b	3	2.33	3.69	3.28	f	5	3.42	2.35	2.67	b	5	3.42	2.35	2.67	b
2	2.94	3.65	3.44	b	18	2.45	3.96	3.51	f	14	3.12	2.78	2.88	b	14	3.12	2.78	2.88	b
3	2.33	3.69	3.28	f	12	2.63	4.81	4.16	c	7	2.81	2.95	2.91	c	7	2.81	2.95	2.91	c
4	4.78	2.98	3.52	a	7	2.81	2.95	2.91	c	1	3.11	2.98	3.02	b	1	3.11	2.98	3.02	b
5	3.42	2.35	2.67	b	17	2.89	3.51	3.32	c	4	4.78	2.98	3.52	a	15	3.05	3.29	3.22	c
6	3.03	4.52	4.07	c	2	2.94	3.65	3.44	b	8	4.36	3.11	3.49	a	3	2.33	3.69	3.28	f
7	2.81	2.95	2.91	c	16	2.95	3.65	3.44	b	15	3.05	3.29	3.22	c	17	2.89	3.51	3.32	c
8	4.36	3.11	3.49	a	11	2.97	3.98	3.68	b	19	3.86	3.44	3.57	b	2	2.94	3.65	3.44	b
9	3.22	4.63	4.21	b	6	3.03	4.52	4.07	c	17	2.89	3.51	3.32	c	16	2.95	3.65	3.44	b
10	3.07	3.78	3.57	c	15	3.05	3.29	3.22	c	2	2.94	3.65	3.44	b	8	4.36	3.11	3.49	a
11	2.97	3.98	3.68	b	10	3.07	3.78	3.57	c	16	2.95	3.65	3.44	b	18	2.45	3.96	3.51	f
12	2.63	4.81	4.16	c	1	3.11	2.98	3.02	b	3	2.33	3.69	3.28	f	4	4.78	2.98	3.52	a
13	3.91	3.71	3.77	a	14	3.12	2.78	2.88	b	13	3.91	3.71	3.77	a	19	3.86	3.44	3.57	b
14	3.12	2.78	2.88	b	9	3.22	4.63	4.21	b	10	3.07	3.78	3.57	c	10	3.07	3.78	3.57	c
15	3.05	3.29	3.22	c	5	3.42	2.35	2.67	b	18	2.45	3.96	3.51	f	11	2.97	3.98	3.68	b
16	2.95	3.65	3.44	b	19	3.86	3.44	3.57	b	11	2.97	3.98	3.68	b	13	3.91	3.71	3.77	a
17	2.89	3.51	3.32	c	13	3.91	3.71	3.77	a	6	3.03	4.52	4.07	c	6	3.03	4.52	4.07	c
18	2.45	3.96	3.51	f	8	4.36	3.11	3.49	a	9	3.22	4.63	4.21	b	12	2.63	4.81	4.16	c
19	3.86	3.44	3.57	b	4	4.78	2.98	3.52	a	12	2.63	4.81	4.16	c	9	3.22	4.63	4.21	b

Fig. 1. T0 is the original information table. T1, T2, T3 are derived from T0 by aligning objects in ascending order of values of attributes a_1, a_2, and a_3, respectively.

Let threshold δ_{a_1} be 0.05. Indiscernibility relation R_{a_1} is:

$$R_{a_1} = \{(o_1, o_1), (o_1, o_{10}), (o_1, o_{14}), (o_2, o_2), (o_2, o_{11}), (o_2, o_{16}), (o_2, o_{17}), (o_3, o_3),$$
$$(o_4, o_4), (o_5, o_5), (o_6, o_6), (o_6, o_{10}), (o_6, o_{15}), (o_7, o_7), (o_8, o_8), (o_9, o_9),$$
$$(o_{10}, o_1), (o_{10}, o_6), (o_{10}, o_{10}), (o_{10}, o_{14}), (o_{10}, o_{15}), (o_{11}, o_2), (o_{11}, o_{11}),$$
$$(o_{11}, o_{16}), (o_{12}, o_{12}), (o_{13}, o_{13}), (o_{13}, o_{19}), (o_{14}, o_1), (o_{14}, o_{10}), (o_{14}, o_{14}),$$
$$(o_{15}, o_6), (o_{15}, o_{10}), (o_{15}, o_{15}), (o_{16}, o_2), (o_{16}, o_{11}), (o_{16}, o_{16}), (o_{17}, o_2),$$
$$(o_{17}, o_{17}), (o_{18}, o_{18}), (o_{19}, o_{13}), (o_{19}, o_{19})\}$$

When $\mathcal{O}$ is characterized by value b of attribute a_4, $\mathcal{O} = \{o_1, o_2, o_5, o_9, o_{11}, o_{14}, o_{16}, o_{19}\}$. Let $\mathcal{O}$ be approximated by objects that are characterized by attributes a_1 with continuous values. Using formulas (3) and (4), lower and upper approximations are:

$$\underline{apr}_{a_1}(\mathcal{O}) = \{o_5, o_9, o_{11}, o_{16}\},$$
$$\overline{apr}_{a_1}(\mathcal{O}) = \{o_1, o_2, o_5, o_9, o_{10}, o_{11}, o_{13}, o_{14}, o_{16}, o_{17}, o_{19}\}.$$

Information table T1 where objects are aligned in ascending order of values of attribute a_1 is derived from information table T0. The above approximations are described as follows:

$$\underline{apr}_{a_1}(\mathcal{O}) = \{o^7, o^8, o^{14}, o^{15}\},$$
$$\overline{apr}_{a_1}(\mathcal{O}) = \{o^5, o^6, o^7, o^8, o^{11}, o^{12}, o^{13}, o^{14}, o^{15}, o^{16}, o^{17}\},$$

where

$$o^5 = o_{17}, \ o^6 = o_2, \ o^7 = o_{16}, \ o^8 = o_{11}, \ o^{11} = o_{10}, \ o^{12} = o_1,$$
$$o^{13} = o_{14}, \ o^{14} = o_9, \ o^{15} = o_5, \ o^{16} = o_{19}, \ o^{17} = o_{13}.$$

From the lower approximation, consistent combined rules are

$$a_1 = [2.95, 2.97] \rightarrow a_4 = b, \ a_1 = [3.22, 3.42] \rightarrow a_4 = b,$$

from collections $\{o^7, o^8\}$ and $\{o^{14}, o^{15}\}$, respectively, where $a_1(o^7) = 2.95$, $a_1(o^8) = 2.97$, $a_1(o^{14}) = 3.22$, and $a_1(o^{15}) = 3.42$. From the upper approximation, inconsistent combined rules are

$$a_1 = [2.89, 2.97] \rightarrow a_4 = b, \ a_1 = [3.07, 3.91] \rightarrow a_4 = b,$$

from collections $\{o^5, o^6, o^7, o^8\}$ and $\{o^{11}, o^{12}, o^{13}, o^{14}, o^{15}, o^{16}, o^{17}\}$, respectively, where $a_1(o^5) = 2.89$, $a_1(o^{11}) = 3.07$, and $a_1(o^{17}) = 3.91$.

Next, we consider the case where $\mathcal{O}$ is characterized by a_3 with a continuous domain. Information table T3 where the objects are aligned in ascending order of values of a_3 is derived from T0. Using lower bound $a_3(o^5) = a_3(o_{15}) = 3.22$ and upper bound $a_3(o^{10}) = a_3(o_8) = 3.49$, $\mathcal{O} = \{o^5, o^6, o^7, o^8, o^9, o^{10}\} = \{o_2, o_3, o_8, o_{15}, o_{16}, o_{17}\}$. We approximate $\mathcal{O}$ by attribute a_2. Information table T2 where the objects are aligned in ascending order of values of a_2 is derived from T0. Let δ_{a_2} be 0.05. Indiscernibility relation R_{a_2} is:

$$R_{a_2} = \{(o_1, o_1), (o_1, o_4), (o_1, o_7), (o_1, o_8), (o_2, o_2), (o_2, o_3), (o_2, o_{16}), (o_3, o_2),$$
$$(o_3, o_3), (o_3, o_{13}), (o_3, o_{16}), (o_4, o_1), (o_4, o_4), (o_4, o_7), (o_4, o_8), (o_5, o_5),$$
$$(o_6, o_6), (o_7, o_1), (o_7, o_4), (o_7, o_7), (o_8, o_8), (o_9, o_9), (o_{10}, o_{10}), (o_{11}, o_{11}),$$
$$(o_{11}, o_{18}), (o_{12}, o_{12}), (o_{13}, o_3), (o_{13}, o_{13}), (o_{14}, o_{14}), (o_{15}, o_{15}), (o_{16}, o_2),$$
$$(o_{16}, o_3), (o_{16}, o_{16}), (o_{17}, o_{17}), (o_{18}, o_{11}), (o_{18}, o_{18}), (o_{19}, o_{19})\}.$$

Using formulas (3) and (4), lower and upper approximations are:

$$\underline{apr}_{a_2}(\mathcal{O}) = \{o_2, o_8, o_{15}, o_{16}, o_{17}\}, \ \overline{apr}_{a_2}(\mathcal{O}) = \{o_1, o_2, o_3, o_4, o_8, o_{13}, o_{15}, o_{16}, o_{17}\}.$$

Using information table T2 where objects are aligned in ascending order of values of attribute a_1, the above approximations are described as follows:

$$\underline{apr}_{a_2}(\mathcal{O}) = \{o^6, o^7, o^9, o^{10}, o^{11}\}, \ \overline{apr}_{a_2}(\mathcal{O}) = \{o^4, o^5, o^6, o^7, o^9, o^{10}, o^{11}, o^{12}, o^{13}\},$$

From the lower approximation, consistent combined rules are

$$a_2 = [3.11, 3.29] \rightarrow a_3 = [3.22, 3.49], \ a_2 = [3.51, 3.65] \rightarrow a_3 = [3.22, 3.49],$$

where $a_2(o^6) = 3.11$, $a_2(o^7) = 3.29$, $a_2(o^9) = 3.51$, and $a_2(o^{11}) = 3.65$. From the upper approximation, inconsistent combined rules are

$$a_2 = [2.98, 3.29] \rightarrow a_3 = [3, 22, 3.49], a_2 = [3.51, 3.71] \rightarrow a_3 = [3.22, 3.49],$$

where $a_2(o^4) = 2.98$ and $a_2(o^{13}) = 3.71$.

For formulas on sets A and B of attributes,

$$R_A = \cap_{a_i \in A} R_{a_i}, \tag{5}$$

$$[o]_A = \{o' \mid (o, o') \in R_A\} = \cap_{a_i \in A} [o]_{a_i}, \tag{6}$$

$$\underline{apr}_A(\mathcal{O}) = \{o \mid \forall o' \in U \ (o, o') \notin R_A \lor o' \in \mathcal{O}\}, \tag{7}$$

$$\overline{apr}_A(\mathcal{O}) = \{o \mid \exists o' \in U \ (o, o') \in R_A \land o' \in \mathcal{O}\}. \tag{8}$$

3 Rough Sets by Indiscernibility Relations in Incomplete Information Systems with Continuous Domains

An information table with incomplete information is called an incomplete information system. In incomplete information systems, $a_i : U \rightarrow s_{a_i}$ for every $a_i \in AT$ where s_{a_i} is a set of values over domain $D(a_i)$ of attribute a_i or an interval on $D(a_i)$. Single value v with $v \in a_i(o)$ or $v \subseteq a_i(o)$ is a possible value that may be the actual one as the value of attribute a_i in object o. The possible value is the actual one if $a_i(o)$ is a single value.

The indiscernibility relation for a_i in an incomplete information system is expressed by using two relations CR_{a_i} and PR_{a_i} from viewpoints of certainty and possibility. CR_{a_i} is a certain indiscernibility relation. When objects (o, o') is an element of CR_{a_i}, o is surely indiscernible with o'. PR_{a_i} is a possible indiscernibility relation. When objects (o, o') is an element of PR_{a_i}, o may be indiscernible with o'.

$$CR_{a_i} = \{(o, o') \mid o = o' \lor (\forall u \in a_i(o) \forall v \in a_i(o') |u - v| \leq \delta_{a_i})\}, \tag{9}$$

$$PR_{a_i} = \{(o, o') \mid o = o' \lor (\exists u \in a_i(o) \exists v \in a_i(o') |u - v| \leq \delta_{a_i})\}. \tag{10}$$

The certain and possible indiscernibility relations are reflexive, symmetric, but are not transitive. We have three patterns. One case is that a pair of objects are not in both certain and possible indiscernibility relations, which means that they are discernible. Another is that they are not in the certain indiscernibility relation, but in the possible one, which means that they are discernible and indiscernible. The other is that they are in both certain and possible indiscernibility relations, which means that they are indiscernible.

We can derive not the actual, but certain and possible approximations from the viewpoint of certainty and possibility, as Lipski obtained in query processing under incomplete information [5,6]. We cannot definitely obtain whether or not an object belongs to approximations, but we can know whether or not the object certainly and/or possibly belongs to approximations. Therefore, we show certain approximations (resp. possible approximations) whose object certainly (resp. possibly) belongs to the actual approximations.

Let $\mathcal{O}$ be a set of objects. According to [11], certain lower approximation $\underline{Capr}_{a_i}(\mathcal{O})$ and possible one $\underline{Papr}_{a_i}(\mathcal{O})$ for a_i are:

$$\underline{Capr}_{a_i}(\mathcal{O}) = \{o \mid \forall o' \in U \ (o,o') \notin PR_{a_i} \vee o' \in \mathcal{O}\}, \tag{11}$$

$$\underline{Papr}_{a_i}(\mathcal{O}) = \{o \mid \forall o' \in U \ (o,o') \notin CR_{a_i} \vee o' \in \mathcal{O}\}. \tag{12}$$

Similarly, Certain upper approximation $\overline{Capr}_{a_i}(\mathcal{O})$ and possible one $\overline{Papr}_{a_i}(\mathcal{O})$ are:

$$\overline{Capr}_{a_i}(\mathcal{O}) = \{o \mid \exists o' \in U \ (o,o') \in CR_{a_i} \wedge o' \in \mathcal{O}\}, \tag{13}$$

$$\overline{Papr}_{a_i}(\mathcal{O}) = \{o \mid \exists o' \in U \ (o,o') \in PR_{a_i} \wedge o' \in \mathcal{O}\}. \tag{14}$$

As with the case of nominal attributes [11], the following proposition holds.

Proposition 3. $\underline{Capr}_{a_i}(\mathcal{O}) \subseteq \underline{Papr}_{a_i}(\mathcal{O}) \subseteq \mathcal{O} \subseteq \overline{Capr}_{a_i}(\mathcal{O}) \subseteq \overline{Papr}_{a_i}(\mathcal{O}).$

Using four approximations denoted by formulae (11)–(14), lower and upper approximations are expressed by interval sets, as is described in [11][1], as follows:

$$\underline{apr}^{\bullet}_{a_i}(\mathcal{O}) = [\underline{Capr}_{a_i}(\mathcal{O}), \underline{Papr}_{a_i}(\mathcal{O})], \tag{15}$$

$$\overline{apr}^{\bullet}_{a_i}(\mathcal{O}) = [\overline{Capr}_{a_i}(\mathcal{O}), \overline{Papr}_{a_i}(\mathcal{O})]. \tag{16}$$

Certain and possible approximations are the lower and upper bounds of the actual approximation.

When objects in $\mathcal{O}$ are characterized by attribute a_j with incomplete information, $\mathcal{O}$ is specified by using an element in domain $D(a_j)$. In the case where $\mathcal{O}$ is specified by value x in $D(a_j)$ of nominal attribute a_j with incomplete information, four approximations: certain lower, possible lower, certain upper, and possible upper ones, are:

$$\underline{Capr}_{a_i}(\mathcal{O}) = \{o \mid \forall o' \in U \ (o,o') \notin PR_{a_i} \vee o' \in CO_{a_j=x}\}, \tag{17}$$

$$\underline{Papr}_{a_i}(\mathcal{O}) = \{o \mid \forall o' \in U \ (o,o') \notin CR_{a_i} \vee o' \in PO_{a_j=x}\}, \tag{18}$$

$$\overline{Capr}_{a_i}(\mathcal{O}) = \{o \mid \exists o' \in U \ (o,o') \in CR_{a_i} \wedge o' \in CO_{a_j=x}\}, \tag{19}$$

$$\overline{Papr}_{a_i}(\mathcal{O}) = \{o \mid \exists o' \in U \ (o,o') \in PR_{a_i} \wedge o' \in PO_{a_j=x}\}, \tag{20}$$

where

$$CO_{a_j=x} = \{o \in \mathcal{O} \mid a_j(o) = x\}, \tag{21}$$

$$PO_{a_j=x} = \{o \in \mathcal{O} \mid a_j(o) \supseteq x\}. \tag{22}$$

Now, $\mathcal{O}$ is characterized by value x of a_j. For rule induction, we can say as follows:

---

[1] Hu and Yao also say that approximations describes by an interval set in information tables with incomplete information [2].

- $o \in \underline{Capr}_{a_i}(\mathcal{O})$ certainly and consistently supports rule $a_i = a_i(o) \to a_j(o) = x$.
- $o \in \overline{Capr}_{a_i}(\mathcal{O})$ certainly and inconsistently supports rule $a_i = a_i(o) \to a_j(o) = x$.
- $o \in \underline{Papr}_{a_i}(\mathcal{O})$ possibly and consistently supports $a_i = a_i(o) \to a_j(o) = x$.
- $o \in \overline{Papr}_{a_i}(\mathcal{O})$ possibly and inconsistently supports $a_i = a_i(o) \to a_j(o) = x$.

We create combined rules from them.

Let $U_{a_i}^C$ and $U_{a_i}^I$ be sets of objects having complete information and incomplete information for a_i. $o \in U_{a_i}^C$ is aligned in ascending order of $a_i(o)$ and is attached the serial superscript with 1 to N_i^C where $|U_{a_i}^C| = N_i^C$. Objects $o \in (\underline{Capr}_{a_i}(\mathcal{O}) \cap U_{a_i}^C)$, $o \in (\overline{Capr}_{a_i}(\mathcal{O}) \cap U_{a_i}^C)$, $o \in (\underline{Papr}_{a_i}(\mathcal{O}) \cap U_{a_i}^C)$, and $o \in (\overline{Papr}_{a_i}(\mathcal{O}) \cap U_{a_i}^C)$ are aligned in ascending order of $a_i(o)$. And then they are expressed by a sequence of collections of objects with a serial superscript like $\{\cdots, o^h, o^{h+1}, \cdots, o^{k-1}, o^k, \cdots\}$ $(h \leq k)$. From collection $(o^h, o^{h+1}, \cdots, o^{k-1}, o^k)$, four types of combined rules expressed with $a_i = [l, u] \to a_j = x$ are derived. For a certain and consistent combined rule,

$$l = \min(a_i(o^h), \min_{CL} e) \text{ and } u = \max(a_i(o^k), \max_{CL} e),$$

$$CL = \begin{cases} e < a_i(o^{k+1}), & \text{for } h = 1 \wedge k \neq N_i^C \\ a_i(o^{h-1}) < e < a_i(o^{k+1}), & \text{for } h \neq 1 \wedge k \neq N_i^C \\ a_i(o^{h-1}) < e, & \text{for } h \neq 1 \wedge k = N_i^C \end{cases}$$

$$\text{with } e \in a_i(o') \wedge o' \in (\underline{Capr}_{a_i}(\mathcal{O}) \cap U_{a_i}^I). \quad (23)$$

For a certain and inconsistent combined rule,

$$l = \min(a_i(o^h), \min_{CU} e) \text{ and } u = \max(a_i(o^k), \max_{CU} e),$$

$$CU = \begin{cases} e < a_i(o^{k+1}), & \text{for } h = 1 \wedge k \neq N_i^C \\ a_i(o^{h-1}) < e < a_i(o^{k+1}), & \text{for } h \neq 1 \wedge k \neq N_i^C \\ a_i(o^{h-1}) < e, & \text{for } h \neq 1 \wedge k = N_i^C \end{cases}$$

$$\text{with } e \in a_i(o') \wedge o' \in (\overline{Capr}_{a_i}(\mathcal{O}) \cap U_{a_i}^I). \quad (24)$$

For a possible and consistent combined rule,

$$l = \min(a_i(o^h), \min_{PL} e) \text{ and } u = \max(a_i(o^k), \max_{PL} e),$$

$$PL = \begin{cases} e < a_i(o^{k+1}), & \text{for } h = 1 \wedge k \neq N_i^C \\ a_i(o^{h-1}) < e < a_i(o^{k+1}), & \text{for } h \neq 1 \wedge k \neq N_i^C \\ a_i(o^{h-1}) < e, & \text{for } h \neq 1 \wedge k = N_i^C \end{cases}$$

$$\text{with } e \in a_i(o') \wedge o' \in (\underline{Papr}_{a_i}(\mathcal{O}) \cap U_{a_i}^I). \quad (25)$$

For a possible and inconsistent combined rule,

$$l = \min(a_i(o^h), \min_{PU} e) \text{ and } u = \max(a_i(o^k), \max_{PU} e),$$

$$PU = \begin{cases} e < a_i(o^{k+1}), & \text{for } h = 1 \wedge k \neq N_i^C \\ a_i(o^{h-1}) < e < a_i(o^{k+1}), & \text{for } h \neq 1 \wedge k \neq N_i^C \\ a_i(o^{h-1}) < e, & \text{for } h \neq 1 \wedge k = N_i^C \end{cases}$$

$$\text{with } e \in a_i(o') \wedge o' \in (\overline{Papr}_{a_i}(\mathcal{O}) \cap U_{a_i}^I). \quad (26)$$

Proposition 4. Let $C\underline{r}$ and $P\underline{r}$ be sets of combined rules obtained from $\underline{Capr}_{a_i}(\mathcal{O})$ and $\underline{Papr}_{a_i}(\mathcal{O})$, respectively. In the case where $\mathcal{O}$ is characterized by value x of attribute a_j, if $(a_i = [l, u] \rightarrow a_j = x) \in C\underline{r}$, then $\exists l' \leq l, \exists u' \geq u \ (a_i = [l', u'] \rightarrow a_j = x) \in P\underline{r}$.

Proposition 5. Let $C\overline{r}$ and $P\overline{r}$ be sets of combined rules obtained from $\overline{Capr}_{a_i}(\mathcal{O})$ and $\overline{Papr}_{a_i}(\mathcal{O})$, respectively. In the case where $\mathcal{O}$ is characterized by value x of attribute a_j, if $(a_i = [l, u] \rightarrow a_j = x) \in C\overline{r}$, then $\exists l' \leq l, \exists u' \geq u \ (a_i = [l', u'] \rightarrow a_j = x) \in P\overline{r}$.

Proposition 6. Let $C\underline{r}$ and $C\overline{r}$ be sets of combined rules obtained from $\underline{Capr}_{a_i}(\mathcal{O})$ and $\overline{Capr}_{a_i}(\mathcal{O})$, respectively. In the case where $\mathcal{O}$ is characterized by value x of attribute a_j, if $(a_i = [l, u] \rightarrow a_j = x) \in C\underline{r}$, then $\exists l' \leq l, \exists u' \geq u \ (a_i = [l', u'] \rightarrow a_j = x) \in C\overline{r}$.

Proposition 7. Let $P\underline{r}$ and $P\overline{r}$ be sets of combined rules obtained from $\underline{Papr}_{a_i}(\mathcal{O})$ and $\overline{Papr}_{a_i}(\mathcal{O})$, respectively. In the case where $\mathcal{O}$ is characterized by value x of attribute a_j, if $(a_i = [l, u] \rightarrow a_j = x) \in P\underline{r}$, then $\exists l' \leq l, \exists u' \geq u \ (a_i = [l', u'] \rightarrow a_j = x) \in P\overline{r}$.

The four types of combined rules are obtained from the following incomplete information table IT in Fig. 2. In the information table $\mathcal{O}$ is characterized by nominal attribute a_4 with incomplete information and $\mathcal{O}$ is approximated by numerical attribute a_1 with incomplete information.

Last, we describe the case where $o \in \mathcal{O}$ is characterized by numerical attribute a_j with incomplete information. $o \in U_{a_j}^C$ is aligned in ascending order of $a_j(o)$ and is attached with the serial superscript with 1 to N_j^C where $|U_{a_j}^C| = N_j^C$. We specify $\mathcal{O}$ by $a_j(o^m) \in U_{a_j}^C$ and $a_j(o^n) \in U_{a_j}^C$ with $m \leq n$.

$$\underline{Capr}_{a_i}(\mathcal{O}) = \{o \mid \forall o' \in U \ (o, o') \notin PR_{a_i} \vee o' \in CO_{[a_j(o^m), a_j(o^n)]}\}, \quad (27)$$

$$\underline{Papr}_{a_i}(\mathcal{O}) = \{o \mid \forall o' \in U \ (o, o') \notin CR_{a_i} \vee o' \in PO_{[a_j(o^m), a_j(o^n)]}\}, \quad (28)$$

$$\overline{Capr}_{a_i}(\mathcal{O}) = \{o \mid \exists o' \in U \ (o, o') \in CR_{a_i} \wedge o' \in CO_{[a_j(o^m), a_j(o^n)]}\}, \quad (29)$$

$$\overline{Papr}_{a_i}(\mathcal{O}) = \{o \mid \exists o' \in U \ (o, o') \in PR_{a_i} \wedge o' \in PO_{[a_j(o^m), a_j(o^n)]}\}, \quad (30)$$

where

$$CO_{[a_j(o^m), a_j(o^n)]} = \{o \in \mathcal{O} \mid a_j(o) \subseteq [a_j(o^m), a_j(o^n)]\}, \quad (31)$$

$$PO_{[a_j(o^m), a_j(o^n)]} = \{o \in \mathcal{O} \mid a_j(o) \cap [a_j(o^m), a_j(o^n)] \neq \emptyset\}. \quad (32)$$

IT

U	a_1	a_2	a_3	a_4
1	$\{3.06, 3.11\}$	2.98	$[3.02, 3.17]$	$\{b, c\}$
2	2.94	$\{3.64, 3.65\}$	3.44	b
3	2.33	$[3.69, 3.72]$	3.28	f
4	4.78	$[2.98, 3.12]$	3.52	a
5	3.42	2.35	2.67	b
6	3.03	4.52	4.07	c
7	2.81	2.95	2.91	c
8	4.36	3.11	3.49	a
9	$\{2.97, 3.22\}$	4.63	4.21	b
10	3.07	3.78	3.57	c
11	$[2.96, 2.97]$	3.98	3.68	b
12	2.63	4.81	4.16	c
13	3.91	3.71	3.77	a
14	3.12	2.78	2.88	b
15	3.05	3.29	3.22	c
16	2.95	$\{3.35, 3.65\}$	3.44	b
17	$[2.89, 2.92]$	3.51	$[3.32, 3.40]$	$\{b, c\}$
18	$[2.45, 2.55]$	3.96	$\{3.49, 3.51\}$	f
19	$[3.86, 3.92]$	3.44	3.57	$\{a, b\}$

Fig. 2. Information table with incomplete information.

$o \in U_{a_j}^C$ is aligned in ascending order of $a_j(o)$ and is attached the serial super-script with 1 to N_j^C. Now, $\mathcal{O}$ is specified by attribute values $a_j(o^m)$ and $a_j(o^n)$ with $o^m \in U_{a_j}^C$ and $o^n \in U_{a_j}^C$. $o \in U_{a_i}^C$ is aligned in ascending order of $a_i(o)$ is attached the serial superscript with 1 to N_i^C. Also, four types of combined rules with $a_i = [l, u] \rightarrow a_j = [a_j(o^m), a_j(o^n)]$ are obtained: certain and consistent, certain and inconsistent, possible and consistent, and possible and inconsistent combined rules.

These types of combined rules are obtained in incomplete information table IT in Fig. 2. In the information table $\mathcal{O}$ is characterized by numerical attribute a_3 with incomplete information and $\mathcal{O}$ is approximated on numerical attribute a_2 with incomplete information.

4   Conclusions

We have described rough sets and rule induction from them in information tables with continuous values. First, we have dealt with complete information tables. Rough sets are obtained from directly using indiscernibility relations. Individual objects that belongs to the rough sets support single rules. The single rules are short of applicability. To improve the applicability of rules, we have combined single rules derived from the rough sets. Combined rules are expressed by using intervals. Second, we have dealt with incomplete information tables. Incomplete information is depicted with a disjunctive set of values or an interval of values. We have dealt with it from viewpoints of certainty and possibility, as was introduced by Lipski in the field of incomplete databases. As a result, four types approximations: certain lower, certain upper, possible lower, and possible upper approximations are obtained, as is so in incomplete information tables with nominal attributes. From these approximations, we have derived four types of combined rules: certain and consistent, certain and inconsistent, possible and consistent, and possible and inconsistent combined rules. Combined rules are more applicable than single ones.

References

1. Grzymala-Busse, J.W.: Mining numerical data – a rough set approach. In: Peters, J.F., Skowron, A. (eds.) Transactions on Rough Sets XI. LNCS, vol. 5946, pp. 1–13. Springer, Heidelberg (2010). https://doi.org/10.1007/978-3-642-11479-3_1
2. Hu, M.J., Yao, Y.Y.: Rough set approximations in an incomplete information table. In: Polkowski, L., Yao, Y., Artiemjew, P., Ciucci, D., Liu, D., Ślęzak, D., Zielosko, B. (eds.) IJCRS 2017. LNCS (LNAI), vol. 10314, pp. 200–215. Springer, Cham (2017). https://doi.org/10.1007/978-3-319-60840-2_14
3. Jing, S., She, K., Ali, S.: A universal neighborhood rough sets model for knowledge discovering from incomplete hetergeneous data. Expert Syst. **30**(1), 89–96 (2013). https://doi.org/10.1111/j.1468-0394.2012.00633.x
4. Kryszkiewicz, M.: Rules in incomplete information systems. Inf. Sci. **113**, 271–292 (1999)
5. Lipski, W.: On semantics issues connected with incomplete information databases. ACM Trans. Database Syst. **4**, 262–296 (1979)
6. Lipski, W.: On databases with incomplete information. J. ACM **28**, 41–70 (1981)
7. Lin, T.Y.: Neighborhood systems: a qualitative theory for fuzzy and rough sets. In: Wang, P. (ed.) Advances in Machine Intelligence and Soft Computing, vol. IV, pp. 132–155. Duke University (1997)
8. Nakata, M., Sakai, H.: Rough sets handling missing values probabilistically interpreted. In: Ślęzak, D., Wang, G., Szczuka, M., Düntsch, I., Yao, Y. (eds.) RSFDGrC 2005. LNCS (LNAI), vol. 3641, pp. 325–334. Springer, Heidelberg (2005). https://doi.org/10.1007/11548669_34
9. Nakata, M., Sakai, H.: Applying rough sets to information tables containing missing values. In: Proceedings of 39th International Symposium on Multiple-Valued Logic, pp. 286–291. IEEE Press (2009). https://doi.org/10.1109/ISMVL.2009.1
10. Nakata, M., Sakai, H.: Twofold rough approximations under incomplete information. Int. J. Gen Syst **42**, 546–571 (2013). https://doi.org/10.1080/17451000.2013.798898
11. Nakata, M., Sakai, H.: Describing rough approximations by indiscernibility relations in information tables with incomplete information. In: Carvalho, J.P., Lesot, M.-J., Kaymak, U., Vieira, S., Bouchon-Meunier, B., Yager, R.R. (eds.) IPMU 2016, Part II. CCIS, vol. 611, pp. 355–366. Springer, Cham (2016). https://doi.org/10.1007/978-3-319-40581-0_29
12. Pawlak, Z.: Rough Sets: Theoretical Aspects of Reasoning about Data. Kluwer Academic Publishers, Dordrecht (1991). https://doi.org/10.1007/978-94-011-3534-4

13. Stefanowski, J., Tsoukiàs, A.: Incomplete information tables and rough classification. Comput. Intell. **17**, 545–566 (2001)
14. Yang, X., Zhang, M., Dou, H., Yang, Y.: Neighborhood systems-based rough sets in incomplete information system. Inf. Sci. **24**, 858–867 (2011). https://doi.org/10.1016/j.knosys.2011.03.007
15. Zenga, A., Lia, T., Liuc, D., Zhanga, J., Chena, H.: A fuzzy rough set approach for incremental feature selection on hybrid information systems. Fuzzy Sets Syst. **258**, 39–60 (2015). https://doi.org/10.1016/j.fss.2014.08.014
16. Zhao, B., Chen, X., Zeng, Q.: Incomplete hybrid attributes reduction based on neighborhood granulation and approximation. In: 2009 International Conference on Mechatronics and Automation, pp. 2066–2071. IEEE Press (2009)

How to Match Jobs and Candidates - A Recruitment Support System Based on Feature Engineering and Advanced Analytics

Andrzej Janusz[1]([✉]), Sebastian Stawicki[1], Michał Drewniak[2],
Krzysztof Ciebiera[1], Dominik Ślęzak[1], and Krzysztof Stencel[1]

[1] Institute of Informatics, University of Warsaw,
ul. Banacha 2, 02-097 Warsaw, Poland
{janusza,s.stawicki,ciebie,slezak,stencel}@mimuw.edu.pl
[2] Toolbox for HR, ul. Dobra 56/66, 00-312 Warsaw, Poland
md@tb4hr.com

Abstract. We describe a recruitment support system aiming to help recruiters in finding candidates who are likely to be interested in a given job offer. We present the architecture of that system and explain roles of its main modules. We also give examples of analytical processes supported by the system. In the paper, we focus on a data processing chain that utilizes domain knowledge for the extraction of meaningful features representing pairs of candidates and offers. Moreover, we discuss the usage of a word2vec model for finding concise vector representations of the offers, based on their short textual descriptions. Finally, we present results of an empirical evaluation of our system.

Keywords: Recommender systems · Feature engineering
Word2vec model · Similarity of job offers · Scoring models

1 Introduction

In recent years, the e-recruitment industry is thriving with an estimated average yearly growth rate of 9% in Europe and the United States. It is expected that in the U.S. alone, its turnover will surpass US$10 billion by 2020 [1]. With the growing value of the industry inevitably comes growing competition among recruitment companies. In order to achieve a competitive advantage in this market, companies invest resources into intelligent systems for data analysis and exploration. One of numerous applications of such systems is in the area of automatic recommendation of suitable job candidates.

Project co-financed by European Union Funds under the Smart Growth Operational Programme, Sub-programme 2.3.2, Innovation Voucher. Under the project: POIR.02.03.02-14-0009/15 European Fund of Regional Development. Data used in this study was provided by the *Toolbox for HR* company.

© Springer International Publishing AG, part of Springer Nature 2018
J. Medina et al. (Eds.): IPMU 2018, CCIS 854, pp. 503–514, 2018.
https://doi.org/10.1007/978-3-319-91476-3_42

In a typical Recruitment Process Outsourcing company (RPO) operating in this industry, sourcers (i.e. people responsible for finding job applicants and convincing them to participate in a recruitment process) contact hundreds of potential candidates in order to find only several who are willing to change their current employer. The success rate is particularly small when the position which the sourcer is proposing requires specialized professional skills and expertise. In such cases, RPO companies are often forced to search for applicants outside a local employment market, which as a consequence, lowers the chances for a successful recruitment even further, since selected candidates need to take into account the necessity of a relocation. Moreover, people who are contacted many times with propositions of unsuitable job offers quickly loose their interest in the correspondence and become more difficult to recruit in future.

One way to improve the efficiency of sourcers and shorten the overall time needed for a single recruitment process is by using a system which provides recommendations regarding the potential candidates. Such recommendations may take a form of scores assigned to pairs of known candidate profiles and descriptions of job offers. Apart from measuring the suitability of applicants to a given position, the scores may express different factors influencing chances of a successful recruitment, such as the likeliness that a person will consider changing his current job or agree on a relocation to a different country. Using these scores, sourcers could modify the order in which they contact the candidates, thus they would find good applicants faster and significantly limit the number of unnecessary emails to people that are unlikely to be interested in changing their current job. Unfortunately, even though there were several attempts at constructing such a system, to our best knowledge, none of them was successful in practice.

In this paper, we describe a recruitment support system (RSS) designed for the purpose of management, analysis and automatic scoring of candidate profiles browsed by sourcers. The system has been deployed in a RPO company operating in Warsaw, Poland, namely Toolbox for HR. This company specializes in recruiting technology teams for fast-growing startups, as well as Fortune 500 companies. Recruiters at Toolbox for HR source tech candidates in over 100 countries and place them in 30 cities globally while reviewing over 1000 profiles every week. We discuss an architecture of the RSS software and explain purpose of modules which we designed. We also give examples of analytic processes which can be conducted using the system. Finally, we present results of an empirical study that demonstrates benefits which the system brings to the sourcers and the company.

2   Related Works

The problem of measuring suitability of job applicants has been in a scope of research since the first Internet job boards appeared in the 1990s. One of the most commonly taken approach can be positioned in the area of recommender systems [2]. Typically, recommender systems predict preferences of users with regard to a set of items of services. They become increasingly popular in a variety of areas, such as selecting movies and music, finding interesting books or research articles, or choosing good restaurants [3].

Researchers designing recommender systems often classify them into two main approaches, namely content-based filtering and collaborative filtering [4]. The content-based methods predict preferences regarding a set of items based on their similarity to items which a user has preferred in the past (i.e. to a user's preference profile). In the collaborative filtering, items are recommended to a user based on preferences of similar users. Therefore, in the first case, preference profiles of users need to be compared to descriptions of items and in the latter case, the similarity between users' profiles has to be determined [5]. In many practical applications hybrid methods are used, where the concepts of content-based and collaborative filtering are applied simultaneously [6]. The system described in this paper can be seen as an example of this particular approach. In our case, a user corresponds to a job offer, whereas the items are candidate profiles. Examples of different recommender systems designed for the purpose of human resource matching and on-line recruitment can be found in [7–9].

One of the most important aspects of any recommender system is the underlying method of measuring the similarity. Depending on the utilized approach, the system needs to evaluate similarity between items or users' preference profiles. In either case, the selection of an appropriate similarity measure is crucial for the task. Numerous measures found their practical use in recommender systems, however, due to a sparsity and high-dimensional nature of typical preference data, the most common choices are Jaccard and cosine similarity [10]. Among other popular measures, there are Pearson correlation and the standard Euclidean distance (treated as a dissimilarity measure).

Moreover, many researchers notice a need for adaptation of the similarity to a particular data domain. This can be achieved by either constructing a custom similarity measure using expert knowledge or employing methods capable of learning the appropriate similarity from available data [11]. As an example of such methods, one can give the rule-based similarity model [12] or the networks of comparators [13]. Alternatively, instead of learning the measure for the considered problem, it is possible to learn a better representation of the data and evaluate the similarity in a standard way. This approach, may require constructing new features to represent items or users. The feature engineering methods have been widely studied in this context [14,15].

Learning a meaningful representation of items in a recommender system can be particularly difficult in situations when they correspond to some complex objects. For instance, in our case the system needs to be able to evaluate the similarity between different job offers and score the suitability of candidates in that context. Positions for the recruitment are usually described by a short text and a small set of categorical features, such as the city and country of the employer. One way to represent this type of data is to combine the available features and a bag-of-words representation of the text [12]. A different possibility, utilized in the presented system, is to compute embeddings of the texts and the categorical feature values into a relatively low-dimensional vector space. It can be done using techniques such as the word2vec [16] and its extensions [17] or the neural probabilistic language model [18].

3   Architecture of the System

In order to facilitate advanced data analytics on recruitment processes, one needs to have an environment allowing for convenient processing and managing information about candidates and job offers. Such environment has to provide a robust access to data, support automation of the data processing chain and allow execution of compound data transformation operations. In this section, we show how those objectives are achieved in the system deployed at Toolbox for HR.

The applicant tracking system deployed at Toolbox for HR is built around a data acquisition and storage module called FERMI. This module was developed by an external company in order to facilitate collecting information about potential job candidates from Internet CV databases and professional networking sites. Its main functionality is to provide sourcers with convenient means for accessing and sharing profiles of potential job candidates. It also enables managing the recruitment process at each of its stages.

We extended FERMI by three different modules aimed at intelligent analysis of data collected by the sourcers:

- BIZON[1] – a browser plug-in which, when enabled, automatically downloads HTMLs of profile pages browsed by sourcers,
- SILO – a centralized service for storing raw HTMLs, along with additional meta-data related to the currently investigated profile and a recruitment process (e.g. ID of the job offer),
- FARM – a relational database containing data extracted from collected profiles, dictionary data manually prepared by domain experts (e.g. lists of possible country names with synonyms, names of different universities with common abbreviations, a list of normalized skill names, etc.) and any additional data that can be an output of analytic processes.

Additionally, the modules are supported by scripts responsible for processing the raw HTML data, extracting relevant information and augmenting it with matched elements from the auxiliary data tables stored in FARM (i.e. from the dictionaries and analytic tables). There are also dedicated scripts for combining data stored in different FARM tables (e.g. joining candidate profiles with corresponding recruitment processes and auxiliary data) and constructing information or decision systems in a tabular format. The auxiliary scripts can also perform analytical tasks, such as constructing and serializing scoring models, finding similar job offers or candidate profiles. They are arranged in an extensible script library stored in a repository and can be used on-demand basis.

---

[1] BIZON is a name of a popular Polish combine harvester. https://en.wikipedia.org/wiki/Bizon_(company).

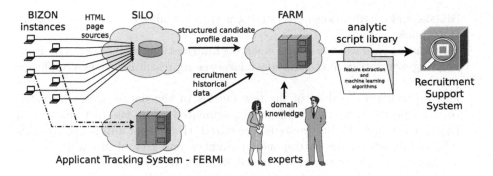

Fig. 1. The architecture and data flow in the described recruitment support system.

Data flow in the system, depicted in Fig. 1, can be described as linear, incremental and buffered. A relevant website, visited using a browser with the BIZON plug-in (the websites are initially filtered using a manually created list of irrelevant domains), is downloaded and stored in the SILO server. Then, information from the website is extracted and augmented using scripts and the auxiliary data from FARM. Finally, the processed data is archived in FARM's database where it can be used for further analysis and constructing scoring models. The whole operation is incremental, since new websites are processed independently from the websites already stored in the system. However, for efficiency reasons, all computations can be conducted in chunks of buffered websites, whose size is dependent on parameter settings.

The BIZON plug-in is an extension of Google Chrome. It has been uploaded to Chrome Web Store and can be installed in a standard way on any computer operating in the company's domain. It uses WebNavigation API provided by Chrome to analyze each page opening request. If the website passes through the excluded domain filter, it downloads the HTML code which is then stored in SILO. Apart from the raw code, SILO stores additional meta-data, such as a browser ID, website address and ID of the recruitment process. The SILO application works in a frame of the client-server architecture of Google App Engine platform. It uses a server operating in Google Cloud in order to secure the communication scalability, regardless of the number of BIZON plug-ins which are working in parallel.

The raw data stored in SILO needs to be transformed into a relational form and augmented by additional information from external sources. Only then it can be utilized for the construction of scoring models and verification of hypothesis describing phenomena related to recruitment processes. The required data transformation consists of several phases:

- *fetch requests* – obtaining information about new websites visited by sourcers; communication with the Google Cloud Datastore NoSQL database, the content of which is systematically updated by the SILO application,
- *parse requests* – unprocessed websites are parsed; relevant information is extracted (e.g. home city, current position, education, employment history,

courses and certifications, etc.) and stored in a consistent form as information chunks,

- *information chunks processing* – emerging information chunks are the basis for creating new entries in the relational database representing the new applicant profiles,
- *enhance content* – values of the new entries in the database are corrected (e.g. misprints in textual fields) and augmented using the dictionary data (e.g. if the name of a university is recognized, the system automatically adds information about the corresponding country and city of studies); data is standardized and entity naming inconsistencies are reduced,
- *fetch FERMI data* – the data corresponding to applicants and recruitment processes from FERMI database is fetched and matched with the corresponding entries in FARM; the relational representation of the new pairs of applicant profiles and recruitment processes is completed.

All these phases were implemented as extensible elements of the FARM module. They operate in the frame of Google App Engine Flexible platform. Planned executions of each phase are scheduled using the App Engine Cron service.

4   Feature Engineering

After storing the augmented representations of the data in FARM, it is possible to run analytic processes defined in the script library. Processes, such as a construction and deployment of a profile scoring model, often require extracting additional characteristics describing the pairs of profiles and recruitment processes. Values of additional features which were defined in a cooperation with recruiters and sourcers working at Toolbox for HR can be computed using a dedicated script. In total, we defined 374 features which can be categorized into three groups:

1. Features of job offers.
2. Relations between candidates and offers.
3. Features of candidates.

The first group contains characteristics of particular job offers. The features in this group are computed based only on information which can be revealed to potential candidates in the initial contact email. They need to be computed only once for a given offer. The group includes features such as the company name, proposed position type, recruitment country and city. We also store a short textual description of the offer, which we embed into a vector space using a word2vec model [16].

The second group contains auxiliary features that express how a given offer fits to a given candidate. Exemplary features from this group include an indicator whether a candidate speaks the required language, a geographical distance between candidate's city and the city from the offer, or a Jaccard similarity value between a short description of the candidate and the description of the offer.

Since information covered by those characteristics is relative to particular candidates and offers, it needs to be computed for every single candidate-offer pair.

The third group corresponds to features of individual candidates. It can be further divided into five subgroups representing different aspects of professional experience: employment history (e.g. number of jobs, average time between switching jobs), skills (e.g. number of certifications, keywords from a list of skills), education (e.g. academic degree, date of the latest education entry), place of residence (e.g. country's GDP, population) and current status (e.g. current employment length, if a candidate has an open code repository). All these features need to be computed only once for a given profile stored in SILO and can be joined with any job offer. However, after some time the profiles may become outdated and sourcers may have to update them using the BIZON module.

5   Example of an Analytic Process in a Recruitment Support System

The main goal of any applicant tracking software is to support recruiters and sourcers in conducting recruitment processes. The RSS described in the previous section enable conducting advanced analysis on historical data collected by sourcers in order to improve efficiency of new recruitments. An exemplary analytical process may consider a problem of identifying historical job offers which are similar to an offer from a new recruitment campaign, finding related candidate profiles, constructing a scoring model and providing scores for new profiles selected by sourcers. In our case, such a process is divided into several phases conducted in a sequence:

1. **Processing of a new job offer.** A recruiter uploads a textual description of a new job offer to the system along with some basic meta-data. The description is standardized (e.g. stop words are removed, common abbreviations are disambiguated), key phrases (from a predefined list) are identified and the text is divided into a list of terms. A word2vec model, trained on descriptions of historical offers, is used to compute an embedding of the offer into a vector space. This vector is stored in a corresponding FARM table.
2. **Fetching relevant profiles.** The computed embedding is compared to embeddings of other offers. The closest k offers are selected and representations of profiles marked as relevant to those offers by sourcers are fetched from FARM (i.e. profiles of people that were contacted in the corresponding recruitment processes).
3. **Constructing a scoring model.** The retrieved data is transformed into a tabular format and (if necessary) augmented using the auxiliary tables from FARM. A scoring model is constructed using a predefined script from the library. Alternatively, a new script can be written and added to the library by a data analyst. Popular scripting languages can be used, e.g. R or Python. The model should take the data table as an input and output scores in a key-value format. After the construction, the model is serialized and stored in a repository.

4. **Computing profile scores.** The profiles fetched in the phase 2 are updated – values of features which depend on a particular job offer are recomputed to reflect the new offer (e.g. if a candidate lives in the same city/country where the job is offered, if a candidate natively speaks the required language, etc.). The updated data corresponding to new candidate-offer pairs becomes an input to a scoring model created for the given offer. The results are presented to a user as a sorted list of candidates with assigned scores and stored in FARM (along with the computed representations of new candidate-offer pairs). Alternatively, on a request from a sourcer, a completely new candidate profile may undergo processing described in Sect. 3 and be assessed by the model.

The architecture of the developed system is flexible and allows to perform different types of data analysis. For instance, a data analyst can easily use it to explore historical data in a search for meaningful dependencies, perform a selection of the most important features or visualize the data for sourcers and recruiters in hope for providing them with useful insights.

6   Experimental Evaluation

We conducted a series of experiments aiming at measuring the performance of the modules described in Sect. 3 in a typical recruitment support task described in Sect. 5. We deployed the modules and asked a group of sourcers to use BIZON in order to feed the FARM database with data. We tracked each profile they contacted between November and December 2017. In total, we collected data on 2288 profiles corresponding to 76 job offers. Each pair in the data was labeled by a tag expressing whether a candidate was interested in the offer and underwent a prescreen procedure. It is worth to notice that the distribution of classes in our data was highly imbalanced, with only about 4% of cases from the positive class. This reflects a typical success rate of sourceres (usually, only a small fraction of contacted candidates is interested in the proposed job offer).

In the first part of the experiment, we checked how the embeddings of offers constructed using the word2vec model reflect their similarity perceived by domain experts. We used the standard skip-gram model with the noise-contrastive training [19], implemented in TensorFlow, to train the model on available textual descriptions of historical offers. The texts were divided into terms and common stop-words were removed. Different embedding sizes were tried, between 5 and 100, but since the number of distinct terms was relatively low (much less than 1000), the final size was set to 12. The embeddings of offers were created from the embeddings of individual terms by averaging vectors corresponding to keywords terms occurring in the descriptions.

In order to measure how well the embeddings reflect similarities between different job offers in our data, we asked experts to manually divide them into categories. They assigned each offer to one of 10 groups (e.g. *BACKEND*, *BIG_DATA*). We performed leave-one-out classification using 1-NN algorithm in the embedding space to predict the category of a give offer based on a category of

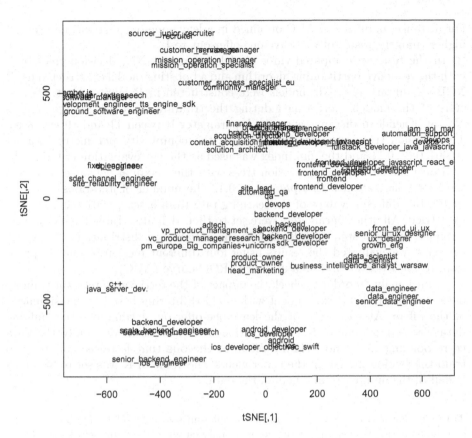

Fig. 2. A visualization of job offers from our data using tSNE technique. The labels correspond to concatenated first few keywords. The colors on the plot correspond to different categories of the offered positions, which were manually assigned by experts. (Color figure online)

the closest neighbor. The average accuracy of predictions was 71% which seems reasonably good considering the relatively high number of categories. Additionally, we visualized the embeddings in a two-dimensional space using the tSNE technique. Figure 2 shows the resulting plot.

In the second part of the experiment, we measured the performance of our profile scoring module. We divided our dataset into chunks corresponding to different offers. For each chunk, we constructed three prediction models using all remaining chunks as the training data and test the performance on the selected chunk. In this paper, we call this method leave-one-offer-out, due to the analogy with the standard leave-one-out technique. It is important to note, however, that using a different validation schema, such as the leave-one-out or 10-fold cross-validation, would not reflect a typical deployment scenario of the discussed system and would be likely to produce over-optimistic performance estimations.

For instance, in our case, AUC obtained in a leave-one-out test was nearly 20% higher than the result of the leave-one-offer-out test.

In the test, we compared three models, namely k-NN, decision trees built with the recursive partitioning algorithm and a boosting model constructed using XGBoost library [20]. All models were constructed in R environment. Parameters of the models were tuned during the training by a grid search. For k-NN, an Euclidean distance with the k parameter between 11 and 19 was used. The decision trees were constructed with the complexity parameter in a set $\{0.001, 0.01, 0.1\}$ and the gini index was used as the cut evaluation criteria. The XGBoost model was using decision trees with the maximal depth set to 6 and the L1 regularization with *alpha* set to 0.1. The number of iterations was fixed to 100 but different values of the learning rate from a set $\{0.001, 0.01, 0.1, 0.3\}$ were tried. All other parameters were set to their defaults. Table 1 shows results obtained by the tested models. Due to the imbalanced distribution of labels in the data, we evaluated the results using four different measures, namely precision, recall, F1-score and area under the ROC curve (AUC).

Additionally, in order to check the impact of the retrieval of relevant training data, we repeated this experiment with enabled filtering based on the similarity of the offers. At each step of the leave-one-offer-out evaluation, the training data was filtered before the construction of a scoring model. Only the data corresponding to K most similar offers to the one that is tested was used to train the scoring model. In the experiments, the value of K was set to 38 which is half of the number of all offers in the data.

Table 1. Comparison of the performance of different scoring models. The values in the column 'experts' were computed based on actual responses to emails sent by sourceres. We assume that Recall of the experts is 1.0. The columns marked as 'filtered' give the results for the models constructed on training data filtered with regard to the similarity of offers (computed in the offer embedding space).

Model measure	k-NN all	k-NN filtered	rpart tree all	rpart tree filtered	XGBoost all	XGBoost filtered	Experts
Precision	0.064	0.055	0.086	0.125	0.118	0.163	0.041
Recall	0.284	0.168	0.147	0.179	0.116	0.158	1.000
F1-score	0.104	0.082	0.109	0.147	0.116	0.160	0.079
AUC	0.551	0.559	0.520	0.615	0.613	0.641	–

The results show that the best performance was achieved by XGBoost which received the highest values of F1-score and AUC among the tested models. Interestingly, filtering of the training data increased results of both tree-based models but at the same time, it slightly degraded the performance of k-NN. To provide a baseline to those results, we investigate the success rate of human sourceres. If we assume that they sent the offer to all potentially relevant candidates which they could find, their recall would be 1.0 (in practice, it does not need to be true).

We can estimate the precision of sourceres by taking the percentage of positive examples in our dataset and thus, we can compute their F1-score. The result of sourceres is shown in the last column of Table 1. The fact that F1-score achieved by XGBoost is over two times greater than that of humans clearly shows the benefits which our system can bring to the company.

7   Conclusions

In the paper, we described a recruitment support system aiming to help recruiters in finding candidates who are likely to be interested in a given job offer. We presented its architecture and explained roles of its main modules. We also presented a data flow schema and gave examples of analytical processes supported by the system. In our example, the system is used to create vector representations of job offers and candidate-offer pairs. The textual descriptions of offers are transformed into vectors using a word2vec model. Then, for a set of new candidate-offer pairs, the system finds examples of cases from similar recruitment processes. It uses them as a training set to compute a scoring model dedicated to the new cases. Our empirical evaluation of the system's performance revealed that its results can be significantly more precise than the judgment of human sourceres.

Nevertheless, our work on the described RSS is by no means complete. Our data acquisition modules are successively storing more and more examples of candidate-offer pairs from a constantly growing number of recruitment processes. We believe that with a sufficient amount of good quality data, we will be able to construct even more reliable scoring models. We are also planning to deploy models based on the deep learning approach and check whether they can bring benefits to our system. Finally, our team is working on further automation of the analytic processes, and at the same time, incorporating more expert knowledge to the system. We hope that in this way, we will make the system even more appealing to its users and help to find a right career path to thousands of people who are looking for their dream jobs.

Acknowledgments. We would like to cordially thank Marcin Smolinski and Anna Czartoszewska-Świerczyńska from Toolbox for HR for cooperation in this project, providing access to the data and supplying us with domain knowledge. We would also like to thank Aleksander Wasiukiewicz and Agnieszka Legut for their administrative support and advise.

References

1. Jobboard Finder: What is the online recruitment market worth today? https://www.jobboardfinder.com/en/e-recruitment
2. Resnick, P., Varian, H.R.: Recommender systems. Commun. ACM **40**(3), 56–58 (1997)
3. Singla, A., Tschiatschek, S., Krause, A.: Actively learning hemimetrics with applications to eliciting user preferences. In: Proceedings of the 33rd International Conference on International Conference on Machine Learning, ICML 2016, vol. 48, pp. 412–420. JMLR.org (2016)

4. Ricci, F., Rokach, L., Shapira, B., Kantor, P.B.: Recommender Systems Handbook, 1st edn. Springer, New York (2010). https://doi.org/10.1007/978-0-387-85820-3
5. Keim, T.: Extending the applicability of recommender systems: a multilayer framework for matching human resources. In: Proceedings of the 40th Annual Hawaii International Conference on System Sciences, HICSS 2007. IEEE Computer Society, Washington, D.C. (2007)
6. Burke, R.: Hybrid recommender systems: survey and experiments. User Model. User-Adap. Inter. **12**(4), 331–370 (2002)
7. Yi, X., Allan, J., Croft, W.B.: Matching resumes and jobs based on relevance models. In: Proceedings of the 30th Annual International ACM SIGIR Conference on Research and Development in Information Retrieval, SIGIR 2007, pp. 809–810. ACM, New York (2007)
8. Singh, A., Rose, C., Visweswariah, K., Chenthamarakshan, V., Kambhatla, N.: PROSPECT: a system for screening candidates for recruitment. In: Proceedings of the 19th ACM International Conference on Information and Knowledge Management, CIKM 2010, pp. 659–668. ACM, New York (2010)
9. Mehta, S., Pimplikar, R., Singh, A., Varshney, L.R., Visweswariah, K.: Efficient multifaceted screening of job applicants. In: Proceedings of the 16th International Conference on Extending Database Technology, EDBT 2013, pp. 661–671. ACM, New York (2013)
10. Bobadilla, J., Ortega, F., Hernando, A., Gutiérrez, A.: Recommender systems survey. Knowl.-Based Syst. **46**(Suppl. C), 109–132 (2013)
11. Lu, J., Wu, D., Mao, M., Wang, W., Zhang, G.: Recommender system application developments: a survey. Decis. Support Syst. **74**(Suppl. C), 12–32 (2015)
12. Janusz, A., Ślęzak, D., Nguyen, H.S.: Unsupervised similarity learning from textual data. Fundam. Inform. **119**(3), 319–336 (2012)
13. Sosnowski, Ł.: Framework of compound object comparators. Intell. Decis. Technol. **9**(4), 343–363 (2015)
14. Han, E.H.S., Karypis, G.: Feature-based recommendation system. In: Proceedings of the 14th ACM International Conference on Information and Knowledge Management, CIKM 2005, pp. 446–452. ACM, New York (2005)
15. Xie, J., Leishman, S., Tian, L., Lisuk, D., Koo, S., Blume, M.: Feature engineering in user's music preference prediction. In: Proceedings of the 2011 International Conference on KDD Cup 2011, KDDCUP 2011, vol. 18, pp. 183–197. JMLR.org (2011)
16. Mikolov, T., Sutskever, I., Chen, K., Corrado, G., Dean, J.: Distributed representations of words and phrases and their compositionality. In: Proceedings of the 26th International Conference on Neural Information Processing Systems, NIPS 2013, vol. 2, pp. 3111–3119. Curran Associates Inc., New York (2013)
17. Le, Q., Mikolov, T.: Distributed representations of sentences and documents. In: Proceedings of the 31st International Conference on Machine Learning, ICML 2014, vol. 32, pp. II-1188–II-1196. JMLR.org (2014)
18. Bengio, Y., Ducharme, R., Vincent, P., Janvin, C.: A neural probabilistic language model. J. Mach. Learn. Res. **3**, 1137–1155 (2003)
19. Mikolov, T., Chen, K., Corrado, G., Dean, J.: Efficient estimation of word representations in vector space. CoRR **abs/1301.3781** (2013)
20. Chen, T., Guestrin, C.: XGBoost: a scalable tree boosting system. In: Proceedings of the 22nd ACM SIGKDD International Conference on Knowledge Discovery and Data Mining, KDD 2016, pp. 785–794. ACM, New York (2016)

Similarity-Based Accuracy Measures
for Approximate Query Results

Agnieszka Chądzyńska-Krasowska[✉]

Polish-Japanese Academy of Information Technology,
Koszykowa 86, 02-008 Warsaw, Poland
honzik@pjwstk.edu.pl

Abstract. We introduce a new approach to empirical evaluation of the accuracy of the select statement results produced by a relational approximate query engine. We emphasize the meaning of a similarity of approximate and exact outcomes of queries from the perspective of practical applicability of approximate query processing solutions. We propose how to design the similarity-based procedure that lets us compare approximate and exact versions of the results of complex queries. We not only offer a measure of the accuracy of query results, but also describe the results of research on users intuition regarding the properties of such a measure, as well as perception query results as similar. The study is supported by theoretical and empirical analyses of different similarity functions and the case study of the investigative analytics over data sets related to network intrusion detection.

Keywords: Approximate Database Engines
Approximate Query Accuracy · Query Result Similarity
Data Granulation and Summarization

1 Introduction

When assessing the classic database engines, their performance is primarily taken into account, which is verified using some previously defined measures. Sample measures can be found, for example, in the benchmark for assessing the performance of TPC-DS decision support systems, prepared by the well-known TPC (Transaction Processing Performance Council) organization [1]. Before starting any assessment, it is necessary to design an appropriate schema, develop procedures that load data according to the rules and prepare a pool of representative queries. In the case of approximate engines [2], in addition to the performance aspect, there is also another aspect of the assessment - the accuracy of query results. Here one should consider two questions: what to test and which assessment measure to use. The answer to the first question is quite obvious - one should perform queries in the approximate engine and compare the obtained results with the results of the same queries obtained on real data. We present

© Springer International Publishing AG, part of Springer Nature 2018
J. Medina et al. (Eds.): IPMU 2018, CCIS 854, pp. 515–527, 2018.
https://doi.org/10.1007/978-3-319-91476-3_43

the proposal for the answer to the second question in the context of summary-based engines, together with research into user intuitions regarding the property of such a measure and the perception of query results as similar.

The focus here is on the measure of the accuracy of analytical query results. The motivation is to aid analysts struggling with the summaries of huge quantities of detailed data, who are the key users of approximate engines. The process of selecting the measure was two-fold. On the one hand, mathematical properties were taken into account, on the other hand, it was remembered that the measurement of the accuracy of the query result will be information for the future user, hence it should be intuitive first of all. It is difficult because different groups of people have different, often unarticulated, expectations towards the measure of the accuracy of query results, which, among other things, is shown in the article. It should be emphasized that all comparisons concerned the engine testing stage, not its final operation, hence the focus was on using only the measure of accuracy, and not, for example, on determining the confidence intervals important at the stage of operation.

This paper presents a new approach to empirical evaluation of the accuracy of the select statement results produced by a relational approximate query engine, emphasizing the meaning of a similarity of approximate and exact outcomes of SQL queries from the perspective of practical applicability of approximate query processing solutions. We propose designing a similarity-based procedure that lets us compare approximate and exact versions of the results of complex queries, simultaneously presenting the results of survey of users intuition regarding the properties of such a measure, as well as perception query results as similar.

The paper is organized as follows. Section 2 provides a proposal for measuring the accuracy of the results of analytical queries. Section 3 presents the process of selecting the proper measure, together with results of survey on users intuition regarding the properties of such a measure. Section 4 presents results of survey on perception query results as similar. In Sect. 5 are conclusions.

2   Similarity of Analytical Queries

The result of any SELECT is a table, which can be denoted as $R = (T, A)$, where T is a set of tuples, and A is a set of attributes. The analytical queries considered in this article are understood as classical queries with the GROUP BY clause, in which the elements of A are S columns reflecting summary functions, and G columns used in the GROUP BY clause. The G columns used in the GROUP BY clause are always treated as categorical attributes, while S columns are treated as numerical or categorical, according to the rules presented below.

When comparing the results of analytical queries, three types of values should be considered, corresponding to three groups compatible with business intuitions in analytics [3]: (1) numerical values corresponding to the measures of the fact table, constituting arguments of any summary functions - will be called measures, according to the naming in [4], (2) values describing the dimensions of the analysis, which should be treated as categorical, e.g. keys from dimensions placed in the fact table that are the basis for creating groups in the GROUP BY clause - will be called categorical dimensions, (3) values describing the dimensions of the analysis, which can be treated as categorical or numerical depending on the context in which they are used in the query, e.g. timestamp - will be called linear dimensions. The following principles of comparison have been adopted: (i) group labels defined in the GROUP BY clause are treated within a given query as categorical attributes, (ii) if the argument of the summary function is a value from the group 1 (measures) or 3 (linear dimensions), within the given query these are treated as numerical attributes, (iii) if the argument of the summary function is a value from the group 2 (categorical dimensions), within the given query these are treated as categorical attributes.

The analytical query may contain WHERE and HAVING clauses imposing conditions for rows and groups, as well as the ORDER BY clause frequently used with the LIMIT directive, which is a limitation of the number of displayed rows. It is assumed that the ORDER BY clause will be treated as a new summary function added to the S. In the general case, T may be a subset of the original rows, G may be a set of columns for which PRIMARY KEY constraints were applied, and S all other columns.

Let $R = (T, A)$ be the result of the Q query obtained on the real data, and $\widetilde{R} = (\widetilde{T}, A)$ be the result of the Q query obtained on the approximate data. We need to determine to what extent T and $\widetilde{T}$ tuples can be matched on G columns and, for matched tuples, determine how their values in S are similar. The Jaccard index [5] is used as a measure of matching the set of tuples T and $\widetilde{T}$: $J(T, \widetilde{T}) = |T \cap \widetilde{T}|/|T \cup \widetilde{T}|$, where the sum and intersection of the tuple sets is understood as the sum and intersection of the truncation of the tuple sets to the set of G columns present in the GROUP BY clause. For simplicity of notation, no additional marks have been added to denote truncation. While constructing the final measure of accuracy of query results, the given formula was enriched with the similarity ratio of matched tuples. A pair of tuples $t \in T$ and $\widetilde{t} \in \widetilde{T}$ are called matched, if for every g belonging to the set of columns G appearing in the clause GROUP BY, $g(t) = g(\widetilde{t})$. Finally, this measure takes the form:

$$Q(R, \widetilde{R}) = \frac{\sum_{t \in T, \widetilde{t} \in \widetilde{T}: g(t) = g(\widetilde{t})} Q(t, \widetilde{t})}{|T \cup \widetilde{T}|} \tag{1}$$

where $Q(t, \widetilde{t})$, which is a measure of the accuracy of a single row, is given by:

$$Q(t, \widetilde{t}) = MIN_{a \in S} Q_a(f(t), f(\widetilde{t})) \tag{2}$$

in which $Q_a(f(t), f(\widetilde{t}))$, which is a measure of the accuracy of a single value (in the case of an analytical query it is the value of the summary function f), for the numerical values is as follows:

$$Q_a(f(t), f(\widetilde{t})) = \begin{cases} 1 - \frac{|f(t) - f(\widetilde{t})|}{|f(t)| + |f(\widetilde{t})|} & \text{for } f(t) \neq 0 \vee f(\widetilde{t}) \neq 0 \\ 1 & \text{otherwise} \end{cases} \tag{3}$$

taking into account formula 4 for the NULL pseudo-value:

$$Q_a(f(t), f(\widetilde{t})) = \begin{cases} 1 \text{ for } f(t) \text{ is null } \wedge f(\widetilde{t}) \text{ is null} \\ 0 \text{ otherwise} \end{cases} \qquad (4)$$

and for categorical values is as follows:

$$Q_a(f(t), f(\widetilde{t})) = \begin{cases} 1 \text{ for } f(t) = f(\widetilde{t}) \\ 0 \text{ otherwise} \end{cases} \qquad (5)$$

The choice of such a strict measure in the case of categorical columns is dictated primarily by the expectations of the business user, for whom even the smallest difference in value in the categorical column can be crucial for the decision-making process. Values for categorical columns, which are certain labels or categories must be accurate, in contrast to numeric columns, for which the order of magnitude is often more important than the specific value.

Figure 1 presents an example of calculating the accuracy of a query result using measure 1 for a query with one column in the GROUP BY clause, while in [4] there is an analogous example for a query with more than one column in the GROUP BY clause. The net_traffic dataset is described in [6]. At the beginning, for each element from the set S for each tuple from the intersection of the truncation of the set of tuples T and $\widetilde{T}$ to the set of columns G, $Q_a(f(t), f(\widetilde{t}))$ is calculated. Then, the minimum function of individual tuple components is used to compare whole tuples. It was considered that the minimum would be a better choice than the product or the arithmetic mean, because it guarantees the invariability of the result in the case of duplication of the same expression in the SELECT list. The total result of the comparison of tuples is divided by the number of all tuples belonging to the sum of the truncation of the sets of tuples T and $\widetilde{T}$ to the set of columns G. Therefore, in the result of the query performed on the approximate data some groups present in the result of the query performed on the real data are missing, or there are extra groups, then the final accuracy deteriorates in the eyes of business users. If an ORDER BY clause appears in the query, it is treated as a new summary function added to the set S. An example of calculation of the accuracy of the query result with the ORDER BY clause can be found in [7].

Both for groups that do not occur in real data and occur in approximate data (called *false presence*), as well as for groups that occur in real data, and do not occur in approximate data (called *false absence*), $Q(t, \widetilde{t}) = 0$ due to the fact that the suggested measure of the accuracy of query results returns 0 if and only if exactly one of $f(t)$ and $f(\widetilde{t})$ is equal to 0 (for count(*)) or NULL (for other summary functions).

3   Selecting the Accuracy Measure of Query Results

The measure 1 proposed in the previous section is of course not the only possible measure of the accuracy of query results. An alternative approach, including any queries, is shown in [8]. Another source of inspiration is [9], which is a review of similarity measures for probabilistic distributions represented using histograms.

```
SELECT P_ELEMENT AS PE, COUNT(*) AS CNT, SUM(BYTES) AS SUM, AVG(BYTES) AS AVG
       FROM NET_TRAFFIC WHERE S_CLASS = 100 GROUP BY P_ELEMENT;
```

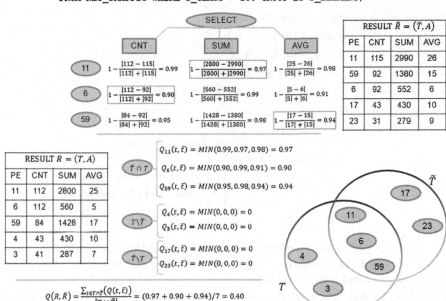

Fig. 1. Example of calculating the accuracy of a query result.

Nevertheless, no reference has been made to the reception of results of a given measure by end users. It focused only on its mathematical properties.

Looking for a measure of the accuracy of query results, the focus was on the comfort of its reception by the engine users. It was assumed that both the set of accepted properties of the measure and the final result of the measurement should be consistent with the intuition of users. Basing on previous practical experience, it was assumed that primarily business analysts and field experts will be the users of the approximate engine, as well as all other people who for some reason will work with data. In view of the above, it was decided to refer to the opinions of various groups of actual and potential users of such an engine. Since it was assumed from the very beginning that the basic building block for the accuracy measure of the query results will be the comparison of two corresponding atomic values, all research has focused only on this aspect. In addition, due to the assumption of absolute treatment of categorical attributes, the focus was only on numerical results.

The search process began with creating a list of properties that such a measure should or could possess. To begin with, referring to the common understanding of the similarity function, it was assumed that the obtained value will belong to the interval $[0, 1]$. It was also assumed that the accuracy calculated with the use of this measure must be monotonically increasing in relation to the generally perceived perception of people who will use the engine in the future. Additionally,

Table 1. Considered properties of measure.

Property name	Property specification
Identity	$Q(x, y, N) = 1 \leftrightarrow x = y$
Symmetry	$Q(x, y, N) = Q(y, x, N)$
Strong zero	$y \neq 0 \rightarrow Q(0, y, N) = 0$
N-monotonicity	$N_1 < N_2 \rightarrow Q(x, y, N_1) < Q(x, y, N_2)$
Complementarity	$Q(x, y, N) = Q(x', y', N)$
Maximum error	$lim_{N \to \infty} Q(x, N, N) = 0$ where z' is a score of complementary query (for count(*) it is assumed $z' = N - z$)
Big data:	$lim_{N \to \infty} Q(x, y, N) = \begin{cases} \text{(a) some kind of function } f(x, y) \\ \text{(b) some kind of constant value} \end{cases}$

by supporting preliminary analyses of selected applications, a handful of possible properties was formulated that could potentially be of interest to users. Those properties are displayed in Table 1.

In theory, each of these properties could be accepted or rejected. The identity of indistinguishable elements is the property of many similarity measures and its adoption seems quite intuitive, however, as argued in [10], the human perception of the same stimuli differs depending on their specifics. In the case of query results, some doubts may arise, particularly whether different results can not be considered as equal. While in the case of 1 and 10 it is rather obvious, in the case of 1000000001 and 1000000010 it is not obvious at all. Symmetry, another frequently accepted property of similarity, is also not entirely obvious. Will the overestimation of the result by the approximate engine be received identically as underestimate by the same value? Strong zero property clearly depends on the context. In same situations, the difference between 0 and 1 is exactly the same as the difference between 1 and 2, in others the difference is huge. Considering it in terms of existence and non-existence, the first case is infinitely worse than the second. N-monotonicity means that the error between 1 and 5 is more significant when the number of all rows is 10 than when it is 100. In other words, the measure would have to depend on N. The opposite of N-monotonicity would be invariability understood as independence from N. Complementarity is a very specific property when comparing query results. Accepting this property would mean that the accuracy of the counting query at N equal to, for example, 1000 would be the same for 990 and 999 as for 10 (1000 − 990) and for 1 (1000 − 999). This is quite problematic, because on the one hand 990 may seem more similar to 999 than 10 to 1, but, on the other hand, the rejection of this property would mean that the implementation of a complementary query (with NOT condition) may give a different accuracy than the initial query, and yet the answer to both of these queries brings equivalent information from the point of view of a decision making process. Maximum error property allows to distribute the value of the

Table 2. Six measures selected for further analysis.

Measure 1	$Q(x,y) = \begin{cases} 1 - \frac{|x-y|}{|x|+|y|} & \text{for } x \neq 0 \vee y \neq 0 \\ 1 & \text{otherwise} \end{cases}$
Measure 2	$Q(x,y,N) = \begin{cases} \dfrac{2}{\frac{1}{MIN(\frac{x}{y},\frac{y}{x})} + \frac{1}{MIN(\frac{N-y}{N-x},\frac{N-x}{N-y})}} & \text{for } x \neq 0 \wedge y \neq 0 \\ 1 & \text{for } x = 0 \wedge y = 0 \\ 0 & \text{for } x = 0 \veebar y = 0 \end{cases}$
Measure 3	$Q(x,y,N) = 1 - \frac{|x-y|}{max(x,N-x)}$
Measure 4	$Q(x,y,N) = 1 - \frac{log(|x-y|+1)}{log(max(x,N-x)+1)}$
Measure 5	$Q(x,y,N) = 1 - \frac{|log(x+1)-log(y+1)|}{log(N+1)}$
Measure 6	$Q(x,y) = \begin{cases} MIN(\frac{x}{y},\frac{y}{x}) & \text{for } x \neq 0 \wedge y \neq 0 \\ 1 & \text{for } x = 0 \wedge y = 0 \\ 0 & \text{for } x = 0 \veebar y = 0 \end{cases}$

similarity function over the entire interval [0,1]. Big data property emphasizes the problem of selecting a measure. With small N the case is relatively simple. But such cases are usually not considered for approximate engines. For N at the level of terabytes or petabytes the constant convergence measure practically ceases to distinguish the results. On the other hand, measures with non-trivial (functional) convergence tend to take a more severe treatment of small results than large ones, which may or may not cause some discomfort to the engine user.

Due to the aforementioned aspect of the users comfort, we did not accept or reject any of the above properties arbitrarily. Three groups of people were consulted - internal experts developing an approximate engine, possessing strong mathematical and analytical skills, real engine users who are experts in detecting network attacks and potential users of such a system - students of Information Management Faculty at the Polish-Japanese Academy of Information Technology.

The question for internal experts was asked directly, that is, each of them received a list of suggested properties with a brief explanation of their consequences for the process of assessing the accuracy of query results and was asked to indicate whether the measure of the accuracy of query results should consider the given property or not. Opinions were divided into three groups: definitely not (which was given (with) value 0), yes/no (which was given the value 1) and deciding yes (which was given the value 2). The summarized results of internal experts were presented in Table 3, from which it can be read that the most important properties for this group are the Identity and Maximum error properties, while the most undesirable is Strong zero property.

Before referring to the intuition of potential users of the approximate engine, it was decided to propose several measures of accuracy of the results of count(*) queries and prepare a survey on this basis. Many different similarity measures have been tested, trying to soften and sharpen them, and to get a compromise

Table 3. On the left summary opinions of internal experts regarding the properties of measures. On the right summary intuitions regarding the properties of the measurements of potential users of the approximate engine. 39 out of 42 respondents completed the survey in accordance with the instructions.

Internal experts opinions				Students opinions			
Property	Yes	No	Yes %	Property	Yes	No	Yes%
Identity	6	0	100%	Identity	20	19	51.28%
Symmetry	2	4	33.33%	Symmetry	16	23	41.03%
Strong zero	0	6	0%	Strong zero	13	26	33.33%
N-monotonicity	3	3	50%	N-monotonicity	16	23	41.03%
Complementarity	4	2	66.67%	Complementarity	0	39	0%
Maximum error	6	0	100%	Maximum error	39	0	100%
Big data	4	2	66.67%	Big data	-	-	-

between the measure that distinguishes the results of a wide range of queries even at large N, and on the other hand did not give the impression of favoring higher results at the expense of smaller ones. Six measures selected for further analysis are displayed in Table 2.

Forty-two students of the last year of the first-degree studies at the Information Technology Faculty were asked to complete a survey identifying intuitions regarding the properties of the measures. These students were considered a very good sample group of potential users of the approximate engine. The survey contained the real and approximate results, the difference module, the ratio and the proposed accuracy calculated in accordance with the six measures - the softened version of the measure 1 ($1 - \frac{|x-y|}{|x|+|y|+1}$), the measures 3–6, and the test measure, the value of which was different from all previous ones. Respondents were asked to select for each pair of results level of the accuracy consistent with their intuition. Among the given pairs of results were those that allowed to check the intuition of almost all of the previously presented properties. Summary of respondents' intuitions regarding the properties of measures is presented in Table 3. It is clearly visible that all respondents indicated intuitively the property of the maximum error, but no one indicated the property of complementarity. It can be presumed, therefore, that the intuition of these two properties is firmly established. It is clearly visible that the property of complementarity, despite a very good logical justification, is completely unintuitive for potential users.

The last group to consult and collect opinions about the property of the measure were real engine users who were experts in detecting network threats. For this group, in contrast to the other two, the Strong zero property is extremely important. It turned out that from the point of view of network traffic diagnostics it is very important to match the actual results with approximate results to detect existence or non-existence of a network attack. This distinction is more important than the nominal number of attacks. Identity and non-trivial Big data

Table 4. Properties of measures 1–6.

Property	M 1	M 2	M 3	M 4	M 5	M 6
Identity	+	+	+	+	+	+
Symmetry	+	+	−	−	+	+
Strong zero	+	+	−	−	−	+
N-monotonicity	−	+	+	+	+	−
Complementarity	−	+	+	+	−	−
Maximum error	+	+	+	+	+	+
Big data	$Q(x,y)$	$\frac{2x}{x+y} for\ x \leqslant y$	1	1	1	$\frac{x}{y} for\ x \leqslant y$

properties were also desirable for this group. The remaining properties were not negated, but they were also not mentioned as important.

Taking all this into consideration, it was decided that the measure of the accuracy of query results must have the following properties: Identity, Strong zero, Maximum error, non-trivial Big data property (dependent on x and y). The other three properties (Symmetry, N-monotonicity and Complementarity) were considered possible but not necessary.

Table 4 presents whether the proposed measures possess several pre-defined properties. It can be read from the table that Measure 2 (harmonic measure) fulfills all properties and has a non-trivial Big data property - with $N \to \infty$ the accuracy of the query result still depends on x and y.

The basic disadvantage of the harmonic measure is its complexity and unintuitive for the business user. However, it can be easily proved that the harmonic measure is asymptotically equal to a much simpler measure 1, i.e.:

$$Q(x,y,N) = \frac{2}{\frac{1}{MIN(\frac{x}{y},\frac{y}{x})} + \frac{1}{MIN(\frac{N-y}{N-x},\frac{N-x}{N-y})}} \to_{N\to\infty} \frac{2\frac{x}{y}}{1+\frac{x}{y}}$$

Measure 1 has no N-monotonicity and Complementarity properties, which were not indicated as necessary properties.

4 Intuitions About the Similarity of Query Results

Since it was assumed from the very beginning that the measurement of the accuracy of the query result should be intuitive to the user, the question was asked concerning the feeling about the similarity of numbers in general and the similarity of the query results in particular. Information on the perception and discrimination of the cardinality of numerical sets can be found in the neurobiological literature, e.g. [11,12], in which two important aspects are emphasized - distance effect and size effect. The distance effect meaning that two numerosities are easier to discriminate when the distance between them is larger. The size effect, on the other hand, meaning that for a given distance comparison is easier

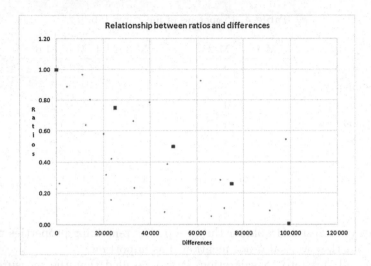

Fig. 2. Relationship between the ratios and the differences between the real and approximate results for the medium version of the survey.

when numerosities are smaller. Both effects point to the fact that it is easier for people to distinguish small cardinality than large. The research was conducted both on non-symbolic sets (sets of dots) and on symbolic sets (numbers). It was noticed that in symbolic sets, and such are numerical query results, the distance and size effects are less pronounced, but they are still there [11]. It could therefore be assumed that the larger the numbers, the greater the differences between them will be accepted, while with small numbers even small deviations will be treated sharply.

We decided to conduct another survey to see how potential engine users will judge the degree of similarity between real and approximate outcomes based on their absolute difference and ratio. As part of this research, the number of pairs of potential query results obtained on real and approximate data was reduced to 25. Respondents were asked to organize the obtained pairs of results from the most similar (marked by number 1) to the least similar ones (marked by number 25). They were to choose a similarity criterion basing on own intuition. A group of respondents, consisting of 110 students of the second-degree studies at the Information Technology Faculty of the Polish-Japanese Academy of Information Technology, was divided into three subgroups. The first subgroup, composed of 34 people, rated small results (in the order of 10^2), the second subgroup of 38 people, rated average results (10^4), and the third subgroup of 38 people rated high results (10^6).

Figure 2 presents the relationship between the ratios and the differences between the real and approximate results for the medium version of the survey. In the small and large versions, the ratios are almost identical, and the differences are respectively smaller in the small version and larger in the large version. Points that were determined as the specific validation test were marked

Table 5. Survey results.

Questionnaire	Small	Medium	Large	In total
Number of respondents	33	38	38	109
Number of accepted surveys	22 (67%)	24 (63%)	20 (53%)	66
The number of results close to the difference	8	8	8	24
The number of results close to the ratio	9	7	6	22
Number of unspecified results	5	9	6	20

with larger squares. If the respondent had not kept monotonicity within these points, his survey was rejected because it was considered insufficiently credible.

At the stage of processing survey results, order numbers were assigned to pairs of query results sorted according to $|x - y|$, where x is the result obtained on real data, and y is the result obtained on approximate data. The pair of results with the smallest difference in each group was individually assigned 1, while the pair of results with the largest difference was assigned to 25. The same was done by sorting the results according to the proportion described by the $MIN(\frac{x}{y}, \frac{y}{x})$ formula (omitting the situation of the query result equal to 0, which was not in the surveys). In this case, the pair of results with the largest quotient in each group were individually assigned 1, while the pair of results with the smallest quotient was assigned 25. Then, for each survey that positively passed the above validation test, the distance L_1 of its ranking was calculated from the difference and the proportion. Due to the fact that respondents sometimes replicated ordinal numbers - deliberately or as a result of a mistake - it was decided to use standardized figures for the sum of all the numbers used.

The main goal of the research was to determine if the intuitions of users are closer to the difference or ratio. It was assumed that the result is close to the difference if the distance between the result and the difference is less than 0.2. An analogous assumption was made for ratios. Table 5 presents the obtained results. It can be noticed that the intuition of about one-third of respondents who passed the validation test is close to the difference, the intuition of one-third of the respondents is close to the ratio, and the intuition of the remaining respondents deviates from both the difference and the proportion.

Before doing the test, differences in the assessment of small and large results were expected, i.e. it was predicted that with small results respondents would more often approach the difference, and with large the ratio. However, this was not confirmed by the analyzed surveys. Analyzing the results, however, it is worth noting that if the larger numbers were assessed, fewer users would pass the validation test, which means that the higher the numbers, the less systematized the accuracy evaluation and the more dependent on the moment of assessment. This observation is consistent with the distance and size effects cited earlier.

5   Conclusions

This paper presents a new approach to empirical evaluation of the accuracy of the select statement results produced by a relational approximate query engine, emphasizing the meaning of a similarity of approximate and exact outcomes of SQL queries from the perspective of practical applicability of approximate query processing solutions. We propose how to design a similarity-based procedure that lets us compare approximate and exact versions of the results of complex queries. Additionally we collected and presented the results of survey on users intuition regarding the considered aspects of similarity.

The surveys clearly show that the intuitions of users about the measure of the accuracy of query results are very different, both in terms of the property of measure and the similarity of approximate and actual results. A very strongly established property of the measure is the maximum error, while a very poorly established property is complementarity. Strong zero property is absolutely necessary for some people, while others reject it. Symmetry and N-monotonicity are important for about one third of users, Identity for more than half.

When it comes to the similarity of numbers, the intuitions of about one third of users are close to the ratios, about one third are close to the difference, while the intuitions of the remaining third part of users are away from both the difference and the ratio. Perhaps this group is guided by a linear combination of difference and ratio, but it is also not excluded that this is the result of incorrect calculations. Determining this would require further research.

References

1. Nambiar, R.O., Poess, M.: The making of TPC-DS. In: Proceedings of 32nd International Conference on Very Large Data Bases, VLDB Endowment, pp. 1049–1058 (2006)
2. Ślęzak, D., Glick, R., Betliński, P., Synak, P.: A new approximate query engine based on intelligent capture and fast transformations of granulated data summaries. J. Intell. Inf. Syst. **50**(2), 385–414 (2018)
3. Kimball, R., Ross, M.: The Data Warehouse Toolkit: The Complete Guide to Dimensional Modeling. Wiley, New York (2011)
4. Ślęzak, D., Chądzyńska-Krasowska, A., Holland, J., Synak, P., Glick, R., Perkowski, M.: Scalable cyber-security analytics with a new summary-based approximate query engine. In: Proceedings of Big Data 2017, pp. 1840–1849 (2017)
5. Deza, M.M., Deza, E.: Encyclopedia of distances. In: Deza, M.M., Deza, E. (eds.) Encyclopedia of Distances, pp. 1–583. Springer, Heidelberg (2009). https://doi.org/10.1007/978-3-642-00234-2_1
6. Chądzyńska-Krasowska, A., Stawicki, S., Ślęzak, D.: A metadata diagnostic framework for a new approximate query engine working with granulated data summaries. In: Polkowski, L., Yao, Y., Artiemjew, P., Ciucci, D., Liu, D., Ślęzak, D., Zielosko, B. (eds.) IJCRS 2017. LNCS (LNAI), vol. 10313, pp. 623–643. Springer, Cham (2017). https://doi.org/10.1007/978-3-319-60837-2_50
7. Chądzyńska-Krasowska, A., Kowalski, M.: Quality of histograms as indicator of approximate query quality. In: Proceedings of FedCSIS 2016, pp. 9–15 (2016)

8. Ioannidis, Y.E., Poosala, V.: Histogram-based approximation of set-valued query-answers. In: Proceedings of VLDB 1999, pp. 174–185 (1999)
9. Cha, S.H.: Comprehensive survey on distance/similarity measures between probability density functions. Int. J. Math. Models Methods Appl. Sci. **4**(1), 300–307 (2007)
10. Tversky, A.: Features of similarity. Psychol. Rev. **84**(4), 327 (1977)
11. Verguts, T., Fias, W.: Representation of number in animals and humans: a neural model. J. Cogn. Neurosci. **16**(9), 1493–1504 (2004)
12. Nieder, A., Miller, E.K.: Coding of cognitive magnitude: compressed scaling of numerical information in the primate prefrontal cortex. Neuron **37**(1), 149–157 (2003)

A Linear Model for Three-Way Analysis of Facial Similarity

Daryl H. Hepting$^{(\boxtimes)}$ (ID), Hadeel Hatim Bin Amer (ID), and Yiyu Yao (ID)

Department of Computer Science, University of Regina, Regina, SK S4S 0A2, Canada
{hepting,yyao}@cs.uregina.ca, binamerh@uregina.ca

Abstract. Card sorting was used to gather information about facial similarity judgments. A group of raters put a set of facial photos into an unrestricted number of different piles according to each rater's judgment of similarity. This paper proposes a linear model for 3-way analysis of similarity. An overall rating function is a weighted linear combination of ratings from individual raters. A pair of photos is considered to be similar, dissimilar, or divided, respectively, if the overall rating function is greater than or equal to a certain threshold, is less than or equal to another threshold, or is between the two thresholds. The proposed framework for 3-way analysis of similarity is complementary to studies of similarity based on features of photos.

Keywords: Similarity · Three-way decision · Card sorting
Linear model

1 Introduction

A basic idea of three-way decisions (3WD) is thinking and problem solving in threes [15]. According to a trisecting-and-acting model of 3WD, we divide a whole into three parts and devise strategies to process the three parts [14,15]. While each part captures a particular aspect of the whole or consists of elements of particular interest, their integration reflects the whole. By thinking in threes, 3WD may provide a simplification of processing the whole through processing three parts. The theory of 3WD has been applied in many fields [6–9,13,16].

In a previous paper [3], we presented some preliminary results on applying the 3WD theory to a card sorting problem. Card sorting has been successfully applied to gain insight about the structure of information in different contexts [1, 2,10,11]. For our card sorting problem, we have a set of facial photos and a group of raters. Each rater was instructed to sort similar photos into the same pile. A pair of photos is similar if both photos are sorted into the same pile. A pair of photos is dissimilar if the photos are sorted into different piles. An analysis of rating results shows that there is a large variance amongst raters in terms of the number of piles and the sizes of those piles (see Fig. 1).

To arrive at an overall rating of similarity of photos by combining judgments from individual raters, it seems unrealistic to consider only two values for each

© Springer International Publishing AG, part of Springer Nature 2018
J. Medina et al. (Eds.): IPMU 2018, CCIS 854, pp. 528–537, 2018.
https://doi.org/10.1007/978-3-319-91476-3_44

rater, i.e., similar or dissimilar. Following the philosophy of 3WD, we take three values: a pair of photos is considered to be similar if the group agrees on their similarity at or above a certain degree and to be dissimilar if the group agrees on their similarity at or below another degree; otherwise, the pair is considered to be divided or undecided.

In our earlier paper [3], we only considered a simple function to synthesize ratings from a group of raters. The main objective of the current study is to propose a general 3-way framework for analyzing facial similarity. We suggest and study a more general function for pooling together individual ratings, in order to arrive at an overall 3-way rating of similar, dissimilar, and divided. The results of such a 3-way analysis would be useful for an in-depth understanding of and further applications of group ratings.

The purpose of this work is to understand what, if any, differences can be reliably extracted from the card sorting data so that raters who may define similarity in different ways can be identified, classified, and perhaps quantified. Three-way classification of similarity may be viewed as a first step towards a more comprehensive model of similarity analysis.

Once we have a 3-way classification, we can attempt to extract features that contribute to the similarity and dissimilarity of photos. A more in-depth analysis of undecided photos may also reveal possible reasons that raters are divided. Similarity analysis based on card sorting by people is complementary to similarity analysis based on photo features. It will be interesting to compare and integrate the two types of approaches. For example, based on photo features, we may be able to ask raters to sort a sample set of photos, rather than the entire set. Alternatively, card sorting by people may provide insights into the design of a feature-based similarity measure. Results from the present study serves as a basis for these future investigations on similarity.

2   A Simple Linear Model of Three-Way Analysis

A group of raters ($N = 25$) were asked to sort a set of facial photographs ($M = 356$) into an unrestricted number of piles based on how they judged similarity of the photos. Through this activity, each rater contributed to the assessment of the similarity amongst the photos. It is not possible for any rater to directly consider the similarity of all $\binom{356}{2} = 63,190$ pairs of photos. Data about which comparisons were made directly (between the photo being sorted and the top photo on each pile) and which indirectly was not recorded. Therefore, a means of analysing the similarity judgements is sought in order to reduce the number of photos under consideration in further studies. If there is more than one strategy being used by different raters to judge similarity, we seek to focus our efforts to understand these different strategies on the photos about which there is possibly disagreement, those photo pairs whose similarity score is between the two thresholds, α and β.

An algorithmic approach such as Eigenfaces, popularized by Turk and Pentland [12], provides a feature-based calculation of facial similarity. In contrast, our card sorting approach to similarity attempts to understand the human perception of facial similarity. Ideally, the piles made by each rater represent equivalence classes. More pragmatically, the boundaries between the piles are likely not so clear. Intuitively, the larger the pile the more difficult it is to maintain the same high threshold for inclusion of photos in that pile.

Raters were instructed to not create any pile with only a single photo, because such a pile conveys no similarity information, only that such a photo is unrelated to all others. A small number of piles with single photos were removed from further consideration. Also, during data entry, a small number of photos were not recorded. Therefore, not all raters made judgements based on all 356 photographs. Table 1 presents the total number of photos considered by each rater.

The stimuli used in the card sorting activity combined two sets of facial photos: one set of 178 Caucasian male subjects and the other set of 178 First Nations male subjects. All photos are identified by a 4 digit code, which is a departure from earlier publications describing the card sorting study (see Hepting et al. [4,5]). The first digit indicates the stimulus set (1: Caucasian, 2: First Nations) followed by 3 digits to indicate the sequence number in that stimulus set (0–177).

Let us begin with an expression for the similarity, S, of two photos A and B, according to a rater, r, who has made n_r piles $P_1, \ldots, P_{n_r}$. For each rater, we can obtain a binary interpretation of similarity in terms of piles, namely, photos in the same pile are similar and photos in different piles are dissimilar.

Let P denotes the set of photos. Formally, we define a function $s_r : P \times P \longrightarrow \{0, 1\}$ for rater r as follows:

$$s_r(A, B) = \begin{cases} 1, & A \text{ and } B \text{ are in the same pile,} \\ 0, & A \text{ and } B \text{ are in two different piles.} \end{cases} \tag{1}$$

In order to obtain an overall evaluation of similarity, we can synthesize ratings from all raters. A simple fusion function is a summation of ratings of individual raters, that is,

$$S(A, B) = \frac{1}{N} \sum_{r=1}^{N} s_r(A, B). \tag{2}$$

It is simply the average of the similarity values given by individual raters. We have $0 \leq S(A, B) \leq 1$, $S(A, B) = 0$ if all raters put A and B in different piles, and $S(A, B) = 1$ if all raters put A and B in the same pile.

Raters may provide different classifications of photos, in terms of piles. It seems reasonable to expect that a pair of photos is similar if a majority of raters view the pair as similar and dissimilar if a majority of raters view the pair as dissimilar. If the raters are divided in the middle, we introduce the case of divided ratings. In this way, we have a 3-way interpretation of similarity. Given a pair of thresholds (α, β) with $0 \leq \beta < \alpha \leq 1$, a 3-way rating of similarity is given by:

Table 1. The number of photos, M_r, that each rater considered. As described in Sect. 2, every rater may not have considered all 356 photos. The photo identifiers listed in the third column are as described in Sect. 2. Of the photos not considered, only 2133 and 1067 appear more than once (respectively 3 times and 2 times).

Rater	M_r: number of photos considered	Identifiers of photos not considered
1	356	
2	356	
3	356	
4	356	
5	356	
6	352	1063, 2043, 2095, 2170
7	353	1085, 2007, 2131
8	354	1164, 2001
9	356	
10	356	
11	352	1067, 2002, 2005, 2036
12	356	
13	356	
14	355	2133
15	350	1018, 1050, 1067, 2055, 2133, 2157
16	356	
17	356	
18	356	
19	356	
20	356	
21	355	2133
22	356	
23	356	
24	356	
25	355	2175

$$\mathbb{S}(A, B) = \begin{cases} \text{Dissimilar,} & S(A, B) \leq \beta, \\ \text{Similar,} & S(A, B) \geq \alpha, \\ \text{Divided,} & \beta < S(A, B) < \alpha. \end{cases} \tag{3}$$

The pair of thresholds reflect our confidence in deciding similarity and dissimilarity. When α approaches 1, we become more confident about similarity, and when β approaches 0, we become more confident about dissimilarity. In previous studies, we used $\beta = 0.4$ and $\alpha = 0.6$.

3   A Generalized Linear Model of 3-Way Analysis

This section looks at the large variance amongst individual raters and suggests a linear function for combining ratings. The linear function takes into consideration the variance amongst the raters.

3.1   Variance of Raters

From the card sorting data, we have two important observations: that the numbers of piles made by different raters are very different (see Fig. 1(a)), and that the sizes of piles made by each rater differ to a large extent (See Fig. 1(b)). The expected inverse relationship between Fig. 1(a) and (b) holds in general. In Fig. 1(c), the histogram of the sizes of piles for all raters illustrates that small pile sizes clearly predominate. It suggests that similarity judgments from different raters are of different strengths.

3.2   A Linear Function of Fusion

The analysis of last subsection show two types of variance amongst raters, one is the number of piles and the other is the sizes of piles. These two types of difference affect the strength of similarity. The simple linear function (2) does not reflect these differences. Accordingly, we introduce a generalized linear model to account for both. More specifically, we propose the following linear function:

$$S(A, B) = \frac{1}{N} \sum_{r=1}^{N} w_r \cdot s_r(A, B). \tag{4}$$

The weights, w_r, reflect the differences of individual raters. The similarity function s_r is generalised as $s_r : P \times P \longrightarrow [0, 1]$, from the set $\{0, 1\}$ to the closed unit interval $[0, 1]$. The values 0 and 1 indicate full dissimilarity and full similarity, respectively. A value between 0 and 1 indicates partial similarity and partial dissimilarity. The generalised function s_r reflects the differences in pile size.

For the weights, we assume that a rater who made more of piles, and consequently, with smaller pile sizes is more informative and confident in assessing similarity. This rater's judgements should be assigned a larger weight. For the similarity function, we assume that pairs in a pile of smaller size are more similar than pairs in piles of larger size. These two assumptions are in fact two different forms of an underlying assumption that, when deciding different piles, a rater implicitly uses a threshold on a perceived degree of similarity. A pair of photos is put into the same pile if the similarity is above the threshold.

3.3   Determining the Weights

Without considering the number of photos in each of the piles that a rater made, a first attempt to quantify the differences between raters can be made by looking

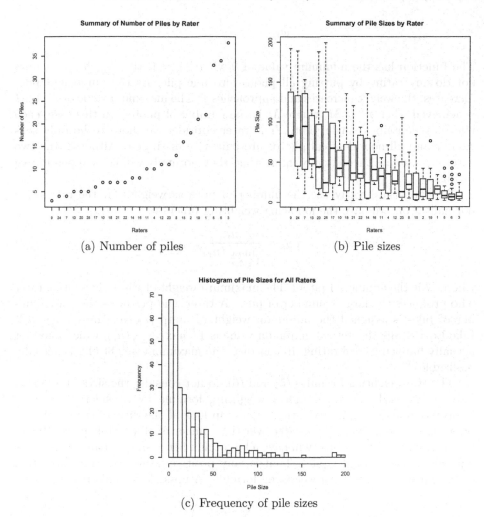

(a) Number of piles

(b) Pile sizes

(c) Frequency of pile sizes

Fig. 1. Raters and the piles they made, organized by increasing number of piles. In (a), the number of piles made by each rater is plotted on the vertical axis. In (b), the sizes of the piles made by each rater is summarized with a boxplot, where the median pile size for each rater is indicated by the bold line. The expected inverse relationship between number of piles and sizes of piles per rater holds in general, but with plenty of variability. In (c), the histogram summarizes the sizes of piles for all raters and small pile sizes clearly predominate.

at the number of piles, n_r, made by rater r. We assume that the more piles made by a rater, the more reliable the rater's ratings. This suggests that any positive monotonic increasing function of n_r will serve the purpose.

A very simple weighting function, based on the number of piles that each rater made, is given by:

$$w_{1r} = 1 - \frac{1}{n_r}. \tag{5}$$

The function has the minimum value of 0 when $n_r = 1$, that is, the rater does not do any rating by putting all photo into one pile. As the number of piles increases, the weight increases and approaches 1. The maximum value of $1 - 1/M$ is achieved when $n_r = M$, where M is the number of photos. In this case, each pile only contain a single photo. The rater considers a photo to be only self-similar and all non-identical photos are pair-wise dissimilar. Although the two extreme cases are theoretically interesting, they normally will not happen in real ratings.

We can also directly use the number of piles as weights for the raters. By normalisation, we have the following weighting function:

$$w_{2r} = \frac{n_r}{\max_{1 \leq i \leq N}(n_i)}, \tag{6}$$

where N is the number of raters. The maximum weight of 1 is assigned to a rater who produces the largest number of piles. A rater who produces the least number of piles is assigned the minimum weight of $\min_{1 \leq i \leq N}(n_i)/\max_{1 \leq i \leq N}(n_i)$. Like Eq. (5), the theoretical minimum value is $1/\max_{1 \leq i \leq N}(n_i)$, which may not actually happen in real rating. In contrast, the maximum weight of 1 is actually realizable.

The two weighting formulas (5) and (6) do not consider the sizes of different piles. In general, we can design a weighting formula by considering detailed information about similarity ratings. One can look at the ratio of the number of photo pairs rated similar by a rater over the number of all possible pairs. Raters who made fewer, larger piles will have a larger number of similar pairs than raters who made more, smaller piles. That is, any positive decreasing transformation of the ratio may serve as a weighting formula. A possible formula is given by:

$$w_{3r} = 1 - \frac{\sum_{i=1}^{n_r} \binom{|P_i|}{2}}{\binom{M_r}{2}}, \tag{7}$$

where M_r is number of photos that rater r actually considered when rating similarity. This formula has the same value as the formula given by Eq. (5) for the two extreme cases. More specifically, the minimum weight 0 is obtained if a rater only has one pile. The maximum weight of 1 is obtained when a rater puts each photo as a separate pile. In general, the fewer the similar pairs, the closer to 1 will be the weight.

For our data set of 356 photos and 25 raters, Fig. 2 summarizes the weights obtained from the three formulas. It can be observed that the three formulae behave similarly: the weights generally increase as the number of piles increases.

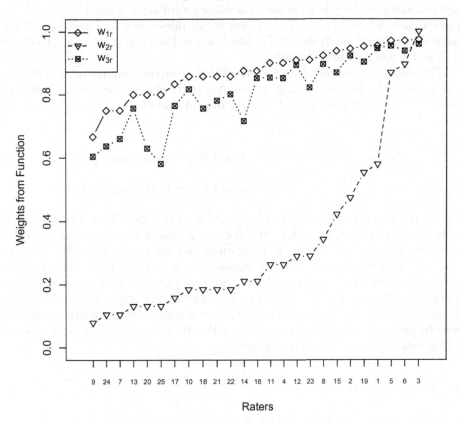

Fig. 2. Plots of the three weighting functions, from Eqs. 5, 6, and 7.

3.4 Determining the Individual Similarity Values

For a specific rater, we can generalize the binary similarity function (1) by taking the sizes of the different piles into consideration.

Let $k_1, \ldots, k_{n_r}$ denote the sizes of the n_r piles, $P_1, \ldots, P_{n_r}$, made by rater r. We assume that the strength of similarity for a pair photos depends on the size of pile into which the pair is sorted. More specifically, the larger the pile, the weaker the degree of similarity. If we assume that the degree of similarity is 1 for pairs in a smallest pile, the similarity will decrease as the size of a pile increases. This immediately leads to the following generalised similarity measure: for two photos A and B,

$$
s_{1r}(A, B) = \begin{cases} \dfrac{\min\limits_{1 \leq i \leq n_r} (k_i)}{k_j}, & A \text{ and } B \text{ are in the same pile } P_j, \\ 0, & A \text{ and } B \text{ are in two different piles.} \end{cases} \tag{8}
$$

It can be seen that s_{1r} has the maximum value of 1 for pairs in a smallest pile and increases as the pile increases. The binary similarity defined by Eq. (1) is a special case. When $k_1 = \ldots = k_{n_r}$, that is, all piles given by rater r are of the same size, we have $s_r(A, B) = 1$ if A and B are in the same pile and $s_r(A, B) = 0$ if A and B are in different piles.

Formula (8) uses the size of a pile. When studying similarity, we consider the number of pairs produced by a pile. Alternatively, we can use the following pair-based similarity measure: for two photos A and B,

$$
s_{2r}(A, B) = \begin{cases} \dfrac{\min\limits_{1 \leq i \leq n_r} \dbinom{k_i}{2}}{\dbinom{k_j}{2}}, & A \text{ and } B \text{ are in the same pile } P_j, \\ 0, & A \text{ and } B \text{ are in two different piles.} \end{cases} \tag{9}
$$

Again, the maximum similarity value is assigned to a pair from a smallest pile. However, similarity value of Eq. (9) will decrease faster than the value of Eq. (8) when the size of a pile increases. Their minimum values are normally different.

Once we introduce weighting formulas for individual raters and generalize similarity value into unit interval $[0, 1]$, we can form different linear models as defined by Eq. (4). Furthermore, we can immediately apply Eq. (3) to obtain 3-way division of similarity, dissimilarity, and divided. An important question is how to interpret and determine a pair of thresholds in a general linear model, which will be a topic of a future paper.

4   Conclusion

Three-way analysis of facial similarity offers a new viewpoint for interpretation. While the similarity or the dissimilarity of a pair of photos is the result of a large degree of agreement of a group of raters, the divided judgment is the result of disagreement. This 3-way classification provides an effective method for dealing with uncertainty involved in facial similarity judgments. This study only considers the step of 3-way division of similarity. As future research, we plan to investigate each of these three divisions with pairwise comparisons in order to obtain a better understanding of facial similarity. Furthermore, it will be interesting to combine the similarity analysis based on human perception with that based on feature calculation to create a more comprehensive model.

Acknowledgements. The authors would like to thank Dominik Ślęzak for his encouragement and the anonymous reviewers for their constructive comments. This work has been supported, in part, by two NSERC Discovery Grants.

References

1. Deibel, K., Anderson, R., Anderson, R.: Using edit distance to analyze card sorts. Expert Syst. **22**(3), 129–138 (2005)
2. Faiks, A., Hyland, N.: Gaining user insight: a case study illustrating the card sort technique. Coll. Res. Libr. **61**(4), 349–357 (2000)
3. Hepting, D.H., Bin Amer, H.H., Yao, Y.: Three-way analysis of facial similarity judgements. In: Proceedings of 2nd International Symposium on Fuzzy and Rough Sets (ISFUROS 2017), October 2017
4. Hepting, D.H., Maciag, T., Spring, R., Arbuthnott, K., Ślęzak, D.: A rough sets approach for personalized support of face recognition. In: Sakai, H., Chakraborty, M.K., Hassanien, A.E., Ślęzak, D., Zhu, W. (eds.) RSFDGrC 2009. LNCS (LNAI), vol. 5908, pp. 201–208. Springer, Heidelberg (2009). https://doi.org/10.1007/978-3-642-10646-0_24
5. Hepting, D.H., Spring, R., Ślęzak, D.: A rough set exploration of facial similarity judgements. In: Peters, J.F., Skowron, A., Sakai, H., Chakraborty, M.K., Slezak, D., Hassanien, A.E., Zhu, W. (eds.) Transactions on Rough Sets XIV. LNCS, vol. 6600, pp. 81–99. Springer, Heidelberg (2011). https://doi.org/10.1007/978-3-642-21563-6_5
6. Hu, B.Q., Wong, H., Yiu, K.C.: On two novel types of three-way decisions in three-way decision spaces. Int. J. Approx. Reason. **82**, 285–306 (2017). http://www.sciencedirect.com/science/article/pii/S0888613X1630319X
7. Li, H., Zhang, L., Zhou, X., Huang, B.: Cost-sensitive sequential three-way decision modeling using a deep neural network. Int. J. Approx. Reason. **85**, 68–78 (2017). http://www.sciencedirect.com/science/article/pii/S0888613X17302086
8. Li, X., Sun, B., She, Y.: Generalized matroids based on three-way decision models. Int. J. Approx. Reason. **90**, 192–207 (2017). http://www.sciencedirect.com/science/article/pii/S0888613X17304784
9. Liang, D., Xu, Z., Liu, D.: Three-way decisions based on decision-theoretic rough sets with dual hesitant fuzzy information. Inf. Sci. **396**, 127–143 (2017). http://www.sciencedirect.com/science/article/pii/S002002551730539X
10. Martine, G., Rugg, G.: That site looks 88.46% familiar: quantifying similarity of web page design. Expert Syst. **22**(3), 115–120 (2005)
11. Soranzo, A., Cooksey, D.: Testing taxonomies: beyond card sorting. Bull. Am. Soc. Inf. Sci. Technol. **41**(5), 34–39 (2015)
12. Turk, M., Pentland, A.: Eigenfaces for recognition. J. Cogn. Neurosci. **3**(1), 71–86 (1991)
13. Yang, X., Li, T., Liu, D., Chen, H., Luo, C.: A unified framework of dynamic three-way probabilistic rough sets. Inf. Sci. **420**, 126–147 (2017). http://www.sciencedirect.com/science/article/pii/S0020025517308939
14. Yao, Y.: An outline of a theory of three-way decisions. In: Yao, J.T., Yang, Y., Słowiński, R., Greco, S., Li, H., Mitra, S., Polkowski, L. (eds.) RSCTC 2012. LNCS (LNAI), vol. 7413, pp. 1–17. Springer, Heidelberg (2012). https://doi.org/10.1007/978-3-642-32115-3_1
15. Yao, Y.Y.: Three-way decisions and cognitive computing. Cogn. Comput. **8**(4), 543–554 (2016). https://doi.org/10.1007/s12559-016-9397-5
16. Yu, H., Jiao, P., Yao, Y., Wang, G.: Detecting and refining overlapping regions in complex networks with three-way decisions. Inf. Sci. **373**, 21–41 (2016). http://www.sciencedirect.com/science/article/pii/S0020025516306703

Empirical Comparison of Distances for Agglomerative Hierarchical Clustering

Shusaku Tsumoto[1(✉)], Tomohiro Kimura[2], Haruko Iwata[3], and Shoji Hirano[1]

[1] Department of Medical Informatics, Faculty of Medicine, Shimane University,
Izumo, Japan
{tsumoto,hirano}@med.shimane-u.ac.jp
[2] General Coordination Division, Faculty of Medicine, Shimane University,
Izumo, Japan
t-kimura@med.shimane-u.ac.jp
[3] Center for Bed-Control, Shimane University Hospital,
89-1 Enya-cho, Izumo 693-8501, Japan
haruko23@med.shimane-u.ac.jp
http://www.med.shimane-u.ac.jp/med_info/tsumoto/index.htm

Abstract. This paper proposes a method for empirical comparison of distances for agglomerative hierarchical clustering based on rough set-based approximation. When a set of target is given, a level of clustering tree where one branch includes all the targets can be traced with the number of elements included. The pair (*#clustersofalevel*, *#elementsofacluster*) can be viewed as indices-pair for a given clustering tree.

Keywords: Agglomerative hierarchical clustering · Distances
Empirical comparison

1 Introduction

Introduction of a hospital information system enables us to store all the clinical information and to apply machine learning and data mining methods to those data [1,7,8]. For example, ordinary statistical methods are applied for hospital management [13,14], temporal data mining are applied for capturing behavior of medical staff and for analyzing disease progression [3,9,11].

Clinical plans can be also extracted from histories of executed clinical actions. Iwata and Tsumoto proposed a construction of clinical pathway from histories of executed nursing orders [6,12].

In these analysis, clustering methods play an important roles because clustering results will capture not only rediscovery of medical knowledge but also some new discovery which are unexpected to domain experts.

In most cases, euclidean distance was used for clustering. However, many types of clustering distances have been proposed [2], and it is not clear whether

© Springer International Publishing AG, part of Springer Nature 2018
J. Medina et al. (Eds.): IPMU 2018, CCIS 854, pp. 538–548, 2018.
https://doi.org/10.1007/978-3-319-91476-3_45

which clustering distances is suitable for construction of a nursing pathway. Also, it is not easy to determine which clustering is better.

This paper proposes a method for empirical comparison of distances for agglomerative hierarchical clustering based on rough set-based approximation. The key point is that a clustering result are represented as a tree. Thus, if a set of target items a given, we can trace the level where all the target items are included in one group. The number of clusters at the level with the number of items included in that level can be used as indices. For example, let us consider the case when the number of target items is 7. If one cluster method will generate a tree where one group includes only all the attributes with total two clusters, let us describe the pair $(2, 7/7)$. Then, if another clustering method has $(5, 20/7)$, we can say that the former clustering method is better than the latter one with respect to the targets.

In this paper six distances and seven clustering methods were compared by using the above pair. The target items are based on nursing cares necessary for surgical operation of cataracta. Empirical results show that euclidean distance is the best for obtaining a suitable clinical pathway for all clustering methods.

The paper is organized as follows. Section 2 gives research background, where an example of a clinical pathway is shown. Section 3 defines a metric pair obtained from a given clustering tree and a given target concept. Section 4 shows an algorithm for empirical comparison. Section 5 shows experimental evaluation in which the above proposed method was applied to the data of cataract operation. Finally, Sect. 6 concludes this paper.

2   Background

2.1   Clinical Pathway

A clinical pathway for a disease describes a schedule of medical care, which is optimized during the hospitalization [4, 15]. It is very important for efficient clinical process management, but usually its construction is manually acquired from doctors or nurses, according to their experiences. Let us give an example. Figure 1 shows a clinical pathway on cataracta used in Shimane University hospital. The hospitalization period consists of three periods: preoperation, operation and post-operation periods. The preoperation date is denoted by -1 day, and operation date is by 0 day.

The pathway will be executed as follows. For the preoperation date, body temperature (BT), pulse rate (PR) and blood pressure (BP) are checked and preoperation instruction will be given. For operation date, BT, PR and BP are checked, and the symptoms of nausea, vomitting and eye pain are inspected. Then, during postoperation period, in addition to nursing orders for the operation date, coaching will be applied. Finally, if the status of a patient is stable, the patient will be discharged five days after the operation.

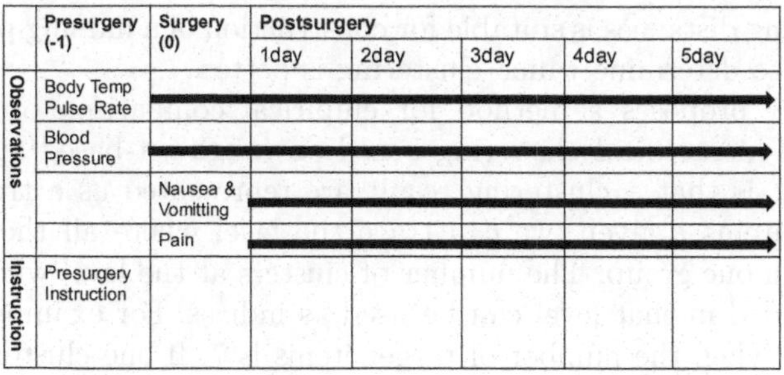

	Presurgery (-1)	Surgery (0)	Postsurgery					
			1day	2day	3day	4day	5day	
Observations	Body Temp Pulse Rate							
	Blood Pressure							
	Nausea & Vomitting							
	Pain							
Instruction	Presurgery Instruction							

Fig. 1. Clinical pathway for cataracta

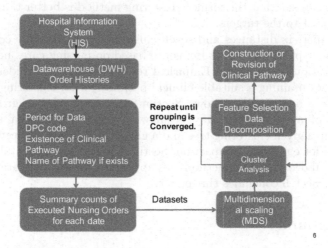

Fig. 2. Construction of clinical pathway (clustering + feature selection)

2.2 Clinical Pathway Construction

This subsection summarizes methods for clinical pathway construction [5,6,10, 12]. Tsumoto and Iwata firstly introduces combination of agglomerative hierarchical clustering and feature selection method for its construction [5,6] (Fig. 2.) Clustering is applied to data on executed nursing orders where rows and columns give nursing orders and date of hospitalization. Then, grouping of nursing actions is extracted. Such groups can be used as classification labels, and information gain is calculated for each attribute. Then, attribute will be grouped by the values of the gains. Since attributes are each date of hospitalization, such grouping of attributes corresponds to clinical schedule (Fig. 4.)

It is notable that the above second step (calculation of information gain) is actually a kind of grouping. Thus, clustering for attributes (attribute clustering) can be applied for this purpose. Tsumoto et al. introduces combination of sample clustering and attribute clustering, called dual clustering [12] and obtains the

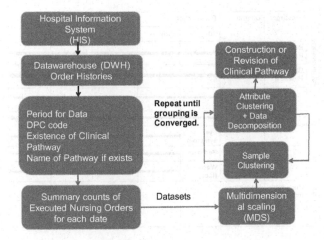

Fig. 3. Construction of clinical pathway

same performance as the former approach [6]. Tsumoto et al. generalize the approach [10] with combination of data decomposition and dual clustering.

The method consists of the following five steps (Fig. 3): first, histories of nursing orders are extracted from hospital information system. Second, orders are classified into several groups by using clustering on the principal components (sample clustering). Third, feature selection method is applied and the dataset is decomposed into subtables. The second and the third process will be repeated until the clustering results are converged.

Figure 4 shows the pathway generated by the above construction algorithm shown in Fig. 3 The induced results show that coaching and wash, whose chronological characteristics are similar to the orders indispensable to the treatment of glaucoma, were not included in the existing pathway. Furthermore, coaching and wash should be treated as postoperation follow-up and routine process, respectively.

The results show that the method is able to construct a clinical pathway for this disease. Furthermore, the temporal intervals suggested the optimal and maximum length of stay.

The above approaches are based on Ward's method [2] with Euclidean distance. However, there exist many types of clustering even for agglomerative types, such as complete-linkage, average-linkage. It is not well known which type of clustering method and which type of distance give the best performance.

3   Definition

Definition 1 (Clustering Tree). *A clustering tree $T(D, strategy, metric)$ is obtained by applying an agglomerative hierarchical clustering with a given strategy and a metric to a dataset D. For example, $T(D, ward, euclidean)$ is a clustering tree of D obtained by Ward's method with euclidean distance.*

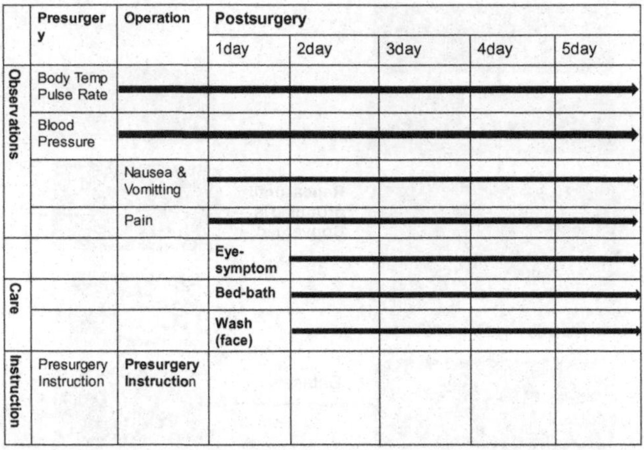

		Presurgery	Operation	Postsurgery				
				1day	2day	3day	4day	5day
Observations	Body Temp Pulse Rate			⟶				
	Blood Pressure			⟶				
			Nausea & Vomitting	⟶				
			Pain	⟶				
Care				Eye-symptom	⟶			
				Bed-bath	⟶			
				Wash (face)	⟶			
Instruction		Presurgery Instruction	Presurgery Instruction					

Fig. 4. Clinical pathway for cataracta obtained from data

Definition 2 (Representation of Clustering Tree: List form). *Let us a Dataset consist of n examples. A clustering tree T consists of* (*level, data_partition*), *such as:*

$$T = \{(1, \{a_1, a_2, \cdots, a_n\}), (2, \{a_1, a_2\}, \cdots), \cdots, (k, \{a_1\}, \{a_2\}, \cdots .\{a_n\})\}$$

It is notable that all the sets of a partition are single, whose number of the partition is n.

For example, Let us consider a clustering tree shown in Fig. 5, consisting of 10 examples.

$$
\begin{aligned}
T = \quad & \{(1, \{1, 2, 3, 4, 5, 6, 7, 8, 9, 10\}), \\
& (2, \{1, 2, 3, 4, 5, 6, 7\}, \{8, 9, 10\}), \\
& (3, \{1, 2, 3, 4\}, \{5, 6, 7\}, \{8, 9, 10\}), \\
& (4, \{1, 2, 3\}, \{4\}, \{5, 6\}, \{7\}, \{8, 9\}, \{10\}), \\
& (5, \{1, 2\}, \{3\}, \{4\}, \{5, 6\}, \{7\}, \{8\}, \{9\}, \{10\}), \\
& (6, \{1\}, \{2\}, \{3\}, \{4, \}, \{5\}, \{6\}, \{7\}, \{8\}, \{9\}, \{10\})\}
\end{aligned}
$$

It is notable that this structure is not the same as a dendrogram obtained by a given agglomerative hierarchical clustering method. In the case of Fig. 5, corresponding dendrogram may be like Fig. 6, where vertical branch shows the strength of the fusion. A dendrogram shows how two or more examples are grouped according to the value of dissimilarities. Here, we focus on how data are partitioned. Thus, we look down the dendrogram from the top level of the tree. Accurate distance values are ignored, but the structure of the dendrogram is conserved.

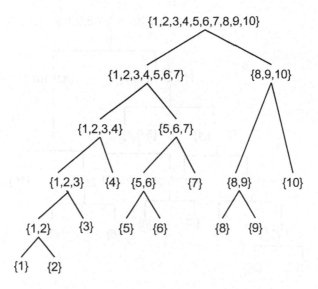

Fig. 5. Example of clustering tree

Definition 3 (Metric pair). *A metric pair of a clustering tree*
$T(D, Strategy, Metric)$ *with a target concept* C *is defined as:*

$$\left(Par(l), \frac{|\min(C \subseteq S|)}{|C|} \right)$$

where l *is the level of* $\min(C \subseteq S)$ *and* $Par(l)$ *denotes the number of a partition of the level* l.

Example 1. Let us consider a clustering tree shown in Fig. 5, consisting of 10 examples. Each horizontal level shows the data partition. For example, the second level has two partitions: $\{1, 2, 3, 4, 5, 6, 7\}$ and $\{8, 9, 10\}$. Let a targe concept have $\{1, 2, 3, 4\}$. Then, we can find this set at the third level where are composed of 4 clusters. Then, the metric pair is given as $(3, 4/4)$. If we take $\{1, 2, 3, 4, 5\}$ as a target concept, $\{1, 2, 3, 4, 5, 6, 7\}$ in the second level is a super set, and since no other small super set is not found, the pair is obtained as $(2, 7/4)$.

4 Algorithm for Clustering Tree

A matching pair can be obtained after a clustering tree is calculated from an original dendrogram. Figure 7 shows an algorithm for a clustering tree. At each hiearchical level, first compare the data partition at the level with a target set. If a target set is found at this level, the value of l is equal to this level. Then, the system checks the lower hierarchical level. If any partition is not equal to the target set at any level, the value of l is equal to 1.

Let us illustrate how the method works. From the dendrogram, the following clustering tree is obtained.

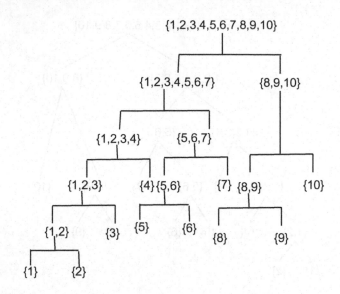

Fig. 6. Example of dendrogram

$[(2, \{BP, BT/PR, Eye\_symptoms, NauseaVomitting, Pain, Coaching, Wash\},$
$\{Preoperation\_Instruction, Shampoo, BS(Ward), Nerve, Others,$
$Psychological, Shower, Bath, SPO2, Skin/Nail, (Ns)Vital_{sign},$
$(Ns)BT/PR, DM, (Ns)BP, Guidance,$
$Fall\_prevention, Fall\_prevention(Bed)\}),$

$(3, \qquad \{BP, BT/PR\},$
$\{Eye\_symptoms, NauseaVomitting, Pain, Coaching, Wash\},$
$\{Preoperation_{I}nstruction, Shampoo, BS(Ward), Nerve, Others,$
$Psychological, Shower, Bath, SPO2, Skin/Nail, (Ns)Vital_{sign},$
$(Ns)BT/PR, DM, (Ns)BP, Guidance,$
$Fall\_prevention, Fall\_prevention(Bed)\}),$

$(4, \qquad \{BP, BT/PR\},$
$\{Eye_{s}ymptoms, NauseaVomitting, Pain, \}$
$\{Coaching, Wash\},$
$\{Preoperation\_Instruction, Shampoo, BS(Ward), Nerve, Others,$
$Psychological, Shower, Bath, SPO2, Skin/Nail, (Ns)Vital_{sign},$
$(Ns)BT/PR, DM, (Ns)BP, Guidance,$
$Fall\_prevention, Fall\_prevention(Bed)\}),$

$(5, \qquad \{BP, BT/PR\},$
$\{Eye_{s}ymptoms, NauseaVomitting, Pain\}$
$\{Coaching, Wash\},$
$\{Preoperation\_Instruction\}$
$\{Shampoo, BS(Ward), Nerve, Others,$
$Psychological, Shower, Bath, SPO2, Skin/Nail, (Ns)Vital_{sign},$
$(Ns)BT/PR, DM, (Ns)BP, Guidance,$
$Fall\_prevention, Fall\_prevention(Bed)\}), ...]$

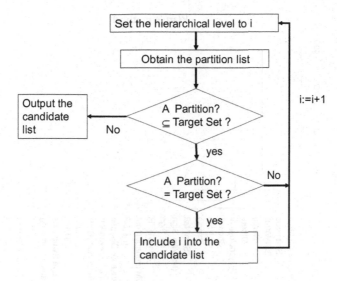

Fig. 7. Algorithm

The target set is $\{BP, BT/PR, Eye\_symptoms, NauseaVomitting, Pain,$ $Coaching, Wash\}$ Then, first, the level is set to 2 ($l = 2$). $P(2)$ is equal to 2. Since the first element of data partition is exactly equal to the target set, this level will include into the candidate list. Then, set $l = 3$. since $\{BP, BT/PR\}$ is separate from the target set, the condition fails and the process is stopped. Since the candidate list is $\{3\}$, $P(3) = 3$. The target set is equal to the first element, so the second element of the pair is 7/7.

If the target set is $\{Eye\_Symptoms, Nausea/Vomitting, Pain\}$, the second element of the data partition is a superset of the target set, not only the level, but also the second level satisfies the condition. Since this partition is observed in lower levels, the candidate list is equal to $\{3, 4, 5, 6 \ldots\}$. The minimum value of the level is equal to 3, and $P(3) = 3$. And the second element of the pair is equal to 3/3.

4.1   How to Compare Metric Pairs

Comparison of metric pairs is two-fold (Fig. 9). First, the numbers of partition $P(l)$ are compared. The smaller pairs are better. Then, secondly, the matched pairs are compared. The pairs whose value is close to 1.0 are better.

5   Experimental Results

Cases of cataracta, lenticular diseases with operation of both eyes and without other operations (DPC code: 020110xx97x0x1), who were admitted to Shimane University Hospital, were extracted from the hospital information system. We selected 65 out of 134 cases where the pathway acquired manually from doctors

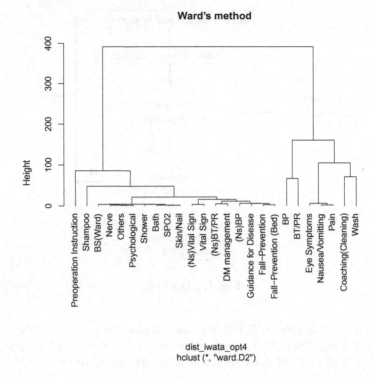

Fig. 8. Sample of dendrogram (cataract)

and nurses was applied an completed and applied the clinical pathway construction algorithm with the following combinations of strategy and clustering dissimilarities. Methods are Ward method, Single Linkage, Complete Linkage, Average Linkage, Mcquittybmethod, Median method and Centroid method. Clustering dissimilarities are Euclidean, Mahalanobis, Manhattan, Chebyshev, Minkowski and Canberra.

Table 1 shows the metric pairs obtained by each strategy and dissimilarity, where seven nursing orders are selected as a target.

Since $(2, 7/7)$ will give the best form of a clinical pathway [6] for this dataset, we can estimate the best combination of a method and a distance. The best ones are: Ward method with Euclidean, Manhattan and Minkowski distance, Single Linkage with Euclidean and Minkowski, Average Linkage with Euclidean, Manhattan and Minkowski, Mcquitty method with Euclidean, Manhattan and Minkowski, Median method with Euclidean, Manhattan and Minkowski and Centroid method with Euclidean, Manhattan and Minkowski.

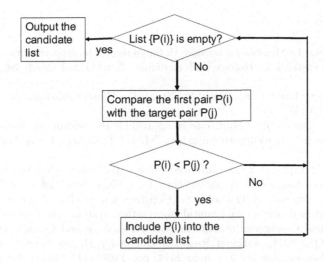

Fig. 9. Empirical comparison of metric pairs

Table 1. Empirical comparison of clustering distances

	Euclidean	Mahalanobis	Manhattan	Chebyshev	Minkowski	Canberra
Ward	(2,7/7)	(3,23/7)	(2,7/7)	(2,8/7)	(2,7/7)	(8,7/7)
Single Linkage	(2,7/7)	(2,24/7)	(1,25/7)	(1,25/7)	(2,7/7)	(11,7/7)
Complete Linkage	(2,7/7)	(1,25/7)	(2,7/7)	(1,25/7)	(2,7/7)	(9,7/7)
Average Linkage	(2,7/7)	(3,23/7)	(2,7/7)	(1,25/7)	(2,7/7)	(5,7/7)
Mcquitty	(2,7/7)	(3,23/7)	(2,7/7)	(1,25/7)	(2,7/7)	(8,7/7)
Median	(2,7/7)	(3,23/7)	(2,7/7)	(1,25/7)	(2,7/7)	(9,7/7)
Centroid	(2,7/7)	(1,25/7)	(2,7/7)	(1,25/7)	(3,7/7)	(9/7/7)

6   Conclusion

This paper proposes a method for empirical comparison of distances for agglomerative hierarchical clustering based on rough set-based approximation. The key
point is that a clustering result are represented as a tree. Thus, if a set of target
items a given, we can trace the level where all the target items are included in
one group. The number of clusters at the level with the number of items included
in that level can be used as indices. Six distances and seven clustering methods
were compared by using the above pair. The target items are based on nursing
cares necessary for surgical operation of cataracta. Empirical results show that
euclidean distance is the best for obtaining a suitable clinical pathway for all
clustering methods.

Acknowledgements. This research is supported by Grant-in-Aid for Scientific
Research (B) 15H2750 from Japan Society for the Promotion of Science(JSPS).

References

1. Bichindaritz, I.: Memoire: a framework for semantic interoperability of case-based reasoning systems in biology and medicine. Artif. Intell. Med. **36**(2), 177–192 (2006)
2. Everitt, B.S., Landau, S., Leese, M., Stahl, D.: Cluster Analysis, 5th edn. Wiley, Hoboken (2011)
3. Hirano, S., Tsumoto, S.: Multiscale comparison and clustering of three-dimensional trajectories based on curvature maxima. Int. J. Inf. Technol. Decis. Mak. **9**(6), 889–904 (2010)
4. Hyde, E., Murphy, B.: Computerized clinical pathways (care plans): piloting a strategy to enhance quality patient care. Clin. Nurse Spec. **26**(4), 277–282 (2012)
5. Iwata, H., Hirano, S., Tsumoto, S.: Construction of clinical pathway based on similarity-based mining in hospital information system. In: Proceedings of 2nd International Conference on Information Technology and Quantitative Management, ITQM 2014, National Research University Higher School of Economics (HSE), Moscow, Russia, 3–5 June 2014, pp. 1107–1115 (2014). https://doi.org/10.1016/j.procs.2014.05.366
6. Iwata, H., Hirano, S., Tsumoto, S.: Maintenance and discovery of domain knowledge for nursing care using data in hospital information system. Fundam. Inform. **137**(2), 237–252 (2015). https://doi.org/10.3233/FI-2015-1177
7. Shortliffe, E., Cimino, J. (eds.): Biomedical Informatics: Computer Applications in Health Care and Biomedicine, 3rd edn. Springer, London (2006). https://doi.org/10.1007/978-1-4471-4474-8
8. Tsumoto, S., Hirano, S.: Risk mining in medicine: application of data mining to medical risk management. Fundam. Inform. **98**(1), 107–121 (2010)
9. Tsumoto, S., Hirano, S.: Detection of risk factors using trajectory mining. J. Intell. Inf. Syst. **36**(3), 403–425 (2011)
10. Tsumoto, S., Hirano, S., Iwata, H.: Construction of clinical pathway from histories of clinical actions in hospital information system. In: 2016 IEEE International Conference on Big Data, BigData 2016, Washington DC, USA, 5–8 December 2016, pp. 1972–1981 (2016). https://doi.org/10.1109/BigData.2016.7840819
11. Tsumoto, S., Hirano, S., Iwata, H., Tsumoto, Y.: Characterizing hospital services using temporal data mining. In: SRII Global Conference, pp. 219–230. IEEE Computer Society (2012)
12. Tsumoto, Y., Iwata, H., Hirano, S., Tsumoto, S.: Construction of clinical pathway using dual clustering. Neurosci. Biomed. Eng. **3**, 49–56 (2015)
13. Tsumoto, Y., Tsumoto, S.: Exploratory univariate analysis on the characterization of a university hospital: a preliminary step to data-mining-based hospital management using an exploratory univariate analysis of a university hospital. Rev. Socionetw. Strateg. **4**(2), 47–63 (2010)
14. Tsumoto, Y., Tsumoto, S.: Correlation and regression analysis for characterization of university hospital (submitted). Rev. Socionet. Strateg. **5**(2), 43–55 (2011)
15. Ward, M., Vartak, S., Schwichtenberg, T., Wakefield, D.: Nurses' perceptions of how clinical information system implementation affects workflow and patient care. Comput. Inform. Nurs. **29**(9), 502–511 (2011)

Soft Computing for Decision Making in Uncertainty

Missing Data Imputation by LOLIMOT and FSVM/FSVR Algorithms with a Novel Approach: A Comparative Study

Fatemeh Fazlikhani$^{(\boxtimes)}$, Pegah Motakefi$^{(\boxtimes)}$, and Mir Mohsen Pedram

Data Mining Lab, Department of Electrical and Computer Engineering,
Kharazmi University, 15719-14911 Tehran, Iran
`fateme.fazlikhani.1990@gmail.com,`
`pegah.motakefi@gmail.com, pedram@khu.ac.ir`

Abstract. Missing values occurrence is an inherent part of collecting data sets in real world's problems. This issue, causes lots of ambiguities in data analysis while processing data sets. Therefore, implementing methods which can handle missing data issues are critical in many fields, in order to providing accurate, efficient and valid analysis.

In this paper, we proposed a novel preprocessing approach that estimates and imputes missing values in datasets by using LOLIMOT and FSVM/FSVR algorithms, which are state-of-the-art algorithms. Classification accuracy, is a scale for comparing precision and efficiency of presented approach with some other well-known methods. Obtained results, show that proposed approach is the most accurate one.

Keywords: Missing data · Imputation
Local linear neuro-fuzzy model (LOLIMOT)
Fuzzy support vector machine (FSVM)
Fuzzy support vector regression (FSVR)

1 Introduction

Nowadays, knowledge discovery is growing up significantly in social, economic and medical application fields. In medical research, diagnosis is usually based on previous patient's information. The diagnosis accuracy of patient's disease like diabetes, breast cancer and others, is greatly depending on expert's experiences [1]. One important issue that is often regarded by many different researchers is missing data occurrence. In practice, it is possible that an analyst cannot have all response variables for any reason, which is called missingness in response. Therefore, missing information draw a statistician's attention to itself. Missing data may cause a lot of problems in processing and analyzing data in data sets. Clearly, inferences that are discovered from complete data are more accurate than the incomplete data, especially when missing rate is high.

F. Fazlikhani and P. Motakefi—First two authors have contributed equally.

© Springer International Publishing AG, part of Springer Nature 2018
J. Medina et al. (Eds.): IPMU 2018, CCIS 854, pp. 551–569, 2018.
https://doi.org/10.1007/978-3-319-91476-3_46

Since the incomplete data are an inherent part of studies and leads a lot of critical conditions, most of researchers are looking for techniques which reduce effects of the missing values in data analysis. Usually, detection of missing data in data sets, is easy and these missing data appears as a null or wrong data. In addition, estimating the missing values in variables which have a dependency with the other variables, is critical. In these cases, estimation of missing values is based on substantial relationship between corresponding variables. Rational solution for dealing with missing data, depends on how the data has missed.

Missing data can be handled by three different kinds of methods [2]:

- Using of deletion methods. In these techniques, a record of data, which contains missing values will be deleted from data set. Eliminating the record of missing data may cause small data sample size.
- Using of means and modes in each feature that contains missing values. Imputing missing values by means, is common in numerical data and also, mode imputation is utilized in nominal data sets.
- Missing value imputation with machine learning and data mining methods. Machine learning imputation techniques seem to be more accurate than the traditional methods [3].

This paper presents a novel preprocessing approach with usage of two state-of-the-art imputation methods based on Local Linear Neuro-Fuzzy (LLNF) and FSVM/FSVR algorithms. The quality of data will improve by applying these efficient imputation methods in incomplete data sets. Then the imputed and completed data is fed to MLP classifier algorithm for comparing imputation accuracy.

The rest of this paper is divided into following sections. Section 2 is completely considering the background study of imputation methods and a review of MLP classifiers. Subsequently, Sect. 3 presents the neuro-fuzzy model and FSVM/FSVR. Evaluation of proposed preprocessing method and usage of two mentioned algorithms is along with in Sect. 4. Eventually, results are shown in Sect. 5 and the conclusion is provided to be described in section.

2 Literature Review

This section presents a brief summary of missing concepts, missing value handling methods, including some statistic and machine learning techniques.

2.1 Missing Data

Date sets can contain missing values which are distributed in all over them. Missing data mechanisms and structures in multivariate data samples are grouped in three modes:

- **Missing At Random (MAR).** When the distribution of missing values, just depends on known values and not depend on attributes which have missing values. In this case missingness is unavoidable [6].

- **Missing Completely At Random (MCAR).** Missing data mechanism is called missing completely at random if the distribution of missing values is independent with other attributes, neither known attributes nor missing values [6].
- **Missing Not At Random (MNAR).** MNAR occurs when the distribution of missing values can depend on the attributes with missing value [7].

This study, only considers the MCAR structure in data. In missing concepts, missing data patterns could be introduced which shows the missing locations among variables of data sets. Figure 1 depicts different types of missing data patterns. The yellow areas indicate the missing data in the data set.

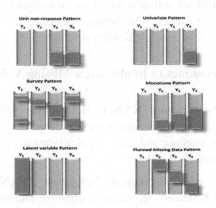

Fig. 1. Different types of missing data patterns [4]

2.2 Missing Value Imputation

One of the known approaches for analyzing and handling missing data are imputation-based methods. In these particular methods, missing values have been filled or imputed by an estimated value, rather than eliminating missing data. Imputation methods appear in a wide range, from simple methods to the most complex ones, but the most important advantage of all imputation techniques is that they may not reduce the sample size [31].

2.3 Missing Value Handling Techniques

Due to analyzing and evaluating the proposed novel models in missing data problems, section below contains a brief look at missing data treating methods as follows:

Deletion methods or Ignore Missing. Excluding all missing units from the data set that can lead to biases and small sample size [8, 11].

Most Common (MC) Value Imputation. Uses the most common value of attributes for imputing missing values, it combines with the mean imputation method for numeric and continuous attributes [8–10, 12].

Event Covering (EC). EC includes 3 steps:
- Detecting statistical interdependency from data patterns.
- Clustering data based on detected interdependency.

- interpret the data patterns for each identified cluster [8].

Singular Value Decomposition Imputation (SVD). Firstly, missing values are estimated with EM algorithm, then SVD will be computed. Ultimately SVD obtains a set of mutually orthogonal expression patterns that can be linearly combined to approximate the values of every features in the data set [8].

Bayesian Principal Component Analysis (BPCA). BPCA consists of three basic steps:

- Principal component (PC) regression
- Bayesian estimation
- A repetitive algorithm like EM [8, 13].

EM Algorithm (EM). EM algorithm is based on an irregular idea formulated to deal with incomplete data. It is named EM, because expected value in each iteration of algorithm, calculates and then a maximization performs [14].

2.4   Data Mining Techniques to Implement a Missing Value Estimator

K-Nearest Neighbor Imputation (KNNI). The missing values are imputed with k-nearest neighbors based on a similarity measure between units. In numerical attributes, it is computed the average and in nominal attributes, the most common unit in neighbors has been chosen [8, 11, 15].

Weighted Imputation with K-Nearest Neighbor (WKNNI). In this method weighted mean of these K nearest neighbors is imputed with missing values. Weights have inverse relation with neighborhood distances [8, 12, 15].

K-means Clustering Imputation (KMI). All the units are clustered with the k-means algorithm and missing values are estimated based on the cluster that belongs to it [8, 12].

Fuzzy K-means Clustering Imputation (FKMI). Data points cannot assign to a specific cluster and each of them belongs to all K clusters with different membership degree. Membership degree is a number between 0 and 1 [8].

3   The Proposed Approach

In this section, the proposed approach and used methods are described. The main novel approach of the study is the type of data preprocessing and modeling. A single model is built for each feature that contains missing values. Data set is preprocessed by eliminating records and features which contain missing values, except a feature that has missing values and will be modeled and imputed. Then this modeling approach will be continued until the missing values imputation is completed.

3.1   Data Preparing and Preprocessing

This section, demonstrates data preparing and preprocessing for two methods (LOLI-MOT and FSVM/FSVR).

Preparing Data Set. Assume a single missing value that is placed in a row and a particular feature for preparing data set these steps are done.

- If the feature contains numerical value, the data preprocessing begins to apply into the models.
- If the feature type is categorical, the values have to convert into numerical values first. In order to do that, a number must be considered for each specified category. But it should be noted that the gained numeric models must replace with missing categorical values and estimated values should be assigned to its own category based on pre-determined threshold at the end.

Preparing Train Data and Test Data. Then, in the next step of data pre-processing:

- If we have enough complete records of data in data set, it is divided to train and test data. At most 2/3 of data is considered as train data and the other part is considered as test data.
- The records which contain missing values are moved to test data part, so depends on data set size, the test data can contain both missing data records and some completed data records.

Data Preprocessing. After preparing data set, data preprocessing is done.

- In this step, all samples that contain missing value, except the sample intended for imputation, are deleted manually or by a generated code.
- In the case which is considered for imputation, if there is more than one variable with missing values, those variables are omitted too.

The model is prepared to estimate the missing values.

3.2   Applied Methods

Local Linear Neuro-Fuzzy Model. The main approach in local linear neuro-fuzzy models are dividing the input space into several sub-partitions which are simpler and linear with validation functions in order to determine the valid area for each LLM. A local linear neuro-fuzzy model structure is displayed in Fig. 2. Each local linear model (LLM) is assigned to a neuron. A validity function is assigned to per neuron.

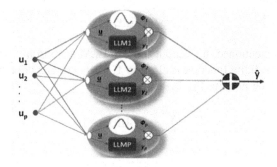

Fig. 2. A local linear neuro-fuzzy model structure [21]

The local output of each local linear model is calculated by the weighted sum of the inputs in their valid region. Then the overall output is calculated through the sum of all local outputs for all neurons in the model, Eq. 1.

$$\hat{y} = \sum_{i=1}^{M} \left(w_{i0} + w_{i1}u_1 + w_{i2}u_2 + \cdots + w_{ip}u_p \right) \Phi_i(\underline{u}) \tag{1}$$

$\Phi_i(u)$ or validity functions are very similar to the basic RBF functions. Validity functions on input vectors are normalized and are defined as Eq. 2.

$$\sum_{i=1}^{M} \Phi_i(\underline{u}) = 1 \tag{2}$$

Validity functions are usually normalized Gaussian functions. If these Gaussian functions also have orthogonal mode, then it is defined as Eq. 3.

$$\Phi_i(\underline{u}) = \frac{\mu_i(\underline{u})}{\sum_{j=1}^{M} \mu_j(\underline{u})} \tag{3}$$

Where $\mu(\underline{u})$ defined in Eq. 4:

$$\mu_i(\underline{u}) = exp\left(-\frac{1}{2} \left(\frac{(u_1 - c_{i1})^2}{\sigma_{i1}^2} + \cdots + \frac{(u_p - c_{ip})^2}{\sigma_{ip}^2} \right) \right) \tag{4}$$

To create a local linear neuro-fuzzy model it will need 3 kinds of parameters. Weight w, Center coordinate C_{ij} and standard deviation σ_{ij} [18, 19, 21].

$$\begin{bmatrix} 1 & u_1(1) & u_2(1) & \cdots & u_p(1) \\ 1 & u_1(2) & u_2(2) & \cdots & u_p(2) \\ \vdots & \vdots & \vdots & & \vdots \\ 1 & u_1(N) & u_2(N) & \cdots & u_p(N) \end{bmatrix} \tag{5}$$

The regression matrices for all LLMs i = 1, 2, ..., M are the same because Xi is independent of i. The output of each neuron is calculated as Eq. 6.

$$\underline{\hat{y}}_i = \underline{X}_i \underline{w}_i \tag{6}$$

As previously mentioned, the output of each LLM is valid in a specific region that the corresponding validity function is close to 1. This action is done by minimizing the loss function for each neuron, Eq. 7.

$$I_i = \sum_{j=1}^{N} \Phi_i(\underline{u}(j)) e^2(j) \tag{7}$$

According to the matrix below, Eq. 8:

$$\underline{Q}_i = \begin{bmatrix} \Phi_i(u(1)) & 0 & \cdots & 0 \\ 0 & \Phi_i(u(1)) & \cdots & 0 \\ \vdots & \vdots & \ddots & \vdots \\ 0 & 0 & \cdots & \Phi_i(u(1)) \end{bmatrix} \tag{8}$$

Optimized weight parameters are calculated as Eq. 9:

$$\hat{\underline{w}}_i = \left(\underline{X}_i^T \underline{Q}_i \underline{X}_i\right)^T \underline{X}_i^T \underline{Q}_i \underline{y} \tag{9}$$

LLNF Non-linear Parameters Estimation. The center coordinate Cij and standard deviation σ_{ij} are related parameters to the validity functions. The input space that has been partitioned into three rectangular areas by taking 3 validity function is displayed in Fig. 3 Using the normal Gaussian validity functions makes center coordinate Cij present center of the rectangle and standard deviations σ_{ij} specifies a rectangular extends in all dimensions. In order to make the relationship between validity functions standard deviations with rectangles extends, the relationship is considered as follows [21], Eq. 10.

$$\sigma_{ij} = k_\sigma \cdot \Delta_{ij} \tag{10}$$

Determining the validity function parameters is a nonlinear optimization problem. There are many techniques to determine these parameters, such as network partitioning, clustering the input space and etc. [22].

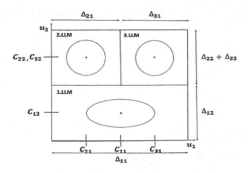

Fig. 3. Partitioning the input space into three rectangular areas [20]

Local Linear Model Tree Algorithm (LOLIMOT). LOLIMOT is an incremental tree-constructional algorithm that divides the input space by axis-orthogonal splits. At each iteration of the algorithm, a new law or local linear model (LLM) is added to the overall model and validity functions which correspond to the current partition of the input space are calculated and model weight parameters are obtained by using the least square technique. The only parameter that must be pre-specified is a proportional factor between rectangles extends and standard deviation. This parameter is usually considered to be equal to 1/3 [23].

LOLIMOT Algorithm. LOLIMOT algorithm contains an external loop for calculating non-linear parameters and an inner loop for calculating weight parameter by applying the local estimation approach [20].

1. Start with a basic model: Create validity functions for partitioning the space and estimating the LLM parameters using the least square algorithm. M is the number of elementary LLMs. If there is no pre-existing partition on the input space, M is set to 1 and starts working with one LLM (because validity function covers whole input space with $\Phi 1(u)$, use global linear model).
2. Choose the worst LLM: Calculate a local loss function for every i = 1, ..., M local linear models. It can be calculated by using the model's weighted square error. Choose the worst LLM according to efficiencies and consider i as the index for the worst LLM. This can be done through max (Ii) equation.
3. Check all the dimensions: Consider the worst LLM for optimization. The hyper-rectangle of this LLM is split into two halves with an axis-orthogonal split. Try division in all dimensions. Then for each division in each dimension dim = 1, ..., P do following steps:

   - Construct μ membership functions for both hyper-rectangle.
   - Construct all the validity functions.
   - Estimate the parameters for new generated LLMs.
   - Calculate the loss function for the overall model.

4. Choose the best division: The best division in the previous step is selected. The validity function and new LLMs will be constructed and the number of LLM or neurons is incremented to M = M + 1.
5. Test the threshold condition: If the threshold is met, then stop, else go to step 2 (Fig. 4).

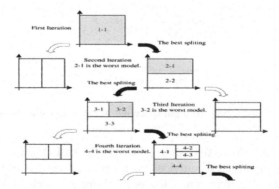

Fig. 4. Operational steps for 4 step LOLIMOT algorithm on a two-dimensional input space [20]

Support Vector Machine (SVM). SVM is one of the supervised methods that provide mapping function from training data, this mapping function can be a classification or regression function. In fact, SVM is a mathematical entity for maximizing a specified math function. For Adjusting SVM learning, considering that there is some unknown and non-linear dependency $y = f(x)$, between the input vector of x with high dimension and a nominal output of y is important [25]. The main idea behind the SVM algorithm needs to use four essential concepts:

Separating the hyperplanes. This rule is about drawing a line between clusters. After separating clusters of data, prediction of unknown elements would be easy because the element would be definitely on one side of the separating line, Fig. 5.

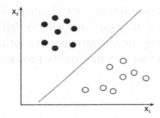

Fig. 5. Separating data classes by hyperplane [26]

The equation of the separating line can be modified with Eq. 11.

$$W_i X_i + b = 0 \tag{11}$$

It is considered that data set is like $\{x_i, y_i | i = 1, 2, \ldots, n\}$ that $x_i \in \mathcal{R}^d$, $y_i \in \{+1, -1\}$ and b is bias parameter (Figs. 6 and 7).

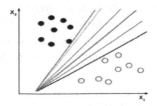

Fig. 6. Existence of multiple separating hyperplanes [26]

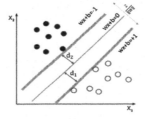

Fig. 7. Choosing a separating hyperplane with maximum margin [26, 37]

Choosing the best margin allows risks or errors between margins. The aim of SVM is finding the maximum margin, Eq. 12.

$$\max(d_1 + d_2) \rightarrow \max\left(\frac{2}{\|W\|}\right) \tag{12}$$

Equations 11 and 12 can lead to Eq. 13 as below [34–38].

$$\min\|w\| \longrightarrow \|w\| = W^T W \min \frac{1}{2} W^T W \tag{13}$$

The soft margin. Many of data sets are not separable with a single straight line. Its causes the SVM dealing with errors and allows falling wrong elements on the wrong side of the separating line. Consequently, for carrying out this issue SVM can add a soft margin without affecting on its final results, Fig. 8.

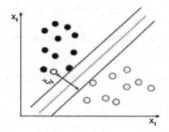

Fig. 8. Soft margin, allowing presence of faulty data among classified data [26, 39]

In addition, we don't want to allow many wrong classified elements. Describing soft margin has to provide a parameter for the user to determine how many samples can break separating hyperplane rule and how far from that margin they can be located. It's obvious the tradeoff between both maximum margin and have a correct classification of samples will be complex [17].

In this case, according to Fig. 8, slack variables (ξ_i) can use in goal function, Eq. 14. $C \sum_i \xi_i$ Specifies maximum errors [34–38].

$$F(x) = \min \frac{1}{2} \|W\|^2 + C \sum_i \xi_i \tag{14}$$

With constraints:

$$S = \{i \,|\, 0 < \alpha_i < C\}$$

The kernel function. Sometimes there are inseparable data set and there is no single point that can separate two classes and even there isn't any separating soft margin [27], Fig. 9.

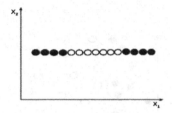

Fig. 9. Linear, non-separable data set [26]

Kernel function can solve this problem by adding an additional dimension to the data. For obtaining new dimension, values of the main function are squared. Kernel functions, map data from a lower dimension to a higher dimension by selecting a suitable function. Thus, data set would be separable in a higher dimension space which is called feature space. The feature space in Fig. 9 converted to higher dimension by kernel function in Fig. 10.

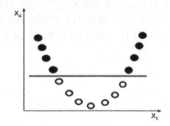

Fig. 10. Non-separable data set with augmenting new dimension [26]

With kernel functions variable x maps to $\varphi(x)$, Fig. 11.

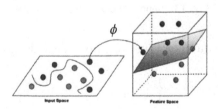

Fig. 11. Mapping data from input space to feature space [40]

It's provable that there is at least one kernel function for each data set which can separate data sets linearly. Although mapping data to a higher dimension can make some problems like increasing the number of values and possible solutions. Data mapping into excessive higher space causes special boundaries shown in Fig. 12, [34–36].

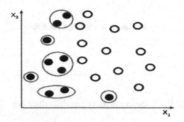

Fig. 12. Transferring training data to the higher dimension [26]

Support Vector Regression (SVR). Support Vector Regression had been used for recognizing patterns, then it has been developed for dealing with non-linear regression problems [28]. SVR model is based on non-linear mapping of main x data to a higher dimension feature space. In fact, SVR is a way of function estimating that maps an input object to a real number base on training data [29]. In SVR, estimating errors are using instead of SVM's margin. Vapnik's epsilon error function determines a ε-cylinder [6]. If predicted values were in the cylinder, the error would be zero, but for all out of cylinder, the error would be equal to difference between predicted value and cylinder ε radius, Fig. 13, [16, 24].

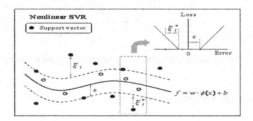

Fig. 13. Support Vector Regression [30]

Vapnik's linear loss function with ε sensitive range defined as Eq. 15, [29].

$$E(x, y, f) = |y - f(x, w)|_\varepsilon = \begin{cases} 0 & \text{for} \quad |y - f(x, w)| \leq \varepsilon, \\ |y - f(x, w)| - \varepsilon & \text{if} \quad |y - f(x, w)| > \varepsilon \end{cases} \quad (15)$$

If SVR algorithm considered soft margin, the Eq. 15, with ξ_i the slack variabale would be as Eq. 16.

$$|y - f(x, w)| \leq \varepsilon + \xi_i \quad , \quad \xi_i \geq 0 \tag{16}$$

Fuzzy Membership. Fuzziness should be used in systems which their information is not precise and certain. A model of a vague phenomenon might be presented as a fuzzy relation that introduced by 'Lotfi zadeh'. A membership function for a fuzzy set 'A', with x statistical population, is $s\_i:x \rightarrow [0,1]$. While each x_i element mapped to a value between 0 and 1. This value is called fuzzy membership, which calculates the amount of element's membership in a fuzzy set [31–33].

Fuzzy Support Vector Methods. Support vector technologies are strong tools for classification and regression, but there are some restrictions in this theory. In SVM, each training element belongs to just one class. In many applications, some of the input points are not assigned to a specific class. Also, some points, are meaningless due to noises and it is better to ignore them. Considering fuzzy membership for support vector methods make them able to reduce the impacts of noises and outlier data [32, 33].

It can be mentioned that in many real-world applications, training data have different effects, also some of them are more important in classification problems. Therefore, in classification algorithms, meaningful training data, must be classified correctly and classifying or not classifying of some of those points like noises, is not important [41, 42].

In standard SV algorithms, the importance of number of errors for all training elements is considered the same, while it should not be like that. The importance of each element can be calculated with fuzzy logic in training phase, and then instead of hard decision in decision phase, a soft decision can be gained [41, 43].

Local Outlier Factor (LOF). One of the algorithms for determining outlier points is LOF. This algorithm by comparing local density of an element with local density of its neighbors, can specify areas with same densities or specify elements which have natural lower density. Thus, this algorithm is able to determine outliers in a data set and fuzzy membership of each element is calculated according to that. In this paper, fuzzy membership of each element is calculated with LOF algorithm [44].

4 Experimental Study

Each applied data set has missing values naturally, therefore our goal is to estimate missing values based on 14 missing value imputation methods. 12 of these methods are based on *Luengo et al. study*. They have been developed a tool called "KEEL" in order to impute and classify incomplete datasets. Our proposed approach is implemented with 2 mentioned methods and also, has been compared with those 12 methods [8].

This section of study, describes the experiments which had been performed for our study. First of all, the incomplete units are imputed in data sets with imputation methods and secondly, the result data sets that are completed are fed to MLP neural network as a classifier. Finally, the classification error on each completed data set which had been imputed by an individual imputation method is compared. This section also included the graphical analysis of these different imputation methods.

4.1 Data Sets

Seven individual data sets had been selected from UCI repository in order to experiment study. The properties of these applied data sets are described in detail in Table 1.

Table 1. Properties of seven chosen individual data sets from UCI repository

Data set	#Records	#Features	#Samples	#Classes	%Missing values	#Records with missing value
Cleveland	303	13	3939	5	0.14%	6
CRX	689	16	11024	2	0.61%	37
Post-Operative	90	9	810	3	0.37	30
Wisconsin	699	10	6990	2	0.23%	16
Breast	286	10	2860	2	0.31%	9
Autos	205	26	5330	6	1.11%	52
Mushroom	8124	23	186852	2	1.33%	2438

In Table 2, used parameters with their amount, is shown for each algorithm. Determined parameters in Table 2, had best results on used data sets.

Table 2. Considered methods in experimental study

MV Imputation Method	Ignore MV	Event Covering	KNN	WKNN	K-mean	Fuzzy K-mean	SVM	EM	SVD	BPCA	Most Common	Concept Most Common	LOLIMOT	FSVM/FSVR
Parameters	-	T=0.05	K=10	K=10	K=10	K=3	Kernel= RBF C = 1	Iteration=30 Inflation Factor=1 Regression Type=Multiple ridge regression	Iteration=30 Inflation Factor=1 Regression Type=Multiple ridge regression Singular Vectors=10	-	-	-	LLM=3	C=10 Kernel Function= linear

4.2 Graphical Analysis of the Classification

Accuracy of All Applied Methods. Two applied algorithms are compared with 12 other algorithms which are mentioned in previous sections. These figures depict results of all compared methods and indicate rate of correctly classified in each dataset. As which have been shown in figures it is obvious that suggested algorithms have higher range of accuracy based on these datasets Figs. 14, 15, 16, 17, 18, 19 and 20. Also, Figs. 21, 22, 23 and 24 show the differences between the target data and the predicted data by the used methods, i.e., LOLIMOT and FSVM/FSVR, on Wisconsin data set.

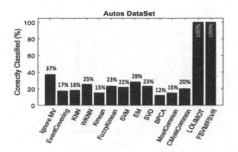

Fig. 14. Classification accuracy of different methods on Autos dataset.

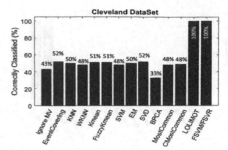

Fig. 15. Classification accuracy of different methods on Cleveland dataset

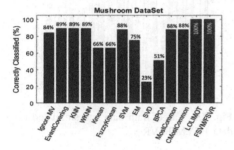

Fig. 16. Classification accuracy of different methods on Mushroom dataset.

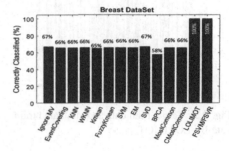

Fig. 17. Classification accuracy of different methods on Breast dataset

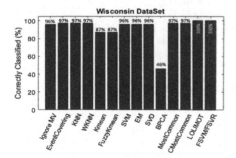

Fig. 18. Classification accuracy of different methods on Wisconsin dataset

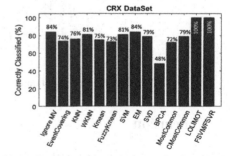

Fig. 19. Classification accuracy of different methods on CRX dataset

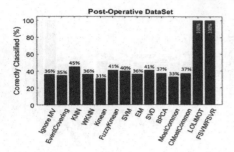

Fig. 20. Classification accuracy of different methods on Post-operative dataset

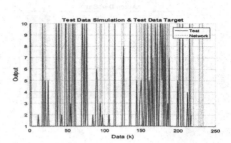

Fig. 21. Target test data and simulated test data by FSVM/FSVR

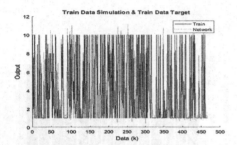

Fig. 22. Target train data and simulated train data by LOLIMOT

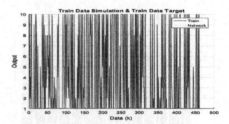

Fig. 23. Target train data and simulated train data FSVM/FSVR

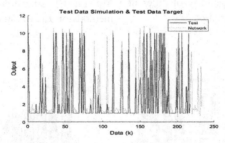

Fig. 24. Target test data and simulated test data by LOLIMOT

5 Conclusion

Although our proposed approach enforces computational burdens, it delivers high accuracy results. Thus, this approach can be recommended in those studies, that computational complexities can be disregarded.

According to obtained results, it has been recognized that used algorithms in missing data imputation, can model the train data and also predict test data with high precision and high accuracy. LOLIMOT can gain more accuracy by applying divide and conquer strategy and local linear models in order to solve a nonlinear problem. The main reason for precise results of FSVM/FSVR is the usage of fuzzy membership in modelling train data. In addition, finding out a better initializing substantial parameters will result in less computation time. Therefore, finding techniques to indicate better initial parameters can help for better sufficiency. Also, using appropriate preprocessing on different datasets will cause higher authenticity in results. As a suggestion, to indicate better initial parameters and find more appropriate kernel functions, usage of meta-heuristics methods can be useful.

References

1. Meesad, P., Yen, G.G.: Combined numerical and linguistic knowledge representation and its application to medical diagnosis. In: IEEE Transactions on Systems, Man, and Cybernetics - Part A: Systems and Humans, pp. 206–222, August 2003
2. Li, D., Deogun, J., Spaulding, W., Shuart, B.: Towards missing data imputation: a study of fuzzy K-means clustering method. In: Tsumoto, S., Słowiński, R., Komorowski, J., Grzymała-Busse, Jerzy W. (eds.) RSCTC 2004. LNCS (LNAI), vol. 3066, pp. 573–579. Springer, Heidelberg (2004). https://doi.org/10.1007/978-3-540-25929-9_70
3. Schafer, J.L.: Analysis of Incomplete Data, pp. 10–13. Chapman & Hall, London (1997)
4. Little, R.J., Rubin, D.B.: Statistical Analysis with Missing Data, 2nd edn, pp. 3–19. Wiley, New York (2002)
5. Wayman, C.: Multiple imputation for missing data: what is it and how can I use it. In: Annual Meeting of the American Educational Research Association, Chicago, IL, pp. 2–16 (2003)
6. Jiri, K.: Dealing with missing values in data. Faculty of Civil Engineering, Czech Technical University, pp. 1–10 (2013)
7. Schafer, L.J., Graham, J.W.: Missing data: our view of the state of the art. Psychol. Methods 7(2), 147–177 (2002)
8. Luengo, J., Garcia, S., Herrera, F.: A study on the use of imputation methods for experimentation with Radial Basis Function Network classifiers handling attribute values: the good synergy between RBFNs and Event Covering method. CITIC-University of Granada, pp. 406–418 (2010)
9. Grzymała-Busse, J.W., Grzymała-Busse, W.J., Goodwin, L.K.: A closest fit approach to missing attribute values in preterm birth data. In: Zhong, N., Skowron, A., Ohsuga, S. (eds.) RSFDGrC 1999. LNCS (LNAI), vol. 1711, pp. 405–413. Springer, Heidelberg (1999). https://doi.org/10.1007/978-3-540-48061-7_49
10. Grzymala-Busse, J.W., Hu, M.: A comparison of several approaches to missing attribute values in data mining. In: Ziarko, W., Yao, Y. (eds.) RSCTC 2000. LNCS (LNAI), vol. 2005, pp. 378–385. Springer, Heidelberg (2001). https://doi.org/10.1007/3-540-45554-X_46
11. Kantardzic, M.: Data Mining-Concepts, Models, Methods, and Algorithms. IEEE, pp. 165–176 (2003)
12. Site dedicated to missing values: http://sci2s.ugr.es/MVDM/index.php#four, Bibliography on missing values: http://sci2s.ugr.es/MVDM/biblio.php
13. Little, R.J., Rubin, D.B.: Statistical analysis with missing data, 2nd edn, pp. 1–409. Wiley, Hoboken (2002)
14. Hand, D.J., Manilla, H., Smyth, P.: Principles of Data Mining, A Bradford Book, pp. 157–160. MIT Press, Cambridge (2001)
15. Gustavo, E., Monard, B., Monard, M.C.: A Study of K-Nearest Neighbour as an Imputation Method. Institute of Mathematics and Computer Science– ICMC, pp. 1–10 (2002)

16. Smola, A.J., Scholkophf, B.: A tutorial on Support Vector Regression. NeuroCOLT2 Technical report Series, NC2-TR-1998-03, pp. 1–73, October 1998
17. Scholkopf, B., Burges, C., Vapnik, V.: Extracting support data for a given task. In: Fayyad, U.M., Uthurusamy, R. (eds.) Proceedings, First International Conference on Knowledge Discovery and Data Mining, pp. 252–257. AAAI Press, Menlo Park (1995)
18. Chen, Z.: Data Mining and Uncertain Reasoning: An Integrated Approach, pp. 1–392. Wiley, Hoboken (2001)
19. Fahlman, S.E., Lebiere, C.: The cascade-correlation learning architecture. In: Touretzky, D.S. (ed.) Advances in Neural Information Processing Systems 2, pp. 1–17. Morgan-Kaufmann, Los Altos (1990)
20. Nelles, O.: Nonlinear System Identification, pp. 69–341. Springer, New York (2001). https://doi.org/10.1007/978-3-662-04323-3
21. Sohani, M., Kermani, K.K.: A neuro-fuzzy approach to diagnosis of neonatal jaundice. In: Proceedings of the 1st International Conference on Bio Inspired Models of Network, Information and Computing Systems Cavalese, Italy, pp. 2–6 (2006)
22. Janghorbani, A., Arasteh, A.: Application of local linear neuro-fuzzy model in prediction of mean arterial blood pressure time series. In: Proceedings of the 17th Iranian Conference of Biomedical Engineering (ICBME 2010), pp. 1–4 (2010)
23. Nikookar, A., Lucas, C.: Artificial bee colony based learning of local linear neuro-fuzzy models. In: IEEE Fuzzy Systems (IFSC), pp. 1–4 (2013)
24. Sen, W., Hong, C., Xiaodong, F.: Clustering algorithm for incomplete data sets with mixed numerical and categorical attributes. Int. J. Database Theory Appl. 6(5), 95–104 (2013)
25. Wang, L. (ed.): Support Vector Machines: Theory and Applications, pp. 1–434. Springer, Heidelberg (2005). https://doi.org/10.1007/b95439
26. Nobel, W.S.: What is a support vector machine? Comput. Biol. 1–3 (2006)
27. Pigott, T.D.: A review of methods for missing data. Educ. Res. Eval. 7(4), 353–383 (2001)
28. Vapnik, V., Golowich, S., Smola, A.: Support vector machine for function approximation regression estimation and signal processing. In: Advances in Neural Information Processing Systems, vol. 9, pp. 281–287 (1996)
29. Yu, H., Kim, S.: SVM tutorial — classification, regression and ranking. In: Rozenberg, G., Bäck, T., Kok, J.N. (eds.) Handbook of Natural Computing, pp. 479–506. Springer, Heidelberg (2012). https://doi.org/10.1007/978-3-540-92910-9_15
30. http://proj.ncku.edu.tw/research/articles/e/20080620/3.html
31. Enders, C.: Applied Missing Data Analysis, pp. 3–55. Guilford Press, New York (2010)
32. Luengo, J., Garcia, S., Herrera, F.: On the choice of the best imputation methods for missing values considering three groups of classification methods. Knowl. Inf. Syst. 32, 77–108 (2011)
33. https://en.wikipedia.org/wiki/Support_vector_machine
34. http://faradars.org/courses/mvrnn9102fh-support-vector-machine-in-matlab-video-tutorial
35. Learning: Support Vector Machines. https://www.youtube.com/watch?v=PwhiWxHK8o
36. 5 Minutes with Ingo: Understanding Support Vector Machines. https://www.youtube.com/watch?v=YsiWisFFruY
37. Bottou, l., et al.: Comparison of classifier methods: a case study in handwritten digit recognition. In: Proceedings of the 12th IAPR International Conference on Pattern Recognition, vol. 2, pp. 77–82, 9–13 October 1994
38. Vapnik, V.: The Nature of Statistical Learning Theory, 2nd edn, pp. 1–314. Springer, New York (2000). https://doi.org/10.1007/978-1-4757-3264-1
39. http://www.saedsayad.com/support_vector_machine.html
40. Liao, R.: Support Vector Machines, pp. 1–33, 10 November 2015

41. Lin, C., Wang, S.: Fuzzy support vector machines. IEEE Trans. Neural Netw. **13**(2), 464–471 (2002)
42. Lin, K., Pai, P.: A fuzzy support vector regression model for business cycle predictions. Expert Syst. Appl. **37**, 5430–5435 (2010)
43. Huang, H., Liu, Y.: Fuzzy support vector machines for pattern recognition and data mining. Int. J. Fuzzy Syst. **4**(3), 826–835 (2002)
44. Local outlier factor. https://en.wikipedia.org/wiki/Local_outlier_factor

Two Modifications of the Automatic Rule Base Synthesis for Fuzzy Control and Decision Making Systems

Yuriy P. Kondratenko[1,2]($\boxtimes$) (iD), Oleksiy V. Kozlov[2] (iD),
and Oleksiy V. Korobko[2] (iD)

[1] Intelligent Information Systems Department, Petro Mohyla Black Sea National
University, 68-th Desantnykiv str. 10, Mykolaiv 54003, Ukraine
y_kondrat2002@yahoo.com
[2] Computerized Control Systems Department, Admiral Makarov National
University of Shipbuilding, Heroes of Ukraine ave. 9, Mykolaiv 54025, Ukraine
{oleksiy.kozlov, oleksii.korobko}@nuos.edu.ua

Abstract. This paper presents two modifications of the method of synthesis
and optimization of rule bases (RB) of fuzzy systems (FS) for decision making
and control of complex technical objects under conditions of uncertainty. To
illustrate the advantages of the proposed method, the development of the RB of
Mamdani type fuzzy controller (FC) for the automatic control system (ACS) of
the reactor temperature of the experimental specialized pyrolysis plant (SPP) is
carried out. The efficiency of the presented method of synthesis and optimization
of the FS RB is investigated and its comparison with the other existing methods
is carried out on the basis of this FC. Analysis of simulation results confirms the
high efficiency of the proposed by the authors method of synthesis and reduction
of the FS RB.

Keywords: Fuzzy controller · Rule base · Synthesis · Optimization
Pyrolysis reactor · Fuzzy control · Decision making systems

1 Introduction

Automation of control and decision making processes in the complex industrial systems gives the opportunity to substantially increase and improve their energy, economic and operational indicators [1, 2]. Among the main components of such industrial
systems can be considered the electric and hydraulic drives, marine and land vehicles,
industrial and mobile robots, floating structures, pyrolysis and thermoacoustic plants
for utilization of secondary energy resources, etc., that are, in turn, complex technological objects with nonstationary and nonlinear parameters [3–5]. Also, such technological objects can operate in the conditions of uncertainty, which include uncertain
parametric and coordinate disturbances, as well as uncertain changes of operating
characteristics and modes. When creating decision making and control systems for
these technological objects, it is advisable to use advanced effective approaches based
on intelligent technologies and soft computing techniques [6–8]. Fuzzy logic is one of

© Springer International Publishing AG, part of Springer Nature 2018
J. Medina et al. (Eds.): IPMU 2018, CCIS 854, pp. 570–582, 2018.
https://doi.org/10.1007/978-3-319-91476-3_47

the most efficient soft computing approach which is widely used for designing of decision making and control systems for the complex nonlinear and nonstationary industrial objects, that operate in the conditions of uncertainty [9–11]. One of the main features of fuzzy control and decision making systems is that they are developed predominantly on the basis of expert assessments, and their efficiency essentially depends on the qualifications and experience of the developers (experts) as well as of a number of subjective factors [12, 13]. Therefore, for the development of highly effective control and decision making systems based on fuzzy logic it is expedient to create and apply effective methods and algorithms for fuzzy systems designing, that will reduce the influence of subjective factors on the design process as well as increase in general the quality indicators of control and decision making processes in the conditions of uncertainty.

2   Problem Statement and Related Works

Many successful examples of development and application of fuzzy control and decision making systems in such areas as: technological processes automation, technical and medical diagnostics, financial management, pattern recognition, etc., are presented in literature [14–16]. Also, quite many studies are devoted to synthesis, as well as structural and parametric optimization of fuzzy systems in order to obtain their best possible performance [11, 15, 17–19]. Works on parametric optimization are given in [19–23], in particular, on optimization of parameters of linguistic terms membership functions (LTMF) [20, 22], preliminary coefficients [23], weights of fuzzy rules [21], etc. In turn, methods and approaches of fuzzy systems structural optimization are presented in papers [11, 20, 21, 24–31]. Among them the selection of defuzzification methods [11, 21], reduction of the rule bases [25, 26] based on combining rule antecedents [25], rule base interpolation [27], linguistic 3-tuple representation [28], evolutionary algorithms [29, 30], multi-objective optimization [31] and other. For successful structural-parametric optimization of fuzzy control and decision making systems it is necessary to have a proper rule base, the compilation of which in some cases in the absence of sufficient experience of experts and in the presence of substantial uncertainty can be a serious problem.

Thus, the main purpose of this work is development and research of the synthesis and optimization method of the rule bases of Mamdani type fuzzy control and decision making systems for their quality indicators and efficiency increasing.

3   Synthesis and Optimization of the Fuzzy Systems Rule Bases

At the synthesis of a Mamdani type fuzzy system the vector of input variables X and output variable Y are selected at the initial stage [11]. Thus, the vector X in the general form can be represented as follows

$$\mathbf{X} = \{\mathbf{X}_i\}, i = \{1, \ldots, n\},\tag{1}$$

where i is a number of FS input variable; n is a total number of FS input variables.

After that, the choice of the number of linguistic terms m_i for each i-th input variable of the vector $\mathbf{X}$ ($i = 1, \ldots, n$) and linguistic terms k for the output variable Y of the FS is performed [13]. Also, the types and parameters of the linguistic terms membership functions for each input and output variables of the FS are previously selected.

The total number of rules s of the FS RB is determined by the number of all possible combinations of the linguistic terms of the FS input variables $\mathbf{X}$ and calculated as follows [13, 21]

$$s = \prod_{i=1}^{n} m_i.\tag{2}$$

In turn, each r-th rule of the RB is a linguistic statement

$$\begin{aligned}\text{IF } "X_1 = a" \text{ AND } "X_2 = b" \text{ AND } \ldots \text{ AND } "X_i = d" \ldots \\ \ldots \text{ AND} \ldots \text{ AND } "X_n = e" \text{ THEN } "Y = z",\end{aligned}\tag{3}$$

where a, b, d, e, z are the corresponding values of linguistic terms.

The consequents for each r-th rule of the RB ($r = 1, \ldots, s$) are selected from a set of possible consequents of rules, that consists of all k linguistic terms of the FS output variable Y.

The task of synthesis and optimization of the FS RB is to find such a vector of RB consequents $\mathbf{Z}$, at which the required quality indicators of the FS will be sufficient [20]. In turn, the vector of RB consequents $\mathbf{Z}$ in general form can be represented as follows

$$\mathbf{Z} = \{\mathbf{Z}_r\}, r = \{1, \ldots, s\}.\tag{4}$$

In many cases, at the synthesis of the FS RB, the vector of consequents $\mathbf{Z}$ is determined on the basis of experts knowledge [11, 21]. At the same time, if the FS quality indicators for this vector of RB consequents are not sufficient, then further parametric adjustment and optimization of the FS can be carried out.

Another approach to the synthesis of the FS RB is the automatic generation of the RB, which is used in the software package FuzzyTECH [32, 33]. In this software package the "Rule Block Wizard" allows to formulate a rule base on the basis of the analysis of the effect of each input variable from the vector $\mathbf{X}$ on the output variable Y. In this case, the task of the expert is to assess the degree of influence of each input variable on the output, which can be: "Very Negative", "Negative", "Not at All", "Positive" or "Very Positive" [33]. Based on this information the "Rule Block Wizard" generates the RB and vector of its consequents $\mathbf{Z}$ in the automatic mode. If for the generated RB the quality indicators are not sufficient, then the vector of its consequents $\mathbf{Z}$ can be further adjusted by the expert in the manual mode.

These two approaches are basically used if there is no full information about the system and specific numerical value of the goal function I can not be calculated.

The proposed by the authors method can be applied in cases, when the goal function I is defined and can be calculated. It provides an iterative search of the optimal vector of the RB consequents $\mathbf{Z}$ at which the goal function I of the processes of control and/or decision making will be optimal ($I = I_{opt}$). The search of the optimal vector of the RB consequents $\mathbf{Z}$ is carried out by means of a sequential search of the consequents of each rule of the FS RB.

In turn, this method has two modified versions: the first one is used for further optimization of the FS RB, which is previously synthesized on the basis of the experience of experts; the second one - for the synthesis of the FS RB in the automatic mode in the absence or non-use of any experts knowledge.

The first modification of this method consists of the following steps.

Step 1. Method initialization. The preliminary synthesis of the FS RB on the basis of expert knowledge and the choice of the goal function I is carried out at the given stage. Synthesis of the structure of the FS RB is carried out on the basis of the previously selected linguistic terms of the input variables $\mathbf{X}$, and the set of possible consequents for each rule is determined on the basis of previously selected linguistic terms of the output variable Y. In turn, the initial vector of the RB consequents $\mathbf{Z}$ is determined on the basis of the usage of experts knowledge and experience and is set in FS RB. Also, at this step, the type of the goal function I is selected for further optimization or efficiency evaluation of the control or decision making process on the basis of the FS.

Step 2. Transition to the 1st rule of the FS RB. The transition to the 1st rule of the RB is carried out at this stage to initiate the iterative procedures for finding the optimal vector of the RB consequents $\mathbf{Z}$.

Step 3. Checking of the Checklist. All vectors of the RB consequents $\mathbf{Z}$, for which the goal function I has been already calculated during the implementation of the method, as well as the corresponding values of the goal function I are entered to the Checklist. If the current vector of the RB consequents $\mathbf{Z}$ is already placed in the Checklist, then the transition is carried out to *Step 6*, and in the opposite case the transition is performed to *Step 4*.

Step 4. Calculation of the value of the goal function for the FS with the current vector of the RB consequents. The calculation of the value of the goal function I for the FS with the current vector of the RB consequents $\mathbf{Z}$ as well as recording of this information to the Checklist is carried out at this stage.

Step 5. Checking of the completion of the previous optimization process of the current rule. The optimization calculations for the current r-th rule are considered complete if the values of the goal function I for each consequent from the set of all k possible consequents for this rule has been calculated. If the checking has given a positive result, then the transition to *Step 7* is carried out. In the opposite case the transition to *Step 6* is performed.

Step 6. Setting of the next consequent from the set of possible consequents in the current rule. After that, the transition to *Step 3* is carried out.

Step 7. Choosing of the consequent for which the value of the goal function is the smallest. The choice of the consequent, for which the value of the goal function I is the smallest among obtained at the optimization calculations for the current r-th rule, and its fixation in this rule is carried out at this stage.

Step 8. Checking of the completion of the RB optimization. Optimization of the RB is considered to be complete, if all the s rules of this FS RB have been optimized in the previous steps of the method implementation. If the checking has given a positive result, then the transition to *Step 10* is carried out. In the opposite case the transition to *Step 9* is performed.

Step 9. Transition to the next rule of the FS RB. After that, the transition to *Step 3* is carried out.

Step 10. Completion of the optimization process of the FS RB. After that, the re-implementation of the proposed method or parametric optimization of the FS and its realization can be carried out for its further application in the decision making and control processes.

The Checklist and its checking at *Step 3* is used to avoid recurring calculations of the goal function I for the FS with the same vector of the RB consequents $\mathbf{Z}$. This allows to get rid of unnecessary iterations, the number of which is equal to $s - 1$.

Thus, the proposed method in this modification allows implementing further optimization of the FS RB, that is previously synthesized on the basis of expert knowledge, to minimize the value of the goal function I of the decision making and control processes. The total number of iterations τ_{max} of this modification of the method is

$$\tau_{max} = ks - (s - 1). \tag{5}$$

At the absence of any experts knowledge for the FS RB synthesis in the automatic mode the second modification of this method can be applied, which differs from the first by the following steps.

The preliminary synthesis of the FS RB structure is carried out on the basis of the previously selected linguistic terms of the input variables $\mathbf{X}$ and the set of possible consequents for each rule is determined based on the previously selected linguistic terms of the output variable Y at *Step 1*. The initial vector of the RB consequents $\mathbf{Z}$, in turn, is determined randomly and is set in the FS. Furthermore, the maximum number of iterations at the implementation of the method τ_{max} is set at this stage. The type of the goal function I of the decision making or control process on the basis of the FS is also selected at this step.

The RB optimization is considered complete if the optimal value of the goal function is achieved ($I = I_{opt}$) or if the maximum number of iterations τ_{max}, previously set at *Step 1*, is carried out at *Step 8*. If the checking, carried out at this stage, has given a positive result, then the transition to *Step 10* is performed. In the opposite case, the transition to *Step 9* is carried out. In the case, when all of the s rules of this FS RB have been optimized at the previous steps of the method, and the checking at *Step 8* does not give a positive result, then the transition to *Step 2* is carried out, and iterative procedures for finding the optimal vector of the RB consequents **Z** continue again, starting from the 1st rule of the RB.

Other steps at this method implementation are the same in both modifications.

Thus, in the second modification of the proposed method the consistent iterative procedures for optimizing the rules of the RB from the 1st to the s-th can be carried out l times before finding the optimal vector of the RB consequents **Z**, at which $I = I_{opt}$, or performing the maximum number of iterations τ_{max}, that is set by the operator at *Step 1*. In this case, the total number of iterations for this modification of the method may be slightly larger than for the first modification.

To study the effectiveness of the proposed method and to compare it with the other existing (presented) methods, the design of the RB of Mamdani type fuzzy controller for the automatic control system of the reactor temperature of the experimental specialized pyrolysis plant [34] is carried out in this paper.

4 Synthesis and Optimization of the Fuzzy Controller RB for the Reactor ACS of the SPP

The operating volume of the pyrolysis reactor and the maximum power of the gas burner of this experimental SPP [34] are 100 L and 25 kW, respectively. The mathematical model of this pyrolysis reactor as a temperature control object can be represented as a transfer function $W_R(s)$ [35]

$$W_R(s) = \frac{T_R(s)}{P_H(s)} = \frac{k_R e^{-\tau s}}{(T_1 s + 1)(T_2 s + 1)^n},\tag{6}$$

where $T_R(s)$ is the image of the reactor heating temperature $T_R(t)$; $P_H(s)$ is the image of the gas burner heat power $P_H(t)$, which is the control action of the temperature ACS; k_R, τ, T_1, T_2, n are the gain, time delay, time constants and order of the inertial link of the transfer function of the SPP reactor, respectively.

The functional structure of the PD-fuzzy-controller for the temperature ACS of the SPP reactor is presented in Fig. 1.

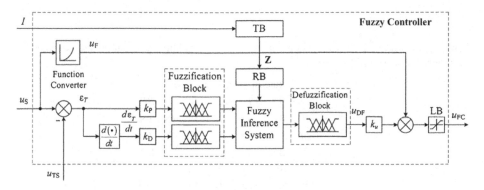

Fig. 1. Functional structure of the PD-FC for the temperature control of SPP reactor

The following notations are accepted in Fig. 1: TB is the tuning block of the RB; LB is the limiting block; u_S is the reactor temperature set signal; u_F, u_{TS}, u_{DF}, u_{FC} are the output signals of the functional converter, temperature sensor, defuzzification block and FC, respectively; ε_T, $d\varepsilon_T/dt$ are the temperature control error and rate of its change; k_P, k_D are the FC preliminary coefficients; k_u is the normalizing coefficient of the output signal. The given FC provides control of the temperature modes of the SPP pyrolysis reactor on the basis of feedforward and feedback. The functional converter implements the dependency $u_F = f(u_S) = f(T_S)$, which corresponds to the inverse static characteristic of the open-loop control system of the reactor temperature.

In this case, the vector of the FC input variables $\mathbf{X}$ consists of temperature control error ε_T, and rate of its change $d\varepsilon_T/dt$.

The signal u_{DF} acts as the output variable Y. In turn, for each input and output variables, 5 linguistic terms ($m_1 = 5$; $m_2 = 5$; $k = 5$) of the triangular type (with parameters in relative units) are chosen: BN (-1.5; -1; -0.5) – big negative; SN (-1; -0.5; 0) – small negative; Z (-0.5; 0; 0.5) – zero; SP (0; 0.5; 1) – small positive; BP (0.5; 1; 1.5) – big positive.

Thus, the total number of the RB rules s of the PD-FC according to formula (2) is equal to 25. Each r-th rule of the given RB ($r = 1,\dots, 25$) is a linguistic statement

$$\text{IF } ``\varepsilon_T = a" \text{ AND } ``\frac{d\varepsilon_T}{dt} = b" \text{ THEN } ``u_{DF} = z". \tag{7}$$

The consequents for each r-th rule of the RB ($r = 1,\dots, 25$), in this case, are chosen from the set of possible consequents of rules, that consists of 5 linguistic terms (BN; SN; Z; SP; BP) of output FC variable u_{DF}. The vector of the RB consequents $\mathbf{Z}$ of the given FC is presented by the expression (4). The gravity center method of defuzzification is chosen for the given FC.

In turn, the synthesis of FC RB of the reactor ACS in this work is carried out in several ways: (a) on the basis of experts knowledge, (b) on the basis of automatic generation of rules in the software package FuzzyTECH and (c) with the help of two modifications of proposed by the authors method.

The FC RBs developed on the basis of the experts knowledge and automatic generation of rules in the software package FuzzyTECH are given in Table 1.

Table 1. FC rule bases developed on the basis of: experts knowledge/automatic generation in the program FuzzyTECH

		Rate of error change, $d\varepsilon_T/dt$				
		BN	SN	Z	SP	BP
Error, ε_T	BN	BN	BN	BN	BN/SN	SN/Z
	SN	BN	SN	SN	SN/Z	Z
	Z	BN	SN	Z	SP	BP
	SP	SP/Z	SP/Z	SP	SP	BP
	BP	Z	SP	BP	BP	BP

At the development of the FC RB on the basis of the first and second modifications of the proposed method, in this case, the goal function I is represented by the equation

$$I(t, \mathbf{Z}) = \frac{1}{t_{max}} \int\limits_0^{t_{max}} \left[(E_T)^2 + k_1 \left(\frac{dE_T}{dt} \right)^2 + k_2 \left(\frac{d^2 E_T}{dt^2} \right)^2 \right] dt, \tag{8}$$

where t_{max} is the total time of the transient of the reactor temperature ACS; k_1, k_2 are the weights; E_T is the deviation of the real transient characteristic of the ACS $T_R(t, \mathbf{Z})$ from the desired transient characteristic of the ACS reference model (RM) $T_D(t)$

$$E_T = T_D(t) - T_R(t, \mathbf{Z}). \tag{9}$$

The RM, in turn, has the transfer function of the aperiodic link of the second order with the set time constant.

The simulation of transients of the reactor temperature ACS is carried out at all possible operation modes (at various types of input and disturbing influences) for efficient optimization of all the rules of the RB, in particular, at calculating the value of the goal function (8) on each iteration in the process of the method implementation.

At the development of the FC RB on the basis of the first modification of the proposed method in its *Step 1* the initial vector of the RB consequents $\mathbf{Z}$ is determined on the basis of the experts knowledge as in the RB, that is given in Table 1. The further optimization of the FC RB is carried out by an iterative search on the basis of *Steps 2– 10* of the first modification of the proposed method. The total number of iterations, calculated by the formula (5), in this case is 101.

At the development of the FC RB on the basis of the second modification of the proposed method in its first step the initial vector of the RB consequents $\mathbf{Z}$ is determined randomly. The maximum number of iterations of the method is set $\tau_{max} = 350$ in order to perform iterative procedures of the RB rules optimization from the 1st to the 25th at least $l = 3$ times. The synthesis of the FC RB is carried out by an iterative search on the basis of *Steps 2–10* of the second modification of the proposed method.

The characters of change of the goal function (8) value in the process of FC RB iterative optimization and synthesis on the basis of first (a) and second (a) modifications of the proposed method are presented in Fig. 2.

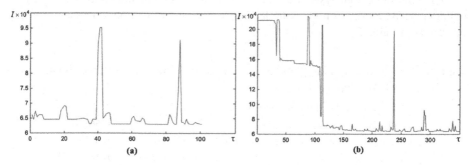

Fig. 2. Goal function value in the processes of FC RB iterative optimization on the basis of the proposed method: (a) 1st modification; (b) 2nd modification

The FC RBs developed on the basis of first and second modifications of the proposed method are presented in Table 2.

Table 2. FC rule bases developed on the basis of 1/2 modifications of the proposed method

		Rate of error change, $d\varepsilon_T/dt$				
		BN	SN	Z	SP	BP
Error, ε_T	BN	BN	BN	BN	BN	Z/SN
	SN	BN	BN	SN/BN	SN	Z/SN
	Z	BN	SN	Z	SP	BP
	SP	Z/SP	SP	SP/BP	BP	BP
	BP	Z/SP	BP	BP	BP	BP

The characteristic surfaces of the FC with RBs, developed on the basis of automatic generation in the program FuzzyTECH (a) and 2 modification of the proposed method (b), are presented in Fig. 3.

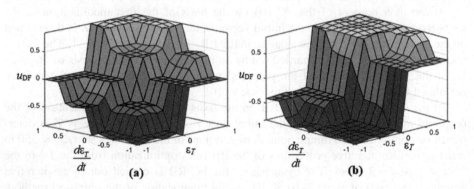

Fig. 3. Characteristic surfaces of the FC developed on the basis of: (a) automatic generation in FuzzyTECH; (b) second modification of the proposed method

The comparative analysis of the goal function values I at simulation of the reactor temperature ACS based on FC with developed RBs in all possible operation modes is presented in Table 3.

Table 3. Comparative analysis of the goal function values for the reactor temperature ACS based on FC with developed RB

ACS based on FC with developed RB	I
On the basis of experts knowledge	$6.568 \cdot 10^4$
On the basis of automatic generation in FuzzyTECH	$6.867 \cdot 10^4$
On the basis of first modification of the proposed method	$6.483 \cdot 10^4$
On the basis of second modification of the proposed method	$6.474 \cdot 10^4$

The results of computer simulation in the form of transients of the reactor initial heating for the temperature ACS on the basis of FC with developed RBs are presented in Fig. 4.

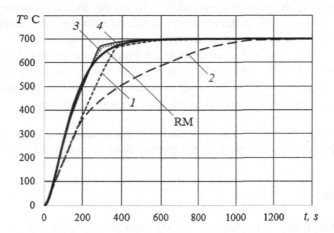

Fig. 4. Transients at $T_S = 700$ °C of SPP reactor's heating temperature for ACS with FC RB developed on the basis of: *1* experts knowledge; *2* automatic generation in FuzzyTECH; *3* first modification of the proposed method; *4* second modification of the proposed method

As can be seen from Table 3 and Fig. 4 the smallest values of the goal function *I* and the best quality indicators of control have temperature ACSs on the basis of FCs with the RBs, developed using the first and second modifications of the proposed by the authors method, that confirms its high efficiency. In addition, for further increasing of the quality indicators of this ACS after the synthesis and optimization of the RB with the help of the proposed method, the further structural and parametric optimization of its FC [17–20] can be carried out, in particular, optimization of the parameters of the LTMF, preliminary coefficients, etc.

5 Conclusions

In this work the authors developed two modifications of the method of automatic synthesis and optimization of the RBs of FSs, that can be used for automation of decision making and control processes in the complex industrial systems under conditions of uncertainty. The proposed method gives the opportunity to implement further optimization of the FS RB, which has been previously developed on the basis of experts knowledge, in order to increase its efficiency as well as to synthesize the FS RB in automatic mode at the absence of any experts knowledge.

It is advisable to use the developed by the authors method for the synthesis and optimization of the RBs of the decision making and management FSs in the following cases: (a) in the presence of incomplete experts knowledge, (b) in the absence of any knowledge of experts and (c) at a sufficiently large number of rules in the FS RB.

To study the effectiveness of the developed method of synthesis and optimization of the FS RB as well as to compare it with the other existing methods the development of the RB of Mamdani type FC for the ACS of the reactor temperature of the SPP is carried out in this work. The analysis of the results of computer simulation shows that the lowest value of the goal function and the best quality indicators of the ACS of the reactor temperature based on the FC can be achieved with the RBs, developed by the proposed by the authors method, which confirms its high efficiency.

In further research it is planned to apply the proposed method of the RB synthesis and optimization in conjunction with gradient, evolutionary and multi-agent methods of parameters optimization for the complex structural and parametric optimization of the decision making and control FS under conditions of uncertainty.

Acknowledgment. Prof. Dr.Sc. Yuriy P. Kondratenko thanks the Fulbright Scholar Program for the possibility to conduct research in USA, Cleveland State University, 2015–2016.

References

1. Mehta, B.R., Reddy, Y.J.: Chapter 7 - SCADA systems. In: Industrial Process Automation Systems, pp. 237–300 (2015)
2. Kondratenko, Y.P., Kozlov, O.V., Korobko, O.V., Topalov, A.M.: Internet of things approach for automation of the complex industrial systems. In: Ermolayev, V. et al. (eds.) Proceedings of the 13th International Conference on Information and Communication Technologies in Education, Research, and Industrial Applications. Integration, Harmonization and Knowledge Transfer, ICTERI 2017, CEUR-WS, Kyiv, Ukraine, vol. 1844, pp. 3–18 (2017)
3. Xiao, Z., Guo, J., Zeng, H., Zhou, P., Wang, S.: Application of fuzzy neural network controller in hydropower generator unit. J. Kybern. **38**(10), 1709–1717 (2009). https://doi.org/10.1108/03684920910994079
4. Hayajneh, M.T., Radaideh, S.M., Smadi, I.A.: Fuzzy logic controller for overhead cranes. Eng. Comput. **23**(1), 84–98 (2006). https://doi.org/10.1108/02644400610638989
5. Topalov, A., Kozlov, O., Kondratenko, Y.: Control processes of floating docks based on SCADA systems with wireless data transmission. In: Perspective Technologies and Methods in MEMS Design: Proceedings of the International Conference MEMSTECH 2016, Lviv-Poljana, Ukraine, pp. 57–61 (2016). https://doi.org/10.1109/memstech.2016.7507520
6. Zadeh, L.A., Abbasov, A.M., Yager, R.R., Shahbazova, S.N., Reformat, M.Z. (eds.): Recent Developments and New Directions in Soft Computing. SFSC, vol. 317. Springer, Cham (2014). https://doi.org/10.1007/978-3-319-06323-2
7. Jamshidi, M., Kreinovich, V., Kacprzyk, J. (eds.): Advance Trends in Soft Computing. Springer, Cham (2013). https://doi.org/10.1007/978-3-319-03674-8
8. Jang, J.-S.R., Sun, C.-T., Mizutani, E.: Neuro-Fuzzy and Soft Computing: A Computational Approach to Learning and Machine Intelligence. Prentice Hall, Upper Saddle River (1996)
9. Zadeh, L.A.: Fuzzy sets. Inf. Control **8**, 338–353 (1965)
10. Zadeh, L.A.: The role of fuzzy logic in modeling, identification and control. Model. Identif. Control **15**(3), 191–203 (1994)
11. Piegat, A.: Fuzzy Modeling and Control, vol. 69. Physica-Verlag, Heidelberg (2013). https://doi.org/10.1007/978-3-7908-1824-6

12. Tanaka, K., Wang, H.O.: Fuzzy Control Systems Design and Analysis: A Linear Matrix Inequality Approach. Wiley, New York (2001)
13. Hampel, R., Wagenknecht, M., Chaker, N. (eds.): Fuzzy Control: Theory and Practice, p. 410. Physica-Verlag, Heidelberg (2000). https://doi.org/10.1007/978-3-7908-1841-3
14. Merigo, J.M., Gil-Lafuente, A.M., Yager, R.R.: An overview of fuzzy research with bibliometric indicators. Appl. Soft Comput. 27, 420–433 (2015)
15. Driankov, D., Hellendoorn, H., Reinfrank, M.: An Introduction to Fuzzy control. Springer Science & Business Media, Berlin (2013). https://doi.org/10.1007/978-3-662-03284-8
16. Suna, Q., Li, R., Zhang, P.: Stable and optimal adaptive fuzzy control of complex systems using fuzzy dynamic model. J. Fuzzy Sets Syst. 133, 1–17 (2003)
17. Oh, S.K., Pedrycz, W.: The design of hybrid fuzzy controllers based on genetic algorithms and estimation techniques. J. Kybern. 31(6), 909–917 (2002)
18. Lodwick, W.A., Kacprzhyk, J. (eds.): Fuzzy Optimization. STUDFUZ, vol. 254. Springer, Heidelberg (2010). https://doi.org/10.1007/978-3-642-13935-2
19. Kondratenko, Y.P., Al Zubi, E.Y.M.: The optimization approach for increasing efficiency of digital fuzzy controllers. In: Annals of DAAAM for 2009 and Proceeding of the 20th International DAAAM Symposium on Intelligent Manufacturing and Automation, pp. 1589–1591 (2009)
20. Kondratenko, Y., Simon, D.: Structural and parametric optimization of fuzzy control and decision making systems. In: Zadeh, L., Yager, R.R., Shahbazova, S.N., Reformat, M.Z., Kreinovich, V. (eds.) Recent Developments and the New Direction in Soft-Computing Foundations and Applications. STUDFUZZ, vol. 361. Springer, Heidelberg (2018). https://doi.org/10.1007/978-3-319-75408-6
21. Rotshtein, A.P., Rakytyanska, H.B.: Fuzzy evidence in identification, forecasting and diagnosis, vol. 275. Springer, Heidelberg (2012). https://doi.org/10.1007/978-3-642-25786-5
22. Simon, D.: H∞ estimation for fuzzy membership function optimization. Int. J. Approx. Reason. 40, 224–242 (2005)
23. Kondratenko, Y., Korobko, V., Korobko, O., Kondratenko, G., Kozlov, O.: Green-IT approach to design and optimization of thermoacoustic waste heat utilization plant based on soft computing. In: Kharchenko, V., Kondratenko, Y., Kacprzyk, J. (eds.) Green IT Engineering: Components, Networks and Systems Implementation. SSDC, vol. 105, pp. 287–311. Springer, Cham (2017). https://doi.org/10.1007/978-3-319-55595-9_14
24. Simon, D.: Design and rule base reduction of a fuzzy filter for the estimation of motor currents. Int. J. Approx. Reason. 25, 145–167 (2000)
25. Cornejo, M.E., Medina, J., Ramírez-Poussa, E.: Attribute and size reduction mechanisms in multi-adjoint concept lattices. J. Comput. Appl. Math. 318, 388–402 (2017). https://doi.org/10.1016/j.cam.2016.07.012
26. Julián-Iranzo, P., Medina, J., Ojeda-Aciego, M.: On reductants in the framework of multi-adjoint logic programming. Fuzzy Sets Syst. 317, 27–43 (2017)
27. Koczy, L.T., Hirota, K.: Size reduction by interpolation in fuzzy rule bases. IEEE Trans. Syst. Man Cybern. Part B: Cybern. 27(1), 14–25 (1997)
28. Alcalá, R., Alcalá-Fdez, J., Gacto, M.J., Herrera, F.: Rule base reduction and genetic tuning of fuzzy systems based on the linguistic 3-tuples representation. Soft. Comput. 11(5), 401–419 (2007). https://doi.org/10.1007/s00500-006-0106-2
29. Pedrycz, W., Li, K., Reformat, M.: Evolutionary reduction of fuzzy rule-based models. In: Tamir, D.E., Rishe, N.D., Kandel, A. (eds.) Fifty Years of Fuzzy Logic and its Applications. SFSC, vol. 326, pp. 459–481. Springer, Cham (2015). https://doi.org/10.1007/978-3-319-19683-1_23
30. Simon, D.: Evolutionary Optimization Algorithms: Biologically Inspired and Population-Based Approaches to Computer Intelligence. Wiley, Hoboken (2013)

31. Ishibuchi, H., Yamamoto, T.: Fuzzy rule selection by multi-objective genetic local search algorithms and rule evaluation measures in data mining. Fuzzy Sets Syst. **141**(1), 59–88 (2004). https://doi.org/10.1016/S0165-0114(03)00114-3
32. Von Altrock, C.: Applying fuzzy logic to business and finance. Optimus **2**, 38–39 (2002)
33. Von Altrock, C.: Fuzzy Logic and Neurofuzzy Applications in Business and Finance. Prentice Hall, NJ (1996)
34. Kondratenko, Y.P., Kozlov, O.V., Gerasin, O.S., Topalov, A.M., Korobko, O.V.: Automation of control processes in specialized pyrolysis complexes based on web SCADA systems. In: Proceedings of the 9th IEEE International Conference on Intelligent Data Acquisition and Advanced Computing Systems: Technology and Applications (IDAACS), Bucharest, Romania, vol. 1, pp. 107–112 (2017). https://doi.org/10.1109/idaacs.2017.8095059
35. Kondratenko, Y.P., Kozlov, O.V.: Mathematic modeling of reactor's temperature mode of multiloop pyrolysis plant. In: Engemann, K.J., Gil-Lafuente, A.M., Merigó, J.M. (eds.) MS 2012. LNBIP, vol. 115, pp. 178–187. Springer, Heidelberg (2012). https://doi.org/10.1007/978-3-642-30433-0_18

Decision Making Under Incompleteness Based on Soft Set Theory

José Carlos R. Alcantud[1,2](✉) ⓘ and Gustavo Santos-García[2] ⓘ

[1] BORDA Research Unit, University of Salamanca, Salamanca, Spain
jcr@usal.es
[2] Facultad de Economía y Empresa and Multidisciplinary Institute of Enterprise (IME), University of Salamanca, Campus Unamuno, E37007 Salamanca, Spain
santos@usal.es
http://diarium.usal.es/jcr, http://diarium.usal.es/santos

Abstract. Decision making with complete and accurate information is ideal but infrequent. Unfortunately, in most cases the available information is vague, imprecise, uncertain or unknown. The theory of soft sets provides an appropriate framework for decision making that may be used to deal with uncertain decisions. The aim of this paper is to propose and analyze an effective algorithm for multiple attribute decision-making based on soft set theory in an incomplete information environment, when the distribution of incomplete data is unknown. This procedure provides an accurate solution through a combinatorial study of possible cases in the unknown data. Our theoretical development is complemented by practical examples that show the feasibility and implementability of this algorithm. Moreover, we review recent research on decision making from the standpoint of the theory of soft sets under incomplete information.

Keywords: Soft sets · Decision making · Incomplete information
Choice value · Combinatorics

1 Introduction

The aim of this paper is to propose an effective algorithm that facilitates a multiple attribute decision-making based on the theory of soft sets under incomplete information, under the general assumption that the distribution of the incomplete data is unknown. Han *et al.* [16], Qin *et al.* [26] and Zou and Xiao [37] laid the foundation of soft-set based decision making under incomplete information. Here we follow a novel and different approach.

Complete and accurate information is ideal for decision making, but this situation is rarely met in practice. In most cases the available information is vague, imprecise, uncertain or unknown.

The second author is grateful to the Spanish projects TRACES TIN2015–67522–C3–3–R and Strongsoft TIN2012–39391–C04–04.

ⓒ Springer International Publishing AG, part of Springer Nature 2018
J. Medina et al. (Eds.): IPMU 2018, CCIS 854, pp. 583–595, 2018.
https://doi.org/10.1007/978-3-319-91476-3_48

Motivated by these concerns, Zadeh [35] marked the beginning of fuzzy sets theory. At its core, partial membership allows that imprecise information about more complex situations can be faithfylly represented and correctly handled. Many extensions of fuzzy sets broaden their scope (v., [4,10] for definitions and relationships). Relatedly, Molodtsov initiated the concept of soft set theory [25], whereas Aktaş and Çağman [1], Ali et al. [9], Maji et al. [24], Sezgin and Atagün [29] and Feng et al. [15] are further essential references. Concerning extensions of soft sets, Maji et al. [22] introduced fuzzy soft sets (v., [2,3,8,20] for decision making criteria in this model), and Wang et al. [30] introduced hesitant fuzzy soft sets. Ma et al. [21] provided a review of decision making methods based on hybrid soft set models.

Maji et al. [23] provides a criteria for selecting an object in a soft set scenario, which consists of maximization of the choice values of the problem. Applications of extended soft set theory include rule mining (Herawan and Deris [18]), data mining processes (Qin et al. [27]), international trade (Xiao et al. [31]), and medical diagnosis (cf., e.g., [7,11,12,17,34]). Recent advances in this field include the parameter reduction problem in soft set based decision making (cf., [36]).

Zou and Xiao [37] observed that in the process of collecting data, the practitioner often encounters unknown, missing or inexistent data, which suggests the concept of incomplete soft sets (v., Han et al. [16], Qin et al. [26], Lin et al. [19] and Alcantud and Santos-García [5,6] for additional analyses). Deng and Wang [14] extended this notion to incomplete fuzzy soft sets in order to predict unknown data in fuzzy soft sets.[1]

However, there are situations where there is perfect uncertainty about the real value of missing data, or we are sure that the alternatives or attributes are independent. In those cases we cannot presume that averages, probabilities or any other specific evaluations produce reliable estimations as in previous solutions. To deal with these situations Alcantud and Santos-García [6] presented a completely redesigned approach to soft set based decision making problems under incomplete information. It relies on the classical Laplacian argument of probability theory and consequently it suggests to examine all completed tables arising from the original incomplete table. All these tables are then evaluated as is standard, i.e., by their respective choice values (cf., [23]). The alternatives are ultimately ranked according to the number or proportion of tables where they are choice value maximizers. The computational costs of this procedure are examined in [6]. The conclusion is that with a large number of missing values, the problem cannot be efficiently solved by bruteforce. Hence in [5] these authors propose two modified algorithms that permit to tackle problems where the number of unknown values is larger.

In this paper we produce an exact solution to that problem based on the application of combinatorics. We provide an algorithm that implements the mathematical solution. The computational performance of the algorithm is then compared with prior solutions in the literature.

---

[1] Although Yang et al. [32] showed some weaknesses of this approach, Deng and Chen [13] successfully resolved these conflicting issues.

The remainder of this paper is organized as follows. Section 2 briefly states some terminology and definitions from soft set theory. Then we review the relevant antecedents, define the technical notions that we need, and recall the *domination sieve* for incomplete soft sets. Section 3 shows the steps in the proposed algorithm, as well as a fully developed example that proves the feasibility and implementability of our proposal. Finally, Sect. 4 concludes the paper.

2 Definitions and Notation

2.1 Definitions: Soft Sets and Incomplete Soft Sets

In this section, we first introduce the standard definitions for complete and incomplete soft sets. There is a fixed universe U of objects, options or alternatives, and there is a universal set E of parameters, characteristics or attributes.

Definition 1 (Molodtsov [25]). A pair (F, A) is a *soft set* over U when $A \subseteq E$ and $F : A \longrightarrow \mathcal{P}(U)$, where $\mathcal{P}(U)$ denotes the set of all subsets of U.

Definition 2 (Han et al. [16]). A pair (F, A) is an *incomplete soft set* over U when $A \subseteq E$ and $F : A \longrightarrow \{0, 1, *\}^U$, where $\{0, 1, *\}^U$ is the set of all functions from U to $\{0, 1, *\}$.

The $*$ symbol in the previous definition represents an unknown data. In other words, if the membership of an element u in the subset of U approximated by e is unknown, then $F(e)(u) = *$. Of course, as in standard soft sets, if an object u is (resp. is not) an element of the subset of U approximated by e, then $F(e)(u) = 1$ (resp., $F(e)(u) = 0$). Thus any soft set can be regarded as an incomplete soft set in a natural way.

Henceforth we assume $U = \{u_1, u_2, \ldots u_N\}$ and $A = \{e_1, e_2, \ldots e_M\}$. Yao [33] explained that under this finiteness assumption, soft sets can be represented either by matrices or in tabular form. Rows correspond to the options, and columns correspond to the parameters. The same is true for incomplete soft sets. Suppose that $T = (t_{ij})_{N \times M}$ is the $N \times M$ matrix associated with the incomplete soft set (F, A). Then element t_{ij} is either one or zero or $*$, depending on whether object i verifies property j, does not verify it, or it is unknown whether i verifies property j, respectively.

Concerning the choice decision mechanism for soft sets, we agree with existing literature in that choice values should be used. As to incomplete soft sets, our proposal is original in that it does not discard any of the possible filled tables or completed soft sets.

2.2 Previous Literature

Table 1 summarizes the main previous approaches to our problem.

Table 1. Summary of research studies about incomplete soft set based decision making. The indexes are named as in the original papers.

Methodology and references		Indexes
Weight-average of all possible choice values of objects		d_i, d_{i-p}, $c_{i(0)}$, $c_{i(1)}$
Weights of each choice value given by distribution of the other objects		
Original approach [37]		
Data filling based on association between parameters		Q_i
Choice: choice values for completed set		
It presumes that objects are cross-related		
Inspired by above approach [26]		
Elicitation criterions for incomplete soft sets generated by restricted intersection		-
This problem is related although different [16]		
Elimination of dominated options	*Random sample*, thus results depend on sample [5]†	s_i^2
Choice mechanism: choice values		
Laplacian argument: equal probability to all completed tables. Suitable when there is no guarantee that objects are related to each other	*Brute force*, thus computationally costly [6]	s_i

2.3   Notation and Fundamentals of Our Algorithm

From the input matrix T we calculate the number of ones and unknown values for every object in the soft set. These are fundamental elements for the analysis of the optimal solution in a fully uncertain environment.

Let v^1 be the vector of numbers of 1's (ones values) by rows in T, i.e., $v^1 = (v_1^1, \ldots v_N^1)$, where v_i^1 is the number of ones in object i. Observe that if $c_{i(0)}$ is i's choice value if all missing data are replaced with 0, then $v_i^1 = c_{i(0)} = |\{e_j \in A : F(e_j)(u_i) = 1\}|$.

The maximum value in vector v^1 is $c_0 = \max\{c_{i(0)} : i = 1, \ldots, N\} = \max\{v_i^1 : i = 1, \ldots, N\}$.

Let v^* be the vector of numbers of $*$'s (unknown values) by rows in T, i.e., $v^* = (v_1^*, \ldots v_N^*)$, where v_i^* is the number of unknown values in object i. This means $v_i^* = |\{e_j \in A : F(e_j)(u_i) = *\}|$. We let m_* denote the maximum value in vector v^*, i.e., $m_* = \max\{v_i^* : i = 1, \ldots, N\}$.

Observe that if $c_{i(1)}$ is i's choice value when all missing data are replaced with 1, then $c_{i(1)} = c_{i(0)} + v_i^*$ for each i.

The number of unknown values is $M^* = \sum_{i=1}^N v_i^*$. A simple combinatorial analysis shows that the number of *combinations* of unknown choice values is:

$$M^c = \prod_{i=1}^N (v_i^* + 1).$$

Example 1 below illustrates these notions.

2.4   Domination Sieve

In order to gain efficiency, our algorithm performs a pre-screening by removing the objects whose choice values can never be maximal. Put shortly, we eliminate dominated alternatives defined as follows:

Definition 3 (Alcantud and Santos-García [6]). Let (F, A) be an incomplete soft set over U. Option i *dominates* option k if and only if $c_{k(1)} < c_{i(0)}$.

Hence in order to check if an option i dominates an option k one needs to verify if $c_{k(0)} + v_k^* < c_{i(0)}$. Intuitively, no matter how we complete the soft set, option i has a choice value that is strictly higher than the choice value of option j. Therefore dominated options should be rejected in any choice-valued approach.

In order to simplify our problem we compute the maximum value $c_0 = 3$ of the choice values $v_i^1 = c_{i(0)}$ for every i. If c_0 is greater than the choice value $c_{k(1)}$ of object u_k, then object u_k can be dropped from the initial matrix because its choice value can never be maximum in any posible completed soft set. In this way, by reducing the number of rows in the matrix form of an incomplete soft set, there are less missing data, which reduces runtime and improves final results.

After dominated options are sieved out, we apply the remaining steps to the new trimmed matrix. To simplify the notation, this reduced matrix is also called T. Its objects and features will be reappointed N (also number of rows of T) and M (also number of columns of T), respectively.

Example 1. Let $T = (t_{ij})$ be the following 5×4 initial matrix, which represents an incomplete soft set over a universe of 5 objects with 4 relevant characteristics for evaluation of the alternatives:

$$T = \begin{pmatrix} 1 & 1 & 1 & 0 \\ 1 & 1 & * & 0 \\ 1 & * & * & 0 \\ 1 & * & 0 & 0 \\ 0 & 0 & 0 & 0 \end{pmatrix}$$

It is easy to compute that $v^1 = (3, 2, 1, 1, 0), v^* = (0, 1, 2, 1, 0)$ therefore $m_1 = 3$, $m_* = 2$, $M^* = 4$, and $M^c = 12$.

We calculate the maximum value c_0 of all choice values $c_{i(0)}$. Observe that $c_{4(0)} + v_4^* = 1 + 1 < c_{1(0)} = c_0 = 3$ and $c_{5(0)} + v_5^* = 0 + 0 < c_{1(0)} = c_0 = 3$. Thus c_0 is unattainable for the choice values $c_{i(1)}$ of objects u_4 and u_5, and these objects can be safely removed from the initial matrix (see Table 2, where the relevant items are underlined). Observe that no matter how the table is completed, the choice values of options u_4 and u_5 will be smaller than the choice value of u_1.

Table 2. Tabular representation of the soft set in Example 1, and noteworthy associated indices.

	e_1	e_2	e_3	e_4	$v_i^1 = c_{i(0)}$	$c_{i(1)}$	v_i^*
u_1	1	1	1	0	$c_0 = \underline{3}$	3	0
u_2	1	1	*	0	2	3	1
u_3	1	*	*	0	1	3	2
(u_4)	1	*	0	0	1	$\underline{2}$	1
(u_5)	0	0	0	0	0	$\underline{0}$	0

After removing objects u_4 and u_5, the new sieved matrix is:

$$T \text{ (sieved)} = \begin{pmatrix} 1 & 1 & 1 & 0 \\ 1 & 1 & * & 0 \\ 1 & * & * & 0 \end{pmatrix}$$

For the new trimmed matrix the vector of numbers of one values v_1 and the vector of numbers of unknown values v^* are:

$$v^1 = (\, 3, \, 2, \, 1 \,); \qquad v^* = (\, 0, \, 1, \, 2 \,)$$

which can be drawn from the corresponding columns in Table 2. And now it is immediate to compute $m_1 = 3$, $m_* = 2$.

Moreover $M^* = 3$ (number of unknown values) and $M^c = 6$ (number of combinations of unknown choice values). This means that the four unknown values of the initial matrix T become three values for the new sieve matrix T, while the number of feasible states (in terms of configurations of choice values) is halved. These states are:

(1) the choice values can be 3 for option u_1, 2 for u_2 and 1 for u_3. This fact happens in exactly one completed table.
(2) the choice values can be 3 for option u_1, 2 for u_2 and 2 for u_3. This fact happens in exactly two completed tables.
(3) the choice values can be 3 for option u_1, 2 for u_2 and 3 for u_3. This fact happens in exactly one completed table.

(4) the choice values can be 3 for option u_1, 3 for u_2 and 1 for u_3. This fact happens in exactly one completed table.

(5) the choice values can be 3 for option u_1, 3 for u_2 and 2 for u_3. This fact happens in exactly two completed tables.

(6) the choice values can be 3 for option u_1, 3 for u_2 and 3 for u_3. This fact happens in exactly one completed table.

We can represent the completed soft sets for cases (1), (2) and (3) as follows:

	e_1	e_2	e_3	e_4	e_1	e_2	e_3	e_4	e_1	e_2	e_3	e_4	e_1	e_2	e_3	e_4
u_1	1	1	1	0	1	1	1	0	1	1	1	0	1	1	1	0
u_2	1	1	0	0	1	1	0	0	1	1	0	0	1	1	0	0
u_3	1	0	0	0	1	1	0	0	1	0	1	0	1	1	1	0

And the completed soft sets for cases (4), (5) and (6) are represented as

	e_1	e_2	e_3	e_4	e_1	e_2	e_3	e_4	e_1	e_2	e_3	e_4	e_1	e_2	e_3	e_4
u_1	1	1	1	0	1	1	1	0	1	1	1	0	1	1	1	0
u_2	1	1	1	0	1	1	1	0	1	1	1	0	1	1	1	0
u_3	1	0	0	0	1	1	0	0	1	0	1	0	1	1	1	0

3   Algorithm: Incomplete Soft Sets for Decision Making Problem

3.1   The Main Elements of the Algorithm

Our algorithm computes the number of tables that produce every possible configuration of choice values. Under the Laplacian assumption that they are equally probable, it chooses the alternative that is best in the highest proportion of completed soft sets. However our algorithm does not compute them explicitly as in Example 1, but instead we use a combinatorial analysis in order to define several auxiliary matrices corresponding to possible values and probabilities for each of the objects that we actually use. Let PV_1 (resp., PPV_1) be the $(N \times (m_* + 1))$-matrices of potential values (resp., proportion by rows of potential values) for each object. For computational purposes, the rows with a number of elements less than $m_* + 1$ are filled up with zero values. Its elements are calculated according to the following formulas:

$$PV_1(i,j) = v_i^1 + j - 1; \qquad PPV_1(i,j) = \frac{\binom{v_i^*}{j}}{2^{v_i^*}},$$

where i varies from 1 to N and j varies from 1 to $v_i^* + 1$.[2] These matrices are shown in Example 2.

Example 2. According to the data of Example 1, now we calculate matrices PV_1 and PPV_1:

$$PV_1 = \begin{pmatrix} 3 & 0 & 0 \\ 2 & 3 & 0 \\ 1 & 2 & 3 \end{pmatrix}, \qquad PPV_1 = \begin{pmatrix} 1.00 & 0 & 0 \\ 0.50 & 0.50 & 0 \\ 0.25 & 0.50 & 0.25 \end{pmatrix}.$$

Hence for example, the fact $PV_1(1,1) = 3$ and $PPV_1(1,1) = 1$ means that in 100% of the 8 filled tables, the choice value of u_1 is 3. The fact $PV_1(2,2) = 3$ and $PPV_1(2,2) = 0.5$ means that in 50% of the 8 filled tables, the choice value of u_2 is 3. And $PV_1(3,2) = 2$ and $PPV_1(3,2) = 0.5$ means that in 50% of all filled tables, the choice value of u_3 is 2.

It is well-known that if two events A and B are independent then their joint probability equals the product of their probabilities, i.e., $P(A \cap B) = P(A) P(B)$. Because choice values are independent events, with probability $1 \cdot 0.5 \cdot 0.5 = 0.25$ the choice values of u_1, u_2 and u_3 are 3, 3, and 2, respectively. And in that case, both u_1 and u_2 are optimal because they are choice value maximizers.

After calculating the auxiliary matrices PV_1 and PPV_1, we analyze all feasible combinations of decision values. As shown in Example 2, we know the probabilities for each of the possible events. Computationally, we build all possible different vectors CV with N rows, in which each element will be a non-zero element of each row in PV_1 matrix. That is, the element CV_i for each object i has a value between $c_{i(0)}$ and $c_{i(0)} + v_i^*$.

For this particular case, we now calculate the probabilities of occurrences from matrix PPV_1. The joint probability XP is the product of all the individual probabilities. When the choice value of that particular case is maximal (may be several maximal), we add that probability XP to the decision values matrix DV.

We repeat the process for all possible vectors CV. The final decision of our Algorithm will consist of the object(s) with greater DV values.

Figure 1 shows a flowchart for the proposal of a decision making procedure under incomplete information that we have described throughout this section.

Example 3. According to the data of Example 1 and 2, now we calculate final matrices CV, XP, and DV. In particular we obtain

$$DV = \begin{pmatrix} 1.00 \\ 0.50 \\ 0.25 \end{pmatrix}$$

which we interpret as follows. In 100% of the randomly filled tables, option 1 achieves the maximum choice value. In 50% of the randomly filled tables, option 2

---

[2] As usual, $\binom{n}{k}$ or "n choose k" returns the binomial coefficient, i.e., the number of combinations of n items taken k at a time, defined as $\dfrac{n!}{(n-k)!k!}$.

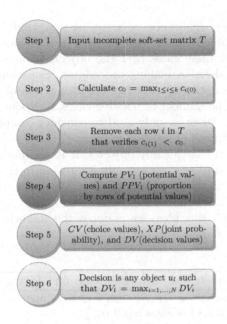

Fig. 1. Step by step procedure for decision making under incompleteness.

achieves the maximum choice value. And in 25% of the randomly filled tables, option 3 achieves the maximum choice value. We recall that ties may happen, which explains why this figures sum up over 100%.

Table 3 contains the elements that produce our solution. In view of our arguments, option 1 should be selected.

Table 3. Combinations of choice values and their respective probabilities in Example 1.

Choice value			Probability	Optimal solutions
u_1	u_2	u_3		
3	2	1	1/8	u_1
3	2	2	1/4	u_1
3	2	3	1/8	u_1, u_3
3	3	1	1/8	u_1, u_2
3	3	2	1/4	u_1, u_2
3	3	3	1/8	u_1, u_2, u_3

3.2 Decision of the Algorithm

Intuitively, our practical implementation of the ideas above is based on the following arguments. According to Laplace's principle of indifference in probability theory, under complete ignorance we must assume that in all tables *'s are replaced in equiprobable manner with either 0 or 1. Hence the best we can do is consider in each of these cases that the objects should be selected according to soft-set based decision making procedures. Consequently we should opt for any object that is optimal in the highest proportion of cases with the completed information. The number of possible cases is exponential on the number of unknown values, however we only need to consider completed soft sets which are equivalent for purposes of decision making.

To do this we note that for any object i, if it has multiple unknown characteristics that are completed, we are only interested in the number, not in the order, of 1's assigned to these unknown values. Thus, the calculation is reduced to a combinatorial analysis in each object.

In accordance with this idea, we endorse the following algorithm for the problems where both U and A are finite:

Algorithm 1. Incomplete Soft Sets Algorithm for Making Decision

---

1: Input an incomplete soft set (F, E) with k objects and l parameters in a matrix form, where $t_{ij} \in \{0, 1, *\}$ denotes a cell (i, j).

2: Calculate $c_0 = \max_{1 \le i \le k} c_{i(0)}$ from the incomplete matrix, where $c_{i(0)}$ is the choice value for i if all missing data are assumed to be 0.

3: Remove each row i from the incomplete matrix that verifies $c_{i(1)} < c_0$, where $c_{i(1)}$ is the choice value for i if all missing data are assumed to be 1.

4: From the new reduced matrix, calculate the auxiliary matrices PV_1 (matrix of potential values) and PPV_1 (matrix of proportion by rows of potential values).

5: Compute the following matrices: choice values CV, joint probability XP, and decision values DV.

6: The result of the decision is any object u_l such that $DV_l = \max_{i=1,...,N} DV_i$.

---

4 Discussion and Conclusions

The works of Han et al. [16], Qin et al. [26] and Zou and Xiao [37] present interesting approaches to incomplete soft set based decision making. These authors used averages, probabilities or other specific evaluations in order to estimate the real value of missing data in a general way and afterwards, they made a decision based on the complementary data.

In this paper we look at the problem from an altogether different perspective. Rather than filling the incomplete data table (see also Khan et al. [28]), we propose a combinatorial study through all possible filled tables that can be produced from the original incomplete table. Then the alternatives are ranked by the proportion of filled tables where they achieve the highest choice value. In other words, a final indicator for each of the objects by our algorithm is defined as the value of this ratio. And our decision making procedure consists of selecting alternatives that maximize this indicator. A classical Laplacian argument from probability theory justifies our research method. In general there is perfect uncertainty about the real value of missing data. Therefore, we cannot support the idea that other aspects would let us faithfully estimate these unknown values. Under Laplace's principle of indifference, due to our complete ignorance we are entitled to assume that all possible tables where the missing data are replaced with either 0 or 1 are equiprobable.

Our novel approach meets the following targets: (1) We do not need to assume any cross-relations among options. (2) We do not attempt to fill the tables with hypothesized values. (3) We adopt a Laplacian position and make use of combinatorics. (4) A unique, computationally tractable solution is provided.

An example illustrates the detailed implementation process of our approach and shows evidence of its potential applications in decision-making problems with incomplete information.

In practice, problems where all parameters describing compared options are equally important rarely exist. Our algorithm has been tightly designed for the exclusive purpose of decision making, hence the introduction of weighted parameters would not be trivial and as such deserves a separate analysis. Domination sieve is simple to read in that instance, though.

References

1. Aktaş, H., Çağman, N.: Soft sets and soft groups. Inf. Sci. **177**, 2726–2735 (2007)
2. Alcantud, J.C.R.: Fuzzy soft set decision making algorithms: some clarifications and reinterpretations. In: Luaces, O., Gámez, J.A., Barrenechea, E., Troncoso, A., Galar, M., Quintián, H., Corchado, E. (eds.) CAEPIA 2016. LNCS (LNAI), vol. 9868, pp. 479–488. Springer, Cham (2016). https://doi.org/10.1007/978-3-319-44636-3_45
3. Alcantud, J.C.R.: A novel algorithm for fuzzy soft set based decision making from multiobserver input parameter data set. Inform. Fusion **29**, 142–148 (2016)
4. Alcantud, J.C.R.: Some formal relationships among soft sets, fuzzy sets, and their extensions. Int. J. Approx. Reason. **68**, 45–53 (2016)
5. Alcantud, J.C.R., Santos-García, G.: Incomplete soft sets: new solutions for decision making problems. Decision Economics, In Commemoration of the Birth Centennial of Herbert A. Simon 1916-2016 (Nobel Prize in Economics 1978). AISC, vol. 475, pp. 9–17. Springer, Cham (2016). https://doi.org/10.1007/978-3-319-40111-9_2
6. Alcantud, J.C.R., Santos-García, G.: A new criterion for soft set based decision making problems under incomplete information. Int. J. Comput. Intell. Sys. **10**, 394–404 (2017)
7. Alcantud, J.C.R., Santos-García, G., Hernández-Galilea, E.: Glaucoma diagnosis: a soft set based decision making procedure. In: Puerta, J.M., Gámez, J.A., Dorronsoro, B., Barrenechea, E., Troncoso, A., Baruque, B., Galar, M. (eds.) CAEPIA 2015. LNCS (LNAI), vol. 9422, pp. 49–60. Springer, Cham (2015). https://doi.org/10.1007/978-3-319-24598-0_5
8. Alcantud, J.C.R., Mathew, T.J.: Separable fuzzy soft sets and decision making with positive and negative attributes. Appl. Soft Comput. **59**, 586–595 (2017)

9. Ali, M.I., Feng, F., Liu, X., Min, W.K., Shabir, M.: On some new operations in soft set theory. Comput. Math. Appl. **57**(9), 1547–1553 (2009)

10. Bustince, H., Barrenechea, E., Pagola, M., Fernandez, J., Xu, Z., Bedregal, B., Montero, J., Hagras, H., Herrera, F., De Baets, B.: A historical account of types of fuzzy sets and their relationships. IEEE Trans. Fuzzy Syst. **24**(1), 179–194 (2016)

11. Çelik, Y., Yamak, S.: Fuzzy soft set theory applied to medical diagnosis using fuzzy arithmetic operations. J. Inequal. Appl. **2013**(1), 82 (2013)

12. Chetia, B., Das, P.K.: An application of interval valued fuzzy soft sets in medical diagnosis. Int. J. Contemp. Math. Sci. **38**(5), 1887–1894 (2010)

13. Deng, T., Chen, Y.: Comments from the author of "An object-parameter approach to predicting unknown data in incomplete fuzzy soft sets". Appl. Math. Model. **39**(23–24), 7744–7745 (2015). [Appl Math Model 37 (2013) 4139–4146]

14. Deng, T., Wang, X.: An object-parameter approach to predicting unknown data in incomplete fuzzy soft sets. Appl. Math. Model. **37**(6), 4139–4146 (2013)

15. Feng, F., Cho, J., Pedrycz, W., Fujita, H., Herawan, T.: Soft set based association rule mining. Knowl.-Based Syst. **111**, 268–282 (2016)

16. Han, B.H., Li, Y., Liu, J., Geng, S., Li, H.: Elicitation criterions for restricted intersection of two incomplete soft sets. Knowl.-Based Syst. **59**, 121–131 (2014)

17. Hassan, N., Sayed, O.R., Khalil, A.M., Ghany, M.A.: Fuzzy soft expert system in prediction of coronary artery disease. Int. J. Fuzzy Syst. **19**(5), 1546–1559 (2017)

18. Herawan, T., Deris, M.M.: A soft set approach for association rules mining. Knowl.-Based Syst. **24**(1), 186–195 (2011)

19. Lin, H., Xiao, Z., Cheng, Y.: New methods for decision making with soft set under incomplete information. J. Comput. Theor. Nanosci. **13**(2), 1247–1252 (2016)

20. Liu, Z., Qin, K., Pei, Z.: A method for fuzzy soft sets in decision-making based on an ideal solution. Symmetry **9**(10), 246 (2017)

21. Ma, X., Liu, Q., Zhan, J.: A survey of decision making methods based on certain hybrid soft set models. Artif. Intell. Rev. **47**, 507–530 (2016)

22. Maji, P., Biswas, R., Roy, A.: Fuzzy soft sets. J. Fuzzy Math. **9**, 589–602 (2001)

23. Maji, P., Biswas, R., Roy, A.: An application of soft sets in a decision making problem. Comput. Math. Appl. **44**, 1077–1083 (2002)

24. Maji, P., Biswas, R., Roy, A.: Soft set theory. Comput. Math. Appl. **45**, 555–562 (2003)

25. Molodtsov, D.: Soft set theory - first results. Comput. Math. Appl. **37**, 19–31 (1999)

26. Qin, H., Ma, X., Herawan, T., Zain, J.M.: Data filling approach of soft sets under incomplete information. In: Nguyen, N.T., Kim, C.-G., Janiak, A. (eds.) ACIIDS 2011. LNCS (LNAI), vol. 6592, pp. 302–311. Springer, Heidelberg (2011). https://doi.org/10.1007/978-3-642-20042-7_31

27. Qin, H., Ma, X., Zain, J.M., Herawan, T.: A novel soft set approach in selecting clustering attribute. Knowl.-Based Syst. **36**, 139–145 (2012)

28. Sadiq Khan, M., Al-Garadi, M.A., Wahab, A.W.A., Herawan, T.: An alternative data filling approach for prediction of missing data in soft sets (ADFIS). Springer-Plus **5**(1), 1–20 (2016)

29. Sezgin, A., Atagün, A.O.: On operations of soft sets. Comput. Math. Appl. **61**(5), 1457–1467 (2011)

30. Wang, F., Li, X., Chen, X.: Hesitant fuzzy soft set and its applications in multi-criteria decision making. J. Appl. Math. **2014**, 10 (2014)

31. Xiao, Z., Gong, K., Zou, Y.: A combined forecasting approach based on fuzzy soft sets. J. Comput. Appl. Math. **228**(1), 326–333 (2009)

32. Yang, Y., Song, J., Peng, X.: Comments on "An object-parameter approach to predicting unknown data in incomplete fuzzy soft sets". Appl. Math. Model. **39**(23–24), 7746–7748 (2015). [Appl Math Modell 37 (2013) 4139–4146]
33. Yao, Y.: A comparative study of fuzzy sets and rough sets. Inf. Sci. **109**, 227–242 (1998)
34. Yuksel, S., Dizman, T., Yildizdan, G., Sert, U.: Application of soft sets to diagnose the prostate cancer risk. J. Inequal.Appl. **2013**(1), 229 (2013)
35. Zadeh, L.: Fuzzy sets. Inf. Control **8**, 338–353 (1965)
36. Zhan, J., Alcantud, J.C.R.: A survey of parameter reduction of soft sets and corresponding algorithms. Artif. Intell. Rev. 1–34 (2018). https://doi.org/10.1007/s10462-017-9592-0
37. Zou, Y., Xiao, Z.: Data analysis approaches of soft sets under incomplete information. Knowl.-Based Syst. **21**(8), 941–945 (2008)

Intelligent Decision Support System for Selecting the University-Industry Cooperation Model Using Modified Antecedent-Consequent Method

Yuriy P. Kondratenko[(✉)] [iD], Galyna Kondratenko [iD],
and Ievgen Sidenko [iD]

Intelligent Information Systems Department, Petro Mohyla Black Sea National
University, 68th Desantnykiv Str., 10, Mykolaiv 54003, Ukraine
{yuriy.kondratenko,halyna.kondratenko,
ievgen.sidenko}@chmnu.edu.ua

Abstract. This work is devoted to the analysis and selection of the most rational model of the university/IT-company cooperation (UIC) using intelligent decision support systems (DSSs) in the conditions of input information uncertainty. The modification of a two-cascade method for reconfiguration of the fuzzy DSS's rule bases is described in details for situations when the volume of input data can be changed. Authors propose an additional observer procedure for checking the fuzzy rule consequents before their final correction. The modified method provides (a) structural reduction of the rule antecedents, (b) correction of the corresponding consequents in an interactive mode and (c) avoiding the results' deformation in the decision making process with variable structure of input data. Special attention is paid to the hierarchically organized DSSs (with variable input vector and discrete logic output) and to design of the web-oriented instrumental tool (WOTFS-1). The simulation results confirm the efficiency and expediency of using (a) the software WOTFS-1 and (b) modified method of fuzzy rule base's antecedent-consequent reconfiguration for the efficient selection of the rational model of academia-industry cooperation.

Keywords: Fuzzy logic · Linguistic term · Rule base · Reconfiguration
Decision support system · University-industry cooperation

1 Introduction

To implement multidimensional fuzzy dependencies, it is expedient to use a hierarchical approach in the synthesis of DSS for automation of decision-making processes based on fuzzy logic output. In the process of developing fuzzy DSS there is a problem of the sharp increase in the rules bases (RBs) at increasing the dimension of the input vector and the number of corresponding linguistic terms (LTs). At present there is a sufficient number of publications [2, 3, 22] on the development and optimization of fuzzy DSSs, including for solving multi-criteria problems in conditions of uncertainty. However, a further solution is required for the methods and technologies of the

© Springer International Publishing AG, part of Springer Nature 2018
J. Medina et al. (Eds.): IPMU 2018, CCIS 854, pp. 596–607, 2018.
https://doi.org/10.1007/978-3-319-91476-3_49

synthesis of hierarchical DSSs based on fuzzy models with variable structure of the input vector, in particular, taking into account the input information from experts and a person, who is a decision maker (DM) and whose estimates are fuzzy [29–32]. It should be noted that the development of fuzzy DSSs to increase the efficiency of decision-making in the conditions of multi-criteria and a priori informational uncertainty, in particular, with the variable structure of the input coordinate's vector, is one of the perspective directions for the creation of intelligent information systems [27].

2   Related Works and Problem Statement

One of the problems of DSSs synthesis based on fuzzy logic output is the complexity of making decisions with variable structure of the input data of the system. This is due to the need to develop approaches to correcting fuzzy RBs in the decision-making process in a priori information uncertainty, in particular when a DM is not able to evaluate and insert specific input coordinates into the system [7, 25, 26, 34, 38].

In fuzzy modeling, Mamdani's algorithm is used most often, according to which the antecedents and the consequents of the rules of fuzzy RBs are given by fuzzy sets (linguistic terms) such as "Low", "Medium", "High", etc. [10, 11, 15–17, 21].

The process of determining the most important input coordinates and the experts' formation of their estimates significantly influences the structure of the hierarchically organized DSS, in particular on the dimensionality of the RBs [9, 14]. Previous studies [18–20, 33, 35–37] show that the decision-making results undergo significant deformation in the application of fuzzy DSSs (with a fixed structure of the knowledge base) under the conditions of the variable structure of the input vector. This is due to the fact that the values of the input coordinates, which do not participate in the simulation of DSSs and a consequent with zero value, due to the corresponding fuzzy rules negatively affect the result. Consequently, the change in the dimension of the input vector in the interactive modes of fuzzy DSSs requires the development of effective methods for the reconfiguration and correction of fuzzy RBs [23, 28].

Among the well-known approaches to correction of fuzzy RBs is the use of weighting coefficients for fuzzy rules [24, 29]. Changing the vector of weighting coefficients for the corresponding rules of fuzzy knowledge bases can reduce the influence of input parameters, which, by the choice of a DM in some situations may not participate in the decision making process, on the result of the system. In addition, there is an approach to the correction of RBs, which consists of identifying non-essential parameters of the model. The number of rules is significantly reduced, which allows to increase the sensitivity of the system to change the values of input signals.

The limited properties of the considered approaches and methods of correction of RBs do not allow them to be used directly to optimize the fuzzy hierarchical DSSs with variable structure of the input vector [9, 11, 33].

The purpose of this paper is to increase the efficiency of the processes of multi-criteria decision-making in fuzzy hierarchical DSSs with discrete logical output under the condition of the variable structure of the input vector and in the formation of incoming information with a high level of uncertainty (with the application of DSS for choosing a cooperation model within the "University – IT-company" consortium).

3   Structure of Fuzzy DSS for Choosing the UIC Model

The problem of choosing one of the models of cooperation between an IT-company and a department (faculty) of the university is relevant today, especially at the beginning of their cooperation and under conditions of a possible change in the direction of joint research. Previous studies and analysis of successful cooperation experience within different types of consortia [1, 4, 8, 9] show that the solution to the task of choosing a model of cooperation between the university and the IT-Company today is to select one of four alternatives $(m = 4)$, such as alternative solution variants $d_i, (i = 1...m)$, where the solution variant d_1 corresponds the model A1; variant d_2 - model A2; variant d_3 - model B; variant d_4 - model C.

The authors developed a fuzzy DSS for choosing a model for cooperation between universities and IT-companies $y = d^*, (d^* \in d, d = \{d_1, d_2, d_3, d_4\})$ in terms of previously proposed and defined indicators (input coordinates). DSS includes 11 fuzzy subsystems and has 27 input coordinates $X = \{x_j\}, j = 1, \ldots, 27$ and one output y, which are interconnected by the fuzzy dependencies $y_k = f_k(x_1, x_2, \ldots, x_{27}), k \in \{1, 2, \ldots, 11\}$ of the appropriate RBs of the 11 subsystems, where f is the functional dependence of the output coordinate y_k to the inputs $x_1, x_2, \ldots, x_{27}$ in the form of a fuzzy RBs [9, 14, 33].

The structure of the corresponding fuzzy DSS provides the choice of one of the four basic (system) solutions $(y \in \{A1, A2, B, C\})$, since 4 LTs were used to describe the output variable y. This structure provides the possibility of choosing additional (program) $\{A1(OR)A2, A1(OR)B, A1(OR)C, A2(OR)B, A2(OR)C, B(OR)C\}$ solutions, in the case where exponents of the degrees of membership $\mu^{d^*}(X^*)$, for example, of two system solutions are of the same value. Program solutions can also be formed for possible (in similar cases) combinations with 3 and 4 system solutions.

To estimate the input $X = \{x_j\}, j = 1, \ldots, 27$, 3 LTs with a triangular form of membership function (MF), in particular "low - L", "medium - M" and "high - H", are chosen. We should represent some of the above-mentioned coordinates: x_4 - level of IT experience of students of the university department; x_5 - participation of students in international exchange programs; x_6 - the level of student co-work with IT-companies; x_7 - students' success in studying. Subsequently, the pre-developed RBs of all subsystems are transformed into a matrixes of knowledge by a combination of rules for the source LT of the corresponding consequent, for example, for the 2nd $y_2 = f_2(x_4, x_5, x_6, x_7)$ fuzzy subsystem $y_2 \in \{L, LM, M, HM, H\}$ (Table 1) [14].

When user inserts the data $X^{(2)*} = (x_4^*, x_5^*, x_6^*, x_7^*)$ to the inputs of the second subsystem, its output result y_2^* is generated based on the corresponding LTs: $y_2^* \in \left\{ LT_j^{(2)} \middle| LT_j^{(2)} \in \{L, LM, M, HM, H\}, j = 1, \ldots, 5 \right\}$. The output of the second subsystem y_2^* is transferred in the form of the accumulated LT to the RB of the next (in terms of the hierarchy) subsystem $y_8 = f_8(y_1, y_2)$.

Implementation of the procedure of fuzzy logical output in more details is represented in researches [9, 14, 31, 33].

Table 1. Partial set of rules of knowledge matrix for subsystem $y_2 = f_2(x_4, x_5, x_6, x_7)$

Number of rule and combination		Coordinates of Subsystem $y_2 = f_2(x_4, x_5, x_6, x_7)$				
		x_4	x_5	x_6	x_7	y_2
1	1,1	L	L	L	L	
...	...		...			L
55	1,8	H	L	L	L	
5	2,1	L	L	M	M	
...	...		...			LM
64	2,19	H	M	L	L	
9	3,1	L	L	H	H	
...	...		...			M
73	3,31	H	H	L	L	
18	4,1	L	M	H	H	
...	...		...			HM
79	4,18	H	H	H	L	
54	5,1	M	H	H	H	
...	...		...			H
81	5,5	H	H	H	H	

4 Antecedent-Consequent Method of Rules Bases Correction of Fuzzy DSS for Choosing the Rational UIC Model

In the process of decision making using fuzzy hierarchical DSSs with a variable structure of the input vector, it is necessary to apply effective approaches for reconfiguring (correction) of fuzzy RBs. This is due to the fact that the values of input coordinates that cannot be estimated at the time of decision making are uncertain and due to the corresponding fuzzy rules negatively affect the result y. The limited properties of existing approaches of reduction of the RBs do not allow them to be used directly to optimize fuzzy hierarchical DSSs with a variable structure of the input vector [29].

To solve this problem, the authors propose: (a) to use the two-stage method of fuzzy RBs correction, which allows for correction of RBs (antecedents and consequents of rules) in hierarchically-organized DSSs with variable structure of the input data vector (N) [9, 11, 13, 14, 20, 33]; (b) to modify this two-stage method by introducing embedded preliminary procedure for additional verification of the consequents before the start of their correction.

Let's discuss the implementation of the first multi-step cascade of the considered two-stage method, which is responsible for correcting the rules' antecedents, for the case when the DM is not able to evaluate some specific input coordinates (N_{NE}).

Step 1. Assessment of the general characteristics of each particular subsystem (dimensionality of input coordinates and rules structure).

Step 2. Checking the state of the input coordinates. If DM is not able to evaluate the value of any of the input coordinates of a particular subsystem, then the output of this subsystem will automatically be excluded from further consideration.

Step 3. Assigning to all LT $\left(LT_i^j, i \in \{1, \ldots, N\}, j \in \{1, \ldots, K\}\right)$ the i-th input coordinate of the j-th rule (K- number of rules in a particular subsystem) of the corresponding numerical values. For example, for three LTs: $LT_i^j = 1$ if $LT_i^j = L$, $LT_i^j = 2$ if $LT_i^j = M$, $LT_i^j = 3$ if $LT_i^j = H$. In the case, for example, for five LTs: $LT_i^j = 1$ if $LT_i^j = L$, $LT_i^j = 1.5$ if $LT_i^j = LM$, $LT_i^j = 2$ if $LT_i^j = M$, $LT_i^j = 2.5$ if $LT_i^j = HM$, $LT_i^j = 3$ if $LT_i^j = H$.

Step 4. Correction of the antecedents of the rules based on the analysis of the inputs of the coordinates, which DM has no opportunity to evaluate (N_{NE}). In this case, all LTs $\left(LT_i^j, i \in \{1, \ldots, N\}, j \in \{1, \ldots, K\}\right)$, for which the input coordinates cannot be assessed $x_i = NE$, are assigned a zero numerical value $LT_i^j = 0$. This means that in the future, the corresponding LT will not influence the decision-making process.

Step 5. Of all the rules that have the same antecedents after correcting them (step 4), there is only one rule, the first in the list in the RB.

Step 6. Formation of a reduced RB with corrected antecedents.

Second cascade [9, 11, 13, 14, 20, 33], *which* is responsible for correction of the rules' consequents (in the reconfigured RB [28] by the first cascade), can be modified on the step 3 by introducing embedded preliminary procedure for additional verification of the consequents before the start of their correction. In this case, the modified second cascade consists of the next steps:

Step 1. Processing of the information on the reduced RB, checking the presence of the LTs with the assigned zero numerical value $LT_i^j = 0$.

Step 2. Processing of corrected antecedents of the rules. In this case, the numerical values of the antecedents for each j-th rule (first cascade, step 3, step 4) of the reduced RB are added. Next, the amount is divided by the number of input coordinates that the DM has the ability to evaluate $(N_e = N - N_{NE})$, for each j-th rule:

$$\text{Result}_j = \sum_{i=1}^{N} LT_i^j \Big/ N_e. \tag{1}$$

Step 3. Using the proposed preliminary procedure which is based on the proposition: if the existing numerical value $(LT_i^j = 1$ for L, $LT_i^j = 1.5$ for LM, $LT_i^j = 2$ for M, $LT_i^j = 2.5$ for HM, $LT_i^j = 3$ for H) of the j-th rule consequent corresponds (is equal to) the calculated value of the result (1), then there is no need to make a correction of the consequent (step 4) of the corresponding j-th rule and in this case we need to go over to the next $(j + 1)$-th rule's consequent correction. This avoids the process of overwriting the value of the program variable (in our case, the value of the consequent of the j-th rule) at the physical level (in RAM). The appropriate procedure for additional verification of the consequents before the beginning of their correction allows increasing the speed of the proposed correction method of fuzzy RBs and reducing its energy intensity by eliminating the process of allocating additional RAM when modifying the consequent. If the existing numerical value of the j-th rule consequent does not correspond to the value of the generated result (1), then it is necessary to correct this consequent (step 4).

Step 4. Determining the new linguistic term $LT_{Result} \subset \{L, LM, M, HM, H\}$ for the j-th rule consequent, using of the value of the calculation result (1) and the scale:

$$Result_j \in \{L \in [1, 1.5), LM \in [1.5, 2), M \in [2, 2.5), HM \in [2.5, 3), H \in [3, 3]\},$$
$$j \in \{1, \ldots, K\}.$$

Step 5. Correcting of the consequent according to the formation of the corresponding modified LTs at the step 4.

Step 6. Formation of reconfigurable RB based on corrected fuzzy rules.

Introduction of additional verification of the consequents before their correction is a necessary procedure for increasing the speed and decreasing the energy intensity of the antecedent-consequent method of reconfiguring fuzzy RBs developed by the authors. This procedure is especially relevant and useful in using the proposed method of correction of the RBs in the subsystems of DSSs with a large dimension of the vector of the input coordinates ($N \geq 5$). This is due to the fact that in the RBs of such subsystems the number of rules is considerably larger compared to other subsystems, which in turn leads to increased energy intensity and slow performance when correcting consequents of rules (second cascade) without the need for additional verification.

Consider the application of an antecedent-consequent method of fuzzy RBs correction using an example of a second subsystem $y_2 = f_2(x_4, x_5, x_6, x_7)$ of the developed DSS for choosing a cooperation model. If, for example, DM is not able to evaluate at the moment two input coordinates $x_5 = NE$, $x_7 = NE$, then when implementing the first cascade, the initial RB (Table 1) is transformed into a reduced RB [28, 37, 38].

From the results of the implementation of the first cascade, it is seen that the number of rules decreased from 81 to 9. But since the initial values of the consequents of the relevant rules do not correspond to the actual expert estimates, the implementation of the second cascade for the correction of consequents is necessary. After realization of the second cascade, the RB has the final corrected form [9, 33].

Table 2 shows some actual set of input coordinates for the second subsystem, which characterize different types of actual situations [13, 19].

Table 2. Some actual sets of the second subsystem $y_2 = f_2(x_4, x_5, x_6, x_7)$

Input coord.	Actual sets of input coordinates							
	I	II	III	IV	V		XIV	XV
x_4	(L,M,H)	NE	NE	NE	NE		(L,M,H)	(L,M,H)
x_5	(L,M,H)	NE	NE	NE	(L,M,H)	...	(L,M,H)	(L,M,H)
x_6	(L,M,H)	NE	(L,M,H)	(L,M,H)	NE		NE	(L,M,H)
x_7	(L,M,H)	(L,M,H)	NE	(L,M,H)	NE		(L,M,H)	NE

After applying the first cascade of the method of fuzzy RBs correction, the RB of the second subsystem is transformed into a reduced RB by number of rules (Table 3).

Table 3. Reduced RB of the second subsystem in accordance with the actual sets

Actual sets	Function of output coordinate	Amount of rules	Number of rules
I	$y_2 = f_2^I(x_4, x_5, x_6, x_7)$	81	1,2,...,81
II	$y_2 = f_2^{II}(x_7)$	3	1,2,3
...	...	...	...
XV	$y_2 = f_2^{XIV}(x_4, x_5, x_6)$	27	1,4,7,10,13,16,19,22,25,28,31,34,37,40, 43,46,49,52,55,58,61,64,67,70,73,76,79

The scheme of the application of the antecedent-consequent method of fuzzy RBs correction for the second subsystem $y_2 = f_2(x_4, x_5, x_6, x_7)$ on the actual sets of II-XV input data (Tables 2, 3) is presented in Fig. 1.

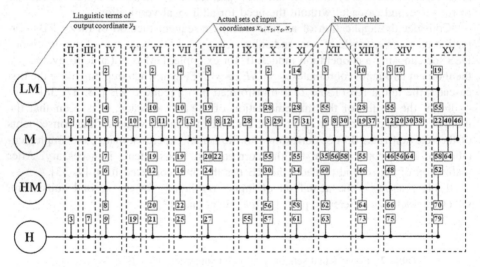

Fig. 1. The implementation scheme of consequents correction

The corresponding scheme (Fig. 1) shows exactly which rules are subject to correction for various types of input data sets. For example, rules with numbers 3, 12, 19, 20, 30, 38, 46, 48, 55, 56, 64, 66, 75 (Table 3) are subject to correction for the current set of input data XIV. In particular, for the rules with numbers 3, 19, 55 the modified consequent corresponds LT "LM", and for the rule No. 75 - "H". In this current set of XIV, this made it possible to increase the speed of the proposed method of correction of fuzzy RBs by 19% and reduce the energy intensity by 8% [13, 14, 20]. The application of the proposed preliminary procedure for additional verification of the consequents before the start of their correction (for the second subsystem in the actual set XIV) allowed avoiding the need for correction of 14 rules (1, 2, 10, 11, 21, 28, 29, 37, 39, 47, 57, 65, 73, 74) from 27 (Table 3). The main conception of the proposed preliminary procedure is described in step 3 of the second cascade.

On all the actual sets of input data of the second subsystem, the number of reduced rules, after the application of the first cascade, is 174 (Table 3, Fig. 1). At the same time, after applying the procedure for additional verification of the consequents before the start of their correction for the method of the correction of fuzzy RBs proposed by the authors, the number of rules, for which the consequent correction needs to be done, decreased from 174 to 99.

The comparative analysis shows that without the application of the proposed method of RBs correction, in particular for the second subsystem ($x_5 = NE$, $x_7 = NE$), there is a deformation of results, since the output of the unmodified subsystem corresponds LT M for which $\mu_2^{y_2^*}(X^{(2)*}) = \max_{j=\overline{1,5}}\left(\mu^{LT_j^{(2)}}(60,0,90,0) \in \{0; 0.1; 0.8; 0; 0\}\right)$.

At the same time, the result of the DSS (y) for choosing a cooperation model deforms to some extent, recommending the choice of a rational **model A2** [31, 33].

When applying the proposed method with $x_5 = NE$, $x_7 = NE$, the output signal of the second subsystem remains unchanged and corresponds LT HM, for which $\mu_2^{y_2^*}(X^{(2)*}) = \max_{j=\overline{1,5}}\left(\mu^{LT_j^{(2)}}(60,80,90,45) \in \{0; 0.1; 0.4; 0.6; 0.2\}\right)$. In this case, the result of the DSS (y) coincides with the result for a complete set of input data – the rational cooperation **model B** [9, 14, 31, 33].

The proposed method of two-stage correction of fuzzy RBs in case of change in the dimension of the input vector, allows in interactive mode to perform automatic correction of fuzzy rules without changing the structure of DSS, which provides an increase in the efficiency and speed of DSS for decision-making in various situations.

5 Design of Fuzzy DSS for Decision Making About Most Rational UIC-Model Based on Developed Web-Oriented Tool

To enhance the effectiveness of the design of DSSs of this class, the authors have developed specialized software and a tool WOTFS-1 (Fig. 2), which has web orientation. The use of WOTFS-1 prevents the formation of the structure of the DSS on choosing a model of cooperation and obtaining an effective result: (a) operatively (including in the absence of time for decision-making), (b) at any given time in the presence of the Internet and (c) without local linking to expert data/knowledge. This feature is important because the ability to evaluate and select a model for collaboration for DM (heads of structural subdivisions of universities and IT-companies) without any restrictions of the time and place of access to DSS is a priority [12, 24].

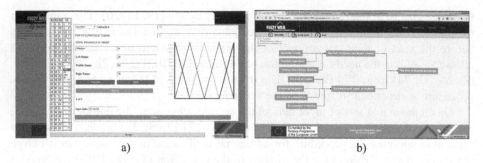

a)   b)

Fig. 2. WOTFS-1 for design of fuzzy DSS: a – inputs and LTs, b – hierarchical structure

A fragmentary (demonstration) structure (3 subsystems: $y_1 = f_1(x_1, x_2, x_3)$, $y_2 = f_2(x_4, x_5, x_6, x_7)$, $y_8 = f_8(y_1, y_2)$) of a fuzzy hierarchical DSS with a discrete logical output for choosing a cooperation model also is presented in Fig. 2.

All data/knowledge from experts are stored on the server side of the web-oriented tool. This avoids making mistakes and increases the system's protection against possible external influences. In addition, the independent global servers provide access to the management of the process of DSSs development (linguistic variables, terms, knowledge bases, etc.) with the limited rights (for example, only for experts). In this way, the expert and client can be in different places and work on creating a DSS and its subsequent testing at different times, which will significantly save time on design and technical resources [10]. The author's web-oriented tool WOTFS-1 is intended for the development of fuzzy hierarchical DSSs with discrete logical output, which allows obtaining the resulting evaluation in the form of a linguistic term, which in turn corresponds to the cooperation model, since for DM it's a more understandable result.

The developed WOTFS-1 tool allows you to develop in real time, without localization of data, your own fuzzy hierarchical DSS with discrete logical output for various purposes, in particular, for solving logistic problems [13, 33], assessing innovation and investment projects [5–7], choosing a cooperation model within a consortium "University – IT-company" [9], optimization in control systems [10, 21] and others. The developed tool is user-friendly and adapted to modern visualization means.

6   Conclusions

The results of the testing of the developed DSS for choosing a model of cooperation within the consortium "University – IT-company" with different variants of the size of the input vector confirm the effectiveness of the proposed two-stage (antecedent-consequent) method of correction of fuzzy RBs and its invariance with respect to (a) the limits on the number of subsystems and (b) the number of input variables of DSS, (c) the number of fuzzy rules and (d) the number of LTs for evaluating input and output coordinates. The application of the procedure for additional verification of the consequents before the beginning of their correction has allowed to increase the speed of the developed by the authors method of the reconfiguration of fuzzy RBs by 8–22% and reduce the energy intensity by 3–12% (depending on the degree of uncertainty of the input information and the size of the current set of input data).

The author's approach and the tool WOTFS-1 can be widely used for implementation of the decision-making processes under uncertainty, in particular when selecting partners for investment objectives in the marine business [33], when choosing cooperation models within academic-industry consortia [9, 14], while optimizing and planning routes [13, 19] etc.

Acknowledgments. The authors thank Tempus Programme of the European Union for support of this research in the framework of the International project TEMPUS- CABRIOLET 544497-TEMPUS-1-2013-1-UK-TEMPUS-JPHES "Model-Oriented Approach and Intelligent Knowledge–Based System for Evolvable Academia-Industry Cooperation in Electronics and Computer Engineering" (2013–2017).

References

1. Bogel, S., Stieglitz, S., Meske, C.: A role model-based approach for modelling collaborative processes. Bus. Process Manag. J. **20**(4), 598–614 (2014)
2. Chen, M.-Y., Linkens, D.A.: Rule-base self-generation and simplification for data-driven fuzzy models. Fuzzy Sets Syst. **142**(2), 243–265 (2004). https://doi.org/10.1016/s0165-0114(03)00160-x
3. Cornejo, M.E., Diaz-Moreno, J.C., Medina, J.: Multi-adjoint relation equations: a decision support system for fuzzy logic. Int. J. Intell. Syst. **32**(8), 778–800 (2017). https://doi.org/10.1002/int.21889
4. Drozd, J., Drozd, A., Maevsky, D., Shapa, L.: The levels of target resources development in computer systems. In: Proceedings of IEEE East-West Design & Test Symposium, pp. 185–189. Kiev, Ukraine (2014). https://doi.org/10.1109/ewdts.2014.7027104
5. Gil-Lafuente, A.M., Merigo, J.M.: The induced generalized OWA operator. Inf. Sci. **179**(6), 729–741 (2005). https://doi.org/10.1016/j.ins.2008.11.013
6. Julián-Iranzo, P., Medina, J., Ojeda-Aciego, M.: On reductants in the framework of multi-adjoint logic programming. Fuzzy Sets Syst. **317**, 27–43 (2017). https://doi.org/10.1016/j.fss.2016.09.004
7. Kacprzyk, J., Zadrozny, S., Tre, G.: Fuzziness in database management systems: half a century of developments and future prospects. Fuzzy Sets Syst. **281**, 300–307 (2015). https://doi.org/10.1016/j.fss.2015.06.011
8. Kazymyr, V.V., Sklyar, V.V., Lytvyn, S.V., Lytvynov, V.V.: Communications management for academia-industry cooperation in IT-engineering: training. In: Kharchenko, V.S. (ed.) Chernigiv-Kharkiv: MESU, ChNTU, NAU "KhAI" (2015). (in Ukrainian)
9. Kondratenko, G., Kondratenko, Y., Sidenko, I.: Fuzzy decision making system for model-oriented academia/industry cooperation: university preferences. In: Berger-Vachon, C., Gil Lafuente, A.M., Kacprzyk, J., Kondratenko, Y., Merigó, José M., Morabito, C.F. (eds.) Complex Systems: Solutions and Challenges in Economics, Management and Engineering. SSDC, vol. 125, pp. 109–124. Springer, Cham (2018). https://doi.org/10.1007/978-3-319-69989-9_7
10. Kondratenko, Y., Gerasin, O., Topalov, A.: A simulation model for robot's slip displacement sensors. Int. J. Comput. **15**(4), 224–236 (2016)

11. Kondratenko, Y., Kondratenko, V.: Soft computing algorithm for arithmetic multiplication of fuzzy sets based on universal analytic models. In: Ermolayev, V., Mayr, H., Nikitchenko, M., Spivakovsky, A., Zholtkevych, G. (eds.) Information and Communication Technologies in Education, Research, and Industrial Applications. ICTERI 2014, vol. 469, pp. 49–77. Springer, Cham (2014). https://doi.org/10.1007/978-3-319-13206-8_3

12. Kondratenko, Y.P., Encheva, S.B., Sidenko, E.V.: Synthesis of intelligent decision support systems for transport logistic. In: Proceedings of the 6th IEEE International Conference on Intelligent Data Acquisition and Advanced Computing Systems: Technology and Applications, Prague, Czech Republic, pp. 642–646 (2011). https://doi.org/10.1109/idaacs.2011.6072847

13. Kondratenko, Y.P., Klymenko, L.P., Sidenko, I.V.: Comparative analysis of evaluation algorithms for decision-making in transport logistics. In: Jamshidi, M., Kreinovich, V., Kacprzyk, J. (eds.) Advance Trends in Soft Computing. SFSC, vol. 312, pp. 203–217. Springer, Cham (2014). https://doi.org/10.1007/978-3-319-03674-8_20

14. Kondratenko, Y.P., Kondratenko, G.V., Sidenko, Ie.V., Kharchenko, V.S.: Models of cooperation between universities and IT-companies: decision making systems based on fuzzy logic. Monograph. In: Kondratenko, Y.P. (ed.). Kharkiv: MESU, PMBSNU, NAU "KhAI" (2015). (in Ukrainian)

15. Kondratenko, Y.P., Kondratenko, N.Y.: Reduced library of the soft computing analytic models for arithmetic operations with asymmetrical fuzzy numbers. Int. J. Comput. Res. 23 (4), 349–370 (2016)

16. Kondratenko, Y.P., Kondratenko, N.Y.: Soft computing analytic models for increasing efficiency of fuzzy information processing in decision support systems. In: Hudson, R. (ed.) Decision Making: Processes, Behavioral Influences and Role in Business Management, pp. 41–78. Nova Science Publishers, New York (2015)

17. Kondratenko, Y.P., Kozlov, O.V., Korobko, O.V., Topalov, A.M.: Synthesis and optimization of fuzzy control system for floating dock's docking operations. In: Santos, W. (ed.) Fuzzy Control Systems: Design, Analysis and Performance Evaluation, Series: Computer Science, Technology and Applications, pp. 141–214. NOVA Science Publishers, Hauppauge (2017)

18. Kondratenko, Y.P., Rudolph, J., Kozlov, O.V., Zaporozhets, Y.M., Gerasin, O.S.: Neuro-fuzzy observers of clamping force for magnetically operated movers of mobile robots. Tech. Electrodyn. 5, 53–61 (2017). (in Ukrainian)

19. Kondratenko, Y.P., Sidenko, I.V.: Decision-making based on fuzzy estimation of quality level for cargo delivery. In: Zadeh, L.A., Abbasov, A.M., Yager, R.R., Shahbazova, S.N., Reformat, M.Z. (eds.) Recent Developments and New Directions in Soft Computing. SFSC, vol. 317, pp. 331–344. Springer, Cham (2014). https://doi.org/10.1007/978-3-319-06323-2_21

20. Kondratenko, Y.P., Sidenko, I.V.: Method of actual correction of the knowledge database of fuzzy decision support system with flexible hierarchical structure. J. Comput. Optim. Econ. Financ 4(2/3), 57–76 (2012)

21. Kondratenko, Y.P., Simon, D.: Structural and parametric optimization of fuzzy control and decision making systems. In: Proceedings of the 6th World Conference on Soft Computing (WCSC), pp. 1–6, Berkeley, USA (2016). http://academic.csuohio.edu/simond/pubs/Kondratenko2016a.pdf

22. Lodwick, W.A., Untiedt, E.: Introduction to fuzzy and possibilistic optimization. In: Lodwick, W.A., Kacprzyk, J. (eds.) Fuzzy Optimization. Studies in Fuzziness and Soft Computing, vol. 254, pp. 33–62. Springer, Heidelberg (2010). https://doi.org/10.1007/978-3-642-13935-2_2

23. Maslovskyi, S., Sachenko, A.: Adaptive test system of student knowledge based on neural networks. In: Proceedings of the 8th IEEE International Conference on Intelligent Data Acquisition and Advanced Computing Systems: Technology and Applications (IDAACS), Warsaw, Poland, pp. 940–945 (2015). https://doi.org/10.1109/idaacs.2015.7341442
24. Meerman, A., Kliewe, T. (eds.): Fostering University-Industry Relationships, Entrepreneurial Universities and Collaborative Innovations. Good Practice Series (2013)
25. Mendel, J.M.: Uncertain Rule-Based Fuzzy Logic Systems: Introduction and New Directions. Prentice Hall PTR, Upper Saddle River (2001)
26. Merigo, J.M., Gil-Lafuente, A.M., Yager, R.R.: An overview of fuzzy research with bibliometric indicators. Appl. Soft Comput. **27**, 420–433 (2015)
27. Messarovich, M.D., Macko, D., Takahara, Y.: Theory of Hierarchical Multilevel Systems. Academic Press, New York (1970)
28. Palagin, A.V., Opanasenko, V.N.: Reconfigurable computing technology. Cybern. Syst. Anal. **43**(5), 675–686 (2007). https://doi.org/10.1007/s10559-007-0093-z
29. Piegat, A.: Fuzzy Modeling and Control. Springer, Heidelberg (2001). https://doi.org/10.1007/978-3-7908-1824-6
30. Prokopenya, A.N.: Motion of a swinging atwood's machine: simulation and analysis with mathematica. Math. Comput. Sci. **11**(3–4), 417–425 (2017)
31. Rotshtein, A.P.: Intelligent Technologies of Identification: Fuzzy Logic, Genetic Algorithms Neural Networks. Universum Press, Vinnitsya (1999). (in Russian)
32. Setnes, M., Babuska, R., Kaymak, U., van Nauta Lemke, H.R.: Similarity measures in fuzzy rule base simplification. IEEE Trans. Syst. Man Cybern. Part B (Cybern.) **28**(3), 376–386 (1998)
33. Solesvik, M., Kondratenko, Y., Kondratenko, G., Sidenko, I., Kharchenko, V., Boyarchuk, A.: Fuzzy decision support systems in marine practice. In: Proceedings of the IEEE International Conference on Fuzzy Systems (FUZZ-IEEE), Naples, Italy (2017). https://doi.org/10.1109/fuzz-ieee.2017.8015471
34. Trunov, A.N.: An adequacy criterion in evaluating the effectiveness of a model design process. East.-Eur. J. Enterp. Technol. **1**(4(73)), 36–41 (2015)
35. Yager, R.R.: Golden rule and other representative values for atanassov type intuitionistic membership grades. IEEE Trans. Fuzzy Syst. **23**(6), 2260–2269 (2015). https://doi.org/10.1109/tfuzz.2015.2417895
36. Yager, R.R.: On the OWA aggregation with probabilistic inputs. Int. J. Uncertain. Fuzziness Knowl.-Syst. **23**(1), 1–14 (2015). https://doi.org/10.1142/s0218488515400115
37. Zadeh, L.A., Abbasov, A.M., Shahbazova, S.N.: Fuzzy-based techniques in human-like processing of social network data. Int. J. Uncertain. Fuzziness Knowl. Syst. **23**(1), 1–14 (2015). https://doi.org/10.1142/s0218488515400012
38. Zadeh, L.A.: Fuzzy sets. Inf. Control **8**(3), 338–353 (1965)

Strategy to Managing Mixed Datasets with Missing Items

Inna Skarga-Bandurova[(✉)], Tetiana Biloborodova,
and Yuriy Dyachenko

Volodymyr Dahl East Ukrainian National University, Severodonetsk, Ukraine
{skarga-bandurova, biloborodova}@snu.edu.ua,
y.dyachenko@i.ua

Abstract. The paper refers to the problem of decision making and choosing appropriate ways for decreasing the level of input information uncertainty related to absence or unavailability some values of mixed data sets. Approaches to addressing missing data and evaluating their performance are discussed. The generalized strategy to managing data with missing values is proposed. The study based on real pregnancy-related records of 186 patients from 12 to 42 weeks of gestation. Three missing data techniques: complete ignoring, case deletion, and random forest (RF) missing data imputation were applied to the medical data of various types, under a missing completely at random assumption for solving classification task and softening the negative impact of input information uncertainty. The efficiency of approaches to deal with missingness was evaluated. Results demonstrated that case deletion and ignoring missing values were the less suitable to handle mixed types of missing data and suggested RF imputation as a useful approach for imputing complex pregnancy-related data sets with missing data.

Keywords: Missing data · Case deletion · Imputation · Classification
Gestation course data

1 Introduction

Missing data is a well-known problem in medicine, particularly for research using routinely collected clinical health records. In many instances, while analyzing and classifying medical data, there is a shortage of clinical health records of patients with target disease, and most of these records are incomplete. The fact is that medical data are represented by a large number of different health screening records, involving a personal and family health history, results of physical examinations, laboratory tests and often include samples with many lost or unknown elements. Missing values may occur by reasons of human errors and misinterpretation during data gathering, equipment failures, loss of data, and troubles in reading a hand-written text, lack of availability due to privacy, legal requirements, and a host of other factors. In these datasets, uncertainty rises from data unavailability, and the adequacy of the interpretation of such data can be questionable.

There exist many techniques for managing datasets with missing values, but there is no all-purpose theoretical solution suited for any given data. Various ways and different models may be practiced for mixed data with missing items, but the problem becomes if it is possible to find a right approach so far as the wrong choice can hardly lead us to any meaningful results [1].

The primary objective of this study is to provide medical researchers with a generalized strategy for handling mixed datasets with missing values, taking into account the missing data mechanisms, their types, initial volume and proportion of missing data.

The motivation for this paper was a research focusing on neonatal hypoxia to be a result of a wide variety the causes. In [2] we noted that identification of the risk factors related to this disease is a challenging task due to the informal nature, missingness, and a wide space of the input data. The diagnosis of fetal hypoxia is based on the information provided by series of pregnancy-related examination, a thorough antenatal screening, ultrasonography observations, interview of the patient and a case history assessment. To set up a profile of pregnant women with a high risk of hypoxia in the newborn, on the previous research we ignored records with missing data. However, in order to obtain better results for classifying and prediction a more detailed study of this question is needed.

The paper is organized as follows: Sect. 2 provides some background on types of missing data, and reviews the main approaches to addressing mostly missing data imputation methods; Sect. 3 presents a generalized strategy for handling data with missing values and describes some performance indicators; Sect. 4 describes a case study that compares various approaches to handling missing data; and Sect. 5 summarizes the main conclusions.

2 Background and Related Work

Imputation is the usual approach for decreasing the level of input information uncertainty and dealing with missing or incomplete data. These days, the most common method is multiple imputation proposed by Rubin [3]. Since the first issue the studies on missing data imputation is not decreasing, on the contrary, the researchers have been developing new and improved methods to cope with missingness. Whereas many statistical methods have been developed for imputing missing data, many of them are weak dealing with complex or mixed data sets. Mixed datasets (i.e., data having both nominal and categorical variables) are often a big challenge for performing imputation [4]. For these reasons, there has been much interest in using machine learning methods for missing data imputation.

The authors of the study [5] developed a two-stage machine learning algorithm, called ensemble of classifiers, for the classification of data containing missing values, without imputation, in real time, to predict the patient's condition in the intensive care unit. In [6, 7] authors analyzed the efficiency of a multiple imputation technique with a broad range machine learning (ML) algorithms including k-nearest neighbor (kNN), decision tree (C4.5) and rule induction (CN2) for recovery missing or unavailable values. When working with data containing missing values, some researchers [6, 8, 9] use special approaches to restore missing values, for example, substitution through average, maximum likelihood methods using the expectation maximization (EM) algorithm. The results of the study [10] showed that the application of the naive Bayesian algorithm is most suitable for the classification of data containing missing values. In [11], the EM algorithm was employed to impute missing data. Authors [12] developed stochastic EM algorithm for obtaining the maximum likelihood estimates of model parameters and mixing proportions. In many cases, EM algorithms showed an excellent power and accuracy of data imputation. A new technique for filling missing values and performing implementation of large datasets classification was proposed in [4]. It is grounded on multilevel cost-sensitive SVM-based algorithms, and imputation method based on expected maximization relied on multiple regression. The study [13] discovered the multiple imputation for categorical and continuous variables using stochastic regression. The authors leveraged this approach to recovery data that missing completely at random and compared the performance of different categorical imputation approaches with the improved continuous method for categorical items. As a result, they recommend a multiple imputation through stochastic regression for categorical data that are missing completely at random. A novel imputation approach dealing with missingness for classifying and prediction of medical records is discussed in [14].

A fertile area for research can be imputation method based on the random forest (RF) [15, 16]. RF imputation is an ML technique which does not rely on distributional assumptions and can adjust nonlinearities and uncertainties in interactions [17]. It is widely used in medical studies. We assume that RF may be useful in multiple imputation of pregnancy-related data sets where there are large numbers of different clinical variables per patient.

2.1 Classes of Missing Data

Missingness and unavailability of some values in data can be due to a lack of documentation, equipment failures, data-entry errors, refusal of answering some questions, etc. Understanding mechanisms of missing data is a crucial point for choosing an efficient technique to manage data with missing items.

The types of missing data, depending on the mechanism of their occurrence, according to [9, 18], are defined as follows.

- Missing completely at random (MCAR) – the absence of values in the data, as well as the probability of missingness, does not depend on any conditions.
- Missing at random (MAR) – the conditional probability of missingness depends only on observed variables. If there is a variable y is missing and there is another variable x, we can define the data as missing at random when $Pr(y \text{ missing} \mid y, x) = Pr(y \text{ missing} \mid x)$.
- Not missing at random (NMAR) – the probability of missingness depends on variables that are incomplete.

In the rest of the paper, we follow the notations used in [18]. Let $Y = (y_{ij})$ denote a data set without missing values, where y_{ij} is the value of variable Y_j for subject i. Missing data indicator matrix $M = (M_{ij})$ defines the pattern of missing data, where

$M_i = 0$ if the value in the observation y_{ij} exists and $M_i = 1$ if y_{ij} is missing. The mechanism of missing data is characterized by a conditional distribution $f(M \mid Y, \phi)$, where ϕ denotes the unknown values. If missingness does not depend on the data values Y, the condition $f(M \mid Y, \phi) = f(M \mid \phi)$ is satisfied then the absence of data does not depend on the data values, and the data fall into category MCAR. That is, if for all Y, ϕ

$$f(M \mid Y, \phi) = f(M \mid \phi). \tag{1}$$

It can be said that missing data mechanism is MAR if for all Y_{mis}, ϕ data missingness depends only on the components Y_{obs}, that are observed, and not on the items that are missing, that is, if

$$f(M \mid Y, \phi) = f(M \mid Y_{obs}, \phi), \tag{2}$$

where Y_{obs} denote the observed components of Y, and Y_{mis} the missing components.

It can be said that missing data mechanism is NMAR if either of the above two classifications is not met.

Suppose that the distribution (y_i, M_i) does not depend on the absence of the value in the observation, so the probability of missingness does not depend on the values of Y or M for other missing values.

$$f(Y, M \mid \theta, \phi) = f(Y \mid \theta)f(M \mid Y, \phi) = \prod_{i=1}^{n} f(y_i \mid \theta) \prod_{i=1}^{n} f(M_i \mid y_i, \phi), \tag{3}$$

where $f(y_i \mid 0)$ denotes the density of y_i indexed by unknown parameters θ. $f(M_i \mid y_i, \phi)$ is the density of a Bernoulli distribution for the binary indicator M_i with probability $\Pr(M_i = 1 \mid y_i, \phi)$ that yi is missing. Thus, if the absence does not depend on Y, i.e. if $\Pr(M_i = 1 \mid y_i, \phi) = \phi$ it does not depend on y_{ij}, then the data missingness are MCAR or MAR. Otherwise, the probability depends on y_{ij} and the data are NMAR.

2.2   Methods for Handling Missing Data

There are different ways to processing missing data: delete the records with missing values, leave them as is, or perform imputation. The main missing data techniques are classified under three headings (Fig. 1):

- Discarding missing values or case deletion and dealing with only complete sets;
- Imputation of missing data and their further use in shared data collection, i.e., complete cases and incomplete cases with imputed values;
- Analysis the full, incomplete data set utilizing model-based likelihood methods. This technique does not impute any data but uses available datasets to compute maximum likelihood estimates.

For making a decision concerning managing data with missing items, it is necessary to take into account the data type (continuous, ordinal, nominal, etc.), the data missingness mechanisms, size of initial datasets and proportion of missing data, method of further analysis, and missing data imputation method.

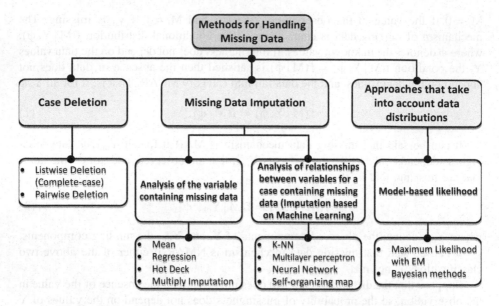

Fig. 1. The main techniques for dealing with missing data (adapted from [19])

Our recommendations for choice of missing data handling approach are summarized in the next section.

3   The Decision Process Governing Strategy for Handling Missing Data and Evaluation Criteria

3.1   Generalized Strategy for Handling Data with Missing Values

In view of the above, a generalized strategy for handling data with missing values considering the different missing data mechanisms, the type of the data, and missing data handling method can be defined as a five-stage process (Fig. 2).

At the first stage, a qualitative assessment of the missing data is carried out. The types of input data are fixed subject to a specific level of measurement (numerical (continuous or discrete), categorical (ordinal or nominal)) and depending on the mechanism of their occurrence (MCAR, MAR, NMAR) which is determined by formulas (1)–(3).

As a first approximation, for missing data containing only numerical type, all methods described above can be applied. For categorical variables, the case deletion or imputation based on machine learning algorithms is recommended.

At the second stage, the missing data in each class is quantitatively evaluated. It is necessary to choose between imputing data and discarding missing values. Usually, if there is no specific goal, for all practical purposes, records with more than 50% missing items should be removed [20] regardless of their type and significance for further analysis.

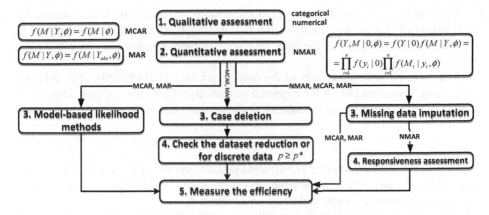

Fig. 2. Generalized strategy for handling mixed datasets with missing values

For datasets with only numerical type the following results for multiple imputation methods, taking into account the number of missing data and the absence mechanism are given [21]:

Under MCAR conditions (when there are up to 50% of the missing values), all methods show a minimal deviation from the complete dataset.

Under MAR conditions (when there are up to 50% of the missing values), only multiple imputation is acceptable. When data is being missing at random are verging towards 50%, then multiple imputation is still possible, but the standard errors will increase a lot, and it will be difficult to draw conclusions.

Under NMAR conditions with 50% missing values, there's nothing will get done well, none of the methods show satisfactory results. At 25% of the missing values, the use of multiple imputations with a 7.8-fold imputation is recommended.

For datasets containing only categorical data with their missingness from 5% to 50%, the imputation using machine learning algorithms is justified [22].

At the third stage, the question of choosing the method for dealing with the missing data should be adjusted.

Model-based likelihood methods work well with MAR data. As it recorded in [23], almost all techniques discussed above show good results for MAR and MCAR data. The case deletion can be used for all data type with any mechanism of missingness. However, it is not recommended to utilize the listwise deletion for MAR or NMAR data. For continuous NMAR data, an imputation with logistic regression is applicable, but case deletion shows poor performance for NMAR data [24]. The general rule for handling NMAR data is that in the majority of cases the same methods suitable for MAR can be applied but with a mandatory responsiveness assessment [25]. In [26] it a multiple imputation and maximum likelihood methods for NMAR data are recommended.

The fourth stage depends on the method being chosen on the previous phase and includes checking the datasets reduction after case deletion or procedures of responsiveness assessment after finishing imputation. After case deletion, the proportion of initial and remaining data should be estimated.

For classification task, the data set reduction range should be carefully evaluated for each subset of classes. It stems from the fact that applying this approach can lead to imbalance and skewness of the data sets with further complications in subsequent analysis.

For data sets with two classes of the output variable, the probability of a particular class value is optimal only if the expected probability of the value of this class is less than or equal to the expected probability of the second class value [27], i.e., only when

$$(1 - p)c_{10} + pc_{11} \leq (1 - p)c_{00} + pc_{01} \tag{4}$$

under $p = P(j = 1 \mid x)$, where P is estimated probability, p – optimal probability, c_{00} – true-positive cases, c_{01} – false-positive cases, c_{10} – false-negative cases, c_{11} – true-negative cases, j – current class of observation x.

If this inequality is actually an equality, the probability of any class is optimal. The threshold for making optimal decisions is p' such that

$$(1 - p')c_{10} + p'c_{11} = (1 - p')c_{00} + p'c_{01} \tag{5}$$

Hence, the probability of a specific class value is optimal only if it is satisfied $p \geq p'$. The threshold can be computed as follows

$$p' = \frac{c_{10} - c_{00}}{c_{10} - c_{00} + c_{01} - c_{11}} \tag{6}$$

If the missing data are MCAR and the result set after case deletion provides adequate power for tests of hypotheses, then case deletion is quite sufficient.

If condition (5) is not fulfilled, then it is necessary to return to the previous stage and apply a multiple imputation or model-based likelihood methods.

At the fifth stage, it is necessary to conduct performance evaluation and assess the efficiency of missing data method.

3.2   Accuracy Metrics and Measuring the Efficiency of Missing Data Imputation

The efficiency of methods for handle missing data can be assessed using the following performance indicators:

1. Classification error rate (CER). For two class output variable, the CER is computed as follows

$$CER = \frac{c_{01} + c_{10}}{c_{00} + c_{01} + c_{10} + c_{11}} \tag{7}$$

2. Root mean square error (RMSE). RMSE estimates standard deviation between imputed and true values. We will assume that $\tilde{x}_i$ denotes an imputed version of the i-th attribute, and $\hat{x}_i$ denotes the true value of the same variable.

$$\text{RMSE} = \sqrt{\frac{\sum_{i=1}^{n} (\hat{x}_i - \tilde{x}_i)^2}{n}} \tag{8}$$

3. Predictive accuracy (PAC). Given that the i-th value of the variable is missing, its imputed version $\tilde{x}_i$ should be close to $\hat{x}_i$ (true value). Pearson's Correlation between $\tilde{x}_i$ and $\hat{x}_i$ is a good measure of imputation efficiency and is defined as follows

$$\text{PAC} \equiv r = \frac{\sum_{n=1}^{N} (\tilde{x}_{i,n} - \bar{\tilde{x}}_i)(\hat{x}_{i,n} - \bar{\hat{x}}_i)}{\sqrt{\sum_{n=1}^{N} (\tilde{x}_{i,n} - \bar{\tilde{x}}_i)^2 (\hat{x}_{i,n} - \bar{\hat{x}}_i)^2}} \tag{9}$$

where $\tilde{x}_{i,n}$ and $\hat{x}_{i,n}$ denotes, respectively, the n-th value of $\tilde{x}_i$ and $\hat{x}_i$; $\bar{\tilde{x}}_i$ and $\bar{\hat{x}}_i$ contains, respectively, N values including $\tilde{x}_i$ and $\hat{x}_i$. A good imputation method will have a correlation coefficient close to 1.

4. Distributional accuracy (DAC). The distance between the empirical distribution function for both imputed and true values can be used as a measure to maintaining the distribution of true values. Empirical distribution functions $F_{\hat{x}_i}$ for cases with true values and $F_{\tilde{x}_i}$ for cases with imputed values

$$F_{\hat{x}_i}(x) = \frac{1}{N} \sum_{n=1}^{N} I(\hat{x}_{i,n} \leq x) \tag{10}$$

$$F_{\tilde{x}_i}(x) = \frac{1}{N} \sum_{n=1}^{N} I(\tilde{x}_{i,n} \leq x) \tag{11}$$

where I is an indicator function. The distance between these functions can be determined using the Kolmogorov-Smirnov distance D_{KS} and is given by formula

$$\text{DAC} \equiv D_{KS} = \max_n (||F_{\hat{x}_i}(x_n) - F_{\tilde{x}_i}(x_n)||) \tag{12}$$

where the values of x_n are both true and imputed values of the variable x_j. A good imputation method gives us a small distance value.

As it mentioned in [19], the CER is one of the most significant accuracy metrics due to the fact that the primary task of imputation in this study is to solve the problem of classifying data with missing values, and the imputation is a secondary task that helps to solve this problem.

4   Case Study

4.1   Data Properties

The study is conducted on real sets of pregnancy data, containing missing values. Our focus in this paper is on case-control studies. The experiments were performed using 186 datasets (medical records), six input variables and one output variable. Variables, selected at the first stage of the study [2], are presented by the results of clinical and laboratory tests of pregnant women and fetus at different terms of pregnancy. Table 1 shows data on the patient's gestation course and the diagnosis of a newborn.

Table 1.   Fragment of the data set used in the experiment

ID	ESR2	ESR3	Prothrombin index	Vertical amniotic size	Placenta maturity	Placenta thickness	Newborn diagnosis
51	43	42	5	146	1	29	Pathology
52	69	46	5	70	3	35	Pathology
56	12	17	6	235	1	36	Pathology
78	55	26	5	64	1	33	Pathology
95	25	47	5	58	3	87	Norm
143	41	37	5	65	3	34	Norm
144	35	22	5	125	1	38	Norm
145	40	40	5	124	2	42	Norm
...	...	...	...	...	...	...	...

Variable "ID" is the patient's ID and is not used for further data analysis, "ESR2" – erythrocyte sedimentation rate (ESR) at 21 weeks of gestation; "ESR3" – ESR index at 38 weeks of gestation, "prothrombin_index" – blood prothrombin index; "vertical_amniotic_size" – amniotic fluid index; "placenta_maturity" – maturation of the placenta at 30 to 38 weeks; "placenta_thickness" – placenta thickness at 30 to 38 weeks of gestation; and variable "newborn_diagnosis." Information on these variables and percentage of missing data in pregnancy-related health records [2] is presented in Table 2.

Table 2.   Percentage of missing data for gestation course variables

Variable	Missing data, %
ESR at 21 weeks of gestation	8, 1
ESR at 30 weeks of gestation	19, 4
Prothrombin index	31, 2
Amniotic fluid index at 30 to 38 weeks of gestation	37, 6
Maturation of the placenta at 30 to 38 weeks of gestation	21, 5
Placenta thickness at 30 to 38 weeks of gestation	21, 5

It can be seen from the table that the indicator of the vertical size of the amniotic fluid on the ultrasound examination at 30–38 weeks of gestation has the most missing data. However, it does not exceed the boundary of 50%. Some variables contain a large number of missing data as well, but the missing elements do not have a systematic picture. All data in the dataset are MCAR. Possible reasons for their missingness are randomness or human factor (loss of biological material, unavailability of physical examinations results and laboratory tests, lack of survey data, i.e., missing data are not related to other data and patient characteristics).

A qualitative assessment of the gestation course data revealed that all missing data are MCAR, type of input variables "ESR2", "ESR3", "vertical_amniotic_size", and "placenta_thickness" is continuous, type of input variables "blood_prothrombin_index" and "placenta_maturity" is ordinal, and type of output variable "diagnosis of newborn" is nominal. Data analysis was performed with the missing value analysis (MVA) module in SPSS.

4.2   Experiments and Results

Having regard to data types, mechanism of their missingness and missing data percentage range there are at least two approaches; they are case deletion and machine learning imputation. As an ML technique, an imputation based on Random Forest algorithm was employed. For classification purposes the experiment with ML imputation was performed in the following manner.

First, we split a full dataset into two sub-datasets: containing only records without missing values and containing only records with missing values. For each complete sub-dataset (records without missing data) a set of decision trees were constructed. For each incomplete sub-dataset (records with missing data) the missing values were imputed. Then all records were combined to form a completed dataset without any missing values, and classification procedure was performed on the new full data set. Random Forest algorithm was applied both for imputing missed data and for classification. Finally, the classification accuracy was evaluated by applying the corresponding classification model, and the classification error rates were analyzed.

Main phase of the experiment has been performed with classifiers from the WEKA tool [28]. The results of the classification error estimation for ignored, deleted and imputed missing data is summarized in Table 3.

Ignoring missing values do not facilitate further data analysis and shows the worst result of the classification error rate. After deleting cases with missing values, the data set became 30% of the original volume. For initially small dataset, this reduction is

Table 3. CER for datasets where data with missed values were ignored, for data without missing values (after case deletion), and on imputed data

Approach	Classification error rate
Ignoring missing values	0.62
Case deletion	0.53
Random forest imputation	0.39

significant and affects the sensitivity estimate. Case deletion shows average results and can be applied with sufficient amounts of input data or low percentage of records containing missing values. The ML imputation based on RF provides a significant improvement compared with the case deletion and ignoring missingness.

As expected, the classification error rates demonstrate that classification on imputed data is more accurate than classification on missing data or classification on data containing only complete cases. Therefore, we conclude that in these conditions, for mixed datasets (i.e., data having both numerical and categorical variables) with amounts of missing values up to 37.6%, imputation improves the classification.

5   Conclusions and Future Work

There is increasing demand for handling missing data as the numerous databases contain missing values. The missing data is an ongoing issue for clinical health records, but so far there is no universal approach performing best results in different situations.

The primary objective of this study was to formulate general strategy and guidelines for medical researchers in choosing among multiple approaches for handling missing data subject to various combinations of data types, mechanisms of missingness, number of missing values, and further analytical task.

The main contributions of the paper are as follows. In this study, we presented a unified approach for decision support concerning managing mixed datasets with missing items. We applied our technique to clinical health records containing missing data to examine its impact on the missing data model selection and subsequently classification. The study was carried out based on the proposed strategy using a real data set with missing values. Following the proposed strategy, three methods such as ignoring missing values, deleting cases with missing values, and missing data imputation using the Random Forest algorithm were applied.

It is still premature to draw any conclusions about the operational suitability this methodological approach and further studies are needed to understand the relationships between different aspects in more detail. However, our results suggest that under various missing data conditions task of model selecting is much-simplified. Our research has been done on a limited number of datasets, and sole missing data imputation method has been tested, so further research required to confirm obtained results.

References

1. Magnani, M.: Techniques for dealing with missing data in knowledge discovery tasks, **15** (01), 2007 (2004). http://magnanim.web.cs.unibo.it/index.html
2. Skarga-Bandurova, I., Biloborodova, T.: Exploratory data analysis to identifying meaningful factors of hypoxic fetal injuries. Inf. Model. **44**(1216), 122–135 (2016). Herald of the NTU "KhPI". NTU "KhPI", Kharkov. https://doi.org/10.20998/2411-0558.2016.44.09
3. Rubin, D.B.: Multiple Imputation for Nonresponse in Surveys, vol. 81. Wiley, Hoboken (2004)

4. Razzaghi, T., Roderick, O., Safro, I., Marko, N.: Multilevel weighted support vector machine for classification on healthcare data with missing values. PLoS ONE **11**(5), e0155119 (2016)
5. Conroy, B., Eshelman, L., Potes, C., Xu-Wilson, M.: A dynamic ensemble approach to robust classification in the presence of missing data. Mach. Learn. **102**(3), 443–463 (2016). https://doi.org/10.1007/s10994-015-5530-z
6. Batista, G.E.A.P.A., Monard, M.C.: An analysis of four missing data treatment methods for supervised learning. Appl. Artif. Intell.: Int. J. **17**(5–6), 519–533 (2003)
7. Schmitt, P., Mandel, J., Guedj, M.: A comparison of six methods for missing data imputation. J. Biom. Biostat. **6**(224), 1 (2015). https://doi.org/10.4172/2155-6180.1000224
8. Ibrahim, J.G., Molenberghs, G.: Missing data methods in longitudinal studies: a review. Test **18**(1), 1–43 (2009)
9. He, Y.: Missing data analysis using multiple imputation. Circ.: Cardiovasc. Qual. Outcomes **3**(1), 98–105 (2010)
10. Oba, S., Sato, M., Takemasa, I., Monden, M., Matsubara, K., Ishii, S.: A Bayesian missing value estimation method for gene expression profile data. Bioinformatics **19**, 2088–2096 (2003). https://doi.org/10.1093/bioinformatics/btg287
11. Calikli, G., Bener, A.: An algorithmic approach to missing data problem in modeling human aspects in software development. In: Proceedings of 9th International Conference on Predictive Models in Software Engineering, p. 10. ACM, New York (2013)
12. Fu, Y.Z.: Stochastic EM algorithm of a finite mixture model from hurdle Poisson distribution with missing responses. Commun. Stat.-Theory Methods **45**(20), 5918–5932 (2016)
13. Finch, W.H.: Imputation methods for missing categorical questionnaire data: a comparison of approaches. J. Data Sci. **8**, 361–378 (2010)
14. Yelipea, U.R., Porikab, S., Gollaa M.: An efficient approach for imputation and classification of medical data values using class-based clustering of medical records. Comput. Electr. Eng. In Press. https://doi.org/10.1016/j.compeleceng.2017.11.030
15. Tang, F., Ishwaran, H.: Random forest missing data algorithms. Stat. Anal. Data Min.: ASA Data Sci. J. **10**, 363–377 (2017). https://doi.org/10.1002/sam.11348
16. Breiman, L., Cutler, A.: Manual on Setting Up, Using, and Understanding Random Forests V3.1. University of California, Berkeley (2002). http://oz.berkeley.edu/users/breiman/Using_random_forests_V3.1.pdf
17. Shah, A.D., Bartlett, J.W., Carpenter, J., Nicholas, O., Hemingway, H.: Comparison of random forest and parametric imputation models for imputing missing data using MICE: a CALIBER study. Am. J.Epidemiol. **179**(6), 764–774 (2014)
18. Little, R.J.A., Rubin, D.B.: Statistical Analysis with Missing Data. Wiley, Hoboken (2014)
19. García-Laencina, P.J., Morales-Sánchez, J., Verdú-Monedero, R., Larrey-Ruiz, J., Sancho-Gómez, J.L., Figueiras-Vidal, A.R.: Classification with incomplete data. In: Handbook of Research on Machine Learning Applications and Trends: Algorithms, Methods, and Techniques: Algorithms, Methods, and Techniques, pp. 147–175 (2009)
20. Hair, J.F., et al.: Multivariate Data Analysis. Prentice Hall, Upper Saddle River (2016)
21. Scheffer, J.: Dealing with missing data. Res. Lett. Inf. Math. Sci. **3**, 153–160 (2002)
22. Farhangfar, A., Kurgan, L., Dy, J.: Impact of imputation of missing values on classification error for discrete data. Pattern Recognit. **41**(12), 3692–3705 (2008)
23. Peugh, J.L., Enders, C.K.: Missing data in educational research: a review of reporting practices and suggestions for improvement. Rev. Educ. Res. **74**(4), 525–556 (2004)
24. Huisman, M.: Imputation of missing network data: some simple procedures. J. Soc. Struct. **10**(1), 1–29 (2009)

25. Doidge, J.C.: Responsiveness-informed multiple imputation and inverse probability-weighting in cohort studies with missing data that are non-monotone or not missing at random. Stat. Methods Med. Res., 1–15 (2016). https://doi.org/10.1177/0962280216628902
26. Cheema, J.R.: Some general guidelines for choosing missing data handling methods in educational research. J. Modern Appl. Stat. Methods 13(2), 53–75 (2014)
27. Elkan, C.: The foundations of cost-sensitive learning. In: Proceedings of the Seventeenth International Joint Conference on Artificial Intelligence (IJCAI 2001), pp. 973–978. Lawrence Erlbaum Associates Ltd. (2001)
28. Frank, E., Hall, M.A., Witten, I.H.: The WEKA Workbench. Online Appendix for "Data Mining: Practical Machine Learning Tools and Techniques", 4 edn. Morgan Kaufmann (2016)

Predicting Opponent Moves for Improving Hearthstone AI

Alexander Dockhorn[✉][iD], Max Frick[iD], Ünal Akkaya[iD], and Rudolf Kruse[iD]

Institute for Intelligent Cooperating Systems, Otto-von-Guericke University,
Universitaetsplatz 2, 39106 Magdeburg, Germany
{alexander.dockhorn,max.frick,uenal.akkaya,rudolf.kruse}@ovgu.de

Abstract. Games pose many interesting questions for the development of artificial intelligence agents. Especially popular are methods that guide the decision-making process of an autonomous agent, which is tasked to play a certain game. In previous studies, the heuristic search method Monte Carlo Tree Search (MCTS) was successfully applied to a wide range of games. Results showed that this method can often reach playing capabilities on par with humans or even better. However, the characteristics of collectible card games such as the online game Hearthstone make it infeasible to apply MCTS directly. Uncertainty in the opponent's hand cards, the card draw, and random card effects considerably restrict the simulation depth of MCTS. We show that knowledge gathered from a database of human replays help to overcome this problem by predicting multiple card distributions. Those predictions can be used to increase the simulation depth of MCTS. For this purpose, we calculate bigram-rates of frequently co-occurring cards to predict multiple sets of hand cards for our opponent. Those predictions can be used to create an ensemble of MCTS agents, which work under the assumption of differing card distributions and perform simulations according to their assigned distribution. The proposed ensemble approach outperforms other agents on the game Hearthstone, including various types of MCTS. Our case study shows that uncertainty can be handled effectively using predictions of sufficient accuracy, ultimately, improving the MCTS guided decision-making process. The resulting decision-making based on such an MCTS ensemble proved to be less prone to errors by uncertainty and opens up a new class of MCTS algorithms.

Keywords: Hearthstone · Monte Carlo Tree Search
Knowledge-base · Ensemble · Uncertainty · Bigrams

1 Introduction

Computational intelligence in games is a thriving research topic, with a growing demand from the video game industry. Especially, applications in video games make it possible to quickly compare agents, and their related methods, by letting them play against each other. The development of artificial intelligence (AI)

© Springer International Publishing AG, part of Springer Nature 2018
J. Medina et al. (Eds.): IPMU 2018, CCIS 854, pp. 621–632, 2018.
https://doi.org/10.1007/978-3-319-91476-3_51

Fig. 1. Elements of the Hearthstone game board: (1) weapon slot (2) hero (bottom: player, top: opponent) (3) opponent's minions, (4) player's minions, (5) hero power, (6) hand cards, (7) mana, (8) decks, (9) history

agents for games often involves solving decision-making tasks, for which heuristic search processes, such as Monte Carlo Tree Search (MCTS) are frequently used [2].

Recent studies showed that the actual skill-level of an MCTS agent is dependent on the quality of performed simulations during the search [7]. This is especially hard in the context of collectible card games, such as the online game Hearthstone, where the set of playable card-combinations is extremely large. Even more critical is the high amount of uncertainty involved in the game. Game-mechanics such as the random card-draw, the unknown hand cards of our opponent, as well as random card effects hinder the accuracy of performed simulations and restrict the simulation depth.

In the context of a 2-player card game, the prediction of our opponent's moves, his hand-cards, as well as the cards in his deck largely influence which moves the player needs to consider. In this work we create a knowledge-base of frequently played card combinations from a database of human player game replays. During a game the knowledge-base is used to predict the opponent deck based on previously played cards. From the estimated deck we sample multiple hand card sets for our opponent. An ensemble of MCTS procedures is initialized based on the hand card samples. The ensemble's best result will be returned as final outcome of the decision-making process.

The developed agent is exemplary for uncertainty handling in MCTS agents. The result of the simulation is less prone to wrong assumptions, due to the included knowledge-base and the search on an ensemble of MCTS procedures Our results indicate that the prediction-based ensemble improves the playing capabilities of our developed agent.

The remainder of this paper is structured as follows: In Sect. 2 we review the game Hearthstone: Heroes of Warcraft and previous research efforts on the development of Hearthstone AIs. Section 3 covers the Flat Monte Carlo and the Monte Carlo Tree Search algorithm. Additionally, we shortly review recently used agents found in the literature in Sect. 4. In Sect. 5 we first discuss our approach to learn frequent card combinations from a large database of replays. We further introduce our roll-out policy, which includes the prediction of our opponent's cards based on our knowledge database. The influence of the new rollout policy and general parameters of MCTS is tested in Sect. 6 followed by a detailed discussion of our results. Finally, in Sect. 7 we highlight the implications of our work and summarize our suggestions for further improvements.

2 Hearthstone: Heroes of Warcraft

Hearthstone is a turn-based digital collectible card game developed and published by Blizzard Entertainment [1]. Players compete in one versus one duels choosing a self-constructed deck and one out of nine available heroes. In these matches players try to beat their opponents by reducing the opponent's health from 30 to 0. This can be achieved by playing cards from the hand onto the game board at the cost of mana. Played cards can be used to inflict damage to the enemies hero or to destroy cards on his side of the game board. The amount of mana available to the player increases every turn (up to a maximum of 10), which also increases the complexity of later turns. At the beginning of a turn the player also draws a new card until his deck is empty. If no cards remain he receives *fatigue*-damage, which is steadily increasing from turn to turn. The standard game board is shown in Fig. 1.

A deck is a self-constructed set of 30 cards, which can be chosen from more than 1000 cards currently included in the game. Each card brings unique effects to influence the current game board. Additionally, heroes can use class-specific hero-powers and cards.

Cards can be of the type minion, spell, or weapon. Figure 2 shows one example of each card type. Minion cards assist and fight on behalf of the hero. They usually have an attack-, health-, and mana cost-value, as well as additional abilities or a special minion type. Once played, they can attack the enemies side of the board every consecutive turn to inflict damage on either enemy minions or the opponent's hero. Attacking also reduces its own health points by the attack points of the defender. In case a minion's health drops to zero or less, it is removed from the board and put into the player's graveyard. Spell cards can be cast using mana to activate various abilities and are discarded after use. They can have a wide range of effects, such as raising a minion's attack, inflicting damage to a hero or minion, etc. Secrets, which are a special kind of spell, can be played without immediately activating their effect. Given a certain event a secret will be activated and for example destroy an attacking minion. Once activated, the secret is removed from the board. Weapon cards are directly attached to the player's hero and also enable him to attack. Their durability value limits the number of attacks till the weapon breaks. Only one weapon can be equipped at the same time.

(a) Minion card   (b) Spell card   (c) Weapon card

Fig. 2. General card types: cards include (1) mana cost, (2) attack damage, (3) health/durability, (4) and special effects.

Hearthstone decks often contain multiple cards which act in synergy, e.g. pirate minion cards of the same type buff each other. Generated decks can be categorized in the three major categories: Aggro, Mid-range, and Control. Aggro decks build on purely offensive strategies, which often include a lot of minions. Control decks try to win on the long run by denying the opponent from executing his strategy and dominating the game situation. Mid-range decks are in-between Aggro and Control decks. They try to counter early attacks to dominate the game board with high cost minions during the mid of the game.

Game length and branching factor can be dependent on the decks being played by both players. The possibility of making multiple moves per turn and the enormous amount of possible decks make Hearthstone a challenging problem for AI research.

3 Flat Monte Carlo and Monte Carlo Tree Search (MCTS)

MCTS is a heuristic search algorithm commonly used in a wide range of computer game AIs [2]. Its exploration of the game tree consists of four major steps (1) selecting a node, (2) expanding the node with any legal move, (3) simulating (random) playouts called rollout, (4) determining the final value of the playout and propagate it back along the path to the root node. Those four phases are commonly grouped into tree policy (selection and expansion) and default policy (simulation and back-propagation). Moves under consideration can be evaluated by repeatedly simulating games to approximate the players chances of winning the game after executing this move. This can either be done by counting the number of won simulated games or rating any intermediate simulated game-state using a scoring function. The number of simulations as well as the quality of the playout are crucial in determining an accurate estimate of the chances of success. Finally, the node with the highest success-rate is played.

Using a random selection and expansion as tree policy can lead the agent to lose a lot of time on the simulation of unpromising nodes. For example if nodes are known, of which the first wins in 80% of the computed simulations and the second in only 10% it would likely be more useful to further analyse the subtree of the first node. This can increase confidence in the approximated chances of winning the game after picking the node. Nevertheless, it would be possible that more simulations on the second node would uncover new and more promising paths. We might also expand a new node, which was not considered yet, that would be even better than the previous nodes. Therefore, the UCB formula (see Eq. (1)) [12], which balances exploration and exploitation, can be used during the selection step.

$$\underbrace{R(s')}_{\text{Exploitation}} + C \underbrace{\sqrt{\frac{log_2(V(s))}{V(s')}}}_{\text{Exploration}} \tag{1}$$

where s' is a child node of s, $R(s')$ is the average success after choosing node s', and $V(s)$ counts the number of visits of state s during previous episodes. The constant C balances both parts of the equation. Using the UCB as a tree policy, the search tree is known to converge to the minimax tree as the number of simulation grows to infinity [12].

4   Previous Work and the Hearthstone AI Competition

Hearthstone is closed source. Thanks to the efforts of an active community, multiple simulators exist as part of the HearthSim project [10].

This work uses a simulator called Sabberstone [5], which tries to remodel each part of the game as close as possible. The C#-programming interface allows researchers to implement algorithms in this rich test environment. Sabberstone currently implements 98% of cards included in the game. Therefore, to the best of our knowledge, Sabberstone represents the most complete simulator currently available. In the future, researchers will be able to directly compare their results in the Hearthstone AI competition [8].

Metastone is another simulator, which was frequently used for research projects in the past [6]. It already includes a greedy optimization agent, which uses a scoring function to choose the best sequence of moves for the current turn. The scoring heuristic takes each player's minions, their number of hand cards, and current health points into account. Each minion receives a score, which is determined by weighting the minion's attack and health values, as well as taking typical abilities, such as taunt, into account. The proposed scoring function results in an aggressive play-style, which heavily relies on reducing the opponent's health, as well as achieving minion dominance on the game board. Other heuristics were developed for the AAIA'17 Data Mining Challenge [11], which inspired multiple research papers on the development of winner-prediction models. Very good results were achieved by Neural Network based methods, which achieved about

80% prediction accuracy of mid-game game-states. Therefore, more enhanced search heuristics than the one provided by Metastone are possible [9].

Metastone also includes an implementation of a Flat Monte Carlo agent, which we use in our evaluation. Based on Metastone, two MCTS agents were developed [13,14], which both performed well against the random and Flat Monte Carlo agent. However, no comparable data is available.

Another well-received work was done on next card prediction [3]. A bag-of-words of card-co-occurrence bi-grams was used for training a prediction system for the next upcoming card. Prediction rates of up to 95% were recorded during the evaluation. The high prediction accuracies inspired our work on enhancements for MCTS.

5   Enhancing MCTS by the Prediction of Opponent Hand Cards

Hearthstone players deduce the best move under the effect of multiple sources of uncertainty. Throughout the game the current game-state cannot be fully accessed by the player. During the player's move it is unknown which card will be drawn next, which cards our opponent currently holds in his hand, and which cards are contained in the opponent's deck. Even if the game-state would be fully accessible, players need to anticipate their opponent s next move(s) for laying out their own strategy. Nevertheless, players can elicit different levels of skill, which is suggesting that those tasks can at least be partially handled.

Applying MCTS will have to face the same sources of uncertainty. In this work we predict our opponent's move from a database of frequently co-occurring cards. Predicting our opponent's move increases the accuracy of performed simulations and let us reduce the number of necessary simulations without loss of quality [7]. Ultimately, this increases the skill level of the MCTS agent with only limited overhead during the simulation.

5.1   Bigram Extraction

Building up a knowledge-base of card-co-occurrences was done by analysing a total of 544.628 replays by human players. The dataset was obtained from an openly available database of replays [4]. On this website players are able to upload replays of past games for statistical tracking. Our knowledge-base is based on replays from June 2016 to October 2017. Each replay consists of the players deck as well as a history of moves for the recorded game. Additionally, decks in the replay file are classified in deck categories such as "pirate warrior" or "token paladin", each consisting of cards which exploit certain card synergies. The amount of games per hero class are summarized in Table 1.

For each card we determined the number of co-occurrences with other cards. Three types of co-occurrences were differentiated

- **isolated:** cards need to be played at the same turn
- **successive:** cards need to be played in successive turns
- **combined:** isolated and successive counts are added

We further studied the influence of taking the games result into account on the final system skill level. For this purpose we either counted only co-occurrences for moves of the winning player, the losing player, or both. The co-occurrence database was stored as compressed .json-file and can be accessed during the simulation phase for to determine likely cards for the next turn.

Table 1. Game statistics

Hero	Avg. length	Avg. actions	#games
Hunter	393.213	12.927	42.740
Druid	433.383	15.889	72.867
Warrior	428.505	14.757	77.526
Priest	525.571	17.156	53.133
Mage	506.471	18.920	63.887
Shaman	443.933	14.706	76.092
Paladin	459.677	14.561	56.395
Rogue	436.423	17.103	58.541
Warlock	449.750	15.681	43.447
All	452.932	15.786	544.628

5.2 MCTS Enhancement

In this work we extend MCTS for creating an agent for Hearthstone. In our adaptations of MCTS we aim for widening the prediction to the next 3 turns determined by consecutive MCTS searches. Extending the simulation can only be allowed by the knowledge-base. The full algorithm is described below. For an easier understanding we recommend comparing the described process with Fig. 3.

Phase 1: During our MCTS simulation we use the current state of the game board as the initial root node. Each action, such as playing a card, attack with a minion, etc., advances the state of the game and represents another node. The next node is selected based on UCT selection using the Metastone score and the visit count of the node (see Eq. (1)). Each transition consists of exactly one move. A player's turn is made of multiple moves and ends with an end-turn move.

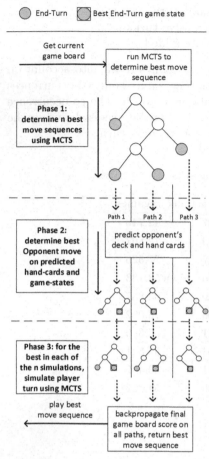

⬤ End-Turn   ▣ Best End-Turn game state

Fig. 3. MCTS with prediction

A simulation uses a greedy selection of moves based on the Metastone scoring function. The last move of a simulation consists of an end-turn move. The final score after ending the player's turn is back-propagated along the simulation path.

Phase 2: After a number of simulations we pick the best n nodes and use the game-state of their best leaf node (end-turn node) for further simulations of the opponent player in phase 2. Based on his previously played cards we determine a likely deck (out of 20 pre-implemented samples). A set of hand-cards is determined using the bigram database. For this we use all previously played cards to determine their most probable follow-up card. Out of all follow-up cards we randomly sample the opponent's hand cards to approximate the real set of hand-cards. Using MCTS we determine the best sequence our opponent can play all resulting in a end-turn move and a final game-board.

Phase 3: In phase 3, we determine the best follow-up turn for each of the returned game-states to rate the resulting game-board. Once more, this is done using MCTS. The scores are back-propagated all the way to our node-rating in Phase 1. Finally, the best rated node will be picked for execution.

Even if this process turns out to be computationally expensive, the thorough simulation of our opponent assures a high quality prediction during phase 2. Deeper simulations can be created by stacking phase 2 and 3 multiple times, before back-propagating the final game-board scores. Nevertheless, we chose a depth of 3, because of the limited prediction accuracy for follow-up turns. Simulating deeper playthroughs would accumulate more uncertainty, due to unknown card-draws on both sides. It turned out to be more effective to increase the number of simulations for a better approximation of the game-state score after a few moves. The number of simulations can be adjusted to fit the 75 seconds time limit of the game.

6   Evaluation

6.1   Experimental Setup

During our evaluation we tested the proposed agent against multiple other agents mentioned in the literature or in previous implementations. We included a random agent (random) as general baseline without any planning capabilities. Flat Monte Carlo (flatMC) was implemented as a basic search strategy. MCTS without further prediction of the opponent's hand cards (plainMCTS) was implemented to test the influence of our bigram prediction. Full observation MCTS (foMCTS) was used to compare our AI versus an optimal prediction of opponent's hand cards. Here, we simulate the optimal prediction by providing the AI with the actual hand cards during the current game-state. Finally, an exhaustive search (exh. s.), which is implemented by the Sabberstone framework, allowed us to choose the best of all possible move sequences for the players turn. However, the method is not allowed to simulate the opponent's turn.

Our proposed MCTS algorithm with prediction of our opponent's hand cards (predMCTS) was tested against each of the described agents. Therefore, we used three by Sabberstone pre-implemented decks for representing the possible deck types and play-styles in Hearthstone. The "Aggro Pirate Warrior" deck (Aggro) consists of low cost pirate minions, offensive weapons, and spell cards of the warrior class. In contrast, the "Mid-Range Jade Shaman" deck (Mid-Range) is made of multiply minion buffs and damaging spells. The third deck-type (Control) is represented by the "Reno Kazaku's Mage" deck, which is mainly based on spell damage, and destroying or taking over control of enemy minions.

For each combination of player deck, opponent deck, and opponent agent we simulated at least 100 games. The multiple random factors of hearthstone lead us to simulate more games for the comparison of predMCTS versus plainMCTS and foMCTS to get stable results. Winning percentages for each match-up are summarized in Table 2.

6.2   Discussion

Reviewing the results in Table 2 indicates the advantages of the implemented AI approach. Looking at all the match-ups in which both agents play the same deck, it becomes clear that our proposed agent is able to beat them in all cases except one. In general the agent performs best when playing the Mid-Range deck. Playing the Aggro deck also leads to very good results. The results of the Control deck against other decks are worse than the results of both other decks, which might be due to the deck being generally worse than the other two variants.

Both random and flat Monte Carlo consistently lose in most match-ups, while the remaining opponents are able to win games more often. It is not surprising that nearly all games ($\approx 98\%$) are won against the random player. Games against the flat Monte Carlo agent show a promising success rate of $\approx 81\%$ on average.

Table 2. Winning chances of predMCTS using various decks against other agents. 100 games were simulated against the Random, flatMC, and exhaustive search agent. Up to 500 games were simulated against predMCTS and foMCTS, because bigger variances were observed during the simulation process. Columns in which both agents play the same deck are highlighted in gray.

Wins in %	Aggro	Mid	Control
Random	95	100	100
flatMC	81	73	94
plainMCTS	59	47	58
foMCTS	46	36	60
exh.s.	65	47	70

(a) predMCTS Aggro Deck

Wins in %	Aggro	Mid	Control
Random	99	98	100
flatMC	88	85	99
plainMCTS	71	55	76
foMCTS	59	50	76
exh.s.	62	70	85

(b) predMCTS Mid-Range Deck

Wins in %	Aggro	Mid	Control
Random	97	97	100
flatMC	73	54	89
plainMCTS	36	31	68
foMCTS	41	16	51
exh.s.	45	20	61

(c) predMCTS Control Deck

The plainMCTS agent was beaten in most games while playing the Mid-Range or the Aggro deck. However, our AI performs worse in the match-ups where it uses the Control deck against other deck types. Nevertheless, we achieved a win-rate of 68% in cases where both agents play the control deck. The overall win-rate is ≈56%. This result and the better performance in the other match-ups where both agents play the same deck shows that the improved prediction of the opponent move results in a better overall performance.

The only agent that has beaten our proposed approach in more than 50% of simulated games is the foMCTS agent. Our overall win-rate is ≈48%. This is not surprising due to the fact, that the foMCTS receives full information of the players hand cards. Therefore, our results indicate that precise knowledge can increase the performance of the MCTS algorithm by a large degree. For this reason, we are motivated to even further increase the performance of our card prediction algorithm. The advantage of knowing the true hand cards is a bit unfair and impractical in a legal game situation.

Our proposed approach was able to beat exhaustive search in 6 out of 9 match-ups. The overall win-rate is ≈58%.

Generally, the player playing the control deck performs worse in most match-ups. This could be due to two possible reasons. First, short-term prediction is not suitable for playing a control deck, because the used deck is in need of long-term planning. Therefore, the implemented approaches may not perform very well. A planning horizon of more than three turns may increase the play-strength on this deck-type. Another reason may be Metastone's scoring function, which puts more weight on the current game-board, than on the remaining utility of the hand cards effects. Playing a spell that kills all the opponent's minions can be played for immediate effect or kept for a later situation. For example killing just one minion would be a waste, if another card could yield the same effect. Either improving Metastone's scoring function or implementing a deck dependent scoring function may increase the play-strength for this deck.

7   Conclusions and Future Works

In this work we proposed a method for handling uncertainty in MCTS. Our method involves the creation of a knowledge-base, which is used to initialize an ensemble of MCTS procedures. Our current sampling approach is based on bi-grams, but can theoretically be exchanged by any probabilistic or heuristic sampling. The resulting decision-making based on such an ensemble proofed to be less prone to uncertainty at the cost of a marginally increased computation time. Our sampling based ensemble opens up a new class of MCTS algorithms, which in its current version already outperforms other agents based on MCTS.

We evaluated our approach based on the collectible online card game Hearthstone. Bigrams, which are fast to process and cheap in memory consumption, are learned from a database game replays. The simulation phase is guided by sampling multiple hand card sets for the opponent player based on the gathered knowledge-base. Due to this sampling multiple game situations are considered during the decision-making phase. This enables our agent to choose better moves than comparable agents depending on the current state of the game. The proposed agent was able to consistently beat those agents in multiple game setups.

Our tests using an MCTS agent with full information on the current game-state show that a perfect prediction would yield slightly better results in some match-ups, suggesting that improving the prediction accuracy would even further increase the play-strength. In the future we plan to further analyse the capabilities of partially informed MCTS agent ensembles.

In the context of creating a Hearthstone AI, we plan to further extend the opponent card and deck prediction. For this purpose, we would like to incorporate other sources of knowledge, such as a deck database of current meta-decks. Further adaptations need to be made for detecting commonly played decks and inferring the opponent's strategy. In this work we limited our deck database to 20 commonly played decks and the deck strategies Aggro, Mid-Range, and Control. This can be further extended by focusing on deck dependent core-mechanics, such as strong board clears, discarding cards, or buffing minions. Detecting which core-mechanic the deck relies on would further improve the opponent prediction. Also, updates of cards or additions of new cards can drastically change the meta-game. Hence, gathered knowledge needs to be frequently revised to stay up to date with currently played strategies. Further work will be put into better scoring functions, which are based on those core-mechanics.

References

1. "Blizzard Entertainment": Hearthstone Webpage. https://playhearthstone.com/en-gb/0. Accessed 06 Mar 2017
2. Browne, C.B., Powley, E., Whitehouse, D., Lucas, S.M., Cowling, P.I., Rohlfshagen, P., Tavener, S., Perez, D., Samothrakis, S., Colton, S.: A survey of monte carlo tree search methods. IEEE Trans. Comput. Intelli. AI Games **4**(1), 1–43 (2012). https://doi.org/10.1109/TCIAIG.2012.2186810
3. Bursztein, E.: I am a legend: hacking hearthstone using statistical learning methods. In: 2016 IEEE Conference on Computational Intelligence and Games (CIG), pp. 1–8. IEEE, September 2016. https://doi.org/10.1109/CIG.2016.7860416
4. "Collect-o-Bot": Hearthstone Replay Database. http://www.hearthscry.com/CollectOBot. Accessed 06 Mar 2017
5. "darkfriend77": Sabberstone Github Repository. https://github.com/HearthSim/SabberStone. Accessed 06 Mar 2018
6. "demilich1": Metastone Github Repository. https://github.com/demilich1/metastone. Accessed 06 Mar 2017
7. Dockhorn, A., Doell, C., Hewelt, M., Kruse, R.: A decision heuristic for Monte Carlo tree search doppelkopf agents. In: 2017 IEEE Symposium Series on Computational Intelligence (SSCI), pp. 1–8. IEEE, November 2017. https://doi.org/10.1109/SSCI.2017.8285181
8. Dockhorn, A., Mostaghim, S.: Hearthstone AI Competition. http://www.is.ovgu.de/Research/HearthstoneAI.html. Accessed 06 Mar 2018
9. Grad, Ł.: Helping AI to play hearthstone using neural networks. In: Ganzha, M., Maciaszek, L., Paprzycki, M. (eds.) Proceedings of 2017 Federated Conference on Computer Science and Information Systems, vol. 11, pp. 131–134, Sepember 2017. https://doi.org/10.15439/2017F561
10. HearthSim Project: Hearthsim Webpage Hearthstone Simulation & AI. https://hearthsim.info/. Accessed 06 Mar 2017
11. Janusz, A., Świechowski, M., Zieniewicz, D., Stencel, K., Puczniewski, J., Mańdziuk, J., Ślęzak, D.: AAIA'17 Data Mining Challenge: Helping AI to Play Hearthstone. https://knowledgepit.fedcsis.org/contest/view.php?id=120. Accessed 06 Mar 2017
12. Kocsis, L., Szepesvári, C.: Bandit based Monte-Carlo planning. In: Fürnkranz, J., Scheffer, T., Spiliopoulou, M. (eds.) ECML 2006. LNCS (LNAI), vol. 4212, pp. 282–293. Springer, Heidelberg (2006). https://doi.org/10.1007/11871842_29
13. Santos, A., Santos, P.A., Melo, F.S.: Monte Carlo tree search experiments in hearthstone. In: 2017 IEEE Conference on Computational Intelligence and Games (CIG), pp. 272–279. IEEE (2017). https://doi.org/10.1109/CIG.2017.8080446
14. Tzourmpakis, G.: Hearthagent, a Hearthstone Agent, Based on the Metastone Project. http://www.intelligence.tuc.gr/~robots/ARCHIVE/2015w/Projects/LAB51326833/. Accessed 06 Mar 2017

A New Generic Framework
for Argumentation-Based Negotiation
Using Case-Based Reasoning

Rihab Bouslama[1(✉)], Raouia Ayachi[2], and Nahla Ben Amor[1]

[1] LARODEC, ISG, University of Tunis, Tunis, Tunisia
rihabbouslama@yahoo.fr, nahla.benamor@gmx.fr
[2] LARODEC, ESSECT, University of Tunis, Tunis, Tunisia
raouia.ayachi@gmail.com

Abstract. The growing use of Information Technology in automated negotiation leads to an urgent need to find alternatives to traditional protocols. New tools from fields such as Artificial Intelligence (AI) should be considered in the process of developing novel protocols, in order to make the negotiation process simpler, faster and more realistic. This paper proposes a new framework based on both argumentation and Case-Based Reasoning (CBR) as means of guiding the negotiation process to a settlement. This paper proposes a new generic domain-independent framework that overcomes the limits of domain-dependent frameworks. The proposed framework was tested in tourism domain using real data.

Keywords: Negotiation · Argumentation
Argumentation-based negotiation · Case-based reasoning · CBR

1 Introduction

Conflicts resolve techniques are very useful to overcome misunderstanding caused by the differences in the opinions, beliefs, and goals of different parties. Negotiation has been one of the most used techniques [1–3]. Further, with the growth of Artificial Intelligence (AI) techniques and their huge success in the last decades, the automation of negotiation where software agents negotiate on behalf of humans was the interest of many researchers [4,5]. Automated negotiation gives software agents with conflicting positions a way to find a beneficial settlement [6].

From a similar perspective, argumentation is another way to reach an agreement between parties that have opposing positions and cannot agree on one decision or action. Back in the mid of nineties, Multi-Agent Systems (MAS) gained a significant attention from several researchers. The argumentation techniques have been used to facilitate the interaction between autonomous agents of MAS that are able to make their own decisions based on their mental states (e.g. preferences, intentions, goals etc.) [7].

While negotiating and arguing, one can refer to his past experiences unconsciously. Case-based reasoning (CBR) is based on the idea that similar problems

© Springer International Publishing AG, part of Springer Nature 2018
J. Medina et al. (Eds.): IPMU 2018, CCIS 854, pp. 633–644, 2018.
https://doi.org/10.1007/978-3-319-91476-3_52

have similar solutions. Thus, its integration in a MAS where agents are in conflict will increase the probability of reaching an agreement [8]. Several researchers proposed to combine these disciplines, namely CBR, argumentation and negotiation. In the literature, several works proposed partial combinations such as [9–12] where argumentation and negotiation have been combined. Moreover, CBR and argumentation have been combined in [8,13] while other works gathered negotiation and CBR [14,15]. Thus, numerous research works were dedicated to partial combinations of these disciplines with object to improve the conflict solving process between different parties. For that reason, we believe that the total combination of the three fields would be more beneficial. However, only few works proposed a total combination of the three domains.

In early ninetees, Sycara proposed the PERSUADER system [16] which is a mediator between a company and its trade union. CBR was used to retrieve past cases representing the arguments used in the past in order to make better negotiation scenarios. Moreover, CBR can help the system to negotiate with an unknown agent based on past experiences with similar agents. But, this proposal is highly domain specific and the negotiation is only ensured by the mediator. Soh et al. [17], proposed another framework where a set of agents collaborate in order to track as many targets as possible. Each agent owns a case-base and a CBR manager which ensures the use of past cases to negotiate in new situations. More recently, in [18] a MAS where agents are able to negotiate through argumentation and use CBR to reach conclusions is presented. The system is for e-commerce domain and used to help users to buy products. CBR was used to reduce system's complexity and hence, users will get the results of the sites selling their desired product more rapidly. However, these frameworks are very domain specific and cannot be easily adapted to other domains. No generic framework has been proposed in the literature that combines CBR, negotiation and argumentation which leads us to the main purpose of this paper which is providing a generic framework for an argumentation-based negotiation using CBR. This work combines the three fields (i.e. negotiation, argumentation and CBR) in a domain independent framework by proposing a generic architecture that defines the basis of the environment and a state machine protocol that specifies the rules of interaction between two argumentative negotiator agents that interact following a set of dialogue games. This framework is called GANC for Generic Argumentation-based Negotiation using CBR.

This paper is organized as follows: Sect. 2 is dedicated to define our new generic argumentation-based negotiation using CBR framework by highlighting its architecture, its main components and the protocol followed by agents in the negotiation process. Section 3 concerns the conducted experiments and the results and Sect. 4 concerns the conclusion of our work and our future perspectives.

2   A New Framework for Argumentation-Based Negotiation Using CBR

This section details our new generic framework for argumentation-based negotiation using CBR (so called GANC for Generic framework for Argumentation-based Negotiation using CBR). Figure 1 illustrates the GANC framework architecture for two agents (A and B).

2.1 GANC Architecture

The GANC architecture is domain-independent and its main components are as follows:

– *Agents:* they are self-motivated entities that can enter and leave a dialogue at any time. These argumentative negotiator agents are not only able to interpret and evaluate incoming locutions from the counter parties, but are also able to evaluate the incoming arguments, update their mental states, generate a set of arguments and propose the appropriate one. Each agent has her own preferences. Based on these preferences, she will accept or reject an offer and generates a proposal. Offers will be proposed in a decreasing order from the best offer to the least favorite one. Indeed, a set of rules that govern the movements (e.g. accept, reject) of an agent are defined. In argumentation theory, each agent has a position that represents its mental state regarding a giving issue. Similarly, an agent has a position regarding a negotiation issue. This position includes her preferences in the form of a set of features, the problem and its offer. Agents follow an argumentation-based negotiation protocol detailed in Sect. 2.2.

– *CBR:* each argumentative negotiator agent has two CBR: (i) *domain CBR* that operates over previous negotiation cases that can be used as arguments in future negotiation situations and (ii) *argument CBR* that operates over past arguments used by agents. Similar cases are generated by comparing the current situation's premises and values with cases' premises in the case-base. This similarity is determined using the normalized Euclidean distance between premises. Proposals and arguments are the main knowledge exchanged between agents. The choice of an argument is so important since it can influence the course of the negotiation process. Later on in this paper,

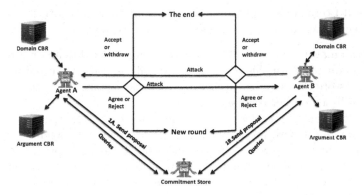

Fig. 1. The proposed architecture of the GANC framework

the structure of the arguments, their types and the way agents use them are discussed. In this framework, agents negotiate through the exchange of arguments. CBR is used to select the appropriate argument that can be sent in a given situation and as a warehouse that stores all past negotiation and argumentation cases which make agents able to take advantage of their past experiences.

- *Commitment store:* it is a particular agent that does not enter in the negotiation process and holds all the information about the negotiation process such as the counter party proposal, the exchanged arguments and all information that concerns the dialogue between the agents. Therefore, an agent queries the commitment store in order to get information about the dialogue (e.g. the offer sent by the opponent).
- *Locutions:* agents have the possibility to send their proposals, accept an offer, refuse an offer by withdrawing the dialogue and to attack, agree or reject an argument. The agreement or the reject of an argument triggers the start of a new round where agents switch roles, the opponent becomes the proponent and vice-versa.

2.2 Argumentation-Based Negotiation (ABN) Protocol Used in GANC

In this section, we discuss the protocol followed by agents during their negotiation process. Several protocols were proposed in negotiation and argumentation communities [19–21]. In negotiation, one of the most used protocols is alternating offers protocol [20]. In argumentation, the protocol can be defined in an *explicit* way (e.g. finite state machine) or in an *implicit* way in the agent's specification [21]. In the GANC framework, the protocol is presented explicitly in a finite state machine format because it is explicit and easily accessible. Agents will follow the alternating offers protocol and exchange offers in an alternating way. The possible moves that an agent can do are presented in Fig. 2 that depicts the ABN protocol used in GANC.

Exchanged Locutions. Agents use locutions in order to negotiate and interact between each other. As some examples, we can mention:

- *Propose:* an agent will enter to the dialogue only if she has an offer to propose to its opponent. With this move an agent will send her proposal to the commitment store that will be checked by the other agent. An agent generates a proposal based on her preferences.
- *Assert:* if an agent receives a "why" locution she has to assert her proposal by supporting it with an argument. After asserting, the agent will wait for the other agent's response that can either be: attack, accept, reject or agree.
- *Attack:* Using this locution, an agent can attack the other agent's supporting argument if she is not convinced or to counter-attack a received attack.
- *Accept:* in our framework an agent accepts a proposal in two cases: (i) if the proposal is the agent's current preferred value. (ii) If the argument respects

the agent's conditions such as the number of similar premises between agent's premises characterizing her offer and the premises sent by the opponent in his argument and based on the number of the received distinguished premises that exists in her CBR (these premises were once a reason to choose a given offer). As an example, an agent can accept the other agent's proposal if their positions are characterized by two common premises with the same value. These conditions differ from agent to another based on their flexibility and how much they are open minded. In case if a counter-example argument is received, the agent will immediately accept the offer since this kind of argument has the highest power of convincing. In GANC framework, based on agents' acceptance conditions we can classify agents into three types: a tolerant agent, a medium strict agent and a very strict agent.

- *Agree:* this means that an agent agrees with the opponent's argument and believes that her offer is not the best. If the opponent agrees on the argument then a new round will start where he will be the proposing agent.
- *Reject:* an agent sends this locution if she has no more arguments to present and she still does not accept the counter party's offer. After this locution a new round will start.
- *Withdraw:* an agent can withdraw the dialogue at any time. The cause of a withdraw can be that an agent does not have any proposals or she does not accept the other agent's proposal neither the argument supporting that offer.

Followed Rules. The dialogue between agents is governed by a set of standard rules as follows:

- A rejected offer or argument cannot be proposed a second time in the same dialogue.
- A new round starts only if one of the agents responds by agree or reject and the other agent accepts that and does not withdraw.
- An agent is free to quit the dialogue: (i) if she can't generate any positions and still refuses its opponent's offer, (ii) if she refuses the offer and refuses to negotiate, (iii) if she receives an invitation to negotiate but she has no offers.
- At each round, only one offer will be discussed and there will be rounds as long as no agreement is reached and the agents still have offers to propose. At each round one agent plays the role of opponent and the other of proponent and then they switch roles in the next round.
- The negotiation ends if one of the agents withdraws the dialogue, if an agreement is reached or if agents exposed all their offers and still no agreement is reached.

At the end of a dialogue, there are two possible outcomes: an *agreement* or *not*. Agents follow an alternating offers protocol in form of a finite state machine inspired from the one in [22,23]. We modified it in order to make it suitable for a negotiation context.

Followed Protocol. Figure 2 depicts the proposed protocol in a finite state machine that presents a round in the negotiation where only one issue is discussed. Every time one of the agents responds by agree or reject, a new round will be started and agents will be back to the state *Enter* but with updated beliefs. It will be rounds as long as no agreement is reached and agents still have offers to propose. For the sake of clarity, the withdraw moves were not presented but an agent can withdraw the dialogue at any step.

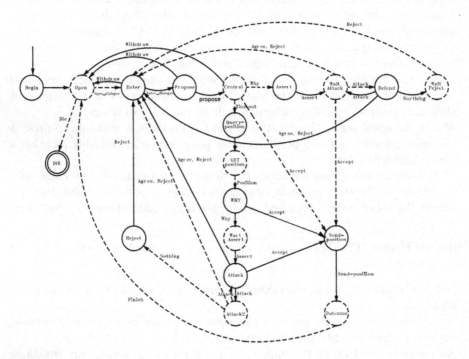

Fig. 2. Proposed argumentation-based negotiation protocol used in GANC (dotted lines indicate wait states and solid lines are send states).

An agent can enter the dialogue only if she has an offer to propose. Thus, agents start by sending their offers and then two different scenarios will happen according to the agent's role.

1. *The agent is a proponent:* after proposing, the agent will move to the central state where she will wait for a response from the opponent. If it is an accept, then the negotiation will end with an agreement on the current offer of the proponent but if it is a challenge then the agent has to assert her offer by a support argument. After asserting, the proposing agent will wait for a reply. If the agent's support argument was attacked, then the two agents will enter in a series of attacks and counter attacks until the opponent accepts the offer or one of them agrees, rejects or has nothing to say about the argument and thus a new round will be started.

2. *The agent is an opponent:* after proposing; the agent will query the commitment store to get the counter party's offer. In this stage, the agent can either accept the offer or challenge it by sending a "why" locution. Same thing, when the opponent receives an assert she can accept the offer, agrees or rejects the argument or attack it. In case of attack, agents will argue until one of them agrees or rejects with the other argument or if the opponent accepts the offer.

2.3   How to Use Arguments in GANC?

An argument is an additional information used by agents to backup their proposals and attacks. It is composed of the agent's promoted value (i.e. offer) and a support set that can be:

- *Premises:* a set of features that the agent used to generate this argument.
- *Negotiation cases:* a list of old negotiations similar to the current scenario.
- *Distinguishing premises:* premises that the other party did not take into consideration and can affect their preferences.
- *Counter example:* arguments that contain same premises as the ones presented by the other party but can give another outcome.

Agents choose their argument's support set based on the type of argument they desire to send. Two types exist in our framework namely:

- *Support argument:* is an argument that supports and explains a proposal. More specifically it is the knowledge used to generate an argument such as previous negotiation and argumentation experiences or agent's premises that defines her proposal.
- *Attack argument:* used to attack an incoming argument. It can contain distinguishing premises or counter examples.

Agents choose the type of argument to send based on the last locution they received. In other words, if an agent is asked to assert her proposal then, she will need an explanatory argument to explain her offer and thus, a support argument will be sent. Otherwise, if the agent's offer or last argument is attacked then she will send a counter attack to defend her position and the offer she proposes.

The argument state changes according to the opponent's response. An argument can have the state *acceptable* if the other agent accepts that argument, it can be in the state *rejected* if it was rejected or *undecided* if the opponent said nothing about it and couldn't neither accept it nor reject it. The structure of an argument will be better explained in an illustrative example.

2.4   Illustrative Example

To illustrate the GANC framework, we propose to instantiate it in a simplified example in the tourism domain. In our example, two agents are negotiating over a holiday destination. Each one, agent 1 and agent 2 has three cases in her case base (see Fig. 3). Agent 1 has two preferences, Maldives and Singapore whilst agent

2 prefers Bali and Kuala Lampur. The negotiation starts when agent 1 proposes Maldives as a destination to agent 2. Since Maldives is not the current preferred offer to agent 2 she has to ask "why" in order to get explanations and arguments. Thus, agent 1 explains her choice by sending an "assert" locution explaining that Maldives is characterized by its hot weather and the availability of three stars hotels with all inclusive pension. However, according to agent 2's case base, the Maldives has hotels with only housing pension. Therefore, she sends an attack containing her argument to agent 1. Hence, agent 1 send a counter example mentioning that according to her experiences, there are hotels with all inclusive pension. Since the counter example is the most convincing argument in our framework and since the only drawback to agent 2 is the pension of the hotel, she accepts the Maldives as a destination. At the end of this negotiation, agent 2 will add a new case to her case-base. The new case is (Maldives; Pension: All inclusive, Nb stars: 3, Weather: hot, Month: august), that contains "Maldives" as a conclusion and four premises. The new case will be used as an argument in future negotiation scenarios.

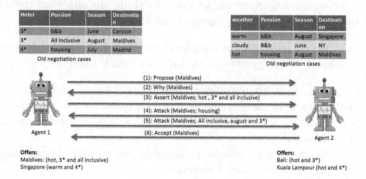

Fig. 3. Bilateral negotiation over travel destination

3   Experimental Study

3.1   Experimental Protocol

In what follows, we explain the experimental protocol by defining the environment of implementation, the data and the experiments that we conducted.

Experiments Environment and Data. All the framework is implemented on the platform Magentix2 [23] using eclipse Luna. Magentix2 is a multi-agent platform that presents argumentative agents that have their own strategies and tactics and interact following a state machine protocol. We modified these agents as well as their strategies and the followed protocol to make their behavior suitable for a negotiation context. Thus, we propose new negotiator argumentative agents based on Magentix2 argumentative agents. For instance, we tested the GANC framework with two agents. In order to implement the framework, we used the Java language. The old argumentation and negotiation experiences (i.e. case bases) are stored in files acceded by the two CBRs.

In order to test the GANC framework, experiments were conducted in the tourism domain where two agents negotiate over a destination. To the best of our knowledge there is no database containing real arguments or real negotiation offers available. Due to this lack of data, we constructed our own database from real data provided by the tourism ministry of Tunisia and collected from many websites such as Trivago[1], HolidayWeather[2], Bandsintown[3] and eDreams[4].

Our database concerns all information that can influence one's choice of a destination. Therefore, data contains information about available hotels and everything that concerns the accommodation (e.g. rooms' prices per night). It also contains the price of means of transport whether it is a bus for destinations in Tunisia or plane for international destinations. Other important features that characterize countries in the world such as their safety rank, their health and hygiene rank and in general their global rank are taken into consideration. These ranks are important in the choice of a holiday destination and were collected from *The Travel and Tourism Competitiveness Report 2017* published by the *World Economic Forum*. We also collected information about concerts and festivals planned in a given holiday period in the destination country and information about its general weather. Moreover, data was collected based on few assumptions. We assume that agents negotiate over a holiday in high season time (i.e. June, July, August) and that the plane tickets are for 10 days duration from the 1st to the 10th of each month. Hotels availability and prices are collected for one night in the first weekend of each month for one person.

The final database contains 505 lines about 126 destinations that are characterized by 29 features cited before (e.g. safety rank, plane ticket's price and the weather). The database is available on line[5].

Protocol. Tests are carried on negotiation scenarios that last 3 rounds. At each round an agent plays the role of an opponent and the other of a proponent and at the end of the round they switch roles to negotiate over the next offer.

In order to highlight the importance of CBR as well as the importance of the exchanged arguments we conducted 120 different experiences in 4 different possible scenarios. Firstly, we supposed that agents are tolerant and thus, they accept an offer if one of the premises characterizing this offer is one of the agent's preferred premises. Let us illustrate this in an example, if we suppose that an agent proposes Djerba for three reasons: hot weather, cheap and holiday period

---

[1] https://www.trivago.fr/.
[2] http://www.holiday-weather.com/.
[3] https://news.bandsintown.com/home.
[4] https://www.edreams.fr/.
[5] https://drive.google.com/drive/folders/1vLTId4KkfkAwq8A64fqo50TkRbxA1821? usp=sharing.

in June and the opponent agent's premises that characterize her choice are hot weather, a given set of means of transport and a set of restaurants then she will accept Djerba as a destination. In the case of tolerant agents, we made 40 runs. First ten runs are done using 10 cases available in the case-base of agents' CBR. Second ten runs are done using 50 cases, the third ten runs are done using case-base with 100 cases and the last ten runs are done using case-base with 500 cases.

Same experiments are conducted using medium strict agents that accept an offer only if there are more than 2 similar premises. Same for highly strict agents that only accept an offer if it is characterized by all agent's preferred premises (i.e. all three premises of each agent are the same).

3.2   Experimental Results

The importance of CBR was highlighted by Conducting these experiments. The first impact was on time that decreases from period nearby 18 s to 10 s. Agents negotiate through the exchange of arguments which make the existence of arguments primary. Therefore in Fig. 4 we can see that in case agents have zero cases in their case base then no agreement is reached. Thus, agents must have arguments at the beginning of a negotiation. The second impact that was highlighted on Fig. 4 is the number of agreements. Agreement is the goal of each agent starting a negotiation and thus, it makes it an important factor to evaluate a negotiation. In the case where agents have only 10 cases in their case-bases, only 10 agreements were reached from 30 negotiation scenarios in each type of agents. The bigger the number of cases is, the higher the probability of reaching an agreement. Negotiation seems to be more successful when the number of cases is more than 100 (see Fig. 4). Thus, negotiation is easier and more beneficial when agents are more experienced.

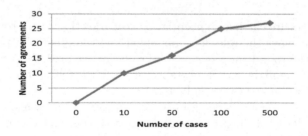

Fig. 4. Number of agreements

Another important feature that influenced the negotiation is the agent type (i.e. tolerant, strict and very strict). In fact, the negotiation process is longer when agents are very strict due to the difficulty of convincing such agent and consequently, the number of agreement decreases.

4   Conclusion

This paper proposes a generic domain-independent framework for argumentation-based negotiation using CBR. GANC is generic thanks to its domain-independent architecture, the structure of the arguments (i.e. attributes- values) and the fact that all domain-dependent information can be instantiated in the domain CBR only by instantiating a set of premises, their values and the outcome. We claim that argumentation presents additional information exchanged by agents that facilitate the negotiation between them and helps to find a settlement. The CBR gives agents the possibility to take advantage of their past experiences and thus, better decisions can be made. Experiments proved that this combination of the three fields (i.e. argumentation, negotiation and CBR) is beneficial and makes the automated negotiation better in terms of time and the reached agreements. Moreover, a benchmark was constructed. It contains information about several travel destinations.

Even though in this paper we focus on bilateral negotiation, we believe, that this work can be the basis for many future works. We plan to extend the GANC framework to support multilateral and multi-issues negotiations. We also plan to take into consideration the uncertainty in such environments and to add the possibility of learning opponent's tactics and the way they reason as well as the moves that they usually apt to in similar situations using CBR.

References

1. Kreps, D.M.: Game Theory and Economic Modelling. Oxford University Press, Oxford (1990)
2. Pruitt, D.G., Carnevale, P.J.: Negotiation in Social Conflict. Thomson Brooks/Cole Publishing Co, Belmont (1993)
3. Su, S.Y., Huang, C., Hammer, J., Huang, Y., Li, H., Wang, L., Lam, H.: An internet-based negotiation server for e-commerce. VLDB J. Int. J. Very Large Data Bases 10(1), 72–90 (2001)
4. Faratin, P.: Automated service negotiation between autonomous computational agents. Unpublished doctoral dissertation, University of London (2000)
5. Kraus, S.: Strategic Negotiation in Multiagent Environments. MIT Press, Cambridge (2001)
6. Beer, M., D'inverno, M., Luck, M., Jennings, N., Preist, C., Schroeder, M.: Negotiation in multi-agent systems. Knowl. Eng. Rev. 14(03), 285–289 (1999)
7. Heras, S., Botti, V., Julián, V.: Challenges for a CBR framework for argumentation in open MAS. Knowl. Eng. Rev. 24(04), 327–352 (2009)
8. Heras, S., JordáN, J., Botti, V., JuliáN, V.: Argue to agree: a case-based argumentation approach. Int. J. Approx. Reason. 54(1), 82–108 (2013)
9. Sycara-Cyranski, K.: Arguments of Persuasion in labor Mediation. In: IJCAI, pp. 294–296 (1985)
10. Amgoud, L., Dimopoulos, Y. and Moraitis, P.: A unified and general framework for argumentation-based negotiation. In: Proceedings of 6th International Joint Conference on Autonomous Agents and Multiagent Systems, p. 158 (2007)

11. Sierra, C., Jennings, N.R., Noriega, P., Parsons, S.: A framework for argumentation-based negotiation. In: Singh, M.P., Rao, A., Wooldridge, M.J. (eds.) ATAL 1997. LNCS, vol. 1365, pp. 177–192. Springer, Heidelberg (1998). https://doi.org/10.1007/BFb0026758
12. Amgoud, L., Prade, H.: Reaching agreement through argumentation: a possibilistic approach. In: KR 2004, pp. 175–182 (2004)
13. Karacapilidis, N., Papadias, D.: Computer supported argumentation and collaborative decision making: the HERMES system. Inf. Syst. **26**(4), 259–277 (2001)
14. Carneiro, D., Novais, P., Andrade, F., Zeleznikow, J., Neves, J.: Using case-based reasoning and principled negotiation to provide decision support for dispute resolution. Knowl. Inf. Syst. **36**(3), 789–826 (2013)
15. Li, H.: Case-based reasoning for intelligent support of construction negotiation. Inf. Manag. **30**(5), 231–238 (1996)
16. Sycara, K.P.: Negotiation planning: an AI approach. Eur. J. Oper. Res. **46**(2), 216–234 (1996)
17. Soh, L.K., Tsatsoulis, C.: Agent-based argumentative negotiations with case-based reasoning. In: AAAI Fall Symposium Series on Negotiation Methods for Autonomous Cooperative Systems, North Falmouth, Massachusetts, pp. 16–25 (2001)
18. Jain, P., Dahiya, D.: An intelligent multi agent framework for ecommerce using case based reasoning and argumentation for negotiation. In: International Conference on Information Systems, Technology and Management, pp. 164–175 (2012)
19. Adnan, M.H.M., Hassan, M.F., Aziz, I., Paputungan, I.V.: Protocols for agent-based autonomous negotiations: a review. In: 2016 3rd International Conference on Computer and information sciences (ICCOINS), pp. 622–626 (2016)
20. Rubinstein, A.: Perfect equilibrium in a bargaining model. Econometrica: J. Econ. Soc. **50**(1), 97–109 (1982)
21. Rahwan, I., Ramchurn, S.D., Jennings, N.R., Mcburney, P., Parsons, S., Sonenberg, L.: Argumentation-based negotiation. Knowl. Eng. Rev. **18**(4), 343–375 (2003)
22. Hadidi, N., Dimopoulos, Y., Moraitis, P.: Argumentative alternating offers. In: International Workshop on Argumentation in Multi-Agent Systems. pp. 105–122 (2010)
23. Pacheco, N.C., et al.: Magentix 2 User's Manual (2015)

Soft Computing in Information Retrieval and Sentiment Analysis

Obtaining WAPO-Structure Through Inverted Indexes

Úrsula Torres-Parejo[1]([✉]) [iD], Jesús R. Campaña[2] [iD], Maria-Amparo Vila[2] [iD],
and Miguel Delgado[2] [iD]

[1] Department of Statistics and Operational Research, University of Cádiz,
Cádiz, Spain
ursula.torres@uca.es

[2] Department of Computer Science and Artificial Intelligence, University of Granada,
Granada, Spain

Abstract. In order to represent texts preserving their semantics, in earlier work we proposed the WAPO-Structure, which is an intermediate form of representation that allows related terms to remain together. This intermediate form can be visualized through a tag cloud, which in turn serves as a textual navigation and retrieval tool. WAPO-Structures were obtained through a modification of the APriori algorithm, which spends a lot of processing time computing frequent sequences, for which it must perform numerous readings on the text until finding the frequent sequences of maximal level.

In this paper we present an alternative method for the generation of the WAPO-Structure from the inverted indexes of the text. This method saves processing time in texts for which an inverted index is already computed.

Keywords: Inverted indexes · Implications · Primary rules
Content representation · Text processing · Semantics
Frequent sequences · Text retrieval

1 Introduction

One of the main problems of information management in textual databases is the amount of unstructured text that is difficult to recover, due to the way of processing it. Many retrieval systems perform only syntactic text processing, which means that much of the content identification capability is lost in the retrieved text.

Frequent ordered itemsets preserve the semantics of the text since they allow related terms to remain ordered and united. This is achieved through the APO (Ordered APriori)-Structure [9], which represents the text through an intermediate form facilitating its processing, representing the content of the information and allowing greater precision and recall with the query results.

The WAPO-Structure introduces weights into the APO-Structure, so that the ordered sets can be visualized through a tag cloud with different font sizes.

© Springer International Publishing AG, part of Springer Nature 2018
J. Medina et al. (Eds.): IPMU 2018, CCIS 854, pp. 647–658, 2018.
https://doi.org/10.1007/978-3-319-91476-3_53

In this way, the tag cloud works as an assistant for the query formulation and as a tool for exploring the contents of the database.

There are many methods to obtain the WAPO-Structure, such as all those for obtaining frequent itemsets [1,2,11] with slight modifications. But if the inverted indexes [4] are available, the processing time is considerably reduced, since it does not have to perform repeated readings on the database to calculate support values.

Hipp et al. [6] compare several of the algorithms in terms of performance for obtaining frequent itemsets, verifying that all have a similar behavior with respect to the execution time, although spending different time depending on tasks. While some of the algorithms compared use most of the time to determine the support of the candidate itemsets below level four, the Apriori algorithm finds the greatest difficulty in calculating support for itemsets of level four or higher.

In previous work [8,9] we obtained the WAPO-Structure from a modification of the Apriori algorithm [1]. In this paper, we propose its generation from a complete inverted index which is more advantageous as the level of the itemsets increases. With the Apriori algorithm, each time we add an item to an itemset, a new reading of the database is required, while with the inverted indexes, the maximum level itemsets are located in a single reading.

To the better understanding of the terminology used in this paper, we establish some previous concept definitions in Table 1.

Table 1. Concept definitions

Expression	Definition
Term	A word or group of words
Mono-term	Single word
Item	An individual article or unit, especially one that is part of a list, collection, or set
Sequence	Ordered list of items

This paper is organized as follows. Section 2 defines APO-Structure and WAPO-Structure. Section 3 gives the definition of Full Inverted Index. Section 4 explains the method for obtaining the WAPO-Structure from a complete inverted index. Section 5 illustrates this process with a practical example and, finally, we end with a brief discussion and conclusions in Sect. 6.

2   APO-Structure and WAPO-Structure

The lack of structure in textual attributes complicates their automatic processing. In [8,9] we see how the APO and WAPO Structures provide the mathematical structure to obtain the semantics inherent to the text, facilitating the processing. These semantics are achieved by allowing the frequently related terms to remain together.

The complete process is carried out in five steps: Selecting the textual attribute, syntactic preprocessing, semantic preprocessing, structure generation and displaying of the structure.

The WAPO-Structure provides weight into the APO-Structure.

2.1 APO-Structure [9]

Definition 1. AP-Seq (AP-Sequences)

Let $X = \{x_1, x_2, \ldots, x_n\}$ be a referential set of items and R a sequence of frequent itemsets. R is then an AP-Seq if and only if:

1.
$$\forall Z = (z_1, z_2, \ldots, z_k) \in R \Rightarrow \begin{cases} (z_1, z_2, \ldots, z_{k-1}) \in R \\ (z_2, z_3, \ldots, z_k) \in R \end{cases} \forall k \in [2, n] \qquad (1)$$

2. $\exists\, Y \in R$ such that:

$$card(Y) = max_{Z \in R}\, (card(Z)) \text{ and } \nexists\, Y' \in R \mid card(Y') = card(Y) \qquad (2)$$
$$\forall Z \in R \Longrightarrow Z \subseteq Y \qquad (3)$$

The sequence Y with higher order is the spanning sequence of R, $R = g(Y)$, in other works $g(Y)$ is the AP-Seq with spanning sequence Y, being the cardinal of Y the level of $g(Y)$.

Example 1. Let $X = \{$intelligence, online, measure, test, partner$\}$.

Let $R = \{<intelligence, test, partner>, <intelligence, test>,$
$<test, partner>, <intelligence>, <test>, <partner>\}$.

Then, R is an AP-Seq with spanning sequence $Y = <intelligence, test, partner>$. All the other sequences are included in it with the same order position between the terms.

Definition 2. APO-Structure

Let $X = \{x_1, \ldots, x_n\}$ be a referential set of items and $S = \{A, B, \ldots\}$ a set of frequent item-seqs with a cardinal higher than or equal to one and $A, B, \ldots$ AP-Seqs such as:

$$\forall\, A, B \in S;\ A \nsubseteq B, B \nsubseteq A \text{ and } B \neq A$$

An APO-Structure generated by S, $E = g(A, B, \ldots)$, is the set of AP-Seqs whose spanning sequences are $A, B, \ldots$

2.2   WAPO-Structure [9]

Definition 3. Frequent weighted item-seq of an AP-Seq
Let $R = g(Y)$ be an AP-Seq with a referential set of items X. It is said that $\widetilde{\alpha}_t \subseteq Y$ is a frequent weighted item-seq from Y if:

$$\widetilde{\alpha}_t = [\alpha_t, \omega_t]. \tag{4}$$

where α_t is a frequent term sequence and ω_t is its weight or frequency.

WAPO-Structures are structures composed of weighted AP-Seqs which are AP-Seq composed of weighted item-seqs.

Definition 4. WAPO-Structure
Let $X = \{x_1, x_2, \ldots, x_n\}$ be a referential set of items and $\widetilde{S} = \{\widetilde{A}, \widetilde{B}, \ldots\}$ a set of frequent weighted item-seqs with a cardinal higher than or equal to one, and $\widetilde{A}, \widetilde{B}, \ldots$ weighted AP-Seqs such as:

$$\forall\, A, B \in S;\ A \nsubseteq B,\ B \nsubseteq A\ \text{and}\ B \neq A. \tag{5}$$

A WAPO-Structure generated by $\widetilde{S}$, $\widetilde{E} = g(\widetilde{A}, \widetilde{B}, \ldots)$ is the set of AP-Seqs whose spanning sequences are $\widetilde{A}, \widetilde{B}, \ldots$

Note 1. We express the spanning sequence $\widetilde{A}$ as well as $\widetilde{g}(A)$.

Example 2. Let us suppose a database containing tuples in Table 2.

Table 2. Tuples in the Example 2

n	Item-seqs
1	<intelligence, test, online, measure>
2	<measure, intelligence, test>
3	<measure, online>

Setting a support of 2 in terms of absolute frequency to consider an item-seq to be frequent, we obtain the following structures:

$$
\begin{aligned}
APO-Structure\ :\ & g(<intelligence, test>, <online>, <measure>) \\
= & (<intelligence, test>, <intelligence>, <test>, \\
& <online>, <measure>) \\
WAPO-Structure\ :\ & \widetilde{g}(<intelligence, test>, <online>, <measure>) \\
= & (<intelligence, test>, 2), (<intelligence>, 2), \\
& (<test>, 2), (<online>, 2), (<measure>, 3)
\end{aligned}
$$

3   Complete Inverted Index

The inverted indexes are widely used in information retrieval [3,7], as well as for other applications [5,10]. In this work we use the definitions given in [4], to understand them, some notations are established in Table 3.

Table 3. Notations

Symbol	Definition
$\sum$	Finite non-empty alphabet
$\sum^*$	Set of all items on $\sum$
λ	Empty word
$\sum^+$	$\sum^* - \{\lambda\}$
S	Finite set of text words $S \subseteq \sum^+$
$Sub(S)$	Set of sub-strings in S

Note 2. If $\omega = xyz$ for the terms $x, y, z \in \sum^* \Rightarrow y$ is a subterm of ω, x is a prefix of ω and z is a suffix of ω.

Definition 5. Complete Inverted Index [4]

Given a finite alphabet $\sum$, a set of terms $k \subseteq \sum^+$ and a set of texts $S \subseteq \sum^+$, a complete inverted index for $(\sum, k, S)$ is an abstract data type that implements the following functions:

1. **find:** $\sum^+ \to k \cup \{\lambda\}$, where find($\omega$) is the largest prefix x of ω with $x \in k \cup \{\lambda\}$ and x occurs in S, that is, x is a subset of terms of a text in S.
2. **freq:** $k \to \mathbb{N}$, where freq(ω) is the number of times that ω occurs is a subset of terms of a text in S.
3. **locations:** $k \to 2^{N \times N}$, where locations(ω) is the number of ordered pairs indicating the number of the text in which ω occurs and its position within the text.

Definition 6. Rule of S [4]

A rule of S (r_S) is a production $x \to_s \gamma x \beta$ where $x \in sub(S), \gamma, \beta \in \sum^*$ that occurs each time that x is preceded by γ and followed by β in S.

Definition 7. Primary rule of S [4]

$t_S : x \to_s \gamma x \beta$ is a primary rule of S if it is a rule S and γ and β are sets of terms of the highest possible order, that is, $\nexists \delta, \tau \in \sum^*$ with $\delta, \tau \neq \lambda$ such as $x \to_s \delta \gamma x \beta \tau$ be a rule of S.

Definition 8. Implication of x in S [4]

If $x \to_s \gamma x \beta$ is a primary rule of S, then $\gamma x \beta$ is called implication of x in S and is denoted $imp_S(x)$: $P(S) = \{imp_S(x) : x \in sub(S)\}$

The members of $P(S)$ are called subsets of primary terms of S.

4   Obtaining the WAPO-Structure Through Complete Inverted Indexes

It is possible to obtain the WAPO-Structure from the APO-Structure through inverted indexes mainly in two ways:

The first is through the Apriori algorithm, in a similar way as the AprioriTid and AprioriHybrid algorithms work [1], with a slight modification to introduce order and weight into the itemsets.

These algorithms construct iteratively the set of frequent terms, using the frequent itemsets found in a step to build the candidate itemsets and check if they are frequent in the next step.

In the first step, the support of elementary items or items of level 1 is calculated and determines which of these items are considered frequent according to the minimum support. In each subsequent step, it starts with a "seed" set consisting of itemsets found in the previous step combined with each other to generate the candidate itemsets deciding which of these are, in turn, frequent.

To do this, the Apriori algorithm requires at each step to re-read data, but the AprioriTid has the property that it is not necessary to go through the entire database to calculate the support of the candidate itemsets after the first step. For this purpose, a codification of the candidate itemsets found in the previous step is created, before deciding whether they are frequent in the subsequent step. In successive steps, the size of this coding is becoming much smaller than that of the database, saving a lot of reading effort.

This coding is the one that we can perform through the inverted index, to later apply the Apriori algorithm, just instead of going through the entire database at each step, only the inverted indexes are read, which indicate the ordered positions of each term in the text, discarding those that do not correspond with a frequent itemset for the later step.

The second way is the one proposed in this article and consists of identifying the implications of x_i^1 with the spanning sets of the APO-Structure, being x_i^1 each of the frequent itemsets of level equal to 1. Obviously, we would have to identify those implications that, in turn, were frequent.

To do this, the primary rules of x_i^1 (tuples containing the term in question) are previously obtained and the maximals are selected, storing the frequent ones. Those not frequent are divided into sub-rules, deciding which of these are, in turn, frequent. Once the set of all the frequent rules is obtained, we eliminate the redundant and not maximal ones and the remaining rules are what we will call "frequent implications of x_i^1", identifying them with the spanning sets of the APO-Structure.

Finally, we use the function **freq** to obtain the weight of the item-seqs in the APO-Structure and generate the WAPO-Structure.

Next, we define the set of frequent implications of x in S, where all the frequent implications of x_i^1 are stored for identifying them with the spanning sets of the APO-Structure.

Definition 9. Set of frequent implications of x in S

Let $P(S) = \{imp_S(x) : x \in sub(S)\}$ be the set of all the implications in S, we call $Pf(S)$ the set of all the frequent implications in S, then $Pf_{x_i^1}(S) = \{imp_S^j(x_i^1) : x \in sub(S)\}$ with $x_i^1 = $ frequent itemset of level 1 and j each of the frequent implications of the itemset i.

Definition 10. Correspondence between frequent implications and the spanning sets of the APO-Structure

Let $E = g(A, B, \ldots, K, \ldots)$ be the APO-Structure generated by the sets $A, B, \ldots, K, \ldots$, then:
$A = imp_S^1(x_1^1), B = imp_S^2(x_1^1), \ldots, K = imp_S^1(x_2^1), \ldots$ removing redundancies.

The following algorithm specifies the process in more detail. For its application it is necessary to have the complete inverted index for all the terms of the base of texts S.

Algorithm

1. *Identify frequent mono-terms according to support:*
   $\omega_i = x_i^1 \ i \in S$
   If $freq(\omega_i) > support \Rightarrow \omega_i$ is frequent (denoted as ω_i^f)
2. *Calculate the primary rules of ω_i^f:*
   If $\omega_i^f \in t_j$ and $t_j = $maximal tuple in S
   $\Rightarrow t_j$ *primary rule of ω_i (denoted as r_k)*
3. *Remove redundancies*
4. *Check if the rules obtained are frequent:*
   If $freq(r_k) > support \Rightarrow r_k$ frequent rule (denoted as r_k^f)
5. *Store the frequent rules in the set of frequent implications:*
   Input r_k^f en $Pf(S)$
6. *Split non-frequent rules into sub-rules:*
   If $freq(r_k) < support \Rightarrow r_k$ no frequent rule (denoted as $\overline{r_k^f}$)
   $\forall \ (\overline{r_k^f}) = \{i_1, \ldots, i_n\}$ *non-frequent rule of S $\Rightarrow$*
   $\{i_i, \ldots, i_{n-1}\}$ y $\{i_2, \ldots, i_n\}$ *rule of S.*
7. *Go back to step 4.*
8. *Remove redundancies in $Pf(S)$*
9. *Identify each of the frequent implications in $Pf(S)$ with the spanning sets of the APO-Structure.*

5   Example of How to Obtain APO-Structure and WAPO-Structure Through Implications

Let us suppose we obtain the item-seqs listed in Table 4 from a database after cleaning the text.

The function **locations(ω)** for all mono-terms is presented in Table 5 and the image of the function **freq(ω)** in Table 6.

Table 4. Item-seqs after text cleaning

n_i	Item-seqs(i)
1	$<pink, yellow, blue>$
2	$<green, pink, yellow>$
3	$<green, blue>$
4	$<orange, pink, yellow, blue>$
5	$<green, pink>$
6	$<yellow, green>$
7	$<green, pink, yellow>$

Table 5. locations(ω)

Term	n_1	n_2	n_3	n_4	n_5	n_6	n_7
pink	(1,1)	(2,2)		(4,2)	(5,2)		(7,2)
yellow	(1,2)	(2,3)		(4,3)		(6,1)	(7,3)
blue	(1,3)		(3,2)	(4,4)			
green		(2,1)	(3,1)		(5,1)	(6,2)	(7,1)
orange				(4,1)			

Table 6. freq(ω)

ω_i	Locations(ω_i)	F_{ω_i}
pink	{(1,1), (2,2), (4,2), (5,2), (7,2)}	5
yellow	{(1,2), (2,3), (4,3), (6,1), (7,3)}	5
blue	{(1,3), (3,2), (4,4)}	3
green	{(2,1), (3,1), (5,1), (6,2), (7,1)}	5
orange	{(4,1)}	1

Let us consider a minimum support for an item-seq to be frequent greater or equal than an absolute frequency of 2. In the current example the frequent item-seqs of cardinality 1 are the mono-terms $<pink>$, $<yellow>$, $<blue>$ and $<green>$.

We compute the implications for each of these frequent item-seqs of cardinality 1. To do it first, it is necessary to compute their rules. The rules for item-seq $<pink>$ along with each rule frequency are shown in Table 7, where r_i represents the rule i and $freq_i$ represents its frequency.

Then, the primary rules are identified. We select only maximal rules. Rule r_4 it is not a primary rule, as it is contained in rule r_2. Rule r_1 it is neither a primary rule as it is contained in rule r_3.

Table 7. Rules for item-seq $<pink>$

i	$r_i(<pink>)$	$freq_i(r_i(<pink>))$
1	$<pink, yellow, blue>$	2
2	$<green, pink, yellow>$	2
3	$<orange, pink, yellow, blue>$	1
4	$<green, pink>$	3

Table 8. Primary rules for item-seq $<pink>$

i	$t_i(<pink>)$	$freq_i(t_i(<pink>))$
1	$<green, pink, yellow>$	2
2	$<orange, pink, yellow, blue>$	1

Table 9. Subrules for (t_2)

i	$r_i'(<t_2>)$	$freq_i(r_i'(<t_2>))$
1	$<orange, pink, yellow>$	1
2	$<pink, yellow, blue>$	2

Table 10. Subrules for (r_1')

i	$r_i''(<r_1'>)$	$freq_i(r_i''(<r_1'>))$
1	$<orange, pink>$	1
2	$<pink, yellow>$	4

Table 8 shows the primary rules for item-seq $<pink>$, where t_i represents the primary rule i and $freq_i$ its frequency. Of these primary rules, only t_1 is frequent regarding the support, so we store it in the set of frequent implications Pf(S).

Since t_2 is not frequent, it is divided into two sub-rules that we can see in Table 9, where r_i' represents the sub-rule i and $freq_i$ its frequency. In this case, r_2' is frequent and comes from a non-frequent primary rule, so r_2' becomes a frequent primary rule (since there is no frequent higher-order rule) and it is stored in the set of frequent implications Pf(S).

Since r_1' is not frequent, it is divided into two sub-rules r_i'' (see Table 10). The rule r_2'' is frequent and it is stored in Pf(S), however it is not a primary rule since there is another maximal rule in Pf(S) containing it, so it will be removed from this set. The r_1'' rule is not frequent, so the operations of dividing into sub-rules, checking the frequencies and storing the frequent rules would be repeated.

Finally, two frequent implications for $<pink>$ in Pf(S) have been obtained. They are shown in Table 11.

The same procedure is applied for the item-seqs $<yellow>$, $<blue>$ and $<green>$, obtaining their frequent implications. They are shown in Tables 12, 13 and 14.

Table 11. Frequent implications for $<pink>$

i	$imp_i(<pink>)$	$freq_i(imp_i(<pink>))$
1	$<green, pink, yellow>$	2
2	$<pink, yellow, blue>$	2

Table 12. Frequent implications for $<yellow>$

i	$imp_i(<yellow>)$	$freq_i(imp_i(<yellow>))$
1	$<pink, yellow, blue>$	2
2	$<green, pink, yellow>$	2

Table 13. Frequent implications for $<blue>$

i	$imp_i(<blue>)$	$freq_i(imp_i(<blue>))$
1	$<pink, yellow, blue>$	2

In total we have determined the next implications:

- $imp_S(x_1^1) = imp(<pink>) \Rightarrow imp^1(x_1^1) = <pink, yellow, blue>$ and $imp^2(x_1^1) = <green, pink, yellow>$
- $imp_S(x_2^1) = imp(<yellow>) \Rightarrow imp^1(x_2^1) = <pink, yellow, blue>$ and $imp^2(x_2^1) = <green, pink, yellow>$
- $imp_S(x_3^1) = imp(<blue>) \Rightarrow imp^1(x_3^1) = <pink, yellow, blue>$
- $imp_S(x_4^1) = imp(<green>) \Rightarrow imp^1(x_4^1) = <green, pink, yellow>$

When duplicate implications are removed, two implications remain: $imp^1(x_1^1) = <pink, yellow, blue>$ and $imp^2(x_1^1) = <green, pink, yellow>$

These implications will be used as the maximal itemseqs in the APO-Structure, which has cardinal 2.

Let E be an APO-Structure, with spanning sequences A and B:
$E = g(A, B) \Rightarrow A = imp^1(x_1^1)$ and $B = imp^2(x_1^1)$
$E = g(<pink, yellow, blue>, <green, pink, yellow>)$
$E = (<pink, yellow, blue>, <green, pink, yellow>, <pink, yellow>, <yellow, blue>, <green, pink>, <pink>, <yellow>, <blue>$ and $<green>)$

In order to compute the weight for the WAPO-Structure, we use the function **freq**(ω_i) with $i =$ item-seq $\in$ APO-Structure. The resulting WAPO-Structure is the following:

Table 14. Frequent implications for $<green>$

i	$imp_i(<green>)$	$freq_i(imp_i(<green>))$
1	$<green, pink, yellow>$	2

$\widetilde{E} = g((<pink, yellow, blue>, 2), (<green, pink, yellow>, 2))$
$\widetilde{E} = ((<pink, yellow, blue>, 2), (<green, pink, yellow>, 2),$
$(<pink, yellow>, 4), (<yellow, blue>, 2), (<green, pink>, 3)$
$(<pink>, 5), (<yellow>, 5), (<blue>, 3)$ and $(<green>, 5))$

6 Conclusions

The inverted index helps us to identify the primary rules of the frequent terms and the frequent sub-rules of the non-frequent primary rules. The set consisting of all frequent maximal rules is called "the set of frequent implications". Each of the rules in this set corresponds to a spanning set of the APO-Structure, so we have the WAPO-Structure from these frequent implications and their frequencies.

The biggest drawback of this method is that when there are many maximal non-frequent item-seqs, a lot of time is lost in the decomposition of these sequences until finding the level in which they are frequent, in order to find the frequent implications. This drawback makes the method more appropriate in a text in which many large repeated sequences are found. In other case, it is preferable to use the Apriori algorithm, which according to Hipp et al. [6] finds the greatest difficulty in calculating support for itemsets of level four or higher and is best known for its ease and simplicity of implementation.

In short, the method proposed in this paper saves reading time by not having to go through the database repeatedly to get the frequent item-seqs as the Apriori algorithm method [1] does. Its application is recommended in databases where most of the frequent sequences are long, but the Apriori algorithm works better in other cases, depending on the characteristics of the text.

Acknowledgements. This work has been partially supported by the "Plan Andaluz de Investigación, Junta de Andalucía" (Spain) under research project P10-TIC6019.

References

1. Agrawal, R., Srikant, R.: Fast algorithms for mining association rules. In: Proceedings of the 20th International Conference in Very Large Data Bases, VLDB, vol. 1215, pp. 487–499. Citeseer (1994)
2. Agrawal, R., Srikant, R.: Mining sequential patterns. In: Proceedings of the Eleventh International Conference on Data Engineering, pp. 3–14. IEEE (1995). https://doi.org/10.1109/ICDE.1995.380415

3. Baeza-Yates, R., Ribeiro-Neto, B., et al.: Modern Information Retrieval. ACM Press, New York (1999)
4. Blumer, A., Blumer, J., Haussler, D., McConnell, R., Ehrenfeucht, A.: Complete inverted files for efficient text retrieval and analysis. J. ACM **34**(3), 578–595 (1987). https://doi.org/10.1145/28869.28873
5. Cutting, D., Karger, D., Pedersen, J., Tukey, J.: Scatter/Gather: a cluster-based approach to browsing large document collections. In: ACM SIGIR Forum, vol. 51, pp. 148–159. ACM (2017). https://doi.org/10.1145/3130348.3130362
6. Hipp, J., Güntzer, U., Nakhaeizadeh, G.: Algorithms for association rule mining - a general survey and comparison. SIGKDD Explor. Newsl. **2**, 58–64 (2000). https://doi.org/10.1145/360402.360421
7. Patil, M., Thankachan, S., Shah, R., Hon, W., Vitter, J., Chandrasekaran, S.: Inverted indexes for phrases and strings. In: Proceedings of the 34th International ACM SIGIR Conference on Research and Development in Information Retrieval, pp. 555–564. ACM (2011). https://doi.org/10.1145/2009916.2009992
8. Torres-Parejo, U., Campaña, J.R., Vila, M.A., Delgado, M.: MTCIR: a multi-term tag cloud information retrieval system. Expert Syst. Appl. **40**(14), 5448–5455 (2013). https://doi.org/10.1016/j.eswa.2013.04.010
9. Torres-Parejo, U., Campaña, J., Vila, M., Delgado, M.: A theoretical model for the automatic generation of tag clouds. Knowl. Inf. Syst. **40**(2), 315–347 (2014). https://doi.org/10.1007/s10115-013-0651-9
10. Vdorhees, E.: The cluster hypothesis revisited. In: ACM SIGIR Forum, vol. 51, pp. 35–43. ACM (2017). https://doi.org/10.1145/3130348.3130353
11. Zaki, M.: SPADE: an efficient algorithm for mining frequent sequences. Mach. Learn. **42**(1), 31–60 (2001). https://doi.org/10.1109/ICDE.2004.1320012

Automatic Expansion of Spatial Ontologies for Geographic Information Retrieval

Manuel E. Puebla-Martínez[1] , José M. Perea-Ortega[2](✉) ,
Alfredo Simón-Cuevas[3] , and Francisco P. Romero[4]

[1] Universidad de las Ciencias Informáticas, La Habana, Cuba
mpuebla@uci.cu
[2] Universidad de Extremadura, Badajoz, Spain
jmperea@unex.es
[3] Universidad Tecnológica de La Habana José Antonio Echeverría, La Habana, Cuba
asimon@ceis.cujae.edu.cu
[4] Universidad de Castilla-La Mancha, Ciudad Real, Spain
franciscop.romero@uclm.es

Abstract. One of the most prominent scenarios for capturing implicit knowledge from heterogeneous data sources concerns the geospatial data domain. In this scenario, ontologies play a key role for managing the totality of geospatial concepts, categories and relations at different resolutions. However, the manual development of geographic ontologies implies an exhausting work due to the rapid growth of the data available on the Internet. In order to address this challenge, the present work describes a semi-automatic approach to build and expand a geographic ontology by integrating the information provided by diverse spatial data sources. The generated ontology can be used as a knowledge resource in a Geographic Information Retrieval system. As a main novelty, the use of OWL 2 as an ontology language allowed us to model and infer new spatial relationships, regarding the use of other less expressive languages such as RDF or OWL 1. Two different spatial ontologies were generated for two specific geographic regions by applying the proposed approach, and the evaluation results showed their suitability to be used as geographic-knowledge resources in Geographic Information Retrieval contexts.

Keywords: Spatial ontologies · Spatial data sources · OWL 2
Geographic information retrieval

1 Introduction and Background

In recent years, research on applications for capturing implicit knowledge from heterogeneous data sources in different real-world scenarios has been intensified. One of the most prominent scenarios concerns the geospatial data domain, in

© Springer International Publishing AG, part of Springer Nature 2018
J. Medina et al. (Eds.): IPMU 2018, CCIS 854, pp. 659–670, 2018.
https://doi.org/10.1007/978-3-319-91476-3_54

which ontologies deals with the totality of geospatial concepts, categories, relations and processes and with their interrelations at different resolutions [1–5]. However, the manual development of geographic ontologies implies an exhausting work due to the rapid growth of the data available on the Internet, thus facilitating the possible occurrence of human mistakes.

In the related literature there are several works that address the problem of the semi-automatic generation of geographic ontologies from Spatial DataBases (SDB) [5–8] and from sources that are not SDB [9,10]. Baglioni et al. [6] and Lima et al. [7] define mapping rules between generic databases and ontologies without modeling table restrictions on the generated ontology. They model the geometric information of the spatial objects by using classes such as *Point*, *Line* and *Polygon*, instead of using their *Minimum Bounding Rectangles* (MBR)[1], which allows to represent the geometry more accurately and thus be able to identify new spatial relations between them. During the GeoKnow EU project (2013–2015)[2] [8], a spatial ontology based on RDF[3] triplets was generated from different SDB and RDF data sources. Hasani et al. [5] describe an approach to integrate spatial data from different SDB into an OWL[4] ontology, facing a high semantic heterogeneity presented in these sources. This method considers only SDB as data sources and no new knowledge is generated from them. Finally, regarding approaches that integrate spatial data from sources that are not SDB, Hahmann and Burghardt [9] propose an integration method based on RDF triplets from LinkedGeoData[5] and Geonames[6], but focusing on residential areas exclusively. More recently, Zaila and Montesi [10] present GeoNW, a spatial ontology generated from three data sources such as GeoNames, WordNet[7] and Wikipedia, although only GeoNames is really used as a spatial data source, since the other two are used to improve the information extracted from GeoNames.

In order to address the challenge of developing spatial ontologies, the present work describes a semi-automatic approach to build and expand a geographic ontology by integrating the information provided by diverse data sources, some of them publicly available on the Internet, such as GeoNames and OpenStreetMap (OSM)[8]. Thus, the generated ontology could be used as a knowledge resource in a Geographic Information Retrieval (GIR) system. The proposed approach starts

---

[1] MBR is an expression of the maximum extents of a 2-dimensional object (e.g. point, line, polygon). MBRs are frequently used as an indication of the general position of a geographic feature.

[2] http://geoknow.eu.

[3] http://www.w3.org/TR/rdf-concepts.

[4] http://www.w3.org/OWL.

[5] http://linkedgeodata.org.

[6] http://www.geonames.org. GeoNames is an open access geographic database that contains more than eight million place names from all countries in the world.

[7] http://wordnet.princeton.edu.

[8] http://www.openstreetmap.org. OSM is a collaborative project inspired by Wikipedia that emerged to create an editable and free world map where, instead of editing articles as in Wikipedia, users edit geographic entities.

from a preliminary ontology generated manually, which is then automatically extended from the information provided by these sources. Besides, the proposed method also allows the integration of data provided by a SDB, which should be normalized in *First Normal Form* (1NF) at least and it could be related to any geographic area of interest. Finally, new spatial relations are automatically generated, considering all the spatial objects managed by the ontology. The main contributions are: *(i)* the description of a semi-automatic method to build an OWL 2 ontology by integrating spatial data provided by heterogeneous resources; *(ii)* the use of OWL 2 as an ontology language has allowed modeling and infer new spatial relationships, regarding the use of other languages such as RDF or OWL 1; *(iii)* the public availability of two spatial ontologies coded in OWL 2 that are related to the geographic areas of Marianao (Cuba) and Acapulco (Mexico). In this sense, several experiments and an in-depth evaluation were performed for demonstration purposes.

The remainder of this paper is organized as follows: the proposed method to build and expand a spatial ontology is described in Sect. 2. Section 3 presents the results and observations obtained for the case study. The paper concludes with a summary and outlook in Sect. 4.

2 Proposed Approach

The proposed method has been designed to generate a geographic ontology through a semi-automatic process from different sources, with the aim that this geographic-knowledge resource can be used to improve performance in the GIR context. The Semantic Web language selected to build the ontology was OWL 2[9], which is considered a more expressive language than OWL 1 or RDF, due to the differences between their object property characteristics [11]. For instance, while OWL 1 allows assertions that an object property is symmetric or transitive, it is impossible to assert that the property is reflexive, irreflexive or asymmetric [12]. Therefore, such expressiveness in the spatial domain enables to define more assertions in the generated ontology.

The proposed approach starts from a preliminary ontology generated manually, which is automatically enriched from the information provided by Geonames, OSM and any SDB, through several knowledge-extension processes supported by Protégé[10] as an ontology editor. It should be noted that the knowledge-extension processes are guided by a Geographical Area of Interest (GAI) specified by the user, so the extraction tasks of concepts, axioms and individuals from these sources, and some procedures for the similarity resolution depend on this GAI. Finally, within the last phase, the gathered information already integrated into the ontology is used for automatic generation of spatial relationships. Figure 1 shows the overview of the proposed approach.

In the preliminary ontology, several common concepts, properties and spatial relationships of the geographical domain are manually represented. The construc-

---

[9] http://www.w3.org/TR/owl2-overview.
[10] http://protege.stanford.edu.

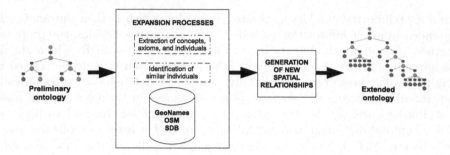

Fig. 1. Overview of the proposed approach

tion process was carried out using *GeoNames Ontology*[11] as a reference source. Other concepts and properties from OSM, as well as object properties commonly used in related work [13,14] such as *isNorthOf*, *contains*, *overlap* or *intersects*, were also represented in the preliminary ontology. Finally, the spatial relations were modeled in a similar way as described in Halimi et al. [15] and Tasic and Porter [16], taking advantage of the high expressiveness that OWL 2 provides with regard to the spatial data integration based on RDF triplets, as explained above. It is noteworthy that all the features of the spatial data sources considered have been represented as a *datatype property* in the preliminary ontology. In this sense, a new *string datatype property* named WKT^{12} (*Well Known Text*) was defined in order to model the geometric information of the spatial objects. This property will enable to represent semantically the geometric fields that SDB usually manage and, therefore, all the geometric information provided by data sources such as GeoNames or OSM. In the GIR context, this is an important aspect regarding the use of the generated ontology because the possible spatial relations between spatial objects will be able to be automatically generated from the *WKT* property. Figure 2 shows the spatial concepts taxonomy formalized for the preliminary ontology.

2.1 Expansion Processes

As shown in Fig. 1, different expansion processes are carried out to extend the preliminary ontology. These processes are guided by a specific GAI, so only the data related to that geographical area are automatically extracted from the information sources.

During the first expansion process, all classes related to the GAI and their OWL annotations are extracted from GeoNames and integrated into the taxonomy of the preliminary ontology. Then, the geographical objects related to the GAI, as well as their properties, are also extracted and integrated, considering the relationships with the classes previously represented on the preliminary ontology. Moreover, spatial relationships between the geographical objects

---

[11] http://www.geonames.org/ontology/ontology_v3.1.rdf.

[12] http://www.opengeospatial.org/standards/wkt-crs.

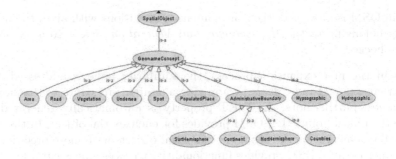

Fig. 2. Spatial concepts taxonomy for the preliminary ontology

(not explicitly represented in GeoNames) are automatically inferred from the extracted information. For instance, *contains* and *inside* are spatial relations included and semantically formalized as *object properties* in the extended ontology. Two main issues were addressed regarding the expansion process from GeoNames. On the one hand, we found spatial objects without a defined MBR property. For these cases, the WKT property was automatically generated from the geographical coordinates of the point that represents the spatial object, with the aim of modeling its geometry. On the other hand, we found similar spatial objects. Since GeoNames is a resource built collaboratively, typographic errors may appear in place names. For this reason, a method to identify similar objects is performed during this process. It is based on the assumption that two spatial objects are similar if the spatial distance among their nearest geographical points is smaller than 1 Km and their names or alternative names are syntactically similar according to the Damerau-Levenshtein function [17]. The similar objects are formalized as equivalent individuals in the extended ontology (*SameIndividual* in OWL 2).

Once the preliminary ontology is extended with the information provided by GeoNames, new spatial data are automatically extracted and inferred from OSM during the second expansion process:

- *Individuals* related to the GAI. Moreover, similar classes that identify any of those *individuals* are defined as *equivalent classes* in the ontology. For instance, the class *Route* from OSM is defined as a *equivalent class* of the class *Road* from GeoNames.
- Descriptions for each class. The descriptions are short texts that describe the OSM classes, and they are stored as OWL *annotations* in the ontology.
- Hierarchical relations between classes. The hierarchical relations are established between classes semantically related. For instance, the first time that a spatial object belonging to the *Building* class from OSM is identified, then such class is added to the ontology and it is also established as a subclass of the *Spot* class from GeoNames.
- Axioms of equivalence between individuals. They are considered when a spatial object meets the similarity conditions of all the individuals belonging to the class. For instance, when a spatial object that belongs to the *Natural* class

from OSM is identified, then the similarity conditions with all of the individuals belonging to the *Hypsographic* and *Vegetation* classes from GeoNames are checked.

As in the first expansion process, the similarity issue is addressed at the level of *classes* and *individuals* by applying a similar procedure to that used for GeoNames. Furthermore, the WKT property is automatically generated from the geographical points that OSM provides for each spatial object. In this sense, OSM allows generating the geometry of spatial objects with more precision than GeoNames because OSM provides functionalities to access the geographical coordinates of them. This process is supported by the use of the *osm4j* Java library[13] developed for working with OSM data.

The data source used for the third expansion process is a SDB, whose only requirement is that it should be normalized in First Normal Form (1NF) at least. Thus, the aim of this process is to extract all the spatial data from the SDB and integrate them into the ontology previously extended with the information provided by GeoNames and OSM, avoiding the possible similarity of individuals and classes. Several issues are addressed in this process:

- Each table from the SDB becomes a new class with the same name in the ontology. First, an automatic search procedure based on the Damerau-Levenshtein function is performed to find existing classes with a similar name. Second, since the proposed method is semi-automatic, the end user decides if the new class is equivalent to any of the candidate ones provided. If so, the class is defined as a *equivalent class* in the ontology or a new class is added otherwise.
- The attributes of each table are mapped as new data properties, except those representing the geometry and the name of the spatial object (they are already included in the preliminary ontology).
- The use of the WKT property (string type) to model the geometry field of the SDB in the ontology. This alternative of representing the geometry field of any SDB as a string format can be considered an advantage for the majority of the tools that manage ontologies, due to the flexibility and easy understand that such format provides for automatically generating spatial relations.
- Relationships between tables are modeled with functional relationships such as *isPartOf* or *isWholeOf*, that are defined previously during the construction of the preliminary ontology. Thus, for instance, if there are two tables named *Continents* and *Countries*, and *Countries* has the primary key of *Continents* as a foreign key, then a new object property is defined in the ontology with the name *Countries-isPartOf-Continents*.
- Four integrity constraints that usually support relational database managers (*max cardinality*, *not null*, *unique* and *primary key*) are also modeled in the ontology. All the data properties from the SDB are restricted with maximum cardinality 1. Then, as described in Mogotlane and Dombeu [18], for each *not null* attribute A of each table T, the minimum cardinality of the data

---

[13] http://jaryard.com/projects/osm4j/.

property representing A in the class corresponding to T, is set to 1. Moreover, the attributes that constitute the *unique* constraint for each table *T* are identified, thus avoiding instances from T with same values in the *unique* field. This was made possible through the use of the *Disjoint Data Properties* axiom provided by OWL 2. Finally, the possible existence of two instances from the same class with identical values for the *primary key* field is avoided by using the *Has Key* axiom provided by OWL 2.

2.2   Generation of New Spatial Relationships

The aim of the last phase of the proposed approach is to infer new spatial relationships between the existing objects in the ontology previously extended during the expansion processes. First, this process automatically identifies all the pairs of individuals that are related by the *spatial* object properties defined in the preliminary ontology. The *spatial* object properties are organized in four groups within the taxonomy defined for the preliminary ontology: *directional* (*isEastOf*, *isNorthOf*, *isSouthOf*, *isWestOf*), *mereological* (*isPartOf*, *isWholeOf*), *proximity* (*isFarOf*, *isNearOf*) and *topological* (*contains*, *crosses*, *disjoint*, *equal*, *inside*, *intersects*, *overlap*, *touches*). Specifically, the existence of several *spatial* object properties is verified for each pair of spatial objects in order to infer new spatial relationships between them.

Second, once the *spatial* object properties are verified for each pair of spatial objects, then the ontology is updated with the new information inferred, only in the case that new spatial relationships have been identified. For instance, if two spatial objects *O1* and *O2* are related by the *disjoint* relationship, and *O2* is related to another spatial object *O3* by the same relationship, then, since the *disjoint* relationship is defined as symmetric and transitive in the preliminary ontology, this process updates the ontological information by adding three new spatial relationships: *O1* is *disjoint* from *O3*, *O2* is *disjoint* from *O1*, and *O3* is *disjoint* from *O2*. For the specific case of the *directional* spatial relationships, the centroid of the two spatial objects is calculated. Then, the four *directional* relationships (*isEastOf*, *isNorthOf*, *isSouthOf*, *isWestOf*) are calculated from the comparison of the coordinates of the centroids.

The inference of new spatial relationships was supported by the use of the *Esri Java Geometry Library*[14]. This library includes methods for spatial operations and topological relationships, and it allows creating simple geometries from supported formats like WKT.

3   Experiments and Evaluation

In this section, we present the case study carried out in order to show the applicability of the proposed approach. Due to the exponential growth of the generated ontology, we have focused on two different GAI like Marianao (Cuba)

---

[14] http://github.com/Esri/geometry-api-java.

and Acapulco (Mexico), although the proposed approach is designed to process information related to any geographical region. Marianao is one of the 15 municipalities in the city of Havana, with a population of around 137,000 inhabitants[15] and a geographic area of $22 \, km^2$. Acapulco is the largest city in Mexico, with a population of around 735,000 inhabitants[16] and a geographic area of $1,880.6 \, km^2$.

The construction process of the preliminary ontology is independent of the GAI to be considered, so an OWL 2-coded ontology was obtained as a result of this initial phase with a total of 15 classes, 14 taxonomical relationships, 21 object properties (non-taxonomic relationships among defined classes) and 44 datatype properties with 80 annotations. Once the preliminary ontology was generated, the proposed approach was applied for both GAI. Only for the GAI of Marianao, a SDB with information about toponyms of Cuba was generated from the Spatial Data Infrastructure of the Republic of Cuba (IDERC)[17] and it was also used as a data source. Specifically, the spatial data were obtained from the *GeoServer* service[18]. Table 1 shows the enrichment of the ontological information during each phase of the proposed approach for different object properties and semantic relations, and for both GAI, Marianao (M) and Acapulco (A).

Table 1. Enrichment of the ontological information for each expansion process and both geographic areas, Marianao (M) and Acapulco (A)

	Preliminar ontology	Expansion processes						Generation of spatial relations	
		GeoNames		OSM			SDB		
		M	A	M	A		M	M	A
Classes	15	667	667	697	693		721	721	693
EquivalentClasses	0	0	0	1	1		12	12	1
SubClassOf relations	14	682	682	744	732		951	951	732
Individuals	0	47	536	2,367	1,890		4,662	4,662	1,890
ObjectPropertyAssertions	0	264	2,398	264	2,398		264	27,763,887	14,618,863
DataPropertyAssertions	0	1,316	15,008	23,220	26,540		26,008	26,017	26,780
SameIndividuals	0	1	7	13,115	634		13,321	13,321	634
Annotations	80	748	748	770	764		770	770	764

As shown in Table 1, the enrichment of information achieved in the preliminary ontology is relevant. It should be noted, for instance, the great increase produced in the total number of *Individuals*, reaching 4,662 and 1,890 new individuals for Marianao and Acapulco, respectively. The enrichment of individuals produced by the spatial data sources goes from 47 to 2,367 (+2,320) by OSM, and from 2,367 to 4,662 (+2,295) by SDB, in the case of Marianao. In the case of Acapulco, the enrichment of individuals achieved was +1,354 (from 536 to

---

[15] http://population.city/cuba/marianao.
[16] http://population.city/mexico/acapulco-de-juarez.
[17] http://www.iderc.cu.
[18] http://www.iderc.cu/geoserver/web.

1,890). Another noteworthy improvement occurred with regard to the semantic relations established between individuals connected by an object property (*ObjectPropertyAssertions*), with a total increase of more than 27.7 million and 14.6 million for Marianao and Acapulco, respectively. Finally, there was also a notable increase (+26,017 and +26,780 respectively) with regard to the semantic relations established between individuals connected by a data property (*DataPropertyAssertions*). These results acquire a special relevance if the generated ontologies were used as a source of knowledge in any GIR system related to the geographic areas considered.

The evaluation of the preliminary ontology was performed by using the OOPS tool[19] [19], which considers 6 dimensions to analyze: *consistency, completeness, conciseness, structural dimension, functional dimension* and *usability-profiling dimension*. No errors were detected in any of the indicators supported by OOPS. Then, we carried out the evaluation of both expanded ontologies: *OntoMarianao* and *OntoAcapulco*.

Since another spatial ontology related to the geographic area of Marianao was not available in order to make a feasible comparison, we decided to apply the *task-based* approach proposed by Raad and Cruz [20] to evaluate *OntoMarianao*. Such approach measures to what extent an ontology helps improve the results of a particular task (GIR in this case). Therefore, 50 queries were performed on the ontology with the aim of retrieving information regarding spatial objects related to the area, obtaining satisfactory results. The results were formalized by using the triplet *<source, object_id, object_name>*, where *source* refers to the information source from which the spatial object was obtained (*G* for Geonames, *O* for OSM and *S* for SDB). For instance, for the query "*hydrographic resources in Marianao*", hydrographic and their descendants were the classes involved and some of the information retrieved was: *<G, 3547581, Arroyo Marinero> <G, 3545352, Arroyo Paila> <G, 3746220, Presa Teresita> <O, 288952829, Río Orengo> <O, 223205922, Río Almendares> <O, 288923290, Río Quibú>...*

Regarding the evaluation of the ontology generated for the geographic area of Acapulco (*OntoAcapulco*), we could perform a comparison with other ontology designed for tourism applications (*TurismoAcapulco*), which was built from the geographic domain ontology *KaabOntology* [21]. The evaluation results obtained for both ontologies by using the 6 dimensions supported by OOPS are shown in Table 2 (*Consistency, Conciseness, Structural Dimension, Functional Dimension* and *Usability-Profiling Dimension*) and Table 3 (*Completeness*). For the *Completeness* dimension, the same object properties and semantic relations used in Table 1 were considered.

As shown in Table 2, no errors were detected during the evaluation of the *OntoAcapulco* ontology for the *Consistency* and *Functional* dimensions. Besides, less and minor errors were detected for the *Conciseness, Structural* and *Usability-Profiling* dimensions with regard to the *TurismoAcapulco* ontology. Finally, the evaluation results shown in Table 3 reveal a greater enrichment of knowledge

---

[19] OOPS! (OntOlogy Pitfall Scanner!). http://oops.linkeddata.es.

Table 2. Evaluation and comparison of the spatial ontologies *TurismoAcapulco* and *OntoAcapulco* with regard to several OOPS dimensions

Dimension	TurismoAcapulco	OntoAcapulco
Consistency	Error P24 (using recursive definitions) in element Ontology1173811255.owl#Playas	No errors
Conciseness	Error P03. Classified as a critical error	Error P02 (creating synonyms as classes). Classified as a minor error
Structural Dimension	Errors P03, P11 (4 issues) P13 (5 issues) and P24	Error P30 (equivalent classes not explicitly declared)
Functional Dimension	Error P04 in element Ontology1173811255.owl# TurismoAcapulco	No errors
Usability-Profiling Dimension	Errors P08 (141 issues), P11 (4 issues), P13 (5 issues), P22 and P41	Errors P02 and P22 (using different naming conventions in the ontology). Classified as minor errors

Table 3. Evaluation and comparison of the spatial ontologies *TurismoAcapulco* and *OntoAcapulco* with regard to the *Completeness* dimension of OOPS

Completeness dimension	TurismoAcapulco	OntoAcapulco
Classes	102	694
EquivalentClasses	0	1
SubClassOf relations	107	735
Individuals	434	1,910
ObjectPropertyAssertions	4	14,618,863
DataPropertyAssertions	2,428	26,780
SameIndividuals	0	601
Annotations	27	765

achieved by the ontology generated by the proposed approach, regarding all the object properties and semantic relations considered.

4   Conclusions and Further Work

We presented an approach to semi-automatically generate a semantic enriched geospatial ontology coded in OWL 2 from heterogeneous spatial data sources. The generated OWL2-coded ontology conceptualizes the place names of any geographical area of interest, and it gives the user a greater expressive power and a semantic view of the geographical data of that area, due to the greater expressiveness that OWL 2 provides in comparison to other ontology languages such as OWL 1 or RDF. Using an ontology evaluation tool and a collection of queries, we could perform an evaluation of the suitability of two generated spatial ontologies for the geographic regions of Marianao (Cuba) and Acapulco (Mexico). Based on the experiments and evaluations carried out, we conclude that modeling spatial relationships using OWL 2 and modeling the geometry of the spatial objects as a *datatype property* of string type (WKT property) can be considered an advantage regarding other related work.

Our future directions include the application of the proposed approach to generate spatial ontologies from other wider geographical regions, as well as the investigation of how to exploit and evaluate the generated ontologies within GIR systems.

Acknowledgments. This work has been partially supported by FEDER and the State Research Agency (AEI) of the Spanish Ministry of Economy and Competition under grant MERINET: TIN2016-76843-C4-2-R (AEI/FEDER, UE).

References

1. Mostafavi, M.A., Edwards, G., Jeansoulin, R.: An ontology-based method for quality assessment of spatial data bases. In: Frank, A.U., Grum, E. (eds.) Third International Symposium on Spatial Data Quality. Geoinfo Series, Department for Geoinformation and Cartography, Vienna University of Technology, pp. 49–66, vol. 1/28a (2004)
2. Fu, G., Jones, C.B., Abdelmoty, A.I.: Building a geographical ontology for intelligent spatial search on the web. In: Hamza, M.H. (ed.) Databases and Applications, pp. 167–172. IASTED/ACTA Press (2005)
3. Hess, G.N., Iochpe, C., Ferrara, A., Castano, S.: Towards effective geographic ontology matching. In: Fonseca, F., Rodríguez, M.A., Levashkin, S. (eds.) GeoS 2007. LNCS, vol. 4853, pp. 51–65. Springer, Heidelberg (2007). https://doi.org/10.1007/978-3-540-76876-0_4
4. Albrecht, J., Derman, B., Ramasubramanian, L.: Geo-ontology tools: the missing link. Trans. GIS **12**(4), 409–424 (2008). https://doi.org/10.1111/j.1467-9671.2008.01108.x
5. Hasani, S., Sadeghi-Niaraki, A., Jelokhani-Niaraki, M.: Spatial data integration using ontology-based approach. ISPRS - Int. Arch. Photogramm. Remote Sens. Spatial Inf. Sc. **XL-1/W5**, 293–296 (2015). https://doi.org/10.5194/isprsarchives-XL-1-W5-293-2015
6. Baglioni, M., Masserotti, M.V., Renso, C., Spinsanti, L.: Building geospatial ontologies from geographical databases. In: Fonseca, F., Rodríguez, M.A., Levashkin, S. (eds.) GeoS 2007. LNCS, vol. 4853, pp. 195–209. Springer, Heidelberg (2007). https://doi.org/10.1007/978-3-540-76876-0_13
7. Lima, D., Mendonça, A., Salgado, A.C., Souza, D.: Building geospatial ontologies from geographic database schemas in peer data management systems. In: Vinhas, L., Davis Jr., C.A. (eds.) GeoInfo, pp. 1–12. MCT/INPE (2011)
8. Lehmann, J., Athanasiou, S., Both, A., Buehmann, L., Garcia-Rojas, A., Giannopoulos, G., Hladky, D., Hoeffner, K., Grange, J.J.L., Ngomo, A.C.N., Pietzsch, R., Isele, R., Sherif, M.A., Stadler, C., Wauer, M., Westphal, P.: The GeoKnow Handbook. Technical report (2015)

9. Hahmann, S., Burghardt, D.: Connecting LinkedGeoData and GeoNames in the spatial semantic web. In: Proceedings of the 6th International GIScience Conference, Zurich (2010)
10. Zaila, Y.L., Montesi, D.: Geographic information extraction, disambiguation and ranking techniques. In: Proceedings of the 9th Workshop on Geographic Information Retrieval, pp. 11:1–11:7. ACM, New York (2015). https://doi.org/10.1145/2837689.2837695
11. Cuenca, B., Horrocks, I., Motik, B., Parsia, B., Patel-Schneider, P., Sattler, U.: OWL 2: the next step for OWL. Web Semant. **6**(4), 309–322 (2008)
12. W3C OWL Working Group: OWL 2 Web Ontology Language Document Overview (Second Edition) - W3C Recommendation 11 December 2012 (2012)
13. Egenhofer, M.J.: On the equivalence of topological relations. Int. J. Geogr. Inf. Syst. **9**, 133–152 (1995)
14. Ping, D., Yong, L.: Building place name ontology to assist in geographic information retrieval. In: 2009 International Forum on Computer Science-Technology and Applications, vol. 1, pp. 306–309, December 2009
15. Halimi, M., Farajzadeh, M., Delavari, M., Takhtardeshir, A., Moradi, A.: Modelling spatial relationship between climatic conditions and annual parasite incidence of malaria in southern part of Sistan&Balouchistan Province of Iran using spatial statistic models. Asian Pac. J. Trop. Dis. **4**, S167–S172 (2014). https://doi.org/10.1016/S2222-1808(14)60434-5
16. Tasic, I., Porter, R.J.: Modeling spatial relationships between multimodal transportation infrastructure and traffic safety outcomes in urban environments. Saf. Sci. **82**, 325–337 (2016). https://doi.org/10.1016/j.ssci.2015.09.021
17. Damerau, F.J.: A technique for computer detection and correction of spelling errors. Commun. ACM **7**(3), 171–176 (1964). https://doi.org/10.1145/363958.363994
18. Mogotlane, K.D., Dombeu, J.V.F.: Automatic conversion of relational databases into ontologies: a comparative analysis of protégé plug-ins performances. Int. J. Web Semant. Technol. **7**(3/4), 21–40 (2016)
19. Poveda Villalón, M.: Ontology evaluation: a pitfall-based approach to ontology diagnosis. Ph.d. thesis, Universidad Politécnica de Madrid, Escuela Técnica Superior de Ingenieros Informáticos, España, February 2016
20. Raad, J., Cruz, C.: A survey on ontology evaluation methods. In: Proceedings of the 7th International Joint Conference on Knowledge Discovery, Knowledge Engineering and Knowledge Management, pp. 179–186 (2015). https://doi.org/10.5220/0005591001790186
21. Torres, M., Quintero, R., Levachkine, S., Moreno, M., Guzmán, G.: Geospatial information integration approach based on geographic context ontologies. In: Popovich, V.V., Schrenk, M., Claramunt, C., Korolenko, K.V. (eds.) IF&GIS. LNGC, pp. 177–191. Springer, Heidelberg (2009). https://doi.org/10.1007/978-3-642-00304-2_12

Using Syntactic Analysis to Enhance Aspect Based Sentiment Analysis

Juan Moreno-Garcia$^{(\boxtimes)}$ (iD) and Jesús Rosado$^{(\boxtimes)}$ (iD)

Escuela de Ingeniería Industrial, University of Castilla La-Mancha, Toledo, Spain
{juan.moreno,jesus.rosado}@uclm.es
http://www.uclm.es/to/eii/investigacion/index_en.asp

Abstract. Many companies/corporations are interested in the opinion that users share about them in different social media. Sentiment analysis provides us with a powerful tool to discern the polarity of the opinion about a particular object or service, which makes it an important research field nowadays. In this paper we present a method to perform the sentiment analysis of a sentence through its syntactic analysis, by generating a code in *Prolog* from the parse tree of the sentence, which is automatically generated using natural language processing tools. This is a preliminary work, which provides encouraging results.

Keywords: Aspect extraction · Sentiment analysis · Parse tree
Natural language processing

1 Introduction

Over the last couple of decades the habit of on-line shopping has become increasingly more popular and with it, the availability of costumer reviews on products and services. Nowadays, with the boom of social networks and specialized forums, the numbers have escalated drastically. Because of that, the development of tools that allow for an automated analysis of all this information has become a matter of interest among researchers [3,4,9]. The goal of this analysis is to be able to decide if a particular item is good or bad from the perspective of the buyer writing the review, which motivates the emphasis in what is called *sentiment analysis*, that is, deciding if the reviewer has a positive or a negative opinion about it. First attempts approached the subject from the perspective of language analysis of the whole comment [2,5], while some authors emphasize the value given to some aspect of the item, like the waiting time in a restaurant or the life of the battery in a laptop [1,8,10].

In this work we want to set the framework for a tool that combine both perspectives. While still acknowledging the relevance of the aspect analysis and focusing on it, we will resort to the grammatical analysis of the sentences to find both, the relevant aspect the review refers to and the opinion of the reviewer about it.

© Springer International Publishing AG, part of Springer Nature 2018
J. Medina et al. (Eds.): IPMU 2018, CCIS 854, pp. 671–682, 2018.
https://doi.org/10.1007/978-3-319-91476-3_55

More precisely, we implement an application that generates a code in *Prolog*, which in turn will perform the sentiment analysis of the review. This application uses the *Python Natural Language Toolkit (NLTK)* [6], with an appropriate grammar, to create the facts and rules needed in *Prolog*. Our goal is to implement a completely autonomous system. Here we present a preliminary work, showing the general scheme of the system that we propose, together with some basic working examples.

The paper is organized as follows. First, in Sect. 2 we give a detailed description of the proposed method. Next, in Sect. 3 we show some examples of the *Prolog* code generated and the output that they provide. Finally, we present the conclusions and discuss the next steps to follow in Sect. 4.

2   Our Approach

As mentioned in the introduction, we propose a method for sentiment analysis that uses additional information, coming from a simplified syntactic analysis of the comment which is being studied, in order to improve its aspect based analysis.

On the one hand, this method processes the sentences through a syntactic analysis based on a grammar and translates it into a new structure which, in turn, goes through a second analysis that will allow us to identify the prevalent aspect and select its polarity. To perform the syntax analysis, several options are available. We have chosen the *Natural Language Toolkit (NLTK)* which allows an easy implementation in *Python* of applications to process data extracted from everyday speech. Among other features, such as providing easy-to-use interfaces to on-line lexical resources, this software provides a collection of text processing libraries for tokenization, parsing, semantic reasoning, etc. We are interested in the syntactic analysis of the sentences, that is, we want to find the parse tree associated to the sentence, which we can obtain in a relatively easy way using the *NLTK*.

On the other hand, each aspect must be labeled. We follow the classification proposed in [7], which uses the labels: *positive, negative, neutral* and *conflict*. In this first approach, we shall focus on comments which fall into the first three categories, while the classification of comments tagged as *conflict* will be addressed in follow up works. In order to assign a label to a comment, we use *Prolog* to process the parse tree obtained in the first step with the *NLTK*. *Prolog* is a programing language developed in the early '70s in the Aix-Marseille University by students Alain Colmerauer and Philippe Roussel. It was the fruit of a project, whose goal was algorithmic classification of natural languages, with both students being tasked with processing natural language. This makes *Prolog* a most fitting choice for our purpose. It looks for all possible combinations of logic solutions based on a system defined by *facts* and *rules*. In our case, the facts correspond to words from our grammar found in the lexical productions, while the rules are those of our grammar.

Figure 1 shows a diagram describing how our method works. It is made up of three parts:

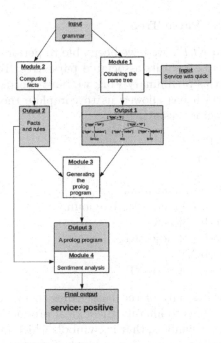

Fig. 1. Scheme of our approach.

1. Preliminary processing: the different structures that will be needed are defined. Specifically, we obtain the parse tree associated to the sentence being studied and define the *facts* and *rules* to be used with *Prolog*. These tasks are carried out by modules 1 and 2 respectively.
2. Generating the final code: a code in *Prolog* is automatically generated using the system of *facts* and *rules* to classify the sentence fed as input. This is done by Module 3.
3. Computing the output: the *Prolog* code generated in the previous step performs the sentiment analysis of the input (Module 4).

The system needs two inputs: the sentence to be analyzed and the definition of a grammar that is adequate for the input language. This grammar will, in turn, be used by the *NLTK* to generate the parse tree, in Module 1, which is described in detail in Sect. 2.1. The provided grammar will be used to generate, automatically, a system of *facts* and *rules* by Module 2, as described in Sect. 2.2.

Next, Module 3 gets the outputs from Module 1 and Module 2 as inputs and creates a *Prolog* code to carry out the sentiment analysis (Sect. 2.3). It should be pointed out that this code, the output of Module 3, is already Module 4, which will provide us with the final output.

The rest of this section is devoted to a detailed description of each module.

2.1   Computing the Parse Tree

Thanks to *Python* and *NLTK* we have been able to perform a syntactic analysis of the input sentence to obtain its associated parse tree. To do this, a grammar must be defined. For this preliminary work we have specified a simplified grammar, shown below, which has allowed us to complete the first tests in simple sentences, with satisfactory results.

Rule 1) S   − > NP VP
Rule 2) NP − > noun
Rule 3) NP − > pronoun
Rule 4) NP − > determinant noun
Rule 5) NP − > determinant adjective noun
Rule 6) VP − > verb adjective
Rule 7) VP − > verb not adjective
Rule 8) VP − > verb NP
Rule 9) VP − > verb not verb NP

The grammar consists only of the basic rules to process simple sentences. We have used low-case tags to identify the lexical productions, and capital case tags to identify non-final symbols, that is, symbols which must be solved using a different rule. This grammar uses a big amount of symbols which lead to lexical productions. The consequence of this is a simpler parse tree, and therefore, an ease of the task to be performed by the following modules.

In addition to these rules, we have to include lexical productions to identify words as noun, adjective, etc.

noun − > 'service' | 'thing' | 'food' | 'price'
adjective − > 'quick' | 'good' | 'right' | 'decent'

Figure 2 shows the parse tree associated to the sentence: *"Service was quick"*. NLTK generates the parse tree according to its own structure. We, then, transform it into a different structure that will make the following steps easier. For instance, we can represent the structure associated to the sentence used as input in the previous example.

('S', ('NP', 'noun', 'service'), ('VP', ('', 'verb', 'was'), ('', 'adjective', 'quick')))

The output of this module is a tuple where the first component is the identifier of the rule and the rest are, either a new tree or leaf. The new tree has the same structure that we have just described, whilst the leaf represents the rule that has been used. Finally, the last symbol is the word that is being represented. For instance, (NP, noun, service) is a leaf that stands for the rule "NP -> noun". We can see here the relation to the one shown in Fig. 2. From now on we will refer to this tuple as Syntactic Tuple.

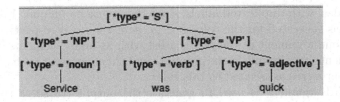

Fig. 2. Parse tree obtained for "Service was quick".

2.2 Calculating the Facts and the Rules

To generate the system of *facts* and *rules* that will be used in our *Prolog* code we have implemented in Module 2 an algorithm to transform each rule in the grammar given as input into either a *fact* or a *rule*. This is the most delicate part of our approach. Albeit it has worked as desired for the grammar described in Sect. 2.1, it must be thoroughly revised in order to adapt it to more complex grammars.

The first step consists in obtaining the *facts* from the lexical productions. These *facts* will be used to tag each word with the labels *negative, neutral* or *positive*. Thinking on how it should work for some specific words, we want to define them using a structure that generalizes examples like *verbSA (was, neutral)*, *adjectiveSA (quick, positive)*, *adjectiveSA (bad, negative)*, etc. It would be desirable that the system had a priory all the words that may be used in the input sentence already labeled and, furthermore, that this dictionary can be easily expanded.

Table 1. From grammar rule NP to Prolog fact or rule (Rules from 2 to 5).

Grammar	Prolog
NP – > noun	np(n,$[N]$,$[[N]]$)
NP – > pronoun	np(p,$[P]$,$[[P]]$)
NP – > determinant noun	np(d_n,$[\_, N]$,$[[N]]$)
NP – > determinant adjective noun	np(d_a_n,$[\_, A, N]$,$[[N, S]]$) :- adjectiveSA(A,S)

Each rule in the given grammar must be automatically translated into a *Prolog fact* or *rule*. We shall describe the process that we have followed to achieve this through examples to clarify each step. Table 1 shows how the rules NP (Rules from 2 to 5) are adapted. In general, we follow these steps:

1. The identifier of the rule, given in lower case, is used as predicate, *np* in this example.
2. Each predicate has a label (first argument) that is used to identify the grammar rule in the *Prolog fact* or *rule*. This label is formed by combining the first letter of each component of the outcome of the grammar rule in an ordered way, separated by the character '_'. In this example there has not been coincidences in the labels, but for further works, it is necessary to design some

rule to deal with duplicated labels, since they must be unique. Table 1 shows the labels generated for the rules NP: n, p, d_n y d_a_n respectively.

3. The second argument of the *fact* is a list with as many elements as different consequents has the rule in the grammar and they represent each of the words in the parse tree associated to this rule.

4. The third argument is used to generate the output (to solve the rule). In the example shown in this table, the *fact* returns the noun or pronoun for rules 2 to 4, and the pair [*noun, word*] for rule 5, where *word* is the label describing the sentiment associated to *noun* (negative, neutral or positive in this case) since it is accompanied by an adjective in the structure. We use a *fact, adjectiveSA*, to get the label. The rules of the input grammar that include some word subject to polarization use a *fact* from a lexical production to get the label that corresponds with this polarization.

We have followed a similar process to classify the rules VP of the grammar (Rules from 6 to 9). For each of this rules, we introduce a *Prolog rule* that evaluates each instance of it. This *rule* has also a label, a list with as many elements as the consequent of the corresponding rule from the grammar and an output list. As an example, the rule "$VP - > verb\ adjective$" translates into de *Prolog rule*:

$$vp(v\_a, [V, A], [[S]]) :-verbSA(V, V1), adjectiveSA(A, A1), solutionSA(V1, A1, S).$$

This rule evaluates the adjective and the verb through the predicates *verbSA* and *adjectiveSA* to generate the output using the *fact solutionSA*, which returns *negative, neutral* or *positive* from the values of $A1$ and $V1$, as illustrated in Table 2.

Table 2. Obtaining the calificative word.

$V1$	$A1$	S
negative	negative	positive
negative	positive	negative
positive	negative	negative
positive	positive	positive
X	neutral	X
neutral	X	X

The *Prolog rule*

$$vp(v\_np, [V, B|NP], NP1) :-verbSA(V, \_), np(B, NP, NP1),$$

solves the rule "$VP - > verb\ NP$" (Rule 8). Again, this *rule* is defined following the same structure as previous ones, with a label, an input list and an output list. It uses the predicate *verbSA* and invokes a rule of type NP (tagged as *np*) which provides the solution $NP1$. Rules 7 and 9 are solved in a similar way to rules 6 and 8 respectively. We skip the details for the sake of brevity.

Finally, the initial rule of the grammar, "$S - > NP\ VP$" is processed using the predicates *np* and *vp* that correspond to the labels A and B. It translates into the *Prolog rule*

$$phrase([A|NP], [B|VP], S, S1) :- np(A, NP, S), vp(B, VP, S1).$$

The rule *phrase*, has four inputs:

1. First, a list with the label of the rule NP followed by all the words in the sentence.
2. The second argument is another list, with the same structure as that of the first argument, but related to VP.
3. Third argument stores the evaluation of NP
4. Fourth argument sotres the evaluation of VP.

To conclude, we want to point out that the structure based on *facts* and *rules* used in *Prolog* naturally fits that of the grammar, which makes of *Prolog* a very appropriate tool for the processing of this kind of data.

2.3   Generating the Prolog Program

To conclude this section, we take a look to the structure of the *Prolog* codes generated by *Module 3*. They use the rule *prhase* described in Sect. 2.2 together with an extension to generate the final output. As an example, the generated rule that solves the sentences "Service was quick" is the following:

$$phrase([n, service], [v\_a, was, quick], S1, S2), aggregateSA(two, S1, S2, S), write(S).$$

It uses the rule *phrase* and the fact *aggregateSA*, joined through the *Prolog* conjuction ','. The predicate *phrase* has already been described. The predicate *aggregateSA* combines the outputs of *prhase* to produce S. It has a similar structure to that of the previous ones: uses a label, two lists to combine and an output list. In this stage, only two *facts* have been necessary:

1. *aggregateSA (two, [[N]], [[W]], [[N, W]]).*: Combines a noun N with a descriptive word W in the output (last parameter).
2. *aggregateSA (new, [[N1]], [[N2, W]], [[N1, W], [B, C]]).*: Combines a noun $N1$ with a descriptive word W which is already combined with another noun $N2$. For instance, in the sentence "*The food was of bad quality*" the word *food* would be combined with the label *negative*, which is already assigned to "*bad quality*".

This has been possible due to the simplicity of the sentences selected for the test, as will be described in the following section. In future research, the definition of a system of *rules* and *facts* that would be able to correctly process complex sentences will be a priority.

3   Tests: Some Examples

The algorithm described in the previous section has been tested on five sentences selected from the url *"http://alt.qcri.org/semeval2014/task4/ index.php?id=data-and-tools"*. This address corresponds to a web page which provides datasets in order to test programs whose goal is the aspect term extraction, aspect term polarity, aspect category detection and aspect category polarity. These datasets consist of customer reviews, to which human-made comments have been added in order to identify the key terms in the sentences and their polarity. We have chosen the following sentences, belonging to the dataset

> *Restaurants trial data*
> *Service was quick*
> *The food was of bad quality*
> *The music is not good*
> *The price is right*
> *I did not like the chicken*

This set is sufficient to test all the rules of the grammar discussed in Sect. 2.1. We want to emphasize that Module 2 generates all the *facts* and *rules* that will be used with all the test sentences, since they are created from the grammar itself, which is the same for all sentences.

Next, we present the outputs that our code has yielded for each of the sentences.

Service was quick

This is a short and simple sentence. It has a noun, a verb and an adjective. There is only one word in the subject (*service*) and only one adjective, which refers to that word (*quick*). When the sentence is processed with the NLTK in Module 1 the output is a single parse tree, shown in Fig. 2. This tree is represented by the tuple:

('S', ('NP', 'noun', 'Service'), ('VP', ('', 'verb', 'was'), ('', 'adjective', 'quick')))

The parse tree is generated by using the grammar rules:

S − > NP VP
NP − > noun
VP − > verb adjective

together with the lexical productions:

noun − > service
verb − > was
adjective − > quick

Using the *facts* and *rules* described in Sect. 4, which we have obtained from *Module 2* and the syntactic tuple provided by *Module 1 Module 3* generates the *Prolog* code:

$p :- phrase([n, service], [v\_a, was, quick], S1, S2), aggregateSA(two, S1, S2, S), write(S).$

When we run it, the output that we get is $S = [[service, positive]]$, which shows that the polarity of the term *service* is *positive*. The sentence has been properly analyzed and the proposed scheme has worked correctly.

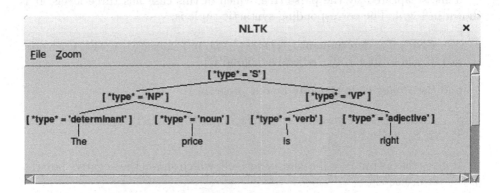

Fig. 3. Parse tree obtained for "The price is right".

The price is right

This sentence is similar to the previous one, but in this case the subject is of the form *determinant noun*. Figure 3 shows the parse tree yielded by our code, which, in this case presents two levels. The tuple that will be passed to module three is the following:

('S', ('NP', (", 'determinant', 'The'), (", 'noun', 'price')), ('VP', (", 'verb', 'is'), (", 'adjective', 'right')))

and the *Prolog* code:

p :- phrase([d_n, the, price],[v_a, is, right],S1,S2),aggregateSA(two,S1,S2,S), write(S).

The output it provides is $S = [[price, positive]]$ which shows that the polarity of the term *price* is positive in this case.

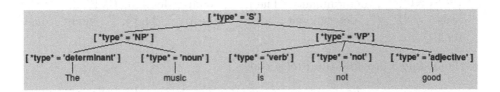

Fig. 4. Parse tree obtained for "The music is not good".

The music is not good

This sentence exhibits a negative sentiment about the key term in the subject, due to the effect of the adverb *not* on the adjective *good*.

This is captured by the parse tree, which in this case has three levels. It is shown in Fig. 4. The corresponding syntactic tuple is:

('S', ('NP', ('', 'determinant', 'The'), ('', 'noun', 'music')), ('VP', ('', 'verb', 'is'), ('', 'not', 'not'), ('', 'adjective', 'good')))

and the *Prolog* code generate by our program is:

p :- phrase([$d\_n, the, music$],[$v\_no\_a, is, no, good$], S1, S2), aggregateSA(two, S1, S2, S), write(S).

It gives the output $S = [[music, negative]]$, which shows the negative polarity of the term *music*.

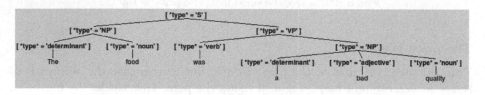

Fig. 5. Parse tree obtained for "The food was a bad quality".

The food was of bad quality

This sentence also shows a negative feeling about the subject, which comes from an *adjective + noun* structure. In this case, the adjective *bad* gives its negative polarity to the noun that it accompanies, which was initially neutral. Then the polarity of the noun in the verb phrase carries over to the noun in the noun phrase.

Figure 5 shows the resulting parse tree, which uses more rules than previous examples. The syntactic tuple generated by Module 1 is:

('S', ('NP', ('', 'determinant', 'The'), ('', 'noun', 'food')), ('VP', ('', 'verb', 'was'), ('NP', ('', 'determinant', 'a'), ('', 'adjective', 'bad'), ('', 'noun', 'quality'))))

while Module 3 provides the *Prolog* code:

p :- phrase([$d\_n, the, food$], [$v\_np, was, d\_a\_n, a, bad, quality$], S1, S2), aggregateSA(new,S1,S2,S), write(S).

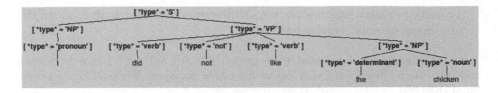

Fig. 6. Parse tree obtained for "I did not like the chicken".

Its output is $S = [[food, negative], [quality, negative]]$, which assigns the polarity *negative* to the terms *quality* and *food*.

I did not like the chicken

The last sentence in our set displays a negative verb form, made of three words (*did not like*). This will test some rules in our grammar that had not yet been used.

Figure 6 shows the generated parse tree. It has four levels due to the complexity of the sentence, which requires the use of a larger amount of rules from the grammar. This tree can be described with the tuple:

('S', ('NP', 'pronoun', 'I'), ('VP', (", 'verb', 'did'), (", 'not', 'not'), (", 'verb', 'like'), ('NP', (", 'determinant', 'the'), (", 'noun', 'chicken')))))

The corresponding *Prolog* code is:

p :- phrase($[p, i]$, $[v\_no\_v\_np, did, no, like, d\_n, the, chicken]$,_,S2), write(S2).

which gives the output $S = [[chicken, negative]]$. This shows the negative feel about the term *chicken* due to the verb *like* (positive) being in negative form.

4   Conclusions and Future Works

This paper presents our first approach to the problem of sentiment analysis through the syntactic analysis of the sentence. We have proposed an algorithm that can be used to process any sentence which can be formed in accordance with the provided grammar. Furthermore, we have used *Python* together with *NLTK* and *Prolog* to do the tests.

The results are promising since our design solves correctly the cases that we study and shows flexibility to be adapted to more complex sentences.

As future research, we contemplate the following:

– Obtaining a large set of basic *facts* since they are the basis upon which the polarity of the terms is computed.
– To improve some aspects of our design. Namely, how *facts* and *rules* are computed in Module 3 and the way the different results are combined.

– To carry out more experiments with more powerful grammar that allows the processing of more complex sentences.

Acknowledgements. Supported by the project TIN2015-64776-C3-3-R of the Science and Innovation Ministry of Spain, co-funded by the European Regional Development Fund (ERDF).

References

1. Kumar Gupta, D., Srikanth Reddy, K., Shweta, Ekbal, A.: PSO-ASent: feature selection using particle swarm optimization for aspect based sentiment analysis. In: Biemann, C., Handschuh, S., Freitas, A., Meziane, F., Métais, E. (eds.) Natural Language Processing and Information Systems 20th International Conference on Applications of Natural Language to Information Systems, NLDB 2015, Passau, Germany, Proceedings, vol. 9103, pp. 220–233. Springer, Heidelberg (2015). https://doi.org/10.1007/978-3-319-19581-0_20
2. Dong, L., et al.: A statistical parsing framework for sentiment classification. Comput. Linguist. **41**(2), 293–336 (2015)
3. Feldman, R.: Techniques and applications for sentiment analysis. Commun. ACM **56**(4), 82–9 (2013)
4. Liu, B.: Sentiment analysis: a multifaceted problem. IEEE Intell. Syst. **25**(3), 76–80 (2010)
5. Negi, S., Buitelaar, P.: INSIGHT galway: syntactic and lexical features for aspect based sentiment analysis. In: Proceedings of the 8th International Workshop on Semantic Evaluation (SemEval 2014), pp. 346–350 (2014)
6. Perkins, J.: Python 3 Text Processing with NLTK 3 Cookbook. Packt Publishing, Birmingham (2014)
7. Pontiki, M. et al.: SemEval-2014 task 4: aspect based sentiment analysis. In: Proceedings of the 8th International Workshop on Semantic Evaluation (SemEval 2014), pp. 27–35 (2014)
8. Pontiki, M. et al.: SemEval-2016 task 5: aspect based sentiment analysis. In ProWorkshop on Semantic Evaluation (SemEval-2016), pp. 19–30. Association for Computational Linguistics (2016)
9. Serrano-Guerrero, J., et al.: Sentiment analysis: a review and comparative analysis of web services. Inf. Sci. **311**, 18–38 (2015)
10. Thet, T.T., Na, J.C., Khoo, C.S.: Aspect-based sentiment analysis of movie reviews on discussion boards. J. Inf. Sci. **36**(6), 823–848 (2010)

A Probabilistic Author-Centered Model for Twitter Discussions

Teresa Alsinet[1]([⊠])[ID], Josep Argelich[1][ID], Ramón Béjar[1][ID], Francesc Esteva[2][ID], and Lluis Godo[2][ID]

[1] INSPIRES Research Center, University of Lleida, Jaume II, 69, 25001 Lleida, Spain
{tracy,jargelich,ramon}@diei.udl.cat
[2] AI Research Institute, IIIA-CSIC, Bellaterra, Spain
{esteva,godo}@iiia.csic.es

Abstract. In a recent work some of the authors have developed an argumentative approach for discovering relevant opinions in Twitter discussions with probabilistic valued relationships. Given a Twitter discussion, the system builds an argument graph where each node denotes a tweet and each edge denotes a criticism relationship between a pair of tweets of the discussion. Relationships between tweets are associated with a probability value, indicating the uncertainty on whether they actually hold. In this work we introduce and investigate a natural extension of the representation model, referred as probabilistic author-centered model. In this model, tweets by a same author are grouped, describing his/her opinion in the discussion, and are represented with a single node in the graph, while edges stand for criticism relationships between author's opinions. In this new model, interactions between authors can give rise to circular criticism relationships, and the probability of one opinion criticizing another is evaluated from the criticism probabilities among the individual tweets in both opinions.

Keywords: Twitter discussions
Probabilistic author-centered model · Argumentation

1 Introduction

In a recent work [2], an argumentative approach has been proposed for discovering relevant opinions in Twitter with probabilistic valued relationships.

Argumentation-based reasoning models aim at reflecting how humans make use of conflicting information to construct and analyze arguments. An argument is an entity that represents some grounds to believe in a certain statement and that can be in conflict with arguments establishing contradictory claims.

This work was partially funded by the Spanish MINECO/FEDER Projects TIN2015-71799-C2-1-P and TIN2015-71799-C2-2-P, by the European H2020 Grant Agreement 723596, and by the 2017 SGR 1537 and 172.

J. Medina et al. (Eds.): IPMU 2018, CCIS 854, pp. 683–695, 2018.
https://doi.org/10.1007/978-3-319-91476-3_56

The most commonly used general argumentation framework is Dung's abstract argumentation model [8].

In abstract argumentation, a graph is used to represent a set of arguments and counterarguments. Each node is an argument and each edge denotes an attack between arguments. Different kinds of semantics for abstract argumentation frameworks have been proposed that highlight different aspects of argumentation (for reviews see e.g. [4,5,16]). Usually, semantics for abstract argumentation frameworks are given in terms of sets of extensions, which are suitable consistent sets of arguments. For a specific extension, an argument is either accepted or rejected and, usually, there is a set of extensions that is consistent with the semantic context.

The analysis of Twitter by means of argumentation frameworks has also been explored by Grosse et al. [13] with the aim of detecting conflicting elements in an opinion tree to avoid potentially inconsistent information. Moreover, in order to mine arguments from Twitter, Bosc et al. [6] proposed a binary classification mechanism (argument-tweet vs. non argument) and Dusmanu et al. [10] applied supervised classification to identify arguments on Twitter and evaluated facts recognition and source identification for argument mining.

Given a Twitter discussion, i.e. a set of tweets generated from a root tweet, the system developed in [2] builds a weighted argument graph where each node denotes a tweet, each edge denotes a criticism relationship between a pair of tweets of the discussion and the weight of nodes models the social relevance of tweets from data obtained from Twitter. In Twitter, a tweet always answers or refers to previous tweets in the discussion, so the obtained underlying argument graph is acyclic. Moreover, when constructing relationships between tweets from informal descriptions expressed in natural language with other attributes such as emoticons, jargon, onomatopoeia and abbreviations, it is often evident that there is uncertainty about whether some of the criticism relationships actually hold. So, to take into account this fact in the model, each edge of an argument graph is associated with a probability value, quantifying such uncertainty on criticism relationships between pairs of tweets. The solution of a weighted argument graph for a Twitter discussion is computed by means of the reasoning system we developed in [1], where the graph is mapped to a valued abstract argumentation framework (VAF) [3] and the so-called ideal semantics [9] is used to evaluate the set of socially accepted tweets in a discussion from the weights assigned to the tweets and the criticism relationships between them.

In this work we introduce a natural extension of our previous representation model for Twitter discussions [2], that will be called *probabilistic author-centered model*. In this new model, tweets within a discussion are grouped by authors, such that tweets of a same author describe his/her opinion in the discussion that is represented by a single node in the graph, and criticism relationships denote controversies between the opinions of Twitter users in the discussion. In this model, the interactions between authors can give rise to circular criticism relationships, and the probability of one opinion criticizing another is evaluated from the individual probabilities of criticism among the tweets that compose both opinions. So, the underlying argument graph can contain cycles and a model for the aggregation of probabilities has to be proposed. Moreover, to compute the set of accepted authors' opinions in a discussion, we also extend our previous reasoning system [1] which is based on the acceptance of tweets of a discussion and not on its authors. This new representation and reasoning model can be of special relevance for assessing Twitter discussions in fields where identifying groups of authors whose opinions are globally compatible or consistent is of particular interest.

The rest of the paper is organized as follows. In Sect. 2, we recall from [2] the formal graph structure to model Twitter discussions. Then, in Sect. 3, we describe the author-centered model for representing discussions in Twitter and, in Sect. 4, we formalize the probabilistic weighting scheme of criticism relationships between authors' opinions. Finally, in Sect. 5 we define the reasoning system to compute the sets of accepted and rejected opinions and, in Sect. 6, we conclude.

2   Twitter Discussion Graph

In this section, we introduce a simplified computational structure of the one proposed in [2] to represent a Twitter discussion with probabilistic valued relationships, that will be called *probabilistic discussion graph*. In such a graph, each node will denote a tweet, each edge will denote an answer relationship between a pair of tweets of the discussion, and each edge will be attached a probability value, indicating the probability that a criticism relationships between the pair of tweets actually holds. We provide more formal definitions next.

Definition 1 *(Twitter Discussion). A Twitter discussion Γ is a non-empty set of tweets. A tweet $t \in \Gamma$ is a triple $t = (m, a, f)$, where m is the up to 140 characters long message of the tweet, a is the author's identifier of the tweet and $f \in \mathbb{N}$ is the number of followers of the author, according to its temporal instant generation during the discussion. Moreover, if t_1 and t_2 are tweets from different authors, We say that t_1 answers t_2 iff t_1 is a reply to the tweet t_2 or t_1 mentions (refers to) tweet t_2.*

Definition 2 *(Discussion Graph). The Discussion Graph (DisG) for a Twitter discussion Γ is the directed graph (T, E) such that for every tweet in Γ there is a node in T and if tweet t_1 answers tweet t_2 there is a directed edge (t_1, t_2) in E. Only the nodes and edges obtained by applying this process belong to T and E, respectively.*

Definition 3 *(Probabilistic Discussion Graph). A probabilistic discussion graph (PDisG) for a Twitter discussion Γ is a triple $\langle T, E, P \rangle$, where*

- *(T, E) is the DisG graph for Γ and*
- *P is a labeling function $P : E \to [0, 1]$ that attaches a probability value $p \in [0, 1]$ to every edge $(t_1, t_2) \in E$, meant as the degree of belief that tweet t_1 is a*

criticism to tweet t_2, i.e. that the message of t_1 does not agree with the claim expressed in the message of t_2. So, $p = 1$ means that it is fully believed that tweet t_1 disagrees with the claim expressed in tweet t_2, while $p = 0$ means that it is fully believed that tweet t_1 agrees with the claim expressed in tweet t_2.

Given a PDisG $\langle T, E, P \rangle$ for a Twitter discussion Γ and two tweets $t_1, t_2 \in \Gamma$, we will say that t_1 *criticizes* t_2, written $t_1 \rightsquigarrow t_2$, iff t_1 answers t_2 and the degree of belief that the message of tweet t_1 is a criticism to the message of tweet t_2 is greater than zero. In other words, $t_1 \rightsquigarrow t_2$ iff $(t_1, t_2) \in E$ and $P(t_1, t_2) > 0$.

In Twitter, every tweet in a discussion can reply at most one tweet, but can mention many tweets, and all of them are prior in the discussion. So, every tweet can answer (and criticize) many prior tweets, either from a same author or from different ones. Given a tweet t_1, we consider the set of tweets $\{t_{1_{a_1}}, \ldots, t_{1_{a_n}}\}$ that t_1 is answering to as those tweets including (i) the tweet that t_1 is replying to, and (ii) all the other previous tweets in the discussion by authors mentioned by t_1.

To check whether a tweet t_1 does not agree with the claim expressed in one of its answered tweets $t_{1_{a_i}}$, the system uses an automatic labeling system based on Support Vector Machines (SVM). The description of the method we used to train the SVM can be found in [2]. The SVM model is built from a set of 582 pairs of tweets (answers) obtained from a discussion set on Spanish politics, and manually labeled with the most probable label: criticism or not criticism. To build the SVM model, for each pair of tweets $(t_1, t_{1_{a_i}})$ we consider different attributes from the tweets of the pair: attributes that count the number of occurrences of relevant words in the tweets and attributes that have to be computed from the message. In particular, for each tweet, we have considered regular words and stop-words, the number of images, the number of URLs mentioned in the tweet, the number of positive and negative emoticons and the sentiment expressed by the tweet. We use LibSVM [7] to train a probabilistic SVM model, that is, a labeling function that assigns a probability value p for each possible label to each answer $(t_1, t_{1_{a_i}})$. The probability estimates can be obtained by using Platt's likelihood method [15]. LibSVM uses the same Platt's method but algorithmically improved [14]. With our SVM model for Spanish politics discussions, we obtain an accuracy of 75% over our training set of tweet pairs. This SVM model, obtained from such small data set, may not be good enough to be used in a final system, but one can always consider training a SVM model with a larger data set.

In Fig. 1 we show the PDisG for a Twitter discussion[1] from the political domain obtained by our discussion retrieval system. Each tweet is represented as a node and each criticism relationship between tweets is represented as an edge (answers with probability values greater than zero). The root tweet of the discussion is labeled with 0 and the other tweets are labeled with consecutive identifiers according to their generation order. The discussion has a simple structure. The root tweet starts the discussion (node 0), the reply (node 1) criticizes

---

[1] The discussion URL is
   https://twitter.com/jordievole/status/574324656905281538.

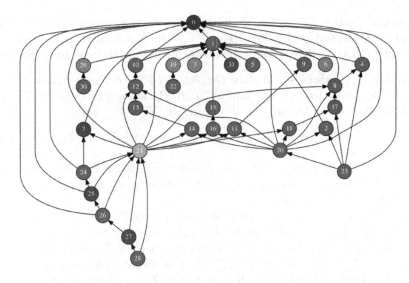

Fig. 1. Tweet-based model for a Twitter discussion. (Color figure online)

the root tweet and the rest of tweets within the discussion criticize mainly node 0 and node 1. The discussion contains 32 tweets of 14 different authors, and 81 criticizes relations between tweets. Nodes are colored in *blue scale*, where the darkness of the color is directly proportional to the number of followers of the authors of the tweets with respect to the maximum value in the discussion. Notice that the graph does not contain cycles, since a tweet only answers previous tweets in the discussion.

3   Author-Centered Model

As we have already pointed out, our goal is to introduce and investigate an author-centered model of Twitter discussions with probabilistic valued relationships. To this end, we group tweets by authors and we consider that criticism relationships between tweets denote controversies at the level of authors.

In this work we consider discussions in which every author's opinion is consistent, discussions in which authors are not self-referenced and do not contradict themselves. That is, for each author a_i and each pair of tweets $t_1 = (m_1, a_i, f_1)$ and $t_2 = (m_2, a_i, f_2)$, we assume that messages m_1 and m_2 do not express neither conflicting nor inconsistent information. Next we define what we will understand by the opinion and the number of followers of an author in a Twitter discussion Γ (with authors' identifiers $\{a_1, \ldots, a_n\}$):

- The opinion of an author a_i in the discussion Γ, denoted T_{a_i}, is the set of tweets of a_i in Γ, i.e. $T_{a_i} = \{(m, a_i, f) \in \Gamma\}$.
- The number of followers of an author a_i in Γ, denoted $f_{a_i} \in \mathbb{N}$, is the mode of the set $\{f \mid (m, a_i, f) \in \Gamma\}$, which provides us with the most frequent number of followers of the author during the discussion.

Given a Twitter discussion, we notice that, in fact, every author a_i can be uniquely represented by his/her opinion T_{a_i}. So, we shall refer to both terms indistinctly. Next we define the probabilistic author graph for a given discussion.

Definition 4 *(Probabilistic Author Graph). Let Γ be a Twitter discussion with authors' identifiers $\{a_1, \ldots, a_n\}$ and let $\langle T, E, P \rangle$ be the PDisG for Γ. The probabilistic author graph (ADisG) for Γ is a triple $\langle \mathcal{T}, \mathcal{E}, \mathcal{P} \rangle$, where*

- *the set of nodes $\mathcal{T}$ is the set of authors' opinions $\{T_{a_1}, \ldots, T_{a_n}\}$, i.e. a node for each author.*
- *the set of edges $\mathcal{E}$ is the set of answers between different authors in the discussion; i.e. there is an edge $(T_{a_i}, T_{a_j}) \in \mathcal{E}$, with $a_i \neq a_j$, iff there is $(t_1, t_2) \in E$ such that $t_1 \in T_{a_i}$ and $t_2 \in T_{a_j}$.*
- *$\mathcal{P}$ is a probabilistic weighting scheme, i.e. a map $\mathcal{P} : \mathcal{E} \to [0,1]$ assigning to every edge $(T_{a_i}, T_{a_j}) \in \mathcal{E}$ a probability value in $[0,1]$, that expresses a degree of belief with which the author a_i actually criticizes the author a_j. For each edge $(T_{a_i}, T_{a_j}) \in \mathcal{E}$, the value $\mathcal{P}(T_{a_i}, T_{a_j})$ is meant to be computed from the set of individual probabilities that tweets in T_{a_i} criticize tweets in T_{a_j}, i.e. from the set*

$$\{P(t_1, t_2) \mid (t_1, t_2) \in E, t_1 \in T_{a_i} \text{ and } t_2 \in T_{a_j}\}.$$

Note that an author can answer several authors in a discussion, and thus criticize several authors. However, if an author criticizes the opinion of another through several tweets, the set of discrepancies is represented with a single edge in $\mathcal{E}$ and with a single probability value, denoting the global belief that one opinion criticizes the other.

The ADisG graph shows discrepancies between authors only if there is some (explicit) criticism relationship between the tweets of the authors, and thus, indirect criticism relations between authors have not been considered yet in our model. For instance, consider a Twitter discussion with tweets $t_1 = (m_1, a_1, f_1)$, $t_2 = (m_2, a_2, f_2)$ and $t_3 = (m_3, a_3, f_3)$, with $a_1 \neq a_2 \neq a_3$. Suppose that $t_1 \rightsquigarrow t_2$ and $t_3 \rightsquigarrow t_1$ i.e. $\{(t_1, t_2), (t_3, t_1)\} \subseteq E$, $P(t_1, t_2) > 0$ and $P(t_3, t_1) > 0$. In our current approach, we restrict ourselves to consider that $t_3 \rightsquigarrow t_2$ iff t_3 answers (replies or mentions) t_2. The reason is that the information contained in a typical tweet, written in natural language and with possibly other attributes, almost never allows us to consider a sound way to assess an indirect criticism relation between two tweets t and t' if t' does not directly reply or mention t.

In the next section we introduce three different probabilistic weighting schemes.

4   Probabilistic Weighting Schemes

In our approach, each node of an ADisG graph denotes an author's opinion, and relationships between nodes are mined from the prevailing sentiment among the aggregated tweets of the opinions. To be more precise, let Γ be a Twitter discussion and let $\langle \mathcal{T}, \mathcal{E}, \mathcal{P} \rangle$ be the probabilistic author graph (ADisG) for Γ. Suppose further we have two authors' opinions or sets of authors' tweets $T_a, T_b \in \mathcal{T}$, with $(T_a, T_b) \in \mathcal{E}$. Our aim is to define a probabilistic weighting scheme $\mathcal{P} : \mathcal{E} \to [0, 1]$ for edges in $\mathcal{E}$, by combining in an appropriate form the individual probabilities values $\{P(t_1, t_2) \mid t_1 \in T_a \text{ and } t_2 \in T_b\}$, where we consider $P(t_1, t_2) = 0$ for pairs of tweets such that $(t_1, t_2) \notin E$. As we will see, the addition of zero values to this set will be harmless.

In the rest of this section we define three possible probabilistic weighting schemes $\mathcal{P}$, depending on the semantics assumed for the criticism relationship between the authors' opinions T_a and T_b.

4.1 Skeptical Scheme

A skeptical notion of criticism between T_a and T_b can be defined as follows: T_a criticizes T_b, written $T_a \rightsquigarrow T_b$, when every tweet in T_b is attacked by some tweet in T_a, i.e. for all $t \in T_b$, there is $t' \in T_a$ such that $t' \rightsquigarrow t$. In logical terms, we can define $T_a \rightsquigarrow T_b$ by the following clause:

$$T_a \rightsquigarrow T_b := \bigwedge_{t \in T_b} \left(\bigvee_{t' \in T_a} t' \rightsquigarrow t \right)$$

Assuming independence of all the $t' \rightsquigarrow t$'s, which is a reasonable assumption in our context,[2] we can easily compute the probability of $T_a \rightsquigarrow T_b$ as

$$\mathcal{P}(T_a \rightsquigarrow T_b) = \prod_{t \in T_b} \left(\bigoplus_{t' \in T_a} P(t', t) \right),$$

where $\oplus$ corresponds to the probabilistic sum operation $x \oplus y = x + y - x \cdot y$. Observe that 0 is a neutral element for $\oplus$ (i.e. $x \oplus 0 = x$), and so having probability values such that $P(t', t) = 0$ does not affect the computation of $\mathcal{P}(T_a \rightsquigarrow T_b)$. Analogously for the next schemes.

4.2 Credulous Scheme

On the other hand, a credulous notion of criticism between T_a and T_b can be defined as follows: T_a criticizes T_b, written $T_a \rightsquigarrow^c T_b$, when there is at least one tweet $t \in T_b$ that is attacked by a tweet $t' \in T_a$, i.e. when there are $t \in T_b$ and $t' \in T_a$ such that $t' \rightsquigarrow t$. In logical terms, $T_a \rightsquigarrow^c T_b$ can be now expressed by the following clause:

$$T_a \rightsquigarrow^c T_b := \bigvee_{t \in T_b} \left(\bigvee_{t' \in T_a} t' \rightsquigarrow t \right).$$

---

[2] This is because in our probabilistic model the label $P(t_1, t_2)$ assigned to an edge (t_1, t_2) is based only on the information inside the tweets t_1 and t_2 and not on other answers from the same authors.

Again, assuming independence of all the $t' \rightsquigarrow t$'s, we can easily compute the probability of $T_a \rightsquigarrow T_b$ as

$$P(T_a \rightsquigarrow^c T_b) = \bigoplus_{t' \in T_a, t \in T_b} P(t', t).$$

4.3   Intermediate Scheme

A more flexible definition of when T_a criticizes T_b is to stipulate that this holds when for *most* of the tweets $t \in T_b$ there is a tweet $t' \in T_a$ such that $t \rightsquigarrow t'$. We denote this notion of attack as $T_a \rightsquigarrow_{most} T_b$.

The question is how we interpret the quantifier *most*. A first option is to understand *most* as a proportion of at least r, for some $r \geq 0.5$ to be chosen. For any set X, let us define $most(X) = \{S \subseteq X \mid \frac{|S|}{|X|} \geq r\}$. Then we can express $T_a \rightsquigarrow_{most} T_b$ as follows:

$$T_a \rightsquigarrow_{most} T_b := \bigvee_{S \in most(T_b)} T_a \rightsquigarrow S.$$

But we can simplify a bit this expression. Indeed, since if $S \subset R$ then $(T_a \rightsquigarrow S) \vee (T_a \rightsquigarrow R) = T_a \rightsquigarrow S$, we can write

$$T_a \rightsquigarrow_{most} T_b := \bigvee_{S \in Min(most(T_b))} T_a \rightsquigarrow S,$$

where $Min(most(X))$ denotes the minimal subsets of X with a proportion of at least r. Then, we can compute:

$$P(T_a \rightsquigarrow_{most} T_b) = P(\bigvee \{T_a \rightsquigarrow S : S \in Min(most(T_b))\}).$$

This can be computationally expensive. However, we can provide a lower approximation taking into account that for any probability we have $P(A \cup B) \geq \max(P(A), P(B))$:

$$P_*(T_a \rightsquigarrow_{most} T_b) = \max\{P(T_a \rightsquigarrow S) : S \in Min(most(T_b))\}.$$

Interestingly enough, there is a simple procedure to compute P_*:

(i) compute, for all $t \in T_b$, the probabilities $P(T_a \rightsquigarrow t) = \bigoplus_{t' \in T_a} P(t', t)$;
(ii) rank them, from higher to lower: $P(T_a \rightsquigarrow t_1) \geq P(T_a \rightsquigarrow t_2) \geq \ldots$;
(iii) let k be the smallest index such that $\frac{k}{|T_b|} \geq r$.

Then, we have $P_*(T_a \rightsquigarrow_{most} T_b) = \prod_{i=1}^{k} P(T_a \rightsquigarrow t_i)$.

5   Mining the Set of Consistent Opinions

Once we have introduced the author-centered model of discussions in Twitter, the next key component is the definition of the reasoning system to compute the set of accepted authors' opinions. To this end, we have extended the reasoning system developed in [1] to deal here with ADisG graphs. The approach, described in the rest of the section, consists of mapping an ADisG graph, with a particular probabilistic weighting scheme, to a valued abstract argumentation framework (VAF) and considering the ideal semantics to compute the (unique) set of consistent authors' opinions of the discussion. Bench-Capon's valued abstract argumentation [3] is an extension of abstract argumentation with a valuation function Val for arguments taking values on a set R equipped with a (possibly partial) preference relation $Valpref$. Ideal semantics [9] guarantees that all opinions in the solution are consistent and that the solution is maximal in the sense that it contains all acceptable arguments.

5.1   The Argumentation-Based Reasoning System

Given an ADisG for a Twitter discussion with a given probabilistic weighting scheme, we build a corresponding VAF where arguments represent authors' opinions and attacks between arguments represent discrepancies between authors' opinions according to an uncertainty threshold α, which characterizes how much uncertainty on probability values we are ready to tolerate.

Definition 5 *(VAF for an ADisG). Let Γ be a Twitter discussion with authors identifiers $\{a_1, \ldots, a_n\}$ and let $\alpha \in [0,1]$ be a threshold on the probability values. If $G = \langle \mathcal{T}, \mathcal{E}, \mathcal{P} \rangle$ is the ADisG graph for Γ with probabilistic weighting scheme $\mathcal{P}$, the Valued Argumentation Framework for G relative to the threshold α, written $VAF(G, \alpha)$, is the tuple $VAF(G, \alpha) = \langle \mathcal{T}, \text{attacks}, R, Val, Valpref \rangle$, where*

- *each node (or author's opinion) T_{a_i} in $\mathcal{T}$ results in an argument,*
- *attacks is an irreflexive binary relation on $\mathcal{T}$ and it is defined according to the threshold α as follows: $\text{attacks} = \{(T_{a_i}, T_{a_j}) \in \mathcal{E} \mid \mathcal{P}(T_{a_i}, T_{a_j}) \geq \alpha\}$,*
- *R is a non-empty set of relevance values,*
- *$Valpref \subseteq R \times R$ is an order relation (transitive, irreflexive and asymmetric) on the set of relevance values R.*
- *$Val : \mathcal{T} \to R$ is a valuation function that assigns relevance values to authors' opinions or arguments,*

An important element of our approach is the use of an uncertainty threshold α. It represents the maximum probability value under which we would be prepared to disregard criticism relationships between authors' opinions. So, the `attacks` relation is interpreted as follows: the opinion of the author a_i is in disagreement with the opinion of the author a_j with at least a probability value α, according to the probabilistic weighting scheme $\mathcal{P}$.

Given such a $VAF(G, \alpha) = \langle \mathcal{T}, \text{attacks}, R, Val, Valpref \rangle$, a *defeat* relation (or effective attack relation) between arguments (authors' opinions) is defined according to the valuation function *Val* and the preference relation *Valpref* as follows:

$$defeats = \{(T_{a_i}, T_{a_j}) \in \texttt{attacks} \mid (\mathit{Val}(T_{a_j}), \mathit{Val}(T_{a_i})) \notin \mathit{Valpref}\}.$$

As we have already pointed out, we consider the ideal semantics for computing the set of consistent authors' opinions of a discussion. The ideal semantics for valued argumentation is defined through the ideal extension (solution) which guarantees that the set of tweets in the solution is the maximal set of tweets that is consistent, in the sense that there are no defeaters among them, and all the tweets outside the solution are defeated by a tweet within the solution. That is, if a tweet outside the solution defeats a tweet within the solution, it is, in turn, defeated by another tweet within the solution. In other words, the solution is the biggest consistent set of tweets that defeats any defeater outside the solution. In [9] the authors prove that the ideal extension is unique.

Formally, given a $\mathrm{VAF}(G, \alpha) = \langle \mathcal{T}, \texttt{attacks}, R, Val, Valpref \rangle$, a set of arguments $S \subseteq \mathcal{T}$ is *conflict-free* iff for all $T_{a_i}, T_{a_j} \in S, (T_{a_i}, T_{a_j}) \notin defeats$. Given a conflict-free set of arguments $S \subseteq \mathcal{T}$, S is *maximally admissible* iff

(i) for all $T_{a_1} \notin S$, $S \cup \{T_{a_1}\}$ is not conflict-free and
(ii) for all $T_{a_1} \notin S$ and $T_{a_2} \in S$, if $(T_{a_1}, T_{a_2}) \in defeats$, there exists $T_{a_3} \in S$ such that $(T_{a_3}, T_{a_1}) \in defeats$.

Accordingly, we define what the solution of a discussion Γ is as follows.

Definition 6 *(Solution of a discussion). Given the* ADisG *graph $G = \langle \mathcal{T}, \mathcal{E}, \mathcal{P} \rangle$ for a discussion Γ and a probabilistic weighting scheme $\mathcal{P}$, the set of accepted authors' opinions of Γ for given a threshold α, or solution of Γ, is the largest admissible conflict-free set of authors' opinions $S \subseteq \{T_{a_1}, \ldots, T_{a_n}\}$ in the intersection of all maximally admissible conflict-free sets in the valued argumentation framework $VAF(G, \alpha)$.*

5.2 Implementation and Analysis of Results

As for the implementation purposes, we have instantiated the set of relevance values R to the set of natural numbers $\mathbb{N}$, and the preference relation *Valpref* to the natural order on $\mathbb{N}$. We have also instantiated the valuation function *Val* to the function $\texttt{followers} : \mathcal{T} \to \mathbb{N}$, with $\texttt{followers}(T_{a_i}) = \lfloor \log_{10}(f_{a_i} + 1) \rfloor$, where $f_{a_i} \in \mathbb{N}$ is the number of followers of the author a_i computed as the mode of the set $\{f \mid (m, a_i, f) \in T_{a_i}\}$ (i.e. the most frequent number of followers of the author during the discussion). This function allows us to quantify authors' relevance from the orders of magnitude of authors' followers, since we want to consider that one author is more relevant than another only if the number of followers is at least ten times bigger for the first author.

To implement the reasoning system, we have used the Answer Set Programming (ASP) approach of the argumentation system ASPARTIX [11]. Actually, we have extended ASPARTIX to deal with VAFs, as the current implementation only works with non-valued arguments. To develop such extension we have modified the manifold ASP program described in [12] to incorporate the valuation function for arguments and the preference relation.

The author-centered approach allows us to perform an analysis of results different from the tweet-based approach proposed in [1]. Aggregating the information by author allows us to identify the set of authors whose opinions are consistent or in agreement in the discussion, the authors involved in a circular argumentative discussion, and the most controversial authors. That is, for instance, we can look for the authors who receive the greatest number of criticisms, the authors who participate in the greatest number of cycles, or the authors that generate the longest argumentative chains.

Figure 2 shows the solution for an ADisG graph instance for the discussion of Fig. 1. To build the ADisG graph, we have used the intermediate probabilistic weighting scheme $\mathcal{P}_*(T_{a_i} \leadsto_{most} T_{a_j})$ with the proportion parameter $r = 0.6$.[3] To find the solution for the ADisG graph (the set of accepted opinions of the discussion according to Definition 6), we have used the uncertainty threshold $\alpha = 0.6$ and the above `followers` valuation function for estimating the authors' relevance in Twitter. According to it, the authors of the discussion are stratified in five levels denoting their relevance, namely: level 0 (lowest level): {11}, level 1: {5, 6, 7, 13}, level 2: {0, 1, 3, 4, 8, 9, 10}, level 3: {12} and level 4: {2}.

The nodes colored in blue are the accepted authors (authors' opinions in the solution) and the nodes colored in gray are the rejected ones, where the darkness of the color is directly proportional to the value of the `followers` function of each author. The edges colored in black are the answers between authors that cannot be classified as attacks, since the criticism probabilities are below the threshold $\alpha = 0.6$, while the edges colored in red are attacks between authors; i.e. answers with a criticism probability of at least the threshold $\alpha = 0.6$. For attack edges, the darkness of the color is directly proportional to the criticism probability with respect to the maximum value. With $r = 0.6$ and $\alpha = 0.6$, 11 answers between authors do not give rise to attacks. The ADisG graph has 13

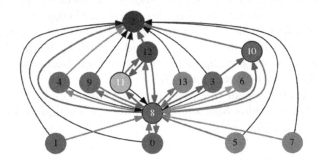

Fig. 2. Author-centered model and its solution. (Color figure online)

---

[3] We plan to implement the other weighting schemes in the near future.

cycles considering all answers among authors, and Authors 8 and 2 seem to be the most controversial ones.

The *solution* contains 11 of the 14 authors and only 3 are rejected (Authors 8, 10 and 11). On the one hand, Author 2 is the owner of the root tweet of the conversation (node 0 in the tweet-based model of Fig. 1), and a total of four other authors (4, 9, 10 and 12) attack him, but he in turn does not reply later to the rest of tweets of the conversation. So, Author 2 is not involved in any cycle in the ADisG graph. Because the weight of Author 2 is greater than the one of any of his attacking authors, Author 2 belongs to the solution of the graph. With respect to the four attackers of Author 2, two of them (4 and 12) are also in the solution, since Author 12 does not defeat Author 2 and his weight is greater than the one of any of his attacking authors. On the other hand, Author 12 defeats Author 8 and this allows Authors 3, 4 and 9 to be in the solution, while in turn, accepting Author 3 causes Author 10 to be rejected. When analyzing the cycles of the graph, we obtain that Author 8 is involved in a total of 8 cycles, considering only attacks answers among authors, and almost all authors involved in cycles with Author 8 are in the solution (0, 6, 13 , 9 and 12). Thus, Author 8 produces a lot of circular discussions, but the weight of Author 12 is high enough to make Author 8 lose the discussion. Observe that in the ideal semantics, authors with a same weight that form a cycle are not accepted if none of the authors in the cycle is attacked by other authors outside of the cycle and accepted in the solution. Hence, in this discussion with high controversy around Author 8 (with a high number of cycles), we end up accepting many of these authors' opinions. Finally, as Authors 1, 5 and 7 only attack Author 8, all of them are also in the solution, while Author 11 is rejected, since it is defeated by Author 12.

6   Conclusions and Future Work

In this paper we have introduced first ideas on a probabilistic author-centered approach to analyze the set of accepted authors' opinions in Twitter discussions. We model discussions with a graph, where nodes represent whole sets of tweets of a single author, and thus representing his opinion, and edges represent criticism relationships between authors. Then, using valued abstract argumentation and ideal semantics, we compute the set of winning authors in the discussion. By comparing the set of accepted opinions with the rejected ones, we can detect the degree of polarization between both sets.

As future work, we plan to extend the author-centered model to also consider support relationships between tweets and also to explore more credulous acceptability semantics.

References

1. Alsinet, T., Argelich, J., Béjar, R., Fernández, C., Mateu, C., Planes, J.: Weighted argumentation for analysis of discussions in Twitter. Int. J. Approx. Reason. **85**, 21–35 (2017). https://doi.org/10.1016/j.ijar.2017.02.004
2. Alsinet, T., Argelich, J., Béjar, R., Fernández, C., Mateu, C., Planes, J.: An argumentative approach for discovering relevant opinions in Twitter with probabilistic valued relationships. Pattern Recognit. Lett. **105**, 191–199 (2018). https://doi.org/10.1016/j.patrec.2017.07.004
3. Bench-Capon, T.J.M.: Value-based argumentation frameworks. In: Proceedings of 9th International Workshop on Non-Monotonic Reasoning, NMR 2002, pp. 443–454 (2002)
4. Bench-Capon, T.J.M., Dunne, P.E.: Argumentation in artificial intelligence. Artif. Intell. **171**(10–15), 619–641 (2007). https://doi.org/10.1016/j.artint.2007.05.001
5. Besnard, P., Hunter, A.: A logic-based theory of deductive arguments. Artif. Intell. **128**(1–2), 203–235 (2001). https://doi.org/10.1016/S0004-3702(01)00071-6
6. Bosc, T., Cabrio, E., Villata, S.: Tweeties squabbling: positive and negative results in applying argument mining on social media. In: Computational Models of Argument - Proceedings of COMMA 2016, pp. 21–32 (2016). https://doi.org/10.3233/978-1-61499-686-6-21
7. Chang, C., Lin, C.: LIBSVM: A library for support vector machines. ACM TIST **2**(3), 27:1–27:27 (2011)
8. Dung, P.M.: On the acceptability of arguments and its fundamental role in non-monotonic reasoning, logic programming and n-person games. Artif. Intell. **77**(2), 321–357 (1995). https://doi.org/10.1016/0004-3702(94)00041-X
9. Dung, P.M., Mancarella, P., Toni, F.: Computing ideal sceptical argumentation. Artif. Intell. **171**(10–15), 642–674 (2007). https://doi.org/10.1016/j.artint.2007.05.003
10. Dusmanu, M., Cabrio, E., Villata, S.: Argument mining on twitter: arguments, facts and sources. In: Proceedings of the 2017 Conference on Empirical Methods in Natural Language Processing, EMNLP 2017, pp. 2317–2322 (2017)
11. Egly, U., Gaggl, S.A., Woltran, S.: ASPARTIX: Implementing argumentation frameworks using answer-set programming. In: Garcia de la Banda, M., Pontelli, E. (eds.) ICLP 2008. LNCS, vol. 5366, pp. 734–738. Springer, Heidelberg (2008). https://doi.org/10.1007/978-3-540-89982-2_67
12. Faber, W., Woltran, S.: Manifold answer-set programs for meta-reasoning. In: Erdem, E., Lin, F., Schaub, T. (eds.) LPNMR 2009. LNCS (LNAI), vol. 5753, pp. 115–128. Springer, Heidelberg (2009). https://doi.org/10.1007/978-3-642-04238-6_12
13. Grosse, K., González, M.P., Chesñevar, C.I., Maguitman, A.G.: Integrating argumentation and sentiment analysis for mining opinions from Twitter. AI Commun. **28**(3), 387–401 (2015). https://doi.org/10.1016/j.artint.2007.05.003
14. Lin, H.T., Lin, C.J., Weng, R.C.: A note on Platt's probabilistic outputs for support vector machines. Mach. Learn. **68**(3), 267–276 (2007). https://doi.org/10.1007/s10994-007-5018-6
15. Platt, J.C.: Probabilistic outputs for support vector machines and comparisons to regularized likelihood methods. In: Advances in Large Margin Classifiers, pp. 61–74. MIT Press, Cambridge (1999)
16. Simari, G.R., Rahwan, I.: Argumentation in Artificial Intelligence. Springer, New York (2009). https://doi.org/10.1007/978-0-387-98197-0

A Concept-Based Text Analysis Approach Using Knowledge Graph

Wenny Hojas-Mazo[1], Alfredo Simón-Cuevas[1] ⓘ,
Manuel de la Iglesia Campos[1] ⓘ, Francisco P. Romero[2(✉)] ⓘ,
and José A. Olivas[2] ⓘ

[1] Universidad Tecnológica de La Habana José Antonio Echeverría, Cujae,
La Habana, Cuba
{Whojas,Asimon,miglesia}@ceis.cujae.edu.cu
[2] Universidad de Castilla-La Mancha, Ciudad Real, Spain
{FranciscoP.Romero,JoseAngel.olivas}@uclm.es

Abstract. The large amounts and growing of unstructured texts, available in
Internet and other scenarios, are becoming a very valuable resource of information and knowledge. The present work describes a concept-based text analysis approach, based on the use of a knowledge graph for structuring the texts
content and a query language for retrieving relevant information and obtaining
knowledge from the knowledge graph automatically generated. In the querying
process, a semantic analysis method is applied for searching and integrating the
conceptual structures from the knowledge graph, which is supported by a disambiguation algorithm and WordNet. The applicability of the proposed
approach was evaluated in the analysis of scientific articles from a Systematic
Literature Review and the results were contrasted with the conclusions obtained
by the authors of this review.

Keywords: Computational text analysis · Graph-based text representation
Knowledge graph querying · Semantic processing

1 Introduction and Background

The large amounts and growing of unstructured texts, available in Internet and other
information-centric application scenarios, are becoming a very valuable resource of
information and knowledge. The effective processing and analysis of those text data
sources to obtain the relevant information and knowledge has been an important
challenge, due to this, the problem of text mining has gained increasing attention in
recent years [2]. In order to address this challenge, the text data has been treated and
processed using different representation levels, in most applications as bag-of-words or
vector-space model, such as in information retrieval system. However, the information
retrieval systems have traditionally focused more on facilitating information access
rather than analyzing information to discover patterns and knowledge, which is the
primary goal of text mining and the concept-based text analysis as specific task. In this
sense, the graph-based text representation is emerging as a promising direction toward
analyzing and exploiting the text structure [13], for example, the conceptual structure

© Springer International Publishing AG, part of Springer Nature 2018
J. Medina et al. (Eds.): IPMU 2018, CCIS 854, pp. 696–708, 2018.
https://doi.org/10.1007/978-3-319-91476-3_57

underlying. The use of graph-based text representation avoids the loss of structural and semantic information and reduced the text data scattering ratio. Once a text is represented as a graph, a variety of tools for graph analysis can be applied to perform quantitative and qualitive analysis of concepts, detecting closely contextually related concepts, identifying the key concepts that produce meaning, and perform integration of several text contents.

The useful of graph models in different text processing task, such as: topic labeling and detection [7, 12], text clustering [1], information retrieval [15, 18], text recommendation [5], and representation of linguistic information [24], have been reported. Several graph models applied to text representation, features and construction methods, have been reviewed in [7, 8, 11, 30]. However, the Concept Maps (CM) [23] is another graph model used to obtain the conceptual structure of a text [19, 25, 31], but very little exploited in the computational analysis of texts content. The CM is a graph-based knowledge representation, composed of concepts and labeled relationship between them that form propositions. The CM is a very useful and intuitive knowledge representation for capturing, representing and organizing the most significant of a topic and a set of conceptual meanings through of propositional structures [23]. Text analytics methods for extracting meaningful keywords and concepts facilitate content analysis, applying various technologies for capturing, processing, analyzing, and visualizing the immense volume, and variety of unstructured data from multiple textual sources [27]. Through CM, the large bodies of text are reduced to a relatively small number of concepts and relationship between them, so that a large corpus can be easily managed and understood via automatic concept mapping.

In this work, a Concept-Based Text Analysis Model (CTAM) is proposed. CTAM is based on the use of CM for structuring the texts content and a query language to retrieve relevant information and obtain knowledge from the constructed CM. CTAM is composed of two fundamental processes: *automatic concept mapping*, and *concept maps querying*. In the first process, a CM is automatically constructed from each text included in the texts collection to be analyzed, using a method based on the reported in [25]. The generated CM are stored in a *CM repository* (CMR). In the second process, an improved version of CMQL (Concept Maps Query Language) [28] is proposed for querying the CMR. CMQL provides the formalization of a set of different types of queries (*union, intersection, sub-map*, and *extension*) for exploring and mining a CMR from different perspectives, and offers more diversity of queries than the reported in [30]. In the search and integration tasks included in the query processing of CMQL, only syntactic aspects have been considered in the similarity analysis of the concepts. This constitutes a weakness due to the knowledge in CM is expressed in natural language, and several problems can be emerged due to the possible ambiguity of the concepts. For example, the not retrieval of useful and interesting information associated to concepts that are not syntactically similar to the included ones in the query, although they can be semantically similar, and the obtain of not appropriate results in the integration of associated information to semantically different concepts. To solve this weakness a semantic analysis method was included in this proposed approach, which is supported in a disambiguation algorithm and WordNet [20]. The word sense disambiguation in unstructured texts has been broadly studied, but there are few works approaching this problem in the CM context [6, 29]. The method reported in [29] improve the disambiguation results respect the reported in [6], through the use of heuristics based on *domain* (according to [4]), *context* and *gloss* and extending the context analysis process of the CM with other relations from WordNet. Nevertheless, the sequential application of these heuristics constitutes a limitation because the sense of the concept was determined according one of them and not taking advantage of the combination of the results obtained from each one. In this sense, a new disambiguation algorithm, based on [29], in which the results obtained for each heuristic are combined (inspired in [21]) to determine the more appropriate sense of the concept is also proposed.

The applicability of the CTAM was evaluated through the development of a case of study, in which 11 scientific articles from the Systematic Literature Review (SLR) reported in [9], were analyzed. The SLR is a means of identifying, evaluating and interpreting all available research relevant to a particular research question, or topic area, or phenomenon of interest [16]. The proposed CTAM offers a new approach of computational support to the analysis phase in the SLR context. In this case of study, different queries were carried out for analyzing the conceptual contents of those articles and to identifying relevant information that facilitated to obtain answer to the research questions outlined in that review. The results obtained were contrasted with the results and conclusions obtained by the authors of the reported review [9].

The rest of the paper is organized as follows: Sect. 2 describes the proposed Concept-Based Text Analysis Model and the defined processes; Sect. 3 presents the results of the developed case of study and the analysis carried out; and conclusions arrived and future works are given in Sect. 4.

2 Concept-Based Texts Analysis Model

CTAM is based on the use of the CM, which are automatically constructed from the texts included in a text collection and stored in a CMR, and the use of different types of queries defined in CMQL [28]. CMQL is applied to retrieve relevant information and obtain knowledge from the CMR. In this sense, two processes were defined: *automatic concept mapping*, and *concept maps querying*. Besides, a semantic analysis method to be applied for searching and integrating information in query processing from CMR, supported by a disambiguation algorithm and WordNet, was include in the last one process. An overview of the proposed approach is shown in Fig. 1.

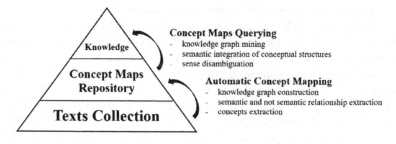

Fig. 1. Graphic overview of the proposed CTAM

2.1 Automatic Concept Mapping from Texts

In the proposed model, the automatic concept mapping is the process in which a CM is automatically constructed from each text included in the texts collection, which are stored in a CMR. This process is carried out through a method based on [25]. The method is conceived in three phases: *preprocessing, concepts extraction,* and *relationships extraction.* In the *preprocessing,* the text is segmented into sentences, and Freeling is used for obtaining the syntactic and grammatical information from the sentences. Next, several tasks are performed on each sentence, such as: tokens extraction, the morpho-syntactic and dependency analysis and the identification of named entities.

The *concepts extraction* phase is based on the identification of simple words or phrases (set of words), that by their composition can constitute a concept, through a set of lexical-syntactic patterns that have been defined for English and Spanish language [25]. The identification of concepts from external knowledge source, such as ontologies, is also included. A concept extracted list is obtained as result. The *relationships extraction* phase allows identifying explicit and implicit links between the previously extracted concepts using the information contained in the text, as well as that represented in an external knowledge source. The explicit relationships are extracted from each sentence using some lexical-syntactic patterns defined for this purpose [25]. The taxonomic relationships (implicit relationships) are extracted from BabelNet [22], and applying the *string matching technic* [14]. Other implicit relationships are extracted evaluating the proximity between two concepts in the text, according to [26]. The extraction of implicit relationships allows linking concepts that are not in the same sentence, which can be a useful information in contextual analysis task. Next, a refining process for eliminating redundancies or inconsistencies resulting from the application of lexical-syntactic patterns is performed. Finally, the concepts and propositions extracted are integrated in a CM. The combined use of the lexical-syntactic patterns for extracting the concepts and relationship between them, the identification of explicit and implicit links between the concepts and the use external knowledge sources allows to achieve a broad coverage of the textual content in the automatic construction of its conceptual representation (CM).

2.2 Concept Maps Querying Process

The CM mining process for retrieving relevant information and knowledge from the CMR is carried out through several types of queries defined in CMQL [28]. Through the different types of queries the system can retrieve information about concepts and propositions, with the results being shown by means of a CM or knowledge graph which is automatically constructed. The knowledge is produced when the captured concepts from different textual sources are integrated as part of the query results. The definition of the queries in CMQL is based on several mechanisms to filter and integrate concepts and propositional structures. This allows obtaining automatically different knowledge views from the CMR. In each query processing, the concepts and propositions including in the search source (set of selected CM) are processed as independent elements and, at the same time, they can be integrated through a semantic analysis process within the queries execution process (described below). Among the defined queries are: *union (CMUnion)*, *intersection (CMInter)*, and *projection (CMProj)*.

The *union* query allows retrieve the knowledge graph that represents the concepts and the relationships between them extracted from a text collection, as a semantically integrated view of the concepts included in these different texts. The *intersection* query allows retrieve the knowledge graph that represents the common concepts included in the texts and the extracted relationships between them, which are represented in a certain amount of CM from the search source, according to the user interest. The minimum percent of CM in which a concept should be represented in the search source is defined as *support value (SV)*. This numeric value is specified by the user when it executes the query. The *projection* queries allow retrieve the knowledge graph that represents the related concepts to a set of interest concepts (previously defined by the user) and the relationship between them from the search source of the query, considering different approaches and a maximum depth of R in the integrated graph of the search source. Three specifications of *projection* queries were included in this model to obtain different knowledge graph views: (1) considering only input link to the interest concept c; (2) considering only output link from c; and (3) combining both types of links. The first two types of queries are very useful to analyze the authority or centrality levels of c with respect to other related concepts. This is a new approach to the application of Kleinberg's concepts [17] in the conceptual analysis of textual contents. In the case of $R = 0$ (when *SV* is not specified by the user), it is assumed that the interest of the user is to identify if the any interesting concepts are included in the search source and if there are any relationship between them.

A refinement of the mathematical formalization of the queries in CMQL is presented in Table 2. This formalization allows to obtain an abstract model, independent of the implementation language of the queries and the storage format of the CM. Before describing the queries, let us consider the following symbolism in Table 1.

Table 1. Symbolism for mathematical formalization of the queries

Symbols	Definitions
SS	Set of CM $\{CM_1, CM_2, ..., CM_n\}$ defined as the search source of a query
CM^q	Concept map obtained as a result of a query $q/q = \{U, I, Proj\}$
IC	Set of *interest concepts* defined by the user (needed in projection query)
P	Set of propositions $p = (c_o; c_d; lp)$
c_o	Origen concept in a proposition p
c_d	Destiny concept in a proposition p
lp	Linked phrase used for labeled the relationship between two concepts
R	Path length between two concepts in a CM
$INC_{CM}^R(c)$	Set of concepts included in all the paths of length R from the concept c in a CM, considering input link to c (relative to the authority level of c)
$OUTC_{CM}^R(c)$	Set of concepts included in all paths of length R from the concept c in a CM, considering output link from c (relative to the centrality or hub level of c).

Table 2. Mathematical formalization of Concept Maps Query Language

Queries	Mathematical formalization
CMUnion	$CMUnion(SS) = (\cup\ CM_i \mid CM_i \in SS), = (C^U, P^U)$ where $C^U = \mid n > 1$ and $P^U = \mid$ $n > 1$.
CMInter	$CMInter^{SV}(SS) = (\cap\ CM_i \mid CM_i \in SS), CM^I = (C^I, P^I)$ where $C^I = \mid n > 1$ and $P^I = \mid (n > 1, p_j \in P^I,$ and $c_o, c_d \in C^I)$
CMProj	$CMProj^R(SS, IC) \subseteq CMUnion(SS)$ and is defined as:
	1. $CMProj^{R,\ IN}(SS, IC) = CM^{Proj} = (C^{Proj}, P^{Proj})$ where $C^{Proj} \mid (c \in IC,$ $CM^U = CMUnion(SS))$, and $P^P = \mid (p_j = (c_o;\ c_d;\ lp)$ and $c_o, c_d \in C^{Proj}$
	2. $CMProj^{R,\ OUT}(SS, IC) = (C^{Proj}, P^{Proj})$ where $C^{Proj} \mid (c \in IC,$ $CM^U = CMUnion(SS))$, and $P^{Proj} = \mid (p_j = (c_o;\ c_d;\ lp)$ and $c_o, c_d \in C^{Proj}$
	3. $CMProj^R(SS, IC) = CM^{Proj} = CMProj^{R,\ IN}(SS, IC) \cup CMProj^{R,\ OUT}(SS, IC)$

In CMQL [28], the information is recovered through identifying syntactic equivalency between the concepts included in the query and the contents in the search source, and the associated semantics to these concepts is not considered. The same thing happens in the process of integration that is carried out as part of the queries processing. Nevertheless, the concepts and propositions in the CMR can be subjected to ambiguity in many cases, because they are expressed in natural language and the ambiguity is an inherent characteristic of the language. Therefore, the effectiveness in the CM querying process can be limited if a semantic analysis task is not included the queries processing; being the identification of the most rational sense of the concepts an important aspect. In the following section, the semantic analysis method that has been proposed in the CTAM is described.

2.3 Semantic Analysis in the Querying Process

The proposed semantic analysis method is based on a process of semantic extension of concepts, and a set of rules in the integration, search and retrieval tasks included in the CMR querying. In this method, the semantic information associated to the concepts is captured from WordNet and a concept sense disambiguation algorithm is used for reducing the ambiguity that may emerged. The semantic extension is applied to all concepts included in a CMR and is defined as the process of associating to one concept other synonym terms identified in WordNet. Initially, the *synsets* in which each concept appears in WordNet are recovered, and then are classified in: *ambiguous - AC -* (those having more than one associated *synsets*), *not ambiguous - NA -* (only one associated *synset*) or *unknown - UC -* (not associated *synset*). Next, a disambiguation algorithm is applied to identify the most appropriated sense (or senses) for the ambiguous concepts. This algorithm, based on [29], improves the disambiguation results, fundamentally through combining the results obtained for each heuristic for determining the sense of the concept. This method is inspired in [21]. After applying the disambiguation algorithm, the lists of *ambiguous* and *not ambiguous* concepts are updated and each one of those concepts are extended with the terms included in their associated *synset*. The algorithm is defined as follows (Table 3):

Table 3. Concept sense disambiguation algorithm

Input: an ambiguous concept (c_a); CM (in which c_a is included); the set of *synsets* of c_a $(S(c_a))$.
Output: the more appropriated *synsets* for disambiguating c_a.

1. Preprocessing: For each concept $c_i | c_i \neq c_a$ and linking phrase (e) of CM
   a. the set of *synsets* of c_i $(S(c_i))$ and e $(S(e))$ are obtained from WordNet.
   b. the n most representative domains associated to the *synsets* included in $(S(c_i))$ and $(S(e))$ are identified and stored in the set D_{mc}, according to their occurrence frequency in those *synsets* and considering $n \leq 5$;
2. Domain analysis: For each $s_i | s_i \in S(c_a)$, the influence degree $(h_d(s_i))$ that each domain $d \in D_{cm}$ exercises on the sense s_i is calculated, through the sum of the occurrence frequencies of each d associated to s_i;
3. Context analysis: For each $s_i | s_i \in S(c_a)$, the influence degree $(h_c(s_i))$ that context (propositional structure in which c_a appear) exercises on the sense s_i is calculated as follows:

$$h_c(s) = w_c * \frac{\sum_{c_i \in C_r} \text{rel}(s, c_i)}{|C_r|} + w_r * \frac{\sum_{r_i \in R_r} \text{rel}(s, r_i)}{|R_r|}$$

   where C_r and R_r are the set of concepts and linking phrases, respectively, included in a vicinity of radius r having c_a as it center, and $rel(s, e)$ is a value indicating the semantic relatedness between the *synset* s and each *synset* associated to the concepts and linking-phrases included in the context. In the formula, w_c and w_r are weights assigned according to the desired relevance degree that information from concepts and linking phrases will have, respectively, for the heuristic. The sum of w_c and w_r should always be 1 to guarantee a maximum possible value of $h_c(s)$ of 1.
4. Gloss analysis: For each $s_i | s_i \in S(c_a)$, the influence degrade $(h_g(s_i))$ of definition (gloss) of the *synset* s_i in WordNet, considering its relation with the elements of context of c_a, is calculated as follows:

$$h_g(s) = w_c * \frac{|C_r \cap G(s)|}{|C_r|} + w_r * \frac{|R_r \cap G(s)|}{|R_r|}$$

   being $G(s)$ the set of words from the gloss of synset s. The weights w_c and w_r have the same purpose described in the previous step.
5. Heuristics combination: For each $s_i | s_i \in S(c_a)$, the global influence of the different heuristics $(h_{dcg}(s_i))$ is calculated as follows:

$$h_{dcg}(s_i) = w_d h_d(s_i) + w_c h_c(s_i) + w_g h_g(s_i)$$

   where w_d, w_c and w_g are weights with values representing the influence degree that each heuristic has in the precision of the full algorithm, which were defined according to precision results obtained by domain, context and gloss heuristics reported in [29]
6. Selection of the resulting *synset*. The more appropriated *synset* for c_a is the *synset* having a higher $h_{dcg}(s_i)$. In case of more than one *synset* having condition, all of them are considered, and the other ones are discarded.

The semantic integration task is aimed at explicitly integrating propositional structures (initially disconnected) through the unification of concepts (in a unique node) represented in different CM and it is applied when the query is performed on more than one CM. The concepts unification process is carried out through the identification of synonymous concepts in the selected CM as the search source of the query, and using several rules (R). Considering that $S(c_i)$ is the set of *synsets* s associated to a concept c_i and c_1 and c_2 are two concepts included in different CM, then c_1 and c_2 are unified if:

- **R1:** $(c_1, c_2 \in NAC) \wedge (S(c_1) = S(c_2))$; or
- **R2:** $(c_1, c_2 \in AC) \wedge (\exists s' | s' \in S(c_1) \wedge s' \in S(c_2))$; or
- **R3:** $((c_1 \in NAC \wedge c_2 \in AC) \vee (c_1 \in AC \wedge c_2 \in NAC)) \wedge (\exists s' | s' \in S(c_1) \wedge s' \in S(c_2))$; or
- **R4:** $(c_1, c_2 \in UC) \wedge (c_1 = c_2)$; *(fundamentally included for unifying not included concepts in WordNet, for example named entities)*

As result, if *R4* was triggered or the labels of c_1 and c_2 (in the case of other triggered rules) the same label of these concepts is used for representing the *unified concept* in the query results. In other cases, the label used for representing the unified concept is constructed with the labels of both concepts separated by a comma and enclosed in [ ] (ex. $[c_1, c_2]$). Finally, the *synset* associated to the *unified concept* is decided according to: (1) *if R1 was triggered, then the synset is the same to the c_1 or c_2*; (2) *if R2 was triggered, then the synsets are the common ones between the associated to c_1 and c_2*; (3) *if R3 was triggered, then the synset is the one associated to the $c_i \in NAC$.*

The semantic analysis is also considered in the proposed retrieval model, specifically in the projection queries (Q). In the process of query specification, the definition of one or more interesting concepts (CQ) by the user is required, besides defining the search source (SS) selecting a set of CM from the CMR. The selected CM are integrated through a union query as internal task in the query processing. Therefore, SS can be formally defined by the tuple (C^{ss}, P^{ss}), where C^{ss} is the set of concepts and P^{ss} is the set of propositions, included in the selected CM. Several rules were defined for identifying if a concept $c_j / c_j \in C^{ss}$ is retrieved or not, from a concept $c_i \in CQ$, where syntactic and semantic analysis are combined. These rules are described below and are executed following the same order in which they appear. However, it is possible to parameterize the combination of the analysis type considered, according to: using the syntactic analysis, using the semantic analysis, or combining both analyses. Being a concept $a/a \in CQ$, a concept $b/b \in C^{ss}$, $T(c_i)$ the set of words included in the label of c_i, and $ST(c_i)$ the set of synonym terms included in $S(c_i)$. The concept b is retrieved from SS if: (R1) $a \equiv b$ *(syntactically equivalent);* or (R2) $a \in T(b)$; or (R3) $a \in ST(b)$.

3 Applicability of CTAM: Case of Study

The evaluation of the proposed model turns out complex because a method for this purpose has not been identified. Nevertheless, in this section we present the case of study carried out in order to show the applicability of CTAM in the SLR context [16]. Much of the SLR processes requires high time-consuming, and several manual tasks [3], implying a great effort when mixing evidence from multiple studies and synthesizing evidence across studies, fundamentally in the analysis phase. On the other hand, some barriers (most of them affects the analysis phase) have been identified for carried out this type of review, such as [3]: lack of support for data extraction and analysis, difficulties of summarizing and aggregating data (especially qualitative data), difficulties of mixing evidence from multiple studies, difficulties for synthesizing evidence across studies, among others. Precisely, the proposed approach contributes to reduce the time consumption and the negative effects of some of those barriers. In order to demonstrate the applicability of our approach, a real SLR reported in [9] was selected. In this review, 11 scientific articles from 1820 primary studies were selected as the most relevant evidences to be analyzed.

The case of study was carried out using a collection with 11 texts, which were constructed using the abstract, introduction and conclusions from those articles analyzed in [9], in a similar way as the reported in [10]. Those texts have an average of 822 words and 33 sentences. Two different queries from CMQL were executed to analyze the content included in those texts and to identify relevant information that facilitated to give answer to the research questions reported in [9]. Specifically, the *CMInter* and *CMProj* were selected for supporting this analysis phase. Through of *CMInter* queries, relevant concepts and relationship between them from the texts can be retrieved, mixing evidence from these multiple studies and synthesizing the contents. In this sense, several *SV* for exploring the evidences, such as: 80% (Q1), 70% (Q2), 60% (Q3) and 50% (Q4), were used in the *CMInter* queries. Through of *CMProj* queries, relevant and useful information for answering the research questions can be retrieved, selecting some identified keywords in these questions as interest concepts.

According the definition of CTAM, initially the concept mapping process was carried out and a CM was automatically constructed from each text. Next, four *CMInter* queries with the different *SV* mentioned were executed on the constructed CMR. The domain specific concepts (DSC) and the generic contextual terms (GCT) identified as relevant keywords in [9] were used to measure the precision, recall and F-measure of retrieved concepts. The results are shown in Table 4, in which the measures were evaluated for the keywords sets: DSC, GCT and DSC + GCT. In Fig. 2, the resultant CM of the *CMInter* (SV = 50%) query is shown. The size of concepts represents the frequency in CMR, therefore [software system, software] and [dependability, reliability] are the most relevant retrieved concepts. The *strong relation* indicates that the concepts [framework, model] and 'ISO' are contextually related, suggesting further analysis by the reviewer. In this example, the results of the proposed semantic analysis method are also illustrated, through the integration of some syntactically different concepts, for example: 'framework'-'model' and 'dependability'-'reliability', because a synonymy relationship was automatically identified among them. This can help reviewers to quickly know the terminologies used in the articles to refer at the same concept.

In addition, the *CMProj*[1] query was executed using the constructed CMR as search source and '*standard*' as the interest concept. The result is shown in Fig. 3. The

Table 4. Results in the keywords identification task using *CMInter* queries

	Precision			Recall			F-measure		
	DSC	GCT	DSC + GCT	DSC	GCT	DSC + GCT	DSC	GCT	DSC + GCT
Q1	100	0	100	37.5	0	16.7	54.5	0	28.6
Q2	80	20	100	37.5	10	22.2	51.1	13.3	36.4
Q3	57.1	28.6	85.7	37.5	20	27.8	45.3	23.5	42
Q4	62.5	25	87.5	37.5	20	27.8	46.9	22.2	42.2
Ave.	**74.9**	**18.4**	**93.3**	**37.5**	**12.5**	**16.8**	**49.5**	**14.8**	**37.3**

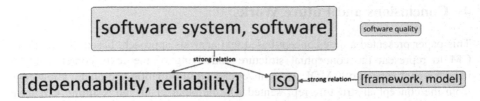

Fig. 2. Result of the *CMInter* query (SV = 50%)

objective of this query is to obtain useful information for answering one of the research questions answered in [9]: *"Which software reliability models have been developed by following the recommendations in International Standards?"*. In this question, *'standard'* is one of the most relevant terms on which is necessary to retrieve information.

Through the selected query it is possible to retrieve those strongly related concepts with the term *'standard'*, including concepts associated to *'reliability models'* and *'International Standards'*. The Fig. 3 shows several retrieved concepts that they represent different international standards, such as: *ISO, ECSS, IEEE, SQuaRE, COSMIC* and *Space Standardization*, most of them (66,6%) were also identified in the manual analysis carried out by Febrero et al. [9]; although all of them are represented in Fig. 3. The analysis of the evidences to answer the research question can be enriched applying others *CMProj* queries, for example, increasing the R value and using others interest concepts, such as: *'reliability'*. As the results of this case of study, several beneficial aspects of the application of CTAM to the exploration, interpretation, and decision-making in the analysis phase of primary studies in a SLR were emerged, such as: obtaining the relevant concepts from the articles that should be reviewed; quickly knowing the terminologies associated to the concepts include in the different articles; assisting reviewers to know which concepts are contextually related to keywords from the research questions; facilitating the qualitative data (concepts and relationship between them) mining and its analysis; and mixing evidences from multiple studies.

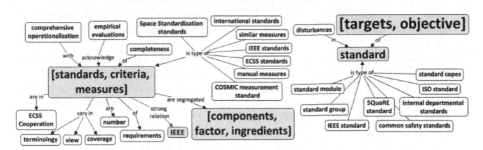

Fig. 3. Result of the *CMProj[1]* query using *standard* as interest concept

4 Conclusions and Future Works

This paper presented a new concept-based text analysis approach, based on the use of CM to represent the conceptual structure underlying of the texts content and an improvement version of CMQL, to retrieve relevant information and obtain knowledge from the conceptual structure represented. On the other hand, the resulting CM from each query provides in CMQL constitutes conceptual and summarized representation views of the content included in the texts. The integration of the proposed semantic analysis method, supported in WordNet and the use of a disambiguation algorithm, to the query processing defined in CMQL allowed improve the results of search and information integration. The results of the case of study carried out to evaluate the applicability of the proposed approach demonstrated several benefits to the conceptual exploration, interpretation, and decision-making in the review of primary studies carried out in a SLR. In future works, others graph operations will be considered to extend the proposed model and increasing the results of the concept-based texts analysis through the automatic detection of frequent patters and topics from the CMR.

Acknowledgments. This work has been partially supported by FEDER and the State Research Agency (AEI) of the Spanish Ministry of Economy and Competition under grant MERINET: TIN2016-76843-C4-2-R (AEI/FEDER, UE)

References

1. Abdulsahib, A.K., Kamaruddin, S.S.: Graph based text representation for document clustering. J. Theor. Appl. Inf. Technol. **76**(1), 1–10 (2015)
2. Aggarwal, C.C., Zhai, C.X. (eds.): Mining Text Data. Springer, New York (2012). https://doi.org/10.1007/978-1-4614-3223-4
3. Al-Zubidy, A., Carver, J.C., Hale, D.P., Hassler, E.E.: Vision for SLR tooling infrastructure: prioritizing value-added requirements. Inf. Softw. Technol. **91**, 72–81 (2017). https://doi.org/10.1016/j.infsof.2017.06.007
4. Bentivogli, L., Forner, P., Magnini, B.; Pianta, E.: Revising WordNet Domains Hierarchy: Semantics, Coverage, and Balancing. In: Proceedings of COLING 2004 Workshop on Multilingual Linguistic Resources, pp. 101–108 (2004). https://doi.org/10.3115/1706238.1706254
5. Benkoussas, C., Bellot, P.: Information retrieval and graph analysis approaches for book recommendation. Sci. World J. **2015**, 1–8 (2015). https://doi.org/10.1155/2015/926418
6. Cañas, A.J., Leake, D.B., Maguitman. A.G.: combining concept mapping with CBR: towards experience-based support for knowledge modeling. In: Proceedings of FLAIRS Conference, pp. 286–290. AAAI Press (2001)
7. Chang, J.Y., Kim, I.M.: Research trends on graph-based text mining. Int. J. Softw. Eng. Appl. **8**(4), 147–156 (2014). https://doi.org/10.14257/ijseia.2014.8.4.16
8. Chen, L., Jose, J.M., Yu, H., Yuan, F.: A semantic graph-based approach for mining common topics from multiple asynchronous text streams. In: Proceedings of the 26th International Conference on World Wide Web, pp. 1201–1209 (2017). https://doi.org/10.1145/3038912.3052630

9. Febrero, F., Calero, C., Moraga, M.A.: Software reliability modeling based on ISO/IEC SQuaRE. Inf. Softw. Technol. **70**, 18–29 (2016). https://doi.org/10.1016/j.infsof.2015.09.006

10. Felizardo, B.K.R., Andery, G.F., Paulovich, F.V., Minghim, R., Maldonado, J.C.: A visual analysis approach to validate the selection review of primary studies in systematic reviews. Inf. Softw. Technol. **54**, 1079–1091 (2012). https://doi.org/10.1016/j.infsof.2012.04.003

11. Hassan, G.S., Abdulsahib, A.K., Kamaruddin, S.S.: Graph-based text representation: a survey of current approaches. Res. J. Appl. Sci. Eng. Technol. **14**(9), 334–340 (2017). https://doi.org/10.19026/rjaset.14.5073

12. Hulpus, I., Hayes, C., Karnstedt, M., Greene, D.: Unsupervised Graph-based Topic Labelling Using Dbpedia. In: Proceedings of the Sixth ACM International Conference on Web Search and Data Mining, pp. 465–474, ACM (2013). https://doi.org/10.1145/2433396.2433454

13. Indurkhya, N.: Emerging directions in predictive text mining. WIREs Data Min. Knowl. Disc. **5**, 155–164 (2015). https://doi.org/10.1002/widm.1154

14. Jiang, X., Tan, A.H.: CRCTOL: a semantic-based domain ontology learning system. J. Am. Soc. Inform. Sci. Technol. **61**(1), 150–168 (2010). https://doi.org/10.1002/asi.21231

15. Karim, G., Mouna, T.K., Lynda, T., Maher, B.J.: Graph-based methods for significant concept selection. Procedia Comput. Sci. **60**, 488–497 (2015). https://doi.org/10.1016/j.procs.2015.08.170

16. Kitchenham, B.A., Charters S.: Guidelines for performing systematic literature reviews in software engineering. Technical report EBSE 2007-001, Keele University and Durham University Joint Report (2007)

17. Kleinberg, J.M.: Authoritative sources in a hyperlinked environment. J. ACM (JACM) **46**(5), 604–632 (1999). https://doi.org/10.1145/324133.324140

18. Koopman, B., Zuccon, G., Bruza, P., Sitbon, L., Lawley, M.: Graph-based concept weighting for medical information retrieval. In: Proceedings of the 17th Australasian Document Computing Symposium, pp. 80–87 (2012). https://doi.org/10.1145/2407085.2407096

19. Kowata, J.H., Cury, D., Silva, M.C.: Concept maps core elements candidates recognition from text. In: Proceedings of the 4th International Conference on Concept Mapping, 1, pp. 120–127 (2010)

20. Miller, G., Fellbaum, C. (eds.): WordNet: An Electronic Lexical Database. The MIT Press, Cambridge (1998)

21. Navigli, R.: Word sense disambiguation: a survey. ACM Comput. Surv. **41**(2), 1–69 (2009). https://doi.org/10.1145/1459352.1459355

22. Navigli, R., Ponzetto, S.P.: BabelNet: The automatic construction, evaluation and application of a wide-coverage multilingual semantic network. Artif. Intell. **193**, 217–250 (2012). https://doi.org/10.1016/j.artint.2012.07.001

23. Novak, J.D., Cañas, A.J.: The theory underlying concept maps and how to construct them. Technical report IHMC CmapTools 2006-01, 32502, USA (2006)

24. Pinto, D., Gómez, H., Vilariño, D., Singh, V.K.: A graph-based multi-level linguistic representation for document understanding. Pattern Recogn. Lett. **41**, 93–102 (2014). https://doi.org/10.1016/j.patrec.2013.12.004

25. Rodríguez, A., Simón, A.: Método para la extracción de información estructurada desde textos. Revista Cubana de Ciencias Inf. **7**(1), 55–67 (2013)

26. Rodríguez, A., Simón, A., Guevara, E., Hojas, W.: Modelo de representación de textos basado en grafo para la minería de texto. Ciencias de la Inf. **46**(1), 63–71 (2015)

27. Sasson, E., Ravid, G., Pliskin, N.: Creation of knowledge-added concept maps: time augmention via pairwise temporal analysis. J. Knowl. Manage. **21**(1), 132–155 (2017). https://doi.org/10.1108/JKM-07-2016-0279
28. Simón, A., Ceccaroni, L., Rosete, A., Suárez, A., Victoria, R.: A support to formalize a conceptualization from a concept maps repository. In: Proceedings of the 3rd International Conference on Concept Mapping, pp. 68–75 (2008)
29. Simón, A., Ceccaroni, L., Rosete, A., Suárez, A., de la Iglesia, M.: A concept sense disambiguation algorithm for concept maps. In: Proceedings of the 3rd International Conference on Concept Mapping, pp. 14–21 (2008)
30. Sonawane, S.S., Kulkarni, P.A.: Graph based representation and analysis of text document: a survey of techniques. Int. J. Comput. Appl. **96**(19), 1–8 (2014). https://doi.org/10.5120/16899-6972
31. Valerio, A., Leake, D., Cañas, A.J.: Using automatically generated concept maps for document understanding: A human subjects experiment. In: Proceedings of 5th International Conference on Concept Mapping, pp. 438–445 (2012)

Tri-partitions and Uncertainty

An Efficient Gradual Three-Way Decision Cluster Ensemble Approach

Hong Yu[✉] and Guoyin Wang

Chongqing Key Laboratory of Computational Intelligence,
Chongqing University of Posts and Telecommunications,
Chongqing 400065, People's Republic of China
{yuhong,wanggy}@cqupt.edu.cn

Abstract. Cluster ensemble has emerged as a powerful technique for combining multiple clustering results. However, existing cluster ensemble approaches are usually restricted to two-way clustering, and they also cannot flexibility provide two-way or three-way clustering result accordingly. The main objective of this paper is to propose a general cluster ensemble framework for both two-way decision clustering and three-way decision. A cluster is represented by three regions such as the positive region, boundary region and negative region. The three-way representation intuitively shows which objects are fringe to the cluster. In this work, the number of ensemble members is increased gradually in each decision (iteration), it is different from the existing cluster ensemble methods in which all available ensemble members join the computing at one decision. It can be ended at a three-way decision final clusters or choose to go on until all the objects are assigned to the positive or negative region of the cluster determinately. The experimental results show that the proposed gradual three-way decision cluster ensemble approach is effective for reducing the running time and not sacrificing the accuracy.

Keywords: Cluster ensemble · Three-way decisions · Gradual decision

1 Introduction

Cluster ensemble has emerged as an important elaboration of the clustering problem [7,8]. Generally speaking, every clustering ensemble method is made up of two steps: (1) generating multiple different clusterings of the data set, also called generation step, and (2) combining these clusterings to obtain a single new clustering, also called consensus step.

From the review of existing studies, we can identify an issue that have not been resolved satisfactorily. That is, they are typically based on two-way (i.e., binary) decisions. A cluster is described by a single set, where every member plays the same role in the cluster. In other words, objects belong to the cluster if they are in the set, otherwise they do not. In this way, it is impossible to indicate which members are fringe members of the cluster through the existing information; in fact, these objects deferred waiting for further information. Therefore,

© Springer International Publishing AG, part of Springer Nature 2018
J. Medina et al. (Eds.): IPMU 2018, CCIS 854, pp. 711–723, 2018.
https://doi.org/10.1007/978-3-319-91476-3_58

Lingras et al. [1,4,5] use an interval set to represent a cluster. Considering to store and compute expediently, Yu [12,13] introduced a framework of three-way cluster analysis (TWC), where a cluster is represented by a pair of sets. Then, the universe is divided into three regions by two sets, namely, positive, boundary and negative regions. The three-way representation brings more insight into the interpretation of cluster. Objects in the positive region certainly belong to the cluster and objects in the negative region definitively do not belong to the cluster based on the already existing information. Objects in the boundary region can not be decided certainly, they might belong to the cluster or not belong to the cluster. They are fringe members, and we need further information to make decisions.

Thus, in this paper, we propose a general framework of gradual three-way cluster ensemble which is suitable for both hard clustering and soft clustering, meanwhile the framework is suitable for both two-way clustering and three-way clustering. In this framework, objects in the boundary region are clustered gradually by the new incremental ensemble members until satisfying the stop conditions. Besides, we present a novel efficient algorithm to obtain the final consensus clustering result. We define cluster cores to reflect the minimal granularity distribution structure agreed by all the ensemble members, thus the original data set can be seen as the union of a set of cores and no-cores instead of objects. The experimental results show that the proposed method is efficient; it uses less resources (ensemble members), the clustering process is quickly while it does not sacrifice the accuracy in comparison to other approaches.

The remainder is organized as follows. Section 2 introduces the framework of gradual three-way decision cluster ensemble. Section 3 presents the gradual three-way decision cluster ensemble algorithm. Section 4 reports experimental results based on an artificial data set and a number of public domain standard data sets. Conclusions are provided in Sect. 5.

2   Framework of Gradual Three-Way Decision Cluster Ensemble

First, let's review the framework of three-way cluster analysis [12,13]. Let $U = \{\mathbf{x}_1, \cdots, \mathbf{x}_n, \cdots, \mathbf{x}_N\}$ be a finite set. $\mathbf{x}_n$ is an object which has D attributes, namely, $\mathbf{x}_n = (x_n^1, \cdots, x_n^d, \cdots, x_n^D)$. x_n^d denotes the value of the d-th attribute of the object $\mathbf{x}_n$, where $n \in \{1, \cdots, N\}$, and $d \in \{1, \cdots, D\}$. In the three-way representation, a cluster is representation by a pair of sets, namely, $C^k = (\text{POS}(C^k), \text{BND}(C^k))$. The objects in $\text{POS}(C^k)$ definitely belong to the cluster C^k, objects in $\text{NEG}(C^k) = U - \text{POS}(C^k) - \text{BND}(C^k)$ definitely do not belong to C^k, and objects in $\text{BND}(C^k)$ might or might not belong to the cluster.

As Yao [11] had pointed out that we can produce the three-way decision rules by building an evaluation function based on a pair of thresholds (α, β), $\alpha \geq \beta$. Although evaluations based on a total order are restrictive, they have a computational advantage. One can obtain the three regions by simply comparing the evaluation value with a pair of thresholds. In the cluster analysis, we usually choose the similarity between an object $\mathbf{x}$ and a cluster C^k as the evaluation function, then we get the following three-way decision rules:

$$\text{RulePOS} : if \; Sim(\mathbf{x}, C^k) \geq \alpha, \; \mathbf{x}_n \in \text{POS}(C^k);$$
$$\text{RuleBND} : if \; \beta < Sim(\mathbf{x}, C^k) < \alpha, \; \mathbf{x}_n \in \text{BND}(C^k); \qquad (1)$$
$$\text{RuleNEG} : if \; Sim(\mathbf{x}, C^k) \leq \beta, \; \mathbf{x}_n \in \text{NEG}(C^k).$$

If the clustering result has K clusters, the result scheme of three-way clustering is represented as: $\mathbf{C} = \{(\text{POS}(C^1), \text{BND}(C^1)), \cdots, (\text{POS}(C^k), \text{BND}(C^k)), \cdots, (\text{POS}(C^K), \text{BND}(C^K))\}$.

Second, we review some concepts about cluster ensemble. Figure 1 depicts our previous proposed cluster ensemble framework based on three-way decisions [14]. Set U^s be the family of H samples, namely, $U^s = \{U_1, U_2, \cdots, U_h, \cdots, U_H\}$, where $U_h \subseteq U$, and $h \in \{1, \cdots, H\}$. Let Φ be a clusterer selection function from a sample to a clustering. Usually, Φ is composed by a set of clustering algorithms and denoted as $\Phi = \{\Phi_1, \cdots, \Phi_l, \cdots, \Phi_L\}$, and a clusterer Φ_l is selected randomly or according to a priori knowledge of the data set. Then, the clustering result of a clusterer Φ_l is called a clustering, also called an ensemble member, denoted by P_h. Let $P = \{P_h | \Phi_l(U_h) \to P_h\}$ be the set of clusterings, where $h \in \{1, \cdots, H\}, l \in \{1, \cdots, L\}$.

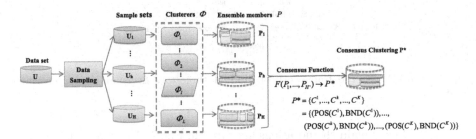

Fig. 1. The framework of three-way decision cluster ensemble

Assume that K^h is the number of clusters of the clustering P_h, where P_h is described as: $P_h = \{C_h^1, \cdots, C_h^k, \cdots, C_h^{K^h}\}$. So far, different clusterings are obtained through the generation step. Actually, the sampling is not a prerequisite; in other words, for $h \in \{1, \cdots, H\}$, we can simply set $U_h = U$. In the consensus step, the labeling clusterings are combined into a single labeling clustering P^* using a consensus function $F : F(P_1, P_2, \cdots, P_{H'}) \to P^*$. Here, $H' \leq H$, because when combining clusterings into a final clustering, we may combine all ensemble members or some of the ensemble members.

The consensus clustering, namely the final result P^*, is represented as: $P^* = \{C^1, \cdots, C^k, \cdots, C^K\} = \{(\text{POS}(C_h^1), \text{BND}(C_h^1)), \cdots, (\text{POS}(C_h^k), \text{BND}(C_h^k)), \cdots, (\text{POS}(C_h^K), \text{BND}(C_h^K))\}$.

Then, we propose a novel framework of gradual three-way decision cluster ensemble, inspired by the idea of sequential decision making [9,10]. It is shown in Fig. 2, abbreviated as the GTWD-CE Model.

The main differences between the new model and the existing models conclude three issues. First, the clustering processing is designed as an iteration step, we can obtain the two-way or three-way results in the first clustering (decision); if some objects cannot be assigned to one cluster definitively, then the model goes to the sequential decision with the different ensemble members. Second, the number of ensemble members H is not constant any more; that is, all of available ensemble members join the computing in other models, but some of ensemble members join the first decision and some join the next decision if needed in the proposed model. Third, the clustering processing is based on cluster cores instead of based on objects; which makes the model efficiently.

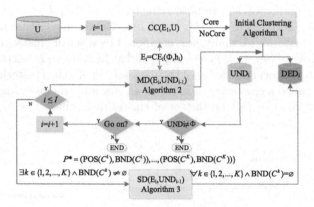

Fig. 2. GTWD-CE: the framework of gradual three-way decision cluster ensemble model

Now, let us explain the notations appeared in Fig. 2. As we have mentioned, we neglect the study of sampling in this work. So, the input data set is denoted by U simply. The time of iterations (decisions) is counted by i; we initialize it to 1 and set the maximal time to I.

The function of $CE_i(\Phi, h_i)$ means choosing h_i ensemble members from the set of all possible ensemble members (clusterings) Φ. The diversity of ensemble members has been proved to be a key factor to improve the quality of the consensus clustering. In fact, the feedback information from the current decision can help the next choice of members, which will be our future work. In this paper, we just choose members randomly, and the amount of members is specific. That is, $H = |\Phi|$ and $\sum h_i \leq H$.

The function of $CC(h_1, U)$ is to obtain cluster cores from U, it divides objects into two kinds of objects such as core objects or no-core objects, the detail will be described in Sect. 3.1. Then, the initial three-way decision clustering algorithm, Algorithm 1, assigns objects into two subsets DED_i and UND_i, and $U = DED_i \cup UND_i$. All objects in positive regions compose DED_i, which means they are decided into one cluster definitively. All objects in boundary regions compose UND_i, which means they are not certainly decided into any clusters in the current clustering/decision. Objects in UND_i might include core objects or no-core objects.

In fact, after Algorithm 1, we really have a consensus clustering result. The result is represented by three-way representations. Of course, if the underlying structure of the original data set is unambiguous, all of the boundary regions are empty. Otherwise, there exist some boundary (fringe) objects, namely, $UND \neq \emptyset$. If we still need to make further decision on these fringe objects, the strategy of gradual decisions starts to work. The model goes to the function $MD(h_i, UND_{i-1})$, which makes further decisions on objects in UND_{i-1} by choosing new cluster ensemble members. The gradual strategy goes on accordingly. When $i = I$ and $UND_{i-1} \neq \emptyset$, the function $SD(h_i, UND_{i-1})$ is invoked. $SD(h_i, UND_{i-1})$ means do a special decision on the UND_{i-1} in order to obtain a finial two-way clustering result. In Sect. 3, we will give the specific strategies for these functions. We need to note that, MD runs based on cluster cores not based on objects as other methods.

The framework is a general cluster ensemble model. It is not only can represent soft clustering, but can also represent hard clustering. If we need two-way clustering (hard clustering) result, we just make the model run on; if we need three-way (soft clustering) result, we keep the boundary regions not empty. Furthermore, the model shows intuitively which objects are fringe in difference from others, and one can develop soft or hard clustering ensemble algorithms based on this framework.

3   Gradual Three-Way Decision Cluster Ensemble Algorithm

In this section, we will introduce the functions in Sect. 2. They are just instances for realizing the model and compose a practical gradual three-way decision cluster ensemble algorithm, shorted by GTWD-CE Algorithm.

3.1   Cluster Core

A cluster core is a subset of objects. A cluster core appears in all the clusterings and reflects the fundamental structure agreed by all clusterings, and a cluster core can not be divided into the smaller subset.

Definition 1. *Let $A \subseteq U$, A is called a cluster core if and only if it satisfies: (C1) for every clustering P_h, there exists a $k \in [1, K^h]$ such that $A = \bigcap_{h \in [1,H]} POS(C_h^k)$; (C2) if A is a cluster core, $B \subset A$, B is not a cluster core.*

Condition (C1) denotes that a cluster core is the intersection of clusters in all clusterings/ensemble members; Condition (C2) denotes that the cluster core is the maximal intersection of these clusters. When a set of objects satisfy these two conditions, we call it a cluster core.

According to Definition 1, we can obtain cluster cores. Here is an example. Supposing there is a $U = \{\mathbf{x}_1, \mathbf{x}_2, \mathbf{x}_3, \mathbf{x}_4, \mathbf{x}_5, \mathbf{x}_6, \mathbf{x}_7, \mathbf{x}_8, \mathbf{x}_9\}$, and there are three different clusterings, namely P_1, P_2 and P_3. The details are shown as follows.

$$P_1 = \{(\{\mathbf{x}_1, \mathbf{x}_2\}, \emptyset), (\{\mathbf{x}_3, \mathbf{x}_4\}, \{\mathbf{x}_7\}), (\{\mathbf{x}_5, \mathbf{x}_6\}, \{\mathbf{x}_7\}), (\{\mathbf{x}_8, \mathbf{x}_9\}, \emptyset)\},$$
$$P_2 = \{(\{\mathbf{x}_1, \mathbf{x}_2\}, \emptyset), (\{\mathbf{x}_3, \mathbf{x}_4\}, \{\mathbf{x}_5, \mathbf{x}_6, \mathbf{x}_7\}), (\{\mathbf{x}_8, \mathbf{x}_9\}, \{\mathbf{x}_5, \mathbf{x}_6, \mathbf{x}_7, \})\},$$
$$P_3 = \{(\{\mathbf{x}_1, \mathbf{x}_2\}, \{\mathbf{x}_7\}), (\{\mathbf{x}_3, \mathbf{x}_4, \mathbf{x}_5, \mathbf{x}_6, \mathbf{x}_8, \mathbf{x}_9\}, \{\mathbf{x}_7\})\}.$$

Then, we have three cluster cores, that is, $\mathbf{Core} = \{Core_1 = \{\mathbf{x}_1, \mathbf{x}_2\}, Core_2 = \{\mathbf{x}_3, \mathbf{x}_4\}, Core_3 = \{\mathbf{x}_8, \mathbf{x}_9\}\}$. That is to say we just need these three cores and three no-core objects for consensus clustering. In this example, the size of computing is reduced from 9 to 6.

3.2   Initial Three-Way Decision Clustering Algorithm

Now, we need to define an appropriate evaluation function in order to obtain the three regions of a cluster. The relationships between cluster cores, between no-core object and cluster core, between no-core objects are defined as evaluation functions used in our framework. The evaluation values for acceptance and rejection are defined by a pair of thresholds α, β.

Definition 2. *The relationship between two cores, $Core_i$ and $Core_j$, is defined by the following equation:*

$$BCC(i, j) = CountCommon(T_i, T_j)/h. \tag{2}$$

T_i and T_j be the cluster mark sequence of $Core_i$ and $Core_j$ respectively, and $CountCommon(T_i, T_j)$ means the number of common marks at the corresponding place in T_i and T_j. h is the number of clusterings (ensemble members). $\mathbf{BCC}_{Z \times Z} = [BCC(i, j)]$ denotes the matrix of relationship between clusters.

Definition 3. *For $\mathbf{x}_i$ and $Core_j$, $\mathbf{x}_i$ is a no-core object, the relationship between $\mathbf{x}_i$ and $Core_j$ is defined by the following equation:*

$$B\overline{C}C(i, j) = CountCommon(T_i, T_j)/h, \tag{3}$$

T_i denotes the cluster mark sequence of $\mathbf{x}_i$, and T_j denotes the cluster mark sequence of $Core_j$. $\mathbf{B\overline{C}C}_{|\overline{\mathbf{Core}}| \times Z} = [B\overline{C}C(i, j)]$ denotes the matrix of relationships between no-core objects and cluster cores.

Definition 4. *For no-core objects $\mathbf{x}_i$ and $\mathbf{x}_j$, the relationship between them is defined by the following equation:*

$$B\overline{CC}(i, j) = |J|/h, \tag{4}$$

$J = \{h \mid \mathbf{x}_i, \mathbf{x}_j \in POS(C_h^k) \cup BND(C_h^k), 1 \le h \le H\}$, and $k = 1, 2, \cdots, K^h$. $B\overline{CC}(i, j)$ means the number of times of $\mathbf{x}_i$ and $\mathbf{x}_j$ occur in the both upper bounds of clusters. $\mathbf{B\overline{CC}}_{|\overline{\mathbf{Core}}| \times |\overline{\mathbf{Core}}|} = [B\overline{CC}(i, j)]$ denotes the matrix of the relationship between no-core objects.

Definition 5. *For two clusters C^z and C^l, the relationship between two clusters is defined as:*

$$S(C^z, C^l) = \frac{1}{|C^z| \cdot |C^l|} \left(\sum_{\substack{Core_u \in POS(C^z) \\ Core_v \in POS(C^l)}} BCC(u,v) + \gamma \left(\sum_{\substack{Core_u \in POS(C^z) \\ Core_v \in BND(C^l)}} BCC(u,v) \right. \right.$$
$$\left. \left. + \sum_{\substack{Core_u \in BND(C^z) \\ Core_v \in POS(C^l)}} BCC(u,v) + \sum_{\substack{Core_u \in BND(C^z) \\ Core_v \in BND(C^l)}} BCC(u,v) \right) \right). \tag{5}$$

Suppose the current no-core object is $\mathbf{x}_n$ and the cluster is C^l. C^l may contain cluster cores and no-core objects. Their relationship is shown as follows.

Definition 6. *For a no-core object $\mathbf{x}_n$ and a cluster C^l, the relationship between no-core object and cluster is defined as:*

$$S(\mathbf{x}_n, C^l) = \frac{1}{|C^l|} \cdot \left(\sum_{Core_u \in POS(C^l)} B\overline{C}C(n,u) + \sum_{\mathbf{x}_m \in POS(C^l)} B\overline{CC}(n,m) \right.$$
$$\left. + \gamma \left(\sum_{Core_u \in BND(C^l)} B\overline{C}C(n,u) + \sum_{\mathbf{x}_m \in BND(C^l)} B\overline{CC}(n,m) \right) \right). \tag{6}$$

To obtain clustering in cluster cores, we can use Eq. (5) to calculate the relationship between clusters. When we need to confirm which clusters no-core objects belong to, we can calculate the relationship between no-core objects and clusters according to Eq. (6).

We can obtain cluster cores **Core** and no-core objects $\overline{\textbf{Core}}$ according to Sect. 3.1; and build the corresponding matrices according to Definitions 2 to 4; then classify cluster cores into big cores and small cores. Furthermore, we initialize every core as a positive region of a cluster, and set the boundary of this cluster to empty. The initial three-way decision clustering algorithm is described in Algorithm 1.

3.3   Strategy of Decision for Pending Data

After running Algorithm 1, the model gets P_1^* and $UND = \bigcup_{k=1}^{K} BND(C^k)$. Then, the model meets two conditions: whether UND_i is $\emptyset$ and whether *Goon* is true. When UND_i is $\emptyset$, there are no objects need to decide; thus the processing will be end. *Goon* is an interactive parameter determined by human. That is to say, when $UND_i \neq \emptyset$ and the user hopes to get further clustering result, then the gradual three-way decision processing will continue.

During the gradual three-way decision processing, namely, $i \geq 2$, the function $E_i = CE_i(\phi, h_i)$ will be called again to obtain the new h_i ensemble members. The number of new ensemble members in the ith decision (clustering) does not require complete equal to the number of ensemble members in the the previous decisions. Then, the model moves to the function $MD(E_i, UND_{i-1})$. Algorithm 2 describes how to cluster the pending objects.

Algorithm 1. IC Algorithm: initial three-way decision clustering

---

Input: **Core**, $\overline{\textbf{Core}}$ and K.
Output: The result of initial clustering.
1 Obtain a matrix of big cores, where the element records the times of the corresponding two cluster cores divided into the same cluster;
2 Find the connected subgraph of the matrix and set the number of subgraphs be SK; and $Noise = \emptyset$.
3 **if** $SK \geq K$ **then** go to Line 7;//K is the real number of clusters
4 **else** the none-zero elements of matrix subtract 1, update the matrix, go to Line 2;
5 **for** *each small core* $Core_z$ **do**
6 | compute the relationships S between $Core_z$ and other big cores;
7 | **if** all relationship values S are no more than β **then** set $Core_z$ be a noise set and go to Line 19;
8 $A = \{m|S \geq \alpha\}$;
9 **if** *cluster cores in A belong to the same cluster* **then**
10 | assign $Core_z$ to the positive region of the cluster;

11 **else**
12 | assign $Core_z$ to the boundary region of those clusters;

13 $B = \{m|\beta < S < \alpha\}$;
14 **for** $Core_z \in B$ **do**
15 | assign $Core_z$ to the boundary region of the corresponding clusters;

16 **for** *each no-core object* $\mathbf{x}_n$ **do**
17 | compute the relationships between $\mathbf{x}_n$ and a cluster;
18 | $A = \{C^l|S(\mathbf{x}_n, C^l) \geq \alpha\}$;
19 | **if** $|A| = 1$ **then** assign $\mathbf{x}_n$ to the positive region of the cluster in the A;
20 | **if** $|A| > 1$ **then**
21 | | $B = \{C^l|max(S(\mathbf{x}_n, C^l), C^l \in A\}$;
22 | | **if** $|B| = 1$ **then** $\mathbf{x}_n$ belongs to the positive region of the cluster in the B;
23 | | **else** $\mathbf{x}_n$ belongs to the boundary region of these clusters in the B;
24 | **if** $|A| = 0$ **then**
25 | | $C = \{C^l|\beta \leq S(\mathbf{x}_n, C^l) < \alpha\}$;
26 | | **if** $C \neq \emptyset$ **then** $\mathbf{x}_n$ belongs to the boundary region of these clusters in the C;
27 | | **else** $Noise = Noise \cup \mathbf{x}_n$;

28 Transfer cores to corresponding objects and obtain a clustering result $P_1^* = UND_1 \cup DED_1$; $DED_1 = \cup POS(C_k)$, and $UND_1 = \cup BND(C_k) \cup Noise$.

Algorithm 2. Making decision algorithm

Input: $P_{i-1}^* = UND_{i-1} \cup DED_{i-1}$, E_i;
Output: The result of clustering P_i^*.
1 Obtain the cluster mark sequences of cores;
2 **for** *every* $\mathbf{x}_a \in UND_{i-1}$ **do**
3 | **for** every $POS(C^k) \in P_{i-1}^*$ **do** calculate $S(\mathbf{x}_a, POS(C^k))$ according to Eq. 7;
4 | Get $A = \{C^k \mid S(\mathbf{x}_a, POS(C^k)) \geq \alpha\}$;
5 | **if** $|A| = 1$ **then** assign $\mathbf{x}_a$ to the positive region of the cluster in A;
6 | **elseif** $|A| > 1$ **then** assign $\mathbf{x}_a$ to the boundary region of these clusters in A;
7 | **else**
8 | | Get $B = \{C^k \mid \beta < S(\mathbf{x}_a, POS(C^k)) < \alpha\}$;
9 | | **if** $B \neq \emptyset$ **then** assign $\mathbf{x}_a$ to the boundary region of these clusters in B;
10 | | **else** union objects in $\mathbf{x}_a$ into $Noise$;

11 Output the updated clustering result P_i^*.

Definition 7. *For* $\mathbf{x}_a \in UND_{i-1}$, $\mathbf{x}_b \in \text{POS}(C^k)$, *i.e.*, $\mathbf{x}_b \in \underline{C}^k$, *and* $C^k \in P^*_{i-1}$, *the relationship between* $\mathbf{x}_a$ *and* $\underline{C}^k$ *is defined by the following equation:*

$$S(\mathbf{x}_a, \text{POS}(C^k)) = \frac{1}{N_k} \sum_{b=1}^{N_k} \frac{\text{CountCommon}(T[\mathbf{x}_a], T[\mathbf{x}_b])}{h_i}. \tag{7}$$

$N_k = |\text{POS}(C^k)|$, $T[\mathbf{x}_a]$ and $T[\mathbf{x}_b]$ denote the cluster mark sequence of $\mathbf{x}_a$ and $\mathbf{x}_b$ respectively. h_i denotes the amount of the ensemble members.

Set $A = \{C^k \mid S(\mathbf{x}_a, \text{POS}(C^k)) \geq \alpha\}$ be the set of clusters whose relationship with $\mathbf{x}_a$ are no less than α.

Now, there exists three cases: Case I: If there only exists one cluster whose relationship with $\mathbf{x}_a$ is no less than α, namely, $|A| = 1$, assign $\mathbf{x}_a$ to the positive region of the cluster in A; Case II: If there exists multiple clusters whose relationship with $\mathbf{x}_a$ are no less than α, namely, $|A| > 1$, which means $\mathbf{x}_a$ has strong relationships with these clusters, assign $\mathbf{x}_a$ to the boundary region of these clusters in A; Case III: If the relationship between $\mathbf{x}_a$ and each cluster is less than α, namely, $|A| = 0$. Suppose $B = \{C^k \mid \beta < S(\mathbf{x}_a, \text{POS}(C^k)) < \alpha\}$. Now there exists two subcases: (1) If $B \neq \emptyset$, $\mathbf{x}_a$ belongs to the boundary region of these clusters in B; (2) If $B = \emptyset$, which means the relationship between $\mathbf{x}_a$ and each cluster is less than β, we regard $\mathbf{x}_a$ is a noise point.

3.4 Strategy for Two-Way Decision

During the gradual decision process, the model decides whether i is bigger than I in each decision. The two-way decision function $SD(h_i, UND_{i-1})$, Algorithm 3, will be invoked when the number of iterations beyond the threshold.

Algorithm 3. Special decision: two-way decision clustering

    Input: $P^*_{i-1} = UND_{i-1} \cup DED_{i-1}$, E_i, $i = I$;
    Output: The result of clustering P^*_I.
1 Obtain the cluster mark sequences of cores;
2 **for** *every* $\mathbf{x}_a \in UND_{i-1}$ **do**
3 | **for** every $\text{POS}(C^k) \in P^*_{i-1}$ **do** calculate $S(\mathbf{x}_a, \text{POS}(C^k))$ according to Eq. 7;
4 | Set the maximal one in $S(\mathbf{x}_a, \text{POS}(C^k))$ be max;
5 | Get $D = \{C^k \mid S(\mathbf{x}_a, \text{POS}(C^k)) = \text{max}\}$;
6 | **if** $max < \beta$ **then** union objects of $\mathbf{x}_a$ into $Noise$;
7 | **elseif** $|D| = 1$ **then** assign objects of $\mathbf{x}_a$ to the positive region of the clusters in D;
8 | **elseif** $|D| > 1$ **then** assign objects of $\mathbf{x}_a$ to the positive region of these clusters in D;
9 Output the updated clustering result P^*_i.

Similarly, Algorithm 3 obtains the cluster mark sequences of cores as Algorithm 2, and it gets the maximal value max in the relationships $S(\mathbf{x}_a, \text{POS}(C^k))$. We set $D = \{C^k \mid S(\mathbf{x}_a, \text{POS}(C^k)) = \text{max}\}$ be the set of clusters whose relationship with $\mathbf{x}_a$ are equal to the maximum.

Now there exist two cases: Case I: If $max < \beta$, which means the maximal relationship between $\mathbf{x}_a$ and each cluster is less than β, we regard objects in

$\mathbf{x}_a$ as noise points; Case II: If $max \geq \beta$, now there exists two subcases: (1) If $|D| = 1$, assign objects in $\mathbf{x}_a$ to the positive region of the cluster in D; (2) If $|D| > 1$, assign objects in $\mathbf{x}_a$ to the positive region of any one cluster in D.

4   Experimental Results

In our experimental framework, we use some of the popular clustering algorithms as ensemble members (clusterers), such as K-means, K-Medoids [3] and RK-Means [5]. We can randomly choose the clustering algorithms as clusterers, or choose an appropriate clustering algorithm according to the prior knowledge. The focus of this paper is the consensus step. Therefore, for each generation step, the conventional K-means algorithm is selected as clusterer Φ in all experiments unless noted otherwise. There are 4 ensemble members for the initial decision of each experiment, i.e. $h_1 = 4$, and the number of clusters of ensemble members is K, unless noted otherwise.

For the experiments described in the following, the results are always averaged over 10 runs, and the standard deviation variances are also reported in results. The quality of the final clustering is evaluated by the accuracy. Due to the uncertainty associated with objects in boundary regions, the accuracy rate is computed based on positive regions.

Test 1: Results of Each Decision. In order to show the gradual decision (clustering) process intuitively, we conduct the experiment on an artificial data AD, and record the result of each decision.

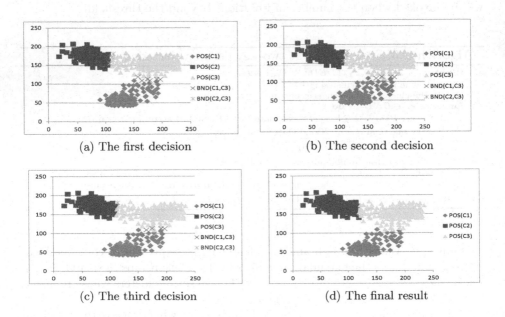

(a) The first decision

(b) The second decision

(c) The third decision

(d) The final result

Fig. 3. The results in the gradual decision process

Figure 3a, b and c are the three-way decision results in the process. Fig. 3d is the final two-way decision result and it is just the original data set. As we can see, our model can discover the underlying structure of the data set, and the data in the fringe are classified gradually during the process.

Figure 3a shows that most of the data has been divided definitely. There are 6 ones in $\text{BND}(C_1, C_3)$, the boundary regions of C_1 and C_3, and there are 3 data points in $\text{BND}(C_2, C_3)$. These points are left to defer decision. In other words, they are put into the boundary region of corresponding clusters and waiting for the next decision. From Fig. 3b, we see that two data points in the boundary region between C_1 and C_3 are divided definitely after the second decision; from Fig. 3c, we see that two data points in the boundary region between C_2 and C_3 are classified after the third decision. Finally, the two-way decision function is used and all the remaining data points are assigned into the specific categories.

Test 2: Results of Comparison Experiments. Table 1 gives a summary of data sets used in our experiments, where "No" denotes serial number of data sets. N, D and K means the number of objects, the number of attributes, the number of ground-truth clusters, respectively. "Distribution" denotes the distribution of clusters in every data set. Take the first data set as an example, Letter AB has two clusters, "766-789" means its first cluster has 766 objects and the second cluster has 789 objects. The first 7 data sets are UCI data sets [2] and the last three ones are artificial data sets. We compared the proposed method with the method presented by Mok et al. [6]. Table 2 shows the results on accuracy and computing time, and time in seconds (s). We record two computing time in Mok Algorithm, the actual computational time is recorded by Time2 and the time when the algorithm just obtains the best result is Time1. We record results of the first decision and the final decision in the proposed method.

Table 1. Datasets

No	Data set, U	N	D	K	Distribution
1	Letter AB	1555	16	2	766-789
2	Image segmentation	2310	19	7	330-330-330-330-330-330-330
3	Pedigits123	3342	16	3	1055-1143-1144
4	Pendigits1469	4398	16	4	1055-1056-1144-1143
5	Pendigits1234	4486	16	4	1055-1143-1144-1144
6	Waveform21	5000	21	3	1647-1657-1696
7	Landsat	6435	36	6	626-703-707-
8	AD5	10000	2	5	2000-2000-2000-2000-2000
9	AD6	15000	2	5	3000-3000-3000-3000-3000
10	AD7	15000	3	5	3000-3000-3000-3000-3000

The results in Table 2 show that the proposed approach is significantly better on both accuracy and computational time. The reason is that the proposed

Table 2. The results of GTWD-CE algorithm and compared algorithm

U	The first decision		The final decision		Mok Algorithm [6]		
	Accuracy	Time(s)	Accuracy	Time (s)	Accuracy	Time1 (s)	Time2 (s)
1	0.85 ± 0.00	0.06 ± 0.02	0.85 ± 0.00	0.07 ± 0.02	0.51 ± 0.00	0.46 ± 0.11	0.56 ± 0.14
2	0.53 ± 0.03	0.12 ± 0.03	0.54 ± 0.02	0.14 ± 0.04	0.51 ± 0.00	0.26 ± 0.07	0.85 ± 0.10
3	0.77 ± 0.09	0.26 ± 0.07	0.79 ± 0.09	0.27 ± 0.05	0.56 ± 0.18	0.60 ± 0.19	1.70 ± 0.68
4	0.88 ± 0.01	0.28 ± 0.04	0.88 ± 0.01	0.30 ± 0.05	0.59 ± 0.00	2.10 ± 0.37	2.41 ± 0.49
5	0.83 ± 0.01	0.28 ± 0.06	0.84 ± 0.01	0.63 ± 0.07	0.34 ± 0.03	0.89 ± 0.26	2.40 ± 0.47
6	0.39 ± 0.00	0.40 ± 0.11	0.39 ± 0.00	0.42 ± 0.13	0.36 ± 0.02	1.21 ± 0.42	2.94 ± 0.48
7	0.68 ± 0.00	0.30 ± 0.04	0.68 ± 0.00	0.32 ± 0.06	0.68 ± 0.00	1.36 ± 1.01	3.82 ± 1.71
8	0.81 ± 0.00	1.20 ± 0.54	0.82 ± 0.00	1.37 ± 0.56	0.73 ± 0.01	5.77 ± 1.05	9.23 ± 0.87
9	0.98 ± 0.01	1.90 ± 0.54	0.99 ± 0.01	2.11 ± 0.36	0.98 ± 0.01	16.78 ± 1.45	19.84 ± 2.10
10	0.99 ± 0.00	1.91 ± 2.69	0.99 ± 0.00	2.13 ± 3.31	0.71 ± 0.02	12.09 ± 3.02	20.02 ± 1.33

method is mainly based on cores not based on objects as the existing methods. The improvement of accuracy just prove the model is effective for clustering because the novel ideas are applied to the model such as the gradual decision and three-way decisions.

5 Conclusion

Cluster ensembles can combine the outcomes of several clusterings to a single clustering. This paper proposed a model of gradual three-way decision cluster ensemble. In the model, the number of ensemble members is increased gradually in each decision (iteration), which is different from the existing cluster ensemble methods. Meantime, the proposed three-way representation for a cluster, objects in the positive region or in the negative region are definitively do or not belong to the cluster respectively. Objects in the boundary region are fringe members and need defer decisions. The representation can describe both hard clustering and soft clustering. To enhance the performance of the proposed method, this paper proposed the concept of cluster core, which reflects the minimal granularity distribution structure agreed by all the ensemble members. Thus, the computing of the proposed algorithms are based on cores instead of on objects as other methods, which makes them high-efficiency. Besides, this paper introduced in detail how to implement the functions in the model. The compared experimental results show that the strategy of gradual three-way decisions is good at consensus clustering. Developing an efficient sampling method is the future work.

Acknowledgment. The authors would like to thank Mr. Lingchao Hu for his help to complete the experimental work. This work was supported in part by the National Natural Science Foundation of China under Grant Nos. 61751312, 61533020 and 61379114.

References

1. Chen, M., Miao, D.Q.: Interval set clustering. Expert Syst. Appl. **38**(4), 2923–2932 (2011)
2. UC Irvine Machine Learning Repository. http://archive.ics.uci.edu/ml/
3. Kaufman, L., Rousseeuw, P.: Clustering by Means of Medoids, pp. 405–416. North-Holland, Amsterdam (1987)
4. Lingras, P., Elagamy, A., Ammar, A., Elouedi, Z.: Iterative meta-clustering through granular hierarchy of supermarket customers and products. Inf. Sci. **257**, 14–31 (2014)
5. Lingras, P., Yan, R.: Interval clustering using fuzzy and rough set theory. In: Proceedings of 2004 IEEE Annual Meeting. Fuzzy Information, June 2004, Banff, Alberta, pp. 780–784 (2004)
6. Mok, P.Y., Huang, H.Q., Kwok, Y.L., Au, J.S.: A robust adaptive clustering analysis method for automatic identification of clusters. Pattern Recogn. **45**(8), 3017–3033 (2012)
7. Peters, G., Crespo, F., Lingras, P., Weber, R.: Soft clustering - fuzzy and rough approaches and their extensions and derivatives. Int. J. Approx. Reason. **54**, 307–322 (2013)
8. Vega-Pons, S., Ruiz-Shulcloper, J.: A survey of clustering ensemble algorithms. Int. J. Pattern Recogn. Artif. Intell. **25**(3), 337–372 (2011)
9. Waldr, A.: Sequential tests of statistical hypotheses. Ann. Math. Stat. **16**(2), 117–186 (1945)
10. Yao, Y.Y.: Granular computing and sequential three-way decisions. In: Lingras, P., Wolski, M., Cornelis, C., Mitra, S., Wasilewski, P. (eds.) RSKT 2013. LNCS (LNAI), vol. 8171, pp. 16–27. Springer, Heidelberg (2013). https://doi.org/10.1007/978-3-642-41299-8_3
11. Yao, Y.Y.: Three-way decisions and cognitive computing. Cogn. Comput. 1–12 (2016)
12. Yu, H.: A framework of three-way cluster analysis. In: Polkowski, L., Yao, Y., Artiemjew, P., Ciucci, D., Liu, D., Ślęzak, D., Zielosko, B. (eds.) IJCRS 2017. LNCS (LNAI), vol. 10314, pp. 300–312. Springer, Cham (2017). https://doi.org/10.1007/978-3-319-60840-2_22
13. Yu, H., Jiao, P., Yao, Y.Y., Wang, G.Y.: Detecting and refining overlapping regions in complex networks with three-way decisions. Inf. Sci. **373**, 21–41 (2016)
14. Yu, H., Zhou, Q.: A cluster ensemble framework based on three-way decisions. In: Lingras, P., Wolski, M., Cornelis, C., Mitra, S., Wasilewski, P. (eds.) RSKT 2013. LNCS (LNAI), vol. 8171, pp. 302–312. Springer, Heidelberg (2013). https://doi.org/10.1007/978-3-642-41299-8_29

Modes of Sequential Three-Way Classifications

Yiyu Yao[(✉)][iD], Mengjun Hu[(✉)][iD], and Xiaofei Deng[(✉)][iD]

Department of Computer Science, University of Regina,
Regina, Saskatchewan S4S 0A2, Canada
{yyao,hu258,deng200x}@cs.uregina.ca

Abstract. We present a framework for studying sequential three-way classifications based on a sequence of description spaces and a sequence of evaluation functions. In each stage, a pair of a description space and an evaluation function is used for a three-way classification. A set of objects is classified into three regions. The positive region contains positive instances of a given class, the negative region contains negative instances, and the boundary region contains those objects that cannot be classified as positive or negative instances due to insufficient information. By using finer description spaces and finer evaluations, we may be able to make definite classifications for those objects in the boundary region in multiple steps, which gives a sequential three-way classification. We examine four particular modes of sequential three-way classifications with respect to multiple levels of granularity, probabilistic rough set theory, multiple models of classification, and ensemble classifications.

Keywords: Sequential · Three-way decision · Classification

1 Introduction

In designing a classification model, we consider two components, namely, a description scheme of objects and an evaluation function of objects based on their descriptions. A description space of a specific format includes all the possible descriptions. An evaluation function evaluates the descriptions and indicates the degree to which a description leads to a positive instance of the given class.

In a two-way classification model, an object is classified as either a positive or negative instance. If the evaluation value of its description is high enough, then the object is classified as a positive instance; otherwise, it is classified as a negative instance. Accordingly, there are two regions: the positive region of those objects classified as positive instances and the negative region of those objects classified as negative instances. One threshold on the evaluation values is used for classification. However, when the information about objects involves

---

Y. Yao—This work is partially supported by a Discovery Grant from NSERC, Canada.

J. Medina et al. (Eds.): IPMU 2018, CCIS 854, pp. 724–735, 2018.
https://doi.org/10.1007/978-3-319-91476-3_59

uncertainty, we may not be able to classify them with confidence as positive or negative instances. In other words, either action may lead to a large error rate.

To resolve the difficulty in dealing with uncertainty, a three-way classification model applies a pair of thresholds on the evaluation function. If the evaluation value of an object is high enough (i.e., higher than or equal to one threshold), then it is classified as a positive instance. If its evaluation value is low enough (i.e., lower than or equal to the other threshold), then it is classified as a negative instance. If the evaluation value is between the two thresholds, a third option is used. The object is neither classified as a positive instance nor a negative one. Instead, a non-commitment decision is made about its classification. The third boundary region consists of those objects that cannot be definitely classified.

Although the boundary region offers a solution to make reasonable classifications on some objects with insufficient information, it introduces non-classification of some objects. A sequential three-way classification model offers an approach to refine the boundary region when more information is available about these objects. Such information may enable us to refine the description space and the evaluation function. Consequently, one may be able to classify the objects in the boundary region as positive or negative instances. This process involves a sequence of description spaces from coarser to finer and a sequence of evaluation functions. A pair of a description space and an evaluation function is applied in each stage of a sequential three-way classification to classify a set of objects in the boundary region from the previous stage. A sequential three-way classification provides an approach to gradually reducing non-commitment about the classification when more information is available. In sequential three-way classifications, an earlier stage involving less detailed information leads to a quick classification on some objects. This may reduce the cost of the classification.

In this paper, we present a framework of sequential three-way classifications by considering two sequences, that is, a sequence of description spaces and a sequence of evaluation functions. We examine four modes of sequential three-way classifications with respect to multiple levels of granularity, probabilistic rough set theory, multiple models of classification, and ensemble classifications.

2 A Framework of Sequential Three-Way Classifications

A three-way classification model provides a third option to meet the challenge of information uncertainty. As more information is obtained gradually, we have a general formulation of sequential three-way classifications.

2.1 Two-Way Classifications v.s. Three-Way Classifications

A classification problem involves classifying a universe of objects into different groups or classes. Based on the available information, the objects can be formally described in various formats. Suppose U is a universe of objects to be classified and DES is a description space that includes all possible descriptions of objects in a certain format, such as all possible logic formulas that are considered to describe the objects. With respect to a specific class, an evaluation function $Eval$: DES $\rightarrow$ [0, 1] is used to evaluate the descriptions and accordingly, the objects with those descriptions are classified. We assume that a greater evaluation value indicates a greater probability or confidence of an object with the description to be in the given class. In the ideal case with perfect knowledge, the evaluation function maps each object under the description into either 0 or 1. An object is classified to be in the given class if its value is 1, and classified to be not in the class if its value is 0. However, due to incomplete information, a description may be mapped to a value between 0 and 1. This leads to uncertainty about the classification of the corresponding objects. In order to classify these objects, it is necessary to specify thresholds on the evaluation values and classify the objects with a tolerance level of error.

A two-way classification model applies one threshold $\gamma \in [0, 1]$ to cut the evaluation values into two parts. Let $Des(x) \in$ DES denote the description of an object $x \in U$. If its evaluation value $Eval(Des(x))$ is high enough (i.e., above or equal to the threshold γ), then x is considered as a positive instance, that is, x is classified to be in the given class C; otherwise, x is considered as a negative instance and classified to be not in C. Accordingly, we can obtain two disjoint positive POS and negative NEG regions as:

$$
\begin{aligned}
\text{POS}(U) &= \{x \in U \mid Eval(Des(x)) \geq \gamma\}, \\
\text{NEG}(U) &= \{x \in U \mid Eval(Des(x)) < \gamma\}.
\end{aligned} \tag{1}
$$

The positive region contains the objects classified as positive instances and the negative region contains those classified as negative instances.

The determination of the threshold γ plays an important role in a two-way classification model. It reflects our confidence or the precision of the classification. However, one threshold cannot lead to a reasonable classification in some cases. If we choose a higher threshold (e.g., $\gamma = 0.9$), an object is more likely to be classified as a negative instance even though its evaluation value is intuitively high enough (e.g., the evaluation value is 0.8). Similarly, a lower threshold tends to classify more objects as positive instances even though their evaluation values are actually very low. The difficulty in determining one optimal and meaningful threshold is also related to the classification of those objects whose evaluation values are around the middle point 0.5. The available information for these objects is not sufficient enough to indicate whether they belong to the class or not. Thus, it is unreasonable to simply classify them as positive or negative instances.

A three-way classification model solves this problem by using a pair of thresholds to cut the evaluation values. It divides U into three pairwise disjoint positive POS, negative NEG, and boundary BND regions:

$$
\begin{aligned}
\text{POS}(U) &= \{x \in U \mid Eval(Des(x)) \geq \alpha\}, \\
\text{NEG}(U) &= \{x \in U \mid Eval(Des(x)) \leq \beta\}, \\
\text{BND}(U) &= \{x \in U \mid \beta < Eval(Des(x)) < \alpha\},
\end{aligned} \tag{2}
$$

where (α, β) is a pair of thresholds with $0 \leq \beta < \alpha \leq 1$. The computation and interpretation of the two thresholds have been discussed by researchers [4,5,18]. In this paper, we assume the two thresholds are given. If $Eval(\mathrm{Des}(x))$ is strong enough to indicate that x belongs to the given class (i.e., $Eval(\mathrm{Des}(x)) \geq \alpha$), then x is put into the positive region. If $Eval(\mathrm{Des}(x))$ is strong enough to indicate that x does not belong to the given class (i.e., $Eval(\mathrm{Des}(x)) \leq \beta$), then x is put into the negative region. Otherwise, we do not have enough information about x and are uncertain about its classification. Consequently, x is put into the boundary region. The actions associated with the three regions are:

(A) If $x \in \mathrm{POS}(U)$, then accept $x \in C$,

(R) If $x \in \mathrm{NEG}(U)$, then reject $x \in C$,

(N) If $x \in \mathrm{BND}(U)$, then neither accept nor reject $x \in C$.

That is, we accept the objects in $\mathrm{POS}(U)$ to be in the given class C, reject those in $\mathrm{NEG}(U)$, and make a non-commitment decision on those in $\mathrm{BND}(U)$. While a two-way classification model forces us to make a definite classification (i.e., either accepting or rejecting) on those objects with uncertain and insufficient information, the boundary region and the non-commitment choice of the classification provide a way to reflect such uncertainty.

2.2   Sequential Three-Way Classifications

The three-way classification discussed in the last section can be iteratively conducted by applying a sequence of evaluation functions on a sequence of description spaces. When more detailed information is obtained, a description space may be refined to include the new details. Accordingly, a sequence of description spaces can be formed from coarser to finer by gradually adding more information. On the other hand, the more detailed information may give us more insights into the problem and enable us to specify a more meaningful and comprehensive evaluation function. Hence, a sequence of evaluation functions can be obtained. A new evaluation function on a finer description space may enable us to classify an object in the previous boundary region as a positive or negative instance.

Suppose we have the following sequence of n pairs of description spaces and evaluation functions from coarser to finer:

$$(\mathrm{DES}_1, Eval_1) \preccurlyeq (\mathrm{DES}_2, Eval_2) \preccurlyeq \cdots \preccurlyeq (\mathrm{DES}_n, Eval_n). \tag{3}$$

The description space $\mathrm{DES}_i (1 \leq i \leq n)$ contains all the possible descriptions of objects in a certain format. The evaluation function $Eval_i : \mathrm{DES}_i \rightarrow [0,1]$ gives the evaluation values of each description. The relationship $(\mathrm{DES}_i, Eval_i) \preccurlyeq (\mathrm{DES}_{i+1}, Eval_{i+1})$ means that $(\mathrm{DES}_{i+1}, Eval_{i+1})$ is finer than $(\mathrm{DES}_i, Eval_i)$ (or, equivalently, $(\mathrm{DES}_i, Eval_i)$ is coarser than $(\mathrm{DES}_{i+1}, Eval_{i+1})$) in the sense that $(\mathrm{DES}_{i+1}, Eval_{i+1})$ is obtained based on more available information. However, the two pairs may not be necessarily different if the extra information does not enable us to refine the pair. It is also possible that either the description space or the evaluation function is refined and the other stays the same. An n-stage sequential three-way classification can be constructed by using these pairs one by one from coarser to finer. Let U_i be the set of objects to be classified in the ith stage and $\mathrm{Des}_i(x) \in \mathrm{DES}_i$ be the description of an object x in the description space DES_i. Given a universe U of objects, the three regions in the ith stage of an n-stage sequential three-way classification are constructed as: let $U_1 = U$ and $U_i = \mathrm{BND}_{i-1}(U_{i-1})(1 < i \leq n)$,

$$\mathrm{POS}_i(U_i) = \{x \in U_i \mid Eval_i(\mathrm{Des}_i(x)) \geq \alpha_i\},$$
$$\mathrm{NEG}_i(U_i) = \{x \in U_i \mid Eval_i(\mathrm{Des}_i(x)) \leq \beta_i\},$$
$$\mathrm{BND}_i(U_i) = \{x \in U_i \mid \beta_i < Eval_i(\mathrm{Des}_i(x)) < \alpha_i\}. \tag{4}$$

The corresponding actions are given by:

(A$_i$) If $x \in \mathrm{POS}_i(U_i)$, then accept $x \in C$,

(R$_i$) If $x \in \mathrm{NEG}_i(U_i)$, then reject $x \in C$,

(N$_i$) If $x \in \mathrm{BND}_i(U_i)$, then defer the classification on x to the next stage.

The pair of thresholds (α_i, β_i) is used in the ith stage which satisfies $0 \leq \beta_i < \alpha_i \leq 1$. It is reasonable to assume that we are more biased towards a deferment decision in an earlier stage where limited information is available [19]. This assumption leads to the following relationships of all the thresholds used:

$$0 \leq \beta_1 \leq \beta_2 \leq \cdots \leq \beta_n < \alpha_n \leq \alpha_{n-1} \leq \cdots \leq \alpha_1 \leq 1. \tag{5}$$

In an earlier stage, the pair of thresholds is more restrictive, that is, α is more closer to 1 and β is more closer to 0. Consequently, an object is more likely to be classified into the boundary region, which indicates a more conservative opinion due to limited information. Another assumption in the above sequential three-way classification is that we do not go back to update the definite classifications made in earlier stages, although those classifications may be inappropriate when finer description space and evaluation function are available in some stage later on. Consequently, in each stage, we only focus on refining the boundary region constructed in the previous stage.

The whole process of the above n-stage sequential three-way classification is illustrated in Fig. 1. In each stage, we adopt a finer pair of a description space and an evaluation function, and attempt to make definite classifications on the objects in the boundary region constructed in the previous stage.

If the boundary region in the last stage $\mathrm{BND}_n(U_n)$ is the empty set, we finally arrive at a two-way classification of all the objects in U. Otherwise, those objects in $\mathrm{BND}_n(U_n)$ are finally associated with a non-commitment decision about their classifications. All available information in the whole process is not sufficient for determining them to be in the given class or not.

3   Four Modes of Sequential Three-Way Classifications

A sequence of pairs of a description space and an evaluation function from coarser to finer is a basis for constructing a sequential three-way classification. The

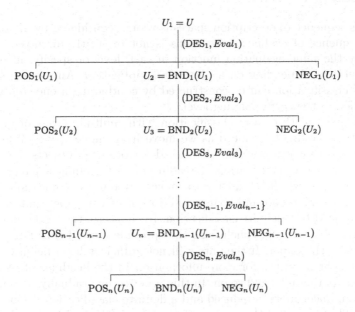

Fig. 1. An n-stage sequential three-way classification model

coarser description spaces and evaluation functions are due to less sufficient or more uncertain information in different stages. The information uncertainty is caused by different reasons in different contexts. In this section, we examine four specific modes of sequential three-way classifications.

3.1 Mode 1: Multiple Levels of Granularity

Granular computing deals with the philosophy, methodology and mechanism of structured thinking, structured problems solving and structured information processing [16]. It is based on a set of granular structures which provide multiple views of a particular application. Each granular structure involves multiple levels of granules in a specific view. Within one granular structure, a higher level granule is much coarser and contains more abstract description of objects. A lower level granule is much finer and offers more concrete description with more details. As we move from higher levels to lower levels, a granular structure naturally gives a sequence of description spaces from coarser to finer.

The construction of sequential three-way decisions in granular computing is examined in [17] with three main components, namely, multiple levels of granularity, multiple descriptions of objects, and three-way decisions at a particular level. Suppose there are n levels of granularity labelled by indices $\{1, 2, \cdots, n\}$ with 1 representing the highest and coarsest level and n the lowest and finest. Let DES_i denote the description space in the ith level of granularity. We can get a sequence of description spaces from coarser to finer as:

$$\mathrm{DES}_1 \preccurlyeq \mathrm{DES}_2 \preccurlyeq \cdots \preccurlyeq \mathrm{DES}_n. \tag{6}$$

While this sequence of description spaces is widely considered by majority studies, the sequence of evaluation functions is not explicitly discussed. One may simply use the same evaluation function in each level or specify different evaluation functions depending on a particular application. An n-stage sequential three-way classification can be constructed by conducting a one-step three-way classification in the corresponding level.

In a sequential three-way classification with multiple levels of granularity, uncertainty is gradually reduced as we move from higher levels of granularity to lower levels that contain more detailed descriptions of objects. For example, in order to make a classification of reading or not reading a paper, a reader may look at the title first and try to make a definite classification. The title alone gives a very broad and coarse description of the paper and may not be sufficient to lead to a definite classification. In this case, the reader may want to examine the abstract and conclusions that provide more details and more finer description of the paper. If they are still not sufficient for a definite classification, the reader may seek for more information in the headings of sections and subsections, paragraphs and so on. In this process, by gradually obtaining more information, uncertainty is reduced and a definite classification is more likely to be made. The sequential three-way classifications with multiple levels of granularity have been discussed with respect to many practical applications, such as face recognition [8,9] and multi-class statistical recognition [14].

3.2   Mode 2: Probabilistic Rough Set Theory

In rough set theory [11], a set of objects is described by a set of attributes. By using a subset of attributes, we may only be able to make definite classifications for some objects. For those objects with identical values on the subset of attributes but different classifications, we cannot make a definite classification. By adding more attributes, we can refine the description of these objects gradually, which may enable us to distinguish them. This process leads to the construction of sequential three-way classifications.

Suppose a universe of objects U is described by a set of attributes AT. By focusing on a subset of attributes $A \subseteq AT$, we can construct a description space DES_A by listing all the logic formulas in the following format:

$$\bigwedge_{a \in A} (a = v_a), \tag{7}$$

where v_a is a value in the domain of attribute a. That is, DES_A contains all the formulas that involve only logic conjunction with each attribute in A appearing exactly once. Given a description or formula $p \in \text{DES}_A$, the set of objects satisfying it is denoted as $m(p)$. Given a class $X \subseteq U$, if a large portion of $m(p)$ is included in X, then p is used as a description of positive instances, that is, all objects satisfying p will be classified as positive instances. If a large portion of $m(p)$ is not included in X, then p is used as a description of negative instances. Otherwise, p cannot be used to classify objects with respect to the class X. Formally, the three regions are constructed as follows:

$$\text{POS}(U) = \bigcup \{m(p) \subseteq U \mid p \in \text{DES}_A, Pr(X|m(p)) \geq \alpha\},$$

$$\text{NEG}(U) = \bigcup \{m(p) \subseteq U \mid p \in \text{DES}_A, Pr(X|m(p)) \leq \beta\},$$

$$\text{BND}(U) = \bigcup \{m(p) \subseteq U \mid p \in \text{DES}_A, \beta < Pr(X|m(p)) < \alpha\}, \qquad (8)$$

where $Pr(X|m(p)) = \frac{|X \cap m(p)|}{|m(p)|}$ and the two thresholds satisfy $0 \leq \beta < \alpha \leq 1$. The three regions are called three-way probabilistic approximations of X in probabilistic rough set theory [18].

A sequential three-way classification can be formed by gradually considering more attributes to refine the description space. Suppose we have the following sequence of subsets of attributes:

$$A_1 \subset A_2 \subset \cdots \subset A_n, \qquad (9)$$

where $A_1 \neq \emptyset$, $A_n \subseteq AT$ and A_i is a proper subset of $A_{i+1}(1 \leq i < n)$. We get the following sequence of description spaces from coarser to finer:

$$\text{DES}_{A_1} \preccurlyeq \text{DES}_{A_2} \preccurlyeq \cdots \preccurlyeq \text{DES}_{A_n}. \qquad (10)$$

A description space $\text{DES}_{A_{i+1}}$ is considered to be finer than DES_{A_i} in the sense that its formulas involve more attributes which may induce smaller and finer sets of objects satisfying these formulas. These finer sets of objects may enable us to transfer some objects in the boundary region into the positive or negative region and, accordingly, make a definite classification. We can construct a sequential three-way classification with respect to a given class X in n stages. The regions in the ith stage are:

$$\text{POS}_i(U_i) = \bigcup \{m(p) \subseteq U_i \mid p \in \text{DES}_{A_i}, Pr(X|m(p)) \geq \alpha_i\},$$

$$\text{NEG}_i(U_i) = \bigcup \{m(p) \subseteq U_i \mid p \in \text{DES}_{A_i}, Pr(X|m(p)) \leq \beta_i\},$$

$$\text{BND}_i(U_i) = \bigcup \{m(p) \subseteq U_i \mid p \in \text{DES}_{A_i}, \beta_i < Pr(X|m(p)) < \alpha_i\}, \qquad (11)$$

where $U_1 = U$ and $U_i = \text{BND}_{i-1}(U_{i-1})(1 < i \leq n)$. This provides an approach to constructing sequential three-way approximations in probabilistic rough set theory.

In sequential three-way classifications with probabilistic rough set theory, the uncertainty is reduced by increasing the number of attributes to form the formulas in a description space. A small number of attributes gives a quick definite classification for some objects (i.e., those in the positive and negative regions) and involves large degree of uncertainty for others (i.e., those in the boundary region). By adding more attributes, we are able to distinguish more objects, reduce the uncertainty for their descriptions and make definite classifications.

3.3   Mode 3: Multiple Models of Classification

Model selection is a fundamental issue in various fields, such as ecology [1], statistics [2], biology [3], economics [7], and social sciences [15]. The main task of model selection is to select one single (or a few) optimal model from a set of candidate models that can solve a problem. For a given classification problem, there may exist multiple candidate models that differ in complexity, cost, efficiency and etc. The selection of these models requires a consideration of tradeoffs between the processing cost and the quality or accuracy of classification results.

A reasonable preference of the models is necessary in model selection. Occam's razor or the principle of parsimony, which gives preferences to simpler models, is widely adopted and argued by researchers in various fields [6]. A simpler classification model uses smaller and simpler description space and an evaluation function with less parameters. In general, it may be able to give a quick classification with a less processing cost. The classification results may have a large error rate due to the limited information used by the model. In contrast, a more complex model may be able to give more accurate classification results with a lower error rate. A larger processing cost may be introduced due to the larger and more complicated description space and extra parameters used in the evaluation function. In the selection of multiple candidate models of classification, we aim at finding a simplest model that can induce satisfactory classification results with a required level of accuracy.

In many classification models, all objects are treated the same way. On the other hand, typical instances of a class usually can be easily and correctly classified by a simpler model. It is not quite advantageous to use a more complex model. If some instances cannot be classified by a simpler model with a required level of accuracy, a more complex model is necessary. In order to balance the processing cost and the accuracy of results, we suggest to adopt the sequential three-way classifications by applying a sequence of classification models from simpler to more complex. If the classification results given by a simpler model are accurate enough, we will classify the objects according to it. Otherwise, a more complex model in the sequence will be applied to get more accurate classification results. Accordingly, a simpler model is adopted whenever its classification results are satisfactory, which has a lower processing cost.

Suppose we have a sequence of classification models from simpler to more complex as:

$$M_1 \preccurlyeq M_2 \preccurlyeq \cdots \preccurlyeq M_n. \tag{12}$$

A classification model is considered to be more complex if its descriptions of objects involve more details or the evaluation function is more complex and costs more to compute. Let $\mathrm{DES}_i(1 \leq i \leq n)$ denote the description space in model M_i and $Eval_i$ denote the evaluation function. Given a universe U of objects to be classified, the models are applied one by one from M_1 to M_n. In the ith stage where M_i is applied, the positive, negative and boundary regions are constructed as: let $U_1 = U$ and $U_i = \mathrm{BND}_{i-1}(U_{i-1})(1 < i \leq n)$,

$$
\begin{aligned}
\mathrm{POS}_i(U_i) &= \{x \in U_i \mid Eval_i(\mathrm{Des}_i(x)) \geq \alpha_i\}, \\
\mathrm{NEG}_i(U_i) &= \{x \in U_i \mid Eval_i(\mathrm{Des}_i(x)) \leq \beta_i\}, \\
\mathrm{BND}_i(U_i) &= \{x \in U_i \mid \beta_i < Eval_i(\mathrm{Des}_i(x)) < \alpha_i\},
\end{aligned}
\tag{13}
$$

where $\mathrm{Des}_i(x)$ is the description of an object x in the description space DES_i and (α_i, β_i) is a pair of thresholds used in the ith stage. One may use a pair of thresholds in a model to control the desired accuracy of classification results. If an object is classified as a positive or negative instance in the ith stage, model M_i is adopted to classify the object. Otherwise, the model M_{i+1} is applied in the next stage. In this way, we use the simplest model in the sequence that gives a desired level of classification accuracy with a minimal processing cost.

3.4   Mode 4: Sequential Ensemble Classifications

Ensemble classification [10,12,13] involves the synthesis of classification results from multiple classification models called a committee. These classification results may be combined by using various ensemble techniques, such as majority voting and weighted average. If there is a large degree of conflict between the classification results from different models, one may have difficulties in combining them into a definite classification. In this case, one may continue with another round of ensemble classification which involves a larger committee with more classification models. This can be iteratively conducted until a general agreement on a definite classification can be made by the committee. The process results in a sequential ensemble classification involving multiple committees.

Suppose we have a sequence of committees from smaller to larger:

$$\mathcal{C}_1 \subseteq \mathcal{C}_2 \subseteq \cdots \subseteq \mathcal{C}_n, \tag{14}$$

where $\mathcal{C}_i (1 \leq i \leq n)$ is a set of classification models. The committee $\mathcal{C}_i$ is used in the ith stage. Given an object x in a universe U, each model in $\mathcal{C}_i$ computes its classification result of x independently. It should be noted that the models in $\mathcal{C}_i$ may use different descriptions of objects and evaluation functions to make the classification. Their individual results are synthesized by using a function. Although there are various functions to do the synthesis, in our discussion, we assume the values of the function evaluate the degree to which the objects are positive instances after the synthesis of the results. Thus, we call it the evaluation function used by the committee $\mathcal{C}_i$ and denote it as $Eval_{\mathcal{C}_i}$. Accordingly, the three regions in the ith stage of a sequential ensemble classification are constructed as: let $U_1 = U$ and $U_i = \mathrm{BND}_{i-1}(U_{i-1})(1 < i \leq n)$,

$$\mathrm{POS}_i(U_i) = \{x \in U_i \mid Eval_{\mathcal{C}_i}(x) \geq \alpha_i\},$$
$$\mathrm{NEG}_i(U_i) = \{x \in U_i \mid Eval_{\mathcal{C}_i}(x) \leq \beta_i\},$$
$$\mathrm{BND}_i(U_i) = \{x \in U_i \mid \beta_i < Eval_{\mathcal{C}_i}(x) < \alpha_i\}, \tag{15}$$

where (α_i, β_i) is a pair of thresholds used in the ith stage. If the synthesized result of the ensemble classification in the ith stage strongly suggests x to be a positive instance (i.e., $Eval_{\mathcal{C}_i}(x) \geq \alpha_i$), then x is classified as a positive instance. If the synthesized result strongly suggests x to be a negative instance (i.e., $Eval_{\mathcal{C}_i}(x) \leq \beta_i$), then x is classified as a negative instance. Otherwise, the classification of x is delayed in the next stage where a larger committee with more classification models is used.

The results of a sequential ensemble classification are affected by the choice of the initial committee, that is, C_1 in the formulation. In order to avoid an easy agreement in the committee, one should include various classification approaches in order to capture different aspects and views of the data set. Moreover, one should also select a reasonable and meaningful size of C_1 to start with. A small size may result in an easy agreement and a large size may not be able to take the advantage of quick decisions in the earlier stages of a sequential approach. The determination of the initial committee is an important issue to be solved.

The sequential ensemble classifications can be generalized with any occasion involving group decision makings. A group of experts may give different opinions of the decisions on a set of objects or entities. These opinions may enable us to make a definite decision on some objects, but not on others. In the latter case, more experts can be engaged until definite decisions can be made on all the objects. One may take the review process of a paper as an example. If a decision of accepting or rejecting the paper can be made by a group of reviewers (e.g., majority of the reviewers strongly accept or reject the paper), then the paper is put into the positive or negative region and a corresponding action can be taken. Otherwise, if a definite decision cannot be made due to large degree of conflict among the opinions from the reviewers (e.g., some reviewers strongly accept the paper and a similar number of reviewers strongly reject the paper), the editor may want to find more reviewers, collect more opinions from different people and try to arrive at a definite decision on the paper.

4   Conclusions

We present a general framework of sequential three-way classifications and, within it, examine four modes. Based on a description space and an evaluation function, a three-way classification model classifies a set of objects into three regions, namely, the positive, negative and boundary regions. While the objects in the positive and negative regions are definitely classified, those in the boundary regions are not. The two sequences of description spaces and evaluation functions from coarser to finer are the basis of a sequential three-way classification. By adopting a finer description space and a finer evaluation function, we may reduce the uncertainty about the boundary region and arrive at definite classifications. Four modes of sequential three-way classifications demonstrate different causes of uncertainty. While current research on granular computing and probabilistic rough sets mainly focuses on the sequence of description spaces, current research on model selection and ensemble classifications mainly focuses on the sequence of evaluation functions. An examination of both sequences in all these four modes may be a direction for future work. Another direction may be a more detailed exploration of the presented framework and the experiments with respect to specific applications.

References

1. Aho, K., Derryberry, D.W., Peterson, T.: Model selection for ecologists: the world-views of AIC and BIC. Ecology **95**, 631–636 (2014). https://doi.org/10.1890/13-1452.1
2. Breiman, L.: Statistical modeling: the two cultures. Stat. Sci. **16**, 199–231 (2001). https://doi.org/10.1214/ss/1009213726
3. Burnham, K.P., Anderson, D.R.: Model Selection and Multimodel Inference: A Practical Information-Theoretic Approach, 2nd edn. Springer, Heidelberg (2002). https://doi.org/10.1007/b97636
4. Deng, X.F., Yao, Y.Y.: An information-theoretic interpretation of thresholds in probabilistic rough sets. In: Li, T., Nguyen, H.S., Wang, G., Grzymala-Busse, J., Janicki, R., Hassanien, A.E., Yu, H. (eds.) RSKT 2012. LNCS (LNAI), vol. 7414, pp. 369–378. Springer, Heidelberg (2012). https://doi.org/10.1007/978-3-642-31900-6_46
5. Herbert, J.P., Yao, J.T.: Game-theoretic rough sets. Fundamenta Informaticae **108**, 267–286 (2011). https://doi.org/10.3233/FI-2011-423
6. Kemeny, J.: The use of simplicity in induction. Philos. Rev. **62**, 391–408 (1953). https://doi.org/10.2307/2182878
7. Keuzenkamp, H.A., McAleer, M.: Simplicity, scientific inference, and econometric modelling. Econ. J. **105**, 1–21 (1995). https://doi.org/10.2307/2235317
8. Li, H.X., Zhang, L.B., Huang, B., Zhou, X.Z.: Sequential three-way decision and granulation for cost-sensitive face recognition. Knowl.-Based Syst. **91**, 241–251 (2016). https://doi.org/10.1016/j.knosys.2015.07.040
9. Li, H.X., Zhang, L.B., Zhou, X.Z., Huang, B.: Cost-sensitive sequential three-way decision modeling using a deep neural network. Int. J. Approx. Reason. **85**, 68–78 (2017). https://doi.org/10.1016/j.ijar.2017.03.008
10. Opitz, D., Maclin, R.: Popular ensemble methods: an empirical study. J. Artif. Intell. Res. **11**, 169–198 (1999). https://doi.org/10.1613/jair.614
11. Pawlak, Z.: Rough classification. Int. J. Man-Mach. Stud. **20**, 469–483 (1984). https://doi.org/10.1016/S0020-7373(84)80022-X
12. Polikar, R.: Ensemble based systems in decision making. IEEE Circ. Syst. Mag. **6**, 21–45 (2006). https://doi.org/10.1109/MCAS.2006.1688199
13. Rokach, L.: Ensemble-based classifiers. Artif. Intell. Rev. **33**, 1–39 (2010). https://doi.org/10.1007/s10462-009-9124-7
14. Savchenko, A.V.: Fast multi-class recognition of piecewise regular objects based on sequential three-way decisions and granular computing. Knowl.-Based Syst. **91**, 252–262 (2016). https://doi.org/10.1016/j.knosys.2015.09.021
15. Weakliem, D.L.: Hypothesis Testing and Model Selection in the Social Science. Guilford Press, New York (2016)
16. Yao, Y.Y.: A triarchic theory of granular computing. Granular Comput. **1**, 145–157 (2016). https://doi.org/10.1007/s41066-015-0011-0
17. Yao, Y.Y.: Granular computing and sequential three-way decisions. In: Lingras, P., Wolski, M., Cornelis, C., Mitra, S., Wasilewski, P. (eds.) RSKT 2013. LNCS (LNAI), vol. 8171, pp. 16–27. Springer, Heidelberg (2013). https://doi.org/10.1007/978-3-642-41299-8_3
18. Yao, Y.Y.: Probabilistic rough set approximations. Int. J. Approx. Reason. **49**, 255–271 (2008). https://doi.org/10.1016/j.ijar.2007.05.019
19. Yao, Y.Y., Deng, X.F.: Sequential three-way decisions with probabilistic rough sets. In: Wang, Y., Celikyilmaz, A., Kinsner, W., Pedrycz, W., Leung, H., Zadeh, L.A. (eds.) ICCI-CC 2011, pp. 120–125 (2011). https://doi.org/10.1109/COGINF.2011.6016129

Determining Strategies
in Game-Theoretic Shadowed Sets

Yan Zhang and JingTao Yao[⊠]

Department of Computer Science, University of Regina,
Regina, Saskatchewan S4S 0A2, Canada
{zhang83y,jtyao}@cs.uregina.ca

Abstract. A three-way approximation of shadowed sets maps the membership grades of all objects into a three-value set with a pair of thresholds. The game-theoretic shadowed sets (GTSS) determine and interpret a pair of thresholds of three-way approximations based on a principle of tradeoff with games. GTSS formulate competitive games between the elevation and reduction errors. The players start from the initial thresholds (1,0) and perform the certain strategies to change the thresholds in the game. The games are repeated with the updated thresholds to gradually reach the suitable thresholds. However, starting from a pair of randomly selected non-(1,0) thresholds is not examined in GTSS. We propose a game approach to make it possible for GTSS starting from a pair of randomly selected thresholds and then determine the strategies associated with them. In particular, given a pair of randomly chosen initial thresholds, we use a game mechanism to determine the change directions that players prefer to make on the initial thresholds. The proposed approach supplements the GTSS, and can be added in the game formulation and repetition learning phases. We explain the game formulation, equilibrium analysis, and the determination of strategies in this paper. An example demonstrates how the proposed approach can supplement GTSS to obtain the thresholds of three-way approximations of shadowed sets when starting from randomly selected thresholds.

Keywords: Game-theoretic shadowed sets
Three-way approximations of shadowed sets · Game theory
Three-way approximations

1 Introduction

A generalized framework of three-way approximations of fuzzy sets maps the membership grades of the objects in U to a set of three values $\{\mathbf{n}, \mathbf{m}, \mathbf{p}\}$ [13,15]. A shadowed set proposed by Pedrycz is constructed from a fuzzy set μ_A and maps the membership grades $\mu_A(x)$ of all objects in U to the set $\{0, [0,1], 1\}$ based on a pair of thresholds (α, β) [7]. The interval set $[0,1]$ represents the uncertainty [7]. We may use a value σ with $0 < \sigma < 1$ to represent the uncertainty in the shadowed sets to obtain a three-way approximation of shadowed sets, i.e.,

© Springer International Publishing AG, part of Springer Nature 2018
J. Medina et al. (Eds.): IPMU 2018, CCIS 854, pp. 736–747, 2018.
https://doi.org/10.1007/978-3-319-91476-3_60

$T : U \rightarrow \{0, \sigma, 1\}$. Therefore, a three-way approximation of shadowed sets can be viewed as a special case of the generalized three-way approximations of fuzzy sets, in which $\mathbf{n} = 0$, $\mathbf{m} = \sigma$, and $\mathbf{p} = 1$ [13]. A three-way approximation of shadowed sets is defined by the membership grades of objects and a pair of thresholds (α, β) while $0 \leq \beta \leq \alpha \leq 1$. The membership grade $\mu_A(x)$ of an object x indicates the degree of the concept A is applicable to x [7, 15]. The elevation and reduction operations change the original membership grades to 1, 0, or σ. These operations produce the elevation and reduction errors which show the difference between the original membership grades and the corresponding elevated or reduced values [3].

One of the fundamental issues of applying the three-way approximations of shadowed sets is the determination and interpretation of the pair of thresholds (α, β) [13]. Yao, Wang and Deng introduce a general optimization-based framework for interpreting and determining the thresholds [13]. Three principles, i.e., a principle of uncertainty invariance, a principle of minimum distance, and a principle of least cost are summarized [13]. Pedrycz uses symmetric $(\alpha, 1 - \alpha)$ model and then computes α by minimizing the difference between the shadowed area and the sum of the elevated and reduced area [8]. Tahayori et al. propose analytical formulas to calculate thresholds when constructing shadowed sets [11]. Grzegorzewski explores the nearest interval approximation of a fuzzy number based on a distance measure [5]. He also proposes an algorithm for fuzzy number approximation that bridge the interval and trapezoidal approximation [6]. Deng and Yao propose a decision-theoretic approach to calculate thresholds by minimizing decision costs, which obtain the thresholds according to the principle of least cost [3, 13]. The error-based $(\alpha, \beta) = (0.75, 0.25)$ model is derived by considering a loss function satisfying additional properties [3]. Game-theoretic shadowed sets (GTSS) determine and interpret the pair of thresholds of three-way approximations according to a principle of tradeoff with games [16].

GTSS gradually reach a balanced threshold pair by repeatedly formulating competitive games between the elevation and reduction errors and finding the tradeoff between these errors [16]. Two game players, the elevation errors and reduction errors, start from the initial thresholds $(\alpha, \beta) = (1, 0)$. The strategies performed by two players are decreasing α and increasing β. Players compete with each other to reach a tradeoff among multiple strategy profiles. The games are repeated with the updated thresholds until the thresholds corresponding to the equilibrium satisfy the stop conditions. In fact, setting the initial thresholds as $(\alpha, \beta) = (1, 0)$ is a very special case. What if the game players start from a pair of randomly chosen thresholds? In this case, how can we determine the strategies performed by two players?

In this paper we formulate a game to solve these problems. Given a pair of randomly chosen initial thresholds, we use a game mechanism to determine the change directions that players prefer to make on these initial thresholds. The game players are elevation errors and reduction errors. The strategies of each player are the possible change directions of two initial thresholds. The equilibrium of the game is a strategy profile on which two players reach a compromise on how to change the initial thresholds. The proposed approach supplements the game-theoretic shadowed set model, and can be added in the game formulation and repetition learning phases. This paper explains the game formulation, equilibrium analysis, and strategy determination in detail. We use an example to show how the proposed approach can supplement GTSS to obtain the thresholds of shadowed sets from a perspective of tradeoff.

2   Background Knowledge

We briefly introduce the background concepts about three-way approximations of shadowed sets and game-theoretic shadowed sets.

2.1   Three-Way Approximations of Shadowed Sets

A three-way approximation of fuzzy sets in the universe U maps the membership grades of all objects in U to a three-value set $\{\mathbf{n}, \mathbf{m}, \mathbf{p}\}$ based on a pair of thresholds (α, β) while $0 \leq \beta \leq \alpha \leq 1$ [3,13],

$$T_{(\alpha,\beta)}(\mu_A(x)) = \begin{cases} \mathbf{p}, & \mu_A(x) \geq \alpha, \\ \mathbf{m}, & \beta < \mu_A(x) < \alpha, \\ \mathbf{n}, & \mu_A(x) \leq \beta. \end{cases} \tag{1}$$

A three-way approximations of shadowed sets can be viewed as a special case of, in which $\mathbf{n} = 0$, $\mathbf{m} = \sigma$, and $\mathbf{p} = 1$ [13]. That is, a three-way approximation of shadowed sets maps the membership grades of all objects in U to a three-value set $\{0, \sigma, 1\}$. The membership grades greater than or equal to α are elevated to 1; the membership grades between α and σ are reduced to σ; the membership grades between σ and β are elevated to σ; the membership grades less than or equal to β are reduced to 0. The three-way approximations of shadowed sets use two operations, elevation operation and reduction operation, to change the original membership grades. The elevated areas or elevation errors, and reduced areas or reduction errors are produced, as the dotted areas and lined areas shown in the Fig. 1(a). Figure 1(b) shows a three-way approximation of shadowed sets after applying the elevation and reduction operations on the membership grades of all objects.

The single value σ represents a situation in which we are far more confident about including an element or excluding an element in the concept A. It shows the most uncertainty in determining elements whose membership grades are around σ. There are different approaches to select the value of σ. Cattaneo and Ciucci use 0.5 to replace the membership grades of the elements in the shadows [1,2]. Deng and Yao use the mean value of the membership grades to represent the shadows [4]. We use $\sigma = 0.5$ to represent a situation of completely uncertainty in this paper. Because 0.5 is in the unit interval, and it is far from either the full membership grade 1 or the null membership grade 0. The vagueness is localized in the shadowed area as opposed to fuzzy sets where

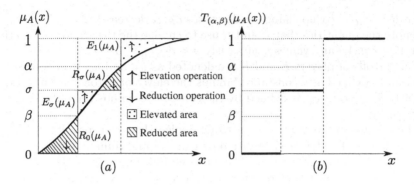

Fig. 1. A three-way approximation of shadowed sets

the vagueness is spread across the entire universe [8]. Shadowed sets have been applied in different research areas, such as granular computing [10,12], image processing [7], clustering analysis [9,14].

If the membership grades are defined by a continuous function, the elevation and reduction errors can be calculated as,

$$E_{(\alpha,\beta)}(\mu_A) = \int_{x_\alpha}^{x_m} (1 - \mu_A(x))dx + \int_{x_\beta}^{x_\sigma} (\sigma - \mu_A(x))dx, \tag{2}$$

$$R_{(\alpha,\beta)}(\mu_A) = \int_0^{x_\beta} \mu_A(x)dx + \int_{x_\sigma}^{x_\alpha} (\mu_A(x) - \sigma)dx. \tag{3}$$

For discrete universe of discourse, we have collection of membership values instead of continuous functions.

No matter which thresholds change and how they change, the elevation errors and reduction errors always change in opposite directions. The decrease of one type of errors inevitably brings the increase of the other type of errors. The balanced shadowed set based three-way approximations are expected to represent a tradeoff between the elevation and reduction errors. The game-theoretic shadowed sets are proposed to find a tradeoff between errors by formulating competitive games.

2.2 Game-Theoretic Shadowed Sets

Game-theoretic shadowed sets (GTSS) use game mechanism to formulate competitive games between the elevation and the reduction errors. The game players are the elevation and reduction errors which are denoted by E and R.

The set of strategy profiles S is made up of the possible strategies or actions performed by two involved players. GTSS use $(\alpha, \beta) = (1, 0)$ as the initial threshold values. All the strategies or actions performed by both players are the changes of the initial thresholds. The player E performs increasing β and the player R performs decreasing α. The strategy set of player E is $S_E = \{\beta \text{ no change}, \beta \text{ increases } c_E, \beta \text{ increases } 2 \times c_E, ...\}$. The strategy set of player R is $S_R = \{\alpha$ no change, α decreases c_R, α decreases $2 \times c_R, ...\}$. c_E and c_R denote the quantities that E and R use to change the thresholds, respectively. c_E and c_R can be any values that satisfy $0 < c_E, c_R \leq 0.1$.

The payoffs of players E and R are denoted as u_E and u_R, respectively. The payoff u_E is defined as a constant C minus the elevation errors shown in Eq. (2). Similarly, the payoff u_R is defined by the constant C minus the reduction errors shown in Eq. (3).

The involved players are trying to maximize their own payoffs in the competitive games. The balanced solution or game equilibrium is a strategy profile from which both players benefit. The game equilibrium represents both players reach a compromise or tradeoff on the conflict. The strategy profile (s_i, t_j) is a Nash equilibrium, if for players E and R, s_i and t_j are the best responses to each other.

GTSS compare the payoffs of both players under the initial thresholds and under the thresholds corresponding to the current equilibrium. The stop condition is set as both players lose their payoffs or the gain of one player's payoff is less than the loss of the other player's payoff.

GTSS attempt to obtain more suitable thresholds with the repetition of thresholds modification. Assuming that the initial thresholds are (α, β), and the thresholds corresponding to the equilibrium are (α^*, β^*). The formulation of the subsequent games depends on if the thresholds (α^*, β^*) satisfy the selected stop conditions. If the thresholds (α^*, β^*) do not satisfy the stop condition, the games are repeated with (α^*, β^*) as the updated thresholds. If the thresholds (α^*, β^*) satisfy the stop condition, the repetition of games is terminated. The thresholds (α, β) are used as the result thresholds.

3   Formulating a Game to Determine the Strategies

In this section, we discuss how to use a game mechanism to determine the strategies for both players in GTSS, given a pair of randomly selected initial thresholds. The game formulation, payoff table, equilibrium analysis, and the determination of strategies are discussed in detail.

3.1   Game Formulation

In the game formulation, we define three elements contained in a game $G = \{O, S, u\}$. O is a set of game players, S is a set of strategy profiles, and u is a set of payoffs. The game players are the elevation errors and the reduction errors which are denoted by E and R, respective, i.e., $O = \{E, R\}$.

Strategies. The set of strategy profiles S is made up of the possible strategies or actions performed by two players E and R. The set of strategy profiles is $S = S_E \times S_R$, where $S_E = \{s_1, s_2, ..., s_{k1}\}$ is a set of possible strategies for player E, and $S_R = \{t_1, t_2, ..., t_{k2}\}$ is a set of possible strategies for player R. k_1 and k_2 denote the numbers of strategies performed by the players E and R, respectively.

All the strategies or actions performed by both players are the changes to the initial thresholds (α, β). Assuming that the randomly selected initial thresholds are (α, β) with $0 \leq \beta \leq 0.5 \leq \alpha \leq 1$, there are five possible changes that can be made on (α, β), i.e., (α, β) no change, increasing α, decreasing α, increasing β, decreasing β. The strategy sets of two players are the same, and they contain five possible strategies ($k_1 = k_2 = 5$ and $s_i = t_i, i = 1, 2, ..., 5$),

$$S_E = S_R = \{\text{no change}, \alpha \uparrow c, \alpha \downarrow c, \beta \uparrow c, \beta \downarrow c\}. \tag{4}$$

c denotes a constant value showing the rate of change made on thresholds, and we set $0 < c < 0.1$. The up arrow $\uparrow$ denotes increasing a value and the down arrow $\downarrow$ denotes decreasing a value. Please note that the above equation is a general strategy set for a pair of initial thresholds. Sometimes, in order to grantee the constraint $0 \leq \beta \leq \sigma \leq \alpha \leq 1$, some strategies in the set may not work. For example, if the initial thresholds $(\alpha, \beta) = (1, 0)$, both players can not increase α and decrease β because α has the maximal value 1 and β has the minimal value 0. The strategy sets of players contain three strategies, i.e., $S_E = S_R = \{\text{no change}, \alpha \downarrow c, \beta \uparrow c\}$.

Payoffs. The payoff set is $u = \{u_E, u_R\}$, and u_E and u_R denote the payoffs of players E and R, respectively. Given a strategy profile $p = (s, t)$ with player E performing s and player R performing t, the payoffs of E and R are $u_E(s, t)$ and $u_R(s, t)$. The strategies s and t performed by E and R are the changes of thresholds (α, β). The payoffs $u_E(s, t)$ and $u_R(s, t)$ are in fact the functions of thresholds (α, β). The thresholds (α, β) are determined by the strategies s and t. That means (α, β) are the result caused by the strategies s and t. The payoff functions $u_E(\alpha, \beta)$ and $u_R(\alpha, \beta)$ are defined by the elevation errors and the reduction errors, respectively. Since we are interested in measuring profits or payoffs in the game-theoretic analysis, we use a constant value C minus the elevation or reduction errors as corresponding payoff functions,

$$u_E(\alpha, \beta) = C - E_{(\alpha, \beta)}(\mu_A),$$
$$u_R(\alpha, \beta) = C - R_{(\alpha, \beta)}(\mu_A). \tag{5}$$

$E_{(\alpha, \beta)}(\mu_A)$ is the elevation error defined in Eq. (2). $R_{(\alpha, \beta)}(\mu_A)$ is the reduction error defined in Eq. (3). The constant value C is defined as the area covered by the function $y = 1$, i.e., $C = \int_x 1 dx$. Please note that the values of thresholds (α, β) in the payoff functions $u_E(\alpha, \beta)$ and $u_R(\alpha, \beta)$ are determined by the strategies performed by both players. The payoff of each player depends on the strategies or actions performed by both game players. The strategy performed by one game player can influence the payoff of the other player.

Payoff Table. We use payoff tables to represent the two-player games. Table 1 shows the payoff table of the formulated game in which the strategy set contains five strategies. The constant c is omitted in the payoff table. $\uparrow$ and $\downarrow$ represent

Table 1. The payoff table of the formulated game

				R		
		No change	$\alpha\uparrow$	$\alpha\downarrow$	$\beta\uparrow$	$\beta\downarrow$
	No change	$\langle u_E(\alpha,\beta), u_R(\alpha,\beta)\rangle$	$\langle u_E(\alpha\uparrow,\beta), u_R(\alpha\uparrow,\beta)\rangle$	$\langle u_E(\alpha\downarrow,\beta), u_R(\alpha\downarrow,\beta)\rangle$	$\langle u_E(\alpha,\beta\uparrow), u_R(\alpha,\beta\uparrow)\rangle$	$\langle u_E(\alpha,\beta\downarrow), u_R(\alpha,\beta\downarrow)\rangle$
	$\alpha\uparrow$	$\langle u_E(\alpha\uparrow,\beta), u_R(\alpha\uparrow,\beta)\rangle$	$\langle u_E(\alpha\uparrow\uparrow,\beta), u_R(\alpha\uparrow\uparrow,\beta)\rangle$	$\langle u_E(\alpha\uparrow,\beta), u_R(\alpha\uparrow,\beta)\rangle$	$\langle u_E(\alpha\uparrow,\beta\uparrow), u_R(\alpha\uparrow,\beta\uparrow)\rangle$	$\langle u_E(\alpha\uparrow,\beta\downarrow), u_R(\alpha\uparrow,\beta\downarrow)\rangle$
E	$\alpha\downarrow$	$\langle u_E(\alpha\downarrow,\beta), u_R(\alpha\downarrow,\beta)\rangle$	$\langle u_E(\alpha,\beta), u_R(\alpha,\beta)\rangle$	$\langle u_E(\alpha\downarrow\downarrow,\beta), u_R(\alpha\downarrow\downarrow,\beta)\rangle$	$\langle u_E(\alpha\downarrow,\beta\uparrow), u_R(\alpha\downarrow,\beta\uparrow)\rangle$	$\langle u_E(\alpha\downarrow,\beta\downarrow), u_R(\alpha\downarrow,\beta\downarrow)\rangle$
	$\beta\uparrow$	$\langle u_E(\alpha,\beta\uparrow), u_R(\alpha,\beta\uparrow)\rangle$	$\langle u_E(\alpha\uparrow,\beta\uparrow), u_R(\alpha\uparrow,\beta\uparrow)\rangle$	$\langle u_E(\alpha\downarrow,\beta\uparrow), u_R(\alpha\downarrow,\beta\uparrow)\rangle$	$\langle u_E(\alpha,\beta\uparrow\uparrow), u_R(\alpha,\beta\uparrow\uparrow)\rangle$	$\langle u_E(\alpha,\beta), u_R(\alpha,\beta)\rangle$
	$\beta\downarrow$	$\langle u_E(\alpha,\beta\downarrow), u_R(\alpha,\beta\downarrow)\rangle$	$\langle u_E(\alpha\uparrow,\beta\downarrow), u_R(\alpha\uparrow,\beta\downarrow)\rangle$	$\langle u_E(\alpha\downarrow,\beta\downarrow), u_R(\alpha\downarrow,\beta\downarrow)\rangle$	$\langle u_E(\alpha,\beta), u_R(\alpha,\beta)\rangle$	$\langle u_E(\alpha,\beta\downarrow\downarrow), u_R(\alpha,\beta\downarrow\downarrow)\rangle$

thresholds increase or decrease c. $\uparrow\uparrow$ and $\downarrow\downarrow$ represent the thresholds increase or decrease $2 \times c$. The threshold values in each cell are determined by the strategies performed by two players. For example, let's look at the cell on the second row and second column in Table 1. Assuming the initial thresholds are (α, β). The player E performs the strategy of increasing α by c and R performs increasing α by c. The strategy profile is $(\alpha\uparrow, \alpha\uparrow)$. The threshold values affected by both players are $(\alpha\uparrow\uparrow, \beta)$. When we set initial values as $(\alpha, \beta) = (0.6, 0.5)$ and $c = 0.05$, the threshold values of the cell on the second row and second column would be $(0.7, 0.5)$.

3.2   Equilibrium Analysis

The pure strategy equilibrium is used as the solution to the formulated game shown in Table 1. The game equilibrium represents both players reach a compromise on how to change the initial thresholds. The strategy profile $(s_i, t_j) \in S = S_E \times S_R$ is a Nash equilibrium if for E and R, s_i and t_j are the best responses to each other, this is,

$$\forall s_k \in S_E, \ u_E(s_i, t_j) \geqslant u_E(s_k, t_j), \quad \text{where } s_i, s_k \in S_E \text{ and } k \neq i, t_j \in S_R,$$
$$\forall t_l \in S_R, \ u_R(s_i, t_j) \geqslant u_R(s_i, t_l), \quad \text{where } t_j, t_l \in S_R \text{ and } l \neq j, s_i \in S_E. \quad (6)$$

This equation can be interpreted as a strategy profile such that no player would like to change his strategy or they would loss benefit if deriving from this strategy profile, provided this player has the knowledge of the other player's strategies.

3.3   The Determination of Strategies

We are able to determine the strategies of both players based on the equilibrium of the game. Assuming that the strategy profile (s_i, t_j) is the equilibrium. s_i and t_j are the changes made on the thresholds, and $s_i, t_j \in \{$no change, $\alpha\uparrow, \alpha\downarrow, \beta\uparrow, \beta\downarrow\}$. The player E will change the thresholds as the strategy s_i shows. Similarly, and R will change the thresholds as the strategy t_j shows. For example, if the equilibrium is $(\alpha\uparrow, \beta\downarrow)$, i.e., E chooses to increase α and R chooses to decrease

β. The strategy set of E would be $S_E = \{$no change, $\alpha \uparrow c_E, \alpha \uparrow 2c_E\}$. The strategy set of R would be $S_R = \{$no change, $\beta \downarrow c_R, \beta \downarrow 2c_R\}$ The constants c_E and c_R denote the quantities that players E and R use to change the thresholds, respectively [16]. c_E and c_R can be any values that satisfy $0 < c_E, c_R \leq 0.1$ When setting $c_E = c_R = 0.05$, we have $S_E = \{\alpha$ no change, $\alpha \uparrow 0.05, \alpha \uparrow 0.1\}$, and $S_R = \{\beta$ no change, $\beta \downarrow 0.05, \beta \downarrow 0.1\}$.

3.4   Using the Proposed Approach in GTSS

The proposed approach to determining the strategies supplements the game-theoretic shadowed set model. Figure 2 shows the flow chart of the game-theoretic shadowed sets, and the two steps inside the dashed line rectangle are the main different part to the original GTSS model.   The revised GTSS are able to

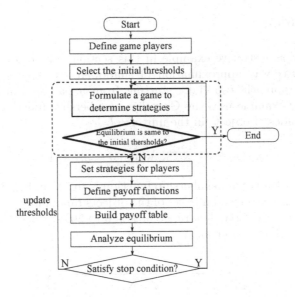

Fig. 2. The flow chart of game-theoretic shadowed sets

randomly select the initial thresholds (α, β). We use the proposed approach to formulate a game and analyze the equilibrium, and then determine the strategies of both players. If the strategy profile corresponding to the equilibrium is (no change, no change) or the strategies of two player cancel out each other, such as $(\alpha \uparrow, \alpha \downarrow)$, we will terminate the search process; otherwise, we set the strategies of players according to the equilibrium. Assuming that the strategy sets of two players are $S_E = \{s_1, s_2, ...\}$ and $S_R = \{t_1, t_2, ...\}$. Next, we define the payoff functions, build the payoff table, and analyze the equilibrium. The thresholds corresponding to the equilibrium are (α^*, β^*). The formulation of the subsequent games depends on if the thresholds (α^*, β^*) satisfy the selected stop conditions.

– If the thresholds (α^*, β^*) do not satisfy the stop conditions, we will update the thresholds in the subsequent games. It means that changing the thresholds by the current strategy sets is able to improve the payoffs of both players. Both players agree on the changes of the thresholds. In this case, we keep the strategy sets same and update the thresholds. The thresholds of the new game will be set as (α^*, β^*), and the strategy sets of both players are same to those used in the old games, i.e., $S_E = \{s_1, s_2, ...\}$ and $S_R = \{t_1, t_2, ...\}$. The strategy sets S_E and S_R are the changes to the updated thresholds (α^*, β^*).
– If the thresholds (α^*, β^*) satisfy the stop conditions, it means (α^*, β^*) are not able to improve the payoffs of both players, and the changes of the thresholds caused by the current strategy sets are not expected. In this case, we may start from the initial thresholds of the current game and use the proposed approach to formulate a game to try other possible strategy sets.

4   Experiment

We present a demonstrative example in this section. The example shows how to obtain a three-way approximation of shadowed sets using game-theoretic shadowed set approach, especially starting from a pair of randomly selected thresholds. The example uses the Gaussian membership function to define the membership grades of objects in the universe U,

$$\mu_A(x) = e^{\frac{-(x-c)^2}{2\theta^2}}, \text{ where } \theta = 3, c = 10. \tag{7}$$

The curve of this Gaussian membership function is shown in Fig. 3. The universe of the objects is formed by a finite set of the objects randomly selected according to the uniform distribution. The range of x is 0 to 20, i.e., $x \in [0, 20]$. We set σ as 0.5 to represent the uncertainty.

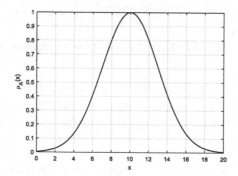

Fig. 3. Gaussian membership

Game-theoretic shadowed sets (GTSS) aim to determine the suitable thresholds of three-way approximations of shadowed sets. The elevation and reduction errors are two game players, i.e., $O = \{E, R\}$. The constant C in the payoff function is the area covered by the function $y = 1$, i.e., $C = \int_0^{20} 1 dx = 20$. The payoff functions of two players are defined as,

$$u_E(\alpha, \beta) = 20 - E_{(\alpha, \beta)}(\mu_A),$$
$$u_R(\alpha, \beta) = 20 - R_{(\alpha, \beta)}(\mu_A). \tag{8}$$

The initial thresholds are set as $(\alpha, \beta) = (0.6, 0.5)$. First, we formulate a game to determine the strategies for both players. The threshold β gets the maximal value 0.5, and both players only are allowed to decrease β. The strategies are $S_E = S_R = \{$no change$, \alpha \uparrow, \alpha \downarrow, \beta \downarrow\}$. The payoff table is a 4×4 matrix. We set $c = 0.05$ and the thresholds decrease or increase 0.05 in each action. The payoffs of players are calculated according to Eq. (8) and the payoff table with the payoff values are shown in Table 2. We analyze the equilibrium of Table 2 according to Eq. (6). The equilibrium is $(\alpha \uparrow, \beta \downarrow)$, which means player E increases α and R decreases β, as the cell at the second row and fourth column shown in Table 2.

Table 2. The payoff table of the game for choosing the strategies

		R			
		No change	$\alpha \uparrow$	$\alpha \downarrow$	$\beta \downarrow$
	No change	$\langle 19.1081, 18.1592 \rangle$	$\langle 19.2939, 18.0973 \rangle$	$\langle 18.8972, 18.1964 \rangle$	$\langle 19.0951, 18.4052 \rangle$
E	$\alpha \uparrow$	$\langle 19.2939, 18.0973 \rangle$	$\langle 19.4569, 18.0095 \rangle$	$\langle 19.1081, 18.1592 \rangle$	$\langle \mathbf{19.2808, 18.3432} \rangle$
	$\alpha \downarrow$	$\langle 18.8972, 18.1964 \rangle$	$\langle 19.1081, 18.1592 \rangle$	$\langle 18.6579, 18.2090 \rangle$	$\langle 18.8842, 18.4424 \rangle$
	$\beta \downarrow$	$\langle 19.0951, 18.4052 \rangle$	$\langle 19.2808, 18.3432 \rangle$	$\langle 18.8842, 18.4424 \rangle$	$\langle 19.0545, 18.6345 \rangle$

When the initial thresholds $(\alpha, \beta) = (0.6, 0.5)$, two players agree to make changes on thresholds by E increasing α and R decreasing β. The constant change steps are $c_E = c_R = 0.05$. The strategy set of E is $S_E = \{\alpha$ no change$, \alpha \uparrow 0.05, \alpha \uparrow 0.1\}$. The strategy set of R is $S_R = \{\beta$ no change$, \beta \downarrow 0.05, \beta \downarrow 0.1\}$. The payoff table with the payoff values are shown in Table 3.

We analyze the equilibria of the game shown in Table 3. The strategy profile $(\alpha \uparrow 0.1, \beta \downarrow 0.1)$ is the equilibrium of the payoff table shown in Table 3.

Table 3. The possible strategy profiles in the payoff table

		R		
		β	$\beta \downarrow 0.05$	$\beta \downarrow 0.1$
	α	$\langle 19.1081, 18.1592 \rangle$	$\langle 19.0951, 18.4052 \rangle$	$\langle 19.0545, 18.6345 \rangle$
E	$\alpha \uparrow 0.05$	$\langle 19.2939, 18.0973 \rangle$	$\langle 19.2808, 18.3432 \rangle$	$\langle 19.2402, 18.5726 \rangle$
	$\alpha \uparrow 0.1$	$\langle 19.4569, 18.0059 \rangle$	$\langle 19.4438, 18.2554 \rangle$	$\langle \mathbf{19.4032, 18.4848} \rangle$

The player E's payoff increases from 19.1081 to 19.4032, and player R's payoff increases from 18.1592 to 18.4848 when the thresholds change from $(0.6, 0.5)$ to $(0.7, 0.4)$.

The stop condition is both players lose their payoffs or the gain of one player's payoff is less than the loss of the other player's payoff in the current game. The games are repeated three times until the stop condition is satisfied. Table 4 shows the three repeated game formulations.

Table 4. The repetition of games

	Initial (α, β)	Strategies	Result (α, β)	Payoffs	Changes
1	(0.6, 0.5)	$(\alpha \uparrow, \beta \downarrow)$	(0.7, 0.4)	<19.4032, 18.4848>	(+0.2951, +0.3256)
2	(0.7, 0.4)	$(\alpha \uparrow, \beta \downarrow)$	(0.8, 0.3)	<19.4875, 18.6334>	(+0.0843, +0.1486)
3	(0.8, 0.3)	$(\alpha \uparrow, \beta \downarrow)$	(0.9, 0.2)	<19.3062, 18.5532>	(−0.1813, −0.0802)

The second column shows the initial thresholds. The third column shows the strategy sets of both players, and the player E performs the first action and the player R performs the second action. The forth column shows the thresholds corresponding to the game equilibrium. The fifth column shows the payoffs of two players. The sixth column shows the changes of two payoffs when the thresholds change from the initial pair to the result pair.

Now we get thresholds $(\alpha, \beta) = (0.8, 0.3)$. Continuing to update the thresholds can not improve the players' payoffs as the third game shown. We need to change the strategies of players. In other words, we may use the proposed approach to choose the other strategies that are different from the previous strategies (E increases α and R decreases β). We formulate a game with players E and R having five strategies as shown in Eq. (4). This game aims to test if both players agree to change the thresholds $(\alpha, \beta) = (0.8, 0.3)$ in other directions. The equilibrium of this formulated game is $(\beta \uparrow, \beta \downarrow)$. The changes made by two players on the thresholds cancel out each other. We terminate the game repetition and the thresholds $(\alpha, \beta) = (0.8, 0.3)$ are the final result. This result is different to that we obtained when setting $(\alpha, \beta) = (1, 0)$ as the initial thresholds in [16].

5   Conclusion

The game-theoretic shadowed sets determine and interpret the thresholds of the three-way approximations of shadowed sets according to a principle of tradeoff with games. GTSS formulate competitive games between the elevation and reduction errors to obtain a tradeoff between these errors. A repetition learning mechanism is adopted to modify the thresholds to reach the balanced thresholds gradually. The limit of GTSS lies in both players start from the fixed initial threshold pair $(1, 0)$ and the strategies of players are also fixed. The proposed approach formulates a game to determine the strategies of both players when

GTSS start from a pair of randomly chosen thresholds. The players are elevation and reduction errors. The strategies are the possible changes that can be made on the initial thresholds. The equilibrium of the game indicates the change directions that two players agree to make on the initial thresholds. Moreover, when a pair of thresholds satisfy the stop conditions after a series of game repetition, the proposed game can be used to examine if it is possible to change these thresholds in other directions. The proposed game approach supplements and extends the game-theoretic shadowed set model.

Acknowledgements. This work is partially supported by a Discovery Grant from NSERC Canada.

References

1. Cattaneo, G., Ciucci, D.: An algebraic approach to shadowed sets. Electron. Notes Theoret. Comput. Sci. **82**(4), 64–75 (2003)
2. Cattaneo, G., Ciucci, D.: Shadowed sets and related algebraic structures. Fundamenta Informaticae **55**(3–4), 255–284 (2003)
3. Deng, X.F., Yao, Y.Y.: Decision-theoretic three-way approximations of fuzzy sets. Inf. Sci. **279**, 702–715 (2014)
4. Deng, X.F., Yao, Y.Y.: Mean-value-based decision-theoretic shadowed sets. In: 2013 Joint IFSA and NAFIPS Annual Meeting, pp. 1382–1387. IEEE (2013)
5. Grzegorzewski, P.: Nearest interval approximation of a fuzzy number. Fuzzy Sets Syst. **130**(3), 321–330 (2002)
6. Grzegorzewski, P.: Fuzzy number approximation via shadowed sets. Inf. Sci. **225**, 35–46 (2013)
7. Pedrycz, W.: Shadowed sets: representing and processing fuzzy sets. IEEE Trans. Syst. Man Cybern. **28**(1), 103–109 (1998)
8. Pedrycz, W.: Shadowed sets: bridging fuzzy and rough sets. In: Rough Fuzzy Hybridization. A New Trend in Decision-Making, pp. 179–199 (1999)
9. Pedrycz, W.: Interpretation of clusters in the framework of shadowed sets. Pattern Recogn. Lett. **26**(15), 2439–2449 (2005)
10. Pedrycz, W., Vukovich, G.: Granular computing with shadowed sets. Int. J. Intell. Syst. **17**(2), 173–197 (2002)
11. Tahayori, H., Sadeghian, A., Pedrycz, W.: Induction of shadowed sets based on the gradual grade of fuzziness. IEEE Trans. Fuzzy Syst. **21**(5), 937–949 (2013)
12. Yao, J.T., Vasilakos, A.V., Pedrycz, W.: Granular computing: perspectives and challenges. IEEE Trans. Cybern. **43**(6), 1977–1989 (2013)
13. Yao, Y.Y., Wang, S., Deng, X.F.: Constructing shadowed sets and three-way approximations of fuzzy sets. Inf. Sci. **412–413**, 132–153 (2017)
14. Zabihi, S.M., Akbarzadeh-T, M.R.: Generalized fuzzy C-means clustering with improved fuzzy partitions and shadowed sets. ISRN Artif. Intell. **2012**, 1–6 (2012)
15. Zadeh, L.A.: Fuzzy sets. In: Fuzzy Sets, Fuzzy Logic, and Fuzzy Systems: Selected Papers by Lotfi A Zadeh, pp. 394–432. World Scientific (1996)
16. Zhang, Y., Yao, J.T.: Game theoretic approach to shadowed sets: a three-way tradeoff perspective. Manuscript

Three-Way and Semi-supervised Decision Tree Learning Based on Orthopartitions

Andrea Campagner and Davide Ciucci[✉][iD]

DISCo, University of Milano–Bicocca, Viale Sarca 336/14, 20126 Milano, Italy
ciucci@disco.unimib.it

Abstract. Decision Tree Learning is one of the most popular machine learning techniques. A common problem with this approach is the inability to properly manage uncertainty and inconsistency in the underlying datasets. In this work we propose two generalized Decision Tree Learning models based on the notion of Orthopair: the first method allows the induced classifiers to abstain on certain instances, while the second one works with unlabeled outputs, thus enabling semi-supervised learning.

Keywords: Orthopair · Three-way decision · Decision tree · Entropy

1 Introduction

Machine Learning has been, in the recent years, one of the most popular research areas in the Computer Science literature, with a variety of proposed models based on different ideas and assumptions (e.g. neural networks, Support Vector Machines, ...). Among these, Decision Tree Learning, both as a standalone technique and as a foundation for more sophisticated ones (e.g. boosting), has been one of the more popular approaches, thanks to its efficiency and the interpretability of the induced models. However, one of the major problems of this approach (and, more in general, of Machine Learning techniques) is the inability to properly represent and cope with information that is uncertain and/or inconsistent. To deal with this information flaws different approaches exist (Fuzzy Set Theory and Rough Set Theory among the others) but their application to traditional Machine Learning approaches have not yet been fully explored.

In this paper, we propose two algorithms, based on Decision Tree Learning and the notion of Orthopair, to properly include uncertainty considerations in the induction of the models. The rest of the paper is organized as follows. In Sect. 2 we provide an introduction to Orthopairs and Orthopartitions focusing, in particular, on uncertainty measures for Orthopartitions. In Sect. 3, we propose two techniques to apply Orthopartitions to Decision Tree Learning. In particular: in Sect. 3.1 we propose an approach to Three-Way Decision Tree Learning, while in Sect. 3.2 we propose an approach to semi-supervised learning using Decision Trees. We notice that our methods are different from the soft decision trees in sense of [5]. Indeed, the "soft" part in soft decision trees regards the way to take a decision, which is still dichotomous. On the contrary, we allow to abstain from a decision, thus we have three possible outcomes.

© Springer International Publishing AG, part of Springer Nature 2018
J. Medina et al. (Eds.): IPMU 2018, CCIS 854, pp. 748–759, 2018.
https://doi.org/10.1007/978-3-319-91476-3_61

2 Orthopartitions: Basic Definitions

In this section, we give the basic definitions of orthopairs and introduce the notion of orthopartition.

2.1 Introduction to Orthopairs

An orthopair on a universe X is a pair of sets (P, N) such that $P \cap N = \emptyset$. Since not necessarily P and N cover the universe, we can define also the set $Bnd = X \setminus (P \cup N)$. Interesting connections with different uncertainty representation frameworks can be put forward (see [1]). We denote with $O(X)$ the set of all orthopairs definable on the universe X. We can define a variety of orderings, and associated algebraic operations, on $O(X)$. For the purposes of this paper we will consider the so-called *truth ordering* $\leq_t$ defined as:

$$O_1 \leq_t O_2 \quad \text{iff} \quad P_1 \subseteq P_2 \quad \text{and} \quad N_2 \subseteq N_1$$

along with the associated *join*, *meet* and *negation* operations:

$$O_1 \sqcap_t O_2 = (P_1 \cap P_2, N_1 \cup N_2)$$
$$O_1 \sqcup_t O_2 = (P_1 \cup P_2, N_1 \cap N_2)$$
$$\neg O = (N, P)$$

We say that a set S is *consistent* with an orthopair O if it holds that

$$x \in P \to x \in S \text{ and } x \in N \to x \notin S.$$

That is, all the positive elements and none of the negative ones of the orthopair are in S, and S can contain also some element in the boundary. We say that two orthopairs O_1, O_2 are *disjoint* if the followings hold:

(Ax O1) $P_1 \cap P_2 = \emptyset$;
(Ax O2) $P_1 \cap Bnd_2 = \emptyset$ and $Bnd_1 \cap P_2 = \emptyset$.

We notice that it is not required that $N_1 \cap N_2 = \emptyset$.

Example 1. Let us consider the universe $U = \{1, 2, \ldots, 10\}$. The two orthopairs $O_1 = (\{1, 2\}, \{9, 10\})$ and $O_2 = (\{9\}, \{1, 2\})$ are disjoint.

2.2 Orthopartitions

The notion of disjoint orthopairs is used in the following to generalize the concept of a partition to an orthopartition. Intuitively, an orthopartition represents an incomplete state of knowledge regarding an underlying, unknown, partition.

Definition 1. *An orthopartition is a set $\mathcal{O} = \{O_1, \ldots, O_n\}$ of orthopairs such that the following axioms hold:*

(Ax O1) $\forall O_i, O_j \in \mathcal{O}$ O_i, O_j *are disjoint;*
(Ax O2) $\bigcap_i N_i = \emptyset$;
(Ax O3) $\forall x \in U$ $(\exists O_i$ *s.t.* $x \in Bnd_i) \to (\exists O_j$ *with* $i \neq j$ *s.t.* $x \in Bnd_j)$;

The rationale behind the axioms is that we suppose that the orthopairs in the collection are mutually exclusive (axiom 1) and that the uncertain elements cannot be only in one of these orthopairs (axiom 3). The axiom 2 expresses a kind of coverage of the universe.

Example 2. Let us consider the two orthopairs O_1 and O_2 in Example 1. They are disjoint but they do not form an orhopartition since $10 \in Bnd_2$ and $10 \notin Bnd_1$. On the other hand, if we also consider $O_3 = (\emptyset, \{1, 2, 9\})$, then the collection $\{O_1, O_2, O_3\}$ is an orthopartition of U.

Definition 2. *We say that a partition π is* consistent *with an orthopartition $\mathcal{O}$ iff $\forall O_i \in \mathcal{O}$, $\exists S_i \in \pi$ s.t. S is consistent with O_i and the S_is are all disjoint. We denote as $\Pi_{\mathcal{O}} = \{\pi | \pi$ is consistent with $\mathcal{O}\}$ the set of all partitions consistent with $\mathcal{O}$.*

Entropy for Orthopartitions. Given the definition of an orthopartition we can give a generalization of the concept of *logical entropy*, given by Ellerman in [2] for classical partitions. At first we recall the classical definition: $h(\pi) = \frac{dit(\pi)}{|U|^2}$ with π a partition and $dit(\pi) = \{(u, u') \in U \times U \,|\, u, u'$ belongs to two different blocks of $\pi\}$.

Definition 3. *Given an orthopartition $\mathcal{O}$, we define a* lower entropy, *an* upper entropy *and a* mean entropy *respectively as:*

$$h_* = min\{h(\pi) | \pi \in \Pi_{\mathcal{O}}\} \tag{1a}$$
$$h^* = max\{h(\pi) | \pi \in \Pi_{\mathcal{O}}\} \tag{1b}$$
$$\hat{h} = \frac{h^*(\mathcal{O}) + h_*(\mathcal{O})}{2} \tag{1c}$$

Moreover, also the Shannon entropy can be generalized to orthopartitions. At first, let us associate with an orthopartition $\mathcal{O}$ the set of probability distributions compatible with $\mathcal{O}$:

$$\mathcal{P}_{\mathcal{O}} = \{\langle p_1, ..., p_n \rangle | p_i \in [\frac{|P_i|}{|U|}, \frac{|P_i \cup Bnd_i|}{|U|}] \text{ and } \sum_{i=1}^n p_i = 1\} \tag{2}$$

Then, we recall the classical definition of Shannon entropy, for a probability distribution $p = \langle p_1, ..., p_n \rangle$, $H_S(p) = \sum_{i=1}^n p_i \log_2(\frac{1}{p_i})$ and use it to define the lower, upper and mean Shannon entropies as:

$$H_{S*} = min\{H_S(p) | p \in \mathcal{P}_{\mathcal{O}}\} \tag{3a}$$
$$H_S^* = max\{H_S(p) | p \in \mathcal{P}_{\mathcal{O}}\} \tag{3b}$$
$$\overset{\wedge}{H_S} = \frac{H_S^*(\mathcal{O}) + H_{S*}(\mathcal{O})}{2} \tag{3c}$$

Remark 1. In the rest of this work, only the mean entropies $\hat{h}$ and $\hat{H}_S$ will be used.

Mutual Information for Orthopartitions. In order to provide a measure of the similarity of two orthopartitions, we can provide a generalized definition of mutual information.

Given two orthopartitions $\mathcal{O}_1, \mathcal{O}_2$ we define a new *meet orthopartition* as:

$$\mathcal{O}_1 \wedge \mathcal{O}_2 = \{O_{i1} \sqcap_t O_{j2} | O_{i1} \in \mathcal{O}_1 \text{ and } O_{j2} \in \mathcal{O}_2\}$$

to which we can associate the set of consistent partitions according to definition 2. This set corresponds also to the meet of the partitions consistent with $\mathcal{O}_1$ and $\mathcal{O}_2$: $\Pi_{\mathcal{O}_1 \wedge \mathcal{O}_2} = \{\pi \wedge_p \sigma | \pi \text{ is consistent with } \mathcal{O}_1 \text{ and } \sigma \text{ is consistent with } \mathcal{O}_2\}$, where $\pi \wedge_p \sigma$ is the standard meet on partitions: for $\pi = \{A_1, \dots A_m\}$ and $\sigma = \{B_1, \dots B_n\}$, then $\pi \wedge_p \sigma = \{A_i \cap B_j | i = 1, \dots, m; j = 1, \dots, n\}$.

Now, the generalized versions of the classical mutual information based on Shannon Entropy and the one proposed by Ellerman in [2] are respectively:

$$I(\mathcal{O}_1, \mathcal{O}_2) = \hat{H}_S(\mathcal{O}_1) + \hat{H}_S(\mathcal{O}_2) - \hat{H}_S(\mathcal{O}_1 \wedge \mathcal{O}_2) \tag{4a}$$

$$m(\mathcal{O}_1, \mathcal{O}_2) = \hat{h}(\mathcal{O}_1) + \hat{h}(\mathcal{O}_2) - \hat{h}(\mathcal{O}_1 \wedge \mathcal{O}_2) \tag{4b}$$

3   Decision Tree Learning

In this section, we are going to introduce two applications of orthopartitions and generalized mutual information to Decision Tree Learning.

Decision Tree Learning is a popular approach in Machine Learning, in which the learned model is represented as a Decision Tree. Let $D = \{x_1, ..., x_d\} \subseteq U$ be a dataset over feature set $\mathbb{A} = \{a_1, ..., a_l\}$.

The classical algorithms for Decision Tree induction (*ID3* [6], *C4.5* [7]) are based on the top-down greedy algorithm 1.

Input: Dataset D
Output: Decision Tree built on D
1 For each feature a compute the mutual information I_a w.r.t. D;
2 Select feature a_{max} with maximum mutual information value and create a decision node on a_{max} (*split attribute*);
3 Recur on the subsets of D determined by the values of a_{max};
           Algorithm 1: Decision Tree Induction

We can extend the Decision Tree Learning to the case of orthopartitions in two different ways:

- In the first generalization, orthopartitions are used to allow induction of Three-way Decision Trees (based on Three-way Decisions, outlined by Yao in [9], and similar in spirit to Three-way Decision Trees proposed by Liu et al. in [4]);
- In the second generalization, orthopartitions are used to allow a form of semi-supervised learning in the context of Decision Tree Learning.

3.1 Three-Way Decision Tree Learning

As regards the first approach, let $D = \{x_1, ..., x_{|D|}\} \subseteq X$ be a given dataset with a set of features $\{a_1, ..., a_m\}$ and a single classification feature C.

We will first consider, for simplicity, that only two classifications are possible, that is $\forall x \in D, \ C(x) \in \{P, N\}$, furthermore we will suppose that any learned model h can classify the instances in three possible ways, that is $\forall x \in X, \ h(x) \in \{P, N, Bnd\}$, where the Bnd decision corresponds to a decision of abstaining from judgement.

Let us define two costs $\epsilon, \alpha \in \mathbb{R}_+$, which represent, respectively, the cost associated with a classification error and the cost corresponding to an abstention and let us suppose that $\alpha < \epsilon$ (otherwise abstaining would not be a meaningful decision). Each feature a, with possible values $v_1^a, ..., v_k^a$, of dataset D (and, thus, each decision node in a corresponding induced Decision Tree) naturally determines an orthopartition on the basis of ϵ and α.

Let $D_i^a = \{x \in D | v_a(x) = v_i^a\}$ be the set of instances that have value v_i^a for feature a. If we associate to D_i^a the classification

$$C_i^a = argmax_{j \in \{P,N\}}\{|\{x \in D_i^a | C(x) = j\}|\}$$

we can compute the *expected classification error cost* as:

$$E(D_i^a | C_i^a) = \epsilon * min_{j \in \{P,N\}}\{|\{x \in D_i^a | C(x) = j\}|\} \tag{5}$$

Similarly we can compute the *expected abstention error cost* as:

$$E(D_i^a | Bnd) = \alpha |D_i^a| \tag{6}$$

Thus, if $E(D_i^a | C_i^a) \leq E(D_i^a | Bnd)$, that is the cost associated with a classification error is less than the cost that we would incur if we were to abstain, we assign to the instances in D_i^a the label C_i^a (that is, $h(x) = C_i^a$); otherwise we assign to the instances in D_i^a the label Bnd.

It is evident that this process of assigning labels determines an orthopair $O_a = (P_a, N_a)$ and, thus, an orthopartition $\mathcal{O}_a = \{O_a, \neg O_a\}$, where:

$$P_a = \bigcup\{D_i^a | C_i^a = P\} \text{ and } N_a = \bigcup\{D_i^a | C_i^a = N\}.$$

We can thus, for each feature a, compute the mutual information m between $\mathcal{O}_a$ and the currently examined dataset D and choose the split attribute as the feature a which gives the greatest value of mutual information.

This process can be synthetically described by the following algorithm:

Input: Dataset D, error cost ϵ, abstention cost α
Output: Three-way Decision Tree built on D

1 For each feature a compute the corresponding orthopartition $\mathcal{O}_a$ using ϵ, α;
2 For each orthopartition $\mathcal{O}_a$ compute the mutual information $m(D, \mathcal{O}_a)$;
3 Select as split attribute the feature a_{max} which gives the greatest mutual information value;
4 Recur on the subsets of D determined by a_{max};

Algorithm 2: Three-way Decision Tree construction

The algorithm is illustrated by the following example.

Example 3. Let us consider the dataset given in Table 1.

Table 1. Dataset on weather conditions.

Temperature	Outlook	Humidity	Windy	Do sport?
Hot	Sunny	High	False	No
Hot	Sunny	High	True	No
Hot	Sunny	High	False	Yes
Cool	Rain	Normal	False	Yes
Cool	Overcast	Normal	True	Yes
Mild	Sunny	High	False	No
Cool	Sunny	Normal	False	Yes
Mild	Rain	Normal	False	Yes
Mild	Sunny	Normal	True	Yes
Mild	Overcast	High	True	Yes
Hot	Overcast	Normal	False	Yes
Mild	Rain	High	True	No
Cool	Rain	Normal	True	No
Mild	Rain	Normal	False	Yes

Let us suppose that $\epsilon = 1$ and $\alpha = 0.4$, thus we get

Temperature	No	Yes	$H_{Shannon}$	Error cost	Abstention cost
Hot	2	2	1	2	1.6
Mild	2	4	0.92	2	2.4
Cool	1	3	0.81	1	1.6

Outlook	No	Yes	$H_{Shannon}$	Error cost	Abstention cost
Sunny	3	3	1	3	2.4
Overcast	0	3	0	0	2.4
Rain	2	3	0.97	2	2

Humidity	No	Yes	$H_{Shannon}$	Error cost	Abstention cost
High	4	3	0.98	3	2.8
Normal	1	6	0.59	1	2.8

Windy	No	Yes	$H_{Shannon}$	Error cost	Abstention cost
False	2	6	0.81	2	3.2
Normal	3	3	1	3	2.4

Comparing the Three-Way Decision Tree Learning algorithm with the *ID3* algorithm (thus using *Information Gain* (IG) as split criterion) we obtain the following values (Table 2):

Table 2. Mutual Information values according to Eq. (4a) and standard Information Gain in ID3.

Feature	I	IG
Temperature	$\frac{14}{196}$	0.03
Outlook	$\frac{24}{196}$	0.165
Humidity	$\frac{25}{196}$	0.155
Windy	$\frac{24}{196}$	0.05

Thus, at the first step, our algorithm will select *Humidity* as the split attribute, while *ID3* would select *Outlook*. The two complete trees will result as in Figs. 1 and 2.

It is easy to see that our algorithm produces a better tree from the point of view of the total error cost: our algorithm incurs a cost of 0.8 while the *ID3*-constructed tree incurs a cost of 1.

This approach can be extended to consider more than two classes, let $C = \{C_1, ..., C_n\}$ be the set of the possible classifications.

In order to extend this approach we have to consider multiple possible abstention decisions. We denote the decision of establishing that a certain instance x belongs to one of the classes $C_i, C_{i+1}, ..., C_{i+k}$ but we abstain to precisely decide which one as $Bnd_{i,i+1,...,i+k}$, with $\{i, i+1, ..., i+k\} \subseteq \{1, ..., n\}$.

By extending decisions in this way the abstention cost can no longer be a constant value α.

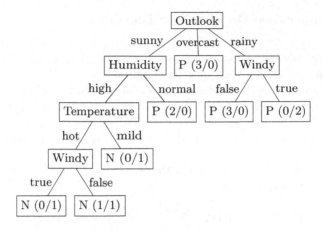

Fig. 1. Decision Tree constructed via the ID3 algorithm (choosing negative classification in case of indecision)

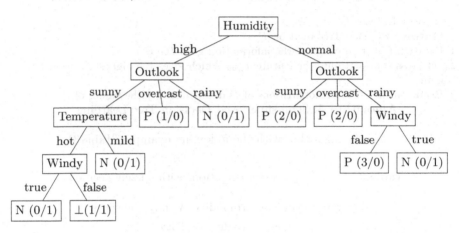

Fig. 2. Decision Tree constructed via the Three-way Decision Tree algorithm

Proposition 1. *If the abstention cost α is a constant, then choosing decision $Bnd_{i,i+1,...,i+k}$ is always costlier than choosing decision $Bnd_{1,...,n}$.*

Proof. $\epsilon * |\{x \in D_i^a | C(x) \notin \{i, i+1, ..., i+k\}\}| +$
$\alpha * |\{x \in D_i^a | C(x) \in \{i, i+1, ..., i+k\}\}| \geq \alpha * |D_i^a|.$

The solution is to define α as a function $\alpha : \{1, ..., |A|\} \to \mathbb{R}_+$ such that, given $A, B \subseteq C$, it holds $|A| \leq |B| \to \alpha(|A|) \leq \alpha(|B|)$.

Remark 2. Note that, since in general every subset of classes should be considered, the complexity of choosing the split attribute is exponential in the number of features $|\mathbb{A}|$, thus, without using heuristics to limit the search space, this approach is applicable only if $\mathbb{A}$ is small.

3.2 Semi-supervised Decision Tree Learning

As regards the second approach, let D be a dataset. The classification, in this case, could be missing for some of the instances, that is $\forall x \in D,\ C(x) \in \{P, N, \perp\}$ where $\perp$ represents a missing classification.

Such a dataset directly represents an orthopartition and we can naturally generalize the classical induction algorithm by considering the mutual information as defined for orthopartitions. For each feature a, with values $v_1^a, ..., v_k^a$, let us denote with D_i^a the (sub)-orthopartition containing the instances $x \in D$ such that $v_a(x) = v_i^a$.

We can associate to each of these orthopartitions D_i^a the entropy $\hat{h}(D_i^a)$ and then compute the information gain as:

$$IG(D, a) = \hat{h}(D) - \sum_{v_i^a} \frac{|\{x : v_a(x) = v_i^a\}|}{|D|} \hat{h}(D_i^a) \tag{7}$$

Thus, the learning process can be described by the following algorithm:

Input: Dataset D
Output: Decision Tree built on D
1 For each feature a compute the information gain $IG(D, a)$;
2 Select as split attribute the feature a_{max} which gives the highest information gain;
3 Recur on the (sub)-orthopartitions of D determined by the values of a;
Algorithm 3: Semi-supervised Decision Tree Learning

We illustrate the algorithm with the following example (Table 3).

Table 3. Dataset on weather conditions with missing decisions.

Temperature	Outlook	Humidity	Windy	Do sport?
Hot	Sunny	High	False	No
Hot	Sunny	High	True	No
Hot	Sunny	High	False	Yes
Cool	Rain	Normal	False	$\perp$
Cool	Overcast	Normal	True	Yes
Mild	Sunny	High	False	No
Cool	Sunny	Normal	False	$\perp$
Mild	Rain	Normal	False	Yes
Mild	Sunny	Normal	True	Yes
Mild	Overcast	High	True	$\perp$
Hot	Overcast	Normal	False	Yes
Mild	Rain	High	True	$\perp$
Cool	Rain	Normal	True	No
Mild	Rain	Normal	False	Yes

Example 4. Consider the following dataset D:

The dataset has a value of $h_* = \frac{20}{49}$, $h^* = \frac{1}{2}$, $\hat{h} = \frac{89}{196}$. We obtain the following values of entropies:

Temperature	h_*	h^*	$\hat{h}$
hot	$\frac{1}{2}$	$\frac{1}{2}$	$\frac{1}{2}$
mild	$\frac{5}{18}$	$\frac{1}{2}$	$\frac{7}{18}$
cool	$\frac{3}{8}$	$\frac{1}{2}$	$\frac{7}{16}$

Outlook	h_*	h^*	$\hat{h}$
sunny	$\frac{4}{9}$	$\frac{1}{2}$	$\frac{17}{36}$
overcast	0	$\frac{4}{9}$	$\frac{2}{9}$
rain	$\frac{8}{25}$	$\frac{12}{25}$	$\frac{10}{25}$

Humidity	h_*	h^*	$\hat{h}$
high	$\frac{5}{18}$	$\frac{1}{2}$	$\frac{7}{18}$
normal	$\frac{7}{32}$	$\frac{15}{32}$	$\frac{11}{16}$

Windy	h_*	h^*	$\hat{h}$
false	$\frac{3}{8}$	$\frac{1}{2}$	$\frac{7}{16}$
normal	$\frac{4}{9}$	$\frac{1}{2}$	$\frac{17}{36}$

Obtaining the following values of information gain: Thus, the algorithm

Feature	Information gain
Temperature	0.024
Outlook	0.064
Humidity	−0.066
Windy	0.004

would select feature *Outlook* as split attribute.

The complete tree would result as follows in Fig. 3, where a label (P, N, Bnd) in the leaf nodes should be interpreted as the number of positive, negative and unknown classifications in Table 3.

The decision given on the leaf nodes can be given with a majority criterion, or by combining this approach with the Three-way Decision Tree approach previously described. In the latter case, the formula for the *expected classification error* is modified to take into account the unlabeled instances:

$$E(D_i^a|C_i^a) = \epsilon * min_{j \in \{P,N\}}\{|\{x \in D_i^a|C(x) = j\}|\}+$$
$$\frac{\epsilon}{2} * |\{|\{x \in D_i^a|C(x) = \bot\}|$$

(8)

We can extend this approach to the multi-class case: if the set of possible classifications is $C = \{C_1, ..., C_n\}$ then each instance is assigned a label in 2^C, where if $|C(x)| > 1$ it means that the exact classification of instance x is unknown.

This approach represents a direct generalization of the one considering only two classes since it determines an orthopartition and we can thus apply the Algorithm 3 described above.

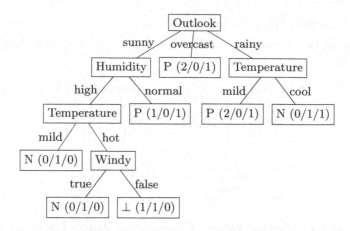

Fig. 3. Decision Tree constructed via the semi-supervised Decision Tree algorithm and the labeling criterion given for Three-Way Decision Trees (with $\epsilon = 1$ and $\alpha = 0.4$).

4   Conclusions

In this work, we proposed two techniques to incorporate uncertainty and inconsistency management in Decision Tree Learning, a traditional and popular Machine Learning approach. The first proposed technique allows the induced models to abstain from judgment on certain instances, thus enabling a trade-off between error and this possibility of abstention. It is similar in spirit to the approach developed in [4]. However, our method is a standalone one with a native uncertainty handling since it builds its own decision tree. The algorithm by Liu et al. requires the construction of a decision tree with ID3 to be subsequently modified in order to classify some elements in the boundary region according to a decision theoretic rough set approach. The second technique, on the other hand, harnesses the correspondence between orthopartitions and datasets with missing classifications to enable semi-supervised learning using Decision Trees. We then showed the application of the two proposed algorithms on two simple datasets. However a number of issues remain to be considered:

- The first, and most important one, is to test the effective applicability of the proposed techniques on real datasets, comparing them with existing Machine Learning techniques;
- As regards the Three-Way Decision Tree Learning algorithm, and specifically its extension to the multi-class case, it is of interest a study of sensible heuristics, in order to reduce the complexity of choosing the split attributes, which in the general case, as argued previously, is exponential in the number of features;
- Studying the applicability of the proposed techniques as a basis for more sophisticated Machine Learning techniques, in a similar way as classical Decision Trees are used to support *boosting* [8] or *Random Forests* [3].

References

1. Ciucci, D.: Orthopairs and granular computing. Granular Comput. **1**, 159–170 (2016)
2. Ellerman, D.: An introduction to logical entropy and its relation to Shannon entropy. Int. J. Semant. Comput. **7**(2), 121–145 (2013)
3. Ho, T.K.: Random decision forests. In: Proceedings of the Third International Conference on Document Analysis and Recognition, ICDAR 1995, vol. 1, pp. 278–282. IEEE Computer Society (1995)
4. Liu, Y., Xu, J., Sun, L., Du, L.: Decisions tree learning method based on three-way decisions. In: Yao, Y., Hu, Q., Yu, H., Grzymala-Busse, J.W. (eds.) RSFDGrC 2015. LNCS (LNAI), vol. 9437, pp. 389–400. Springer, Cham (2015). https://doi.org/10.1007/978-3-319-25783-9_35
5. Nguyen, H.S.: A soft decision tree. In: Kłopotek, M.A., Wierzchoń, S.T., Michalewicz, M. (eds.) Intelligent Information Systems 2002. AINSC, vol. 17, pp. 57–66. Physica, Heidelberg (2002). https://doi.org/10.1007/978-3-7908-1777-5_6
6. Quinlan, J.R.: Induction of decision trees. Mach. Learn. **1**(1), 81–106 (1986)
7. Quinlan, J.R.: C4.5: Programs for Machine Learning. Morgan Kaufmann Publishers Inc., San Francisco (1993)
8. Schapire, R.E.: The boosting approach to machine learning: an overview. In: Denison, D.D., Hansen, M.H., Holmes, C.C., Mallick, B., Yu, B. (eds.) Nonlinear Estimation and Classification. LNS, pp. 149–171. Springer, New York (2003). https://doi.org/10.1007/978-0-387-21579-2_9
9. Yao, Y.: An outline of a theory of three-way decisions. In: Yao, J.T., Yang, Y., Słowiński, R., Greco, S., Li, H., Mitra, S., Polkowski, L. (eds.) RSCTC 2012. LNCS (LNAI), vol. 7413, pp. 1–17. Springer, Heidelberg (2012). https://doi.org/10.1007/978-3-642-32115-3_1

Author Index